Handbook of Spine Technology

Boyle C. Cheng
Editor

Handbook of Spine Technology

Volume 2

With 425 Figures and 94 Tables

Editor
Boyle C. Cheng
Neuroscience Institute, Allegheny Health Network
Drexel University, Allegheny General Hospital Campus
Pittsburgh, PA, USA

ISBN 978-3-319-44423-9   ISBN 978-3-319-44424-6 (eBook)
ISBN 978-3-319-44425-3 (print and electronic bundle)
https://doi.org/10.1007/978-3-319-44424-6

© Springer Nature Switzerland AG 2021
All rights are reserved by the Publisher, whether the whole or part of the material is concerned, specifically the rights of translation, reprinting, reuse of illustrations, recitation, broadcasting, reproduction on microfilms or in any other physical way, and transmission or information storage and retrieval, electronic adaptation, computer software, or by similar or dissimilar methodology now known or hereafter developed.
The use of general descriptive names, registered names, trademarks, service marks, etc. in this publication does not imply, even in the absence of a specific statement, that such names are exempt from the relevant protective laws and regulations and therefore free for general use.
The publisher, the authors, and the editors are safe to assume that the advice and information in this book are believed to be true and accurate at the date of publication. Neither the publisher nor the authors or the editors give a warranty, expressed or implied, with respect to the material contained herein or for any errors or omissions that may have been made. The publisher remains neutral with regard to jurisdictional claims in published maps and institutional affiliations.

This Springer imprint is published by the registered company Springer Nature Switzerland AG.
The registered company address is: Gewerbestrasse 11, 6330 Cham, Switzerland

This book is dedicated to my parents, Samuel and Ruth, who were inspiring from an early age and, moreover, instilled within me the power of harnessing personal talent combined with a strong work ethic to achieve the best of all possible outcomes. In turn, I hope this to be a legacy for my wife, Judy and my two sons, Cooper and Jonathan who have in their own ways, motivated, encouraged, and supported this effort.

Preface

Historically, the excitement generated by a new technology for spine or an innovation in spinal interventions has been followed in relatively short order by sobering patient complications or broad-spectrum failures. Such catastrophes often necessitate salvage procedures and ultimately a dramatic decline in interest that ends in a failed pile of debris with a ruinous perception for the categorical technology that may last a generation or more. The disastrous scenarios have been repeated to the point of becoming frequent in spine technologies with very little evidence of slowing. Accordingly, the spine landscape is littered with the burned-out wreckages of abandoned technologies.

This book documents the fundamentals of spinal treatments, original design intent for spinal devices and the clinical outcomes of spine technologies. Often there is little more than tribal knowledge and even less information documenting the history of surgical approaches and supporting hardware. This is evident by the repeated failure modes for similar devices in databases and registries. The goal of this handbook is to provide a repository of information for both successful spine technologies as well as those with poor clinical outcomes and, moreover, the root cause of the failed spinal implants that contributed to the unsatisfactory results. If nothing else, this will serve the healthcare community by memorializing the history of spine technologies and prevent a repeat of the technological cycle that contributed to problematic patient outcomes. The specific aim of this book is to record from both a clinical and a scientific point of view what we have learned about the human spine and the influence of spine technologies that have contributed to the attempted treatment.

Important devices include those that failed catastrophically, for example, nucleus augmentation devices, to those that were ahead of the times but not necessarily a commercial success including motion preservation devices. Beyond the necessary surgical skills, patient selection has often been cited as essential to the commercialization of a device. Patient reported outcomes are a good proxy to the success of the device as well as a relative metric for regulatory purposes. The systematic diagnosis for patients presenting with back pain requires the most appropriate technology for the patient's symptoms. In one patient, immediate fixation and stabilization is the best solution and, in combination with the appropriate adjuncts to fusion, affords the patient the best opportunity for success. In other segments of the population with different

sets of symptoms, the solution may be a motion-preserving technology. The guidance and design rationale will help the audience understand the premise of each technology and, ultimately, how best the technology may be applied and in which patient population.

It is without fail that the lessons of the past can help mitigate potential disasters and even prevent another timely consuming and financially draining iteration. From engineers to clinical scientists, publications frequently discuss the most successful technologies or the most popular techniques. However, the most valuable lessons may be in failure. Failure may be attributable to the design of the device. It may also be a materials limitation. One failure often not discussed is surgeon error. Regardless, understanding the origin of the failure or root cause analysis is essential to future refinements.

Technologies that cover new materials, design failures, and even technologies developed for the sole purpose of finding an indication are presented. Often, there is no prior art, no reported case studies nor hints of potential complications attributable to a technology. Discovery of this information requires the courage to reflect on such mistakes and the willingness to share them. Documentation can lead to helpful preventative warnings resulting in the potential for reduced iterations and, ideally, failed products resulting in voluntary, or worse, mandated product market withdrawals. It is the goal of this handbook to put on display such failures so that we may advance technology for the patients benefit.

Boyle C. Cheng, Ph.D.

Acknowledgments

I want to acknowledge my professors and students that have planted the seeds throughout my career and, in particular, recognize those that have nurtured the growth through the fertile bed of their own experience and wisdom. This would include my colleagues, mentors, co-authors, and Michele Birgelen who worked tirelessly alongside me in completing the handbook and who epitomizes the definition of dedication.

Contents

Volume 1

Part I   Low Back Pain Is a Point of View   1

1   Low Back Patterns of Pain: Classification Based on Clinical Presentation   3
Hamilton Hall

2   Back Pain: The Classic Surgeon's View   27
Neil Berrington

3   Back Pain: Chiropractor's View   37
I. D. Coulter, M. J. Schneider, J. Egan, D. R. Murphy, Silvano A. Mior, and G. Jacob

4   Chiropractors See It Differently: A Surgeon's Observations   67
John Street

5   Medical Causes of Back Pain: The Rheumatologist's Perspective   93
Stephanie Gottheil, Kimberly Lam, David Salonen, and Lori Albert

6   Psychosocial Impact of Chronic Back Pain: Patient and Societal Perspectives   109
Y. Raja Rampersaud

Part II   Biomaterials and Biomechanics   125

7   Implant Material Bio-compatibility, Sensitivity, and Allergic Reactions   127
Nadim James Hallab, Lauryn Samelko, and Marco Caicedo

8   Mechanical Implant Material Selection, Durability, Strength, and Stiffness   151
Robert Sommerich, Melissa (Kuhn) DeCelle, and William J. Frasier

9   Material Selection Impact on Intraoperative Spine Manipulation and Post-op Correction Maintenance 163
Hesham Mostafa Zakaria and Frank La Marca

10   Biological Treatment Approaches for Degenerative Disc Disease: Injectable Biomaterials and Bioartificial Disc Replacement 171
Christoph Wipplinger, Yu Moriguchi, Rodrigo Navarro-Ramirez, Eliana Kim, Farah Maryam, and Roger Härtl

11   Bone Grafts and Bone Graft Substitutes 197
Jae Hyuk Yang, Juliane D. Glaeser, Linda E. A. Kanim, Carmen Y. Battles, Shrikar Bondre, and Hyun W. Bae

12   Mechanobiology of the Intervertebral Disc and Treatments Working in Conjunction with the Human Anatomy 275
Stephen Jaffee, Isaac R. Swink, Brett Phillips, Michele Birgelen, Alexander K. Yu, Nick Giannoukakis, Boyle C. Cheng, Scott Webb, Reginald Davis, William C. Welch, and Antonio Castellvi

13   Design Rationale for Posterior Dynamic Stabilization Relevant for Spine Surgery 293
Ashutosh Khandha, Jasmine Serhan, and Vijay K. Goel

14   Lessons Learned from Positive Biomechanics and Poor Clinical Outcomes 315
Deniz U. Erbulut, Koji Matsumoto, Anoli Shah, Anand Agarwal, Boyle C. Cheng, Ali Kiapour, Joseph Zavatsky, and Vijay K. Goel

15   Lessons Learned from Positive Biomechanics and Positive Clinical Outcomes 331
Isaac R. Swink, Stephen Jaffee, Jake Carbone, Hannah Rusinko, Daniel Diehl, Parul Chauhan, Kaitlyn DeMeo, and Thomas Muzzonigro

16   The Sacroiliac Joint: A Review of Anatomy, Biomechanics, Diagnosis, and Treatment Including Clinical and Biomechanical Studies (In Vitro and In Silico) 349
Amin Joukar, Hossein Elgafy, Anand K. Agarwal, Bradley Duhon, and Vijay K. Goel

Part III   Considerations and Guidelines for New Technologies 375

17   Cyclical Loading to Evaluate the Bone Implant Interface 377
Isaac R. Swink, Stephen Jaffee, Daniel Diehl, Chen Xu, Jake Carbone, Alexander K. Yu, and Boyle C. Cheng

| 18 | **FDA Premarket Review of Orthopedic Spinal Devices** | 401 |

Katherine Kavlock, Srinidhi Nagaraja, and Jonathan Peck

19 Recent Advances in PolyArylEtherKetones and Their In Vitro Evaluation for Hard Tissue Applications 423
Boyle C. Cheng, Alexander K. Yu, Isaac R. Swink, Donald M. Whiting, and Saadyah Averick

20 Selection of Implant Material Effect on MRI Interpretation in Patients 439
Ashok Biyani, Deniz U. Erbulut, Vijay K. Goel, Jasmine Tannoury, John Pracyk, and Hassan Serhan

21 Metal Ion Sensitivity 459
William M. Mihalko and Catherine R. Olinger

22 Spinal Cord Stimulation: Effect on Motor Function in Parkinson's Disease 473
Nestor D. Tomycz, Timothy Leichliter, Saadyah Averick, Boyle C. Cheng, and Donald M. Whiting

23 Intraoperative Monitoring in Spine Surgery 483
Julian Michael Moore

24 Oncological Principles 505
A. Karim Ahmed, Zach Pennington, Camilo A. Molina, and Daniel M. Sciubba

25 Bone Metabolism 523
Paul A. Anderson

Part IV Technology: Fusion 539

26 Pedicle Screw Fixation 541
Nickul S. Jain and Raymond J. Hah

27 Interspinous Devices 561
Douglas G. Orndorff, Anneliese D. Heiner, and Jim A. Youssef

28 Kyphoplasty Techniques 573
Scott A. Vincent, Emmett J. Gannon, and Don K. Moore

29 Anterior Spinal Plates: Cervical 593
A. Karim Ahmed, Zach Pennington, Camilo A. Molina, C. Rory Goodwin, and Daniel M. Sciubba

30 Spinal Plates and the Anterior Lumbar Interbody Arthrodesis 603
Zach Pennington, A. Karim Ahmed, and Daniel M. Sciubba

31 Interbody Cages: Cervical 633
John Richards, Donald R. Fredericks Jr., Sean E. Slaven, and Scott C. Wagner

**32 Anterior Lumbar Interbody Fusion and Transforaminal
Lumbar Interbody Fusion** 645
Tristan B. Fried, Tyler M. Kreitz, and I. David Kaye

33 Scoliosis Instrumentation Systems 657
Rajbir Singh Hundal, Mark Oppenlander, Ilyas Aleem, and
Rakesh Patel

34 SI Joint Fixation 675
J. Loewenstein, W. Northam, D. Bhowmick, and E. Hadar

35 Lateral Lumbar Interbody Fusion 689
Paul Page, Mark Kraemer, and Nathaniel P. Brooks

36 Minimally Invasive Spine Surgery 701
Bilal B. Butt, Rakesh Patel, and Ilyas Aleem

37 Cervical Spine Anatomy 717
Bobby G. Yow, Andres S. Piscoya, and Scott C. Wagner

38 Thoracic and Lumbar Spinal Anatomy 737
Patricia Zadnik Sullivan, Michael Spadola, Ali K. Ozturk, and
William C. Welch

Volume 2

Part V Technology: Motion Preservation 747

**39 Cervical Total Disc Replacement: FDA-Approved
Devices** 749
Catherine Miller, Deepak Bandlish, Puneet Gulati, Santan
Thottempudi, Domagoj Coric, and Praveen Mummaneni

**40 Cervical Total Disc Replacement: Next-Generation
Devices** 761
Tyler M. Kreitz, James McKenzie, Safdar Khan, and Frank M.
Phillips

41 Cervical Total Disc Replacement: Evidence Basis 771
Kris E. Radcliff, Daniel A. Tarazona, Michael Markowitz, and
Edwin Theosmy

42 Cervical Total Disc Replacement: Biomechanics 789
Joseph D. Smucker and Rick C. Sasso

**43 Cervical Total Disc Replacement: Technique – Pitfalls and
Pearls** 807
Miroslav Vukic and Sergej Mihailovic Marasanov

44 Cervical Total Disc Replacement: Expanded Indications ... 823
Pierce D. Nunley

45 Cervical Total Disc Replacement: Heterotopic Ossification and Complications 829
Michael Paci and Michael Y. Wang

46 Lumbar TDR Revision Strategies 837
Paul C. McAfee and Mark Gonz

47 Posterior Lumbar Facet Replacement and Interspinous Spacers 845
Taylor Beatty, Michael Venezia, and Scott Webb

48 Cervical Arthroplasty: Long-Term Outcomes 857
Thomas J. Buell and Mark E. Shaffrey

49 Adjacent-Level Disease: Fact and Fiction 885
Jonathan Parish and Domagoj Coric

50 Posterior Dynamic Stabilization 893
Dorian Kusyk, Chen Xu, and Donald M. Whiting

51 Total Disc Arthroplasty 899
Benjamin Ebben and Miranda Bice

Part VI International Experience: Surgery 923

52 The Diagnostic and the Therapeutic Utility of Radiology in Spinal Care 925
Matthew Lee and Mario G. T. Zotti

53 Surgical Site Infections in Spine Surgery: Prevention, Diagnosis, and Treatment Using a Multidisciplinary Approach 949
Matthew N. Scott-Young, Mario G. T. Zotti, and Robert G. Fassett

54 Lumbar Interbody Fusion Devices and Approaches: When to Use What 961
Laurence P. McEntee and Mario G. T. Zotti

55 Stand-Alone Interbody Devices: Static Versus Dynamic 997
Ata G. Kasis

56 Allograft Use in Modern Spinal Surgery 1009
Matthew N. Scott-Young and Mario G. T. Zotti

57 Posterior Approaches to the Thoracolumbar Spine: Open Versus MISS 1029
Yingda Li and Andrew Kam

58 Lateral Approach to the Thoracolumbar Junction: Open and MIS Techniques 1051
Mario G. T. Zotti, Laurence P. McEntee, John Ferguson, and Matthew N. Scott-Young

59 Surgical Approaches to the Cervical Spine: Principles and Practicalities 1067
Cyrus D. Jensen

60 Intradiscal Therapeutics for Degenerative Disc Disease 1091
Justin Mowbray, Bojiang Shen, and Ashish D. Diwan

61 Replacing the Nucleus Pulposus for Degenerative Disc Disease and Disc Herniation: Disc Preservation Following Discectomy 1111
Uphar Chamoli, Maurice Lam, and Ashish D. Diwan

62 Spinal Fusion Evaluation in Various Settings: A Summary of Human-Only Studies 1131
Jose Umali, Ali Ghahreman, and Ashish D. Diwan

63 Effects of Reimbursement and Regulation on the Delivery of Spinal Device Innovation and Technology: An Industry Perspective 1149
Emma Young

64 Anterior Lumbar Spinal Reconstruction 1165
Matthew N. Scott-Young, David M. Grosser, and Mario G. T. Zotti

Part VII Challenges and Lessons from Commercializing Products 1209

65 Approved Products in the USA: AxiaLIF 1211
Franziska Anna Schmidt, Raj Nangunoori, Taylor Wong, Sertac Kirnaz, and Roger Härtl

66 Spine Products in Use Both Outside and Inside the United States 1217
Tejas Karnati, Kee D. Kim, and Julius O. Ebinu

67 Trauma Products: Spinal Cord Injury Implants 1229
Gilbert Cadena Jr., Jordan Xu, and Angie Zhang

68 Biologics: Inherent Challenges 1251
Charles C. Lee and Kee D. Kim

69 Robotic Technology 1269
Kyle J. Holmberg, Daniel T. Altman, Boyle C. Cheng, and Timothy J. Sauber

Index 1283

About the Section Editors

Part I: Low Back Pain Is a Point of View
Hamilton Hall Department of Surgery, University of Toronto, Toronto, ON, Canada

Part II: Biomaterials and Biomechanics
Hassan Serhan I.M.S. Society, Easten, MA, USA

Tony Tannoury Department of Orthopedics, Boston University Medical Center, Boston, MA, USA

Part III: Considerations and Guidelines for New Technologies
Boyle C. Cheng Neuroscience Institute, Allegheny Health Network, Drexel University, Allegheny General Hospital Campus, Pittsburgh, PA, USA

Vijay K. Goel University of Toledo, Engineering Center for Orthopaedic Research Excellence (E-CORE), Toledo, OH, USA

Departments of Bioengineering and Orthopaedic Surgery, Colleges of Engineering and Medicine, University of Toledo, Toledo, OH, USA

Part IV: Technology: Fusion
Don K. Moore Department of Orthopaedic Surgery, University of Missouri Health Care, Columbia, OH, USA

William C. Welch Department of Neurosurgery, University of Pennsylvania, Philadelphia, PA, USA

Part V: Technology: Motion Preservation
Domagoj Coric Department of Neurological Surgery, Carolinas Medical Center and Carolina Neurosurgery and Spine Associates, Charlotte, NC, USA

Part VI: International Experience: Surgery
Matthew N. Scott-Young Gold Coast Spine, Southport, QLD, Australia

Faculty of Health Sciences and Medicine, Bond University, Varsity Lakes, QLD, Australia

Part VII: Challenges and Lessons from Commercializing Products
R. Douglas Orr S40 Cleveland Clinic, Center for Spine Health, Cleveland, OH, USA

Michael Y. Oh Department of Neurosurgery + Academic Services, Allegheny Health Network, Pittsburgh, PA, USA

Contributors

Anand K. Agarwal Engineering Center for Orthopaedic Research Excellence (E-CORE), University of Toledo, Toledo, OH, USA

Anand Agarwal Department of Orthopaedic Surgery and Bioengineering, School of Engineering and Medicine, University of Toledo, Toledo, OH, USA

A. Karim Ahmed Department of Neurosurgery, The Johns Hopkins School of Medicine, Baltimore, MD, USA

Lori Albert Rheumatology Faculty, University of Toronto, Toronto, ON, Canada

Ilyas Aleem Department of Orthopaedic Surgery, University of Michigan, Ann Arbor, MI, USA

Daniel T. Altman Department of Orthopaedic Surgery, Allegheny Health Network, Pittsburgh, PA, USA

Paul A. Anderson Department of Orthopedic Surgery and Rehabilitation, University of Wisconsin, Madison, WI, USA

Saadyah Averick Department of Neurosurgery, Neuroscience Institute, Allegheny Health Network, Pittsburgh, PA, USA

Hyun W. Bae Surgery, Department of Orthopaedics, Cedars-Sinai Medical Center, Los Angeles, CA, USA

Board of Governors Regenerative Medicine Institute, Cedars-Sinai Medical Center, Los Angeles, CA, USA

Department of Surgery, Cedars-Sinai Spine Center, Los Angeles, CA, USA

Deepak Bandlish Department of Neurological Surgery, SBKS Medical College, Vadodara, India

Carmen Y. Battles Surgery, Department of Orthopaedics, Cedars-Sinai Medical Center, Los Angeles, CA, USA

Department of Surgery, Cedars-Sinai Spine Center, Los Angeles, CA, USA

Taylor Beatty Orthopaedic Surgery Resident PGY5, Largo Medical Center, Largo, FL, USA

Neil Berrington Section of Neurosurgery, University of Manitoba, Winnipeg, MB, Canada

D. Bhowmick Department of Neurosurgery, University of North Carolina, Chapel Hill, NC, USA

Miranda Bice University of Wisconsin School of Medicine and Public Health, Madison, WI, USA

Michele Birgelen Department of Neurosurgery, Neuroscience Institute, Allegheny Health Network, Pittsburgh, PA, USA

Ashok Biyani ProMedica Physicians Biyani Orthopaedics, Toledo, OH, USA

Shrikar Bondre Chemical Engineering, Prosidyan, Warren, NJ, USA

Nathaniel P. Brooks Department of Neurological Surgery, University of Wisconsin, Madison, WI, USA

Thomas J. Buell Department of Neurological Surgery, University of Virginia Health System, Charlottesville, VA, USA

Bilal B. Butt Department of Orthopaedic Surgery, University of Michigan, Ann Arbor, MI, USA

Gilbert Cadena Jr. Department of Neurological Surgery, University of California Irvine, Orange, CA, USA

Marco Caicedo Department of Orthopedic Surgery, Rush University Medical Center, Chicago, IL, USA

Jake Carbone Louis Katz School of Medicine, Temple University, Philadelphia, PA, USA

Department of Neurosurgery, Allegheny Health Network, Pittsburgh, PA, USA

Antonio Castellvi Orthopaedic Research and Education, Florida Orthopaedic Institute, Tampa, FL, USA

Uphar Chamoli Spine Service, Department of Orthopaedic Surgery, St. George & Sutherland Clinical School, University of New South Wales, Kogarah, NSW, Australia

School of Biomedical Engineering, Faculty of Engineering and Information Technology, University of Technology Sydney, Sydney, NSW, Australia

Parul Chauhan Department of Neurosurgery, Neuroscience Institute, Allegheny Health Network, Pittsburgh, PA, USA

Boyle C. Cheng Neuroscience Institute, Allegheny Health Network, Drexel University, Allegheny General Hospital Campus, Pittsburgh, PA, USA

Domagoj Coric Department of Neurological Surgery, Carolinas Medical Center and Carolina Neurosurgery and Spine Associates, Charlotte, NC, USA

I. D. Coulter RAND Corporation, Santa Monica, CA, USA

Reginald Davis BioSpine, Tampa, FL, USA

Melissa (Kuhn) DeCelle Research and Development, DePuy Synthes Spine, Raynham, MA, USA

Kaitlyn DeMeo Department of Neurosurgery, Allegheny Health Network, Pittsburgh, PA, USA

Daniel Diehl Department of Neurosurgery, Neuroscience Institute, Allegheny Health Network, Pittsburgh, PA, USA

Ashish D. Diwan Spine Service, Department of Orthopaedic Surgery, St. George & Sutherland Clinical School, University of New South Wales, Kogarah, NSW, Australia

Bradley Duhon School of Medicine, University of Colorado, Denver, CO, USA

Benjamin Ebben University of Wisconsin, Madison, WI, USA

Julius O. Ebinu Department of Neurological Surgery, University of California, Davis, Sacramento, CA, USA

J. Egan Southern California University of Health Sciences, Whittier, CA, USA

Hossein Elgafy Engineering Center for Orthopaedic Research Excellence (E-CORE), University of Toledo, Toledo, OH, USA

Deniz U. Erbulut Departments of Bioengineering and Orthopaedic Surgery, Colleges of Engineering and Medicine, University of Toledo, Toledo, OH, USA

Robert G. Fassett Faculty of Health Sciences and Medicine, Bond University, Gold Coast, QLD, Australia

Schools of Medicine and Human Movement and Nutrition Sciences, The University of Queensland, St Lucia, QLD, Australia

John Ferguson Ascot Hospital, Remuera, Auckland, New Zealand

William J. Frasier Research and Development, DePuy Synthes Spine, Raynham, MA, USA

Donald R. Fredericks Jr. Department of Orthopaedics, Walter Reed National Military Medical Center, Bethesda, MD, USA

Tristan B. Fried Sidney Kimmel Medical College, Thomas Jefferson University, Philadelphia, PA, USA

Emmett J. Gannon Department of Orthopaedic Surgery and Rehabilitation, University of Nebraska Medical Center, Omaha, NE, USA

Ali Ghahreman Department of Neurosurgery, St. George Hospital and Clinical School, Kogarah, NSW, Australia

Nick Giannoukakis Institute of Cellular Therapeutics, Allegheny Health Network, Pittsburgh, PA, USA

Juliane D. Glaeser Surgery, Department of Orthopaedics, Cedars-Sinai Medical Center, Los Angeles, CA, USA

Board of Governors Regenerative Medicine Institute, Cedars-Sinai Medical Center, Los Angeles, CA, USA

Department of Surgery, Cedars-Sinai Spine Center, Los Angeles, CA, USA

Vijay K. Goel Engineering Center for Orthopaedic Research Excellence (E-CORE), University of Toledo, Toledo, OH, USA

Departments of Bioengineering and Orthopaedic Surgery, Colleges of Engineering and Medicine, University of Toledo, Toledo, OH, USA

Mark Gonz Vascular Surgery Associates, Towson, MD, USA

C. Rory Goodwin Department of Neurosurgery, Duke University Medical Center, Durham, NC, USA

Stephanie Gottheil University of Toronto, Toronto, ON, Canada

David M. Grosser Southern Queensland Cardiovascular Centre, Southport, QLD, Australia

Puneet Gulati Department of Neurological Surgery, Maulana Azad Medical College and Lok Nayak Hospital, New Delhi, India

E. Hadar Department of Neurosurgery, University of North Carolina, Chapel Hill, NC, USA

Raymond J. Hah Department of Orthopaedic Surgery, University of Southern California, Los Angeles, CA, USA

Hamilton Hall Department of Surgery, University of Toronto, Toronto, ON, Canada

Nadim James Hallab Department of Orthopedic Surgery, Rush University Medical Center, Chicago, IL, USA

Roger Härtl Department of Neurological Surgery, Weill Cornell Brain and Spine Center, New York–Presbyterian Hospital, Weill Cornell Medicine, New York, NY, USA

Anneliese D. Heiner Penumbra, Inc., Alameda, CA, USA

Kyle J. Holmberg Department of Orthopaedic Surgery, Allegheny Health Network, Pittsburgh, PA, USA

G. Jacob Southern California University of Health Sciences, Whittier, CA, USA

Stephen Jaffee College of Medicine, Drexel University, Philadelphia, PA, USA

Allegheny Health Network, Department of Neurosurgery, Allegheny General Hospital, Pittsburgh, PA, USA

Nickul S. Jain Department of Orthopaedic Surgery, University of Southern California, Los Angeles, CA, USA

Cyrus D. Jensen Department of Trauma and Orthopaedic Spine Surgery, Northumbria Healthcare NHS Foundation Trust, Newcastle upon Tyne, UK

Amin Joukar Engineering Center for Orthopaedic Research Excellence (E-CORE), University of Toledo, Toledo, OH, USA

Andrew Kam Department of Neurosurgery, Westmead Hospital, Sydney, NSW, Australia

Linda E. A. Kanim Surgery, Department of Orthopaedics, Cedars-Sinai Medical Center, Los Angeles, CA, USA

Board of Governors Regenerative Medicine Institute, Cedars-Sinai Medical Center, Los Angeles, CA, USA

Department of Surgery, Cedars-Sinai Spine Center, Los Angeles, CA, USA

Tejas Karnati Department of Neurological Surgery, University of California, Davis, Sacramento, CA, USA

Ata G. Kasis Northumbria NHS Trust, UK and Nuffield Hospital, Newcastle-upon-Tyne, UK

Katherine Kavlock Center for Devices and Radiological Health, Food and Drug Administration, Silver Spring, MD, USA

I. David Kaye Rothman Institute, Thomas Jefferson University Hospital, Philadelphia, PA, USA

Safdar Khan Division of Spine Surgery, Department of Orthopedic Surgery, Ohio State University, Columbus, OH, USA

Ashutosh Khandha Department of Biomedical Engineering, College of Engineering, University of Delaware, Newark, DE, USA

Ali Kiapour Departments of Bioengineering and Orthopaedic Surgery, Colleges of Engineering and Medicine, University of Toledo, Toledo, OH, USA

Eliana Kim Department of Neurological Surgery, Weill Cornell Brain and Spine Center, New York–Presbyterian Hospital, New York, NY, USA

Kee D. Kim Department of Neurological Surgery, UC Davis School of Medicine, Sacramento, CA, USA

Sertac Kirnaz Department of Neurological Surgery, Weill Cornell Brain and Spine Center, Weill Cornell Medicine, New York, NY, USA

Mark Kraemer Department of Neurological Surgery, University of Wisconsin, Madison, WI, USA

Tyler M. Kreitz Thomas Jefferson University Hospital, Philadelphia, PA, USA

Dorian Kusyk Department of Neurosurgery, Neuroscience Institute, Allegheny Health Network, Pittsburgh, PA, USA

Frank La Marca Department of Neurosurgery, Henry Ford Hospital, Detroit, MI, USA

Department of Neurosurgery, Henry Ford Allegiance Hospital, Jackson, MI, USA

Kimberly Lam University of Toronto, Toronto, ON, Canada

Maurice Lam Spine Service, Department of Orthopaedic Surgery, St. George & Sutherland Clinical School, University of New South Wales, Kogarah, NSW, Australia

Charles C. Lee Department of Cell Biology and Human Anatomy, School of Medicine, University of California, Davis, Davis, CA, USA

Matthew Lee Western Imaging Group, Blacktown, NSW, Australia

Timothy Leichliter Department of Neurology, Neuroscience Institute, Allegheny Health Network, Pittsburgh, PA, USA

Yingda Li Department of Neurosurgery, Westmead Hospital, Sydney, NSW, Australia

Department of Neurological Surgery, University of Miami, Miami, FL, USA

J. Loewenstein Department of Neurosurgery, University of North Carolina, Chapel Hill, NC, USA

Sergej Mihailovic Marasanov Department of Neurosurgery, University Hospital Rebro, Zagreb, Croatia

Michael Markowitz Department of Orthopaedics, Rowan University School of Osteopathic Medicine, Stratford, NJ, USA

Farah Maryam Department of Neurological Surgery, Weill Cornell Brain and Spine Center, New York–Presbyterian Hospital, New York, NY, USA

Koji Matsumoto Department of Orthopedic Surgery, Nihon University School of Medicine, Tokyo, Japan

Paul C. McAfee Spine and Scoliosis Center, University of Maryland St Joseph Medical Center (UMSJMC), Towson, MD, USA

Laurence P. McEntee Gold Coast Spine, Southport, QLD, Australia

Bond University, Varsity Lakes, QLD, Australia

James McKenzie Department of Orthopaedic Surgery, Thomas Jefferson University, Philadelphia, PA, USA

William M. Mihalko Campbell Clinic Department of Orthopaedic Surgery and Biomedical Engineering, University of Tennessee Health Science Center, Memphis, TN, USA

Catherine Miller Department of Neurological Surgery, University of California-San Francisco, San Francisco, CA, USA

Silvano A. Mior Department of Research, Canadian Memorial Chiropractic College, Toronto, ON, Canada

Centre for Disability Prevention and Rehabilitation, Ontario Tech University and Canadian Memorial Chiropractic College, Toronto, ON, USA

Camilo A. Molina Department of Neurosurgery, The Johns Hopkins School of Medicine, Baltimore, MD, USA

Don K. Moore Department of Orthopaedic Surgery, University of Missouri, Columbia, OH, USA

Julian Michael Moore School of Kinesiology, University of Michigan, Ann Arbor, MI, USA

Department of Neurology, University of Michigan, Ann Arbor, MI, USA

Yu Moriguchi Department of Neurological Surgery, Weill Cornell Brain and Spine Center, New York–Presbyterian Hospital, New York, NY, USA

Hesham Mostafa Zakaria Department of Neurosurgery, Henry Ford Hospital, Detroit, MI, USA

Department of Neurosurgery, Henry Ford Allegiance Hospital, Jackson, MI, USA

Justin Mowbray Spine Service, Department of Orthopaedic Surgery, St George and Sutherland Clinical School, The University of New South Wales, Kogarah, NSW, Australia

Praveen Mummaneni Department of Neurological Surgery, University of California-San Francisco, San Francisco, CA, USA

D. R. Murphy Department of Family Medicine, Alpert Medical School of Brown University, Providence, RI, USA

Department of Physical Therapy, University of Pittsburgh, Cranston, RI, USA

Thomas Muzzonigro Department of Neurosurgery, Neuroscience Institute, Allegheny Health Network, Pittsburgh, PA, USA

Srinidhi Nagaraja Center for Devices and Radiological Health, Food and Drug Administration, Silver Spring, MD, USA

Raj Nangunoori Department of Neurological Surgery, Weill Cornell Brain and Spine Center, Weill Cornell Medicine, New York, NY, USA

Rodrigo Navarro-Ramirez Department of Neurological Surgery, Weill Cornell Brain and Spine Center, New York–Presbyterian Hospital, New York, NY, USA

W. Northam Department of Neurosurgery, University of North Carolina, Chapel Hill, NC, USA

Pierce D. Nunley Spine Institute of Louisiana, Shreveport, LA, USA

Catherine R. Olinger Campbell Clinic Department of Orthopaedic Surgery and Biomedical Engineering, University of Tennessee Health Science Center, Memphis, TN, USA

Mark Oppenlander Department of Orthopaedic Surgery, University of Michigan, Ann Arbor, MI, USA

Douglas G. Orndorff Spine Colorado, Durango, CO, USA

Ali K. Ozturk Department of Neurosurgery, University of Pennsylvania, Philadelphia, PA, USA

Michael Paci Department of Neurological Surgery, University of Miami Miller School of Medicine, Miami, FL, USA

Paul Page Department of Neurological Surgery, University of Wisconsin, Madison, WI, USA

Jonathan Parish Department of Neurological Surgery, Carolinas Medical Center, Charlotte, NC, USA

Carolina Neurosurgery and Spine Associates, Charlotte, NC, USA

Rakesh Patel Department of Orthopaedic Surgery, University of Michigan, Ann Arbor, MI, USA

Jonathan Peck Center for Devices and Radiological Health, Food and Drug Administration, Silver Spring, MD, USA

Zach Pennington Department of Neurosurgery, Johns Hopkins Hospital, The Johns Hopkins School of Medicine, Baltimore, MD, USA

Brett Phillips Institute of Cellular Therapeutics, Allegheny Health Network, Pittsburgh, PA, USA

Frank M. Phillips Division of Spine Surgery, Rush University Medical Center, Chicago, IL, USA

Andres S. Piscoya Department of Orthopaedic Surgery, Walter Reed National Military Medical Center, Bethesda, MD, USA

John Pracyk DePuy Synthes Spine, Raynham, MA, USA

Kris E. Radcliff Department of Orthopaedic Surgery, The Rothman Institute, Thomas Jefferson University, Philadelphia, PA, USA

Y. Raja Rampersaud Arthritis Program, Toronto Western Hospital, University Health Network (UHN), Toronto, ON, Canada

Department of Surgery, Division of Orthopaedic Surgery, University of Toronto, Toronto, ON, Canada

John Richards Department of Orthopaedics, Walter Reed National Military Medical Center, Bethesda, MD, USA

Hannah Rusinko Neuroscience Institute, Allegheny Health Network, Pittsburgh, PA, USA

David Salonen University of Toronto, Toronto, ON, Canada

Lauryn Samelko Department of Orthopedic Surgery, Rush University Medical Center, Chicago, IL, USA

Rick C. Sasso Indiana Spine Group, Carmel, IN, USA

Timothy J. Sauber Department of Orthopaedic Surgery, Allegheny Health Network, Pittsburgh, PA, USA

Franziska Anna Schmidt Department of Neurological Surgery, Weill Cornell Brain and Spine Center, Weill Cornell Medicine, New York, NY, USA

M. J. Schneider School of Health and Rehabilitation Sciences, Clinical and Translational Science Institute, University of Pittsburgh, Pittsburgh, PA, USA

Daniel M. Sciubba Department of Neurosurgery, Johns Hopkins Hospital, The Johns Hopkins School of Medicine, Baltimore, MD, USA

Matthew N. Scott-Young Faculty of Health Sciences and Medicine, Bond University, Gold Coast, QLD, Australia

Gold Coast Spine, Southport, QLD, Australia

Jasmine Serhan Department of Biological Sciences, Bridgewater State University, Bridgewater, MA, USA

Hassan Serhan I.M.S. Society, Easten, MA, USA

Mark E. Shaffrey Department of Neurological Surgery, University of Virginia Health System, Charlottesville, VA, USA

Anoli Shah Engineering Center for Orthopaedic Research Excellence (E-CORE), Departments of Bioengineering and Orthopaedic Surgery, Colleges of Engineering and Medicine, University of Toledo, Toledo, OH, USA

Bojiang Shen Spine Service, Department of Orthopaedic Surgery, St George and Sutherland Clinical School, The University of New South Wales, Kogarah, NSW, Australia

Rajbir Singh Hundal Department of Orthopaedic Surgery, University of Michigan, Ann Arbor, MI, USA

Sean E. Slaven Department of Orthopaedics, Walter Reed National Military Medical Center, Bethesda, MD, USA

Joseph D. Smucker Indiana Spine Group, Carmel, IN, USA

Robert Sommerich Research and Development, DePuy Synthes Spine, Raynham, MA, USA

Michael Spadola Department of Neurosurgery, University of Pennsylvania, Philadelphia, PA, USA

John Street Division of Spine, Department of Orthopedics, University of British Columbia, Vancouver, BC, Canada

Patricia Zadnik Sullivan Department of Neurosurgery, University of Pennsylvania, Philadelphia, PA, USA

Isaac R. Swink Department of Neurosurgery, Neuroscience Institute, Allegheny Health Network, Pittsburgh, PA, USA

Jasmine Tannoury Boston University, Boston, MA, USA

Daniel A. Tarazona Department of Orthopaedic Surgery, The Rothman Institute, Thomas Jefferson University, Philadelphia, PA, USA

Edwin Theosmy Department of Orthopaedics, Rowan University School of Osteopathic Medicine, Stratford, NJ, USA

Santan Thottempudi University of California – Santa Cruz, Santa Cruz, CA, USA

Nestor D. Tomycz Department of Neurosurgery, Neuroscience Institute, Allegheny Health Network, Pittsburgh, PA, USA

Jose Umali Spine Service, Department of Orthopaedic Surgery, St. George Hospital and Clinical School, Kogarah, NSW, Australia

Michael Venezia Orthopaedic Specialists of Tampa Bay, Clearwater, FL, USA

Scott A. Vincent Department of Orthopaedic Surgery and Rehabilitation, University of Nebraska Medical Center, Omaha, NE, USA

Miroslav Vukic Department of Neurosurgery, University Hospital Rebro, Zagreb, Croatia

Scott C. Wagner Department of Orthopaedic Surgery, Walter Reed National Military Medical Center, Bethesda, MD, USA

Michael Y. Wang Department of Neurological Surgery, University of Miami Miller School of Medicine, Miami, FL, USA

Scott Webb Florida Spine Institute, Clearwater, Tampa, FL, USA

William C. Welch Department of Neurosurgery, University of Pennsylvania, Philadelphia, PA, USA

Donald M. Whiting Neuroscience Institute, Allegheny Health Network, Pittsburgh, PA, USA

Christoph Wipplinger Department of Neurological Surgery, Weill Cornell Brain and Spine Center, New York–Presbyterian Hospital, New York, NY, USA

Department of Neurosurgery, Medical University of Innsbruck, Innsbruck, Austria

Taylor Wong Department of Neurological Surgery, Weill Cornell Brain and Spine Center, Weill Cornell Medicine, New York, NY, USA

Chen Xu Department of Neurosurgery, Neuroscience Institute, Allegheny Health Network, Pittsburgh, PA, USA

Jordan Xu Department of Neurological Surgery, University of California Irvine, Orange, CA, USA

Jae Hyuk Yang Surgery, Department of Orthopaedics, Cedars-Sinai Medical Center, Los Angeles, CA, USA

Korea University Guro Hospital, Seoul, South Korea

Emma Young Prism Surgical Designs Pty Ltd, Brisbane, QLD, Australia

Jim A. Youssef Spine Colorado, Durango, CO, USA

Bobby G. Yow Department of Orthopaedic Surgery, Walter Reed National Military Medical Center, Bethesda, MD, USA

Alexander K. Yu Department of Neurosurgery, Neuroscience Institute, Allegheny Health Network, Pittsburgh, PA, USA

Joseph Zavatsky Spine & Scoliosis Specialists, Tampa, FL, USA

Angie Zhang Department of Neurological Surgery, University of California Irvine, Orange, CA, USA

Mario G. T. Zotti Orthopaedic Clinics Gold Coast, Robina, QLD, Australia

Gold Coast Spine, Southport, QLD, Australia

Part V

Technology: Motion Preservation

Secure-C Cervical Disc	757
M6-C Artificial Cervical Disc	758
Conclusion	758
References	758

Abstract

Cervical total disc replacement is a routinely used treatment for radiculopathy due to degenerative disease of the cervical spine. The procedure originated to avoid some of the complications seen with the traditional anterior cervical discectomy and fusion. Appropriate patient selection is paramount to obtain acceptable patient outcomes, with particular indications and contraindications for these procedures. As the procedure gained more acceptance, several cervical artificial discs have been developed and, subsequently, approved by the US Food and Drug Administration (FDA). Each of the eight FDA-approved devices is briefly reviewed in this chapter including outcomes from device-specific studies.

Keywords

Artificial disc · Cervical · Disc replacement · FDA-approved · Outcomes

Introduction

Anterior cervical decompression and fusion (ACDF) is one of the most common surgeries done worldwide to decompress the cervical canal, provide stabilization, and restore the normal lordosis of the cervical spine (Cloward 2007; Smith and Robinson 1958). It has been utilized in the treatment of degenerative disc disease, cervical radiculopathy, myelopathy, instability, and segmental deformity. Fusion rates for ACDFs are reported above 95%, and this fusion has been shown to cause a change in motion characteristics of the adjacent segments (DiAngelo et al. 2003; Eck et al. 2002). The change in the kinematics of adjacent levels may be responsible for increased risk of adjacent segment degeneration (Hilibrand et al. 1999; Dohler et al. 1985).

Cervical total disc replacement was developed to avoid some of these complications seen with ACDF. Preservation of the motion segments after total disc replacement surgery may reduce or delay the progression of adjacent segment disease by maintaining motion as well as normal segmental lordosis and anatomic disc space height (Fuller et al. 1998). Indications and contraindications are reviewed in this chapter as not all patients who are candidates for ACDF are candidates for total disc replacement.

Acceptance of this procedure has led to the development of numerous artificial disc designs. Eight devices have been approved by the US Food and Drug Administration (FDA) after thorough investigation through investigation device exemption (IDE) studies. Each of the eight FDA-approved devices (Prestige ST, Bryan, ProDisc-C, Secure-C, PCM, Mobi-C, Prestige LP, M6-C) and their outcomes are briefly discussed in this chapter. All eight artificial discs are FDA-approved for one-level use from C3-7. Only two artificial discs, Mobi-C and Prestige LP, are FDA-approved for two-level use. Most of the artificial disc designs have either uni- or biarticulating surfaces, although the newest FDA-approved device, M6-C, has a non-articulating, compressible core. Most of these discs are metal-on-polymer (M-o-P), although the Prestige ST and Prestige LP represent metal on metal (M-o-M) designs. Despite the controversy surrounding M-o-M total hip arthroplasty implants, there have been no widespread reports of M-o-M cervical artificial discs causing complications such as elevated serum metal ion levels, osteolysis or pseudotumor formation (Coric et al. 2011).

Rationale for Total Disc Replacement

As mentioned, despite long-term clinical success of ACDF, it has been associated with the development of adjacent segment degeneration. This degeneration can be associated with symptoms such as radiculopathy, myelopathy, or neck pain and may necessitate additional interventions. Due to loss of motion at the fused segment, the kinematics are changed at the levels above and below the fused segment. This has been shown in biomechanical studies to cause increase in intradiscal pressure and motion at the adjacent levels (Eck et al. 2002). It is still unclear whether this degeneration is a result of the natural progression seen with aging or a result of the change in biomechanical stresses seen with ACDF.

In contrast, biomechanical studies have reported that total disc replacement does not disrupt the kinematics at adjacent levels and allow for restoration of more normal load transfer (DiAngelo et al. 2003). Additional studies report that there are reduced stresses at adjacent levels in total disc replacement when compared to levels adjacent to a fusion (Pickett et al. 2005).

Indications/Contraindications

Cervical spondylosis is a common condition and can result in radiculopathy and myelopathy. Patients presenting with these symptoms should undergo appropriate work-up including radiographic evaluation and nonsurgical management. Radiographic evaluation, including MRI and CT imaging, can reveal single versus multilevel disease, presence of facet arthropathy, overall cervical spine alignment, kyphotic deformity, instability, and the location of compressive pathology (anterior, posterior, or both). The results of radiographic evaluation are crucial in determining whether a patient is an appropriate candidate for a total disc replacement.

In the setting of normal cervical alignment and mobility with failure of medical management, appropriate indications for total disc replacement include:

- Radiculopathy due to paracentral or central disc pathology or foraminal stenosis
- Myelopathy due to anterior compression by herniated disc

Contraindications for total disc replacement include:

- Significant multiple level degenerative disc disease (> two levels) with baseline motion abnormalities or advanced degeneration of the facet joints
- Abnormal global spinal alignment
- Cervical Instability (translation >3 mm and/or >11° rotational difference to that or either adjacent level)
- Active or prior discitis
- Osteoporosis (T-score < -1.5)
- Traumatic instability (ligament disruption or facet injury)
- Ossified posterior longitudinal ligament (OPLL) or the presence of bridging osteophytes
- Known allergy to implant materials

FDA-Approved Devices

Starting in 2006, eight cervical artificial disc devices have become available in the United States for one-level use, two of which, Mobi-C and Prestige LP, are approved for two-level use (Table 1). These devices vary in size, shape, materials, and articulating surfaces. They can be categorized based on biomechanical design, biomaterials, and type of fixation (Fig. 1) (Mummaneni and Haid 2004; Mummaneni et al. 2007).

Bryan Cervical Disc

Device Description

The Bryan cervical disc was developed in the early 1990s by neurosurgeon Vincent Bryan. The device is made of two titanium alloy shells with a polyurethane nucleus, which makes it a biarticulating contained bearing design. This is a non-modular disc. Fixation is achieved via milled vertebral end plates, and it allows end plate bony ingrowth through a porous end plate design (Fig. 2).

Table 1 Comparison of the eight FDA-approved artificial disc devices

Name	Design	Modular	Articulating method	Implant composition	Primary fixation	Manufacturer
Bryan	Metal-on-polyurethane, Biarticulating contained bearing	No	Biarticulating	Titanium, polyurethane core	Milled vertebral end plates	Medtronic
PCM	Metal-on-polyethylene Ball-and-socket	No	Uniarticulating	Cobalt-chromium, UHMWPE	Ridged, V-tooth design	NuVasive
ProDisc-C	Metal-on-polyethylene, ball-and-socket	Yes	Uniarticulating	Cobalt-chromium, UHMWPE	Central keel	DePuy Synthes (recently sold to Paradigm Spine)
Prestige ST	Metal-on-metal Ball-and-trough	No	Uniarticulating	Stainless steel	Locked vertebral body screws	Medtronic
Prestige LP	Metal composite, ball-and-trough	No	Uniarticulating	Titanium/ceramic composite	Dual rails	Medtronic
Mobi-C	Metal-on-polyethylene, mobile core	Yes	Biarticulating	Cobalt-chromium, UHMWPE	Lateral self-retaining teeth	LDR
Secure-C	Metal-on-polyethylene, mobile core	No	Biarticulating	Cobalt-chromium, UHMWPE	Ridged central keel	Globus Medical
M6-C	Metal on polyurethane	No	Nonarticulating Compressible	Titanium/Polyurethane UHMWPE	Triple fins	Spinal Kinetics

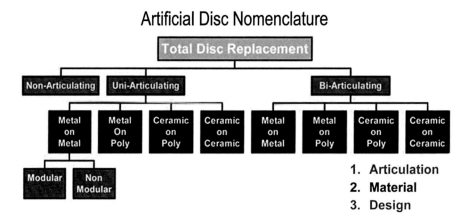

Fig. 1 Nomenclature for artificial disc implants based on design, articulation, and materials. (Permission for the reprint of figure obtained from Journal of Neurosurgery: Spine)

Outcomes

Recently, Goffin et al. reported results in 89 patients treated with the Bryan disc. Ten-year follow-up was available for 72 cases (81%). Maintenance or improvement of the neurological state was seen in 89% of patients. SF-36 patient reported scores improved significantly at all follow-up points. Mean angular motion of the

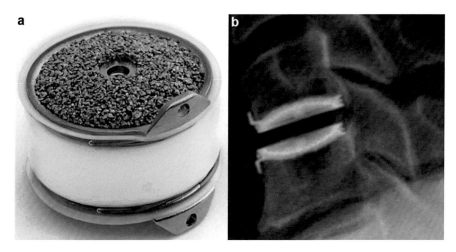

Fig. 2 Bryan artificial disc: (**a**) device; (**b**) lateral X-ray after implantation

prosthesis at 10-year follow-up was 8.6°. Mobility of the device, defined as >2° of angular motion, was reached in 81% of patients. During their study period, 21 patients (24%) developed new or recurrent radiculopathy or myelopathy; the majority of these patients were treated conservatively. Seven patients (8%) required 8 additional spine surgeries to treat persistent or recurrent symptoms. Of these, two patients (2%) were reoperated on at the index level, and five (6%) patients underwent surgery at an adjacent level (Goffin et al. 2003).

Heller et al. presented the results of a randomized controlled multicenter clinical study in 2009 with 242 patients in the investigational group (Bryan arthroplasty) and 221 patients in the control group (single-level ACDF). They showed statistically significant favorable results in the investigational group in various parameters like NDI, neck pain, and return to work and comparable results in other parameters like arm pain and SF 36 physical and mental components. At 24 months, overall success was achieved in 82.6% of the patients in the investigational group and 72.7% in the control group. This difference of 9.9% was statistically significant ($P = 0.010$), and a similar difference was noted at the 12-month follow-up interval ($P = 0.004$) (Heller et al. 2009).

Porous Coated Motion (PCM) Prosthesis

Device Description

The porous coated motion prosthesis is designed to have a metal-on-polyethylene articular surface. This device is a uniarticulating design, which is not modular. It is made up of cobalt-chromium-molybdenum alloy end plates with a TiCaP porous coating for bony ingrowth. Fixation is achieved with a central V-tooth design in a "press fit" fashion (Fig. 3).

Outcomes

In 2015, Philips et al. published the long-term outcomes of the FDA IDE prospective, randomized controlled trial, which compared the PCM prosthesis to anterior cervical discectomy and fusion. The total patient pool of 293 patients (163 PCM, 130 ACDF) was evaluated at 5-year follow-up, and 110 patients had 7-year follow-up. They reported that at 5-year follow-up, all patient-reported outcomes – neck and arm pain visual analogue scale score, neck disability index, and general health (36-Item Short Form Health Survey physical and mental component scores: physical component summary, mental component summary) – were significantly improved from baselines in both groups. Mean scores were

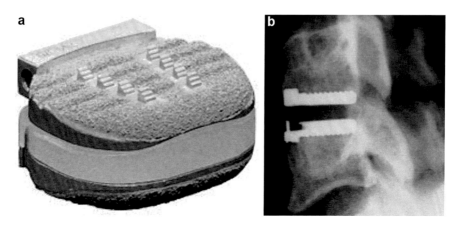

Fig. 3 Porous coated motion prosthesis: (**a**) device; (**b**) lateral X-ray after implantation

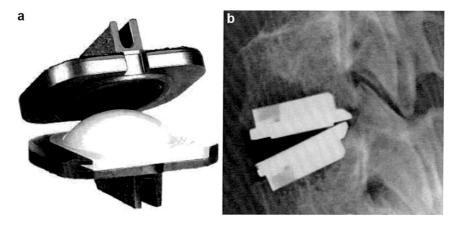

Fig. 4 ProDisc-C cervical disc: (**a**) device; (**b**) lateral X-ray after implantation

significantly better in the PCM group for neck disability index, neck pain, general health, and patient satisfaction. PCM patients trended toward fewer 2- to 7-year device-related serious adverse events and secondary surgical procedures. Adjacent-level degeneration was radiographically more frequent after ACDF and was the primary indication for the increase in late-term secondary surgical procedures after ACDF (Phillips et al. 2015).

ProDisc-C Cervical Disc

Device Description

The ProDisc-C cervical disc is similar in its design to the ProDisc lumbar disc prosthesis. It is a modular ball-and-socket type uniarticulating design. It consists of two cobalt-chromium-molybdenum end plates with an ultrahigh molecular weight polyethylene (UHMWPE) core. Fixation is achieved via a central keel (Fig. 4).

Outcomes

In 2016, Loumeau et al. published data from a randomized controlled trial comparing 7-year clinical outcomes of one-level symptomatic cervical disc disease following ProDisc-C total disc arthroplasty versus ACDF. A total of 22 patients were randomized to each arm of the trial. The authors reported that neck disability index (NDI) scores improved with the ProDisc-C greater than with ACDF. Total range of motion and neck and arm pain improved more in the ProDisc-C group compared to the ACDF group. Patient satisfaction remained higher in the ProDisc-C group at 7 years. Six additional operations (two at the same level; four at an adjacent level) were performed in the ACDF group; however, no reoperations were performed in the ProDisc-C group. They concluded that ProDisc-C implants appear to be safe and effective for the treatment of cervical disc disease and had a lower reoperation rate than those patients treated with an ACDF (Loumeau et al. 2016).

Prestige ST Cervical Disc

Device Description
The Prestige ST was designed by Mr. Brian Cummins and was the first cervical total disc replacement to receive FDA approval in 2006. It is a stainless steel disc, which has a ball-and-trough design with biarticulating surfaces. It is secured to the vertebral body with screws. The superior and inferior surfaces, which contact the end plates, are treated to promote bone integration (Fig. 5).

Outcomes
A FDA IDE randomized controlled study reported by Mummaneni et al. compared cervical disc replacement using the Prestige ST device versus a single-level ACDF. Two-, 5-, and 7-year results have been published. Out of the 541 total patients in the study, 395 patients (212 Prestige ST, 183 ACDF) completed a 7-year follow-up. They found significantly improved NDI scores and neurological improvement scores in the investigational group as compared to the control group. Additionally, rates for subsequent surgical procedures that involved adjacent levels were significantly lower in the Prestige ST group (4.6% vs. 11.9%). They concluded that cervical disc arthroplasty using the Prestige ST cervical disc had the potential for preserving motion at the operated level while providing biomechanical stability and global neck mobility and could result in a reduction in adjacent segment degeneration (Burkus et al. 2014).

Prestige LP Artificial Disc

Device Description
The Prestige LP artificial disc has the same ball-and-trough articulation as the Prestige ST disc. However, the Prestige LP is made from a titanium ceramic composite material. It is anchored to the vertebral bodies via dual rails on the superior and inferior end plates. It also has a porous titanium spray coating to facilitate fixation and bone ingrowth (Fig. 6).

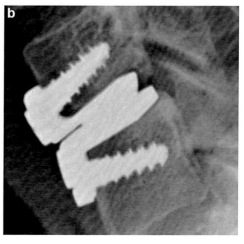

Fig. 5 Prestige ST cervical disc: (**a**) device; (**b**) lateral X-ray after implantation

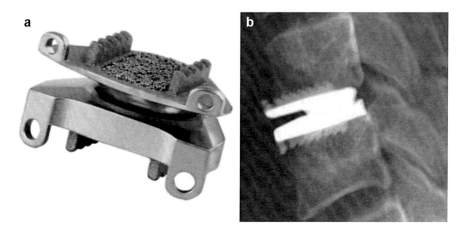

Fig. 6 Prestige LP cervical disc: (**a**) device; (**b**) lateral X-ray after implantation

Outcomes

The results of a randomized control study, investigating the Prestige LP device, were published by Gornet et al. in 2017. They assessed the long-term clinical safety and effectiveness in patients undergoing total disc replacement using the Prestige LP prosthesis to treat degenerative cervical spine disease at 2 adjacent levels compared with ACDF. The study was conducted at 30 centers in the United States with a total of 397 patients (209 Prestige LP, 188 ACDF). At 84 months, the Prestige LP demonstrated statistical superiority over fusion in overall success, NDI improvement, and neurological success. There was no statistically significant difference in the overall rate of implant-related or implant/surgical procedure-related adverse events up to 84 months. The Prestige LP group had fewer serious (Grade 3 or 4) implant- or implant/surgical procedure-related adverse events (3.2% vs. 7.2%,). Patients in the Prestige LP group also underwent statistically significantly fewer second surgical procedures at the index levels (4.2%) than the fusion group (14.7%). Angular range of motion at the superior- and inferior-treated levels on average was maintained in the Prestige LP group up to 84 months (Gornet et al. 2017).

Mobi-C Cervical Disc

Device Description

The Mobi-C cervical disc was first implanted in 2004. This device has a biarticulating design of metal plates articulating on a polyethylene modular core. It has lateral self-retaining teeth on the superior and inferior metal plates, which are pressed into the bone for fixation. The plates are coated with hydroxyapatite to enhance bone integration (Fig. 7).

Outcomes

Hisey et al. published their results in 2016 of a prospective, randomized, controlled study which was conducted as a FDA IDE trial across 23 centers with 245 patients randomized (2:1) to receive total disc replacement with Mobi-C cervical disc or ACDF. The 60-month follow-up rate was 85.5% for the Mobi-C group and 78.9% for the ACDF group. The composite overall success was 61.9% with Mobi-C vs. 52.2% with ACDF, demonstrating statistical non-inferiority. Improvements in NDI, VAS neck and arm pain, and SF-12 scores were similar between groups and were maintained from earlier follow-up through 60 months. There was no significant difference between Mobi-C and ACDF in adverse events or

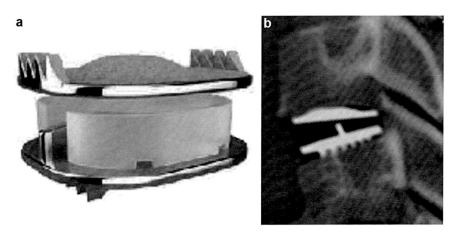

Fig. 7 Mobi-C cervical disc: (**a**) device; (**b**) lateral X-ray after implantation

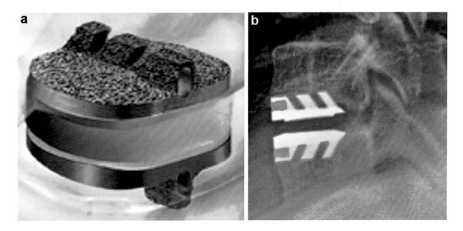

Fig. 8 Secure-C cervical disc: (**a**) device; (**b**) lateral X-ray after implantation

major complications. Range of motion was maintained with Mobi-C through 60 months. Device-related subsequent surgeries (Mobi-C 3.0%, ACDF 11.1%) and adjacent segment degeneration at the superior level (Mobi-C 37.1%, ACDF 54.7%) were significantly lower for Mobi-C cohort. They concluded that total disc replacement with Mobi-C is a viable alternative to single-level ACDF (Hisey et al. 2016).

Secure-C Cervical Disc

Device Description

The Secure-C device is a selectively constrained anterior articulating intervertebral device comprised of two end plates and a central core. The superior and inferior cobalt-chrome alloy end plates have multiple serrated keels for short-term fixation and titanium plasma spray coating on bone contacting surfaces for long-term bony ingrowth. The sliding central core is composed of ultrahigh molecular weight polyethylene, with a spherical superior interface (Fig. 8).

Outcomes

Vaccaro et al. published results of a prospective, multicenter, randomized controlled IDE trial to compare the clinical safety and effectiveness of the Secure-C device versus ACDF. A total of 380 patients from 18 investigational sites were randomized and evaluated. Overall, the study

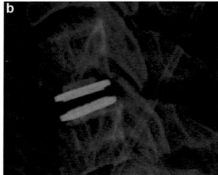

Fig. 9 M6-C artificial disc: (**a**) device; (**b**) lateral X-ray after implantation

demonstrated the statistical superiority of the Secure-C group compared with the ACDF group at 24 months. At 24 months, the Secure-C cohort demonstrated clinically significant improvement in pain and function in terms of NDI scores, VAS scores, and 36-Item Short Form Health Survey. At 24 months, the percentage of patients experiencing secondary surgical interventions at the index level was statistically lower for the Secure-C group (2.5%) than the ACDF group (9.7%). This type of disc has also proven to be a viable alternative to ACDF in appropriately selected patients (Vaccaro et al. 2013).

M6-C Artificial Cervical Disc

Device Description

The M6-C disc is an unconstrained disc with a polyethylene weave (designed to mimic the annulus fibrosus) which houses a compressible viscoelastic polyurethane core (designed to mimic the nucleus pulposus). The end plates are titanium with a plasma spray coating, and fixation is achieved with three rows of "fins" on the upper and lower end plates (Fig. 9).

Outcomes

Lauryssen et al. published results of a prospective, multicenter, non-controlled IDE pilot study to evaluate the clinical safety and effectiveness of the M6-C disc. A total of 30 patients from 3 investigational sites were evaluated and demonstrated significantly improved clinical outcomes (NDI, VAS neck and arm scores) compared to baseline at 2-year follow-up (Lauryssen et al. 2012).

Conclusion

Eight cervical artificial disc devices have been approved by the FDA dating back to 2006. These devices have a sound evidence basis as safe and viable alternatives to ACDF in properly selected patients. Patient selection is key to ensure appropriate patient outcomes as seen in these FDA IDE studies. Further long-term investigations will be necessary to ensure the longevity of these devices.

References

Burkus JK, Traynelis VC, Haid RW Jr et al (2014) Clinical and radiographic analysis of an artificial cervical disc: 7-year follow-up from the Prestige prospective randomized controlled clinical trial: clinical article. J Neurosurg Spine 21(4):516–528. https://doi.org/10.3171/2014.6.SPINE13996

Cloward RB (2007) The anterior approach for removal of ruptured cervical disks. 1958. J Neurosurg Spine 6(5):496–511. https://doi.org/10.3171/spi.2007.6.5.496

Coric D, Nunley P, Guyer RD, Musante D, Carmody C, Gordon C, Lauryssen C, Ohnmeiss D, Boltes MO (2011) Prospective, randomized, multicenter study of cervical arthroplasty: 269 patients from the Kineflex®| C artificial disc IDE study with minimum two year follow-up. J Neurosurg-Spine 15:348–358

DiAngelo DJ, Roberston JT, Metcalf NH et al (2003) Biomechanical testing of an artificial cervical joint and an anterior cervical plate. J Spinal Disord Tech 16(4):314–323

Dohler JR, Kahn MR, Hughes SP (1985) Instability of the cervical spine after anterior interbody fusion. A study on its incidence and clinical significance in 21 patients. Arch Orthop Trauma Surg 104(4):247–250

Eck JC, Humphreys SC, Lim TH et al (2002) Biomechanical study on the effect of cervical spine fusion on adjacent-level intradiscal pressure and segmental motion. Spine (Phila Pa 1976) 27(22):2431–2434. https://doi.org/10.1097/01.BRS.0000031261.66972.B1

Fuller DA, Kirkpatrick JS, Emery SE et al (1998) A kinematic study of the cervical spine before and after segmental arthrodesis. Spine (Phila Pa 1976) 23(15):1649–1656

Goffin J, Van Calenbergh F, van Loon J et al (2003) Intermediate follow-up after treatment of degenerative disc disease with the Bryan Cervical Disc Prosthesis: single-level and bi-level. Spine (Phila Pa 1976) 28(24): 2673–2678. https://doi.org/10.1097/01.BRS.0000099392.90849.AA

Gornet MF, Lanman TH, Burkus JK et al (2017) Cervical disc arthroplasty with the Prestige LP disc versus anterior cervical discectomy and fusion, at 2 levels: results of a prospective, multicenter randomized controlled clinical trial at 24 months. J Neurosurg Spine 26(6): 653–667. https://doi.org/10.3171/2016.10.SPINE16264

Heller JG, Sasso RC, Papadopoulos SM et al (2009) Comparison of BRYAN cervical disc arthroplasty with anterior cervical decompression and fusion: clinical and radiographic results of a randomized, controlled, clinical trial. Spine (Phila Pa 1976) 34(2):101–107. https://doi.org/10.1097/BRS.0b013e31818ee263

Hilibrand AS, Carlson GD, Palumbo MA et al (1999) Radiculopathy and myelopathy at segments adjacent to the site of a previous anterior cervical arthrodesis. J Bone Joint Surg Am 81(4):519–528

Hisey MS, Zigler JE, Jackson R et al (2016) Prospective, randomized comparison of one-level Mobi-C cervical total disc replacement vs. anterior cervical discectomy and fusion: results at 5-year follow-up. Int J Spine Surg 10:10. https://doi.org/10.14444/3010

Loumeau TP, Darden BV, Kesman TJ et al (2016) A RCT comparing 7-year clinical outcomes of one level

symptomatic cervical disc disease (SCDD) following ProDisc-C total disc arthroplasty (TDA) versus anterior cervical discectomy and fusion (ACDF). Eur Spine J 25 (7):2263–2270. https://doi.org/10.1007/s00586-016-4431-6

Lauryssen C, Coric D, Dimmig T, Musante D, Ohnmeiss DD, Stubbs HA (2012) Cervical total disc replacement using a novel compressible prosthesis: results from a prospective Food and Drug Administration-regulated feasibility study with 24-month follow-up. Int J Spine Surg 6:71–77. https://doi.org/10.1016/j.ijsp.2012.02.001

Mummaneni PV, Haid RW (2004) The future in the care of the cervical spine: interbody fusion and arthroplasty. Invited submission from the Joint Section Meeting on Disorders of the Spine and Peripheral Nerves, March 2004. J Neurosurg Spine 1(2):155–159. https://doi.org/10.3171/spi.2004.1.2.0155

Mummaneni PV, Burkus JK, Haid RW et al (2007) Clinical and radiographic analysis of cervical disc arthroplasty compared with allograft fusion: a randomized controlled clinical trial. J Neurosurg Spine 6(3):198–209. https://doi.org/10.3171/spi.2007.6.3.198

Phillips FM, Geisler FH, Gilder KM et al (2015) Long-term outcomes of the US FDA IDE prospective, randomized controlled clinical trial comparing PCM cervical disc arthroplasty with anterior cervical discectomy and fusion. Spine (Phila Pa 1976) 40(10):674–683. https://doi.org/10.1097/BRS.0000000000000869

Pickett GE, Rouleau JP, Duggal N (2005) Kinematic analysis of the cervical spine following implantation of an artificial cervical disc. Spine (Phila Pa 1976) 30(17):1949–1954. https://doi.org/10.1097/01.brs.0000176320.82079.ce

Smith GW, Robinson RA (1958) The treatment of certain cervical-spine disorders by anterior removal of the intervertebral disc and interbody fusion. J Bone Joint Surg Am 40-A(3):607–624

Vaccaro A, Beutler W, Peppelman W et al (2013) Clinical outcomes with selectively constrained SECURE-C cervical disc arthroplasty: two-year results from a prospective, randomized, controlled, multicenter investigational device exemption study. Spine (Phila Pa 1976) 38(26):2227–2239. https://doi.org/10.1097/BRS.0000000000000031

Cervical Total Disc Replacement: Next-Generation Devices

40

Tyler M. Kreitz, James McKenzie, Safdar Khan, and Frank M. Phillips

Contents

Introduction	762
Physiologic Kinematics	762
Design Considerations	763
Elastomeric Implants	764
M6-C Artificial Cervical Disc	764
Freedom Cervical Disc	765
CP-ESP Cervical Disc	765
Cadisc-C	766
Next-Generation Pivot/Ball Type Artificial Discs	766
Synergy Cervical Disc	766
Baguera C	767
Simplify Cervical Disc	767
Conclusions	767
Cross-References	767
References	768

T. M. Kreitz (✉)
Thomas Jefferson University Hospital, Philadelphia, PA, USA
e-mail: tyler.m.kreitz@gmail.com

J. McKenzie
Department of Orthopaedic Surgery, Thomas Jefferson University, Philadelphia, PA, USA
e-mail: mckenzie.jamesc@gmail.com

S. Khan
Division of Spine Surgery, Department of Orthopedic Surgery, Ohio State University, Columbus, OH, USA
e-mail: Safdar.Khan@osumc.edu

F. M. Phillips
Division of Spine Surgery, Rush University Medical Center, Chicago, IL, USA
e-mail: frank.phillips@rushortho.com

© Springer Nature Switzerland AG 2021
B. C. Cheng (ed.), *Handbook of Spine Technology*,
https://doi.org/10.1007/978-3-319-44424-6_72

Abstract

Cervical disc arthroplasty techniques were developed as an alternative to fusion in order to preserve natural motion and reduce the risk of adjacent segment degeneration in the appropriately selected patients with cervical myeloradiculopathy. These arthroplasty implants must provide stability, preserve physiologic motion, and replicate the kinematic signature of the natural disc. There are currently eight cervical arthroplasty implants approved by the Food and Drug Administration (FDA) for use in the United States. The majority of approved implants follow a metal on polyethylene ball-in-socket or saddle-type design. Over the past decade, there has been an explosion of cervical arthroplasty implant designs each with their own advantages and disadvantages. The purpose of this chapter is to review the biomechanics and kinematics of the natural cervical disc. We will also review available in vivo and ex vivo literature on novel elastomeric compression, hydraulic, and next-generation ball-in-socket cervical arthroplasty designs.

Keywords

Cervical spondylosis · Arthroplasty · Radiculopathy · Myelopathy · Novel implants

---

Introduction

Motion-preserving cervical disc arthroplasty implants were developed as an alternative to fusion in order to preserve natural motion and cervical biomechanics and reduce the risk of adjacent segment disease for patients with cervical radiculopathy or myelopathy (Hilibrand et al. 1999). The goals of cervical disc arthroplasty are to restore disc and foraminal height, preserve physiologic motion, and provide long-term stability (Cepoiu-Martin et al. 2011; McAfee 2004; Mummaneni et al. 2007). There are currently eight cervical arthroplasty implants approved by the Food and Drug Administration (FDA) for clinical use. These include the Prestige ST and LP (Medtronic), Bryan (Medtronic), ProDisc-C (Centinel Spine), SECURE-C (Globus Medical), Porous Coated Motion (PCM) (NuVasive), Mobi-C (LDR), and the recently approved M6-C Artificial Cervical Disc (Orthofix). The majority of currently approved designs involve a bi-articulating ball-in-socket type design with a polyethylene core and metal (titanium or cobalt chrome) endplate. Most endplate designs include a keel for initial stability and textured surface to promote long-term bony ingrowth (Staudt et al. 2018).

Physiologic Kinematics

A healthy cervical intervertebral disc is viscoelastic and allows for three-dimensional motion in sagittal, coronal, and axial planes. Physiologic motion of a normal cervical segment allows for 15° of flexion-extension, 4° of lateral bending, and 5° of axial rotation in each direction (Holmes et al. 1994; Iai et al. 1994; Ishii et al. 2006). Additionally, there is a linear coupling of ipsilateral lateral bending and axial rotation resulting from facet and uncinate process orientation in each cervical segment (Bogduk and Mercer 2000; Patwardhan et al. 2012; Senouci et al. 2007). The physiologic sagittal center of rotation (COR) varies by cervical segment. The flexion-extension COR at the C5–C6 segment occurs at the midpoint of the superior endplate of the C6 vertebrae. The COR occurs at a point more caudad and dorsal in upper cervical segments and more cephalad in lower segments (Bogduk and Mercer 2000; Hwang et al. 2008; Patwardhan et al. 2012). An arthroplasty device should replicate both physiologic range of motion (ROM) and maintain a natural COR. An arthroplasty device that alters segments of physiologic COR may result in abnormal translations of the adjacent vertebrae during motion, unnatural forces across the segment including the facet joints and uncinate impingement. These abnormal forces may result in limited motion, pain, or ultimately facet joint or adjacent segment degeneration (Bogduk and Mercer 2000; Patwardhan et al. 2012; Pickett et al. 2006).

The viscoelastic cervical disc demonstrates nonlinear flexion-extension load-displacement curve. This characteristic allows for motion with minimal energy expenditure around the neutral zone, termed high flexibility zone. Increasing stiffness outside this high flexibility zone prevents damaging motion beyond the physiologic range. This graded resistance to angular motion also allows for energy dissipation, thereby reducing forces across index and adjacent segments under physiologic load (Panjabi 1992; Patwardhan et al. 2012). Additionally, the viscoelastic nature of the nucleus pulposus allows the disc to conform under compressive loads and act as a shock absorber, thereby reducing force across adjacent segments and facets (Lazennec et al. 2016). The ideal cervical arthroplasty implant allows for compressibility and graded resistance to motion. Replicating this kinematic signature will reduce shear stresses across the facet joints and adjacent segments and improve implant longevity. First-generation ball-in-socket designs do not allow for compressibility or graded resistance to motion. Elastomeric cervical disc implants were developed as an alternative with these physiologic biomechanical characteristics in mind. Elastomeric compression devices are primarily designed with a polyurethane core that theoretically allows for motion under compression and graded resistance mimicking that of the native disc. To date, there are only a few cervical disc replacement designs that claim to fit this description. These include the M6-C Artificial Cervical Disc (Orthofix), Freedom Cervical Disc (FCD, AxioMed LLC), Cadisc-C (Rainier), and CP-ESP (FH Orthopedics) (Chin et al. 2017; Staudt et al. 2018).

Design Considerations

Multiple characteristics should be considered when designing and evaluating cervical arthroplasty devices. These include articulating surface design, mono or multipiece implant, constraint, materials, and fixation methods. The majority of cervical devices contain a mono or bi-articulating surface. First-generation implants use a ball-in-socket or saddle articulation design. Next-generation implants take advantage of these traditional designs but also include elastomeric and hydraulic-type designs. Implants may exist as a single monoblock or multipiece design. Experience with hip and knee arthroplasty would suggest that monoblock designs may predispose to increase stress across the implant/bone interface leading to early failure. Modular multipiece implants may reduce stress across adjacent interfaces and provide flexibility with sizing, though multipiece implants with more articulating surfaces inherently have more methods of failure. Certain first-generation ball-in-socket designs have highly congruent articulations with a resulting fixed COR. As a result, precise implant position is necessary for restoration of physiologic COR, which varies by cervical segment (Bogduk and Mercer 2000; Hwang et al. 2008; Patwardhan et al. 2012). Other designs that allow for some translation will have a mobile COR and theoretical flexibility in implant position and may accommodate segmental differences. Constraint is defined by the amount of motion in all directions allowed by the implant. Implants may be constrained, unconstrained, or semiconstrained. Constrained designs provide greater stability but may prevent physiologic motion thereby increasing stress on the implant/bone interface, adjacent segments, and facet joints. On the other hand, unconstrained designs may be unstable under physiologic loads. The majority of cervical arthroplasty designs are semiconstrained providing stability with physiologic motion.

Most currently approved cervical arthroplasty implants are made of a metal endplate (titanium, chrome/cobalt, stainless steel) with a polyethylene or polyurethane center. This basic design was born from hip and knee arthroplasty experience, providing a low-friction bearing surface with stable bone interface. Endplate metals offer different advantages and disadvantages based on modulus, stress shielding, biocompatibility, corrosion resistance, and advanced imaging metal artifact. Newer designs are taking advantage of polyetheretherketone (PEEK) and ceramic materials thereby improving MRI compatibility. Articulating surfaces, whether they are metal on metal, metal on polyethylene, or metal on polyurethane, have different wear debris profiles. Wear debris may result in osteolysis, bone loss, loosening, and ultimate implant failure as seen in hip and knee arthroplasty. Metal on metal articulations have been largely abandoned due to concerns for metal wear debris. Overall, the long-term wear profiles of polyethylene and polyurethane devices in the cervical spine are largely unknown. Finally, the majority of devices contain metal spikes or keels for initial fixation into the adjacent vertebral endplates. Long-term fixation is achieved by bony ingrowth into porous-coated (calcium phosphate, hydroxyapatite, and plasma-sprayed titanium) surfaces (Staudt et al. 2018).

Elastomeric Implants

M6-C Artificial Cervical Disc

The M6-C Artificial Cervical Disc Implant (Orthofix) is a next-generation non-constrained viscoelastic compression-type implant. The nucleus core is made of viscoelastic polyurethane surrounded by ultrahigh-molecular weight polyethylene fiber designed to mimic the nucleus and annulus, respectively, and mimic the physiologic properties of the natural disc (Fig. 1). This physiologic core is attached to two titanium endplates and surrounded by a sheath to prevent wear debris elution and tissue ingrowth. Both titanium endplates contain three fins for provisional fixation and titanium plasma spray coating to promote bony ingrowth. Biomechanical analysis of the M6-C design has demonstrated physiologic ROM, COR, and stability in cadaveric specimens. Patwardhan et al. evaluated the biomechanics of an implanted the M6-C artificial disc at the C5–C6 segment in 12 cadaveric specimens. ROM in flexion-extension, lateral bending, axial rotation, coupled motion, stiffness, and COR was evaluated using digital video fluoroscopic images under 1.5 Nm force moments and compared to control segments. They demonstrated implantation of the M6-C prosthesis within 1 mm of the disc-space midline closely replicated control segment COR and ROM in flexion-extension. Additionally, implantation in a more posterior position did not significantly affect ROM, coupling, or stiffness, suggesting an advantage to and flexibility of implant insertion associated with this novel elastomeric implant (Patwardhan et al. 2012). An initial multicenter FDA-regulated feasibility study evaluated 24-month clinical and radiographic outcomes of 30 patients undergoing one- or two-level M6-C prosthesis implantation with 24-month follow-up. They demonstrated improvement in Neck Disability Index (NDI) and Visual Analog Scale (VAS) neck and arm scores at all time points. No patients experienced surgical or neurologic complication. Radiographic disc height increased in all patients, while global and segment ROM in flexion-extension and lateral bending was maintained (Lauryssen et al. 2012). The results of the feasibility study suggested that the M6-C produces excellent results similar to current approved implants and suggested further prospective studies are necessary to determine the motion provided by the elastomeric compression design improves long-term clinical outcomes and reduces adjacent segment disease (Lauryssen et al. 2012). A recent retrospective study by Thomas et al. in Belgium evaluated clinical outcomes of 33 patients who underwent M6-C arthroplasty for spondylotic radiculopathy or myelopathy with mean 17.1-month follow-up. All patients demonstrated improvement in NDI, VAS arm and back, and

Fig. 1 (a) Cutaway schematic of the M6-C Artificial Cervical Disc Implant. It demonstrates viscoelastic polyurethane nucleus core surrounded by ultrahigh-molecular weight polyethylene fiber mimicking the nucleus and annulus of the natural disc. (b) Exterior schematic of M6-C Artificial Cervical Disc Implant demonstrating physiologic core attached to two titanium endplates. The core is surrounded by an external sheath to prevent tissue ingrowth and elution of wear debris. Each titanium endplate contains three fins for provisional fixation and titanium plasma spray coating to promote bony ingrowth

SF-36 scores. Four patients experienced device-related complications, two with endplate subsidence, one with implant loosening after motor vehicle collision, and one with immobility due to heterotopic ossification. All four of these patients had a history of previous cervical surgery. They concluded that the M6-C prosthesis is a good addition to the cervical arthroplasty options, though should be avoided in patients with history of previous cervical surgery (Thomas et al. 2016). Early reports on the FDA Investigational Device Exemption (IDE) outcomes data demonstrated favorable outcomes of 83 patients who underwent M6-C implantation at 12 (Phillips et al. 2017) and 24 months (Sasso et al. 2018) follow-up. There was significant improvement in the mean VAS neck and arm scores and index level lordosis. Mean index level ROM increased slightly from 7.8° preoperatively to 8.1 at 2 years. There was radiographic evidence of subsidence in three cases, no evidence of migration, and no revision procedures in the follow-up period (Phillips et al. 2017; Sasso et al. 2018). Further long-term studies with larger patient cohorts are needed to determine the effects on development of adjacent segment disease and long-term wear properties. The M6-C implant has recently received FDA approval for use.

Freedom Cervical Disc

The Freedom Cervical Disc (AxioMed) is a monoblock viscoelastic design consisting of an elastomeric core fixed to two titanium plates. The elastomeric polymer core consists of a silicone polycarbonate urethane copolymer. This polymer is molded and bonded to two titanium-retaining plates. Both titanium plates have a porous bead coating designed to engage and allow for bony ingrowth between the cephalad and caudad endplates. The Freedom Cervical disc is created with 8 degrees of lordosis and available in heights ranging from 5.7 to 6.9 mm. The prosthesis is designed to mimic a normal physiologic cervical disc by establishing appropriate alignment and lordosis, viscoelasticity to mimic load sharing, and stable range of motion in flexion, extension, lateral bending, and rotation.

Surprisingly there are no biomechanical studies published to confirm the kinematic features claimed by the manufacturer. Specifically, there is no data regarding the stiffness of this mono-block polymer prosthesis and concerns for resultant high bone-implant forces.

The Freedom Cervical disc has undergone previous pilot studies outside the United States but is not currently approved for use within the United States (Chin et al. 2017; Staudt et al. 2018). One study by Chin et al. reported on the 2-year post market clinical outcomes of the Freedom Cervical Disc in Europe. A total of 39 patients with cervical radiculopathy at 5 institutions underwent one- or two-level cervical disc arthroplasty using the Freedom Cervical Disc. At 2 years clinical follow-up, all patients demonstrated improvement in NDI and VAS neck and arm pain scores. There were no new neurologic symptoms or device-related complications. ROM was surprisingly not evaluated in this study. They concluded that the Freedom Cervical Disc performed as expected in the appropriately selected patients with one- and two-level degenerative disc disease (Chin et al. 2017). This single study is limited by the number of patients and lack of long-term follow-up. Criticisms of this implant design include concerns regarding a single polymer of unknown compressibility matching physiologic properties of the native disc (Staudt et al. 2018).

CP-ESP Cervical Disc

A similar design, the CP-ESP cervical disc prosthesis (FH Orthopedics) is an evolution of the LP ESP lumbar prosthesis that has been implanted in Europe for over 10 years. The CP-ESP disc is a monoblock elastomeric implant with a central polycarbonate urethane (PCU) core fixed to two titanium endplates. Both endplates contain anchoring pegs, textured titanium, and hydroxyapatite layers to provide preliminary fixation and allow for bony ingrowth. The PCU core demonstrates resistance to oxidation both in vivo and ex vivo (Kurtz et al. 2007; Lazennec et al. 2016). The core is attached to the endplates via adhesion molding with peg and groove design without the use of adhesives avoiding the risk of fluid infiltration and fatigue fractures. This design also allows the implant to replicate the anisotropy of a healthy disc, allowing for controlled compression while avoiding shear in flexion and extension. Mechanical analysis demonstrates a physiologic flexion/extension arc of 14°, lateral bending of 12°, and rotation of 8°. The CP-ESP implant is available in 5, 6, and 7 mm heights with various anterior-posterior and lateral dimensions.

A biomechanical assessment of wear debris and fatigue measured using a three-axis motion simulator over the course of ten million cycles demonstrated loss of height ranging from 0.02 to 0.12 mm and no detectable wear debris. Lazennec et al. prospectively evaluated 1- and 2-year clinical and radiographic outcomes of 62 patients who underwent one- or two-level cervical disc arthroplasty using the CP-ESP prosthesis. At both time points, all patients demonstrated improvement in NDI and VAS neck and arm scores. They also demonstrated improved radiographic range of motion at the index levels. No patients experienced implant-related complications or revision procedures during follow-up (Lazennec et al. 2016). Though this design is available for use in Europe, it is not currently under FDA review (Staudt et al. 2018).

Cadisc-C

The Cadisc-C (Ranier Technology) is an evolution of the Cadisc-L design for lumbar disc disease. This unique monoblock elastomeric design consists of polycarbonate-polyurethane nucleus with calcium phosphate coating without an associated metal endplate. The polycarbonate-polyurethane implant contains a lower modulus "nucleus" integrated into a surrounding higher modulus "annulus" allowing it to more accurately mimic the biomechanics of the natural intervertebral disc (McNally et al. 2012; Rieger 2014). The lack of metallic endplate and articulating surfaces is theorized to reduce potential for wear debris (McNally et al. 2012). Though, concern exists regarding the all polymer monoblock design lack of fixation and potential for migration. There is also no published data regarding wear debris profile of this design (Staudt et al. 2018). Currently, prospective trials are underway evaluating clinical outcomes of the Cadisc-C design in Germany (Rieger 2014).

Next-Generation Pivot/Ball Type Artificial Discs

Synergy Cervical Disc

The Synergy (Synergy Disc Replacement) Cervical Disc prosthesis is a next-generation ball-in-socket cervical disc comprised of bi-articulating titanium endplates with an ultrahigh-molecular weight polyethylene core. Bony fixation is augmented by six plasma-sprayed titanium "teels" (a combination of teeth and keels) on each articulating surface. Its three-piece design is MRI compatible and is available in 5 or 6 mm height options. The Synergy device also has a proprietary geometry which incorporates 0° or 6° of cervical lordosis (Staudt et al. 2018). Synergy was compared to two similar constrained ball and pivot arthroplasty designs (Bryan and ProDisc-C) in a retrospective study of 60 patients undergoing single-level cervical disc arthroplasty for cervical radiculopathy. Pre- and postoperative ROM along with dynamic lateral cervical spine imaging were assessed for each group. The Synergy cohort showed the least variability in change of sagittal alignment, achieving six degrees of lordosis on average with maintenance of cervical ROM achieved in all groups (Lazaro et al. 2010). A recent retrospective cohort study compared both the clinical outcomes and postoperative sagittal alignment of patients undergoing single-level surgery for cervical radiculopathy or myelopathy. Forty patients in the arthroplasty group were compared to 33 patients in the single-level fusion group with a minimum follow-up of 24 months. Both the arthroplasty and ACDF groups showed significant improvement in NDI and VAS neck and arm scores. The arthroplasty group maintained an average cervical lordosis of 6 +/− 2.7°, while the ACDF group demonstrated an average of 4 +/− 2.4° of lordosis. The authors concluded that the Synergy system demonstrated comparable outcomes and improved sagittal alignment in comparison to cervical fusion (Yucesoy and Yuksel 2017). While it has undergone various stages of testing and pilot studies, the Synergy arthroplasty system lacks FDA approval and is not currently available in the US market.

Baguera C

The Baguera C (Spineart) is a novel ball-in-socket implant with a mobile core designed as a shock absorber. The mobile core is made of ultrahigh-molecular weight polyethylene (UHMWPE) nucleus that articulates with two titanium endplate components. The titanium endplates contain a bioceramic internal coating in contact with the UHMWPE nucleus and a porous titanium exterior intended for endplate ingrowth. Each endplate contains three fins intended to provide initial stability. The nucleus allows to 0.3 mm anterior to posterior translation, 2° rotation, and 0.15 mm elastic deformation mimicking that of the physiologic disc. One biomechanical analysis demonstrated reduced core contact pressures and liftoff throughout ROM compared to ProDisc-C (Centinel Spine) and Discocerv (Alphatec) using a cervical spine finite element model (Lee et al. 2016). Fransen et al. performed a retrospective registry analysis of 99 patients at 5 European investigational centers undergoing one- or two-level cervical arthroplasty for radiculopathy or myelopathy using the Baguera C implant. They demonstrated a decreased range of motion from 10.2° preoperatively to 8.7° for single-level procedures and from 9.8° to 9.1° for two levels at 2 years radiographic follow-up. They also demonstrated evidence of heterotopic ossification in 54% of patients. None demonstrated radiographic evidence of subsidence, kyphosis, or degeneration of the adjacent disc (Fransen 2016). While lack of radiographic evidence of adjacent segment disease is encouraging, larger long-term studies are needed to determine the efficacy of the implant. Additionally, the mobile nucleus design may theoretically predispose to long-term wear debris and potential for osteolysis as seen in hip and knee arthroplasty.

Simplify Cervical Disc

The Simplify disc has completed one- and two-level IDE study but is not yet received FDA approval. It is a semiconstrained design with titanium plasma-sprayed PEEK endplates with a retention ring housing a mobile ceramic core. Simplify is a modern generation disc with novel biomaterials (PEEK and ceramic) which provide for positive imaging characteristics.

Conclusions

The goal of motion-preserving cervical arthroplasty devices is to restore natural kinematics and motion under physiologic load and prevent degeneration of adjacent segments. Traditional, first-generation cervical arthroplasty devices contain ball-in-socket type designs and do not allow for physiologic coupled motion and compressible graded resistance. As a result, these designs may predispose to adjacent segment and facet stress predisposing to facet degeneration, pain, reduced motion, and degeneration. Early biomechanical evidence suggests that next-generation elastomeric compression devices may better replicate physiologic coupled motion and graded resistance. Further studies are necessary to determine the wear properties, durability, and long-term outcomes of these novel implants.

Cross-References

▶ Cervical Total Disc Replacement: Biomechanics
▶ Cervical Total Disc Replacement: Evidence Basis
▶ Cervical Total Disc Replacement: Expanded Indications
▶ Cervical Total Disc Replacement: Heterotopic Ossification and Complications

▶ Cervical Total Disc Replacement: Technique – Pitfalls and Pearls

References

Bogduk N, Mercer S (2000) Biomechanics of the cervical spine. I: normal kinematics. Clin Biomech Bristol Avon 15:633–648

Cepoiu-Martin M, Faris P, Lorenzetti D, Prefontaine E, Noseworthy T, Sutherland L (2011) Artificial cervical disc arthroplasty: a systematic review. Spine 36: E1623–E1633. https://doi.org/10.1097/BRS.0b013e318 2163814

Chin KR, Lubinski JR, Zimmers KB, Sands BE, Pencle F (2017) Clinical experience and two-year follow-up with a one-piece viscoelastic cervical total disc replacement. J Spine Surg 3:630–640. https://doi.org/10.21037/jss.2017.12.03

Fransen P (2016) Search results – radiographic outcome and adjacent segment evaluation two years after cervical disc replacement with the BAGUERA C prosthesis as treatment of degenerative cervical disc disease – National Library of medicine. J Spine 5:1–7. https://www.omicsonline.org/open-access/radiographic-outcome-and-adjacent-segment-evaluation-two-years-after-cervical-disc-replacement-with-the-baguerac-prosthesis-as-tre-2165-7939-1000298.pdf

Hilibrand AS, Carlson GD, Palumbo MA, Jones PK, Bohlman HH (1999) Radiculopathy and myelopathy at segments adjacent to the site of a previous anterior cervical arthrodesis. J Bone Joint Surg Am 81:519–528

Holmes A, Wang C, Han ZH, Dang GT (1994) The range and nature of flexion-extension motion in the cervical spine. Spine 19:2505–2510

Hwang H, Hipp JA, Ben-Galim P, Reitman CA (2008) Threshold cervical range-of-motion necessary to detect abnormal intervertebral motion in cervical spine radiographs. Spine 33:E261–E267. https://doi.org/10.1097/BRS.0b013e31816b88a4

Iai H, Goto S, Yamagata M, Tamaki T, Moriya H, Takahashi K, Mimura M (1994) Three-dimensional motion of the upper cervical spine in rheumatoid arthritis. Spine 19:272–276

Ishii T, Mukai Y, Hosono N, Sakaura H, Fujii R, Nakajima Y, Tamura S, Iwasaki M, Yoshikawa H, Sugamoto K (2006) Kinematics of the cervical spine in lateral bending: in vivo three-dimensional analysis. Spine 31:155–160

Kurtz S, Ong K, Lau E, Mowat F, Halpern M (2007) Projections of primary and revision hip and knee arthroplasty in the United States from 2005 to 2030. J Bone Joint Surg Am 89:780–785. https://doi.org/10.2106/JBJS.F.00222

Lauryssen C, Coric D, Dimmig T, Musante D, Ohnmeiss DD, Stubbs HA (2012) Cervical total disc replacement using a novel compressible prosthesis: results from a prospective Food and Drug Administration-regulated feasibility study

with 24-month follow-up. Int J Spine Surg 6:71–77. https://doi.org/10.1016/j.ijsp.2012.02.001

Lazaro BCR, Yucesoy K, Yuksel KZ, Kowalczyk I, Rabin D, Fink M, Duggal N (2010) Effect of arthroplasty design on cervical spine kinematics: analysis of the Bryan disc, ProDisc-C, and synergy disc. Neurosurg Focus 28:E6. https://doi.org/10.3171/2010.3.FOCUS1058

Lazennec J-Y, Aaron A, Ricart O, Rakover JP (2016) The innovative viscoelastic CP ESP cervical disk prosthesis with six degrees of freedom: biomechanical concepts, development program and preliminary clinical experience. Eur J Orthop Surg Traumatol Orthop Traumatol 26:9–19. https://doi.org/10.1007/s00590-015-1695-1

Lee JH, Park WM, Kim YH, Jahng T-A (2016) A biomechanical analysis of an artificial disc with a shock-absorbing Core property by using whole-cervical spine finite element analysis. Spine 41:E893–E901. https://doi.org/10.1097/BRS.0000000000001468

McAfee PC (2004) The indications for lumbar and cervical disc replacement. Spine J Off J North Am Spine Soc 4:177S–181S. https://doi.org/10.1016/j.spinee.2004.07.003

McNally D, Naylor J, Johnson S (2012) An in vitro biomechanical comparison of Cadisc^{TM}-L with natural lumbar discs in axial compression and sagittal flexion. Eur Spine J 21:612–617. https://doi.org/10.1007/s00586-012-2249-4

Mummaneni PV, Burkus JK, Haid RW, Traynelis VC, Zdeblick TA (2007) Clinical and radiographic analysis of cervical disc arthroplasty compared with allograft fusion: a randomized controlled clinical trial. J Neurosurg Spine 6:198–209. https://doi.org/10.3171/spi.2007.6.3.198

Panjabi MM (1992) The stabilizing system of the spine. Part II. Neutral zone and instability hypothesis. J Spinal Disord 5:390–396. Discussion 397

Patwardhan AG, Tzermiadianos MN, Tsitsopoulos PP, Voronov LI, Renner SM, Reo ML, Carandang G, Ritter-Lang K, Havey RM (2012) Primary and coupled motions after cervical total disc replacement using a compressible six-degree-of-freedom prosthesis. Eur Spine J Off Publ Eur Spine Soc Eur Spinal Deform Soc Eur Sect Cerv Spine Res Soc 21(Suppl 5):S618–S629. https://doi.org/10.1007/s00586-010-1575-7

Phillips F, Sasso R, Coric D, Guyer R, Sama A, Cammisa F, Blumenthal S, Albert T, Zigler J (2017) Clinical and radiographic outcomes for the M6-C artificial cervical disc: 1 year follow-up at five IDE investigation centers for a novel design six degree of freedom prosthesis. Int Soc Adv Spine Surg. https://www.isass.org/abstracts/isass17-oral-posters/isass17-660-Clinical-and-Radiographic-Outcomes-for-the-M6-C-Artificial-Cervical-Di.html

Pickett GE, Sekhon LHS, Sears WR, Duggal N (2006) Complications with cervical arthroplasty. J Neurosurg Spine 4:98–105. https://doi.org/10.3171/spi.2006.4.2.98

Rieger B (2014) Comparison of software-assisted implantation of cadisc-C with the disc prosthesis rotaio in the therapy of cervical disc disease in terms of postoperative changes in Neck Disability Index (NDI) a prospective, controlled and randomized study. NeurochirugieUniversitätsklinikum Köln

Sasso RC, Phillips F, Coric D, Guyer R, Sama A, Cammisa A, Blumenthal S, Albert T, Zigler J (2018) M6-C artificial cervical disc 24 month follow up: clinical and radiographic outcomes from five investigational centers involved in a US FDA approved IDE study. Int Soc Adv Spine Surg. http://www.isass.org/abstracts/isass18-oral-posters/isass18-437-M6-C-Artificial-Cervical-Disc-24-Month-Follow-up-Clinical-and-Radiogr.html

Senouci M, FitzPatrick D, Quinlan JF, Mullett H, Coffey L, McCormack D (2007) Quantification of the coupled motion that occurs with axial rotation and lateral bending of the head-neck complex: an experimental examination. Proc Inst Mech Eng H 221:913–919. https://doi.org/10.1243/09544119JEIM265

Staudt MD, Das K, Duggal N (2018) Does design matter? Cervical disc replacements under review. Neurosurg Rev 41:399–407. https://doi.org/10.1007/s10143-016-0765-0

Thomas S, Willems K, Van den Daelen L, Linden P, Ciocci M-C, Bocher P (2016) The M6-C cervical disk prosthesis: first clinical experience in 33 patients. Clin Spine Surg 29:E182–E187. https://doi.org/10.1097/BSD.0000000000000025

Yucesoy K, Yuksel K (2017) Can cervical arthroplasty impact alignment? A comparison of the synergy disc with cervical fusion. J Spine 6:1–4. https://www.omicsonline.org/open-access/can-cervical-arthroplasty-impact-alignment-a-comparison-of-the-synergy-disc-with-cervical-fusion-2165-7939-1000400.pdf

Cervical Total Disc Replacement: Evidence Basis

41

Kris E. Radcliff, Daniel A. Tarazona, Michael Markowitz, and Edwin Theosmy

Contents

Introduction	772
Biomechanics	773
Constraint	773
Anatomical Considerations	774
Biomaterials	774
End Plate Materials	774
Bearing Designs	775
Surgical Procedure and Technical Pearls	776
Indications and Contraindications	776
Positioning	776
Exposure	776
Technique	777
Postoperative Course	780
Complications	780
Anterior Approach	781
Heterotopic Ossification	781
Adjacent Segment Disease	782
Postoperative Sagittal Imbalance	782
Implant Migration and Subsidence	782
Evidence	783
Single-Level Disc Arthroplasty Outcomes	783
Multilevel Disc Arthroplasty Outcomes	783

K. E. Radcliff (✉) · D. A. Tarazona
Department of Orthopaedic Surgery, The Rothman
Institute, Thomas Jefferson University,
Philadelphia, PA, USA
e-mail: radcliffk@gmail.com;
daniel.tarazona@rothmaninstitute.com

M. Markowitz · E. Theosmy
Department of Orthopaedics, Rowan University School of
Osteopathic Medicine, Stratford, NJ, USA
e-mail: mmarkowitz22@gmail.com; etheosmy@gmail.com

© Springer Nature Switzerland AG 2021
B. C. Cheng (ed.), *Handbook of Spine Technology*,
https://doi.org/10.1007/978-3-319-44424-6_73

Reoperations	784
Cost-Effectiveness	784
Patient Satisfaction	785
Pain Medication Usage	785
Return to Work	785
Conclusion	786
Cross-References	786
References	786

Abstract

Anterior cervical decompression and fusion has long been the gold standard for cervical degenerative disc disease, but concerns about the deleterious effects of fusion on adjacent segments have led to the development of cervical total disc replacement (TDR). While many TDR designs have been evaluated, metal-on-polymer and metal-on-metal designs are the most commonly used today. Different types of metals and surface modification have been introduced in attempt to improve osseous integration and decrease failure of implant. Correct positioning, adequate exposure, and thorough decompression and end plate preparation are necessary to ensure proper disc placement. Patients benefit in the postoperative period from early mobilization, improved range of motion, and often return to work earlier, with a lower risk of reoperations than with fusion. Long-term outcomes from many of the IDE trials consistently demonstrate to be comparable and even superior to fusion with cost-effective analysis further supporting financial feasibility.

Keywords

Cervical disc replacement · Degenerative disc disease · Intervertebral disc · Cervical spine · Arthroplasty · Cervical radiculopathy · Cervical myelopathy

Introduction

Joint arthroplasty has been an evolving treatment modality for degenerative joint disease for decades. Its evolution has permitted improved prostheses, better outcomes, and the inception of arthroplasty being used in numerous areas of the body. Utilization of arthroplasty for cervical spine pathology has been a topic of recent discussion. While prior treatment has been centered on hip and knee arthrodesis, cervical total disc replacement (TDR) displays promise for the maintenance of native spine biomechanics and kinematics while minimizing the progression of adjacent segment disease seen in arthrodesis.

Currently, the most commonly used treatment of cervical spine spondylosis and disc disease centers on Smith-Robinson, who pioneered the widely known and used anterior cervical decompression and fusion (ACDF) surgery in the 1950s with multiple studies reporting remarkably good success with this procedure (Bohlman et al. 1993). However, adjacent segment degeneration, or the appearance of degenerative changes at a level above or below a fused segment, has been seen during long-term follow-up, and efforts toward artificial joint prostheses became a potential alternative. Symptomatic adjacent segment disease (ASD) affects approximately a quarter of all patients who undergo an ACDF by 10 years with more than two-thirds of patient failing nonoperative therapy and requiring addition operative interventions (Hilibrand et al. 1999).

The first attempt to create an artificial intervertebral replacement appeared in the 1960s under Ulf Fernström. The initial implant utilized was a stainless steel ball bearing prosthesis trialed in both the cervical and lumbar spine (Baaj et al. 2009). Unfortunately, little success was obtained, and nearly 90% of the cases demonstrated implant migration and subsidence of the prosthesis (Fernström 1966). Its failure briefly dampened enthusiasm for TDR and shifted the focus back to ACDF procedures. However, B. H. Cummins developed the first modern model of the cervical disc prosthesis in 1989 at the Frenchay Hospital in Bristol. This new device was a metal-on-metal ball-and-socket design with screws anchoring anteriorly, fixing it to the vertebral body. Again, early implants demonstrated a high degree of screw cut out, dysphagia, and implant mobilization (Cummins et al. 1998). Attempting to avoid this complication, Cummins developed a second-generation device where the anterior portion of the device, articulating surface, and locking screw device were all redesigned. This device became known as the Prestige I Disc, which was continually modified to a fifth-generation product ultimately called the Prestige LP made of a titanium-ceramic composite that could be more aggressively anchored to the vertebral body (Nasto and Logroscino 2016). In July 2007, the Prestige ST disc was approved by the FDA. Since then numerous models or varying materials were to follow such as the Bryan, which consists of two titanium alloy end plates articulating with a polyurethane core. This device is fixed to the bone by a porous titanium layer that maintains a tight fit between the vertebrae. The ProDisc-C with cobalt-chrome-molybdenum end plates and a polyethylene articulating surface is a ball-and-socket constrained prosthesis that has a central keel for initial fixation to the bone (Smucker and Sasso 2011).

Over the past 30 years, there have been increased efforts toward the development of TDR. While currently not widely accepted as a substitute for ACDF, the concept of joint arthroplasty in the spine remains in clinical practice and as a topic of evolving research. In a recent international survey of spine surgeons, only 7% used TDR as a standard treatment, while 84% used ACDF (Chin-See-Chong et al. 2017). The development of TDR has the potential to not only minimize adjacent segment disease but to ultimately mimic healthy spine kinematics and biomechanics. Additional studies are required to help gain insight into the better use of these prostheses.

Biomechanics

The goal of cervical total disc replacement is to restore native cervical spine biomechanics. While ACDF remains the standard for many cervical spine pathologies, numerous studies have found that it decreases motion at fused levels and transfers increased motion and stresses to adjacent levels, creating higher intradiscal pressure and ultimately increasing the incidence of adjacent level disease when compared to arthroplasty (Smucker and Sasso 2011; Dmitriev et al. 2004).

Constraint

The different types of movements in the cervical spine can be grouped into rotational and translational. They occur in the sagittal, coronal, and axial planes, making up the six degrees of freedom. When one degree of freedom is limited, that movement is constrained, and a fully constrained device would indicate no movement at all in the six degrees of freedom. There is a lack of consistency in the literature when defining and classifying a device based on constraint, but they have been described as unconstrained, semi-constrained, and fully constrained. The degree of constraint typically refers to translational movements since rotation is unconstrained in all devices; otherwise a fully constrained device would technically be a fusion. An unconstrained prosthesis allows for more motion, but does not resist shearing forces which are shifted on to the facet joints. A more constrained prosthesis limits motion but assumes more of the shearing forces, thus relieving the facet joints.

Kinematic studies of the cervical spine after artificial disc implantation have shown promising results in retaining native biomechanics. TDR has demonstrated preserved postoperative sagittal rotation, translation, center of rotation, and disc heights when compared to preoperative measurements (Pickett et al. 2005). When comparing biomechanical data of different implants in vivo and in vitro, DiAngelo et al. reviewed the biomechanics of different devices and noted the motion of two different prostheses in cadaveric models.

The designs in question were a semiconstrained device allowing anterior-posterior translation and a constrained device with minimal to no anterior-posterior translation. The semiconstrained device, the Prestige ST disc, more closely replicated native cervical spine kinematics at all ranges of motion, whereas the constrained disc, ProDisc-C, failed to reproduce native motion (DiAngelo et al. 2003, 2004).

Anatomical Considerations

The anatomical relationships in the cervical spine are complex, and correct alignment is essential for efficient biomechanics. Relevant bony anatomy with regard to cervical biomechanics includes intervertebral disc, uncovertebral joints, and the facet joints. Each plays a role in the overall range of motion and stability of the spine through the range of motion. During flexion, the center of rotation is located in the anterior region of the inferior vertebra, and forces are relieved off the facets, where as in extension the facets and spinous process are engaged. The two uncovertebral joints aid in lateral bending by providing resistance to shear forces. The significant changes in stability after uncovertebral joint resection highlight the importance of preserving them during decompression (Kotani et al. 1998). During lateral bending, the center of rotation moves to the superior vertebral body, and the lateral uncovertebral joints are engaged like a rail to limit translation.

Sasso et al. further evaluated the biomechanical properties of the BryanTDR versus ACDF preoperatively and postoperatively. After 24 months, TDR slightly improved flexion and extension range of motion from 6.4 to 8° without implant complication. Meanwhile, ACDF only retained about 1° of flexion and extension, which gradually decreased over a 24-month period (Nasto and Logroscino 2016; Sasso et al. 2008). Janssen et al. reported significantly higher flexion-extension range of motion after disc replacement with the ProDisc-C compared to the ACDF group at 7 years follow-up (Janssen et al. 2015). Similarly, Hisey et al.

reported no loss of segmental range of motion in the disc replacement group but did note an increase in overall flexion-extension range of motion at the 5-year follow-up (Hisey et al. 2016). Conversely, others have also reported no difference in cervical range of motion between ACDF and TDR at long-term follow-up, but in general range of motion is maintained or improved with TDR (Radcliff et al. 2016b).

Biomaterials

An ideal arthroplasty device is one that is biocompatible, has superior biomechanical properties, produces minimal wear debris, and can achieve strong fixation. Immediate fixation and stability are based on screw fixation or physical stops, while long-term fixation relies on osseous integration between the prosthesis and vertebral body. Another important consideration is the wear properties of the different materials since the generated particles create an inflammatory response, resulting in osteolysis and eventually failure of implant. To address several of these factors, different materials have been used for the implant design including titanium alloy, stainless steel, and cobalt-chrome (CoCr) alloy. Also, polymers such as ultrahigh molecular weight polyethylene (UHMWPE) and polyurethane (PU) are used as articular surfaces.

End Plate Materials

Titanium alloy is a biocompatible metal that contains titanium, vanadium, and aluminum. Its modulus of elasticity is closer to cortical bone than steel allowing for better osseous integration. This property has led to the use of titanium as part of a porous coating on the outer surface of cervical implants to improve fixation at the implant-bone interface. Calcium phosphate has also been added to titanium resulting in increased osseointegration (Cunningham et al. 2009). Titanium is also inert and forms an oxide film making it highly resistant to corrosion. It also produces less imaging artifact than steel or cobalt-chrome, and both the index and adjacent levels can be easily visualized on MRI (Fayyazi et al. 2015). However, its poor wear properties and propensity to generate more wear debris make it less useful as a bearing surface.

Stainless steel is primarily an iron-carbon alloy that can be combined with other metals to alter its properties. It is a widely available and inexpensive metal that is strong, stiff, and resistant to fatigue. However, its high elasticity modulus leads to stress shielding of bone. It also tends to be more corrosive than other metals, but when used in combination with chromium, it is less susceptible to corrosion. The biggest limitation with stainless steel is the artifact produced with MRI which does not allow for visualization of either the index or adjacent levels such as with Prestige ST (stainless steel end plates) (Fayyazi et al. 2015).

Cobalt-chrome is a strong alloy that is biocompatible and resistant to corrosion. It has demonstrated superior wear characteristics making it a more reliable metal for bearing surfaces. It also produces less debris compared to titanium alloy and is relatively resistant to fatigue. All these properties together have made cobalt-chrome a popular choice for cervical arthroplasty devices. Cobalt-chrome devices are MRI compatible, but imaging artifact lies somewhere in between that of stainless and titanium where adjacent levels can be visualized but the index level is typically obscured.

Until recently, implant designs have been solely evaluated based on the mechanical properties of the materials used, but now the focus has shifted to surface topography. The three most commonly used surface modifications in TDR include titanium spray coating, hydroxyapatite (HA) coating, and porous surfaces. Titanium spray coating has been shown to improve osseous integration on Co-Cr implants through enhanced cellular attachment, proliferation, and differentiation (Pham 2014). Hydroxyapatite accounts for a majority of inorganic bone, and HA coatings provide an osteoconductive environment. Surface porosity is another technique that is often utilized in multiple orthopedic implants due to its ability to promote bony ingrowth and stable interlocking fixation while reducing stress shielding of metals

by lowering the modulus of elasticity (Dabrowski et al. 2010). These techniques function in different ways but ultimately work to enhance the osseous integration at the bony-implant interface in an attempt to achieve long-term fixation.

Bearing Designs

The most commonly used bearing surfaces are metal on metal and metal on polymer. Despite ceramics superior wear and corrosion properties, its brittleness, inability to absorb shock, and potential for catastrophic implant failure have limited its use in arthroplasty.

Metal-on-metal articulations were the initial design for total hip replacements. This was in large part due to in vitro studies that showed lower wear rates compared to polymer and clinical studies which showed a low mechanical failure rate and no osteolysis at long-term follow-up (Walter 1992; Dorr et al. 2000). The wear rates in simulator testing of metal-on-metal cervical arthroplasty have lower rates than that of metal-on-metal or metal-on-polymer total hip arthroplasty (Traynelis 2004). However, studies using metal on metal begun to show elevated levels of metal ions in urine, serum, and several organs (Jacobs et al. 1996; Urban et al. 2000). Concerns about ion toxicity and metal hypersensitivity began to mount, and the design lost some of its initial traction.

The metal on polymer design is similar in principle to modern total hip replacement devices. It is composed of two metallic vertebral end plates along with a core polymer. This was introduced as an alternative to metal on metal designs, due to concerns about systemic toxicity. UHMWPE is a thermoplastic polyethylene polymer with extremely long hydrocarbon chains. The long chains allow for more efficient load transfer giving it a high-impact strength. Cobalt-chromium on UHMWPE has a long track record of success in THA, but osteolysis is a known issue with this combination.

A new alternative currently under investigation is a PEEK-on-ceramic bearing. One of the advantages of this design is the absence of artifact produced on imagining as well as MRI compatibility. This allows for accurate postoperative radiographic monitoring and reduces the exposure to ionizing radiation from CT scans which is often utilized with metallic TDR. Also, the wear rates, particle size, and morphology produced by a PEEK-on-ceramic bearing indicate a potential alternative to the commonly utilized CoCr-on-polyethylene and metal-on-metal bearings (Siskey et al. 2016).

Surgical Procedure and Technical Pearls

Indications and Contraindications

Well-defined patient selection criteria are critical for successful outcomes. Indications for cervical total disc replacement include radiculopathy or myelopathy resulting from one- or two-level disease, primarily anterior or disk-related pathology, preserved segmental motion, preserved disc space height, minimal facet arthropathy, and maintained sagittal alignment. Contraindications to performing TDR are tumor, trauma, infection, known allergy to implant metal, segmental instability, osteoporosis, collapsed disk space, circumferential pathology, severe cervical spondylosis, and prior cervical spine surgery at targeted level. Only about 40% of patients with symptomatic single-level cervical radiculopathy will be candidates for TDR (Auerbach et al. 2008).

Positioning

First, the patient is placed under general anesthesia and positioned supine on a radiolucent table. A bolster is placed between the shoulders and the neck is placed in neutral position. In contrast, for ACDF, the neck is often hyperextended. Tape is placed across the forehead to prevent rotation of the head intraoperatively. The arms are placed along the patient's side and taped with gentle traction, allowing for better access to the lower cervical spine. All bony protuberances should be well padded. The patient is then prepped and draped in the typical sterile fashion. Fluoroscopy should be utilized for intraoperative guidance. Rotational alignment can be assessed on AP fluoroscopy by confirming that the spinous process bisects the pedicles. The use of head weights is not recommended as this will cause a false impression of disc height through distraction of adjacent levels.

Pearls
- Poor head positioning can lead to sagittal malalignment and incorrect implant placement (Buchowski et al. 2009).
- Hyperextension leads to excessive removal of the posterior end plates, resulting in kyphosis.
- Hyperflexion leads to excessive removal of the anterior end plates, resulting in excessive lordosis.

Exposure

Prior to skin incision, fluoroscopy is used to confirm the approximate level of the incision. A standard Smith-Robinson approach is performed to access the anterior cervical spine. First, a transverse incision is made on the neck, either right- or left-sided, and sharp dissection is used down to the platysma. The platysma is dissected from the underlying fascia bluntly and split with electrocautery. Then the interval between medial portion of the sternocleidomastoid muscle (SCM) and the strap muscles is bluntly dissected. The carotid pulse should be palpated to locate the artery. Blunt dissection is continued medially to the SCM and carotid sheath. This is carried down through the pretracheal fascia and into the retropharyngeal space. Overlying the spine are the prevertebral fascia, anterior longitudinal ligament, and longus colli muscle. The longus colli is elevated in a subperiosteal manner. Retractors are then inserted and exposure of the anterior cervical spine is complete. The targeted intervertebral disc is confirmed on lateral fluoroscopy. The midline is located halfway between both uncinate processes, and a mark is made both in the superior and inferior vertebral bodies for implantation.

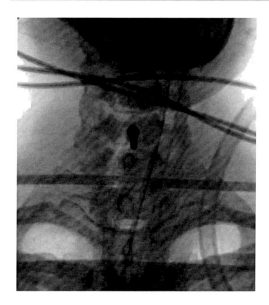

Fig. 1 AP fluoroscopy demonstrating midline placement of Caspar pin to avoid axial rotation

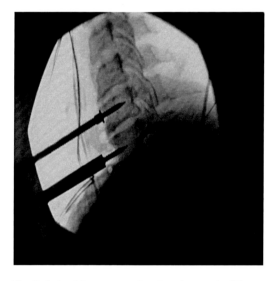

Fig. 2 Lateral fluoroscopy showing placement of Caspar pins parallel to vertebral end plates

Pearl
- Expose the lateral boundaries of the uncinate processes so they can be visually identified in order to ensure the implant is properly centered in the coronal plane.

Technique

After confirming the intervertebral disc, distraction pins are inserted midline in the superior and inferior vertebral bodies (Fig. 1).

Pearls
- Place the distractor pins parallel to the vertebral end plates so that the disc is symmetrically opened (Fig. 2).
- Ensure that the distractor pins are parallel to the true sagittal plane and do not deviate in the axial plane. Otherwise, a rotational displacement will be created once the pins are distracted.
- Use a Cobb elevator to carefully distract the disc space by levering on the end plates. Use the distractor pins to maintain the amount of distraction achieved from end plate leverage. Do not spread the vertebral bodies apart directly with the distractor pins.
- Take a preoperative X-ray to identify the facet joint height prior to distraction.

Depending on the implant used, this part of the procedure may sometimes follow the discectomy. Lateral fluoroscopy is used to confirm pin placement is parallel to the disc space. A locking distractor is attached to the pins and a Cobb elevator, and distracting forces are applied (Figs. 3 and 4). An annulotomy and discectomy are carried out using curettes and rongeurs.

Pearl
- Perform a wide, symmetric annulotomy.

This is completed between uncinate processes and from ventral to dorsal while using progressively smaller curettes. For complete decompression, meticulous removal of all osteophytes is critical. We recommend minimal use of a high-speed burr. However, if a one is used, copious amounts of irrigation and bone wax should be applied to reduce the risk of heterotopic ossification. When preparing the end plates, only the cartilaginous portion should be removed. Special attention should be paid to osteophytes located by

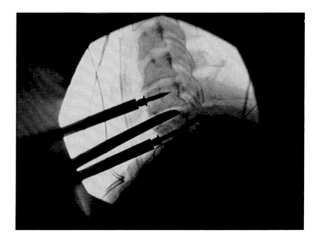

Fig. 3 Lateral image of same patient as Fig. 2 showing use of a Cobb to distract disc space

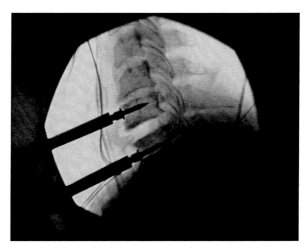

Fig. 4 Lateral image after use of Cobb and distraction forces applied to distract disc space

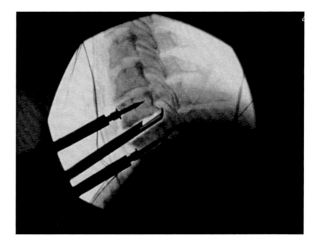

Fig. 5 Lateral fluoroscopy showing use of a Kerrison rongeur to remove posterior osteophytes for a complete decompression

the uncinate process or posterior vertebral body (Fig. 5). Partial resection of the uncinate process can be performed if necessary, but complete resection should be avoided as this could lead to cervical instability. If resection of the uncinates is necessary, try to perform bilateral symmetric

Fig. 6 Trial implant placed following decompression

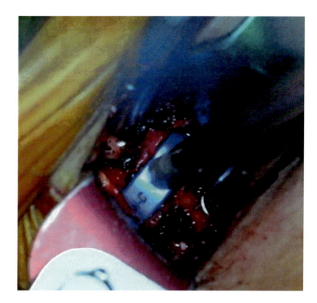

Fig. 7 Lateral image showing compression of the Caspar pins onto the disc replacement. Notice that facets are not distracted compared to other disc levels

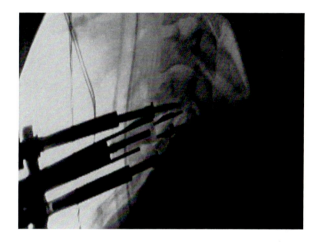

resection. Bilateral neuroforaminal decompression is performed with rongeurs. Once the decompression has been completed, trial implants and fluoroscopy are used to determine the optimal implant size and position (Fig. 6).

Pearls
- Ensure that the trial implant is as far posterior as possible as the center of rotation of the cervical spine is near the posterior vertebral body border.
- Select as wide of a trial as possible to ensure maximal medial-lateral surface area coverage and reduce surface stresses.
- Select as deep of a trial as possible to ensure maximal anterior-posterior surface area coverage and reduce surface stresses.
- Study the adjacent level facet joints when trialing and after disc replacement is placed to ensure that you are not over distracting the segment (Fig. 7).
- Remove all distraction while trialing.
- After trialing, cover all cut bone surfaces (e.g., uncovertebral joints) with bone wax to reduce the rate of heterotopic ossification.
- The trial should fit snugly into the disc space, but ideally some cranial-caudal toggle motion

Fig. 8 Final placement of disc replacement with distractor pins removed

of the trial handle should occur to indicate that the disc is not overstuffed.
- If the surgeon is having difficulty visualizing the precise posterior vertebral body margin, then take a lateral fluoroscopy shot with a nerve hook behind the vertebral body.
- For devices with a keel, ensure that the keel cut goes as far posteriorly as the intended implant. Otherwise, the implant could fracture a piece of posterior vertebral body cortex if it is attempted to be placed more posteriorly than the extent of keel cut.

The device is then inserted according to manufacturer's standards while confirming placement visually (Fig. 8) and radiographically in the coronal and sagittal planes.

Pearls
- Occasionally, the cranial and caudal end plates will not advance uniformly. If you notice that one end plate is more anterior than the other end plate, use the single end plate tamp. Otherwise, the implant will have a kyphotic appearance.
- Performing a thorough discectomy, decompression and preparation of the end plates are critical for proper implant placement and to optimize postoperative range of motion. Care should be taken to not violate the vertebral end plates when using the high-speed burr as this could result in implant subsidence.
- Use bone wax on all cut bone surfaces (e.g., anterior osteophytes, distractor pin holes) to prevent the egress of marrow and reduce the rate of heterotopic ossification.

Postoperative Course

Patients are routinely discharged same day for single-level arthroplasty and on postoperative day 1 or 2 for multilevel disc arthroplasty. Postoperatively, a soft collar is not generally used. Instead, the patient should be mobilized out of bed on postoperative day 0 and can immediately start gentle neck range of motion, but extremes should be avoided. Follow-up radiographs are useful to assess for postoperative kyphosis, implant subsidence, and early and delayed fusion. However, imaging artifact may limit radiographic accuracy. Physical therapy should be instituted as tolerated, generally 2–6 weeks postoperative. One of the early touted benefits of TDR is the restoration of motion at the diseased segment and less strain on adjacent levels. This has clinically manifested in improved postoperative range of motion and earlier return to work (Burkus et al. 2014; Zhu et al. 2016a).

Complications

Complications associated with cervical total disc replacement is one of the main reasons cited for not offering it as an option to patients in a survey of spine surgeons (Chin-See-Chong et al. 2017) However, a recent meta-analysis comparing the adverse events between ACDF and TDR indicated no difference in terms of dysphagia/dysphonia, hardware-related complications, heterotopic ossification, neurological deterioration, overall neurologic adverse events, or mortality. There were three types of adverse events that did show a significant difference. First, there was a small increase in minor wound-related adverse events for TDR, but none required a second procedure for deep wound infection or removal of infected implant. There was a high variation in rates of infection among the different studies, and this was attributed to inclusion of a general category of infection instead of specifying wound-related infections in some studies. Also, ACDF was associated with a higher incidence in surgical-related neurologic adverse events and secondary surgeries which generally occurred late. The definition of neurologic adverse events was not consistent and leads to variable rates from 2.8% to 73.8%. Overall, there does not appear to be a major difference in complication rates between TDR and ACDF, but future studies will need to use uniform definitions in order to make accurate comparison.

Anterior Approach

Cervical total disc replacement shares some of the same complications that are associated with an ACDF by virtue of a similar anterior approach. The anterior approach can be associated with dysphagia due to injury to the esophagus or surrounding nerve plexus. Transient dysphonia or dysarthria can result from injury to recurrent laryngeal, superior laryngeal, and hypoglossal nerve. Periodic release of retractors and a good understanding of the anatomy can reduce the risk of these injuries. Postoperative hematoma should be closely monitored for as it could compromise the airway. Kato and colleagues conducted a propensity score matching analysis of a cohort of CSM patients which showed that there is no difference in outcomes or overall complication rates between anterior and posterior cervical approaches (Kato et al. 2017). However, dysphagia and dysphonia were more common in an anterior approach, while surgical site infections and C5 palsy are more common with a posterior approach.

Heterotopic Ossification

Heterotopic ossification (Fig. 9) is an abnormal deposition of the bone in soft tissue and was first described in total joint arthroplasty procedures. One of the main advantages of a TDR over fusion is its ability to preserve motion, and HO could hamper this by restricting range of motion and potentially affect outcomes. The clinical significance of HO remains questionable. Several predisposing factors have been identified including age, gender, degree of preoperative spondylosis, and implant type. Heterotopic ossification has

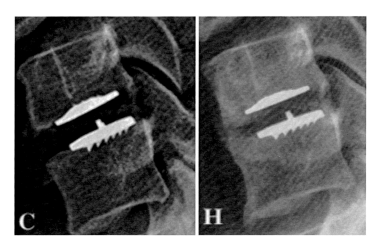

Fig. 9 Normal postoperative X-ray following cervical disc replacement (C) versus same patient after developing heterotopic ossification (H). (Lee et al. 2012)

been found to be more common in two-level TDR compared to single-level TDR (Wu et al. 2012). The incidence of HO is highly variable, ranging from 17.8% to 77.3%, and was proportional to the duration of follow-up (Leung et al. 2005; Lee et al. 2012). Although Lee et al. showed that cervical range of motion was limited by high-grade HO, it has not been found to affect clinical outcomes following TDR (Wu et al. 2012). Treatment of HO with NSAIDs is largely based on literature from total joint arthroplasty, but strong evidence to support its use is lacking. Additionally, in order to attempt to reduce the rate of HO, it is recommended to bone wax all cut bone surfaces (e.g., anterior osteophytes, distractor pin holes) to minimize the exposure of marrow.

Adjacent Segment Disease

Adjacent segment disease is a broad term that includes the degenerative changes at vertebral levels neighboring surgically treated segments (Fig. 10) and is associated with signs and symptoms such as radiculopathy, myelopathy, or instability. ASD is one of the major influences that promoted further investigations of TDR. The motion-preserving properties could theoretically reduce the risk of ASD and subsequent reoperations. A meta-analysis of 14 randomized controlled trials comparing TDR and ACDF confirmed that TDR was associated with a lower rate of ASD and fewer adjacent segment reoperations (Zhu et al. 2016b). The use of TDR for ASD following fusion is controversial with only smaller studies indicating it is potentially a safe option (Rajakumar et al. 2017) (Fig. 11).

Postoperative Sagittal Imbalance

One of the relative contraindications for TDR is the presence kyphosis or positive cervical sagittal balance (Johnson et al. 2004). Sagittal balance determines the load distribution on the device, and any imbalance could lead to abnormal wear and worsening kyphosis. The clinical significance of segmental kyphosis can be profound, resulting in segmental instability, adjacent segment degeneration, axial neck pain, early hardware failure, and poorer functional outcomes (Cao et al. 2011). However, different techniques have been developed to prevent kyphosis, but caution should be used. Some techniques can lead to overcorrection of lordosis, anterior migration of prosthesis, restricted range of motion, and neck pain (Lei et al. 2017).

Implant Migration and Subsidence

Migration and subsidence are generally uncommon. Goffin et al. reported a single case of both migration and subsidence (Goffin 2006; Goffin et al. 2003). He proposed techniques such as maintaining the integrity of the end plates, using the widest possible device to improve load distribution, and avoiding TDR in patients with poor bone quality (i.e., osteoporosis, metabolic bone disorders). Also, the addition of a keel to the arthroplasty device, which lies up against the anterior vertebral body, has been utilized by some devices to reduce the risk of posterior migration.

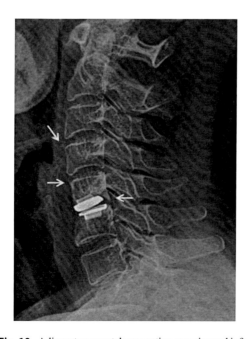

Fig. 10 Adjacent segment degeneration superior and inferior to disc replacement, as well as posterior osteophyte at index level (Kim et al. 2016)

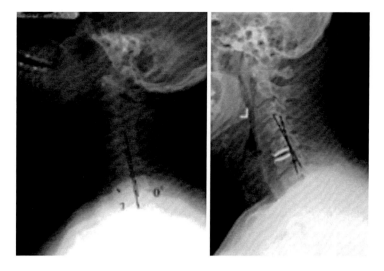

Fig. 11 Preoperative x-ray (right) showing neutral alignment. Postoperative x-ray (right) showing increase in segmental kyphosis. (Johnson et al. 2004)

Evidence

Single-Level Disc Arthroplasty Outcomes

Several long-term randomized prospective trials have been conducted to evaluate patients undergoing cervical total disc replacement. A meta-analysis comparing TDR versus fusion for single-level cervical disc disease concluded that TDR improved neck and arm pain and had a higher neurological and overall success rate (Xing et al. 2013). Later studies have continued to demonstrate comparable and even superior clinical and radiographic outcomes (Janssen et al. 2015; Rožanković et al. 2017).

In a study with longest follow-up to date, Dejaegher et al. reported a 10-year follow-up on patients undergoing disc replacement with the Bryan prosthesis (Dejaegher et al. 2017). Eighty-nine patients underwent single-level disc replacement, and patients were assessed every 2 years. Neurologic success was achieved in more than 80% of patients, and they saw significant improvement in terms of level of disability, neck and arm pain, and functional status. Twenty-four percent of patients developed new or recurrent neurologic symptoms, similar to previously published rates for ACDF.

Multilevel Disc Arthroplasty Outcomes

Similar studies have investigated outcomes for patients who underwent multilevel TDR, which have been largely limited to two and three levels. Gornet et al. investigated outcomes for those patients undergoing adjacent level disc replacement with a Prestige LP device in comparison to ACDF with cortical ring allograft and anterior plating (Gornet et al. 2017). Success for this study was defined by four criteria: Neck Disability Index (NDI) score improvement of ≥ 15 points, maintenance or improvement in neurological status, number of serious adverse events caused by the implant or surgery, and need for additional surgeries. Results showed that patients had better success with multilevel disc replacement compared to a multilevel fusion (81% vs. 69%).

Radcliff et al reported results from a 5-year study comparing two-level TDR to fusion. Patients undergoing disc replacement demonstrated significantly improved Neck Disability Index scores compared to the fusion group (Radcliff et al. 2016b). Similarly, Lanman et al. reported greater overall success of the Prestige LP TDR over fusion at 84-month follow-up (Lanman et al. 2017). Patients undergoing disc replacement also demonstrated improved NDI scores and higher neurological success when compared to fusion.

In comparison to single-level TDR, Huppert and colleagues showed that multilevel TDR had similar levels of improvement in outcomes, satisfaction, and range of motion compared to single-level TDR (Huppert et al. 2011). Although the multilevel cohort had a higher rate of dysphagia and dysphonia, all but one resolved spontaneously. In addition, there was no significant difference in the number complications or reoperations. Likewise, a meta-analysis comparing single- and multilevel TDR revealed no differences in outcomes, functional recovery, or reoperation rates (Zhao et al. 2014).

Several studies have compared multilevel ACDF and TDR, and a recent meta-analysis concluded that both groups had similar clinical outcomes (Wu et al. 2017). However, patients who underwent TDR did experience lower rates of ASD and complications while achieving greater overall range of motion. Altogether multilevel TDR appears to be as effective as single-level TDR and multilevel ACDF.

Reoperations

There are several reasons why patients have to undergo additional surgery after their index cervical total disc replacement. The most common reasons included removal of device with conversion to fusion, recurrence of symptoms, and procedures to address ASD.

Janssen et al. reported that after 7 years, 18% of ACDF patients underwent secondary procedures, while only 7% of the ProDisc-C group needed additional procedures (Janssen et al. 2015). The risk for secondary surgery was approximately 3.7 times higher for single-level ACDF. Mostly reoperations were performed at the index level for both groups; however, adjacent level procedures were more commonly done in the ACDF group. Hisey and colleagues reported similar trends, with the Mobi-C group requiring a higher number of additional procedures compared to the ACDF group (Hisey et al. 2016).

For multilevel disc replacements, Radcliff et al. found that the disc replacement cohort underwent fewer secondary surgeries and adjacent level procedures than the fusion group (Radcliff et al. 2016b). The reasons for reoperations for the Mobi-C group included neck pain and radiculopathy (most common), hematoma, poor device attachment, and inferior level end plate migration. For the Prestige LP, Lanman et al. reported a reoperation rate of 4.2% through 84 months, compared to 14.7% for the ACDF group (Lanman et al. 2017). In this study patients had their device removed mostly due to radicular arm pain, cervical kyphosis and sagittal imbalance, foraminal stenosis and other degenerative changes, "failed arthroplasty," and loosening of hardware. Likewise, Gornet et al. reported on the Prestige LP device, showing reoperation rates of 2.4% compared to 8% with ACDF (Gornet et al. 2017).

Overall, all of these studies and even a meta-analysis have demonstrated that TDR has a superior advantage in regard to reoperations rates (Gao et al. 2015). Reoperation rates for TDR have been consistently lower than 10%. This is particularly important due to the heavy burden that reoperations place on the cost-effectiveness, outcomes, and patient satisfaction.

Cost-Effectiveness

Due to the rapid expansion of spine surgery and increasing healthcare expenditure, the costs of a procedure are an important consideration. In addition, due to the similar outcomes between ACDF and TDR, a cost-effective analysis can offer an economic perspective and potentially shed light on which procedure is superior. Numerous studies have analyzed the cost-effectiveness of the two procedures with some studies indicating no difference (Overley et al. 2017) and others indicating that TDR is more cost-effective (Ament et al. 2016). Clinical outcomes are typically converted to health state utility when performing a cost-effective analysis. In general, utility scores for TDR are either similar (Qureshi et al. 2013) or superior to ACDF (Ament et al. 2015). However, cost has been more variable with some studies indicating lower cost with TDR (Radcliff et al. 2015, 2016a) and others showing higher costs when compared to ACDF (Overley et al. 2017). In comparison to the often-cited willingness-to-pay threshold of $50,000, TDR is a cost-effective procedure in its own right and is comparable to ACDF.

Patient Satisfaction

With a growing emphasis on improving patient satisfaction and linking it to reimbursements, many of the IDE trials have incorporated it into their studies. Overall patient satisfaction is very high for TDR and is comparable to fusion with a more recent meta-analysis indicating that it is higher for TDR (Hu et al. 2016).

Murrey and colleagues asked patients whether or not they would undergo the same surgery again. Results demonstrated that 96% of ProDisc-C patients would choose to have another disc replacement, which was similar to patients who were fused (Murrey et al. 2009). Similarly, Hisey et al. reported high levels of satisfaction among the Mobi-C and ACDF groups at all time points up to a 5-year follow-up (Hisey et al. 2016). Of note, 97.1% of patients stated they would recommend the Mobi-C single-level disc replacement to a friend, compared to 91.1% for patients who had an ACDF.

For multilevel cervical disc replacement, Radcliff et al. reported 96% satisfaction rate at 5 years with the Mobi-C device versus 89.5% satisfaction for the ACDF group. Ninety-five percent of patients receiving the Mobi-C device stated that they would recommend the surgery to a friend, while only 84% of patients in the ACDF group would recommend it (Radcliff et al. 2016b). For the Prestige LP, 95% of patients receiving the disc replacement stated they were "definitely" or "mostly" satisfied, and the same percentage of patients stated they would undergo the procedure again (Lanman et al. 2017). Gornet et al. surveyed patient's responses and reported three findings: (1) patient satisfaction was greater with TDR (94.5% vs. 89%), (2) more TDR patients felt they were helped by their surgery (94% vs. 85.5%), and (3) a larger percent of TDR patients were willing to have the surgery again (93% vs. 89%) (Gornet et al. 2017).

Pain Medication Usage

Due to widespread concerns about narcotic abuse, the impact on cost of care, and the fact that pain medication acts as a surrogate for pain control, the amount of pain medication used is another factor that has been monitored. Janssen et al. reported that the use of narcotic pain medications and muscle relaxants for TDR and ACDF was similar preoperatively and postoperatively (Janssen et al. 2015). However, Murrey et al. found that at 24 months a lower percentage of TDR patients used narcotics and muscle relaxants compared to those who had an ACDF (Murrey et al. 2009).

Return to Work

The ability to return to work is a reflection on functional outcomes and influences patient satisfaction and cost-effectiveness. One of the initially proposed advantages of TDR over fusion was that early mobilization and maintenance of normal cervical kinematics could result in earlier return to work. A systematic review comparing TDR and ACDF indicated that an equivalent rate of patients ultimately returned to work at 6 months, but those who underwent TDR resumed work sooner (Traynelis et al. 2012).

In two separate studies, Gornet and colleagues evaluated return to work following single-level and two-level TDR compared to ACDF (Gornet et al. 2015, 2017). Preoperatively, 67% of single-level and 70% of two-level disc replacement were working, and after 2 years both groups had a return-to-work rate of 73%, indicating they retained their preoperative work status well. For single-level procedures, they found that TDR returned to work on average 20 days earlier than the ACDF group even after adjusting for preoperative work status and propensity scores. With regard to two-level TDR, there was a trend to earlier return, but no statistical difference was noted. Similarly, Malham et al. reported a return to work rate of 74% after 2-year follow-up for patients undergoing disc replacement, with a median return to work time of 39 days.

In general, patients who undergo a TDR maintain their preoperative work status well, have similar return-to-work rates compared to fusion, and do appear to allow for earlier return to work. Return-to-work rates also appear to be roughly similar between the different TDR devices with most returning around 40–50 days after surgery, but no direct comparisons have been made.

Conclusion

The advent of joint arthroplasty has created many avenues in the management of degenerative joint disease; in particular, TDR has challenged the most commonly used treatment modality for cervical spine disease. While ACDF has proven to be an effective and reliable procedure, long-term data demonstrates an inevitably high incidence of adjacent segment disease. In contrast, treatment with TDR aims to eliminate strain on adjacent levels through its ability to recreate cervical spine biomechanics and preserve motion. Many biomaterials including titanium, cobalt-chromium, stainless steel, and polymers have been explored, along with different surface topographic modifications, but the ideal construct has yet to be perfected.

TDR has demonstrated many attractive advantages. It permits patients the ability to initiate range of motion immediately in the postoperative period. This eliminates periods of immobilization and permits quicker recovery to baseline status. Patients who underwent TDR have consistently expressed a very high level of satisfaction. Additionally, in comparison to ACDF, TDR has similar or even superior clinical outcomes, cost-effectiveness, and time to return to work, along with a decreased need for pain medication and lower reoperations rates. However, most of the data has resulted from the initial IDE trials which raises concerns about publication bias, external validity, confirmation bias, and financial conflict of interest (Radcliff et al. 2017). Results have been promising, but future independent research efforts are needed if TDR is to gain acceptance as a reliable alternative to ACDF.

Cross-References

▶ Cervical Total Disc Replacement: FDA-Approved Devices
▶ Cervical Total Disc Replacement: Next-Generation Devices
▶ Cervical Total Disc Replacement: Biomechanics
▶ Cervical Total Disc Replacement: Technique – Pitfalls and Pearls
▶ Cervical Total Disc Replacement: Expanded Indications
▶ Cervical Total Disc Replacement: Heterotopic Ossification and Complications

References

Ament JD et al (2015) A novel quality-of-life utility index in patients with multilevel cervical degenerative disc disease: comparison of anterior cervical discectomy and fusion with total disc replacement. Spine 40(14):1072–1078

Ament JD et al (2016) Cost utility analysis of the cervical artificial disc vs fusion for the treatment of 2-level symptomatic degenerative disc disease: 5-year follow-up. Neurosurgery 79(1):135–145

Auerbach JD et al (2008) The prevalence of indications and contraindications to cervical total disc replacement. Spine J 8(5):711–716

Baaj AA et al (2009) History of cervical disc arthroplasty. Neurosurg Focus 27(3):E10

Bohlman HH et al (1993) Robinson anterior cervical discectomy and arthrodesis for cervical radiculopathy. Long-term follow-up of one hundred and twenty-two patients. J Bone Joint Surg Am 75(9):1298–1307

Buchowski JM et al (2009) Cervical disc arthroplasty compared with arthrodesis for the treatment of myelopathy. Surgical technique. J Bone Joint Surg Am 91(Suppl 2):223–232

Burkus JK et al (2014) Clinical and radiographic analysis of an artificial cervical disc: 7-year follow-up from the Prestige prospective randomized controlled clinical trial: clinical article. J Neurosurg Spine 21(4):516–528

Cao J-M et al (2011) Clinical and radiological outcomes of modified techniques in Bryan cervical disc arthroplasty. J Clin Neurosci 18(10):1308–1312

Chin-See-Chong TC et al (2017) Current practice of cervical disc arthroplasty: a survey among 383 AOSpine International members. Neurosurg Focus 42(2):E8

Cummins BH, Robertson JT, Gill SS (1998) Surgical experience with an implanted artificial cervical joint. J Neurosurg 88(6):943–948

Cunningham BW et al (2009) Bioactive titanium calcium phosphate coating for disc arthroplasty: analysis of

58 vertebral end plates after 6- to 12-month implantation. Spine J 9(10):836–845

Dabrowski B et al (2010) Highly porous titanium scaffolds for orthopaedic applications. J Biomed Mater Res B Appl Biomater 95(1):53–61

Dejaegher J et al (2017) 10-year follow-up after implantation of the Bryan cervical disc prosthesis. Eur Spine J 26(4):1191–1198

DiAngelo DJ et al (2003) Biomechanical testing of an artificial cervical joint and an anterior cervical plate. J Spinal Disord Tech 16(4):314–323

DiAngelo DJ et al (2004) In vitro biomechanics of cervical disc arthroplasty with the ProDisc-C total disc implant. Neurosurg Focus 17(3):E7

Dmitriev AE et al (2004) 3. Adjacent level intradiscal pressures following a cervical total disc replacement arthroplasty: an in vitro human cadaveric model. Spine J 4(5):S4

Dorr LD et al (2000) Total hip arthroplasty with use of the metasul metal-on-metal articulation. J Bone Joint Surg Am 82(6):789–798

Fayyazi AH et al (2015) Assessment of magnetic resonance imaging artifact following cervical total disc arthroplasty. Int J Spine Surg 9:30

Fernström U (1966) Arthroplasty with intercorporal endoprothesis in herniated disc and in painful disc. Acta Chir Scand Suppl 357:154–159

Gao F et al (2015) An updated meta-analysis comparing artificial cervical disc arthroplasty (CDA) versus anterior cervical discectomy and fusion (ACDF) for the treatment of cervical degenerative disc disease (CDDD). Spine 40(23):1816–1823

Goffin J (2006) Complications of cervical disc arthroplasty. Semin Spine Surg 18(2):87–98

Goffin J et al (2003) Intermediate follow-up after treatment of degenerative disc disease with the Bryan cervical disc prosthesis: single-level and bi-level. Spine 28(24):2673–2678

Gornet MF et al (2015) Cervical disc arthroplasty with Prestige LP disc versus anterior cervical discectomy and fusion: a prospective, multicenter investigational device exemption study. J Neurosurg Spine 23(5):558–573

Gornet MF et al (2017) Cervical disc arthroplasty with the Prestige LP disc versus anterior cervical discectomy and fusion, at 2 levels: results of a prospective, multicenter randomized controlled clinical trial at 24 months. J Neurosurg Spine 26(6):653–667

Hilibrand AS et al (1999) Radiculopathy and myelopathy at segments adjacent to the site of a previous anterior cervical arthrodesis. J Bone Joint Surg Am 81(4):519–528

Hisey MS et al (2016) Prospective, randomized comparison of one-level Mobi-C cervical total disc replacement vs. anterior cervical discectomy and fusion: results at 5-year follow-up. Int J Spine Surg 10:10

Hu Y et al (2016) Mid- to long-term outcomes of cervical disc arthroplasty versus anterior cervical discectomy and fusion for treatment of symptomatic cervical disc disease: a systematic review and meta-analysis of eight prospective randomized controlled trials. PLoS One 11(2):e0149312

Huppert J et al (2011) Comparison between single- and multi-level patients: clinical and radiological outcomes 2 years after cervical disc replacement. Eur Spine J 20(9):1417–1426

Jacobs JJ et al (1996) Cobalt and chromium concentrations in patients with metal on metal total hip replacements. Clin Orthop Relat Res 329:S256–S263

Janssen ME et al (2015) ProDisc-C total disc replacement versus anterior cervical discectomy and fusion for single-level symptomatic cervical disc disease: seven-year follow-up of the prospective randomized U.S. Food and Drug Administration Investigational Device Exemption Study. J Bone Joint Surg Am 97(21):1738–1747

Johnson JP et al (2004) Sagittal alignment and the Bryan cervical artificial disc. Neurosurg Focus 17(6):E14

Kato S et al (2017) Comparison of anterior and posterior surgery for degenerative cervical myelopathy: an MRI-based propensity-score-matched analysis using data from the prospective multicenter AOSpine CSM North America and International Studies. J Bone Joint Surg Am 99(12):1013–1021

Kim SW et al (2016) The impact of coronal alignment of device on radiographic degeneration in the case of total disc replacement. Spine J 16(4):470–479

Kotani Y et al (1998) The role of anteromedial foraminotomy and the uncovertebral joints in the stability of the cervical spine. A biomechanical study. Spine 23(14):1559–1565

Lanman TH et al (2017) Long-term clinical and radiographic outcomes of the Prestige LP artificial cervical disc replacement at 2 levels: results from a prospective randomized controlled clinical trial. J Neurosurg Spine 27(1):7–19

Lee SE, Chung CK, Jahng TA (2012) Early development and progression of heterotopic ossification in cervical total disc replacement. J Neurosurg Spine 16(1): 31–36

Lei T et al (2017) Anterior migration after Bryan cervical disc arthroplasty: the relationship between hyperlordosis and its impact on clinical outcomes. World Neurosurg 101:534–539

Leung C et al (2005) Clinical significance of heterotopic ossification in cervical disc replacement: a prospective multicenter clinical trial. Neurosurgery 57:759–763

Murrey D et al (2009) Results of the prospective, randomized, controlled multicenter Food and Drug Administration investigational device exemption study of the ProDisc-C total disc replacement versus anterior discectomy and fusion for the treatment of 1-level symptomatic cervical disc disease. Spine J 9(4):275–286

Nasto LA, Logroscino C (2016) Cervical disc arthroplasty. In: Menchetti P. (eds) Cervical spine. Springer, Cham, pp 193–206

Overley, S.C. et al., 2017. The 5-year cost-effectiveness of two-level anterior cervical discectomy and fusion or cervical disc replacement: a Markov analysis. Spine J. https://doi.org/10.1016/j.spinee.2017.06.036

Pham V-H (2014) Improving osseointegration of Co-Cr by nanostructured titanium coatings. Springerplus 3:197

Pickett GE, Rouleau JP, Duggal N (2005) Kinematic analysis of the cervical spine following implantation of an artificial cervical disc. Spine 30(17):1949–1954

Qureshi SA et al (2013) Cost-effectiveness analysis: comparing single-level cervical disc replacement and single-level anterior cervical discectomy and fusion: clinical article. J Neurosurg Spine 19(5):546–554

Radcliff K, Zigler J, Zigler J (2015) Costs of cervical disc replacement versus anterior cervical discectomy and fusion for treatment of single-level cervical disc disease: an analysis of the Blue Health Intelligence database for acute and long-term costs and complications. Spine 40(8):521–529

Radcliff K, Lerner J et al (2016a) Seven-year cost-effectiveness of ProDisc-C total disc replacement: results from investigational device exemption and post-approval studies. J Neurosurg Spine 24(5):760–768

Radcliff K, Coric D, Albert T (2016b) Five-year clinical results of cervical total disc replacement compared with anterior discectomy and fusion for treatment of 2-level symptomatic degenerative disc disease: a prospective, randomized, controlled, multicenter investigational device exemption clinical trial. J Neurosurg Spine 25(2):213–224

Radcliff K et al (2017) Bias in cervical total disc replacement trials. Curr Rev Musculoskele Med 10(2):170–176

Rajakumar DV et al (2017) Adjacent-level arthroplasty following cervical fusion. Neurosurg Focus 42(2):E5

Rožanković M, Marasanov SM, Vukić M (2017) Cervical disk replacement with discover versus fusion in a single-level cervical disk disease: a prospective single-center randomized trial with a minimum 2-year follow-up. Clin Spine Surg 30(5):E515–E522

Sasso RC et al (2008) Motion analysis of Bryan cervical disc arthroplasty versus anterior discectomy and fusion: results from a prospective, randomized, multicenter, clinical trial. J Spinal Disord Tech 21(6):393–399

Siskey R et al (2016) Are PEEK-on-ceramic bearings an option for total disc arthroplasty? An in vitro tribology study. Clin Orthop Relat Res 474(11):2428–2440

Smucker JD, Sasso RC (2011) Cervical disc replacement. In: Rothman Simeone the spine. pp 808–825

Traynelis VC (2004) The Prestige cervical disc replacement. Spine J 4(6 Suppl):310S–314S

Traynelis VC, Leigh BC, Skelly AC (2012) Return to work rates and activity profiles: are there differences between those receiving C-ADR and ACDF? Evid Based Spine Care J 3(S1):47–52

Urban RM et al (2000) Dissemination of wear particles to the liver, spleen, and abdominal lymph nodes of patients with hip or knee replacement*. J Bone Joint Surg Am 82(4):457–477

Walter A (1992) On the material and the tribology of alumina-alumina couplings for hip joint prostheses. Clin Orthop Relat Res 282:31–46

Wu J-C et al (2012) Differences between 1- and 2-level cervical arthroplasty: more heterotopic ossification in 2-level disc replacement: clinical article. J Neurosurg Spine 16(6):594–600

Wu T-K et al (2017) Multilevel cervical disc replacement versus multilevel anterior discectomy and fusion. Medicine 96(16):e6503

Xing D et al (2013) A meta-analysis of cervical arthroplasty compared to anterior cervical discectomy and fusion for single-level cervical disc disease. J Clin Neurosci 20(7):970–978

Zhao H et al (2014) Multi-level cervical disc arthroplasty (CDA) versus single-level CDA for the treatment of cervical disc diseases: a meta-analysis. Eur Spine J 24(1):101–112

Zhu Y, Tian Z et al (2016a) Bryan cervical disc arthroplasty versus anterior cervical discectomy and fusion for treatment of cervical disc diseases. Spine 41(12): E733–E741

Zhu Y, Zhang B et al (2016b) Cervical disc arthroplasty versus anterior cervical discectomy and fusion for incidence of symptomatic adjacent segment disease: a meta-analysis of prospective randomized controlled trials. Spine 41(19):1493–1502

Cervical Total Disc Replacement: Biomechanics

42

Joseph D. Smucker and Rick C. Sasso

Contents

Introduction 790

Background 790

General Cervical Spine Biomechanics 791

History of Disc Arthroplasty Design Kinematics 794

Current Kinematic Studies 795
The BRYAN® Disc 795
The PRODISC-C® 797
Combined PRODISC-C®/PRESTIGE® ST/LP Studies 798
The PRESTIGE® Disc 800
The PCM® Disc 800
Computer Simulation and Finite Element (FE) Modeling Studies 801
Multidisc Studies 803

Future Kinematic Design Principles 804

Conclusions 804

References 805

Abstract

Cervical disc arthroplasty is an evolving surgical concept designed to treat certain pathological conditions of the cervical spine. The introduction of arthroplasty devices has stimulated novel studies aimed at understanding motion in the cervical spine and has also driven investigators to examine the consequences that result from surgical alteration of pathological structures. The study of cervical "biomechanics" and "kinematics" has evolved from basic analysis of flexion/extension radiographs to complex, computer-assisted modeling that aides investigators in understanding concepts such as center of rotation (COR), functional spinal unit (FSU)

J. D. Smucker (✉) · R. C. Sasso
Indiana Spine Group, Carmel, IN, USA
e-mail: joe@smuckermd.com;
rsasso@indianaspinegroup.com

© Springer Nature Switzerland AG 2021
B. C. Cheng (ed.), *Handbook of Spine Technology*,
https://doi.org/10.1007/978-3-319-44424-6_74

translation, and coupled motion. In recent years kinematic studies have contributed to our understanding of adjacent level degeneration and index-level facet loading. We review the young science of cervical arthroplasty biomechanics.

Keywords

Cervical spine · Arthroplasty · Biomechanics · Kinematics · Finite Element · Motion

Introduction

The design of arthroplasty devices for the human cervical disc has brought about a renewed interest in the biomechanics of the cervical spine. Modern techniques of assessment and measurement are currently being employed parallel to traditional outcome measurements in the hope that such information may advance the collective understanding of disc arthroplasty on cervical motion.

Concepts of cervical arthroplasty have undergone a dramatic evolution since the development of the original Bristol/Cummins disc arthroplasty device. At a basic level, motion retention/preservation is a primary kinematic measure of device success in this procedure, though the current indications for the procedure are typically of neurological origin. Retention of motion or "motion sparing" in cervical arthroplasty has quickly evolved in device design over the past 20–30 years. Materials used in disc arthroplasty have also changed. The evolution of metal-on-metal implants has occurred in parallel with the development of novel bearing concepts incorporating metal alloys, polyethylene, and ceramics.

Currently the term "cervical arthroplasty" is applied to the procedure of "disc arthroplasty" or "disc replacement." A number of these devices are in the process of early use or are involved in US Food and Drug Administration (FDA) trials. While the early data from clinical trials is encouraging, there remains a need to demonstrate the biomechanical properties of these devices and techniques in the intermediate and long term. Cervical arthroplasty of the disc alone is not intended to address the posterior elements at the index surgical level – leaving open the option for future modifications of the concept of cervical arthroplasty and kinematic motion sparing.

Background

The cervical spine consists of vertebral bodies with intervening discs and soft tissue structures that support motion and protect the neural and vascular elements. From a biomechanical perspective, these discs and their corresponding facets function in load bearing and motion transfer allowing for flexion/extension, lateral bending, and rotation as well as complex coupled motions. In addition to its biomechanical functions in motion, the cervical spine serves as the protective passage for the spinal cord and vertebral arteries.

Cervical spondylosis is the process by which the cervical spine most frequently loses motion and is occasionally to blame for ensuing neurological phenomena which have been the traditional indication for surgical interventions. Disc degeneration is well documented as the transition from mild degenerative disc disease to multilevel cervical spondylosis progresses. For many years, the surgical treatment for pathology in the cervical intervertebral disc has been limited to procedures which remove pathologic disc material and address the bony and neurologic pathology in the region of the excised disc.

Anterior cervical discectomy and fusion (ACDF) is a proven intervention for patients with radiculopathy and myelopathy (Bohlman et al. 1993). It has served as the standard by which other cervical and spinal disorders may be judged as the result of its high rate of success. The success of this technique is often judged based upon its consistent ability to relieve symptoms related to neurological dysfunction. In this sense, the clinical results with regard to the patient's index complaint are outstanding. The radiographic results of this technique are also initially predictable with a high rate of fusion. Plating techniques have diminished the need for postoperative immobilization or eliminated them entirely (Campbell et al. 2009). However, because of limitations specific to this procedure, investigators have developed surgical alternatives to fusion that attempt to address the kinematic and biomechanical issues inherent in it.

A major concern related to the treatment of cervical degenerative disc disease (DDD) and spondylosis with ACDF are the issues of adjacent segment degeneration and adjacent segment disease (ASD). Adjacent segment degeneration is manifest as the radiographic appearance of degenerative change at a level directly above or below a level treated with a surgical intervention – typically being associated with degeneration of a level adjacent to a fused level. Adjacent segment disease (ASD) is defined as adjacent segment degeneration causative of clinical symptoms (pain and/or neurological disorders) severe enough to lead to patient complaint and/or require operative intervention (Hilibrand et al. 1999). Adjacent segment degenerative change has been reported to be as high as 92% by Goffin et al. who wrote a long-term follow-up on patients after treatment with anterior interbody fusion (2004). While there remains some debate as to the causation of adjacent segment degeneration – with a mix of postsurgical (altered biomechanics) and naturally determined aging (genetics) cited as root causes – there is little debate as to the existence of this phenomenon. A number of studies have made a consistent point of distinguishing between radiographic "degeneration" and symptomatic "disease" (Goffin et al. 2004; Robertson et al. 2005).

There is clinical evidence to support the postsurgical nature of ASD with respect to kinematics. In patients previously treated with fusion, adjacent segment disease has been documented at a rate of 2.9% of patients per annum by Hilibrand et al., and 25% of patients undergoing cervical fusion will have new onset of symptoms within 10 years of that fusion (Hilibrand et al. 1999). This study has received a great deal of attention and has led to further investigations as to kinematic and biomechanical causation. Other reports have focused on the recurrence of neurological symptoms and degenerative changes adjacent to fused cervical levels (Goffin et al. 1995, 2004). The concept that adjacent levels need to kinematically compensate for loss of motion in the fused segment may also be valid. Segments adjacent to a fusion have an increased range of motion and increased intradiscal pressures (Eck et al. 2002; Fuller et al. 1998).

Total intervertebral disc replacement (TDR) is intended to preserve motion, minimize limitations of fusion, and may allow patients to quickly return to routine activities. The primary goals of the procedure in the cervical spine are to restore disc height and segmental motion after removing local pathology that is deemed to be the source of a patient's index complaint. A secondary intention is the preservation of normal kinematics at adjacent cervical levels, which may be theorized to prevent later adjacent level degeneration. Cervical TDR avoids the morbidity of bone graft harvest (Silber et al. 2003; St. John et al. 2003). It also may avoid complications such as pseudarthrosis, issues caused by anterior cervical plating, and cervical immobilization side effects.

General Cervical Spine Biomechanics

Motion in the cervical spine implies a direct interaction between two or more cervical vertebrae and their supporting structures. A motion segment of the cervical spine, often analyzed as a functional spinal unit (FSU), is complex. The cervical spine is much more than a single FSU, and investigators have found that much more complex kinematic relationships exist as they seek to understand not only the effects of various treatments on a single ("index") FSU but also the effects of that same treatment on adjacent or remote FSUs.

Each FSU consists of three compartments (the disc and two facets) and multiple supporting ligamentous and soft tissue structures. The normal cervical spine exhibits complex coupled motions in addition to the traditionally understood independent kinematic motions such as anterior-posterior translation during flexion and extension. An implant designed to replace the cervical disc should consider the effect of all three compartments and the multiple ligamentous and soft tissue structures present in this complex environment.

One of the primary goals of cervical disc replacement is to reproduce "normal kinematics" after implantation. Fortunately, numerous kinematic studies of various designs have been undertaken parallel to US FDA (IDE) studies. Collectively, these studies may be classified by device and/or study design criteria. Some investigators have taken advantage of novel finite element (FE)-based techniques, while others have used more traditional in vivo or in vitro means. Review of these studies is instructive in understanding the current state of kinematic knowledge with regard to cervical TDR. Over time, similar studies may suggest which type of implant design will provide "kinematically accurate" motion.

Early device designs made use of ball-in-socket articulations within the device. A ball-in-socket (constrained design) does not allow for natural translation. The complexity of the cervical spine requires a "balance" of all the significant structures including facets and ligaments. A ball-in-socket, by its design, dictates the kinematics of motion irrespective of traditional FSU behaviors and eliminates the normal anterior/posterior translation that the facets provide. A number of studies describe the increased forces born by these facets – a phenomenon sometimes described as "kinematic conflict."

The most significant effect of this change in facet loading is in extension. During flexion the facets "un-shingle" and reduce their involvement in constraining the motion of the functional spine unit. However, when the spine goes into extension, the facets "shingle" and become more involved in constraining the motion. Thus, with a constrained facet joint and a constrained arthroplasty device, one would expect to see binding or limited motion as one joint works against the other in the FSU. For this reason device designers have introduced less constraint in more recently designed devices.

There are a number of methods by which kinematic data may be derived. In vivo measurements in the human are often made through review of flexion and extension radiographs that are digitized and subsequently measured with software packages (Sasso and Best 2008) (Figs. 1 and 2). Alternatively, nonhuman in vivo measurements may occur in translational projects wherein the spine is tested via histological and radiographic means as well as benchtop environments with mechanical loading devices, optical tracking (Fig. 3), and pressure sensors. In vitro testing of human cadaveric specimens occurs via similar benchtop testing protocols with the obvious exclusion of histological means (Figs. 3 and 4).

Computer-assisted finite element (FE) modeling is a technique by which a computer-generated

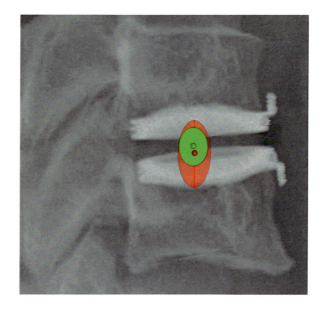

Fig. 1 The BRYAN® Cervical Disc Prosthesis is demonstrated in vivo in this lateral cervical radiograph. The center of rotation (COR) has been calculated pre- and post-placement of the arthroplasty prosthesis at the index surgical level. Software allows for in vivo analysis of kinematic changes in humans via radiographic means over time. Changes in COR may correlate to long-term kinematic outcomes, device survival, and adjacent level changes. (© Courtesy of Rick Sasso, Indianapolis, IN)

Fig. 2 The BRYAN® Cervical Disc Prosthesis is demonstrated in vivo in this lateral cervical radiograph. The center of rotation (COR) has been calculated pre- and post-placement of the arthroplasty prosthesis at the adjacent surgical level. (© Courtesy of Rick Sasso, Indianapolis, IN)

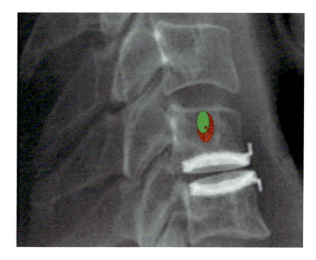

Fig. 3 Explanted spinal specimens may be tested in a number of ways. Optical tracking allows for real-time tracking of motion and is commonly used in conjunction with forces applied to the cervical spine in a controlled, monitored environment. Cameras on this OptiTrack™ Device (NaturalPoint® Inc., Corvallis, Oregon) follow the motion of rigid bodies. (© Courtesy Nicole Grosland, PhD and Joseph D. Smucker, MD – The University of Iowa and Indiana Spine Group)

model of the cervical spine is modified to include surgical procedures such as ACDF or arthroplasty techniques and principles (Ahn and DiAngelo 2008; Kallemeyn et al. 2009). Specimen-specific modeling is a more refined method of testing such principles (Kallemeyn et al. 2009) (Fig. 5). FE modeling has the potential advantage of providing investigators with a more flexible testing environment given the assumption of model-specific limitations.

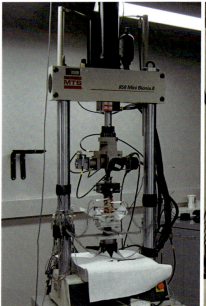 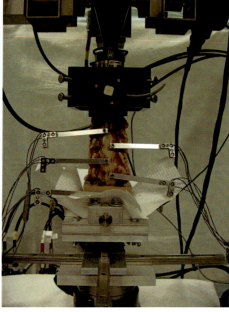

Fig. 4 Controlled application of force within the defined degrees of freedom in the cervical spine is applied to create motion in an ex vivo environment. This MTS™ 858 Mini Bionix II system (MTS Systems Corp., Eden Prairie, MN) applies precise force via computer-controlled hydraulic mechanisms. Optical tracking via the OptiTrack™ system is combined with this controlled application of force to track and analyze simple and coupled motions created in this multi-FSU spinal specimen – allowing for real-time tracking of motion. (© Courtesy Nicole Grosland, PhD and Joseph D. Smucker, MD – The University of Iowa and Indiana Spine Group)

History of Disc Arthroplasty Design Kinematics

An understanding of the evolution of cervical TDR serves as an important lesson in the concepts of kinematic device design properties and articular constraint. In the late 1980s, Cummins et al. (1998) developed a metal-on-metal ball-and-socket cervical disc replacement comprised of 316 L stainless steel. With the acquisition of this technology and the later development of new metal-on-metal devices, a rapid transition evolved to the most recent device, the PRESTIGE® LP (Medtronic Sofamor Danek, Memphis, TN). A predecessor of this device, the PRESTIGE® ST (Medtronic Sofamor Danek, Memphis, TN), is currently approved for human use by the US FDA.

A number of devices have evolved parallel to the metal-on-metal implants and include the BRYAN® Disc (Medtronic Sofamor Danek, Memphis, TN), the Porous Coated Motion Prosthesis (PCM®, NuVasive, San Diego, CA), the SECURE-C® (Globus Medical, Audobon, PA), and the MOBI-C® (Zimmer Biomet, Parsippany, NJ). To date, several such devices have obtained approval for use in the US market: the PRODISC-C® (Centinel Spine, West Chester, PA) and the BRYAN® Disc. Each of the other devices is in the process of limited human trials and/or US FDA-IDE submission and represents an alternative to metal-on-metal bearing surfaces which have the potential for metal debris and systemic concentration of metal ions.

While the ideas of bearing surfaces, wear debris, and constraint are not new to discussions with regard to arthroplasty in general, they are relatively young in the spine. In fact, a full understanding of the term "constraint" with regard to cervical kinematics post-disc arthroplasty has not been agreed upon – as constraint may arise within the device or as a result of the local anatomy

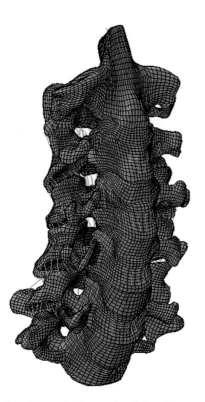 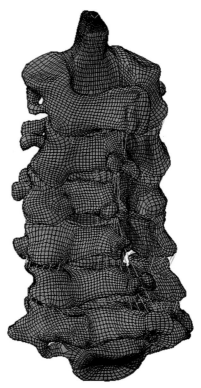

Fig. 5 Dorsal and ventral views of a finite element (FE) model of the human cervical spine (C2-C7) are presented. Multiblock analysis occurs after biomechanical properties are assigned to bony and soft tissue structures. Initial specimen-specific models are created from computed tomograpic (CT) analysis of the human cervical spine. The specimen may then be analyzed in a computer environment with simulation of motion via computer applied forces to the model. The model may be further modified via implantation of spinal devices such as disc arthroplasty devices. Facet forces, intradiscal forces, and other kinematic measurements such as COR may be calculated. (© Courtesy Nicole Grosland, PhD and Joseph D. Smucker, MD – The University of Iowa and Indiana Spine Group)

(facets, PLL, etc.). As the knowledge base in spine TDR increases, intelligent investigations and discussions will include many of these concepts and may redefine our understanding of them.

It is relevant to understand that the load born by devices in the cervical spine is dissimilar to that born in the lumbar spine. The biomechanical environment of the cervical spine has been taken into account in the design of the current generation of these devices. As intermediate- and long-term studies on individual devices become available, the design concepts of these initial devices will have the opportunity for continued examination in their in vivo environment.

Current Kinematic Studies

The BRYAN® Disc

Galbusera et al. published their review in March 2006 of the biomechanics and kinematics at the C5–C6 spinal unit both before and after placement of a BRYAN® Cervical Prosthesis (Galbusera et al. 2006). In this study, the authors produced a finite element (FE) model of the functional spinal unit at C5–C6. The model employed reconstruction of both the vertebral bodies at C5 and C6 and representations of the vertebra, ligaments, and discs at this level. The authors applied motion through the intact FSU to assess several kinematic measures with a compression preload. The kinematic measures studied included flexion/extension moments, pure lateral bending moments, and a pure torsion moment. They reviewed their results comparing this to known data from prior publications. The FE model was then modified to include the placement of the BRYAN® Arthroplasty Device with repeat stimulations.

The authors noted that they were able to calculate the instantaneous center of rotation of C5 with respect to C6 throughout flexion/extension. In general, FSU rotation curves post-arthroplasty were comparable to those obtained from the intact FSU with the exception of a slightly greater stiffness that was noted to be "induced by the artificial disc" (Galbusera et al. 2006). Pre- and post-arthroplasty data suggested that the position of the instantaneous center of rotation was similar in both models and was stable throughout flexion and extension – being confined to a small area "corresponding to the physiological region in both models" (Galbusera et al. 2006).

Galbusera et al. later published a more detailed finite element model from C4 to C7 expanding upon their 2006 study (Galbusera et al. 2008). In this study the group produced a finite element model including functional spinal units and appropriate soft tissue structures from C4 to C7 for kinematic testing in flexion and extension. Once again, a BRYAN® Disc Prosthesis was inserted at the C5–C6 level. Pre- and post-placement motions were analyzed. Once again, in both flexion and extension, placement of the BRYAN® Disc Prosthesis showed that there was a "general preservation of the forces transmitted through the facet joints" and that "calculated segmental motion was preserved after disc arthroplasty" (Galbusera et al. 2008). Similar to the prior study, the instantaneous centers of rotation (ICR) in flexion and extension showed preservation pre- and post-placement of the BRYAN® Disc.

This study did suggest some post-placement asymmetry in flexion and extension that the authors summarized may be secondary to lack of the anterior longitudinal ligament post-prosthesis

placement. However, they were able to conclude that disc arthroplasty with the BRYAN® Disc in this multi-FSU model reproduced "near physiological motion" at the C5–C6 level (Galbusera et al. 2008).

Pickett et al. have also described the kinematics of the cervical spine following implantation of the BRYAN® Cervical Disc (Pickett et al. 2005). In this prospective cohort study, the authors described a total of 20 patients who underwent single- or two-level implantation of the BRYAN® Disc. Each of these patients was treated per protocol for a degenerative condition of the cervical discs that was producing neurologic symptoms including radiculopathy and/or myelopathy. From a kinematic standpoint, this study examined pre- and postsurgical plain radiographs including neutral lateral as well as flexion and extension radiographs at prescribed intervals. Kinematic parameters including rotation, horizontal translation, change in disc height, and center of rotation at each spinal level were evaluated using quantitative motion analysis software produced by Medical Metrics Corporation (Houston, Texas).

The authors demonstrated a postsurgical preservation of range of motion at the operated spinal segment with a mean postsurgical range of motion of 7.8° at the 24-month postsurgical follow-up. They noted that disc placement "either placed at C5–6 or C6–7" seemed to change the "relative contribution of each spinal segment to overall sagittal rotation (DiAngelo et al. 2004)." They also noted that total overall cervical motion as measured from C2 to C7 was increased at late follow-up intervals. There were no significant changes in sagittal rotation, anterior-posterior disc height, translation, or center of rotation following placement of the BRYAN® Arthroplasty Device at the follow-up intervals. The authors concluded that placement of BRYAN® Artificial Disc for cervical radiculopathy and or myelopathy appears to "reproduce the preoperative kinematics of the spondylotic disc (Pickett et al. 2005)." This in vivo study tends to support the finite element studies noted earlier as published by Galbusera et al. (2006, 2008).

Rick Sasso and Natalie Best published a novel BRYAN® Disc article in February 2008 analyzing radiographic data from patients who had undergone either ACDF with allograft and plating or placement of a single-level BRYAN® Cervical Disc (Sasso and Best 2008). In this single-level study, all patients had radiographic follow-ups immediately preoperatively as well as postoperatively at regular intervals up to a 24-month endpoint. The study represents data from a subset of patients involved in the randomized prospective BRYAN® Cervical Disc Arthroplasty study for the US FDA. The authors evaluated flexion/extension and neutral lateral radiographs at the prescribed intervals and analyzed motion using Medical Metrics software similar to that described in the prior chapter by Pickett et al. (2005). They quantified functional spinal unit motion, translation, and center of rotation.

As expected, there was significantly more motion in flexion and extension in the disc replacement group than in the fusion group at the index surgical level. In this study, the arthroplasty FSUs were able to retain an average range of motion of 6.7° at the 24-month follow-up interval. This was in contrast to the range of motion of the fusion group which was initially 2.0° at the 3-month follow-up, decreasing overtime to 0.6° at the final 24-month follow-up. The authors also noted that flexion/extension both above and below the operative level was not statistically different in those groups having undergone cervical arthroplasty versus fusion. An interesting finding, however, is that mobility overall increased for both groups over time. At levels above the fusion, there was an increase in translation in comparison to the arthroplasty device which showed no evidence of an increase in translation at the adjacent level. The finding of increased translation was only statistically significant at the 6-month follow-up interval. The authors concluded that the BRYAN® Disc appeared to preserve preoperative kinematics at adjacent levels in comparison to fusion which showed some changes overall in the kinematics (Sasso and Best 2008). This did support the postulation that arthroplasty has the potential to preserve cervical kinematics at adjacent levels postoperatively.

Sasso et al. also reported upon the motion analysis/kinematic properties of all patients enrolled in a prospective randomized multicenter trial for the BRYAN® Cervical Artificial Disc Prosthesis (Sasso et al. 2008). Their overall objective in this study was to analyze the entire set of patients in a prospective fashion similar to the subset which was previously reported (Sasso and Best 2008). In this study, all patients received either a single-level ACDF or a single-level disc arthroplasty with the BRYAN® Cervical Disc Prosthesis. A total of 221 patients received fusion, whereas 242 received a single-level arthroplasty. Operative segments could include the C3–4 disc space down to the C6–7 disc space. Similar to the previous subset, the authors analyzed flexion/extension and neutral lateral radiographs obtained at prescribed intervals postoperatively in comparison to the preoperative interval. This study examined patients up to and including the 24-month interval. Medical Metrics software was once again used to track the cervical vertebral bodies at the index FSU looking at flexion and extension range of motion as well as translation.

Similar to the prior subset, the arthroplasty group retained statistically significant increases in motion at the index FSU in comparison to the ACDF group. The arthroplasty group had an average of 7.95° of motion at the 24-month follow-up. The preoperative range of motion at the same FSUs was 6.43° with no significant evidence of degeneration of motion at the same FSU following arthroplasty at the 24-month interval. As expected, average range of motion in the fusion group slowly diminished to the point of being 0.87° at 24 months. Preoperatively this group had a range of motion of 8.39°. Also noted was no evidence of BRYAN® Disc migration or subsidence at the 24-month follow-up – suggesting that the arthroplasty device was functioning as designed at this early follow-up interval and reproducing the kinematics of the degenerative disc space at the index FSU in comparison to fusion of those same levels.

The PRODISC-C®

DiAngelo et al. have examined the in vitro biomechanics of the PRODISC-C (DiAngelo et al. 2004). Their study was designed to compare disc arthroplasty to ACDF in cervical spine biomechanics in a multilevel human cadaveric model. This study employed three spinal conditions: intact harvested specimens alone, single-level arthroplasty specimens, and single-level fusion specimens. The study incorporated a total of six fresh human cadaveric specimens harvested from C2 to T1. All specimens were treated according to the group assigned at the C5–6 level following testing in their intact condition. This study simulated fusion in a unique way. Fusion was accomplished across the treated spinal level via custom designed fixtures similar to an external fixation system. Following surgical treatment according to protocols, kinematic principals were tested under biomechanical loading devices. This was done with a programmable testing apparatus that "replicated physiologic flexion/extension, lateral bending, and axial rotation (DiAngelo et al. 2004)." The authors then measured vertebral motion via applied load and bending moments.

As expected, the simulated fusion was successfully able to diminish motion at the treated level relative to the harvested untreated as well as disc arthroplasty conditions. The authors noted that adjacent segment motion increased in those specimens following the reduction of motion at the simulated fusion segment. This study noted that in all modes of testing, the PRODISC-C arthroplasty device "did not alter the motion patterns at either the instrumented level or adjacent segments compared with the harvested condition except in extension (DiAngelo et al. 2004)."

Puttlitz et al. have examined post-disc arthroplasty kinematics using the PRODISC-C in a human cadaveric model (Puttlitz et al. 2004). This study utilized a total of six fresh frozen human cadaveric spines to evaluate two different spinal conditions including both the intact and post-disc arthroplasty condition at the C4–C5 level. Prior to testing, compression and a follower load were applied, as well as pure moment loading to the specimens to evaluate treatment kinematics and pretreatment kinematics. Range of motion (ROM) kinematics was then measured using an optical tracking system, and data was reported.

The results of this limited cadaveric study suggest that the PRODISC-C was able to retain "approximate" intact motion in all three rotation planes "flexion/extension, rotation, and lateral bending (Puttlitz et al. 2004)." They also examined coupled rotations including lateral bending during axial rotation and axial rotation during lateral bending – noting no significant difference in these two tested conditions following arthroplasty. They concluded that ball-and-socket devices such as the PRODISC-C can "replicate physiologic motion at the affected and adjacent levels (Puttlitz et al. 2004)." This is the only study on the PRODISC-C that examines a motion coupling from a kinematic standpoint and suggests maintenance of the coupled motions following cervical arthroplasty. It is possible that a larger in vitro study could provide further insight into the coupling motions examined in this study that were novel to it.

Combined PRODISC-C®/PRESTIGE® ST/LP Studies

Chang et al. have looked at both the PRODISC-C® and PRESTIGE® Artificial Devices compared with ACDF in a cadaveric model (Chang et al. 2007a). The object of the authors' investigation was to examine cervical kinematics at surgically treated levels as well as adjacent segments in a cadaveric model – evaluating two different types of cervical artificial disc devices in comparison to the intact spine and a fusion model. For the purposes of this study, a total of 18 cadaveric human spines were tested in their intact state with kinematic modes including flexion/extension, axial rotation, and lateral bending. These three groups of specimens were then subjected to a surgical intervention including placement of a PRODISC-C®, a PRESTIGE II® Artificial Disc, or ACDF. All specimens were operated at the C6–7 level. This study simulated ACDF with placement of a 7 mm tapered cortical allograft followed by placement of a rigid anterior cervical plate and screws "to maintain lordosis at the treated level (Chang et al. 2007a)." Placement of either the PRESTIGE® or the PRODISC®

device was performed according to the manufacturers' recommended surgical technique at the C6–7 level.

Range of motion was noted to increase after arthroplasty in comparison with the intact spine in extension in both the PRODISC-C® and PRESTIGE® groups as well as in flexion in both arthroplasty groups. With respect to bending, the post-arthroplasty ROMs were greater than those of the intact spine in both arthroplasty groups; this was also similar for rotation. Adjacent level ROM was noted to decrease in all specimens that underwent implantation of a cervical arthroplasty device for all tested kinematic modes. With respect to ROM adjacent to the fusion-treated spines, it was noted to diminish in all motion modes at the treated level but increase at all adjacent levels with a reported range of 3–20%. Adjacent level range of motion diminished in all modes post-arthroplasty with the exception of extension in those patients who underwent a total disc arthroplasty.

This study lends additional credence to the idea of adjacent level disease as a result of surgery as noted by the increased range of motion kinematics at adjacent levels in those cadaveric specimens undergoing ACDF in comparison to the diminished range of motion noted in those patients undergoing cervical disc arthroplasty.

Chang et al. have also evaluated adjacent level disc pressure and facet joint forces after cervical arthroplasty with the PRODISC-C®/PRESTIGE® devices in comparison to ACDF in an in vitro human cadaveric model (Chang et al. 2007b). In this study, the authors examined intradiscal pressures at adjacent levels, as well as facet joint stress following both arthroplasty and cervical spine fusion in 24 human cadaveric spines obtained from C3 to T2. This study examined a surgical intervention at C6–7 in 18 of these specimens. Six specimens were excluded from the original 24 in the study based upon pre-procedural radiographic studies suggesting bone abnormalities. This study examined intradiscal pressures with pressure transducer needles. The forces in the facets, however, were indirectly measured.

The specimens were then divided into three groups with six specimens per group – each receiving either an artificial disc implantation (PRODISC-C® or PRESTIGE®) or in the case of the third group an ACDF. With respect to the PRODISC-C® group, a 7 mm height disc was chosen, and with respect to the PRESTIGE® group, an 8 mm height disc was chosen. These were determined to be "adequate for the cadaveric specimens (Chang et al. 2007b)." The fusion groups, as per a previous study reported by Chang et al. (Rousseau et al. 2008), underwent fusion with a 7 mm lordotic tapered allograft fixed with a rigid plate and screw.

Biomechanical testing ensued with flexion/extension, lateral bending, and axial rotation modes measured. In the arthroplasty-treated specimens, the intradiscal pressure was not significantly different in comparison to the intact spine at adjacent levels proximal and distal to the arthroplasty FSU. However, in those specimens treated with fusions, the intradiscal pressures increased at the location of the posterior annulus fibrosus in extension and at the location of anterior annulus in flexion at the cranial adjacent level. At the caudal adjacent level intradiscal pressure change was not noted to be significant. Indirect measurements of facet forces were computed in this study and were noted to be minimal in flexion, bending, and rotation modes in both arthroplasty- and fusion-treated spines. In extension the arthroplasty models exhibited an increase in facet forces at the treated FSUs in comparison to the fusion model where the facet forces decreased at the treated FSU and increased at the adjacent segments (Chang et al. 2007b).

Rousseau et al. undertook an in vivo analysis of two types of ball-and-socket cervical disc devices which they classified as "two-piece implants (Rousseau et al. 2008)." The authors of this study considered three-piece implants to be those with a mobile nucleus between two metal implants. They examined a total of 26 patients who had been implanted with the PRESTIGE® LP Device and compared them to 25 patients who had been implanted with the PRODISC-C® Device. Investigational specimens were then referenced against the measurements of 200 healthy cervical discs in vivo. Spineview™ software (Surgiview, Paris, France) was used to calculate the intervertebral range of motion and the mean center of rotation kinematic variables. The authors also calculated the center of rotation between full flexion and extension for range of motion.

In comparison to the normal non-implanted vertebral discs, the range of motion kinematics in flexion and extension were noted to be significantly reduced with both types of arthroplasty. Comparing the two arthroplasty groups head to head, range of motion was similar, and the location of the center of rotation with full flexion and extension appeared to be "influenced by the type of intervertebral disc despite interindividual variability (Rousseau et al. 2008)." Specifically, the authors noted that there was a trend toward a "more anterior and superior" location of the center of rotation in full flexion and extension with the prosthetic devices then observed in normal nonoperated control discs (Rousseau et al. 2008). This comparison of two-piece ball-and-socket-type prosthesis was notable for the fact that neither cranial nor caudal types of device designs were able to fully restore flexion and extension kinematics to normal mobility in the kinematic measurements described in the study including range of motion and center of rotation.

The PRESTIGE® Disc

DiAngelo et al. have described an in vitro biomechanical study comparing non-fusion (intact specimen) to ACDF and cervical arthroplasty in a multilevel human cadaveric model (DiAngelo et al. 2003). The study was conducted using a programmable testing apparatus that allowed for replication of physiologic flexion/extension and lateral bending. The authors measured vertebral motion applied load and bending moments. The authors used the PRESTIGE® ST cervical joint for arthroplasty and an Orion® (Medtronic Sofamor Danek, Memphis, TN) plate to simulate fusion in this small cadaveric study. Included were a total of four fresh human cadaveric specimens harvested to include C2–T1.

Following their measurements, they reported findings. The application of an anterior cervical plate significantly decreased the motion across the fusion site relative to the native or artificial joint conditions. The placement of a PRESTIGE® artificial cervical joint "did not alter the motion patterns at either the instrumented level or the adjacent segments compared with the harvested condition (DiAngelo et al. 2003)." This study of kinematics is novel not only in the maintenance of normal range of motion at the implanted FSU but also with regard to maintenance of normal motion at all segments of the spine status post-placement of a PRESTIGE® cervical disc prosthesis. Unfortunately, this small in vitro study did not have the power ability to make large in vitro analyses.

The PCM® Disc

Several novel kinematic studies have been performed with regard to the PCM® Device. The device has undergone basic testing from a kinematic standpoint (Hu et al. 2006) in addition to studies that add to the basic kinematic studies in novel ways (McAfee et al. 2003; Dmitriev et al. 2005). These have included studies that examine the role of the posterior longitudinal ligament (PLL) and those that measure adjacent level intradiscal pressures following placement of the PCM® Device (Hu et al. 2006; McAfee et al. 2003; Dmitriev et al. 2005).

Hu et al. have examined the PCM® arthroplasty device, evaluating biomechanical as well as other factors, in a caprine animal model (Hu et al. 2006). The PCM® Disc was tested in vivo and ex vivo in 12 goats divided into 2 distinct groups. These two groups differed in their survival periods – 6 and 12 months, respectively. Each specimen underwent an anterior discectomy at the level C3–C4 followed by implantation of the PCM® Device. Outcomes of the study were based upon examination of the prosthesis by computerized tomography, multi-directional post-sacrifice flexibility testing, decalcified histology, and histomorphometric and immunochemical analyses.

With regard to postoperative survival, there was no evidence of prosthesis loosening at the two examined survival periods. Multidirectional flexibility testing from a kinematic standpoint was performed in all standard measures. Under axial rotation and lateral bending, there was no significant difference in the range of motion of the operated FSU in comparison to nonoperative controls. The authors concluded that intervertebral range of motion was preserved under axial rotation and lateral bending at the two examined postsurgical time frames in this animal mode (Hu et al. 2006).

McAfee et al. established that the posterior longitudinal ligament (PLL) may provide a stabilizing influence to the cervical spinal segment (McAfee et al. 2003). Biomechanical testing was performed using human cadaveric spines and a six-degree-of-freedom spine simulator with additional optoelectronic motion measurement. The major finding was that biomechanical stability may be restored following complete anterior cervical discectomy with resection of the PLL via implantation of an arthroplasty device such as the PCM$^®$ Device.

Dmitriev et al. have looked at intradiscal pressure and segmental kinematics following cervical disc arthroplasty with a PCM$^®$ Device (Dmitriev et al. 2005). This in vitro human cadaveric study examined a total of ten spines. Each spine underwent intact analysis with subsequent reconstruction at C5–C6 with a total disc replacement, an allograft dowel, or an allograft dowel and an anterior cervical plate. The authors then tested the specimens in displacement control under axial rotation, flexion/extension, and lateral bending kinematic modes. They recorded intradiscal pressure at levels adjacent to the C5–6 space including C4–5 and C6–7 FSUs. Range of motion was monitored at the operative FSU (C5–C6).

The authors noted that the intradiscal pressures recorded at adjacent levels were similar to the intact (nonoperated) condition in those patients who had undergone a total disc replacement with the PCM$^®$ Device. However, the intradiscal pressures at C4–5 in flexion/extension for both types of simulated fusions were noted to be significantly higher than the mean intradiscal pressures measured at these same levels in the intact and disc replacement groups. Similar findings were noted at C6–7, where significantly increased intradiscal pressures were achieved in all three loading methods including axial rotation, flexion/extension, and lateral bending. As expected, both types of simulated fusions at C5–6 produced a significantly diminished range of motion during flexion/extension testing. The authors concluded that the PCM$^®$ Disc has the ability to maintain adjacent level intradiscal pressure in comparison to increased intradiscal adjacent level pressures noted with simulated fusions. This study lends some support to the concept of adjacent level disease as a result of the modified kinematic environment adjacent to a fusion.

Computer Simulation and Finite Element (FE) Modeling Studies

In addition to numerous disc-specific kinematic studies that have been published in recent years, several authors have contributed to the collective understanding of finite element (FE) modeling with respect to artificial cervical disc replacements. Ahn et al. published such a study, noting as background that there was a need for further simulation studies to understand common design themes for restoration of motion as the result of numerous types of cervical disc designs (Ahn and DiAngelo 2008). They cited the numerous examples of both constrained and semi-constrained devices. The study proposed to expand upon the limited number of in vitro studies previously discussed herein.

The study incorporated a three-dimensional graphics-based computer model of the subaxial cervical spine that had previously been developed. This model was used to study the kinematics and mechanics of an arthroplasty device placed at the C5–6 disc space – the validation for which had been described in a previous study by the same group (Ahn and DiAngelo 2008). The basic computer model incorporated the geometry of cervical vertebrae as established from the computer tomographic images of a 59-year-old woman, linking the adjacent vertebrae at C5 and C6 as a "triple joint complex comprised of the intervertebral disc joints in the anterior region and 2 facet joints in the posterior region and the surrounding ligament structure (Ahn and DiAngelo 2008)."

The authors modeled intervertebral discs as nonlinear elements having a total of six degrees of freedom. With this model, they studied three different theoretical prosthetic disc devices. The first device tested was a disc with the center of rotation of a spherical joint located in the midportion of the C5–6 disc, the second device being with the center of rotation of the cervical joint located 6.5 mm below the midportion of the C5–6 disc, and the third being the center of rotation of the cervical joint in a plane located at the C5–6 disc level. The authors simulated removal of the anterior longitudinal ligament and the anterior portion of the annulus as well as the nucleus pulposus for placement of the disc prosthesis. They then tested the three disc implantation designs throughout the six degrees of freedom allowed by the computer model.

With the three types of disc devices, the authors noted that a constrained spherical joint (device design #1 with the joint placed at the midportion of the disc) significantly increased facet loads during cervical spine extension kinematics. Tested design #2 lowered the rotational axis of the spherical joint toward the subjacent body, and this was noted to kinematically cause a "marginal increase in facet loading during flexion, extension, and lateral bending (Ahn and DiAngelo 2008)." Unconstraining the device (device design #3) minimized facet loading buildup during all loading modes by placing the center of rotation of the spherical joint in a plane located at the C5–6 disc level.

The authors concluded that a finite element model was able to demonstrate simple design changes that may have effects on the kinematic behavior of cervical discs placed in human spines at the C5–6 disc space. They were able to predict facet loads calculated from their computer model but noted that the computer model still needs to have validation with regard to in vitro experimental studies. This model does add credence to kinematic principles of device design and goes one step beyond some of the in vitro

research in its theoretical device design principles.

Liu et al. have described a fluoroscopic kinematic study looking at the kinematics of the anterior cervical discectomy fusion versus cervical artificial disc replacement at the C5–6 joint (Liu et al. 2007). In this novel study, the investigators used a controlled group of ten normal subjects as well as ten patients treated with ACDF in comparison to ten patients treated with cervical artificial disc replacement. Both types of surgical procedures were performed at the C5–6 level. Radiographic data was collected with the patient performing a full flexion and extension motion under fluoroscopy surveillance with kinematic data collection obtained from these fluoroscopic images. The data were derived based on the "inverse dynamic model of the entire cervical spine (Liu et al. 2007)." This custom model was created based on "KANE'S Dynamics and the Reduction Modeling Technique (Liu et al. 2007)." The authors then calculated kinematic data using software and reported the results.

The ACDF group had notable increases in intersegmental rotation at adjacent disc spaces (C6–7 and C4–5 levels) in comparison to the intact normal specimen. Also notable was the fact that the intact spine (no surgical intervention) had a greater range of motion than that observed in ACDF despite these increases of adjacent segment rotations in the ACDF population. The authors noted that the kinematic measurements in the cervical arthroplasty group were similar to those in the normal group and postulated (by their measurement principles) that cervical artificial disc arthroplasty has the potential to restore "normal dynamic motion of the cervical spine (Liu et al. 2007)."

This study provides a novel approach for analysis of in vivo contact forces and expands upon basic kinematic measurements that have been reported in disc arthroplasty studies. It also suggests that cervical arthroplasty has the potential to maintain adjacent segment kinematics, although it is difficult to make predictions with respect to adjacent segment degeneration as a result of this motion analysis study.

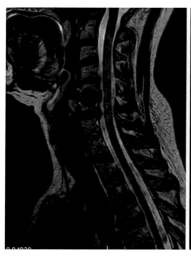 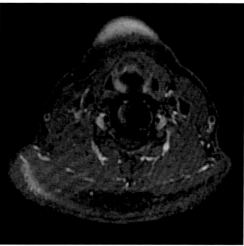

Fig. 6 The BRYAN® Cervical Disc Prosthesis is visualized on these postoperative MR sagittal and axial images. Titanium alloy devices such as the BRYAN® device may have less MRI artifact that similar devices constructed with CoCr or stainless steel. These images demonstrate the imaging characteristics of this device at the index and adjacent surgical levels. (© Courtesy of Rick Sasso, Indianapolis, IN)

Multidisc Studies

Lin et al. created a novel in vivo study to evaluate bone/implant stresses at the C5–6 disc space with placement of BRYAN®, PRESTIGE® LP, and PRODISC® Cervical Disc prostheses (Lin et al. 2009). Their image-based finite element modeling technique was designed to predict stress patterns at the interface between the prosthesis and the lower vertebral endplate – an effort to elucidate possible mechanisms of subsidence and describe load transfers of disc designs. The group built a three-dimensional finite element model of the C5–6 functional spinal unit based on computed tomographic (CT) images acquired from a patient who had previously been identified as a candidate for cervical disc arthroplasty.

The modeling process included facet joints, uncovertebral joints, and specific artificial disc designs that could be placed within the intervertebral disc space. The authors evaluated the discs and endplates in flexion/extension and lateral bending with compression applied. The authors noted that the PRODISC-C® and PRESTIGE® LP Discs caused "high stress concentrations around their central fins or teeth, which may initiate bone absorption (Lin et al. 2009)." With respect to the BRYAN® Disc, the prosthesis appeared to recover the highest range of motion secondary to what the authors described as the "high elastic nucleus" which was notable for diminishing the stresses at the superior endplate of C6 (Lin et al. 2009). The authors also noted that the PRESTIGE® LP Disc, with its rear positioned metal-metal joint, may be a concern for a mechanism of possible subsidence in the posterior aspect of this arthroplasty device.

The authors concluded that the rigidity of the nucleus/core in both the PRESTIGE® LP and the PRODISC-C® prostheses is capable of maintaining initial disc height at the consequence of high contract stresses at the bone endplate interface with either "improper placement or under sizing (Lin et al. 2009)." The BRYAN® Device differs in its core rigidity creating a much larger displacement during motion allowing for "more variation in disc height that may theoretically increase the load sharing of facet and uncovertebral joints compared to more rigid artificial disks (Lin et al. 2009)." This in vivo finite element study goes beyond typical center of rotation and flexion/extension

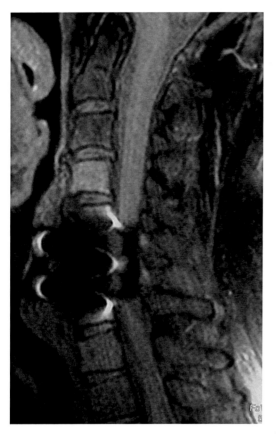

Fig. 7 The MOBI-C® is visualized on this sagittal MRI (T1/FS technique). These images demonstrate the imaging characteristics of this device at the index and adjacent surgical levels. Significant artifact is present at the index and adjacent levels making diagnostic interpretation challenging. (© Courtesy of Rick Sasso, Indianapolis, IN)

kinematics in looking at one of the major causes for implant failure, subsidence. The study is only predictive of the stresses caused by device design and does not predict ultimate subsidence mechanisms. It goes beyond prior studies in elucidating possible areas of increased device/endplate mechanical stresses that are the result of normal device kinematics.

Future Kinematic Design Principles

With respect to basic device design principles, kinematic modeling will likely have an effect on patient outcomes and adjacent segment disease in the long-term. Future design work will continue to make heavy use of preclinical modeling, FE modeling, biomechanical testing, and translational nonhuman testing. Currently implanted cohorts from US FDA trials will alter our understanding of device kinematics over the intermediate and long-term. At the time of this writing, US follow-up of these devices has been published up to 10 years (Sasso et al. 2017). Wear debris caused by device design and kinematic conflicts may play a role in device construction materials and constraint properties as we understand long-term outcomes beyond this interval. Postoperative imaging limitations will also affect future device design as in vivo human studies will continue to make heavy use of imaging techniques and measurements in lieu of biomechanical and histological techniques (Figs. 6 and 7).

Current arthroplasty designs restore only the anterior and middle columns of the cervical spine. They rely on posterior column preservation at the index surgery and over time. Future device designs may include techniques that modify not only structures at the level of the disc but also facets.

Conclusions

We sought to review the basic cervical kinematics that exist and correlate the early data reported from in vivo, in vitro, and finite element (computer-based) studies on disc arthroplasty. Device design with respect to the modified center of rotation at an FSU, device fixation to the vertebral endplates, and flexibility of the articulating nucleus all appear to play a role in reproduction of normal cervical kinematics after cervical disc arthroplasty. A number of these studies also begin to suggest kinematic means of surgical contribution to adjacent level degeneration. It is extremely encouraging to see that many kinematic studies that have been undertaken coincide with the results of US FDA-IDE trials of these devices.

Little data currently exists on how reproduction (or lack of reproduction) of normal kinematics affects intermediate- and long-term patient outcomes and adjacent segment degeneration. Abnormal kinematics may contribute to early

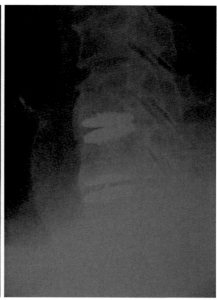

Fig. 8 The MOBI-C® is visualized on these lateral radiographic views of the cervical spine. This patient presented with loss of motion and radiographic evidence of heterotopic ossification at the index surgical levels (C5-C6 and C6-C7). These images demonstrate the imaging characteristics of this device at the index and adjacent surgical levels. Significant artifact is present at the index and adjacent levels making diagnostic interpretation challenging. (© Courtesy of Rick Sasso, Indianapolis, IN)

subsidence in some of these devices; however, other than descriptive subsidence complications in a number of clinical series, the abnormal kinematics of the devices themselves have not clearly been suggested to be at fault for such events. Several studies have suggested that cervical disc arthroplasty causes an early-term risk of heterotopic ossification (Mehren et al. 2006; Leung et al. 2005; Heidecke et al. 2008) (Fig. 8). The authors of this publication are not aware of any current kinematic studies that demonstrate or further elucidate either the biomechanical or kinematic mechanisms that may result in heterotopic ossification. Indeed, it may be that device placement/ implantation techniques place patients at more risk of heterotopic ossification than properties intrinsic to the arthroplasty devices. This is supported by indirect experiential evidence of a diminished rated of heterotopic ossification in patients who have been treated with NSAIDS in some randomized prospective studies (Sasso et al. 2007a, b; Heller et al. 2009).

As cervical device design continues to proceed, it will be critical for both device designers and study investigators to understand the kinematics in the short-, intermediate-, and long-term phases of the various devices. Modified kinematics as the result of improper placement of arthroplasty devices must also be investigated. Such understanding will likely contribute to increased knowledge with respect to the long-term wear and survival of the devices and may possibly alter the patient outcomes in a positive manner.

References

Ahn HS, DiAngelo DJ (2008) A biomechanical study of artificial cervical discs using computer simulation. Spine (Phila Pa 1976) 33(8):883–892

Bohlman HH et al (1993) Robinson anterior cervical discectomy and arthrodesis for cervical radiculopathy. Long-term follow-up of one hundred and twenty-two patients. J Bone Joint Surg Am 75(9):1298–1307

Campbell MJ et al (2009) Use of cervical collar after single-level anterior cervical fusion with plate: is it necessary? Spine 34(1):43–48

Chang UK et al (2007a) Range of motion change after cervical arthroplasty with ProDisc-C and prestige artificial discs compared with anterior cervical discectomy and fusion. J Neurosurg Spine 7(1):40–46

Chang UK et al (2007b) Changes in adjacent-level disc pressure and facet joint force after cervical arthroplasty compared with cervical discectomy and fusion. J Neurosurg Spine 7(1):33–39

Cummins BH, Robertson JT, Gill SS (1998) Surgical experience with an implanted artificial cervical joint. J Neurosurg 88(6):943–948. [see comment]

DiAngelo DJ et al (2003) Biomechanical testing of an artificial cervical joint and an anterior cervical plate. J Spinal Disord Tech 16(4):314–323

DiAngelo DJ et al (2004) In vitro biomechanics of cervical disc arthroplasty with the ProDisc-C total disc implant. Neurosurg Focus 17(3):E7

Dmitriev AE et al (2005) Adjacent level intradiscal pressure and segmental kinematics following a cervical total disc arthroplasty: an in vitro human cadaveric model. Spine (Phila Pa 1976) 30(10):1165–1172

Eck JC et al (2002) Biomechanical study on the effect of cervical spine fusion on adjacent-level intradiscal pressure and segmental motion. Spine 27(22):2431–2434

Fuller DA et al (1998) A kinematic study of the cervical spine before and after segmental arthrodesis. Spine 23(15):1649–1656

Galbusera F et al (2006) Biomechanics of the C5-C6 spinal unit before and after placement of a disc prosthesis. Biomech Model Mechanobiol 5(4):253–261

Galbusera F et al (2008) Cervical spine biomechanics following implantation of a disc prosthesis. Med Eng Phys 30(9):1127–1133

Goffin J et al (1995) Long-term results after anterior cervical fusion and osteosynthetic stabilization for fractures and/or dislocations of the cervical spine. J Spinal Disord 8(6):500–508; discussion 499

Goffin J et al (2004) Long-term follow-up after interbody fusion of the cervical spine. J Spinal Disord Tech 17(2):79–85

Heidecke V et al (2008) Intervertebral disc replacement for cervical degenerative disease – clinical results and functional outcome at two years in patients implanted with the Bryan cervical disc prosthesis. Acta Neurochir 150(5):453–459

Heller JG et al (2009) Comparison of BRYAN cervical disc arthroplasty with anterior cervical decompression and fusion: clinical and radiographic results of a randomized, controlled, clinical trial. Spine 34(2):101–107

Hilibrand AS et al (1999) Radiculopathy and myelopathy at segments adjacent to the site of a previous anterior cervical arthrodesis. J Bone Joint Surg (Am Vol) 81(4):519–528

Hu N et al (2006) Porous coated motion cervical disc replacement: a biomechanical, histomorphometric, and biologic wear analysis in a caprine model. Spine 31(15):1666–1673

Kallemeyn NA et al (2009) An interactive multiblock approach to meshing the spine. Comput Methods Prog Biomed 95(3):227–235

Leung C et al (2005) Clinical significance of heterotopic ossification in cervical disc replacement: a prospective multicenter clinical trial. Neurosurgery 57(4):759–763

Lin CY et al (2009) Stress analysis of the interface between cervical vertebrae end plates and the Bryan, Prestige LP, and ProDisc-C cervical disc prostheses: an in vivo image-based finite element study. Spine (Phila Pa 1976) 34(15):1554–1560

Liu F et al (2007) In vivo evaluation of dynamic characteristics of the normal, fused, and disc replacement cervical spines. Spine (Phila Pa 1976) 32(23):2578–2584

McAfee PC et al (2003) Cervical disc replacement-porous coated motion prosthesis: a comparative biomechanical analysis showing the key role of the posterior longitudinal ligament. Spine 28(20):S176–S185

Mehren C et al (2006) Heterotopic ossification in total cervical artificial disc replacement. Spine 31(24):2802–2806

Pickett GE, Rouleau JP, Duggal N (2005) Kinematic analysis of the cervical spine following implantation of an artificial cervical disc. Spine 30(17):1949–1954

Puttlitz CM et al (2004) Intervertebral disc replacement maintains cervical spine kinetics. Spine 29(24):2809–2814

Robertson JT, Papadopoulos SM, Traynelis VC (2005) Assessment of adjacent-segment disease in patients treated with cervical fusion or arthroplasty: a prospective 2-year study. J Neurosurg Spine 3(6): 417–423

Rousseau MA et al (2008) In vivo kinematics of two types of ball-and-socket cervical disc replacements in the sagittal plane: cranial versus caudal geometric center. Spine Phila Pa 1976 33(1):E6–E9

Sasso RC, Best NM (2008) Cervical kinematics after fusion and Bryan disc arthroplasty. J Spinal Disord Tech 21(1):19–22

Sasso RC et al (2007a) Artificial disc versus fusion: a prospective, randomized study with 2-year follow-up on 99 patients. Spine 32(26):2933–2940

Sasso RC et al (2007b) Clinical outcomes of BRYAN cervical disc arthroplasty: a prospective, randomized, controlled, multicenter trial with 24-month follow-up. J Spinal Disord Tech 20(7):481–491

Sasso RC et al (2008) Motion analysis of Bryan cervical disc arthroplasty versus anterior discectomy and fusion: results from a prospective, randomized, multicenter, clinical trial. J Spinal Disord Tech 21(6):393–399

Sasso WR et al (2017) Long-term clinical outcomes of cervical disc arthroplasty: a prospective, randomized, controlled trial. Spine (Phila Pa 1976) 42(4):209–216

Silber JS et al (2003) Donor site morbidity after anterior iliac crest bone harvest for single-level anterior cervical discectomy and fusion. Spine 28(2):134–139

St. John TA et al (2003) Physical and monetary costs associated with autogenous bone graft harvesting. Am J Orthop (Chatham, Nj) 32(1):18–23

Cervical Total Disc Replacement: Technique – Pitfalls and Pearls

43

Miroslav Vukic and Sergej Mihailovic Marasanov

Contents

Introduction 808

History of Cervical Arthroplasty 808

Rationale for Cervical Disc Arthroplasty 809

Contraindications, Disadvantages, and Specific Complications Related to Cervical Total Disc Replacement (TDR) 810

Principles of Artificial Cervical Disc Designs 811
Prestige Cervical Disc System (Medtronic, Memphis, TN, USA) 811
Bryan Cervical Disc (Medtronic, Memphis, TN, USA) 812
ProDisc-C (Centinel Spine, USA) 812
Porous Coated Motion (PCM) (Nuvasive, San Diego, CA, USA) 813
Mobi-C Cervical Disc (Zimmer, USA) 813
Discover (Centinel Spine, USA) 813

Preoperative Planning 813

Surgical Technique 814
Positioning 814
Approach 815
Midline Determination 816
Placement of Pins 816
Discectomy 817
End Plate Preparation 817
Footprint Size 818
Disc Trial 818
Implant Insertion 819
Wound Closure 820
Postoperative Care and Follow-Up 820

Conclusions 820

References 821

M. Vukic (✉) · S. M. Marasanov
Department of Neurosurgery, University Hospital Rebro,
Zagreb, Croatia
e-mail: vukicmd@gmail.com; smmarasanov@gmail.com

© Springer Nature Switzerland AG 2021
B. C. Cheng (ed.), *Handbook of Spine Technology*,
https://doi.org/10.1007/978-3-319-44424-6_75

Abstract

Anterior cervical discectomy and fusion (ACDF) is still considered the gold standard for surgical management of cervical spondylosis. The discovery of the impact of fusion on other functional spinal units in the form of adjacent segment disease has led to the development of motion-sparing techniques in cervical spine surgery, such as cervical arthroplasty. A substantial number of different cervical artificial disc implants have been approved for clinical use, some with long-term follow-up data demonstrating the safety and efficacy of the implants in maintaining motion of the index level. However, a significant number of prostheses failed to retain the desired mobility, mostly due to heterotopic ossification and unintentional fusion. Choosing the appropriate implant design, along with meticulous surgical technique, is the most important prerequisite for good surgical results and longevity of implant integrity and function. In this chapter we discuss the evolution of key characteristics of implant design crucial for successful surgery, as well as surgical tips and techniques related to cervical arthroplasty.

Keywords

Cervical arthroplasty · Total disc replacement · ACDF · History · Adjacent segment · Implant design · Implant selection · Discectomy · End plate preparation · Heterotopic ossification

Introduction

Since its introduction in the mid-1950s, anterior cervical discectomy and fusion (ACDF), as first described by Robinson and Smith (1955) and Smith and Robinson (1958), has become the gold standard for the treatment of single- and multilevel cervical disc disease and cervical spondylosis. Although many different additions and minor changes have modified the ACDF approach over time, such as by Cloward (1958), the approach itself and the basic surgical technique implied have conceptually remained the same, testimony to the rationale and technical simplicity of this approach. ACDF is still justifying its position as a straightforward technique that yields unequivocally excellent clinical results.

It was not until the late 1980s and 1990s that awareness emerged of the impact of fusion of diseased cervical segments on adjacent functional spinal units in a clinically significant proportion of patients. Although the existence of adjacent segment degeneration has been reported before, it was only after the landmark paper by Hilibrand et al. (1999) in which the authors reported on a small percentage (2.9%) of new-onset adjacent segment symptomatic radiculopathy cases per year, but a significant cumulative rate of 25% in a 10-year period after the index ACDF surgery that cervical spine motion preservation techniques have come again under focus by the spine surgical community. The purported mechanism for the non-negligible incidence of adjacent segment disease following ACDF has been speculated to be a higher, nonphysiological degree of stress imposed upon functional spinal units next to the fused segments, leading to their accelerated degeneration. Various reports have shown that intradiscal pressure in segments adjacent to those fused by surgery increased after surgical immobilization. This begs the question of whether the incidence of adjacent segment disease could be lowered by preservation of motion in the operated segment after surgical decompression.

History of Cervical Arthroplasty

The idea of arthroplasty as opposed to arthrodesis has a long history in orthopedic surgery, specifically in the field of hip and knee surgery (Wiles 1958; McKee and Watson-Farrar 1966).

While loss of function of a hip or knee joint creates a debilitating condition for the individual, the fusion and loss of function of a single functional spinal unit, or even multiple segments in the cervical spine, are surprisingly well tolerated. Nevertheless attempts to retain mobility in operated cervical levels have also a long history dating from the 1960s, with the first reported implantation of a cervical

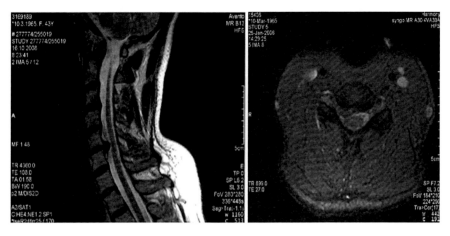

Fig. 1 MRI showing a typical indication for cervical disc arthroplasty – a single level soft disc herniation causing radiculopathy refractory to conservative treatment

arthroplasty device by Fernström (1966). These implants consisted of a stainless steel ball implanted in the intervertebral space after discectomy. Because of a high proportion of implant subsidence and implant migration, the placement of these devices was quickly abandoned.

Spinal arthroplasty for arthrodesis fell out of favor until the newly emerged success of lumbar arthroplasty in the 1980s. It was the invention of the SB Charité lumbar disc prosthesis with excellent results in trials that renewed interest in spinal arthroplasty (Cinotti et al. 1996; Lemaire et al. 1997; Zeegers et al. 1999; Guyer and Ohnmeiss 2003). Different other lumbar devices, such as the ProDisc, have been widely implanted, and all of these have survived to see different design changes and improvements. Because of this trend, renewed interest in cervical spine arthroplasty also reemerged.

Rationale for Cervical Disc Arthroplasty

With the introduction of modern cervical total disc replacement devices, the initial indications for surgery included patients with cervical degenerative disc disease confined to one level causing radiculopathy and/or myelopathy with radiological evidence of only soft disc herniation or mild spondylosis. Cervical arthroplasty was therefore considered only for a selected subgroup of patients with single-level disease between C3 and C7 with no evidence of pathological changes to facet joints and posterior elements of the spine (Fig. 1). Intuitively, this raises questions regarding possible selection bias with regard to surgical outcome. Counterintuitively, and interestingly enough, in the paper by Hilibrand et al., a negative correlation between the number of levels included in the ACDF construct and onset of adjacent segment disease was reported. Two- and three-level ACDF somehow appear to be correlated with a lower incidence of adjacent segment disease than single-level arthrodesis.

After initial reports (Cummins et al. 1998; Goffin et al. 2002), mainly case series in Europe followed by the reports of US Food and Drug Administration (FDA) investigational device exemption (IDE) studies that the spectrum of indications for cervical arthroplasty has expanded. The indications beyond radiculopathy due to soft disc herniations include axial neck pain, myelopathy and foraminal stenosis, and nerve root entrapment due to narrowed disc space.

In contrast to the initial limitation to single-level disease, the indications now are widely accepted for two-level disease (Fig. 2) and in cases where a symptomatic degenerative disc disease (DDD) is addressed in conjunction with an adjacent level to a previously fused level (Fig. 3).

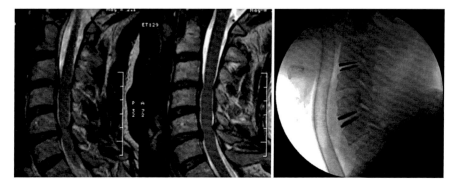

Fig. 2 Extended indication for cervical TDR – two-segment disease

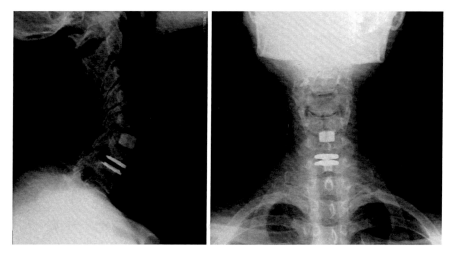

Fig. 3 Extended indication for cervical TDR – arthroplasty adjacent to previous fusion

Patients in whom a cervical arthroplasty procedure is considered must have failed a course of conservative treatment of at least 6–8 weeks.

Contraindications, Disadvantages, and Specific Complications Related to Cervical Total Disc Replacement (TDR)

A number of spinal conditions strongly preclude the use of cervical TDR devices. These include conditions with predominantly posterior compression of neural elements, such as cases with facet joint or yellow ligament hypertrophy and cases of congenitally narrow spinal canal that cannot be adequately addressed only from anteriorly. These also include cases where there is no motion preserved in the index level, such as cases of ossification of the posterior longitudinal ligament (OPLL), for instance. A vast number of metabolic conditions including osteoporosis or severe osteopenia and bone metabolic diseases (such as Paget's) are a contraindication.

When comparing the indications, surgical goals, and expectations in the long run for the two procedures (ACDF vs TDR), different factors come into play with respect to the surgical procedure undertaken. The main goal of both ADCF and TDR is the adequate surgical decompression of neural structures from an anterior approach. In the case of ACDF, the secondary goal after decompression is the solid fusion and restoration of ideal vertebral body alignment. This is achieved by proper end plate preparation and implantation of adequate allograft material with or without an anterior plating system. After achievement of bony fusion, further formation of bony spurs toward the spinal canal in the operated segment(s) is prevented, and sometimes the resorption of previously formed ones can be detected. Basically, outpatient follow-up of these patients in the majority of cases comes to an end, or patients tend to be lost to follow up after achievement of bony fusion because they are no longer symptomatic.

The goals of arthroplasty somewhat differ in that after surgical decompression a different milieu is set up. The surgeon must take into account the need for adequate end plate preparation, which differs depending on the prosthesis to be implanted, but also be minimally disruptive to adjacent tissues. It has been shown that meticulous surgical technique is of paramount importance for the long-lasting success of cervical arthroplasty, both in optimizing clinical results and in complication avoidance. This of course also true with ACDF, but in TDR the impact of surgical technique is even more pronounced. For example, trauma to the longus colli muscle has been connected to the formation of heterotopic ossification, a new complication specific for artificial disc surgery. Unlike with the ACDF procedure, there is no termination to follow-up of these patients since there is a need to follow the functionality of the prosthesis itself.

Principles of Artificial Cervical Disc Designs

A vast number of different artificial cervical disc prostheses are available on the market worldwide. However, some crucial key points of design, biomechanical parameters, construction elements, and materials differentiate these various TDR designs. Although there may not be an ideal artificial disc implant, and due to the variability of individual anatomy and pathology, a single ideal implant likely cannot exist, and the history of implant development has brought to light some features that are clear improvements. An artificial

disc has to be easy to implant, available in different sizes in terms both of height and footprint size (endplate coverage), and to have an incorporated lordotic angle of approximately 7°. Ideally, it should be MRI compatible with as few MRI-related artifacts as possible and has to have radiopaque markers for safe implantation. It also needs to be able to resist the stress imposed by millions of cycles of movement without mechanical failure and at the same time produce virtually no shear material debris. A key function is to have an adequate instantaneous axis of rotation, unrestrained near-to-physiological range of motion in all directions coupled with a small but crucial amount of translation while at the same time allowing for some degree of motion present in the facet joints. Ideally, the prosthesis must be confined to the intervertebral space to minimize dysphagia, while at the same time precluding prosthesis migration and subsidence. And lastly, revision surgery, if required, must not be complicated.

As one can see from the list of characteristics, these implants are subjected to a very rigorous group of requirements. From the history of the evolution of their design, one can also see the evolution of our understanding of the requirements of a functional cervical arthroplasty device.

Prestige Cervical Disc System (Medtronic, Memphis, TN, USA)

The first modern cervical artificial disc device was designed by B. H. Cummins in the Department of Medical Engineering at the Frenchay Hospital in Bristol, UK, in 1989 (Cummins et al. 1998). The first design was a stainless steel two-piece metal-on-metal, ball-on-socket device with anteriorly placed anchoring screws. This so-called Bristol-Cummins artificial cervical joint has subsequently experienced several design improvements. Later on named the Frenchay cervical disc, the device was eventually bought by Medtronic and redesigned to "Prestige" in 1998. The major design improvement was a change of the concave articulating surface from a

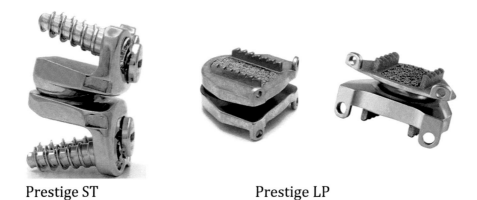

Fig. 4 The Prestige ST and Prestige LP artificial discs

Fig. 5 The Bryan disc

hemispheric cup to a more ellipsoid saucer, allowing for additional freedom of movement, in particular a small amount of anteroposterior translation coupled with flexion and extension. The screw locking mechanism was also changed to a lower profile with a locking mechanism designed to prevent screw pullout. With the fifth generation called the Prestige LP, it has become one of the most widely and extensively studied artificial discs. It has evolved from a stainless steel ball-on-socket device with anterior plate/screw fixation to a titanium ceramic composite using two small keels for low-profile fixation (Fig. 4).

Bryan Cervical Disc (Medtronic, Memphis, TN, USA)

Contrary to the Prestige artificial disc, the Bryan cervical disc (Fig. 5) that was invented in the late 1990s by Vincent Bryan was initially designed as a composite metal-on-plastic design. More adherent to low-friction principles of tribology, it consists of a polyurethane core between two titanium alloy shells, allowing for unrestrained motion and shock absorption. The titanium shells are covered by a porous surface that promotes bony ingrowth. The whole implant is contained in a polyurethane membrane designed to prevent shedding of wear debris. Specific to the Bryan system is the surgical implantation technique that necessitates a specific milling of vertebral end plates that guarantees initial implant stability. As with the Prestige disc system, substantial clinical data with the use of the Bryan disc are available due to the large number of devices implanted and long follow-up (Goffin et al. 2003; Anderson et al. 2004.).

ProDisc-C (Centinel Spine, USA)

Analogous to the lumbar ProDisc-L, the ProDisc-C (Fig. 6) is a uniarticulating ultrahigh-molecular-weight polyethylene (UHMWPE)-on-cobalt chrome ball-in-socket design. Short-term fixation stability is achieved by a protruding midline keel on the end plate surfaces of the device.

Fig. 6 The ProDisc-C disc

Fig. 7 The PCM disc

Porous Coated Motion (PCM) (Nuvasive, San Diego, CA, USA)

Different from the previous designs, the PCM (Fig. 7) is a uniarticulating two-piece non-constrained device. The initial fixation to the vertebrae is achieved by the curved design and serrations of the implant end plates, while long-term fixation is provided by bony ingrowth into its titanium and calcium phosphate surface. The articulation of this implant is UHMWPE-CoCr combination. Specific to this device is the lack of inherent range of motion limitations due to its design which relies on gliding motion. The motion limitation of this unconstrained device is dependent upon the surrounding soft tissues and facet joints.

Mobi-C Cervical Disc (Zimmer, USA)

The Mobi-C cervical disc (Fig. 8) is a combination of two titanium end plates and a polyethylene core in a semi-constrained design. The specific characteristic of this device is the presence of two peripheral stops incorporated in the construct of the inferior titanium plate that limits the mobility of the polyurethane insert.

Discover (Centinel Spine, USA)

The Discover artificial cervical disc (Fig. 9) has a fixed core ball-in-socket joint with articulating surfaces of titanium alloy and a cross-linked UHMWPE core. It offers an inherent 7° of structural lordosis. The immediate fixation is provided by six 1mm fixation teeth, while the titanium plasma and hydroxyapatite coating provides for better bony ingrowth and long-term fixation of the device.

Preoperative Planning

Irrespective of the device to be used, and consistent with good surgical practice, a thorough history and clinical examination and preoperative assessment of the patient's symptoms and signs, as well as a preoperative radiological work-up, must be undertaken. In preparation for cervical arthroplasty surgery, apart from standard C-spine X-rays showing the patient in the neutral standing position (particularly potentially useful during the operative procedure – described below) and MRI scans delineating the localization and characteristics of pathology as well as of neural structures, flexion and extension X-rays are necessary. If available, a CT scan is of use for evaluation of facet joint degenerative changes.

Knowledge of the normal lordotic curvature of the patient's cervical spine, as assessed by the preoperative neutral standing lateral X-ray, is of paramount importance when positioning the patient on the operating table. Failure to position the patient correctly with insufficient extension or overextension of the neck is a crucial mistake that

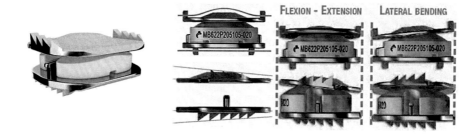

Fig. 8 The Mobi-C cervical disc

Fig. 9 The Discover disc

can occur even before initiating the surgical procedure itself. Measuring the purported implant size on preoperative X-rays can give the surgeon an idea of whether adequate end plate preparation has been made at the completion of neural decompression.

Surgical Technique

Positioning

Since indications for artificial disc replacement surgery differ from the standard ACDF indications that encompass a much larger spectrum of cervical spine degenerative diseases, the surgical nuances that differ are numerous and present in each step of the procedure starting with patient positioning. When planning total disc replacement surgery, it is absolutely mandatory to try and achieve as neutral a position as possible to maintain a midline position and natural lordotic curvature. Midline position is required for accurate coronal alignment and midline placement of the implant in order to minimize adjacent segment stresses and guarantee maximum duration of implant function. The head of the patient must not be rotated to either side. One way of assessing the degree of flexion/extension is to superimpose the preoperative X-ray of the patient taken when standing to the one taken intraoperatively during positioning. Hyperextension during surgery, for example, can lead to unintended AP translation inside the device resulting in suboptimal postoperative implant range of motion.

The head is maintained in the desired position by a self-retaining tape placed over the patient's forehead with care taken not to leave the eyes unprotected (avoided by applying ointment) should they unintentionally open under the drapes during surgery. The neck must not be left unsupported; usually an appropriate pad or a roll of cotton compresses is put under the neck to match its curvature, thus offering support during surgery (Fig. 10).

Positioning Pearls
- Ensuring a centered head position and midline cervical spine alignment is crucial for successful arthroplasty device implantation.
- The head should be taped to the surgical table to prevent any movement during surgery.
- Final positioning before draping can be checked by X-ray to ensure optimal alignment and lordotic curvature have been achieved.

Positioning Pitfalls
- Inadvertent rotation of the cervical spine during surgery can lead to improper implant

Fig. 10 Proper patient positioning for performing cervical disc arthroplasty. Note the placement of the mechanical ventilation tube in the opposite corner of the mouth, the position of the tape used to immobilize the head, and the roll of cotton pads that provide support for the neck in a neutral position

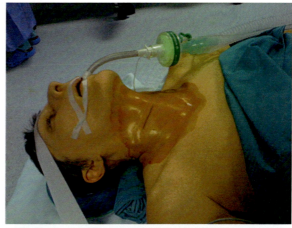

positioning and incorrect stress on the device resulting in possible suboptimal functionality of the prosthesis, neck pain, and even implant migration.

Approach

Depending upon the implant used, different dedicated sets of surgical instruments are present.

As with classical ACDF, the side of approach depends on the preference of the surgeon but also upon patient characteristics. A classical anterolateral approach to the cervical spine is undertaken. Since the vast majority of arthroplasty surgeries performed are for single-level disc disease, a classical approach through a horizontal incision concealed in a skin crease at the level of the conic ligament is undertaken. Care must be taken to minimize any unnecessary tissue trauma and bleeding, especially in the plane of the prevertebral fascia in order to minimize formation of heterotopic ossification. Instead, only mobilization of the medial margins of the longus colli muscle is undertaken with a Penfield No. 4 instrument or a similar dissector in order to visualize the uncovertebral joints and establish the midline. A marker is inserted into the disc, and fluoroscopy is performed to confirm the surgical level.

Approach Pearls
- For patients with short neck and for surgery on lower levels (C6–C7), shoulders should be gently pulled down and taped to the surgical table in a craniocaudal direction to ensure operative level X-ray visualization.
- The skin incision should be made using natural skin crests if visible, to ensure a better cosmetic outcome.

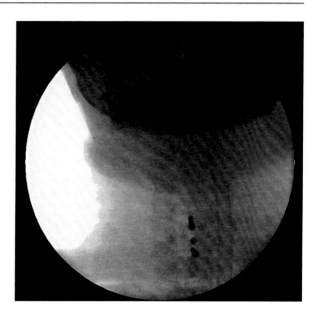

Fig. 11 Diagram and intraoperative fluoroscopy showing the placement of midline pin(s). Placement of the midline pin either in the disc or vertebra should be done equidistant from both uncovertebral joints. Radiographic confirmation of this position is mandatory

- Deep cervical fascia should be widely dissected in respect to the craniocaudal direction to minimize retractor compression to adjacent structures.
- Meticulous hemostasis must be achieved at all steps during surgery to avoid placement of drainage at the time of closure.

Approach Pitfalls
- Extensive electrocautery of the longus colli muscle must be avoided in order to minimize tissue trauma and scar formation, which can predispose to heterotopic ossification.

Midline Determination

Irrespective of the implant type, establishing the midline is of crucial importance for the long-term success of surgery. As previously stated, this is done by identifying the middle of the distance from the two uncovertebral joints while inflicting minimal trauma to surrounding soft tissue. Radiographic verification of the midline position of pins on adjacent vertebrae is performed by fluoroscopy after positioning a midline pin either in the disc itself or in the body of the adjacent vertebrae. Care must be taken not to misinterpret the midline alignment on fluoroscopy, and this is verified when both the pin and spinous process are aligned in a projection perpendicular to the intervertebral space (Fig. 11).

Midline Determination Pearls
- Longus colli muscle dissection should be performed with the use of a dissector to the level where both uncovertebral joints are fully visible, facilitating the determination of the true midline.
- When assessing the midline position of pins by fluoroscopy, it is necessary for the pins to be in line with the superimposed spinous processes of the corresponding vertebrae.

Placement of Pins

Retractor pins are then placed in the adjacent vertebral bodies (Fig. 12). This can be done with the aid of a guiding system. In this manner parallel placement and therefore an even and parallel retraction of both distractor pins are facilitated. If a dedicated guidance system is not available, care of must be taken to assure the placement of the vertebral body pins as close as parallel to one another as possible.

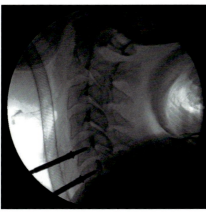
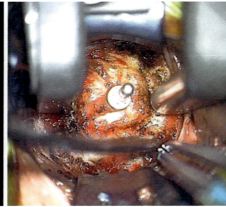

Fig. 12 Intraoperative fluoroscopy and intraoperative photograph showing positioning of the distractor pins

Pin Placement Pearls
- Parallel placement of pins will ensure equal distraction forces to be applied on both vertebrae.
- Both pins must be of same length and diameter.
- Pins must be placed in the mid-portion of the vertebral bodies, parallel to the end plates.

Pin Placement Pitfalls
- It is important not to overdistract, because overdistraction can put additional stress on facet joints leading to either postoperative neck pain or possible degeneration of the facets which at the end can diminish functioning of the prosthesis.

Discectomy

Initial partial discectomy is performed with the pins in place and before their distraction. The discectomy is then finished in a stepwise fashion, with minimal increments in vertebral body distractions over the pins as the discectomy is completed. Overdistraction must be avoided, since this can lead to implantation of an oversized device, leading to excessive distractive stress on the facet joints and postoperative pain and failure of motion preservation.

It is important to remove all residual disc material to the posterior longitudinal ligament and laterally to both uncovertebral joints. Removal of the posterior longitudinal ligament is always necessary, as well as removal of any osteophytes protruding into the spinal canal. However, all unnecessary bone drilling should be avoided and the operative field thoroughly rinsed with saline in order to flush away any bone debris.

Discectomy Pearls
- Removal of the posterior longitudinal ligament is necessary, in order to visualize adequately the decompression of neural structures.
- Additional foraminotomy is usually not required because most cases are limited to soft disc herniation; however, it is crucial to see the entire foramen and the nerve within it and to check for any residual compression using a probe or a nerve hook.

End Plate Preparation

Nuances of end plate preparation differ based on the type of implant used, and manufacturers' advice should be followed. Generally, the end plates should be prepared flat and shaped to match as closely as possible the curvature or angulation of the implant end plate geometry. Implants using keels as means for immediate anchoring have the inherent risk of provoking vertebral body fractures especially in the setting of performing arthroplasty surgery on two adjacent levels or next to a previous fused segment,

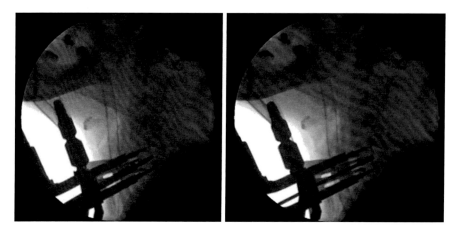

Fig. 13 Intraoperative fluoroscopic visualization when choosing the appropriate footprint size

so care must be taken not to be overly aggressive in preparation for their positioning. Newer generation devices use small spurs or teeth for immediate implant fixation which is generally less traumatic to the bony end plates.

Final end plate preparation is achieved with the use of appropriately sized rasps, with which any residual cartilaginous tissue impeding bony ingrowth into the implant end plates can be removed. Care must be taken not to impose injury to the subchondral bone. The superior end plate usually is concavely shaped, and this surface can be carefully flattened both anteriorly and posteriorly in order to accommodate the implant surface. The end plate of the lower vertebra should be usually flattened only posteriorly. At the end of the endplate preparation phase, both vertebral surfaces should be parallel to one another. A thorough irrigation is performed again.

End Plate Preparation Pearls
- When using prosthesis featuring a predetermined lordotic angle, achieving absolute parallelism during end plate preparation is mandatory; failure to do so may lead to loss of lordotic shape of the prosthesis.

Footprint Size

Choosing the right size of the implant in terms of endplate coverage is the best way to avoid implant subsidence as well as heterotopic ossification. The greater the coverage area, the lesser the chance that the implant will injure the adjacent bone and migrate. In the setting of various implant sizes available, the largest implant that still fits into the intervertebral space without protruding should be used.

Footprint Size Pearls
- Checking the appropriate footprint size is done exclusively under fluoroscopy (Fig. 13).
- The largest footprint size that does not enter the uncovertebral joints is to be chosen.
- When checking for footprint size, the template itself must be positioned parallel to the distraction pins (Fig. 14).

Disc Trial

To further optimize disc implant selection, several sets include disc trials that mimic the final disc implant and can be assessed under fluoroscopy in terms of height, coverage, and positioning.

Disc Trial Pearls
- It is of primary importance that the midline of the trial prosthesis matches the midline of the vertebral body, because this in the end will affect the center of rotation of the implant; again, this can be achieved only by fluoroscopic guidance.

Fig. 14 Intraoperative image showing the proper position of the probe when checking adequacy of footprint size

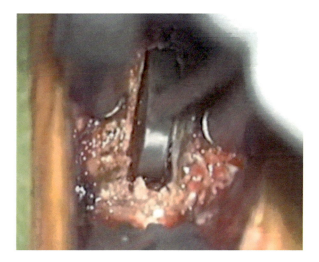

Fig. 15 Intraoperative photograph showing implantation of the Discover disc

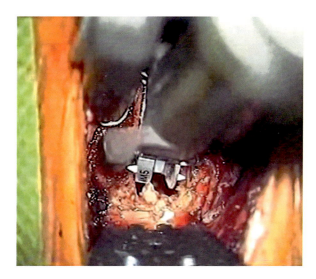

Implant Insertion

Implant placement (Fig. 15) steps must be followed as described by the manufacturer. Any unnecessary manipulation of the implant should be avoided. Also, the simultaneous use of a burr and any metallic suction devices or similar metallic instruments during any surgical step should be performed with care. Contact of the burr with such an instrument can provoke formation of metal debris which can over the course of years lead to implant scratching and accelerated wear.

Final position (Fig. 16) must be checked by AP and LL fluoroscopy and any further positioning adjustment made accordingly. Since these implants are designed to last for millions of cycles of flexion and extension, translational movement, and rotation, any suboptimal placement should not be tolerated.

Implant Insertion Pearls
- Malpositioning of the prosthesis leads to prosthesis malfunction, and should such a malpositioning be verified at implantation, a reinsertion should be undertaken. For prosthesis using keels, this often cannot be done without significant injury to the cortical bone. Therefore prostheses using lower profile anchoring features, such as teeth or spurs, are advantageous, particularly for surgeons with less experience in cervical arthroplasty.

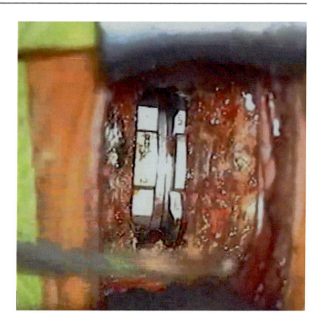

Fig. 16 Intraoperative photograph showing final position of the Discover disc

Wound Closure

After the device insertion is completed, all pins and retractors should be removed and fluoroscopy undertaken to check the final implant positioning. If the implant position is optimal, the wound can be irrigated and inspected for any bleeding. Hemostasis should be meticulously performed, and drainage for one-level surgery is usually not required. Only the platysma is sutured, as well as the subcutaneous layer using a 4–0 resorbable suture. Dermabond is applied on the skin, so that no particular restrictions regarding the skin incision site are necessary.

Postoperative Care and Follow-Up

Patients are encouraged to be upright and walk 3–4 h after surgery and are usually discharged the same or the following day. Administering pain medication as needed as well as NSAIDs for 1–2 weeks after surgery is advised. Wearing of a cervical collar is not recommended.

Follow-up in our practice is usually scheduled at 3, 6, and 12 months after surgery, with dynamic X-rays only.

Patients are also advised to come for further postoperative follow-up 24, 48, 72, and 96 months. The purpose is to check for possible heterotopic ossification that may occur during the patient's postoperative course and to follow and understand long-term outcome of both the patient and the device.

Conclusions

Modern cervical disc arthroplasty surgery started in 2001 in Belgium with the work of Goffin. Since then, cervical disc arthroplasty (CDA) has become a standard procedure in many orthopedic and neurosurgical departments all over the world. Several meta-analyses and randomized control trials showed that CDA is an effective and safe surgical procedure for the treatment of single-level cervical disc disease. Furthermore, in some studies, CDA was found to be superior to ACDF in terms of neck and arm pain, neurological success, and range of motion at the operated level and when assessing for secondary surgical procedures.

CDA is a procedure that has its own history in terms of both technological and surgical development. It has a significant role in our present spinal surgery armamentarium and hopefully a bright future. Evaluation of new technology is an ongoing process, and we have to be careful and prudent in its implementation, but we also have a reason to be optimistic.

References

Anderson PA, Sasso RC, Rouleau JP et al (2004) The Bryan cervical disc: wear properties and early clinical results. Spine J 4:303S–309S

Cinotti G, David T, Postacchini F (1996) Results of disc prosthesis after a minimum follow-up period of 2 years. Spine 21:995–1000

Cloward RB (1958) The anterior approach for removal of ruptured cervical disks. J Neurosurg 15:602–617

Cummins BH, Robertson JT, Gill SS (1998) Surgical experience with an implanted artificial cervical joint. J Neurosurg 88:881–884

Fernström U (1966) Arthroplasty with intercorporal endoprothesis in herniated disc and in painful disc. Acta Chir Scand Suppl 357:154–159

Goffin J, Casey A, Kehr P et al (2002) Preliminary clinical experience with the Bryan cervical disc prosthesis. Neurosurgery 51:840–845

Goffin J, Van Calenbergh F, van Loon J et al (2003) Intermediate follow-up after treatment of degenerative disc disease with the Bryan cervical disc prosthesis: single-level and bi-level. Spine 28:2673–2678

Guyer RD, Ohnmeiss DD (2003) Intervertebral disc prostheses. Spine 28(Suppl 15):S15–S23

Hilibrand AS, Carlson GD, Palumbo MA et al (1999) Radiculopathy and myelopathy at segments adjacent to the site of a previous anterior cervical arthrodesis. J Bone Joint Surg Am 81:519–528

Lemaire JP, Skalli W, Lavaste F et al (1997) Intervertebral disc prosthesis. Results and prospects for the year 2000. Clin Orthop Relat Res 337:64–76

McKee GK, Watson-Farrar J (1966) Replacement of arthritic hips by the McKee-Farrar prosthesis. J Bone Joint Surg (Br) 48:245–259

Robinson RA, Smith GW (1955) Anterolateral cervical disc removal and interbody fusion for cervical disc syndrome. Bull Johns Hopkins Hosp 96:223–224

Smith GW, Robinson RA (1958) The treatment of certain cervical spine disorders by anterior removal of the intervertebral disc and interbody fusion. J Bone Joint Surg Am 40:607–624

Wiles P (1958) The surgery of the osteoarthritic hip. Br J Surg 45:488–497

Zeegers WS, Bohnen LM, Laaper M et al (1999) Artificial disc re- placement with the modular type SB Charite III: 2-year results in 50 prospectively studied patients. Eur Spine J 8:210–217

Cervical Total Disc Replacement: Expanded Indications

44

Pierce D. Nunley

Contents

Brief Worldwide History of cTDR 824

Development of Indications 824

The USA 824

OUS 824

US Current Indications 825

Evidence to Expand US Indications 825

Conclusions 826

References 827

Abstract

Cervical total disc replacement (cTDR) was first used in the 1960s with little success. Widespread use began in the early 2000s, with multiple devices approved in the United States (US) and outside the US (OUS). The US regulatory process, the Investigational Device Exemption (IDE) trial, slows the adoption of some technology in the USA including cTDR. The design, including strict inclusion/exclusion, of IDE trials results in the most compelling, and highest level of evidence data in support of cTDR. The strict patient selection for the IDE trials continues to impact clinical use, as the US indications for use of cTDR remain restrictive compared to OUS. Literature supporting expanded indications of cTDR is limited to low level of evidence and small patient populations from mostly OUS sources. OUS surgeons have many years (10+) of experience with expanded indications, so we will also explore their experiences with cTDR. The reality in the USA is that approval of expanded indications will likely not occur without another IDE trial. What evidence is necessary for US surgeons to adopt expanded indications in their practice?

P. D. Nunley (✉)
Spine Institute of Louisiana, Shreveport, LA, USA
e-mail: pnunley@louisianaspine.org

© Springer Nature Switzerland AG 2021
B. C. Cheng (ed.), *Handbook of Spine Technology*,
https://doi.org/10.1007/978-3-319-44424-6_76

Keywords

Cervical total disc replacement · Cervical arthroplasty · Cervical indications

Brief Worldwide History of cTDR

The first cervical total disc replacement (cTDR) was implanted in 1966 by the Swedish surgeon, Ulf Fernstom. The stainless steel ball bearing, known as the Fernstrom ball, was plagued with high failure rates, so approximately 75 were implanted before use was discontinued (Fernström 1966; Fisahn et al. 2017). In 1989, Cummins (Bristol, UK) developed a stainless-steel artificial disc consisting of metal on metal ball and socket device with anchoring screws. One more redesign, following high failure rates of the original Cummins design, produced the Frenchay cervical disc. The first with promising clinical outcomes, the Frenchay disc, was purchased by Medtronic (Medtronic, Minneapolis, MN, USA) and became the Prestige ST cervical disc (Cummins et al. 1998; Wigfield et al. 2002). The Bryan disc was designed in the USA by Vincent Bryan in 1992. The Bryan disc design was markedly different than Prestige, with two titanium alloy endplates and a polyurethane core filled with saline (Basho and Hood 2012). During this time period, another disc, ProDisc-C, was designed by a French surgeon, Dr. Thierry Marnay. ProDisc-C brought a third unique design to the cTDR market. The device was composed of cobalt-chromium-molybdenum and ultra-high molecular weight polyethylene (UHMWPE) articulating surface, with two keels to facilitate anchoring to the vertebral endplates (Baaj et al. 2009).

The 1990s and early 2000s experienced gaining momentum for design of new cTDR devices, but design was only a beginning for cTDR technology. Regulatory approval is required for cTDR devices in the country of distribution, although these approval requirements vary widely. In addition to approval of the device, many regulatory agencies will also approve the specific indications for use of the device.

Development of Indications

The development of medical device patient indications differs widely throughout the world. We will address the US indications and the OUS indications separately.

The USA

In the USA, the FDA regulates cTDR as a Class III medical device, requiring a large-scale (multi-center, prospective, controlled, and randomized) investigational device exemption (IDE) clinical trial for approval. The high-quality, controlled nature of these trials renders level 1 evidence comparing cTDR to anterior cervical discectomy and fusion (ACDF). During these trials, patient eligibility criteria are strict. Upon approval the manufacturer in conjunction with the FDA issues a document, "Instructions for Use" (IFU) that includes indications, contraindications, warnings, and precautions specific to the cTDR device. For purposes of this chapter, we will refer to indications, contraindications, warnings, and precautions, simply as "indications." The basis for the cTDR indications is the patient population included in the clinical trial, supported as level 1 evidence. Among the seven approved cTDR devices in the USA, the study populations and therefore the indications remain relatively homogeneous. In the USA, surgeons are allowed to operate on patients "off-label" (outside of the indications), but many surgeons choose to respect the indications based on the extensive data and high level of evidence that supports them. Additionally, US reimbursement remains a challenge even for on-label use of cTDR, so insurance approval of off-label use is rare.

OUS

OUS the regulatory approval process varies between countries, but largely these do not require highly controlled or high level of evidence trials for approval. Therefore, post-approval indications are at the discretion of the surgeon and tend to be more expansive. In general, the OUS reimbursement landscape is also less restrictive, allowing the surgeon and patient to determine the best treatment option without coverage as a factor.

The focus of this chapter will be the current US indications and what evidence exists to expand these indications. Many countries OUS already accept broader indications for cTDR, but should and will the USA adopt these expanded indications?

US Current Indications

US cTDR patient indications remain narrow, but highly supported with level 1 evidence. While each cTDR device has slight variation in the approved indications, the following are common among the cTDR devices:

- Single or two contiguous levels between C3 and C7 for conditions
  - Intractable radiculopathy (with or without neck pain)
  - or myelopathy
  - and at least one of the following:
    herniated nucleus pulposus
    spondylosis (defined by osteophytes)
    visible loss of disc height compared to adjacent levels
- Failure of 6 weeks of conservative therapy
- Skeletally mature patients

Additionally, the following relevant (not fully inclusive listing) contraindications and/or warnings and precautions exist for guidelines against the use of cTDR:

- Prior cervical spine surgery, including prior surgery at the index level
- More than two disease levels requiring surgery
- Severe facet joint degeneration
- Segmental instability (translation >3.5 mm and/or >11° angular difference to that of either adjacent level)
- Disc height less than 3 mm measured from the center

- Significant kyphotic deformity or significant reversal of lordosis
- Neck pain alone

The US data and indications exclude large subsets of the population, and the most heavily debated include hybrid treatment (adjacent to a prior or concurrent fusion) and treatment for greater than two levels.

Evidence to Expand US Indications

For the FDA to officially expand US indications, level 1 evidence collected as part of an IDE would be required with the current regulatory cTDR classification. But for surgeons, what evidence is compelling enough to expand indications within your practice?

The published literature on expanded indications is a lessor level of evidence than the data published from the IDE studies. However, several of these studies are robust and provide valuable insight into expanded use of cTDR.

An analysis conducted by Auerbach used the US indications/contraindications of cTDR for Prestige, ProDisc-C, PCM, and Bryan to group patients requiring cervical spine surgery. Patients were grouped into three categories: (1) patients with direct contraindications, (2) qualified cTDR patients, and (3) qualified patient if indications were expanded to include clinical adjacent segment pathology (CASP). Of 167 patients, 95 were contraindicated, with 7/95 that would have qualified if CASP were included. It is noteworthy that the most common exclusion criteria were greater than two operative levels requiring surgery (47 patients). Only the remaining 72/167 patients were qualified to receive a cTDR (Auerbach et al. 2008). Adjacent to a previous fusion and required surgical levels greater than 2, remain a contraindication in the USA, and it could impact over 50% of patients presenting for cervical spine surgery.

Twenty patients were prospectively enrolled in a study in China for treatment at one or two levels with Bryan cTDR. Unlike in the US trials, the two-level treatment was not restricted to contiguous levels. Clinical outcomes were favorable through 4 years, with no reported serious complications (Zhang et al. 2013). Another Chinese study prospectively enrolled 48 patients treated with Bryan at one, two, and three levels. Clinical outcomes remain favorable through 10 years, although reported heterotopic ossification (HO) rates are high, 69.0%. These HO rates are not categorized into grades, as is typically in the USA, so the true magnitude and impact are unclear (Zhao et al. 2016).

PCM cTDR has been analyzed and reported with expanded indications in several Brazilian studies. A 2007 study included patients treated at one ($n = 72$), two ($n = 53$), three ($n = 12$), and four ($n = 4$) cervical levels (not required to be contiguous) between C3 and T1. There was no exclusion for prior cervical fusion, resulting in 11 one-level and 9 multilevel patients treated with PCM as a revision to a failed fusion, and 12 one-level patients and 9 multilevel patients were treated with PCM adjacent to a prior fusion. Results were significantly improved for both groups, with significantly more improvement in the multilevel cohort (Pimenta et al. 2007). The authors found that the incidence of HO in this population was low, 7.7% through 6 years. While higher HO did correlate to loss of motion, clinical outcomes were not impacted (Pimenta et al. 2013).

A similar study in Brazil used CT results to analyze facet degeneration through 5 years postoperatively. Study enrollment criteria were similar to the previous study, prior fusions were not excluded, and patients were operated at one ($n = 72$), two ($n = 67$), three ($n = 17$), and four ($n = 6$) cervical levels (not required to be contiguous) between C3 and T1. Results indicate there is facet joint degeneration, although minimal, 14% with grade 3 or 4 degeneration (Oliveira et al. 2011).

In France, Mobi-C cTDR has been studied in a large prospective, non-controlled population using expanded indications. Patients were treated at one ($n = 175$), two ($n = 51$), three ($n = 4$), and four ($n = 1$) levels from C3 to T1 with outcomes reported through 2 years. Similar to the PCM studies, prior fusions, even at the index level,

were not excluded. Of one-level patients, 21 (with 28 fused levels) had prior cervical fusions, with 18 of these adjacent to the index level. Of multilevel patients (2, 3, and 4 levels) 5 (with 5 fused levels) had prior cervical fusions, with 3 adjacent to the index level and 2 at the index level. Outcomes for both groups were favorable, with no significant difference between the one-level and multilevel groups (Huppert et al. 2011).

Forty-eight patients treated with Bryan cTDR at one-level were retrospectively reviewed in Korea. Twelve of the 48 patients were treated with Bryan adjacent to a fusion (hybrid). Although postoperative kyphotic changes were noted radiographically, the differences were not significant and clinical outcomes remain improved through the 11.8 month mean follow-up (Yoon et al. 2006).

The literature on expanded indications, while compelling, is of lessor evidence than the robust and highly controlled trials that are more common in the USA.

The evidence to expand US indications is not all based on published literature; it is also based upon surgeon experience. Many OUS and some US surgeons are regularly performing cTDR operations outside of the US indications including patients older than 75, greater than two operative levels, replacement of a failed fusion, adjacent to a fusion, and noncontiguous or "skip" levels.

Conclusions

The use of cTDR as an alternative to cervical fusion has gained momentum since the early 2000s and has seen even more widespread support in the last 10 years. Although still considered a novel technology, the number of implantations and supporting literature is abundant. However, the worldwide indications for use of cTDR differ widely. The typical US surgeon remains conservative in patient selection for cTDR, while OUS surgeons commonly expand the patient selection to include more conditions. Continued research, particularly level 1 and 2 evidence studies, focused on expanded indications will help to appropriately advance cTDR, in an evidence-based fashion.

References

Auerbach JD, Jones KJ, Fras CI, Balderston JR, Rushton SA, Chin KR (2008) The prevalence of indications and contraindications to cervical total disc replacement. Spine J 8:711–716. https://doi.org/10.1016/j.spinee.2007.06.018

Baaj AA, Uribe JS, Vale FL, Preul MC, Crawford NR (2009) History of cervical disc arthroplasty. Neurosurg Focus 27: E10. https://doi.org/10.3171/2009.6.FOCUS09128

Basho R, Hood KA (2012) Cervical total disc arthroplasty. Global Spine J 2:105–108. https://doi.org/10.1055/s-0032-1315453

Cummins BH, Robertson JT, Gill SS (1998) Surgical experience with an implanted artificial cervical joint. J Neurosurg 88:943–948. https://doi.org/10.3171/jns.1998.88.6.0943

Fernström U (1966) Arthroplasty with intercorporal endoprosthesis in herniated disc and in painful disc. Acta Chir Scand Suppl 357:154–159

Fisahn C, Burgess B, Iwanaga J, Chapman JR, Oskouian RJ, Tubbs RS (2017) Ulf Fernström (1915–1985) and his contributions to the development of artificial disc replacements. World Neurosurg 98:278–280. https://doi.org/10.1016/j.wneu.2016.10. 135

Huppert J et al (2011) Comparison between single- and multi-level patients: clinical and radiological outcomes 2 years after cervical disc replacement. Eur Spine J 20:1417–1426. https://doi.org/10.1007/s00586-011-1722-9

Oliveira L, Coutinho E, Marchi L, Abdala N, Pimenta L (2011) Cervical facet degeneration after total disc replacement: 280 levels in 162 patients 5-year follow-up. Neurosurg Q 21:17–21

Pimenta L, McAfee PC, Cappuccino A, Cunningham BW, Diaz R, Coutinho E (2007) Superiority of multilevel cervical arthroplasty outcomes versus single-level outcomes: 229 consecutive PCM prostheses. Spine (Phila Pa 1976) 32:1337–1344. https://doi.org/10.1097/BRS.0b013e318059af12

Pimenta L, Oliveira L, Coutinho E, Marchi L (2013) Bone formation in cervical total disk replacement (CTDR) up to the 6-year follow-up: experience from 272 levels. Neurosurg Q 23:1–6

Wigfield CC, Gill SS, Nelson RJ, Metcalf NH, Robertson JT (2002) The new Frenchay artificial cervical joint: results from a two-year pilot study. Spine (Phila Pa 1976) 27:2446–2452. https://doi.org/10.1097/01.BRS.0000032365.21711.5E

Yoon DH, Yi S, Shin HC, Kim KN, Kim SH (2006) Clinical and radiological results following cervical arthroplasty. Acta Neurochir 148:943–950. https://doi.org/10.1007/s00701-006-0805-6

Zhang Z, Gu B, Zhu W, Wang Q, Zhang W (2013) Clinical and radiographic results of Bryan cervical total disc replacement: 4-year outcomes in a prospective study. Arch Orthop Trauma Surg 133:1061–1066. https://doi.org/10.1007/s00402-013-1772-z

Zhao Y, Zhang Y, Sun Y, Pan S, Zhou F, Liu Z (2016) Application of cervical arthroplasty with Bryan cervical disc: 10-year follow-up results in China. Spine (Phila Pa 1976) 41:111–115. https://doi.org/10.1097/BRS.0000000000001145

Cervical Total Disc Replacement: Heterotopic Ossification and Complications

45

Michael Paci and Michael Y. Wang

Contents

Introduction	830
Heterotopic Ossification	830
Definition	830
Grading	830
Occurrence of HO After CTDR	830
Risk Factors for Developing HO After CTDR	831
Clinical Significance of HO	832
Prevention of HO	832
Other Complications Related to CTDR	833
Complications Related to the Anterior Approach	833
Complications Related to the Prosthesis	833
Adjacent Segment Disease	834
Conclusion	835
References	835

Abstract

Cervical total disc replacement (CTDR) can be complicated by the occurrence of heterotopic ossification (HO). HO occurs when bone formation happens in tissues where it is not normally present. It is graded radiographically and develops on a spectrum, from ossification anterior to the cervical spine to ossification involving the articulating surfaces and causing an effective fusion. The true incidence of HO after CTDR is still under debate, with rates reported in the literature varying from approximately 20–90% of patients. The exact causes of HO are still unknown, but associations have been found with male gender, older age, type of prosthesis used, multilevel surgery, and surgical technique. The presence of HO after CTDR has not been correlated with worse clinical outcomes. There is no evidence to date to support a specific strategy to prevent HO, but prescribing a short course of nonsteroidal

M. Paci · M. Y. Wang (✉)
Department of Neurological Surgery, University of Miami Miller School of Medicine, Miami, FL, USA
e-mail: michael.paci@jhsmiami.org;
MWang2@med.miami.edu

© Springer Nature Switzerland AG 2021
B. C. Cheng (ed.), *Handbook of Spine Technology*,
https://doi.org/10.1007/978-3-319-44424-6_77

anti-inflammatory drugs (NSAIDs) is often done. CTDR is also associated with other complications, including complications that can occur during the anterior approach to the cervical spine, complications related to the prosthesis used and adjacent segment disease.

Keywords

Cervical disc replacement · Heterotopic calcification · Anterior cervical approach · Prosthesis-related complication · Adjacent segment disease

Introduction

Cervical total disc replacement (CTDR) has become a procedure of choice in select patients with one-level and two-level, symptomatic cervical spondylosis for whom motion preservation is an important goal. As detailed elsewhere in this book, this procedure has been shown to have benefits when compared to the more traditional anterior cervical discectomy and fusion (ACDF). It also carries some risks and these will be explored in the current chapter. In particular, we will discuss heterotopic ossification (HO), including its definition, its grading, prevalence after CTDR, risk factors for developing it, and methods to prevent it. We will also discuss more general risks associated with the anterior approach needed to perform a CTDR, as well as specific risks related to the different types of implants use in CTDR.

Heterotopic Ossification

Definition

Heterotopic ossification is a process by which bone and calcifications are deposited in tissues in which they are normally not present. This is known to occur in the context of trauma (such as brain trauma and spinal cord injury), as well as in the setting of arthroplasty, including hip replacement and CTDR (Shehab et al. 2002). In the setting of arthroplasty, the amount of bone formed and its location in relation to the articulating surfaces can threaten the range of motion of the joint involved. The exact pathophysiology of HO is still not well understood. It is clear however that a process of bone deposition can be triggered when bony fragments or shavings fall into contact with certain mesenchymal tissues such as muscle or fascia. It is thought that the presence of bony fragments in such tissues, with their accompanying osteoblasts and bone morphogenic proteins, can lead to the abnormal differentiation of mesenchymal cells into osteoblasts (Balboni et al. 2006). This in turn can lead to new bone formation.

Grading

A well-known grading system for heterotopic ossification following hip arthroplasty was classified in 1973 by Brooker et al. This system was based on the radiographic appearance of HO in relation to the hip joint. McAfee et al. (2003) adapted this grading system for HO in the setting of lumbar disc replacement. Mehren et al. (2006) further adapted this classification for HO in the setting of CTDR.

HO in CTDR is classified based on its radiographic appearance and the extent to which it involves the disc space and its movement (Mehren et al. 2006). Grade 0 is given when no HO is visible. Grade 1 is given when some HO is seen anterior to the vertebral column but not in the disc space. Grade 2 is given when HO starts to appear within the disc space, but no bridging osteophytes are visible yet. Grade 3 is given when bridging osteophytes have formed across the disc space, but some movement is still possible. Finally, grade 4 is given when the level treated by CTDR is entirely fused because of HO, and no movement is possible.

Occurrence of HO After CTDR

The exact incidence of HO after CTDR is still not well established. Different studies have presented widely varying data on this phenomenon. Leung et al. (2005) observed a cohort of 90 patients who underwent CTDR. They found that 17.8% of patients developed HO at 12 months postoperatively, with 6.7% showing grade 3 or 4 HO. Heidecke et al. (2008) presented a cohort of 54 patients treated with CTDR (59 levels treated). Of these, 29% showed evidence of HO at 2 years of follow-up. Similarly, Lee et al. (2010) present a cohort of 48 patients treated with one-level CTDR. They reported that 27.1% of patients had developed HO at a mean follow-up time of 14 months. On the other end of the spectrum, Mehren et al. (2006) showed a much higher level of HO formation after CTDR. In their group of 54 patients treated with CTDR (77 levels treated), 66.2% of levels demonstrated evidence of HO at 1 year of follow-up. Park et al. (2013) present even higher figures with a 94.1% incidence of HO at 24 months in a cohort of 75 patients treated by CTDR.

It is clear that the rates of reported HO after CTDR vary significantly. The cohort studies presented are all relatively small, and the patient populations from each group likely differ significantly from one another as well.

Interestingly, it seems that the rate of HO after CTDR may increase with time after the initial intervention. This seems intuitively logical as it mirrors the natural history of cervical spondylosis in which osteophyte formation progresses over time at a diseased segment until the segment is fused. Suchomel et al. (2010) present data that support this idea. They followed a group of 54 patients (65 levels) treated with CTDR over a period of 4 years. At 6 months, 9% of levels showed grade 3 HO, and none showed grade 4 HO. By 4 years, 45% showed grade 3 HO, and 18% showed grade 4 HO.

Risk Factors for Developing HO After CTDR

Risk factors associated with the development of HO after CTDR remain under investigation. Older age, male gender, two-level CTDR, surgical technique, and type of prosthesis have been associated with HO in various studies that have examined this phenomenon.

Yi et al. (2013) presented a cohort of 170 patients who underwent CTDR with a minimum follow-up of 12 months. They reported a 40.6% rate of HO at follow-up. They found that male gender conferred an odds ratio of 2.117 of developing HO when compared to female gender. They also found that prosthesis type was associated with HO in their cohort. Compared to patients who received a Bryan disc (Medtronic, USA), patients who received a Mobi-C (LDR Medical, France) or ProDisc-C (Synthes, USA) had a significantly increased rate of HO (odds ratios of 5.262 and 7.449).

Leung et al. (2005) also examined possible risk factors for the development of HO in their cohort. They found a significant association between the development of HO and male gender. Additionally, they reported an association between age and HO, with older patients having an odds ratio of 1.10 of developing it as compared to younger ones.

Wu et al. (2012) studied whether two-level CTDR was a risk factor for HO as compared to one-level CTDR. Their hypothesis was that patients needing treatment at two levels were more likely to have more advanced spondylosis in their cervical spine and would thus be more at risk for HO. They presented a cohort of 70 patients who were followed for an average of approximately 46 months. Forty-two patients underwent a one-level procedure, whereas 28 underwent a two-level procedure. The authors found that 75.0% of the patients treated at two level developed HO, as compared to 40.5% for those treated at one level, and this was statistically significant ($p = 0.009$). They argue that this confirms their hypothesis and that patients treated at two-levels had more advanced cervical spondylosis that continued to progress postoperatively.

As previously noted, Park et al. (2013) presented a very high rate of HO in their cohort of patients treated with CTDR (94.1%). They argue that surgical technique can play a role in the development of HO as patients in their cohort were treated by two surgeons with a slightly different technique and one was associated with a significantly higher level of HO. Indeed, one surgeon was described as using a ball-type burr, whereas the second was described as using a diamond-tipped burr. Patients who underwent the procedure with the second surgeon had an odds ratio of 3.33 of developing HO as compared to those who were operated by the first one. The authors discuss that the differing burr types and techniques may have led to a different amount of bony exposure, thus leading to differing HO rates.

Clinical Significance of HO

Given that HO occurs in a significant proportion of patients who undergo CTDR, and that it seems to progress with time, many authors have questioned whether it defies the purpose of the procedure itself. Indeed, the choice of CTDR over a more traditional ACDF is usually made with the intent to preserve motion. However, higher grades of HO are known to restrict the implanted prosthesis' motion, thus resulting into something more akin to a fused level from an ACDF.

As previously discussed, Lee et al. (2010) presented a small cohort of 48 patients who underwent CTDR, with a rate of HO of 27.1%. They examined the clinical outcomes of patients with HO and compared them to those without. They did not find any significant differences in pain as quantified by the visual-analog scale (VAS), in function as quantified by the Oswestry Disability Index (ODI) or in range of motion (ROM) as quantified radiographically. These results are however limited by the overall short patient follow-up and the fact that they did not have any patients who presented with grade 4 HO.

Tu et al. (2011) also compared the clinical outcomes of patients with and without HO after CTDR. They present a cohort of 36 patients (52 levels treated) followed for approximately 27 months. They report a rate of HO of 50%. They did not find a significant difference in VAS scores or in Odom outcome criteria in patients who developed HO compared to those who didn't. These results are however also of limited value given the small group of patients presented, the short follow-up and that there was only one patient who developed grade 4 HO.

More studies are needed to better evaluate the effect of HO on patient outcomes. It is however not surprising that patient-reported outcomes such as the VAS score or the Odom criteria have not been shown to differ between patients with and without HO. HO usually does not lead to neural compression. The relief of arm pain achieved during surgery should therefore not be affected by HO. Further, as CTDR was shown to be as effective as ACDF for treatment of such symptoms, even patients with grade 4 HO would be expected to report overall good outcomes. It remains to be seen whether the gradual loss of motion at a CTDR level would eventually lead to clinical deterioration, in particular, in an increase of mechanical neck pain. There is no available evidence to support this at this time.

Prevention of HO

There are currently no evidence-based methods to treat or prevent HO after CTDR. The current practice in our department is to prescribe patients a 1-week course of low-dose indomethacin, discourage the use of any neck collar or brace post-operatively, and encourage a regimen of physiotherapy and light exercise.

Most studies examining possible methods to prevent HO in arthroplasty were carried out in the setting of hip arthroplasty. Radiation therapy and nonsteroidal anti-inflammatory drugs (NSAIDs) have both been shown to be effective methods of preventing HO in patients undergoing hip arthroplasty (D'Lima et al. 2001).

Given the high morbidity of radiation when it is given to the soft tissues of the neck, it is not a reasonable therapeutic option for prevention of HO after CTDR. Indomethacin, which can be contraindicated if patients are known to have a history of renal disease or peptic ulcer disease, nevertheless has a much lower-risk profile. Tu et al. (2015) present a retrospective review of a cohort of 75 patients (107 levels) treated with CTDR. Patients were followed up for a mean of approximately 38 months. The authors examined the rate of HO between those who were given NSAIDs post-op and those who were not.

Although they found a lower rate of HO in the NSAIDs taking group (47.2% vs. 68.2% for those who did not take them), this was not statistically significant. They also found a lower rate of prostheses that had undergone arthrodesis in the NSAIDs taking group (13.2% vs. 22.7% in those who did not take them), but this was not significant either. Both groups had similar VAS scores for neck and arm pain at follow-up, as well as similar Neck Disability Index (NDI) scores. This data therefore does not seem to indicate any clinical benefit from NSAIDs after CTDR. As most of the data presented in this chapter, this data comes from a small retrospective cohort, and this limits its value.

One prevention strategy which is advocated by many authors is the copious irrigation of the wound during surgery. This is done to clear out bone fragments and shavings from the surgical site in order to limit the induction of bone formation that could be triggered by such remnants. Leung et al. (2005) and Tu et al. (2011) advocate for this strategy. Although it makes sense based on our current understanding of the pathophysiology of HO, it is not supported by any evidence to date. Further, as discussed by Lee et al. (2010), most surgeons already irrigate the wound extensively to reduce the rate of infection. This step may therefore not be a factor in the development of HO.

Other Complications Related to CTDR

Complications Related to the Anterior Approach

Although there are technical differences between CTDR and ACDF, the initial approach and neural decompression are essentially the same. As such, complications related to these steps are also similar. CTDR starts with an anterior neck dissection through the natural planes between the trachea and esophagus medially and the sternocleidomastoid and the carotid sheath laterally. This dissection is taken down to the anterior surface of the cervical vertebral column, at which point the proper disc level is identified and the discectomy and neural decompression can commence.

Many structures can be injured from the time of incision to the end of the neural decompression. We usually quote a rate of serious complications (such as neural injury, paralysis, vertebral artery injury) of less than 1% and a combined rate of 3–5% for all other possible complications.

Possible complications that are regularly discussed with patients preoperatively include infection, hematoma, hoarseness from recurrent laryngeal nerve injury, dysphagia from retraction on the esophagus, soft tissue injury in the neck (including esophagus and airway), cerebrospinal fluid leak, nerve injury, paralysis, and complications from anesthesia. More unusual complications can include Horner's syndrome from a too lateral dissection, superior laryngeal nerve injuries from high cervical dissection resulting in a risk of aspiration, and vertebral artery injury. C5 nerve root palsies are also possible from an anterior approach.

In a retrospective review of a cohort of 1015 patients who underwent ACDF at their institution, Fountas et al. (2007) reported the incidence of complications relating to the anterior approach to the cervical spine. They found the overall rate of morbidity to be of 19.3%. The main postoperative complications included dysphagia (9.5%, mostly transient), hematoma (5.6%), laryngeal nerve palsy (3.1%), dural tear (0.5%), esophageal perforation (0.3%), worsening myelopathy (0.2%), wound infection (0.1%), and mortality (0.1%). These are all complications that can also be associated with CTDR.

Complications Related to the Prosthesis

Although the rate of complications associated with the implanted prosthesis in CTDR seems to be very low, several issues have been reported. As each different type of prosthesis has a different design, the issues that have been reported vary from one type to another.

In their preliminary report on CTDR using the Bryan disc (Medtronic, USA) in a group of 60 patients, Goffin et al. (2002) reported that they identified possible device migration of more than 2 mm but less than 3 mm in two patients. In their follow-up report, the same group reported a third case of device migration (Goffin et al. 2003). The authors attributed this phenomenon to inadequate initial milling of the vertebral body endplates when inserting the device. Indeed, one of the steps to implanting a Bryan disc involves inserting a milling/drilling device between both endplates and compressing them over this drill so that they can acquire the proper concavity to accept the prosthesis. Issues with this step of the surgery can therefore lead to improper placement and subsequent migration.

Some prostheses have a metal keel that must be implanted in the vertebral body, and this has been a source of complications as well. The ProDisc-C (Synthes, USA) is composed of two articulating surfaces which are each connected to a keel that must be inserted in the vertebral bodies above and below the disc space being treated. Shim et al. (2007) report a case of CTDR during which the vertebral bodies involved both fractured posteriorly when the authors were using a chisel to prepare them to receive the device's superior and inferior keels. They identified this intraoperatively and were able to remove any bone fragments that were compressing the thecal sac. Tu et al. (2012) present a similar situation in which the superior vertebral body was found to have a sagittal split fracture after the insertion of a ProDisc-C device. The patient was treated conservatively. See Fig. 1 for an example of a keel-based prosthesis.

We found two reports of outright device failure because the hardware's material fissured or cracked. Fan et al. (2012) present a case of Bryan disc failure, in which the prosthesis itself was found to have developed a fissure 8 years after implantation. Similarly, Nguyen et al. (2011) present a case of CTDR in which a prosthesis with a ceramic surface was found to have cracked leading to recurrent symptoms.

Interestingly, we also found one report in which a patient developed an inflammatory reaction to the implant, and this was assumed to be because of intolerance or allergy to the material used. Cavanaugh et al. (2009) report that a patient who underwent a one-level CTDR returned

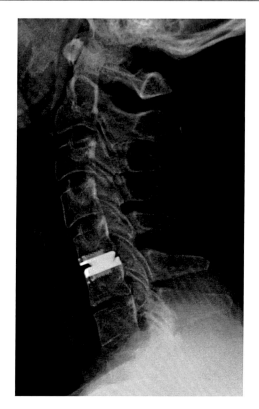

Fig 1 This lateral cervical radiograph demonstrates a keel-based prosthesis, the ProDisc-C (Synthes, USA). This patient did not suffer any complications during surgery and has not developed any HO

6 months postoperatively with recurrence of symptoms. Imaging revealed a mass behind the implant which turned out to be inflammatory tissue at reoperation. The authors conclude that the patient was likely hypersensitive to one of the components of the implant.

Adjacent Segment Disease

The main goal of CTDR is to preserve motion at the operated level. In theory, this preserved motion is also supposed to decrease the incidence of adjacent segment disease (ASD). This issue remains controversial, however, with differing reports of long-term outcomes when CTDR is compared to ACDF.

Robertson et al. (2005) compared the incidence of ASD at 24 months postoperatively in a group of patients who underwent CTDR to that of a group of patients who underwent ACDF in a prospective clinical trial. This trial included 74 patients undergoing CTDR and 158 patients undergoing ACDF. The authors found that at 2 years of follow-up, the CTDR group showed a significantly lower rate of ASD compared to the ACDF group (17.5% vs. 34.6%, $p = 0.009$).

A different conclusion was however reached by the group of Nunley et al. (2012). They analyzed the data from three prospective randomized trials comparing CTDR to ACDF. Their pooled data resulted in a cohort of 113 patients who underwent ACDF and 57 patients who underwent CTDR. The authors found that at a median follow-up of 42 months, both groups showed a similar rate of ASD (14.3% vs. 16.8%, annual rate of 3.23% in the ACDF group and 3.77 in the CTDR group).

A more recent paper by Janssen et al. (2015) presents longer follow-up data of a randomized controlled trial of CTDR versus ACDF. They present data at 7 years of follow-up for 79 patients who underwent CTDR and 73 patients who underwent ACDF. The authors found that significantly more patients in the ACDF group had to undergo revision surgery for ASD than in the CTDR group (13 vs. 6 patients).

Although it remains controversial as to whether CTDR leads to lower rates of ASD than ACDF, ASD is a phenomenon that does occur with CTDR, and we believe it should be discussed with patients.

Conclusion

CTDR can be complicated by HO, as well as by issues related to the anterior approach to the cervical spine, by issues with the prosthesis used and by ASD. Nonetheless, the data presented in this chapter demonstrate that CTDR is an overall safe procedure and that HO has a limited effect on clinical outcomes. More research is needed to better delineate the exact pathophysiology of HO as well as its natural history. We believe that it is only once the phenomenon is better understood that effective strategies to prevent it will be found.

References

Balboni TA, Gobezie R, Mamon HJ (2006) Heterotopic ossification: pathophysiology, clinical features, and the role of radiotherapy for prophylaxis. Int J Radiat Oncol Biol Phys 65:1289–1299

Brooker AF, Bowerman JW, Robinson RA et al (1973) Ectopic ossification following total hip replacement. Incidence and a method of classification. J Bone Joint Surg Am 55:1629–1632

Cavanaugh DA, Nunley PD, Kerr EJ 3rd et al (2009) Delayed hyper-reactivity to metal ions after cervical disc arthroplasty: a case report and literature review. Spine (Phila Pa 1976) 34:E262–E265

D'lima DD, Venn-Watson EJ, Tripuraneni P et al (2001) Indomethacin versus radiation therapy for heterotopic ossification after hip arthroplasty. Orthopedics 24:1139–1143

Fan H, Wu S, Wu Z et al (2012) Implant failure of Bryan cervical disc due to broken polyurethane sheath: a case report. Spine (Phila Pa 1976) 37:E814–E816

Fountas KN, Kapsalaki EZ, Nikolakakos LG et al (2007) Anterior cervical discectomy and fusion associated complications. Spine (Phila Pa 1976) 32:2310–2317

Goffin J, Casey A, Kehr P et al (2002) Preliminary clinical experience with the Bryan cervical disc prosthesis. Neurosurgery 51:840–845; discussion 845–847

Goffin J, Van Calenbergh F, Van Loon J et al (2003) Intermediate follow-up after treatment of degenerative disc disease with the Bryan Cervical Disc Prosthesis: single-level and bi-level. Spine (Phila Pa 1976) 28:2673–2678

Heidecke V, Burkert W, Brucke M et al (2008) Intervertebral disc replacement for cervical degenerative disease – clinical results and functional outcome at two years in patients implanted with the Bryan cervical disc prosthesis. Acta Neurochir 150:453–459; discussion 459

Janssen ME, Zigler JE, Spivak JM et al (2015) ProDisc-C total disc replacement versus anterior cervical discectomy and fusion for single-level symptomatic cervical disc disease: seven-year follow-up of the prospective randomized U.S. Food and Drug Administration Investigational Device Exemption Study. J Bone Joint Surg Am 97:1738–1747

Lee JH, Jung TG, Kim HS et al (2010) Analysis of the incidence and clinical effect of the heterotopic ossification in a single-level cervical artificial disc replacement. Spine J 10:676–682

Leung C, Casey AT, Goffin J et al (2005) Clinical significance of heterotopic ossification in cervical disc replacement: a prospective multicenter clinical trial. Neurosurgery 57:759–763; discussion 759-763

Mcafee PC, Cunningham BW, Devine J et al (2003) Classification of heterotopic ossification (HO) in artificial disk replacement. J Spinal Disord Tech 16:384–389

Mehren C, Suchomel P, Grochulla F et al (2006) Heterotopic ossification in total cervical artificial disc replacement. Spine (Phila Pa 1976) 31:2802–2806 adhesion formation, which prevent accurate identification of the great vessels. The anterior lumbar spine therefore remains a relatively facile approach as an index procedure but is fraught with potential complications in any revision situation. In our experience, many failures of lumbar disc replacement could have been avoided as they can be traced to surgeon-specific factors (as opposed to patient-specific factors) such as incomplete discectomy, improper device insertion, or inappropriate indications.

Our approach for revisions is to over-prepare – assume you will have suboptimal visualization between fascial planes. This means we place ureteral stents to palpate the ureters. In addition, in the event of inadvertent entering of the ureter, the stent facilitates suture repair by acting as a conduit in the early postoperative period. Occasionally for high-risk cases, we also prophylactically cannulate the femoral vessels. This allows faster intraoperative endovascular passage of balloons intraoperatively to assist hemostasis. In addition we can pass covered vascular stents up from the femoral vein and artery in an endovascular technique in the event of friable vessels with limited exposure in a deep retroperitoneal revision.

Keywords

Revision disc replacement · Spinal deformity · Spinal reconstruction

Key Points

1. Revision lumbar anterior retroperitoneal approaches are perhaps the most risky procedures in spinal surgery due to adhesions and difficult visualization of the very friable lumbar veins.
2. It behooves the spine surgeon to have a great working relationship with a vascular specialist and to over-prepare for repeat anterior retroperitoneal approaches.
3. Removal of the prosthesis requires being able to distract the disc space to gain working room between the vertebral endplates.

Introduction

One of the key considerations in approach for surgery in failed total disc replacement is to be able to differentiate between "pilot error" (suboptimal surgical technique) and inherent problems with the technology (metal-on-metal wear debris, osteolysis, poor ingrowth fixation, etc.). The best method for distinguishing inherent shortcomings of the technology is to present the immediate postoperative radiographs to a consensus group of experts – if the experts can predict the subsequent mode of failure, then one is dealing with a complication in surgical technique. However, if a consensus group of experts cannot predict the eventual mode of failure, then the complication is due to an inherent problem in the prosthetic design or biomaterials (Fig. 1).

Implant Failure: UHMWPE Core

Polyethylene core fractures, core dislocations, or implant breakage have been more common with UHMWPE implants. Instances of excessive polyethylene wear and osteolysis have been quite rare and have uniformly occurred in patients implanted before US FDA approval. David reported a case of polyethylene failure due to oxidation 9.5 years following implantation. Similarly, as reported in the largest series of TDR failures reported to date, there was one case of detectable polyethylene core wear noted 12 years postoperatively. Since 1997, an industry-wide enhancement to the sterilization process of polyethylene, whereby irradiation occurs in an inert gas such as nitrogen rather than air, has resulted in dramatically improved wear characteristics due to a reduction in the incidence of oxidation. This results in an increase in cross-linking or highly cross-linked UHMWPE – this improves wear characteristics but increases the brittleness of the core implant. This corrective preventative action of avoiding polymerization and reducing the UHMWPE wear rate was born out of the total hip implant experience on partially cross-linked UHMWPE. In contrast to the above cases, analysis of a retrieved polyethylene core from a

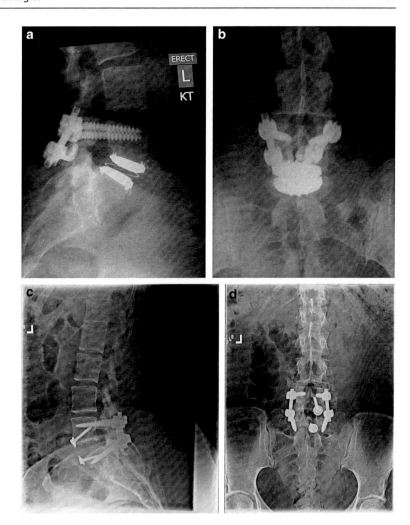

Fig. 1 These are the lateral (a) and the anteroposterior (b) radiographs of a 56-year-old woman who demonstrated a serum-confirmed nickel allergy shortly after implantation of this L5–S1 cobalt-chrome alloy arthroplasty. She also had undergone posterior fixation of an L5 spondylolysis with instrumentation from Elsewhere General. We performed anterior removal of the prosthesis and anterior lumbar interbody fusion at L5–S1 using allograft with buttress screws. The posterior instrumentation was revised into L5–S1 conventional pedicle instrumentation for increased stability. The lateral (c) and anteroposterior (d) radiographs demonstrate a stable 360 fusion and instrumentation. We advocate a combined front and back approach for most revisions following arthroplasty revision. The patients' metal allergy resolved

revision surgery performed for bone-implant loosening 1.6 years postoperatively in a patient implanted post-1997 demonstrated low levels of oxidation with mechanical properties that were not substantially degraded. At the time of this writing, there are only two published cases of anterior dislocation of the polyethylene inlay of a ProDisc artificial disc replacement. We have had over ten cases of posterior core dislocations presented from Elsewhere General. These are some of the most difficult revisions as the patients present with severe pain, pressure of the core on the cauda equina, and leg weakness. The key surgical step in the revision technique is to be able to insert a skeletal distraction device between the failed vertebral bodies to be able to distract the disc space. This is analogous to a Caspar distractor in the cervical spine. One has to distract the lumbar disc space to increase the working space in order to atraumatically remove the failed implant. Invariably we remove the core first and then reach vertebral endplate, in succession. With a keeled prosthesis, the loosening of the metal-bone ingrowth is achieved with a narrow osteotome.

Porous Ingrowth Failure and Loosening

Metal-bone interface complications account for the greatest number of failures of lumbar disc replacement, and the mode of failure depends on the type of device. Sagittal vertebral body fractures can occur with keeled devices because the keel slot creates a stress riser in the bone. Two-level implantations in which keel slots are introduced into the superior and inferior aspects of the intercalated vertebra are at highest risk for this complication. To reduce this risk, the MAVERICK keel has been reduced from 11 mm to 7 mm[19]. In contrast, implants with smaller "anchors" at the bone-implant interface, such as the CHARITE, in which six 3 mm teeth engage the vertebral endplate, exhibit a greater tendency to migrate or dislodge if improperly placed. In the CHARITE US IDE study, there were 15 of 347 implantations that required removal, and none of these had been inserted in the "ideal" position[3]. Regardless of the design, TDR placement anterior to the center of rotation, especially in a hyperlordotic segment, will have a tendency to migrate or dislocate anteriorly due to excessive shear. Spondylolysis and spondylolisthesis present a biomechanically less stable environment for TDR and explain some of the early failures by migration and should be considered an absolute contraindication to lumbar TDR. Damage to the vertebral endplate during insertion, placement of a TDR in an osteoporotic spine, inaccurate positioning, and insertion of an implant that is too small are all factors that can contribute to TDR subsidence. In van Ooij and coworkers' report of 27 complications of the CHARITE device, there were 16 cases of subsidence. The cause, prevalence, or incidence of these failures could not be determined as the study was retrospective without mention of the total number of cases. Strict adherence to indications, surgeon education, and DEXA with preoperative correction of osteoporosis should theoretically reduce the incidence of this complication (Fig. 2).

Spinal instability has a greater tendency to occur in multilevel implantations. Preoperative scoliosis is a relative contraindication to TDR; however, subtle coronal and sagittal plane deformities may not be taken into account. TDR insertions in these scenarios will likely exacerbate rather than reduce any deformity because stabilizing structures such as the anterior and posterior longitudinal ligament as well as the majority of the annulus are removed during the

implantation. Even in the well-aligned spine, incomplete discectomy and "off-axis" TDR placement at one level can create an "off-axis" situation at the next level (a so-called Z deformity). Thus when more than one vertebral level is to be implanted, device placement errors will tend to be compounded at sequential levels of insertion resulting in a "Z deformity." The principle concept is that cervical disc replacement adds inherent rotational stability to the cervical spine – unfortunately lumbar TDR adds to rotational instability to the lumbar spine. This is well documented by McAfee et al. in our laboratory (McAfee et al. 2006b) due to the posterior position of the lumbar facets, the relatively small cross-sectional area of the lumbar facets, and the relative importance of the anterior longitudinal ligament (ALL) in the lumbar spine body. For these three reasons, multiple levels of cervical TDR increase the cervical spine stability, whereas multilevel lumbar TDR decreases lumbar spine stability postoperatively.

Wear Debris and Cytokine Reaction: Prosthetic Inflammatory Response

Particulate debris generated from virtually any articulating biomaterial is characterized by macrophage recruitment and pro-inflammatory cytokine release. The culmination of this cascade is matrix metalloproteinase activation, which leads to bone resorption and osteolysis. Osteolysis is a well-documented complication of total joint replacement; however, most investigators agree that it will not be prevalent following TDR. One reason for the potentially reduced risk in the spine is the fact that the intervertebral disc lacks a surrounding synovial space, the key source for macrophages, and other inflammatory cells. The second reason is the relatively reduced range of motion and, hence, reduced volumetric wear in TDR compared with typical diarthrodial joint replacement. With over 20 years of global experience and over 10,000 TDR implantations worldwide, there are only anecdotal reports of osteolysis, and to our knowledge, only one has been documented histologically. Heterotopic

Fig. 2 (a) This 51-year-old executive underwent L5–S1 CHARITE disc replacement without incident. (b) Unfortunately 6 months following the surgery, he bent forward to pick up a heavy object, and he displaced the UHMWPE core of the prosthesis anteriorly. (c) A venogram was performed as part of the diagnostic workup of the anticipated revision surgical procedure. It demonstrated complete occlusion of the left iliac vein. This required insertion of an IVC umbrella at the start of the anterior revision procedure to avoid pulmonary embolism. (d) Following anterior and posterior arthrodesis and instrumentation at L5–S1, the patient also required CABG. In total the patient required five operative procedures. The illustration indicates no further migration of spinal instrumentation in this complicated patient requiring anticoagulation

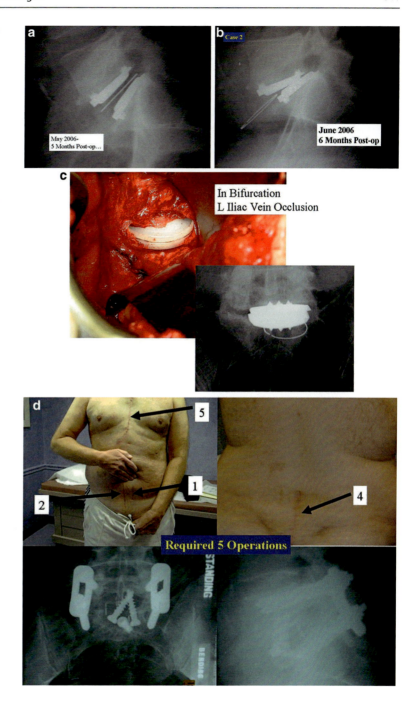

ossification (HO) surrounding lumbar TDR implants has been classified previously by McAfee and colleagues in preparation for the CHARITE US IDE trial. In this trial, the incidence of HO was 4.3%; however, the presence of HO did not have any impact on flexion-extension range of motion or clinical outcome. We know of no other lumbar TDR trial in which the presence of HO has been systematically evaluated; however, isolated cases of periannular ossifications have been reported. To be fair, however, the imaging characteristics of the lumbar devices, which are typically cobalt-chrome or stainless steel alloys, do not allow for identification of all but the larger amounts of periannular bone formation.

Nerve Root Compression from Implant Migration

Nerve root compression is occasionally found in cases of lumbar TDR when compared with fusion controls in prospective randomized evaluation. Most cases of revision surgery in the CHARITE US IDE trial involved reoperation with laminectomy and foraminotomy for new neurologic symptoms in the lower extremities. After any anterior discectomy and instrumentation, patients should be asked if they have any "new" pain in their legs that was not present preoperatively, and a thorough neurologic examination should be performed in the postanesthesia care unit. If there are any new neurological complaints or physical exam findings (i.e., motor or sensory loss), a CT scan with myelographic contrast should be performed to assess positioning of the implant and ensure that a hematoma, bone fragment, or disc material is not present in the spinal canal or neuroforamina. MRI is not particularly helpful postimplantation due to metal artifact obscuring the regions of greatest interest. By rapidly diagnosing the cause for new neurologic symptoms with appropriate imaging, early revision such as implant repositioning is possible. Neurologic injury without radiographic abnormalities is a complication that typically presents with left-sided leg pain in the L4 or L5 dermatome. First reported by Thalgott and *colleagues*, this complication has been thought by some to be due to excessive retraction of the lumbar plexus during the retroperitoneal exposure[26]. The lumbar plexus is at highest risk when exposing the L4–L5 disc space. To mobilize the left common iliac vein to the patient's right side, the iliolumbar vein(s) needs to be identified and ligated as it courses dorsally between the lumbar nerve roots located under the psoas muscle. Alternatively, such symptoms have been thought to be sympathetic mediated and result from the vascular exposure,

as similar symptoms may be seen following non-spinal vascular dissections. In either case, whatever the etiology, the dysesthetic pain pattern typically resolves after 6–8 weeks; however, corticosteroid selective nerve root blocks may help minimize the symptoms. Notwithstanding, the burden of proof remains to demonstrate that the acquired symptoms are not due to new nerve root compression, which may be remedied with further surgery.

Biomechanical Instability

Excessive motion is a phenomenon that is not a prevalent mode of failure for TDR but likely underdiagnosed. It is defined as motion exceeding the natural motion arc in six degrees of freedom from the native motion in a specific patient's functional spinal unit. This motion results in subclinical instability that the body senses and in turn attempts to combat. This can present with postoperative muscle spasms and unexplained and chronic postoperative pain. It is a diagnosis of exclusion. The TDR is typically in excellent position without signs of dislodgement, dislocation, or subsidence. The hallmark is significant hypertrophy of the facet joints. Because the workup algorithm follows closely to ruling out spondylolysis, a CT scan can be very helpful in characterizing this hypertrophy. To determine surgical candidacy following exhaustive conservative management which typically includes long-acting narcotics, local facet blocks are the major objective tool to assess pain relief. Care must be taken to counsel the patient to give feedback following the injections to pay close attention to the duration of relief that may ensue corresponding to the half-life of Marcaine. If the patient notices a significant decrease in pain, a stabilizing procedure in the form of a posterolateral fusion with instrumentation can be discussed as an option of treatment. There is presently no way of predicting which patients will be affected by this phenomenon. Some surgeons have pointed to facet orientation as a predisposition. Especially at L5–S1, more sagittally oriented facets in conjunction with sacral inclination can predispose to increased facet forces and this mode of failure. When diagnosed properly with facet blocks, patients have had resolution of pain and spasms and likewise weaned off of narcotics completely.

Vascular Revision Strategies

Revision total disc replacement is a potentially life-threatening procedure and should be reserved for indications that justify the increased risk.

Revision anterior exposure within 2 weeks of TDR incurs relatively little additional morbidity because adhesion formation is minimal. For this reason, surgeons should have a low threshold for revising implants that are clearly malpositioned or exhibit early migration within this 2-week time frame. If the prosthesis can be repositioned or revised to another TDR, without damage to host bone, it seems reasonable to do so. Beyond this period of time, a revision strategy must be individualized to the particular clinical situation. A posterior instrumented fusion with or without a decompressive laminectomy is currently the most effective salvage procedure. Preoperative planning is critical to a successful anterior revision. The authors analyze the cause for failure to establish individual goals for the revision. Corrections such as polyethylene replacement, size, or position changes are relatively easy to anticipate. Appropriate patient counseling regarding the increased risks, potential for changes, or need to abort the original plan for safety considerations is also critical. Reviewing the initial operative reports, clinic notes, and original indications can be particularly enlightening if a revision to a TDR is contemplated. If a patient failed to meet indications for a TDR at the index procedure, it is unlikely that they will meet indications at the revision. Patients with significant host bone loss, deformity, or instability should be revised to an interbody arthrodesis.

In our experience, major vascular injury, deep venous thrombosis, and potential ureteral injury are at particularly increased risk in the anterior revision scenario, and our operating room preparation strives to mitigate these risks (McAfee et al. 2003, 2006a, 2006b). Ureteral stenting is easily performed preoperatively and is valuable not only for intraoperative identification of the ureter but also for maintaining ureteral patency during the healing phase if an injury were to occur. Significant reduction in the somatosensory-evoked potentials of one leg may be the first indicator of excessive arterial retraction, and for this reason, we routinely utilize this form of somatosensory (SSEP) monitoring. Placing a pulse oximeter probe on the great toes is another form of monitoring that can assist in detecting excessive arterial retraction or occlusion. Often temporary relaxation of the retractors will result in normalization of the SSEP or pulse oximeter readings. We currently insert inferior vena cava (IVC) filters preoperatively as postoperative lower extremity duplex evaluations have missed a deep venous thrombosis (DVT) in the pelvic great vessels which progressed to a postoperative pulmonary embolism (PE) (Tortolani et al. 2006). Finally, we prepare the inguinal region into the operative field through which percutaneous vascular access wires are placed into the right and left femoral veins. If an injury to one of the venous great vessels occurs, balloon catheters can be inflated to tamponade and control bleeding in a timely fashion.

The specific revision surgical approach depends on the reason for failure and the original surgical approach; however, the currently available options include posterior decompression, posterior decompression and instrumented fusion, anterior TDR removal and fusion, or anterior TDR removal and reinsertion. Usually anterior TDR removal and conversion to fusion is the safest salvage strategy because gaining the exposure necessary to implant a new disc replacement is rarely possible due to adhesion formation. Our attitude is we perform anterior and posterior arthrodesis for L4–L5 revisions because we want this to be the absolute last time the great vessel dissection should ever be required. Anterior interbody fusion devices typically require less exposure and can be inserted obliquely across the disc space. This can be extremely handy when the only accessible area to the disc is to one side. Revision to a TDR is no longer a consideration in our experience in the lumbar spine.

Revision anterior exposures should always be performed with an experienced vascular access surgeon, and gaining access via a virgin territory is desired but rarely possible. At L5–S1, if a left-sided retroperitoneal approach was utilized at the index procedure, then transperitoneal or right-sided retroperitoneal approach can be considered. Conversely, if a transperitoneal approach was used at the index procedure, then a right- or left-sided retroperitoneal exposure can be considered. L4–L5 and higher is always more challenging because the left-sided retroperitoneal approach is virtually always utilized during the initial exposure. Transperitoneal and right-sided approaches are technically far more demanding at L4–L5 because of the central to right-sided position of the IVC. For this reason, anterior revisions to L4–L5 should be performed via a transperitoneal or through the same left-sided retroperitoneal approach. By identifying the left psoas muscle, one can generally palpate the lateral border of the lumbar spine, and, from this point, a subperiosteal dissection can facilitate safe exposure toward the midline and beyond. In the end, the experience and comfort level of the access surgeon will be as, if not more, important than the type of approach used during the index procedure.

Given the aforementioned technical challenges, the most common salvage procedure for failed TDR is posterior instrumented fusion with pedicle screws. This technique essentially locks the prosthesis from any further movement and allows for bone grafting in the posterolateral intertransverse region. Cunningham et al. found that pedicle screws alone combined with a lumbar disc replacement were not found to be statistically different biomechanically from pedicle screws and a femoral ring allograft. Posterior hemilaminectomy or laminotomy without fusion is an alternative for focal disc or bone displacements into the spinal canal; however, extreme care must be taken to avoid destruction of stabilizing structures like the facet joints.

Conclusions

Lumbar TDR is a safe and effective treatment option for appropriately selected patients with lumbar degenerative disc disease. Bone-implant failures can be prevented by strict adherence to FDA indications and accurate placement of the device. For at least one device, suboptimal positioning has been correlated with worse patient outcome, a finding which will likely be borne out for other TDR designs. Device failures for current designs are rare and are characterized by polyethylene fracture or dislocation. Anterior revision surgery with an experienced access surgeon and preoperative placement of ureteral stents, vena cava filters, percutaneous vascular access wires, and spinal cord monitoring can reduce the risks of major vessel injury or thrombosis.

References

McAfee PC, Cunningham BW, Devine JD, Williams E, Yu-Yahiro J (2003) Classification of Heterotopic Ossification (HO) in artificial disk replacement. J Spinal Disord Tech 16:384–389

McAfee PC, Geisler FH, Saiedy SS, Moore SV, Regan JJ, Guyer RD, Blumenthal SL, Feder IL, Tortolani PJ, Cunningham B (2006a) Revisibility of the Charite artificial disc. Analysis of 688 patients enrolled in the US IDE study of the Charite artificial disc. Spine 31(11):1217–1226

McAfee PC, Cunningham BW, Hayes V, Sidiqi F, Dabbah M, Sefter JC, Hu N, Beatson H (2006b) Biomechanical analysis of rotational motions after disc arthroplasty. Implications for patients with adult deformities. Spine 31(19):S152–S160

Tortolani PJ, McAfee PC, Saiedy S (2006) Failures of lumbar disc replacement. Semin Spine Surg 18(2):78–86

Posterior Lumbar Facet Replacement and Interspinous Spacers

47

Taylor Beatty, Michael Venezia, and Scott Webb

Contents

Introduction 846

Facet Anatomy and Biomechanics 846

Rationale and Biomechanics of Facet Arthroplasty Systems 848

History of Facet Arthroplasty Systems 849

Total Facet Arthroplasty System (TFAS) 849

ACADIA Facet Replacement System (AFRS) 851

Total Posterior Arthroplasty System (TOPS) 852

Interspinous Devices (ISD) 854

Conclusions 855

References 855

Abstract

Motion preservation technology in the lumbar spine is not confined to lumbar disc arthroplasty. Pathology involving the posterior elements of the lumbar spine is often the source of pain generation and stenotic symptoms. Posterior-based motion preservation systems fall under two categories: facet arthroplasty and posterior dynamic stabilization (PDS). Several devices have gone through clinical trials since initial introduction in the early 1990s for PDS systems and mid-2000s for facet arthroplasty systems presenting viable alternatives to lumbar fusion. Understanding the anatomy and the biomechanical forces acted on the lumbar region has led to the creation of these devices with goals of symptomatic relief, motion preservation, and prevention of adjacent segment disease.

T. Beatty (✉)
Orthopaedic Surgery Resident PGY5,
Largo Medical Center, Largo, FL, USA
e-mail: dr.beatty.ortho@gmail.com

M. Venezia
Orthopaedic Specialists of Tampa Bay, Clearwater, FL, USA
e-mail: mveneziado@gmail.com

S. Webb
Florida Spine Institute, Clearwater, Tampa, FL, USA
e-mail: spinecutter@yahoo.com

© Springer Nature Switzerland AG 2021
B. C. Cheng (ed.), *Handbook of Spine Technology*,
https://doi.org/10.1007/978-3-319-44424-6_79

Keywords

Facet replacement · Posterior dynamic stabilization · TOPS · TFAS · ACADIA · Coflex · X-Stop

Introduction

There has been significant advancement in spinal implants and the treatment of spinal pathology over the last few decades, including a resurgence of motion preservation and joint arthroplasty. While we still have much to learn concerning the pain generators and correlating pathology of the lumbar spine, procedures such as total facet arthroplasty and posterior dynamic stabilization allow for additional tools in a surgeon's armamentarium.

Historically, the standard of care for degenerative pathology with instability of the lumbar spine has been fusion with or without instrumentation. Dynamic stabilization has emerged as a viable alternative to the standard of care. To begin addressing some of the concerns with fusion, posterior dynamic stabilization was introduced beginning with the development of the Graf ligament in 1989 and subsequently the Dynesys (Zimmer Spine, Minneapolis, MN) in 1994. The facet joints have also been targeted as another potential pain generator of the lumbar spine. Total facet arthroplasty has arisen as a treatment option for spinal stenosis as well as degenerative spondylolisthesis, allowing for an alternative to spinal fusion and the ability to address the facet joints as possible lumbar pain generators.

Throughout this chapter, we will review the biomechanical basis as well as clinical literature supporting the role of posterior-based dynamic stabilization (PDS).

Facet Anatomy and Biomechanics

The facet joint is critical to the proper motion of the functional spine unit which consists of the disc, the facet joints, and the ligaments. The motion at each segment is determined by the health of each of these components. Each vertebral segment interacts with the adjacent vertebral segment by means of five articulating joints; the disc and the four facet joints at each level. These components share the loads as they are transmitted through the spine. The facet joint acts like a cam with a multi-radius arc of motion that is engaged as the spine moves. This assists with proper spinal motion, protection of neurologic structures, and transferring of load through the spinal column (White and Panjabi 1978). These diarthrodial, synovial joints bear weight in both compression and shear providing functional range of motion (ROM) while limiting excessive ROM. They help to protect the lumbar disc from excessive stress. Due to this mutualistic nature of disc and associated facets, the degenerative changes in the disc lead to increased loads and progression of degeneration in the facets (Webb and Holen 2008) (Fig. 1). The loads are transferred through each component, and as the spine ages, their interaction with each other changes. The loads transferred are going to vary depending on the health of the segment.

Lumbar facets require high load transmission and therefore are significantly larger than cervical facets. ROM is limited in the lumbar segment with the focus in flexion and extension. The ability of the facets to share the load with the intervertebral disc is position dependent with a load sharing of 0% in full flexion to 33% in full extension (Panjabi et al. 1989). The facets are located more centrally with an adducted positioning to prevent hyperextension and rotational torsion (Webb and Holen 2008). The amount of sagittal angulation increases as you move caudally in the lumbar spine. In childhood, the facets are angled more posteriorly with a transition to a more sagittal position with age; this can be a factor in many of the facet-related issues (Scoles et al. 1988). The lumbar facets main function is to prevent anterior shear which has been proven to damage the intervertebral disc (Reily 2011). Resistance increases toward the endpoint of motion. This gradual increased resistance protects the joint and the adjacent structures by

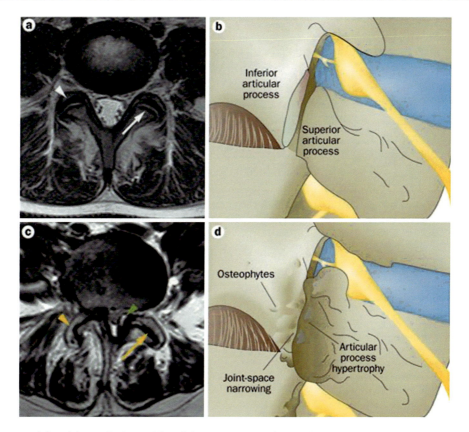

Fig. 1 Normal facet joints and advanced facet joint osteoarthritis. (**a**) T2-weighted axial MRI image of normal facet joints with no joint-space narrowing (white arrow), and no osteophytes or articular process hypertrophy (white arrowheads). (**b**) Normal facet joints. (**c**) T2-weighted axial MRI image of osteoarthritic facet joints with joint-space narrowing (yellow arrow), osteophytes and articular process hypertrophy (yellow arrowhead). Facet joint osteoarthritis, disc-bulging, and a facet joint synovial cyst (green arrowhead) in combination lead to stenosis of the central canal and lateral recesses. (**d**) Osteoarthritic "facet joints." (Gellhorn et al. 2013)

providing a soft stop. When anterior shear load is applied suddenly, the facets carry 1/3 of the load, while the disc carries 2/3 of the load; when the load is applied over time, the majority of the load is carried by the facets (White and Panjabi 1978). Nachemson (1960) used pressure transducers and determined that the facets carry approximately 15% of vertical load in the lumbar spine. Yang and King (1984) established the mechanism of facet transmission of axial load in cadaveric specimens by the use of an intervertebral load cell with results demonstrating a normal facet carrying 3–25% of the load, while an arthritic joint could carry as much as 47%.

The facets are surrounded by capsules that are innervated by type I, II, and III mechanoreceptors. This neural input allows for positional feedback for postural control. Destruction of this innervation may also be a key role in the disease process (Webb and Holen 2008). Cavanaugh et al. (1996) concluded that these nerves are activated by capsular stretch and by neurogenic and non-neurogenic modulators of inflammation, including substance P, bradykinin, and phospholipase A2. Facet joint surfaces are covered with hyaline cartilage that undergoes degenerative changes to its mechanical properties similar to the processes that occur in other joints. The loss of articular cartilage leads to spur formation thought to stabilize the joint. The degenerative process also leads to loss of control of anterior shear forces, facet subluxation, and increased anterior-posterior translation (Reily 2011). This begins early in the degenerative process with the loss of 1 mm of cartilage in the facet joint being shown to significantly increase the translational motion (Reily 2011). Degeneration and inflammation lead to pain with joint motion that causes restriction and thus deconditioning.

It is important to understand how each component of the functional spinal unit performs and how each component interacts and the impact when they begin to fail. The functional spinal unit has a neutral zone where the axial force transmitted by the spine is stiffer at the extreme range of motion and less stiff near the neutral position. In other words, it requires more force to move the functional spinal unit from the neutral zone to the endpoints of motion. When one looks at the facet joint, it is apparent that degenerative changes as well as surgical intervention such as in a fusion or disc arthroplasty changes the way the facet joints at the index level and adjacent levels behave. Facet joints can be overloaded if the surgical intervention substantially changes the way the disc behaves as in a disc replacement. With a fairly non-constrained lumbar disc arthroplasty, one might expect increased facet stresses. In a fusion, the adjacent segment will experience increased loads across the facet joint. It is important to consider the impact of any surgical procedure on the overall motion of each lumbar segment and the long-term potential impact.

ing with adjacent segments, slow the degeneration of these adjacent segments compared to fusion.

Artificial facet joints should provide stability similar to the native facet joint. Ideally the design will include measures to cause the artificial facet joint to behave similar to the native joint to include resistance to flexion and extension forces. Wear of the articular surfaces should also be minimal. Robust fixation to the spine is also critical. In the case of facet replacement, there have been different types of fixation. The Total Facet Arthroplasty System (Archus Orthopedics, Redmond, WA) was a cemented implant in an attempt to enhance the fixation to the vertebral body in a potentially osteoporotic patient. Subsequent implants ACADIA (Globus Medical, Audubon, PA), and TOPS (Premia Spine, Philadelphia, PA) are screw-based implants with coatings or texturing of the screws.

Kinematically all the facet joint replacement devices have done testing that indicates that if a functional spinal unit is destabilized with the removal of the facet joints the kinematics of that segment is restored with the utilization of a facet arthroplasty implant (Fig. 2). When the facet arthroplasty implants were tested in vivo and compared to fusion constructs with loads applied, the facet arthroplasty group experienced much lower implant stresses than the fusion implant (Webb and Holen 2008).

Rationale and Biomechanics of Facet Arthroplasty Systems

The potential benefit of posterior motion preservation devices is load sharing with the anterior and other posterior structures. Additionally, by removing pathologic structures, the symptoms associated with the facet joints including local pain symptoms can be reduced while decompressing the neural structures. The goal of posterior motion preservation devices should be to preserve motion and, by load shar-

Kinematics

- The TFAS™ effectively
  - stabilizes motion in flexion and lateral bending
  - restores the motion in extension
  - limits the motion in axial rotation
- TFAS™ restored motion of an unstable FSU to that of an intact FSU allowing considerable range of motion in all directions when compared to the intact condition

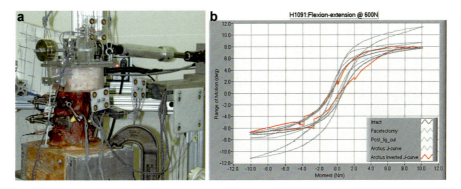

Fig. 2 Kinematic testing (**a**) demonstrated that TFAS effectively stabilized motion in flexion and restores motion in extension of an unstable functional spine unit to that of an intact unit (**b**)

History of Facet Arthroplasty Systems

The goal of posterior lumbar facet replacement systems is to use an alternative to fusion for facet degeneration and stenosis that allows for full decompression and stability while maintaining near physiologic motion (Zhu et al. 2007). The earliest facet replacements can be traced back to the 1980s, with the majority of these implants focused on articular surface replacement similar to the implants for peripheral joints (Serhan et al. 2011).

The TFAS was the initial facet replacement system to enter US investigational device exemption (IDE) in the mid-2000s. Other facet replacements soon followed including the Acadia by Facet Solutions and TOPS by Implant. They were all uniquely designed with different features to address the replacement of a facet joint in a wide decompression necessary to treat significant lumbar stenosis.

Total Facet Arthroplasty System (TFAS)

The first patient in the US IDE clinical trial of the Total Facet Arthroplasty System (TFAS) was performed on August 26, 2005 by Dr. Scott Webb (Figs. 3 and 4). This was the first time a total facet replacement had ever been performed in the United States. The TFAS implant was indicated in patients with lumbar spinal stenosis presenting predominantly with neurogenic claudicatory symptoms that required a wide decompression and stabilization.

The Total Facet Arthroplasty System (TFAS) is metal-on-metal joint prosthesis intended to provide stabilization of lumbar spinal segments in skeletal mature patients as an adjunct to a laminectomy/laminotomy/neural decompression and facetectomy in the treatment of single-level stenosis. Additional indications include degenerative facets at the index level with or without instability. Up to a Grade 1 degenerative spondylolisthesis could be present.

Due to the typical age group of patients with spinal stenosis and neurogenic claudication, the implant was designed to be implanted with cement fixation.

The implant's modularity allowed for a great deal of potential variability (Fig. 5).

Initial midterm clinical data of the US IDE trial demonstrated successful restoration of motion and clinically significant reduction of preoperative symptoms. ZCQ symptom and ZCQ function scores improved by 84% and 81% compared to preoperative scores in the 79 TFAS IDE patients. VAS back and leg pain scores improved in 73 of 79 and 75 of 79, respectively. Radiographic analysis showed all devices to be intact and functioning at time of follow-up (Sachs et al. 2008).Unfortunately,

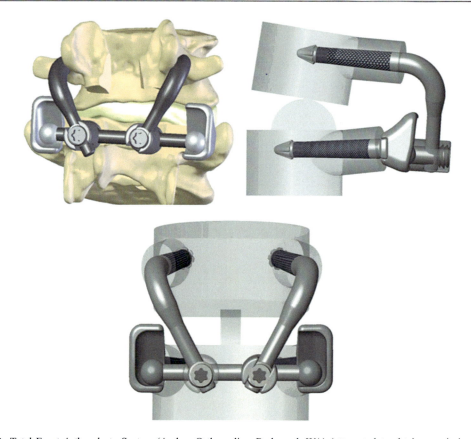

Fig. 3 Total Facet Arthroplasty System (Archus Orthopedics, Redmond, WA) (attempted to obtain permission from Archus Orthopedics but company no longer exists)

Fig. 4 Intraoperative imaging of TFAS (Scott Webb)

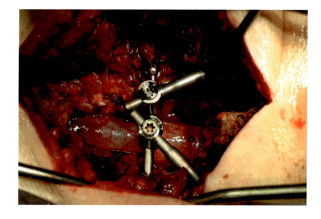

volume wear of metal debris in the TFAS implant was found to be comparable to metal-on-metal hip (M-o-M) implants in regard to particle size and distribution. The fear of similar outcomes despite significantly different biomechanics led to the removal of the product from the market. TFAS pioneered facet arthroplasty and demonstrated safety and efficacy in a small number of patients with limited follow-up.

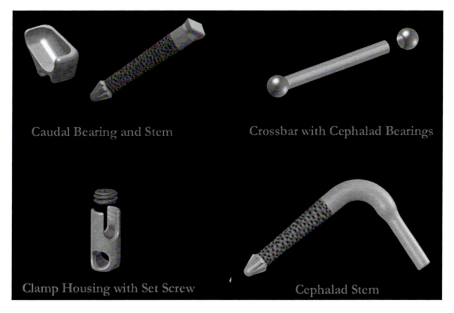

Fig. 5 TFAS components allowed for significant variability creating a custom fit for patient's anatomy

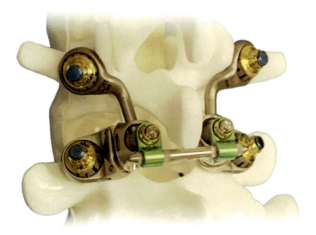

Fig. 6 ACADIA Facet Replacement System (AFRS) (Globus Medical, Audubon, PA)

ACADIA Facet Replacement System (AFRS)

The ACADIA Facet Replacement System (AFRS) (Globus Medical, Audubon, PA) formerly Anatomic Facet Replacement System (Facet Solutions, Logan, UT) evolved under the basic principles of other articulating synovial joint implant systems with a focus on restoring normal ROM with a uniform distribution of forces (Carl et al. 2008) (Fig. 6). International clinical evaluations began in São Paulo, Brazil, in 2005 with approval in the United States for FDA IDE study coming in 2006. AFRS was also intended to provide stabilization of spinal segments in skeletally mature patients as an adjunct to laminectomy, laminotomy, and facetectomy in the treatment of single level, L3–4 or L4–5, degenerative disease of the facets with or without instability including Grade I degenerative

Fig. 7 Postoperative extension radiograph of ACADIA in a patient with a Grade 1 spondylolisthesis (Scott Webb)

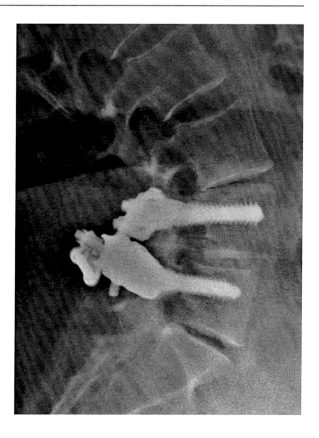

spondylolisthesis with objective evidence of neurologic impairment, central or lateral spinal stenosis (Carl et al. 2008).

The device, like the TFAS system, is a pedicle-screw-based construct with superior and inferior facet implants with articulating surfaces made of cobalt-chromium-molybdenum. Unlike the spherical bearing TFAS system, AFRS creates metal-on-metal articulation that resembles the anatomic structure of the facet joint. A crossbar links the left and right inferior facet implants providing construct stability (Carl et al. 2008). Pedicle screws are made of titanium alloy with hydroxyapatite coatings that allow for improved bone-implant interface (Fig. 7). The polyaxial junction of the implant accommodates for ±15° of variability in pedicle screw placement (Goel et al. 2007).

Preliminary outcomes of the US IDE study involving 162 ACADIA patients demonstrate improvement in ZCQ, ODI, and VAS scores at 2 and 4 years post-op similar to posterior lateral fusion patients. Surgical intervention at subsequent levels occurred in 8% of ACADIA patients and 7.4% of PLF patients (Myer et al. 2014). Ultimately, the IDE trial was terminated in 2017, also due M-o-M wear debris concerns.

Total Posterior Arthroplasty System (TOPS)

The Total Posterior Arthroplasty System (TOPS) (Premia Spine, Philadelphia, PA) is a dynamic facet arthroplasty prosthesis designed to restore segmental stability following extensive posterior decompression, including facet joint resection, while maintaining near anatomic motion (Anekstein et al. 2009) (Fig. 8). The device replaces the degenerated facets and bony elements removed during decompression of the stenotic level without sacrificing motion. The TOPS device differs from previous facet replacement systems as it does not resemble an anatomic

Fig. 8 Intraoperative imaging of Total Posterior Arthroplasty System (TOPS) (Premia Spine, Philadelphia, PA) (Scott Webb)

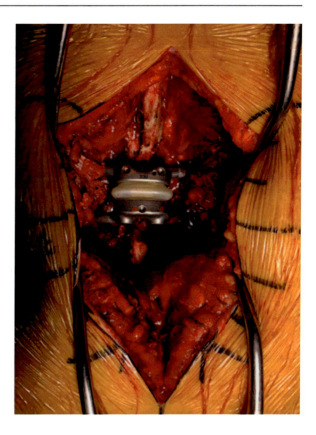

facet joint. The primary indication is symptomatic lumbar stenosis secondary to Grade 1 degenerative spondylolisthesis.

TOPS uses a cannulated pedicle-screw-based implant with superior and inferior titanium constructs fixed to a flexible, polycarbonate urethane (PCU) articulating core. The double horizontal crossbar connects pedicle screws of the same vertebrae creating better load sharing and eliminates screw head torque (Myers et al. 2008). The internal configuration of the bumper limits motion by dissipating energy during load sharing (McAfee et al. 2007). The central core is attached to all four pedicles stabilizing rotation and lateral bending as well as significantly decreasing the load transfer to adjacent segments (McAfee et al. 2007) The PCU bumper acts as a limiter of motion but also absorbs shock in the vertical axis; it incorporates a PEEK ribbon that acts as a restraint for excessive flexion. The boot creates a closed compartment that prevents wear debris from entering the spinal canal which was a major fear of product designers after the issues surrounding M-o-M hip prosthetics (Anekstein et al. 2009) (Fig. 9).

Initial investigational studies with the TOPS device began internationally in 2005, with the first US IDE study under FDA investigation beginning shortly afterward in 2006. Initial results of the study were very promising. McAfee et al. (2007) reported on 29 patients in the initial international study who had met criteria for single-level decompression and fusion of the lumbar spine underwent facet replacement. Outcomes were based on VAS, ODI, and ZCQ scores at 6 weeks, 3 months, 6 months, and 1 year. Preliminary data demonstrated improvement in all three scores and no device-related events or hardware failure. CT scans at 3 months and 1 year demonstrated no signs of subsequent disc height loss at treatment level or adjacent levels. The TOPS device has undergone some minor changes in the structure of the system and is currently undergoing a second IDE study which began in January 2017 (ClinicalTrials.gov 2017).

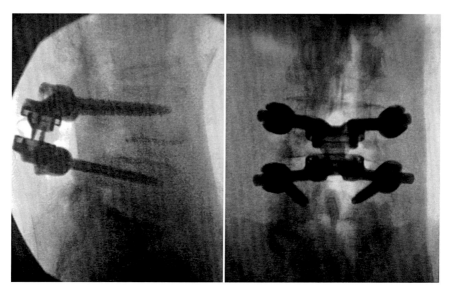

Fig. 9 Intraoperative fluoroscopic imaging (lateral and PA) of second generation of TOPS device (Scott Webb)

The TOPS device is the only posterior facet replacement device still in active US IDE study.

Interspinous Devices (ISD)

Interspinous devices have been primarily used to treat lumbar central stenosis with concomitant neurogenic claudication. These devices result in indirect decompression of the canal and neuroforamen. These devices are typically used in conjunction with decompressive surgery. Earliest documentation off interspinous devices was in the 1950's at which time metal "plugs" were placed between the spinous processes. Senegas et al. described one of the early interspinous spacers in 1988. This was a titanium device that was held to the spinous processes with Dacron tape. This was later redesigned as the Wallis Implant (Abbott Spine, Austin, TX) which used a PEEK material instead of the previous titanium. One of the more popular devices, the X-Stop (Medtronic, Minneapolis, MN) device received FDA premarket approval in 2005, initial studies were very promising, but as later data came out there appeared to be high rates of complications. Medtronic discontinued this device in 2015. The DIAM system (Device for Intervertebral Assisted Motion) (Medtronic Sofamor Danek, Memphis, TN) is another abandoned interspinous device. It is an "H-"shaped device secured by a cord to the spinous processes. The latest device to gain traction in the US market is the Coflex device (Paradigm Spine, New York, New York) (Fig. 10). The FDA has approved Coflex for 1–2 level stenosis in the lumbar spine, with some clinical data promising compared to traditional fusion. Superion (Vertiflex, Carlsbad, CA) is an interspinous device that is delivered percutaneously, similar to devices such as the X-Stop (Fig. 11). In 2015, a randomized controlled trial comparing the Superion to the X-Stop demonstrated a statistically significant benefit to the Superion over the X-Stop at 36 months.

Interspinous devices were primarily developed to treat lumbar spinal stenosis resulting in neurogenic claudication. Patients with these symptoms tend to improve with flexion; therefore these devices place the spinal segment into slight flexion or kyphosis (Richards et al. 2005). There is not compelling data to support ISD over traditional open laminectomy decompression, but it does offer an alternative to an open surgical procedure (Senegas et al. 1988). This may be an attractive alternative for elderly patients with medical comorbidities and higher surgical risk.

Conclusions

The role for motion preservation by means of facet arthroplasty and dynamic stabilization in lumbar spine surgery continues to evolve. Posterior facet remains in active FDA IDE study with promising early results for the treatment of spinal stenosis secondary to degenerative spondylolisthesis. Pedicle-screw-based posterior dynamic stabilization systems have largely fell out of favor. Interspinous spacers have developed a niche in the minimally invasive treatment of spinal stenosis. Overall, the use of posterior arthroplasty for the treatment of lumbar spine pathology remains in its relative infancy, and further study is warranted in defining its role in the future.

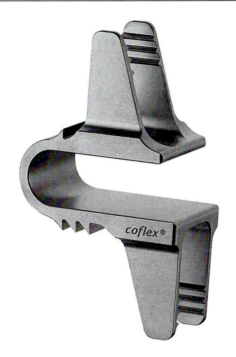

Fig. 10 Coflex device (Paradigm Spine, New York, New York)

References

Anekstein Y, Floman Y, Smorgick Y et al (2009) The total posterior arthroplasty system. In: Bhave A (ed) Emerging techniques in spine surgery, 1st edn. Jaypee Brothers Medical, New Delhi, pp 119–130

Carl A, Oliviera CE, Hoy RW et al (2008) Anatomic facet replacement system (AFRS). In: Yue et al (eds) Motion preservation surgery of the spine: advanced techniques and controversies. Elsevier, Philadelphia, pp 577–580

Cavanaugh JM, Ozaktay AC, Yamashita HT et al (1996) Lumbar facet pain: biomechanics, neuroanatomy and neurophysiology. J Biomech 29:1117–1129

ClinicalTrials.gov (2017) National Library of Medicine. Bethesda. NCT03012776, A pivotal study of the premia spine TOPS System. https://clinicaltrials.gov/ct2/show/NCT03012776?term=TOPS+facet&rank=1. Accessed 2 Feb 2019

Gellhorn AC, Katz JN, Suri P (2013) Osteoarthritis of the spine: the facet joints. Nat Rev Rheumatol 9:216–224

Goel VK, Mehta A, Jangra J et al (2007) Anatomic facet replacement system (AFRS) restoration of lumbar segment mechanics to intact: a finite element study and in vitro cadaver investigation. Int J Spine Surg 1:46–54

McAfee PC, Khoo LT, Pimenta L et al (2007) Treatment of lumbar spinal stenosis with a total posterior arthroplasty prosthesis: implant description, surgical technique, and a prospective report on 29 patients. Neurosurg Focus 22:1–11

Myer J, Youssef JA, Rahn KA et al (2014) ACADIA® facet replacement system IDE study: preliminary outcomes at two and four years postoperative. Spine J 14:160S–161S

Myers K, Tauber M, Sudin Y et al (2008) Use of instrumented pedicle screws to evaluate load sharing in posterior dynamic stabilization systems. Spine J 8:926–932

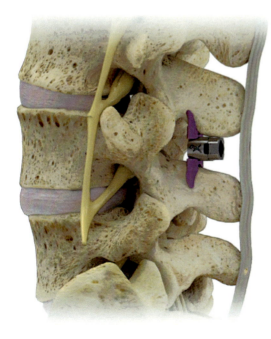

Fig. 11 Superion Image provided courtesy of Boston Scientific. ©2020 Boston Scientific Corporation or its affiliates. All rights reserved

Nachemson A (1960) Lumbar intradiscal pressure: experimental studies on post-mortem material. Almqvist & Wiksells Boktyckeri, Uppsala

Panjabi MM, Yamamoto I, Oxland TR et al (1989) How does posture affect coupling in the lumbar spine? Spine 14:1002–1011

Reily M (2011) Total facet replacement. In: Herkowitz et al (eds) Rothman-Simeone: the spine, 6th edn. Elsevier, Philadelphia, pp 986–997

Richards JC, Majumdar S, Lindsey DP et al (2005) The treatment mechanism of an interspinous process implant for lumbar neurogenic intermittent claudication. Spine 30:744–749

Sachs B, Webb S, Brown C et al (2008) The total faces arthroplasty system (TFAS®) in the treatment of degenerative lumbar spinal stenosis: midterm results of US IDE trial with longest follow-up of 24 months. Spine J 8:60S–61S

Scoles PV, Linton AE, Latimer B et al (1988) Vertebral body posterior element morphology: the normal spine in middle life. Spine 13:1082–1086

Senegas J, Etchevers JP, Vital JM et al (1988) Widening of the lumbar vertebral canal as an alternative to laminectomy in the treatment of lumbar stenosis. Rev Chir Orthop Reparatrice Appar Mot 74:15–22

Serhan H, Mhatre D, Defossez H et al (2011) Motion-preserving technologies for degenerative lumbar spine: the past, present and future horizons. SAS J 5:75–89

Webb S, Holen G (2008) Total facet arthroplasty system (TFAS). In: Yue et al (eds) Motion preservation surgery of the spine: advanced techniques and controversies. Elsevier, Philadelphia, pp 565–576

White AA, Panjabi MM (1978) Clinical biomechanics of the spine. J. B. Lippincott, Philadelphia

Yang KH, King AI (1984) Mechanism of facet load transmission as a hypothesis for low-back pain. Spine 9:557–565

Zhu Q, Larson CR, Sjovold SG et al (2007) Biomechanical evaluation of the Total Facet Arthroplasty System™: 3-dimensional kinematics. Spine 32:55–62

Cervical Arthroplasty: Long-Term Outcomes

48

Thomas J. Buell and Mark E. Shaffrey

Contents

Introduction	858
Bryan Cervical Disc	860
Long-Term Outcomes for Single-Level Bryan CDA Versus ACDF	862
Cummins/Bristol and Prestige Cervical Discs	863
Long-Term Outcomes for Single-Level Prestige CDA Versus ACDF	867
Long-Term Outcomes for Two-Level Adjacent Prestige LP CDA Versus ACDF	867
Porous Coated Motion (PCM) Cervical Disc	868
Long-Term Outcomes for Single-Level PCM CDA Versus ACDF	868
ProDisc-C Cervical Disc	868
Long-Term Outcomes for Single-Level ProDisc-C CDA Versus ACDF	870
Mobi-C Cervical Disc	870
Long-Term Outcomes for Single-Level Mobi-C CDA Versus ACDF	870
Long-Term Outcomes for Two-Level Adjacent Mobi-C CDA Versus ACDF	872
Kineflex-C Cervical Disc	872
Long-Term Outcomes for Single-Level Kineflex-C CDA Versus ACDF	872
Secure-C Cervical Disc	873
Long-Term Outcomes for Single-Level Secure-C CDA Versus ACDF	873
Discover Cervical Disc	873
Long-Term Outcomes for Single- and Multilevel Discover CDA Versus ACDF	874
Summary of Complications Associated with Cervical Disc Arthroplasty	874
Metal Ion Toxicity	876

T. J. Buell (✉) · M. E. Shaffrey
Department of Neurological Surgery, University of
Virginia Health System, Charlottesville, VA, USA
e-mail: tjb4p@hscmail.mcc.virginia.edu; tjb4p@virginia.
edu; mes8c@hscmail.mcc.virginia.edu

© Springer Nature Switzerland AG 2021
B. C. Cheng (ed.), *Handbook of Spine Technology*,
https://doi.org/10.1007/978-3-319-44424-6_80

Patient Selection	876
Cost Efficacy	877
Conclusions	877
References	878

Abstract

Cervical disc arthroplasty (CDA) attempts to preserve normal motion at adjacent segments and in doing so may decrease the incidence of adjacent segment degeneration in comparison with cervical arthrodesis. Since 2006, the United States Food and Drug Administration (FDA) has approved seven CDA prosthetic devices for surgical management of symptomatic cervical spondylosis and disc herniation (seven for 1-level disease and two for two-level disease). Motion-preserving CDA has showed great promise with equivalent quality-of-life outcomes in many long-term comparative studies. Currently, follow-up duration of up to 10 years is available from some of the FDA trials comparing CDA to arthrodesis. In general, study findings have consistently demonstrated that both techniques result in significant clinical improvement by roughly 3 months post-op and that improvement may be maintained at final follow-up. Overall, there exists robust data to support CDA as a viable alternative to arthrodesis in select patients. However, complications such as heterotopic ossification have been reported. In this chapter, we review CDA, with an emphasis on highlighting the published long-term outcomes and complications for this motion-preserving operation in comparison with arthrodesis.

Keywords

Degenerative disc disease · Cervical spondylosis · Disc herniation · Anterior cervical discectomy and fusion · Cervical disc arthroplasty · Heterotopic ossification · Artificial cervical disc · Motion preservation · Adjacent segment degeneration or disease · Bryan cervical disc · Prestige cervical disc ·

Prestige ST · Prestige LP · Porous Coated Motion · ProDisc-C · Mobi-C · Kineflex-C · Secure-C · Discover artificial cervical disc

Introduction

Degenerative disc disease involving the cervical spine is part of the normal aging process (Traynelis 2004). When degenerative changes occur gradually, they may be asymptomatic; however, in a subset of patients, cervical spondylosis or disc herniation may result in compression of nerve roots or the spinal cord, resulting in radiculopathy, myelopathy, or both (Traynelis 2004). A common surgical treatment for patients with symptomatic cervical spondylosis or disc herniation is anterior cervical discectomy and fusion (ACDF) (Alvin et al. 2014). The ACDF procedure, first described over 50 years ago, has been shown to be safe and clinically efficacious (Alvin et al. 2014; Cloward 1959; Smith and Robinson 1958). However, there is debate about further degeneration at adjacent segments after fusion surgery (Gao et al. 2013; McAfee et al. 2012; Xing et al. 2013; Yin et al. 2013). Specifically, it is currently unclear if adjacent segment degeneration is part of the natural history of cervical spondylosis or whether it is related to the adjacent fused levels. Some studies have shown an average 3% reoperation rate, while other studies report revision rates exceeding 10% after 2 years to treat complications related to the index fusion operation (Hilibrand et al. 1999; Yin et al. 2013). By preserving physiologic cervical motion, one of the goals of CDA is to reduce the incidence of adjacent segment degeneration while maintaining the highly effective results of ACDF in maintenance or improvement of neck pain, arm pain, and myelopathy (Alvin et al. 2014).

The initial clinical experience with CDA began in the 1960s with Ulf Fernstrom, a Swedish surgeon, implanting stainless steel ball bearing prosthetic devices following laminectomy (Fernstrom 1966; Fisahn et al. 2017). A high failure rate and concern for hypermobility and device migration into adjacent vertebral cancellous bone ultimately led the industry back to favoring ACDF (Bertagnoli et al. 2005; Fernstrom 1966; Fisahn et al. 2017). Then later in the 1980s, CDA returned with a design by Cummins, who was developing an artificial disc to address the shortcomings of ACDF regarding motion preservation and adjacent segment degeneration. Cummins' artificial cervical disc was developed in collaboration with the Department of Medical Engineering at Frenchay Hospital in 1989, and resulted in improved clinical outcomes after implantation in appropriately selected patients (Cummins et al. 1998; Traynelis 2004; Wigfield et al. 2002b). After performing the index decompression or discectomy, the main advantage of CDA in comparison to ACDF is the possibility for postsurgical segmental motion preservation, which may prevent the occurrence of adjacent segment degeneration and disease (Alvin et al. 2014).

Since 2006, the United States Food and Drug Administration (FDA) has approved seven CDA prosthetic devices for surgical management of symptomatic cervical spondylosis and disc herniation at a single level (Coric et al. 2018; Gornet et al. 2016). In 2007, the Prestige ST (Medtronic Inc.), a metal-on-metal device made from stainless steel, was the first CDA device to receive FDA approval (Mummaneni et al. 2007). A later version with a low profile modification, the Prestige LP (Medtronic Inc.), was made from a titanium ceramic composite and received FDA approval in 2014 (Gornet et al. 2015). The other five FDA-approved devices are metal (cobalt-chrome or titanium alloy)-on-polymer (polyethylene or polyurethane) designs and include (by order of FDA approval for single-level CDA) ProDisc-C (2008; Synthes Spine), Bryan (2009; Medtronic Inc.), Porous Coated Motion (2012; Cervitech), Secure-C (2012; Globus Medical), and Mobi-C (2013; LDR Medical) (Heller et al. 2009; Hisey et al. 2014; Murrey et al. 2009; Phillips et al. 2013; Vaccaro et al. 2013). Two of these devices, the Prestige LP and Mobi-C, have since received FDA approval for CDA at two adjacent levels.

More recently, long-term studies have been published for these FDA-approved artificial discs and suggest CDA is safe and clinically efficacious in appropriately selected patients. Currently, follow-up duration of up to 10 years is available from some of the FDA trials comparing CDA to ACDF. In general, study findings have consistently demonstrated that both CDA and ACDF result in significant clinical improvement by roughly 3 months post-op and that improvement may be maintained at final follow-up (Davis et al. 2015; Heller et al. 2009; Hisey et al. 2016; Phillips et al. 2013; Radcliff et al. 2016a; Vaccaro et al. 2013; Zigler et al. 2013). CDA was found to produce noninferior results in all the studies for certain outcome variables and even demonstrated statistical superiority for some outcome measures. For single-level CDA, four of the seven discs (Prestige ST, Prestige LP, Bryan, and Secure-C) demonstrated superiority in overall success. Prestige ST showed superiority in three of four outcome variables (neurological success, revision surgery, and overall success), while the other discs showed superiority in ≤2 variables (Prestige LP, neurological and overall success; Bryan, Neck Disability Index [NDI] and overall success; Secure-C, revision surgery and overall success; and Pro-Disc C, revision surgery). The Porous Coated Motion (PCM) and Mobi-C discs demonstrated noninferiority for all outcome variables. For two-level (adjacent) CDA, Prestige LP and Mobi-C demonstrated superiority in three outcome variables (NDI, secondary surgery, and overall success), but not neurological success (Turel et al. 2017).

Although the aforementioned devices have met rigorous outcome requirements for FDA approval, there have been reports of complications such as heterotopic ossification (HO; abnormal bone formation around or within the intervertebral disc space) and/or implant migration (Gao et al. 2013; McAfee et al. 2012; Xing et al. 2013; Yin et al. 2013). Therefore, in this review, in addition to summarizing long-term outcomes, we also report complications associated with CDA for the FDA-approved discs. Although not FDA-approved, we also report outcomes and complications for the Discover artificial cervical disc (DePuy Spine) due to its widespread use outside the United States (OUS). At the end of each section, we specifically report outcome variables (NDI, Visual Analogue Scale [VAS] neck score, VAS arm score, Short Form-36 Health Survey Physical and Mental component scores [SF-36 PCS; SF-36 MCS]) comparing CDA and ACDF, when available. Outcomes for two-level adjacent CDA are also summarized for the FDA-approved Prestige LP and Mobi-C discs.

Bryan Cervical Disc

Vincent Bryan designed the Bryan cervical disc (Medtronic Inc.) in the United States in 1992 (Basho and Hood 2012). The Bryan cervical disc (Fig. 1a) is a non-constrained device consisting of a low-friction, wear-resistant, polyurethane nucleus housed between titanium plates (Bryan 2002). These titanium plates have convex porous ingrowth surfaces that function to support bony fixation of adjacent vertebral end plates. Consistent with the goal of motion preservation, the Bryan disc was designed to allow normal or physiologic range of motion, as well as coupled motion in cervical flexion/extension, lateral bending, rotation, and translation (Bryan 2002). Several studies have reported significant improvement in postoperative standardized outcomes scores (NDI, VAS scores, and SF-36 scores) for Bryan CDA in comparison with ACDF, for both single- and two-level procedures in patients with discogenic cervical radiculopathy and/or myelopathy (although the Bryan disc is not FDA-approved for multilevel CDA) (Cheng et al. 2009, 2011; Coric et al. 2006, 2013; Garrido et al. 2010; Goffin et al. 2003, 2010; Hacker 2005; Heidecke et al. 2008; Heller et al. 2009; Lafuente et al. 2005; Leung et al. 2005; Quan et al. 2011; Robertson et al. 2005; Sasso et al. 2007a, b, 2011; Tu et al. 2011; Walraevens et al. 2010; Yang et al. 2008; Zhang et al. 2012). Table 1 summarizes Bryan CDA outcomes data (Alvin et al. 2014; Anderson et al. 2004; Bhadra et al. 2009; Bryan 2002; Cheng et al. 2009; Coric et al. 2006, 2010, 2013; Ding et al. 2012; Duggal et al. 2004; Garrido et al. 2010; Goffin et al. 2003, 2010;

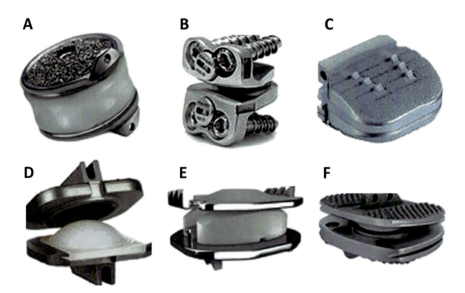

Fig. 1 (**a**) Bryan disc, (**b**) Prestige disc, (**c**) Porous Coated Motion (PCM) disc, (**d**) ProDisc-C disc, (**e**) Mobi-C disc, and (**f**) Kineflex-C disc. Recreated from Alvin et al. Cervical arthroplasty: a critical review of the literature. The Spine Journal. (Alvin et al. 2014)

Table 1 Summary of single- and multilevel Bryan disc arthroplasty outcomes for symptomatic cervical spondylosis or disc herniation. Recreated and modified from Alvin et al. Cervical arthroplasty: a critical review of the literature. The Spine Journal. (Alvin et al. 2014; Anderson et al. 2004; Bhadra et al. 2009; Bryan 2002; Cheng et al. 2009; Coric et al. 2006, 2010, 2013; Ding et al. 2012; Duggal et al. 2004; Garrido et al. 2010; Goffin et al. 2003, 2010; Hacker 2005; Heidecke et al. 2008; Heller et al. 2009; Kim et al. 2008, 2009; Lafuente et al. 2005; Lee et al. 2010; Leung et al. 2005; Pickett et al. 2004, 2006; Quan et al. 2011; Ren et al. 2011; Robertson et al. 2005; Ryu et al. 2010; Sasso et al. 2007a, b, 2011, 2017; Sekhon 2003; Sekhon et al. 2005; Shim et al. 2006; Tu et al. 2011; Walraevens et al. 2010; Wang et al. 2008; Yang et al. 2008; Yoon et al. 2006; Zhang et al. 2012)

Author	Design	Study size	Follow-up duration	NDI	VAS neck/ arm scores	SF-36 PCS	SF-36 MCS	Design LoE
Sasso 2017	RCT	47 (single level)	120 mos	Imp	Imp/Imp			Ib
Coric 2013	RCT	74 (single level)	72 mos	Imp	Imp/Imp			Ib
Zhang 2012	RCT	120 (single level)	24 mos	Imp	Imp/Imp			Ib
Sasso 2011	RCT	463 (single level)	48 mos	Imp	Imp/Imp	Imp	Imp	Ib
Cheng 2009	RCT	65 (all multilevel)	24 mos	Imp	Imp/Imp	Imp	Imp	Ib
Garrido 2010	RCT	47 (single level)	48 mos	Imp	Imp/Imp	Imp	Imp	Ib
Heller 2009	RCT	463 (single level)	24 mos	Imp	Imp/Imp	Imp	Imp	Ib
Sasso 2007	RCT	115 (single level)	24 mos	Imp	Imp/Imp	Imp	Imp	Ib
Hacker 2005	RCT	46 (single level)	12 mos	Imp	Imp/Imp	Imp	Imp	Ib
Coric 2006	RCT	33 (single level)	24 mos	Imp	Imp/Imp	Imp	Imp	Ib
Quan 2011	PC	21 (6 multilevel)	96 mos		Imp/Imp			IIb
Ren 2011	PC	45 (6 multilevel)	35 mos	Imp				IIb
Coric 2010	PC	98 (13 multilevel)	24 mos	Imp				IIb
Goffin 2010	PC	98 (9 multilevel)	72 mos	Imp	Imp/Imp	Imp	Imp	IIb
Ryu 2010	PC	36 (single level)	24 mos	Imp	Imp/Imp			IIb
Walraevens 2010	PC	89 (single level)	48 mos					IIb
Bhadra 2009	PC	60 (single level)	31 mos		Imp/Imp			IIb
Heidecke 2008	PC	54 (5 multilevel)	24 mos					IIIb
Kim 2008	PC	47 (8 multilevel)	24 mos	Imp	Imp/Imp			IIb
Wang 2008	PC	59 (single level)	24 mos	Imp	Imp/Imp			IIb
Yang 2008	PC	19 (3 multilevel)	24 mos		Imp/Imp			IIb
Pickett 2006	PC	74 (21 multilevel)	24 mos	Imp	Imp/Imp	Imp	Imp	IIb
Robertson 2005	PC	74 (single level)	24 mos		Imp/Imp	Imp	Imp	IIb
Sekhon 2005	PC	15 (5 multilevel)	24 mos		Imp/Imp			IIb

(continued)

Table 1 (continued)

Author	Design	Study size	Follow-up duration	NDI	VAS neck/ arm scores	SF-36 PCS	SF-36 MCS	Design LoE
Lafuente 2005	PC	46	12 mos	Imp	Imp/Imp	Imp	Imp	IIb
Pickett 2004	PC	14 (1 multilevel)	24 mos	Imp		Imp	Imp	IIb
Duggal 2004	PC	26 (4 multilevel)	12 mos	Imp		Imp	Imp	IIb
Anderson 2004	PC	136 (30 multilevel)	12 mos			Imp	Imp	IIb
Goffin 2003	PC	143 (43 multilevel)	12 mos			Imp	Imp	IIb
Bryan 2002	PC	97 (single level)	24 mos			Imp	Imp	IIb
Ding 2012	R	32 (included multilevel)	49 mos	Imp	Imp/Imp	Imp	Imp	IIb
Tu 2011	R	36 (16 multilevel)	27 mos		Imp/Imp			IIb
Lee 2010	R	48 (single level)	14 mos		Imp/Imp			IIb
Kim 2009	R	51 (12 multilevel)	19 mos	Imp	Imp/Imp			IIb
Shim 2006	R	47 (8 multilevel)	6 mos	Imp	Imp/Imp			IIb
Yoon 2006	R	46 (single level)	12 mos	Imp	Imp/Imp			IIb
Leung 2005	R	90	12 mos			Imp	Imp	IIb
Sekhon 2003	R	7 (2 multilevel)	6 mos	Imp	Imp/Imp			IIb

Imp improved, *LoE* level of evidence, *MCS* mental component score, *NDI* neck disability index, *PC* prospective cohort, *PCS* physical component score, *R* retrospective, *RCT* randomized controlled trial, *SF-36* short form-36, *VAS* visual analogue scale

Hacker 2005; Heidecke et al. 2008; Heller et al. 2009; Kim et al. 2008, 2009; Lafuente et al. 2005; Lee et al. 2010; Leung et al. 2005; Pickett et al. 2004, 2006; Quan et al. 2011; Ren et al. 2011; Robertson et al. 2005; Ryu et al. 2010; Sasso et al. 2007a, b, 2011, 2017; Sekhon 2003; Sekhon et al. 2005; Shim et al. 2006; Tu et al. 2011; Walraevens et al. 2010; Wang et al. 2008; Yang et al. 2008; Yoon et al. 2006; Zhang et al. 2012). The vast majority of these studies had follow-up duration of up to 2 years; however, some had over 6 years of clinical and radiographic follow-up (Pointillart et al. 2017; Quan et al. 2011).

Long-Term Outcomes for Single-Level Bryan CDA Versus ACDF

In 2012, Zhang and colleagues reported 24-month outcomes for Bryan CDA versus ACDF. Study results demonstrated no significant differences between treatment groups based on mean NDI or median VAS scores (Zhang et al. 2012). These results are consistent with a study by Coric and colleagues (with average follow-up 72 months) that also demonstrated no significant differences between groups based on mean NDI or median VAS scores (Coric et al. 2013). In contrast, Sasso and colleagues reported significantly greater improvement in the CDA cohort based on NDI, VAS neck and arm pain scores, and SF-36 PCS and MCS scores at 48 months post-op (Sasso et al. 2011). Sasso and colleagues also reported an advantage for CDA in comparison with ACDF as measured by 7- and 10-year NDI scores (Sasso et al. 2017). The same authors reported CDA having an advantage over ACDF based on 7-year VAS neck and arm pain scores; however, the comparison was no longer significant at final 10-year follow-up (Sasso et al. 2017).

The data from these studies suggests that Bryan CDA is at least a viable alternative to ACDF for symptomatic cervical spondylosis and/or disc prolapse. The data also suggests that there is a lower incidence of secondary surgery after CDA (Cheng et al. 2009, 2011; Coric et al. 2006, 2013; Garrido et al. 2010; Goffin et al. 2003, 2010; Hacker 2005; Heidecke et al. 2008; Heller et al. 2009; Lafuente et al. 2005; Leung et al. 2005; Quan et al. 2011; Robertson et al. 2005; Sasso et al. 2007a, b, 2011; Tu et al. 2011; Walraevens et al. 2010; Yang et al. 2008; Zhang et al. 2012). A study by Shang and colleagues that focused on "skip" cervical spondylosis provided more evidence for the benefits of Bryan CDA over ACDF (Shang et al. 2017). Also, in a study that utilized a workers' compensation patient cohort, a greater number of CDA patients returned to work at 6 weeks and 3 months after surgery compared to ACDF (Steinmetz et al. 2008).

However, despite the demonstrated benefits of Bryan CDA over ACDF, complications were still reported (Cheng et al. 2009, 2011; Coric et al. 2006, 2013; Garrido et al. 2010; Goffin et al. 2003, 2010; Hacker 2005; Heidecke et al. 2008; Heller et al. 2009; Lafuente et al. 2005; Leung et al. 2005; Quan et al. 2011; Robertson et al. 2005; Sasso et al. 2007a, b, 2011; Tu et al. 2011; Walraevens et al. 2010; Yang et al. 2008; Zhang et al. 2012). For example, new anterior osteophyte formation or enlargement, increased narrowing of the intervertebral interspace, new adjacent degenerative disc disease, and calcification of the anterior longitudinal ligament were reported radiological findings indicative of post-CDA adjacent-level disease (Robertson et al. 2005; Yi et al. 2009). The incidence of heterotopic ossification (HO) causing restricted range of movement of the artificial disc prosthesis appears to increase with time, especially in multilevel (bilevel) CDA. Longer follow-up duration after CDA, gender, and age were noted to be risk factors in the development of HO after CDA (Leung et al. 2005).

Preoperative cervical kyphosis is a contraindication to CDA; therefore, post-CDA alignment has been an important topic of interest (Leven et al. 2017; Nunley et al. 2018). Using Bryan CDA for patients with single- and/or two-level symptomatic disc disease, Kim and colleagues studied postsurgical sagittal alignment of the functional spinal unit (FSU), as well as overall sagittal balance of the cervical spine (Kim et al. 2008). Their results demonstrated that Bryan CDA resulted in preserved motion of the FSU, and although the preoperative lordosis (or kyphosis) of the FSU could not always be maintained at during follow-up, the overall sagittal balance of the cervical spine was usually preserved (Kim et al. 2008). Pickett and colleagues reported similar results. Specifically, they also demonstrated preserved motion of the FSU after CDA. Although both the end plate angle of the treated disc space and the angle of the FSU became kyphotic after CDA, overall cervical spine sagittal alignment was preserved (Pickett et al. 2004). Other authors have found that cervical spine sagittal alignment became kyphotic after surgery, but overall lordosis was restored at a later time on follow-up imaging (Yoon et al. 2006). Possible causes of kyphotic changes included "over-milling" at the dorsal end plate, suboptimal angle of disc insertion, structural absence of lordosis in the Bryan disc prosthesis, removal of the posterior longitudinal ligament, and preexisting cervical kyphosis (Yoon et al. 2006).

Cummins/Bristol and Prestige Cervical Discs

Authors of both single- and multicenter studies have reported statistically significant improved postoperative outcomes for Prestige CDA (Fig. 1b), as well as reduced rates of secondary surgery compared to ACDF (Burkus et al. 2010, 2014; Lanman et al. 2017; Mummaneni et al. 2007; Peng et al. 2011; Porchet and Metcalf 2004; Riina et al. 2008; Robertson and Metcalf 2004). In addition to improved neurological success and outcomes, some studies have also demonstrated that Prestige CDA may restore segmental lordosis and preserve segmental motion (Peng et al. 2011). The follow-up duration for many of these studies was 2 years, but up to 7 years of follow-up data was reported (Lanman

Table 2 Summary of single-level Prestige disc outcomes for symptomatic cervical spondylosis or disc herniation. Recreated and modified from Alvin et al. Cervical arthroplasty: a critical review of the literature. The Spine Journal. (Alvin et al. 2014; Burkus et al. 2010, 2014; Gornet et al. 2015, 2016, 2017; Lanman et al. 2017; Mummaneni et al. 2007; Peng et al. 2011; Porchet and Metcalf 2004; Riew et al. 2008; Riina et al. 2008; Robertson and Metcalf 2004)

Author/Device	Design	Study size	Follow-up duration	NDI	VAS neck/ arm scores	SF-36 PCS	SF-36 MCS	Design LoE
Gornet 2016 Prestige LP	RCT	545	84 mos	Imp	Imp/Imp	Imp	Imp	Ib
Gornet 2015 Prestige LP	RCT	545	24 mos	Imp	Imp/Imp	Imp	Imp	Ib
Burkus 2014 Prestige ST	RCT	541	84 mos	Imp	Imp/Imp	Imp		Ib
Burkus 2010 Prestige ST	RCT	541	60 mos	Imp	Imp/Imp	Imp		Ib
Riew 2008 Prestige ST	RCT	199	24 mos	Imp	Imp/Imp	Imp	Imp	Ib
Mummaneni 2007 Prestige ST	RCT	541	24 mos	Imp	Imp/Imp	Imp	Imp	Ib
Porchet 2004 Prestige II	RCT	49	24 mos	Imp	Imp/Imp	Imp	Imp	Ib
Peng 2011 Prestige LP	PC	115 (includes 1–3 levels)	24 mos	Imp	Imp/Imp	Imp	Imp	IIb
Riina 2008 Prestige ST	PC	19	24 mos	Imp	Imp/Imp			IIb
Robertson & Metcalf 2004 Prestige I	PC	14	48 mos	Imp	Imp/Imp	Imp	Imp	IIb

Imp improved, *LoE* level of evidence, *MCS* mental component score, *NDI* neck disability index, *PC* prospective cohort, *PCS* physical component score, *R* retrospective, *RCT* randomized controlled trial, *SF-36* short form-36, *VAS* visual analogue scale

Table 3 Summary of two-level (adjacent) Prestige LP disc outcomes for symptomatic cervical spondylosis or disc herniation. (Gornet et al. 2017; Lanman et al. 2017)

Author	Design	Study size	Follow-up duration	NDI	VAS neck/arm scores	SF-36 PCS	SF-36 MCS	Design LoE
Lanman 2017	RCT	397	84 mos	Imp	Imp/Imp	Imp	Imp	Ib
Gornet 2017	RCT	397	24 mos	Imp	Imp/Imp	Imp	Imp	Ib

Imp improved, *LoE* level of evidence, *MCS* mental component score, *NDI* neck disability index, *PC* prospective cohort, *PCS* physical component score, *R* retrospective, *RCT* randomized controlled trial, *SF-36* short form-36, *VAS* visual analogue scale

et al. 2017). Tables 2 and 3 summarize Prestige CDA outcomes data (Burkus et al. 2010, 2014; Gornet et al. 2015, 2016, 2017; Lanman et al. 2017; Mummaneni et al. 2007; Peng et al. 2011; Porchet and Metcalf 2004; Riew et al. 2008; Riina et al. 2008; Robertson and Metcalf 2004). Below, we highlight key design steps in the history of Prestige CDA and then summarize one- and two-level outcomes data for Prestige CDA versus ACDF.

In the late 1980s, Cummins introduced a simple ball-and-socket prosthetic cervical joint in an attempt to address some of the problems associated with ACDF (Cummins et al. 1998; Wigfield et al. 2002b). His efforts, in collaboration with the Department of Medical Engineering at Frenchay

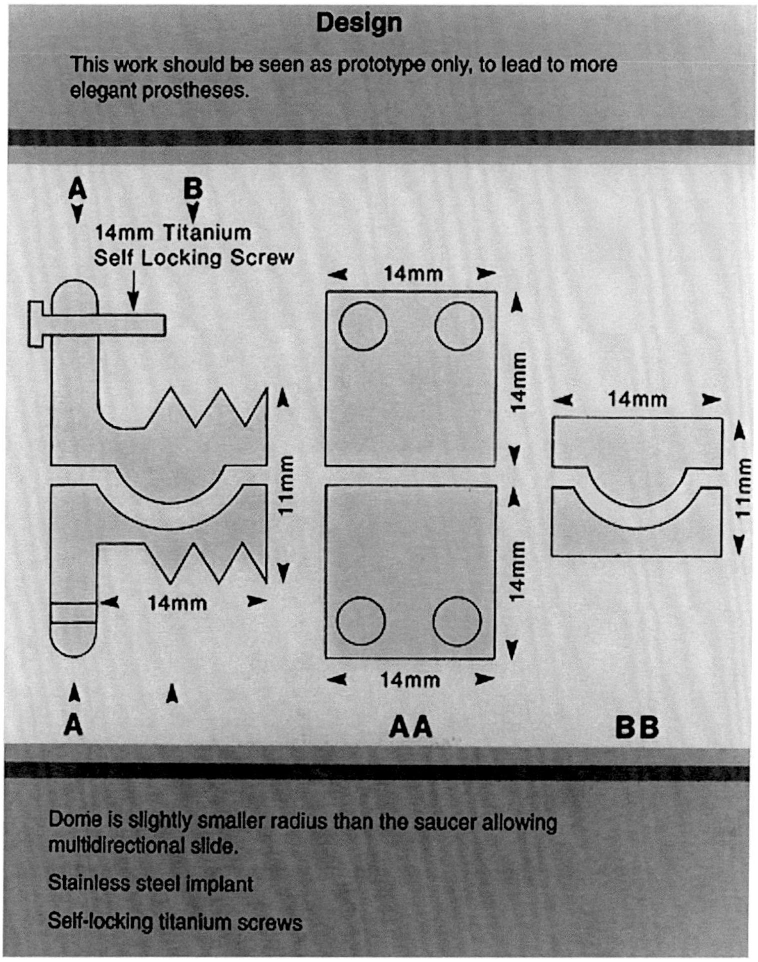

Design

This work should be seen as prototype only, to lead to more elegant prostheses.

14mm Titanium Self Locking Screw

14mm · 14mm · 11mm · 14mm · 14mm · 14mm · 14mm · 11mm

A · B · A · AA · BB

Dome is slightly smaller radius than the saucer allowing multidirectional slide.

Stainless steel implant

Self-locking titanium screws

Fig. 2 Prototype design of the Prestige artificial cervical disc composed of stainless steel (made by Mr. Colin Walker at Frenchay Hospital). (Recreated from Cummins et al. Surgical experience with an implanted artificial cervical joint. Journal of Neurosurgery. (Cummins et al. 1998))

Hospital, led to the development of a prosthetic cervical disc constructed entirely of stainless steel with congruent surfaces and no point loading (Fig. 2) (Cummins et al. 1998; Traynelis 2004; Wigfield et al. 2002b). The Cummins disc occupied 11 mm of the intervertebral space and was secured to the vertebral bodies above and below the index level with screws (Traynelis 2004). Between 1991 and 1996, 22 Cummins discs were implanted in 20 "end-stage" patients who lacked motion over multiple cervical levels because of congenital block vertebrae or prior surgical fusion. On follow-up, two patients lacked motion at the index level. This was attributed to the relatively large implant size which may have caused over-distraction of the facet joints (Cummins et al. 1998). Although there were implant problems such as screw breakages, patients experienced clinical improvement (those with radiculopathy improved, and those with myelopathy improved or stabilized) (Cummins et al. 1998).

The work of Cummins set the foundation for the development of the next generation of artificial cervical discs. The next CDA device was developed in 1998 and was referred to as the Frenchay artificial cervical joint (Fig. 3) (Traynelis 2004; Wigfield et al. 2002b).

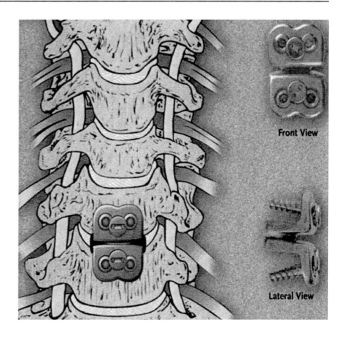

Fig. 3 The two articulating components of the Frenchay artificial cervical joint (or Prestige I) are shown with the bone and locking screws. (Recreated from Wigfield et al. The New Frenchay Artificial Cervical Joint: Results From a Two-Year Pilot Study. Spine (Phila Pa 1976). (Wigfield et al. 2002b))

Medtronic ultimately purchased the Frenchay disc, and it was renamed as Prestige (Medtronic Inc.) (Nunley et al. 2018). This device had some similarities to the prior Cummins joint but was redesigned with a trough rather than a ball-and-socket for articulation. Also, the lower component of the joint was redesigned for translation within three degrees of freedom for both translation and rotation (Wigfield et al. 2002b). Together, these design changes allowed more physiologic motion (anterior-posterior translation coupled with flexion/extension) (Traynelis 2004; Wigfield et al. 2002b). Wigfield and colleagues prospectively evaluated the Frenchay artificial joint in a cohort of 15 patients with cervical radiculopathy or myelopathy from cervical disc herniation or posterior vertebral body osteophytes (Wigfield et al. 2002b). Over the duration of their 2-year study, the Frenchay CDA maintained motion and intervertebral height at the index levels, there were no cases of dislocation screw backout, and clinical outcomes scores improved (Wigfield et al. 2002b).

The next iteration of Prestige CDA, Prestige II, was developed in 1999 (Traynelis 2004). This device had roughened end plate surfaces to promote bony ingrowth for long-term stability (Traynelis 2004). The Prestige II was the first artificial cervical disc to be compared to ACDF (non-instrumented arthrodesis with autograft) in a prospective randomized trial of patients with symptomatic single-level primary cervical disc disease (Porchet and Metcalf 2004; Traynelis 2004). Data after 2 years of follow-up demonstrated improvement in most outcome measurements that favored CDA over ACDF (Porchet and Metcalf 2004). Also, motion analysis demonstrated favorable results in the CDA cohort (motion was maintained in the CDA cohort compared to ACDF patients who displayed no significant motion) (Porchet and Metcalf 2004)

The next Prestige disc, Prestige ST, became available in 2002 (Traynelis 2004). The surfaces of the device contacting the end plates were grit-blasted to promote bone osteointegration (Traynelis 2004). In comparison with its predecessor, there was a 2 mm reduction in the height of the device's anterior flanges (Traynelis 2004). The Prestige ST ball-and-trough articulation design, combined with its angulation between the base and anterior portions of the device, allowed more physiologic motion comparable to normal cervical vertebrae (Traynelis 2004). Mummaneni and colleagues performed a multicenter, prospective, randomized, non-inferiority clinical trial comparing the Prestige ST to ACDF (Mummaneni et al. 2007). The Prestige CDA patients maintained physiological segmental motion and had improved clinical outcomes (summarized below) and reduced rates of secondary surgery compared to ACDF (Mummaneni et al. 2007). Burkus and colleagues demonstrated that the Prestige ST disc maintained improved clinical outcomes (summarized below) and segmental motion after implantation after 5 years post-op (Burkus et al. 2010). Rates of reoperations for adjacent segment degeneration trended lower in the CDA cohort in comparison with the ACDF group, but the differences were not statistically significant (Burkus et al. 2010).

The Prestige LP is the latest generation in the Prestige family of cervical discs (Traynelis 2004). The FDA-approved Prestige LP disc (for both single- and two-level symptomatic cervical spondylosis or disc herniation) is a non-constrained ball-in-trough, metal-on-metal articulation made of a titanium ceramic composite. The unique titanium ceramic composite material is highly durable and results in less artifact during CT and MRI scans (Traynelis 2004). Also, the porous titanium plasma spray coating on the end plate surface facilitates bone ingrowth and long-term fixation (Traynelis 2004). Long-term outcomes for the Prestige family of discs are summarized below.

Long-Term Outcomes for Single-Level Prestige CDA Versus ACDF

Prestige LP: In 2015, Gornet and colleagues reported 24-month outcomes for Prestige LP CDA versus ACDF: NDI and VAS neck and arm scores were noninferior, and SF-36 MCS was noninferior as well as statistically superior (Gornet et al. 2015). Gornet and colleagues reported continued success for Prestige LP CDA versus ACDF at 84 months: NDI and VAS scores were still noninferior, SF-36 PCS was noninferior, and SF-36 MCS was noninferior as well as statistically superior (Gornet et al. 2016).

Prestige ST: Outcomes at 24 months for Prestige ST CDA versus ACDF demonstrated no differences in NDI, VAS neck score (which had been significantly better for the CDA group at 12 months), VAS arm score, and SF-36 PCS and MCS (Mummaneni et al. 2007; Riew et al. 2008). Burkus and colleagues reported outcomes at 60 months for Prestige ST CDA versus ACDF: NDI was significantly better, VAS neck score was significantly better, VAS arm score had no significant difference, SF-36 PCS had no significant difference, and SF-36 MCS comparison was not reported (Burkus et al. 2010, 2014). Later in 2014, Burkus and colleagues reported 84-month outcomes for Prestige ST CDA versus ACDF, and the results were similar to previously reported 60-month outcomes except the SF-36 PCS score for the CDA group was now significantly improved compared to the ACDF treatment group (Burkus et al. 2014).

Prestige II: In 2004, Porchet and colleagues reported outcomes at 24 months for Prestige II CDA versus ACDF: NDI was statistically equivalent, VAS neck score statistical equivalence could not be shown between treatment groups, VAS arm score was statistically equivalent, and no significant differences were demonstrated for SF-36 PCS and MCS (Porchet and Metcalf 2004).

Long-Term Outcomes for Two-Level Adjacent Prestige LP CDA Versus ACDF

Gornet and colleagues reported 24-month outcomes for Prestige LP CDA versus ACDF at two levels: NDI was statistically superior, VAS neck score was noninferior, VAS arm score was noninferior, SF-36 PCS was noninferior, and SF-36 MCS was not reported (Gornet et al. 2017). Lanman and colleagues reported similar results for 84-month CDA outcomes: VAS neck score was statistically superior, SF-36 PCS was statistically superior, and SF-36 MCS was noninferior (Lanman et al. 2017). Although there was no statistically significant difference in the overall rate of implant- or procedure-related adverse events for up to 84 months post-op, the trend favored the CDA treatment cohort (Lanman et al. 2017).

Porous Coated Motion (PCM) Cervical Disc

The Porous Coated Motion (PCM) device (Cervitech) is a non-constrained artificial cervical disc that was originally invented by McAfee and then was improved upon by Helmut Link and Arnold Keller (Fig. 1c) (Pimenta et al. 2004). It has a unique biomechanical design feature that incorporates a large radius ultrahigh-molecular-weight polyethylene bearing surface attached to the inferior vertebrae. This allows the device more physiologic translational motion in an arc, which is consistent with the natural motion of the cervical spine (Pimenta et al. 2004). The porous ingrowth material is composed of two ultra-thin layers of titanium with electrochemically coated calcium phosphate (Pimenta et al. 2004). The pore size was designed to match the bony trabecular architecture of the cervical vertebra (Pimenta et al. 2004).

Pimenta and colleagues reported the results of a pilot study performed between December 2002 and October 2003 in which 82 PCM devices were implanted in 53 patients. Significant improvements in all scores were seen postoperatively (NDI, VAS pain scores, and Treatment Intensity Gradient Test). One device migration of 4 mm was seen at 3 months and was observed (no reoperation). Eighty percent of patients had a good or excellent result at 1 week, improving to 90% of patients having a good or excellent result by 1 month (Odom's criteria), and this result remained stable 3 months after surgery (Pimenta et al. 2004). Later in 2007, Pimenta and colleagues published the first prospective CDA study to show significantly improved clinical outcomes for multilevel compared to single-level CDA (PCM disc) (Pimenta et al. 2007). Table 4 is a summary of PCM CDA outcomes data (Alvin et al. 2014; Delamarter et al. 2010; Phillips et al. 2009, 2013, 2015; Pimenta et al. 2004, 2007).

Long-Term Outcomes for Single-Level PCM CDA Versus ACDF

The FDA randomized controlled trials comparing PCM CDA vs. ACDF were performed by Phillips and colleagues (Phillips et al. 2013, 2015). The study cohort consisted of patients 18–65 years of age with single-level symptomatic cervical spondylosis (radiculopathy and/or myelopathy) unresponsive to nonoperative treatment. This included patients with prior non-adjacent or adjacent single-level fusion operations. The 24-month outcomes demonstrated that NDI was significantly better, VAS neck and arm scores were not significantly different, and SF-36 PCS and MCS were not significantly different for PCM CDA compared to ACDF (Phillips et al. 2013). The patients with PCM CDA had lower rates of prolonged dysphagia, greater patient satisfaction, and superior overall success compared to ACDF (Phillips et al. 2013).

In 2015, Phillips and colleagues reported 60-month outcomes for PCM CDA vs. ACDF: NDI was significantly better, VAS neck score was significantly better, VAS arm score was not significantly different, and SF-36 PCS and MCS were significantly better (Phillips et al. 2015). PCM CDA patients also had a lower rate of radiographical adjacent-level degeneration and a trend toward fewer secondary surgeries (Phillips et al. 2015). The authors interpreted the results of these studies to support PCM CDA as a viable and sustainable alternative to ACDF in appropriately selected patients (Phillips et al. 2015).

ProDisc-C Cervical Disc

The ProDisc-C (Synthes Spine) is an artificial cervical disc designed with these principles in mind: implant stability, ease and safety of insertion, minimal end plate disruption, and optimization of functional range of motion (Fig. 1d). These principles and design characteristics were investigated in several studies, and clinical outcomes are summarized in

Table 4 Summary of single- and multilevel Porous Coated Motion (PCM) disc outcomes for symptomatic cervical spondylosis or disc herniation. Recreated from Alvin et al. Cervical arthroplasty: a critical review of the literature. The Spine Journal. (Alvin et al. 2014; Phillips et al. 2009, 2013, 2015; Pimenta et al. 2004, 2007)

Author	Design	Study size	Follow-up duration	NDI	VAS neck/arm scores	SF-36 PCS	SF-36 MCS	Design LoE
Phillips 2015	RCT	110	60 mos	Imp	Imp/Imp	Imp	Imp	Ib
Phillips 2013	RCT	342	24 mos	Imp	Imp/Imp	Imp	Imp	Ib
Phillips 2009	PC	152	12 mos	Imp	Imp/Imp			IIb
Pimenta 2007	PC	140 (69 multilevel)	NR	Imp	Imp/Imp	NR	NR	IIb
Pimenta 2004	PC	53 (25 multilevel)	NR	Imp	Imp/Imp	NR	NR	IIb

Imp improved, *LoE* level of evidence, *MCS* mental component score, *NDI* neck disability index, *PC* prospective cohort, *PCS* physical component score, *R* retrospective, *RCT* randomized controlled trial, *SF-36* short form-36, *VAS* visual analogue scale

Table 5 Summary of single- and multilevel ProDisc-C disc outcomes for symptomatic cervical spondylosis or disc herniation. Recreated and modified from Alvin et al. Cervical arthroplasty: a critical review of the literature. The Spine Journal. (Alvin et al. 2014; Bertagnoli et al. 2005; Chin et al. 2017; Delamarter et al. 2010; Janssen et al. 2015; Kelly et al. 2011; Kesman et al. 2012; Mehren et al. 2006; Murrey et al. 2009; Nabhan et al. 2007; Peng et al. 2009; Suchomel et al. 2010; Zigler et al. 2013)

Author	Design	Study size	Follow-up duration	NDI	VAS neck/arm scores	SF-36 PCS	SF-36 MCS	Design LoE
Janssen 2015	RCT	209	84 mos	Imp	Imp/Imp	Imp	Imp	Ib
Zigler 2013	RCT	209	60 mos	Imp	Imp/Imp	Imp	Imp	Ib
Kesman 2012	RCT	44	84 mos	Imp	Imp/Imp	Imp	Imp	Ib
Kelly 2011	RCT	199	24 mos					Ib
Delamarter 2010	RCT	345	48 mos	Imp	Imp/Imp	Imp	Imp	Ib
Murrey 2009	RCT	209	24 mos	Imp	Imp/Imp	Imp	Imp	Ib
Nabhan 2007	RCT	49	12 mos		Imp/Imp			Ib
Suchomel 2010	PC	54 (10 multilevel)	48 mos		Imp/Imp			IIb
Mehren 2006	PC	54 (20 multilevel)	12 mos	Imp	Imp/Imp			IIb
Bertagnoli 2005	PC	16 (4 multilevel)	12 mos	Imp	Imp/Imp			IIb
Peng 2009	R	166	24 mos	Imp	Imp/Imp			IIb
Chin 2017	R	110	24 mos	Imp	Imp/Imp			III

Imp improved, *LoE* level of evidence, *MCS* mental component score, *NDI* neck disability index, *PC* prospective cohort, *PCS* physical component score, *R* retrospective, *RCT* randomized controlled trial, *SF-36* short form-36, *VAS* visual analogue scale

Table 5 (Alvin et al. 2014; Bertagnoli et al. 2005; Chin et al. 2017; Delamarter et al. 2010; Janssen et al. 2015; Kelly et al. 2011; Kesman et al. 2012; Mehren et al. 2006; Murrey et al. 2009; Nabhan et al. 2007; Peng et al. 2009; Suchomel et al. 2010; Zigler et al. 2013). The specific advantages of the ProDisc-C device include the absence of anterior plate fixation hardware, preservation of osseous end plates, immediate keel fixation stability, and the possibility of multilevel application. Biomechanically, the ProDisc-C implant is considered to represent a ball-and-socket/semi-constrained design with a fixed axis of rotation (Bertagnoli et al. 2005). DiAngelo and colleagues performed an in vitro biomechanical study to compare the effects of ProDisc-C CDA and ACDF in a multilevel human cadaveric model. Their results demonstrated that ACDF decreased motion at the index level in comparison with CDA (DiAngelo et al. 2004). The reduced motion at the index level was compensated at adjacent segments by an increase in motion. ProDisc-C CDA did not alter the motion patterns at either the index or adjacent levels compared with control (except in extension) (DiAngelo et al. 2004). Long-term outcomes from the FDA trials comparing ProDisc-C CDA to ACDF are summarized below.

Long-Term Outcomes for Single-Level ProDisc-C CDA Versus ACDF

In 2007, Nabhan and colleagues reported no significant difference in 12-month VAS neck and arm scores for ProDisc-C CDA versus ACDF (Nabhan et al. 2007). Later in 2009, Murrey and colleagues reported no significant differences in all outcome variables (NDI, VAS neck and arm scores, SF-36 PCS and MCS) at 24 months post-op (Murrey et al. 2009). This trend continued in 2010 with Delamarter and colleagues reporting 48-month outcomes for ProDisc-C CDA versus ACDF: NDI and VAS neck and arm scores were still not significantly different (Delamarter et al. 2010). However, in 2013, Zigler and colleagues reported 60-month outcomes for ProDisc-C CDA vs. ACDF and found that NDI and VAS neck scores were significantly better (VAS arm score, SF-36 PCS, and SF-36 MCS were not significantly different) (Zigler et al. 2013). Then at 84 months post-op,

two studies demonstrated no significant difference in all outcome variables for ProDisc-C CDA versus ACDF (Janssen et al. 2015). For these two studies, the VAS and SF-36 scores showed noninferiority of the Prodisc-C group, which trended toward statistical superiority (Kesman et al. 2012).

Mobi-C Cervical Disc

The Mobi-C cervical artificial disc (LDR Medical) is a semi-constrained, bone-sparing prosthetic device (Fig. 1e) (Davis et al. 2015; Kim et al. 2007). The implant is composed of two cobalt-chromium-molybdenum alloy shells with an ultrahigh-molecular-weight polyethylene mobile insert facilitating five independent degrees of freedom (Davis et al. 2015; Kim et al. 2007). The mobility of the polyethylene insert decreases the transmission of the constraints on the bone-implant interface and reduces the constraints of the posterior facet joints (Kim et al. 2007). The implant has lateral self-retaining, incline-shaped teeth that were designed to support reliable vertebral end plate anchorage and stability (Kim et al. 2007). Tables 6 and 7 summarize Mobi-C CDA outcomes data (Bae et al. 2015; Beaurain et al. 2009; Davis et al. 2013, 2015; Guerin et al. 2012; Hisey et al. 2014, 2015, 2016; Huppert et al. 2011; Kim et al. 2007; Lee et al. 2012; Park et al. 2008, 2013; Radcliff et al. 2016a). The Mobi-C disc has FDA approval for both single- and two-level symptomatic cervical spondylosis and/or disc disease. Long-term outcomes from the FDA trials are summarized below.

Long-Term Outcomes for Single-Level Mobi-C CDA Versus ACDF

Hisey and colleagues reported 24-, 48-, and 60-month outcomes in multicenter, prospective, randomized, controlled FDA investigational device exemption clinical trials comparing Mobi-C CDA to ACDF in the treatment of

Table 6 Summary of FDA single-level (and other multi-level) Mobi-C cervical disc outcome studies. Recreated from Alvin et al. Cervical arthroplasty: a critical review of the literature. The Spine Journal. (Alvin et al. 2014; Bae et al. 2015; Beaurain et al. 2009; Davis et al. 2013, 2015; Guerin et al. 2012; Hisey et al. 2014, 2015, 2016; Huppert et al. 2011; Kim et al. 2007; Lee et al. 2012; Park et al. 2008, 2013; Radcliff et al. 2016a)

Author	Design	Study size	Follow-up duration	NDI	VAS neck/arm scores	SF-36 PCS	SF-36 MCS	Design LoE
Hisey 2016	RCT	245	60 mos	Imp	Imp/Imp			Ib
Hisey 2015	RCT	245	48 mos	Imp	Imp/Imp			Ib
Hisey 2014	RCT	245	24 mos	Imp	Imp/Imp			Ib
Lee 2012	PC	28 (9 multilevel)	24 mos	Imp	Imp/Imp			IIb
Huppert 2011	PC	231 (56 multilevel)	24 mos	Imp	Imp/Imp	Imp	Imp	IIb
Beaurain 2009	PC	76 (9 multilevel)	24 mos	Imp	Imp/Imp	Imp	Imp	IIb
Park 2013	R	75 (16 multilevel)	40 mos	Imp	Imp/Imp			IIb
Park 2008	R	53	20 mos	Imp	–/Imp			IIb
Kim 2007	R	23 (7 multilevel)	6 mos		Imp/Imp			IIb

Imp improved, *LoE* level of evidence, *MCS* mental component score, *NDI* neck disability index, *PC* prospective cohort, *PCS* physical component score, *R* retrospective, *RCT* randomized controlled trial, *SF-36* short form-36, *VAS* visual analogue scale

Table 7 Summary of FDA two-Level Mobi-C cervical disc outcomes for symptomatic cervical spondylosis or disc herniation. Recreated from Alvin et al. Cervical arthroplasty: a critical review of the literature. The Spine Journal. (Alvin et al. 2014; Bae et al. 2015; Beaurain et al. 2009; Davis et al. 2013, 2015; Guerin et al. 2012; Hisey et al. 2014, 2015, 2016; Huppert et al. 2011; Kim et al. 2007; Lee et al. 2012; Park et al. 2008, 2013; Radcliff et al. 2016a)

Author	Design	Study size	Follow-up duration	NDI	VAS neck/arm scores	SF-36 PCS	SF-36 MCS	Design LoE
Radcliff 2016	RCT	330	60 mos	Imp	Imp/Imp			Ib
Bae 2015	RCT	413 (225 multilevel)	48 mos	Imp	Imp/Imp			Ib
Davis 2015	RCT	291	48 mos	Imp	Imp/Imp			Ib
Davis 2013	RCT	330	24 mos	Imp	Imp/Imp			Ib
Guerin 2012	PC	40	24.3 mos	Imp	Imp/Imp	Imp	Imp	IIb

Imp improved, *LoE* level of evidence, *MCS* mental component score, *NDI* neck disability index, *PC* prospective cohort, *PCS* physical component score, *R* retrospective, *RCT* randomized controlled trial, *SF-36* short form-36, *VAS* visual analogue scale

symptomatic degenerative disc disease in the cervical spine (single level). The results demonstrated similar findings at each of these time points, namely, there were no significant differences in NDI or VAS neck and arm scores (Gornet et al. 2015; Hisey et al. 2014, 2015).

Table 8 Summary of single-level Kineflex-C cervical disc outcomes for symptomatic cervical spondylosis or disc herniation. Recreated from Alvin et al. Cervical arthroplasty: a critical review of the literature. The Spine Journal. (Alvin et al. 2014; Coric et al. 2011, 2013, 2018)

Author	Design	n	Follow-up	NDI	VAS neck/arm	SF-36 PCS	SF-36 MCS	Design LoE
Coric 2018	RCT	269	60 mos	Imp	Imp/Imp			Ib
Coric 2013	RCT	74	48 mos	Imp	Imp/Imp			Ib
Coric 2011	RCT	269	24 mos	Imp	Imp/Imp			Ib

Imp improved, *LoE* level of evidence, *MCS* mental component score, *NDI* neck disability index, *PCS* physical component score, *R* retrospective, *RCT* randomized controlled trial, *SF-36* short form-36, *VAS* visual analogue scale

Table 9 Single-level Secure-C cervical disc outcomes for symptomatic cervical spondylosis or disc herniation (Vaccaro et al. 2013)

Author	Design	Study size	Follow-up duration	NDI	VAS neck/arm scores	SF-36 PCS	SF-36 MCS	Design LoE
Vaccaro 2013	RCT	380	24 mos	Imp	Imp/Imp	Imp	Imp	Ib

Imp improved, *LoE* level of evidence, *MCS* mental component score, *NDI* neck disability index, *PCS* physical component score, *R* retrospective, *RCT* randomized controlled trial, *SF-36* short form-36, *VAS* visual analogue scale

Long-Term Outcomes for Two-Level Adjacent Mobi-C CDA Versus ACDF

In 2013, Davis and colleagues reported 24-month outcomes for Mobi-C CDA versus ACDF at two adjacent levels: NDI was significantly better, and although VAS neck score was significantly improved at 3 and 6 months postoperatively, there were no statistically significant differences at any other time point. Also, there were no significant differences between treatment groups for VAS arm scores at any time point (Davis et al. 2013). Later in 2015, Davis and colleagues reported similar results at 48 months post-op: NDI was significantly better, but there were no significant differences in VAS neck and arm scores between treatment groups (Davis et al. 2015). For 60-month outcomes, Radcliff and colleagues reported that NDI was significantly better, and although there was more improvement in VAS neck and arm scores for the CDA group, the difference was not statistically significant (Radcliff et al. 2016a).

Kineflex-C Cervical Disc

The Kineflex-C artificial cervical disc (SpinalMotion Inc.) is a cobalt-chrome on cobalt-chrome alloy (metal-on-metal) semi-constrained device (Fig. 1f) (Coric et al. 2011). It is composed of three pieces (two end plates and a mobile center that translates within a retention ring). There is a midline keel on the device's end plate that provides immediate fixation, and the end plates are coated with a titanium plasma spray to promote bony ingrowth for long-term fixation (Coric et al. 2011). Table 8 summarizes Kineflex-C CDA outcomes data (Coric et al. 2011, 2013, 2018). Long-term outcomes for the FDA trials comparing Kineflex-C CDA and ACDF are summarized below.

Long-Term Outcomes for Single-Level Kineflex-C CDA Versus ACDF

Coric and colleagues reported 24- and 48-month outcomes for Kineflex-C CDA versus ACDF and found no significant differences between treatment groups based on NDI or VAS scores (Coric et al. 2011, 2013). However, clinical success (maintenance or improvement in neurological exam, minimum of 20% improvement in NDI, no device failure, no reoperation at the index level, no major device-related adverse event) was significantly higher in the Kineflex-C group compared to ACDF (Coric et al. 2011). Recently, Coric and colleagues reported clinical

48 Cervical Arthroplasty: Long-Term Outcomes

Table 10 Summary of single- and multilevel Discover disc outcomes for symptomatic cervical spondylosis or disc herniation. Recreated from Alvin et al. Cervical arthroplasty: a critical review of the literature. The Spine Journal. (Alvin et al. 2014; Du et al. 2011; Fang et al. 2013; Li et al. 2013; Miao et al. 2014; Rozankovic et al. 2017; Shi et al. 2016; Skeppholm et al. 2015)

Author	Design	Study size	Follow-up duration	NDI	VAS neck/ arm scores	SF-36 PCS	SF-36 MCS	Design LoE
Rozankovic 2017	RCT	105	24 mos	Imp	Imp/Imp			Ib
Skeppholm 2015	RCT	137 (43 multilevel)	24 mos	Imp	Imp/Imp			Ib
Shi 2016	PC	128	24 mos	Imp				IIb
Miao 2014	PC	79 (23 multilevel)	31.6 mos		Imp/Imp			IIb
Li 2013	PC	55	24 mos	Imp	Imp/Imp			IIb
Du 2011	PC	25 (1 multilevel)	15 mos	Imp	Imp/Imp			IIb
Fang 2013	R	18	15 mos		Imp/Imp			IIb

Imp improved, *LoE* level of evidence, *MCS* mental component score, *NDI* neck disability index, *PCS* physical component score, *R* retrospective, *RCT* randomized controlled trial, *SF-36* short form-36, *VAS* visual analogue scale

success was significantly improved for the Kineflex-C CDA group compared to ACDF at 60 months post-op (Coric et al. 2018). Also, the results demonstrated there were no significant differences between treatment groups in terms of reoperation/revision surgery or device-/surgery-related adverse events during the 5 years of follow-up (Coric et al. 2018).

Secure-C Cervical Disc

The selectively constrained Secure-C artificial cervical disc (Globus Medical) is an anterior articulating intervertebral device comprised of two cobalt-chrome alloy serrated end plates and a sliding polyethylene central core. The end plates have a titanium plasma spray coating on its bone-contacting surface to promote long-term bony ingrowth (Vaccaro et al. 2013). The Secure-C artificial cervical disc is designed for motion in flexion/extension up to $30 \pm 15°$, lateral bending up to $20 \pm 10°$, and sagittal translation of up to ± 1.25 mm (Vaccaro et al. 2013). There is less available FDA trial outcomes data (compared to the aforementioned discs) comparing Secure-C CDA to ACDF (Table 9) (Vaccaro et al. 2013).

Long-Term Outcomes for Single-Level Secure-C CDA Versus ACDF

Overall success results (improvement of at least 25% in baseline NDI, no device failure requiring revision, and absence of major complications [major vessel injury, neurological damage, or nerve injury]) demonstrated statistical superiority of the randomized Secure-C group compared with the randomized ACDF group at 24 months post-op (Vaccaro et al. 2013). There was non-inferiority of the randomized Secure-C group at all postoperative time points (up to 24 months) for both (1) 25% or more and (2) 15-point or more improvement in NDI (Vaccaro et al. 2013). Also, the study demonstrated statistical noninferiority of Secure-C compared to ACDF for VAS neck and arm pain scores (and also statistical superiority for VAS neck pain) (Vaccaro et al. 2013).

Discover Cervical Disc

The non-constrained Discover artificial cervical disc (DePuy Spine) is an MRI-compatible ball-and-socket design consisting of two end plates manufactured from titanium alloy and a polyethylene core (Du et al. 2011; Shi et al. 2016). The inferior end plate is a two-piece design with an ultrahigh-molecular-weight polyethylene insert and features a spherical bearing surface that allows motion in all rotational directions (Du et al. 2011). The Discover disc has a 7° lordotic angle split evenly between the superior and inferior end plates for restoration of lordosis at the index level (Du et al. 2011). Table 10 summarizes Discover CDA outcomes data (Du et al. 2011; Fang et al. 2013; Li et al. 2013; Miao et al. 2014; Rozankovic et al. 2017; Shi et al. 2016; Skeppholm et al. 2015). In contrast to the aforementioned artificial cervical discs, the Discover disc is not approved by the FDA; however, its widespread use for CDA warrants a brief summary of its outcomes.

Long-Term Outcomes for Single- and Multilevel Discover CDA Versus ACDF

In 2017, Rozankovic and colleagues reported 24-month outcomes for Discover CDA vs. ACDF (single level): NDI and VAS neck and arm scores were significantly improved compared to ACDF (Rozankovic et al. 2017). In contrast, Skeppholm and colleagues did not find significantly better 24-month outcomes for CDA compared to ACDF based on NDI scores. In contrast to the Rozankovic study, the Skeppholm study included patients with multilevel cervical disc degeneration who received CDA at adjacent levels, which could explain the difference in results (Skeppholm et al. 2015).

Summary of Complications Associated with Cervical Disc Arthroplasty

Biomechanical and clinical studies suggest that the rate of adjacent segment degeneration (ASDG; *radiographic* evidence of degeneration at the adjacent level) is significantly higher for ACDF compared to CDA (Baba et al. 1993; Chang et al. 2007; Coric et al. 2010; DiAngelo et al. 2003; Dmitriev

et al. 2005; Eck et al. 2002; Matsunaga et al. 1999; Nunley et al. 2018; Park et al. 2011; Puttlitz et al. 2004; Reitman et al. 2004; Wigfield et al. 2002a). However, rates of adjacent segment disease (ASDI; development of new *clinical* symptoms correlating with adjacent segment degeneration) between CDA and ACDF continue to be debated. Jawahar and colleagues found no difference in the incidence of ASDI between CDA and ACDF. On the contrary, there has been growing evidence from other long-term follow-up studies and meta-analyses that suggest CDA may reduce ASDI and reoperation rates in comparison with ACDF (Gao et al. 2013; Ishihara et al. 2004; Jawahar et al. 2010; McAfee et al. 2012; Robertson et al. 2005; Upadhyaya et al. 2012).

Other adverse outcomes associated with CDA include heterotopic ossification (HO), delayed fusion around cervical disc prosthesis, asymmetric end plate preparation resulting in postoperative kyphosis, and reduction in caudal vertebral body height (Yi et al. 2010). Rates of HO with the FDA investigational device exemption publications have been reported, and grade 4 HO rates are as high as 13% (Gornet et al. 2016; Hisey et al. 2016; Janssen et al. 2015; Nunley et al. 2018; Radcliff et al. 2016a). Table 11 is a summary of the commonly reported complications associated with CDA in the literature (Alvin et al. 2014; Anderson et al. 2004; Beaurain et al. 2009; Bertagnoli et al. 2005; Bhadra et al. 2009; Bryan 2002; Cheng et al. 2011; Coric et al. 2006, 2011; Ding et al. 2012; Du et al. 2011; Duggal et al. 2004; Garrido et al. 2010; Goffin et al. 2003, 2010; Hacker 2005; Heidecke et al. 2008; Heller et al. 2009; Huppert et al. 2011; Kelly et al. 2011; Kesman et al. 2012; Kim et al. 2007, 2008, 2009; Lee et al. 2010; Leung et al. 2005; Li et al. 2013; Mehren et al. 2006; Mummaneni et al. 2007; Murrey et al. 2009; Nabhan et al. 2007; Park et al. 2008, 2013; Peng et al. 2009, 2011; Phillips et al. 2013; Pickett et al. 2004; Pimenta et al. 2004, 2007; Porchet and Metcalf 2004; Quan et al. 2011; Ren et al. 2011; Riew et al. 2008; Riina et al. 2008; Robertson and Metcalf 2004; Robertson et al. 2005; Ryu et al. 2010; Sasso et al. 2011; Sekhon 2003; Sekhon et al. 2005; Shim et al. 2006; Suchomel et al. 2010; Tu et al. 2011; Walraevens

Table 11 Summary of cervical disc arthroplasty complications. Recreated from Alvin et al. Cervical arthroplasty: a critical review of the literature. The Spine Journal. (Alvin et al. 2014; Anderson et al. 2004; Beaurain et al. 2009; Bertagnoli et al. 2005; Bhadra et al. 2009; Bryan 2002; Cheng et al. 2011; Coric et al. 2006, 2011; Ding et al. 2012; Du et al. 2011; Duggal et al. 2004; Garrido et al. 2010; Goffin et al. 2003, 2010; Guerin et al. 2012; Hacker 2005; Heidecke et al. 2008; Heller et al. 2009; Huppert et al. 2011; Kelly et al. 2011; Kesman et al. 2012; Kim et al. 2007; 2008, 2009; Lee et al. 2010, 2012; Leung et al. 2005; Li et al. 2013; Mehren et al. 2006; Mummaneni et al. 2007; Murrey et al. 2009; Nabhan et al. 2007; Park et al. 2008, 2013; Peng et al. 2009, 2011; Phillips et al. 2013; Pickett et al. 2004; Pimenta et al. 2004, 2007; Porchet and Metcalf 2004; Quan et al. 2011; Ren et al. 2011; Riew et al. 2008; Riina et al. 2008; Robertson and Metcalf 2004; Robertson et al. 2005; Ryu et al. 2010; Sasso et al. 2011; Sekhon 2003, 2005; Shim et al. 2006; Suchomel et al. 2010; Tu et al. 2011; Walraevens et al. 2010; Wang et al. 2008; Yang et al. 2008; Yoon et al. 2006; Zigler et al. 2013)

Author	Disc	HO (%)	ASDI (%)[a]	ASDG (%)[a]	Other (%)[a]
Cheng 2011	Bryan	2.4	None	None	Dysphagia (2.4)
Tu 2011	Bryan	50	None	None	None
Lee 2010	Bryan	27	None	None	None
Ryu 2010	Bryan	52.8	None	None	None
Yang 2008	Bryan	None	None	None	None
Shim 2006	Bryan	None	None	None	Op failure (17)
Hacker 2005	Bryan	None	(4.6)	None	Dysphonia (4.5)
Lafuente 2005	Bryan	None	None	None	Dysphonia (7)
Leung 2005	Bryan	17.8	None	None	None
Ding 2012	Bryan	None	None	23	None
Quan 2011	Bryan	47.6	19	19	None
Ren 2011	Bryan	4.4	None	None	None
Garrido 2010	Bryan	None	5	None	Reoperation (6.7)
Bhadra 2009	Bryan	13	None	None	None
Kim 2009	Bryan	None	None	None	None
Heidecke 2008	Bryan	29	None	None	None
Kim 2008	Bryan	None	None	None	None
Wang 2008	Bryan	None	None	None	None
Sasso 2007	Bryan	None	5.4	None	Reoperation (3.5)
Coric 2006	Bryan	None	None	None	None
Yoon 2006	Bryan	None	None	None	None
Duggal 2004	Bryan	None	None	None	None
Pickett 2004	Bryan	None	None	None	None
Sekhon 2003	Bryan	None	None	None	None
Zhang 2012	Bryan	12.5	1.6	None	Reoperation (1.6)
Sasso 2011	Bryan	None	4.1	None	Reoperation (3.7)
Coric 2010	Bryan	5.6	1.7	None	Reoperation (7.5)
Goffin 2010	Bryan	None	4.1	None	Reoperation (8.2)
Walraevans 2010	Bryan	34	None	None	None
Heller 2009	Bryan	None	None	None	Reoperation (2.5)
Pickett 2006	Bryan	2.7	None	None	Reoperation (5.4)
Robertson 2005	Bryan	None	1.3	17.5	None
Sekhon 2005	Bryan	None	None	None	None
Anderson 2004	Bryan	None	None	None	Reoperation (2.2)
Goffin 2003	Bryan	None	None	None	Reoperation (2.0)
Bryan 2002	Bryan	None	None	None	None
Peng 2011	Prestige	None	None	None	None
Riina 2008	Prestige	None	None	None	None
Burkus 2010	Prestige	3.2	2.9	None	Reoperation (10.5)

(continued)

Table 11 (continued)

Author	Disc	HO (%)	ASDI (%)[a]	ASDG (%)[a]	Other (%)[a]
Riew 2008	Prestige	None	None	None	Reoperation (1.9)
Mummaneni 2007	Prestige	None	1.1	None	Reoperation (1.8)
Porchet 2004	Prestige	None	None	None	None
Robertson 2004	Prestige	None	None	None	None
Phillips 2013	PCM	38	39.1	None	None
Pimenta 2007	PCM	0.7	None	None	Reoperation (2.2)
Pimenta 2004	PCM	None	None	None	None
Suchomel 2010	ProDisc-C	88	None	None	None
Peng 2009	ProDisc-C	None	None	None	None
Nabhan 2007	ProDisc-C	None	None	None	None
Mehren 2006	ProDisc-C	57	None	None	None
Bertagnoli 2005	ProDisc-C	None	None	None	None
Zigler 2013	ProDisc-C	None	None	None	Reoperation (2.9)
Kesman 2012	ProDisc-C	None	None	None	None
Kelly 2011	ProDisc-C	None	None	None	None
Murrey 2009	ProDisc-C	2.9	None	None	Reoperation (1.9)
Guerin 2012	Mobi-C	27.7	None	None	None
Lee 2012	Mobi-C	77.3	None	None	None
Park 2013	Mobi-C	94.1	None	None	None
Beaurain 2009	Mobi-C	67	None	9.1	Dysphagia (10.5)
Park 2008	Mobi-C	None	None	None	None
Kim 2007	Mobi-C	None	None	None	None
Huppert 2011	Mobi-C	62	None	None	Reoperation (2.6)
Coric 2011	Kineflex-C	None	None	9	Reoperation (5)
Li 2013	Discover	18	None	7.2	None
Du 2011	Discover	None	9	None	None

[a]Complication rate reported for the arthroplasty investigational cohort

ASDI adjacent segment disease, *ASDG* adjacent segment degeneration, *HO* heterotopic ossification, *PCM* porous coated motion

et al. 2010; Wang et al. 2008; Yang et al. 2008; Yoon et al. 2006; Zigler et al. 2013).

Metal Ion Toxicity

Articulating prosthetic implants are subject to wear and corrosion following implantation. An advantage of metal-on-metal bearings is the substantially lower volumetric wear debris when compared with conventional metal-on-polyethylene bearing couples. A concern regarding any metal-on-metal CDA (e.g., Prestige LP CDA) is that patients may have increased serum metal ion concentrations after surgery since implant wear can lead to local and systemic transport of metal debris (Coric et al. 2018; Gornet et al. 2016). Toxicology-related sequelae

from chronically elevated metal ion levels have not been determined. In support of CDA, a 5-year randomized control trial (comparing single-level Kineflex-C CDA with ACDF) demonstrated that serum ion levels (cobalt and chromium) were significantly lower than the levels that merit monitoring (Coric et al. 2018). However, several case studies have reported some early local effects of wear debris (Cavanaugh et al. 2009; Gornet et al. 2016; Hacker et al. 2013).

Patient Selection

CDA is associated with high success rates when performed for appropriately selected patients. However, complications may occur with improper patient selection, technical errors, or progression of underlying cervical disease (Leven et al. 2017; Nunley et al. 2018; Nunley et al. 2012). Current indications for CDA in the United States (largely dictated by FDA approval of the various prosthetic devices) include skeletally mature patients with cervical radiculopathy and/or myelopathy at a single or two adjacent levels without severe facet joint degeneration, instability, malalignment or kyphosis, or severe neck pain only (Leven et al. 2017; Nunley et al. 2018). Other contraindications include retrovertebral compression (i.e., congenital stenosis or ossification of the posterior longitudinal ligament) and spondyloarthropathies (ankylosing spondylitis) (Leven et al. 2017; Nunley et al. 2018). Patients with a severe axial neck pain due to facet degeneration should be counseled appropriately since these symptoms may not improve after CDA (Leven et al. 2017). Also, some authors have recommended a disc height of 3 mm or greater for adequate disc space access and removal (Ding and Shaffrey 2012). Placing an oversized implant into a collapsed disc space can potentially place excessive forces through the facet joints and lead to worsening of axial neck pain (Ding and Shaffrey 2012).

Cost Efficacy

Although many studies have demonstrated successful treatment with CDA, economic analysis and health costs are also important determinants for obtaining insurance coverage in the United States (Nunley et al. 2018). Therefore, recent studies have focused on analyzing the incremental cost-effectiveness of CDA in comparison with the ACDF. Ament and colleagues reported the incremental cost-effectiveness ratio of CDA compared to ACDF at 2 years post-op for two-level disease was $24954/quality-adjusted life year (QALY). This value is considered to be well within the commonly accepted threshold of $50000/QALY (Ament et al. 2014). Ament and colleagues updated their cost utility analysis at 5 years post-op and reported that the incremental

cost-effectiveness ratio for CDA continued to remain below this $50000/QALY threshold (Ament et al. 2016).

In 2014, McAnany and colleagues analyzed 5-year outcomes data and reported cost benefits of CDA compared to ACDF (McAnany et al. 2014). The CDA cost-effectiveness ratio was $35976/QALY compared to $42618/QALY for ACDF (McAnany et al. 2014). In two studies by Radcliff and colleagues, the results suggested that CDA was also the more cost-effective treatment over ACDF (Radcliff et al. 2015, 2016b). Using 3-year data, they found that the total costs paid by insurers for CDA were $34979 compared to $39829 for ACDF. This difference may have been from readmissions and reoperations, which were higher for the ACDF cohort (Radcliff et al. 2015). In another study which analyzed 7-year data, Radcliff and colleagues reported continued cost benefits of CDA over a range of scenarios (Radcliff et al. 2016b).

In 2016, Ghori and colleagues performed a Markov analysis to evaluate the societal costs of ACDF versus CDA in a theoretical cohort of 45–65-year-old patients (Ghori et al. 2016). Their results demonstrated that the long-term costs for CDA were less expensive throughout the model's age range (Ghori et al. 2016). Factors driving lower costs included lower perioperative costs, earlier return to work, and lower reoperation rates (Ghori et al. 2016).

Conclusions

Total cervical disc replacement attempts to preserve normal motion at adjacent segments and in doing so may decrease the incidence of adjacent segment degeneration and disease. Motion-preserving CDA has showed great promise with equivalent quality-of-life outcomes to ACDF in many long-term comparative studies. However, complications such as heterotopic ossification have been reported to occur with some frequency, but the ultimate clinical consequences or implications (in comparison with ACDF) are yet to be determined. Overall, there exists robust data to support CDA as a viable alternative to ACDF in select patients, but further investigation and continued long-term comparison between CDA and ACDF is warranted.

References

Alvin MD et al (2014) Cervical arthroplasty: a critical review of the literature. Spine J 14:2231–2245. https://doi.org/10.1016/j.spinee.2014.03.047

Ament JD, Yang Z, Nunley P, Stone MB, Kim KD (2014) Cost-effectiveness of cervical total disc replacement vs fusion for the treatment of 2-level symptomatic degenerative disc disease. JAMA Surg 149:1231–1239. https://doi.org/10.1001/jamasurg.2014.716

Ament JD, Yang Z, Nunley P, Stone MB, Lee D, Kim KD (2016) Cost utility analysis of the cervical artificial disc vs fusion for the treatment of 2-level symptomatic degenerative disc disease: 5-year follow-up. Neurosurgery 79:135–145. https://doi.org/10.1227/NEU.0000000000001208

Anderson PA, Sasso RC, Rouleau JP, Carlson CS, Goffin J (2004) The Bryan cervical disc: wear properties and early clinical results. Spine J 4:303S–309S. https://doi.org/10.1016/j.spinee.2004.07.026

Baba H, Furusawa N, Imura S, Kawahara N, Tsuchiya H, Tomita K (1993) Late radiographic findings after anterior cervical fusion for spondylotic myeloradiculopathy. Spine (Phila Pa 1976) 18:2167–2173

Bae HW et al (2015) Comparison of clinical outcomes of 1- and 2-level total disc replacement: four-year results from a prospective, randomized, controlled, multicenter IDE clinical trial. Spine (Phila Pa 1976) 40:759–766. https://doi.org/10.1097/BRS.0000000000000887

Basho R, Hood KA (2012) Cervical total disc arthroplasty. Global Spine J 2:105–108. https://doi.org/10.1055/s-0032-1315453

Beaurain J et al (2009) Intermediate clinical and radiological results of cervical TDR (Mobi-C) with up to 2 years of follow-up. Eur Spine J 18:841–850. https://doi.org/10.1007/s00586-009-1017-6

Bertagnoli R, Yue JJ, Pfeiffer F, Fenk-Mayer A, Lawrence JP, Kershaw T, Nanieva R (2005) Early results after ProDisc-C cervical disc replacement. J Neurosurg Spine 2:403–410. https://doi.org/10.3171/spi.2005.2.4.0403

Bhadra AK, Raman AS, Casey AT, Crawford RJ (2009) Single-level cervical radiculopathy: clinical outcome and cost-effectiveness of four techniques of anterior cervical discectomy and fusion and disc arthroplasty. Eur Spine J 18:232–237. https://doi.org/10.1007/s00586-008-0866-8

Bryan VE Jr (2002) Cervical motion segment replacement. Eur Spine J 11(Suppl 2):S92–S97. https://doi.org/10.1007/s00586-002-0437-3

Burkus JK, Haid RW, Traynelis VC, Mummaneni PV (2010) Long-term clinical and radiographic outcomes of cervical disc replacement with the Prestige disc: results from a prospective randomized controlled clinical trial. J Neurosurg Spine 13:308–318. https://doi.org/10.3171/2010.3.SPINE09513

Burkus JK, Traynelis VC, Haid RW Jr, Mummaneni PV (2014) Clinical and radiographic analysis of an artificial cervical disc: 7-year follow-up from the Prestige prospective randomized controlled clinical trial: clinical article. J Neurosurg Spine 21:516–528. https://doi.org/10.3171/2014.6.SPINE13996

Cavanaugh DA, Nunley PD, Kerr EJ 3rd, Werner DJ, Jawahar A (2009) Delayed hyper-reactivity to metal ions after cervical disc arthroplasty: a case report and literature review. Spine (Phila Pa 1976) 34:E262–E265. https://doi.org/10.1097/BRS.0b013e318195dd60

Chang UK, Kim DH, Lee MC, Willenberg R, Kim SH, Lim J (2007) Changes in adjacent-level disc pressure and facet joint force after cervical arthroplasty compared with cervical discectomy and fusion. J Neurosurg Spine 7:33–39. https://doi.org/10.3171/SPI-07/07/033

Cheng L, Nie L, Zhang L, Hou Y (2009) Fusion versus Bryan cervical disc in two-level cervical disc disease: a prospective, randomised study. Int Orthop 33:1347–1351. https://doi.org/10.1007/s00264-008-0655-3

Cheng L, Nie L, Li M, Huo Y, Pan X (2011) Superiority of the Bryan((R)) disc prosthesis for cervical myelopathy: a randomized study with 3-year followup. Clin Orthop Relat Res 469:3408–3414. https://doi.org/10.1007/s11999-011-2039-z

Chin KR, Pencle FJR, Seale JA, Pencle FK (2017) Clinical outcomes of outpatient cervical total disc replacement compared with outpatient anterior cervical discectomy and fusion. Spine (Phila Pa 1976) 42:E567–E574. https://doi.org/10.1097/BRS.0000000000001936

Cloward RB (1959) Vertebral body fusion for ruptured cervical discs. Am J Surg 98:722–727

Coric D, Finger F, Boltes P (2006) Prospective randomized controlled study of the Bryan cervical disc: early clinical results from a single investigational site. J Neurosurg Spine 4:31–35. https://doi.org/10.3171/spi.2006.4.1.31

Coric D, Cassis J, Carew JD, Boltes MO (2010) Prospective study of cervical arthroplasty in 98 patients involved in 1 of 3 separate investigational device exemption studies from a single investigational site with a minimum 2-year follow-up. Clinical article. J Neurosurg Spine 13:715–721. https://doi.org/10.3171/2010.5.SPINE09852

Coric D et al (2011) Prospective, randomized, multicenter study of cervical arthroplasty: 269 patients from the Kineflex|C artificial disc investigational device exemption study with a minimum 2-year follow-up: clinical article. J Neurosurg Spine 15:348–358. https://doi.org/10.3171/2011.5.SPINE10769

Coric D, Kim PK, Clemente JD, Boltes MO, Nussbaum M, James S (2013) Prospective randomized study of cervical arthroplasty and anterior cervical discectomy and fusion with long-term follow-up: results in 74 patients from a single site. J Neurosurg Spine 18:36–42. https://doi.org/10.3171/2012.9.SPINE12555

Coric D et al (2018) Prospective, randomized multicenter study of cervical arthroplasty versus anterior cervical discectomy and fusion: 5-year results with a metal-on-metal artificial disc J Neurosurg Spine 1–10. https://doi.org/10.3171/2017.5.SPINE16824

Cummins BH, Robertson JT, Gill SS (1998) Surgical experience with an implanted artificial cervical joint. J Neurosurg 88:943–948. https://doi.org/10.3171/jns.1998.88.6.0943

Davis RJ et al (2013) Cervical total disc replacement with the Mobi-C cervical artificial disc compared with anterior discectomy and fusion for treatment of 2-level symptomatic degenerative disc disease: a prospective, randomized, controlled multicenter clinical trial: clinical article. J Neurosurg Spine 19:532–545. https://doi.org/10.3171/2013.6.SPINE12527

Davis RJ et al (2015) Two-level total disc replacement with Mobi-C cervical artificial disc versus anterior discectomy and fusion: a prospective, randomized, controlled multicenter clinical trial with 4-year follow-up results. J Neurosurg Spine 22:15–25. https://doi.org/10.3171/2014.7.SPINE13953

Delamarter RB, Murrey D, Janssen ME, Goldstein JA, Zigler J, Tay BK, Darden B 2nd (2010) Results at 24 months from the prospective, randomized, multicenter investigational device exemption trial of ProDisc-C versus anterior cervical discectomy and fusion with 4-year follow-up and continued access patients. SAS J 4:122–128. https://doi.org/10.1016/j.esas.2010.09.001

DiAngelo DJ, Roberston JT, Metcalf NH, McVay BJ, Davis RC (2003) Biomechanical testing of an artificial cervical joint and an anterior cervical plate. J Spinal Disord Tech 16:314–323

DiAngelo DJ, Foley KT, Morrow BR, Schwab JS, Song J, German JW, Blair E (2004) In vitro biomechanics of cervical disc arthroplasty with the ProDisc-C total disc implant. Neurosurg Focus 17:E7

Ding D, Shaffrey ME (2012) Cervical disk arthroplasty: patient selection. Clin Neurosurg 59:91–97. https://doi.org/10.1227/NEU.0b013e31826b6fbe

Ding C, Hong Y, Liu H, Shi R, Hu T, Li T (2012) Intermediate clinical outcome of Bryan cervical disc replacement for degenerative disk disease and its effect on adjacent segment disks. Orthopedics 35:e909–e916. https://doi.org/10.3928/01477447-20120525-33

Dmitriev AE, Cunningham BW, Hu N, Sell G, Vigna F, McAfee PC (2005) Adjacent level intradiscal pressure and segmental kinematics following a cervical total disc arthroplasty: an in vitro human cadaveric model. Spine (Phila Pa 1976) 30:1165–1172

Du J, Li M, Liu H, Meng H, He Q, Luo Z (2011) Early follow-up outcomes after treatment of degenerative disc disease with the Discover cervical disc prosthesis. Spine J 11:281–289. https://doi.org/10.1016/j.spinee.2011.01.037

Duggal N, Pickett GE, Mitsis DK, Keller JL (2004) Early clinical and biomechanical results following cervical arthroplasty. Neurosurg Focus 17:E9

Eck JC, Humphreys SC, Lim TH, Jeong ST, Kim JG, Hodges SD, An HS (2002) Biomechanical study on the effect of cervical spine fusion on adjacent-level intradiscal pressure and segmental motion. Spine (Phila Pa 1976) 27:2431–2434. https://doi.org/10.1097/01.BRS.0000031261.66972.B1

Fang LM, Zhang YJ, Zhang J, Li Q (2013) Efficacy evaluation of treating cervical spondylopathy with the Discover artificial cervical disc prosthesis. Zhonghua Yi Xue Za Zhi 93:2965–2968

Fernstrom U (1966) Arthroplasty with intercorporal endoprosthesis in herniated disc and in painful disc. Acta Chir Scand Suppl 357:154–159

Fisahn C, Burgess B, Iwanaga J, Chapman JR, Oskouian RJ, Tubbs RS (2017) Ulf Fernstrom (1915–1985) and his contributions to the development of artificial disc replacements. World Neurosurg 98:278–280. https://doi.org/10.1016/j.wneu.2016.10.135

Gao Y, Liu M, Li T, Huang F, Tang T, Xiang Z (2013) A meta-analysis comparing the results of cervical disc arthroplasty with anterior cervical discectomy and fusion (ACDF) for the treatment of symptomatic cervical disc disease. J Bone Joint Surg Am 95:555–561. https://doi.org/10.2106/JBJS.K.00599

Garrido BJ, Taha TA, Sasso RC (2010) Clinical outcomes of Bryan cervical disc arthroplasty a prospective, randomized, controlled, single site trial with 48-month follow-up. J Spinal Disord Tech 23:367–371. https://doi.org/10.1097/BSD.0b013e3181bb8568

Ghori A, Konopka JF, Makanji H, Cha TD, Bono CM (2016) Long term societal costs of anterior discectomy and fusion (ACDF) versus cervical disc arthroplasty (CDA) for treatment of cervical radiculopathy. Int J Spine Surg 10:1. https://doi.org/10.14444/3001

Goffin J et al (2003) Intermediate follow-up after treatment of degenerative disc disease with the Bryan cervical disc prosthesis: single-level and bi-level. Spine (Phila Pa 1976) 28:2673–2678. https://doi.org/10.1097/01.BRS.0000099392.90849.AA

Goffin J, van Loon J, Van Calenbergh F, Lipscomb B (2010) A clinical analysis of 4- and 6-year follow-up results after cervical disc replacement surgery using the Bryan cervical disc prosthesis. J Neurosurg Spine 12:261–269. https://doi.org/10.3171/2009.9.SPINE09129

Gornet MF, Burkus JK, Shaffrey ME, Argires PJ, Nian H, Harrell FE, Jr. (2015) Cervical disc arthroplasty with PRESTIGE LP disc versus anterior cervical discectomy and fusion: a prospective, multicenter investigational device exemption study. J Neurosurg Spine 1–16. https://doi.org/10.3171/2015.1.SPINE14589

Gornet MF, Burkus JK, Shaffrey ME, Nian H, Harrell FE Jr (2016) Cervical Disc arthroplasty with Prestige LP disc versus anterior cervical discectomy and fusion: seven-year outcomes. Int J Spine Surg 10:24. https://doi.org/10.14444/3024

Gornet MF et al (2017) Cervical disc arthroplasty with the Prestige LP disc versus anterior cervical discectomy and fusion, at 2 levels: results of a prospective, multicenter randomized controlled clinical trial at 24 months. J Neurosurg Spine 26:653–667. https://doi.org/10.3171/2016.10.SPINE16264

Guerin P, Obeid I, Gille O, Bourghli A, Luc S, Pointillart V, Vital JM (2012) Sagittal alignment after single cervical disc arthroplasty. J Spinal Disord Tech 25:10–16. https://doi.org/10.1097/BSD.0b013e31820f916c

Hacker RJ (2005) Cervical disc arthroplasty: a controlled randomized prospective study with intermediate follow-up results. Invited submission from the joint section meeting on disorders of the spine and peripheral nerves, March 2005. J Neurosurg Spine 3:424–428. https://doi.org/10.3171/spi.2005.3.6.0424

Hacker FM, Babcock RM, Hacker RJ (2013) Very late complications of cervical arthroplasty: results of 2 controlled randomized prospective studies from a single investigator site. Spine (Phila Pa 1976) 38:2223–2226. https://doi.org/10.1097/BRS.0000000000000060

Heidecke V, Burkert W, Brucke M, Rainov NG (2008) Intervertebral disc replacement for cervical degenerative disease – clinical results and functional outcome at two years in patients implanted with the Bryan cervical disc prosthesis. Acta Neurochir 150:453–459; discussion 459. https://doi.org/10.1007/s00701-008-1552-7

Heller JG et al (2009) Comparison of BRYAN cervical disc arthroplasty with anterior cervical decompression and fusion: clinical and radiographic results of a randomized, controlled, clinical trial. Spine (Phila Pa 1976) 34:101–107. https://doi.org/10.1097/BRS.0b013e31818ee263

Hilibrand AS, Carlson GD, Palumbo MA, Jones PK, Bohlman HH (1999) Radiculopathy and myelopathy at segments adjacent to the site of a previous anterior cervical arthrodesis. J Bone Joint Surg Am 81:519–528

Hisey MS et al (2014) Multi-center, prospective, randomized, controlled investigational device exemption clinical trial comparing Mobi-C cervical artificial disc to anterior discectomy and fusion in the treatment of symptomatic degenerative disc disease in the cervical spine. Int J Spine Surg 8. https://doi.org/10.14444/1007

Hisey MS et al (2015) Prospective, randomized comparison of cervical total disk replacement versus anterior cervical fusion: results at 48 months follow-up. J Spinal Disord Tech 28:E237–E243. https://doi.org/10.1097/BSD.0000000000000185

Hisey MS, Zigler JE, Jackson R, Nunley PD, Bae HW, Kim KD, Ohnmeiss DD (2016) Prospective, randomized comparison of one-level Mobi-C cervical total disc replacement vs. Anterior cervical discectomy and fusion: results at 5-year follow-up. Int J Spine Surg 10:10. https://doi.org/10.14444/3010

Huppert J et al (2011) Comparison between single- and multilevel patients: clinical and radiological outcomes 2 years after cervical disc replacement. Eur Spine J 20:1417–1426. https://doi.org/10.1007/s00586-011-1722-9

Ishihara H, Kanamori M, Kawaguchi Y, Nakamura H, Kimura T (2004) Adjacent segment disease after anterior cervical interbody fusion. Spine J 4:624–628. https://doi.org/10.1016/j.spinee.2004.04.011

Janssen ME, Zigler JE, Spivak JM, Delamarter RB, Darden BV 2nd, Kopjar B (2015) ProDisc-C total disc replacement versus anterior cervical discectomy and fusion for single-level symptomatic cervical disc disease: seven-year follow-up of the prospective randomized U.S. Food and Drug Administration Investigational Device Exemption Study. J Bone Joint Surg Am 97:1738–1747. https://doi.org/10.2106/JBJS.N.01186

Jawahar A, Cavanaugh DA, Kerr EJ 3rd, Birdsong EM, Nunley PD (2010) Total disc arthroplasty does not affect the incidence of adjacent segment degeneration in cervical spine: results of 93 patients in three prospective randomized clinical trials. Spine J 10:1043–1048. https://doi.org/10.1016/j.spinee.2010.08.014

Kelly MP, Mok JM, Frisch RF, Tay BK (2011) Adjacent segment motion after anterior cervical discectomy and fusion versus Prodisc-c cervical total disk arthroplasty: analysis from a randomized, controlled trial. Spine (Phila Pa 1976) 36:1171–1179. https://doi.org/10.1097/BRS.0b013e3181ec5c7d

Kesman T, Murrey D, Darden B (2012) Single-center results at 7 years of prospective, randomized ProDisc-C total disc arthroplasty versus anterior cervical discectomy and fusion for treatment of one level symptomatic cervical disc disease. Evid Based Spine Care J 3:61–62. https://doi.org/10.1055/s-0032-1328144

Kim SH, Shin HC, Shin DA, Kim KN, Yoon DH (2007) Early clinical experience with the mobi-C disc prosthesis. Yonsei Med J 48:457–464. https://doi.org/10.3349/ymj.2007.48.3.457

Kim SW, Shin JH, Arbatin JJ, Park MS, Chung YK, McAfee PC (2008) Effects of a cervical disc prosthesis on maintaining sagittal alignment of the functional spinal unit and overall sagittal balance of the cervical spine. Eur Spine J 17:20–29. https://doi.org/10.1007/s00586-007-0459-y

Kim SW et al (2009) Comparison of radiographic changes after ACDF versus Bryan disc arthroplasty in single and bi-level cases. Eur Spine J 18:218–231. https://doi.org/10.1007/s00586-008-0854-z

Lafuente J, Casey AT, Petzold A, Brew S (2005) The Bryan cervical disc prosthesis as an alternative to arthrodesis in the treatment of cervical spondylosis: 46 consecutive cases. J Bone Joint Surg (Br) 87:508–512. https://doi.org/10.1302/0301-620X.87B4.15436

Lanman TH, Burkus JK, Dryer RG, Gornet MF, McConnell J, Hodges SD (2017) Long-term clinical and radiographic outcomes of the Prestige LP artificial cervical disc replacement at 2 levels: results from a prospective randomized controlled clinical trial. J Neurosurg Spine 27:7–19. https://doi.org/10.3171/2016.11.SPINE16746

Lee JH, Jung TG, Kim HS, Jang JS, Lee SH (2010) Analysis of the incidence and clinical effect of the heterotopic ossification in a single-level cervical artificial disc replacement. Spine J 10:676–682. https://doi.org/10.1016/j.spinee.2010.04.017

Lee SE, Chung CK, Jahng TA (2012) Early development and progression of heterotopic ossification in cervical total disc replacement. J Neurosurg Spine 16:31–36. https://doi.org/10.3171/2011.8.SPINE11303

Leung C et al (2005) Clinical significance of heterotopic ossification in cervical disc replacement: a prospective multicenter clinical trial. Neurosurgery 57:759–763; discussion 759–763

Leven D, Meaike J, Radcliff K, Qureshi S (2017) Cervical disc replacement surgery: indications, technique, and technical pearls. Curr Rev Musculoskelet Med 10:160–169. https://doi.org/10.1007/s12178-017-9398-3

Li J, Liang L, Ye XF, Qi M, Chen HJ, Yuan W (2013) Cervical arthroplasty with Discover prosthesis: clinical outcomes and analysis of factors that may influence postoperative range of motion. Eur Spine J 22:2303–2309. https://doi.org/10.1007/s00586-013-2897-z

Matsunaga S, Kabayama S, Yamamoto T, Yone K, Sakou T, Nakanishi K (1999) Strain on intervertebral discs after anterior cervical decompression and fusion. Spine (Phila Pa 1976) 24:670–675

McAfee PC, Reah C, Gilder K, Eisermann L, Cunningham B (2012) A meta-analysis of comparative outcomes following cervical arthroplasty or anterior cervical fusion: results from 4 prospective multicenter randomized clinical trials and up to 1226 patients. Spine (Phila Pa 1976) 37:943–952. https://doi.org/10.1097/BRS.0b013e31823da169

McAnany SJ, Overley S, Baird EO, Cho SK, Hecht AC, Zigler JE, Qureshi SA (2014) The 5-year cost-effectiveness of anterior cervical discectomy and fusion and cervical disc replacement: a Markov analysis. Spine (Phila Pa 1976) 39:1924–1933. https://doi.org/10.1097/BRS.0000000000000562

Mehren C et al (2006) Heterotopic ossification in total cervical artificial disc replacement. Spine (Phila Pa 1976) 31:2802–2806. https://doi.org/10.1097/01.brs.0000245852.70594.d5

Miao J, Yu F, Shen Y, He N, Kuang Y, Wang X, Chen D (2014) Clinical and radiographic outcomes of cervical disc replacement with a new prosthesis. Spine J 14:878–883. https://doi.org/10.1016/j.spinee.2013.07.439

Mummaneni PV, Burkus JK, Haid RW, Traynelis VC, Zdeblick TA (2007) Clinical and radiographic analysis of cervical disc arthroplasty compared with allograft fusion: a randomized controlled clinical trial. J Neurosurg Spine 6:198–209. https://doi.org/10.3171/spi.2007.6.3.198

Murrey D, Janssen M, Delamarter R, Goldstein J, Zigler J, Tay B, Darden B (2009) Results of the prospective, randomized, controlled multicenter Food and Drug Administration investigational device exemption study of the ProDisc-C total disc replacement versus anterior discectomy and fusion for the treatment of 1-level symptomatic cervical disc disease. Spine J 9:275–286. https://doi.org/10.1016/j.spinee.2008.05.006

Nabhan A, Ahlhelm F, Shariat K, Pitzen T, Steimer O, Steudel WI, Pape D (2007) The ProDisc-C prosthesis: clinical and radiological experience 1 year after surgery. Spine (Phila Pa 1976) 32:1935–1941. https://doi.org/10.1097/BRS.0b013e31813162d8

Nunley PD et al (2012) Factors affecting the incidence of symptomatic adjacent-level disease in cervical spine after total disc arthroplasty: 2- to 4-year follow-up of 3 prospective randomized trials. Spine (Phila Pa 1976) 37:445–451. https://doi.org/10.1097/BRS.0b013e31822174b3

Nunley PD, Coric D, Frank KA, Stone MB (2018) Cervical disc arthroplasty: current evidence and real-world application. Neurosurgery. https://doi.org/10.1093/neuros/nyx579

Park JH, Roh KH, Cho JY, Ra YS, Rhim SC, Noh SW (2008) Comparative analysis of cervical arthroplasty using mobi-c(r) and anterior cervical discectomy and fusion using the solis(r) -cage. J Korean Neurosurg Soc 44:217–221. https://doi.org/10.3340/jkns.2008.44.4.217

Park DK, Lin EL, Phillips FM (2011) Index and adjacent level kinematics after cervical disc replacement and anterior fusion: in vivo quantitative radiographic analysis. Spine (Phila Pa 1976) 36:721–730. https://doi.org/10.1097/BRS.0b013e3181df10fc

Park JH, Rhim SC, Roh SW (2013) Mid-term follow-up of clinical and radiologic outcomes in cervical total disk replacement (Mobi-C): incidence of heterotopic ossification and risk factors. J Spinal Disord Tech 26:141–145. https://doi.org/10.1097/BSD.0b013e31823ba071

Peng CW, Quirno M, Bendo JA, Spivak JM, Goldstein JA (2009) Effect of intervertebral disc height on postoperative motion and clinical outcomes after Prodisc-C cervical disc replacement. Spine J 9:551–555. https://doi.org/10.1016/j.spinee.2009.03.008

Peng CW et al (2011) Intermediate results of the Prestige LP cervical disc replacement: clinical and radiological analysis with minimum two-year follow-up. Spine (Phila Pa 1976) 36:E105–E111. https://doi.org/10.1097/BRS.0b013e3181d76f99

Phillips FM et al (2009) Cervical disc replacement in patients with and without previous adjacent level fusion surgery: a prospective study. Spine (Phila Pa 1976) 34:556–565. https://doi.org/10.1097/BRS.0b013e31819b061c

Phillips FM et al (2013) A prospective, randomized, controlled clinical investigation comparing PCM cervical disc arthroplasty with anterior cervical discectomy and fusion. 2-year results from the US FDA IDE clinical trial. Spine (Phila Pa 1976) 38:E907–E918. https://doi.org/10.1097/BRS.0b013e318296232f

Phillips FM, Geisler FH, Gilder KM, Reah C, Howell KM, McAfee PC (2015) Long-term outcomes of the US FDA IDE prospective, randomized controlled clinical trial comparing PCM cervical disc arthroplasty with anterior cervical discectomy and fusion. Spine (Phila Pa 1976) 40:674–683. https://doi.org/10.1097/BRS.0000000000000869

Pickett GE, Mitsis DK, Sekhon LH, Sears WR, Duggal N (2004) Effects of a cervical disc prosthesis on segmental and cervical spine alignment. Neurosurg Focus 17:E5

Pickett GE, Sekhon LH, Sears WR, Duggal N (2006) Complications with cervical arthroplasty. J Neurosurg Spine 4:98–105. https://doi.org/10.3171/spi.2006.4.2.98

Pimenta L, McAfee PC, Cappuccino A, Bellera FP, Link HD (2004) Clinical experience with the new artificial cervical

PCM (Cervitech) disc. Spine J 4:315S–321S. https://doi.org/10.1016/j.spinee. 2004. 07.024

Pimenta L, McAfee PC, Cappuccino A, Cunningham BW, Diaz R, Coutinho E (2007) Superiority of multilevel cervical arthroplasty outcomes versus single-level outcomes: 229 consecutive PCM prostheses. Spine (Phila Pa 1976) 32:1337–1344. https://doi.org/10.1097/BRS.0b013e318059af12

Pointillart V, Castelain JE, Coudert P, Cawley DT, Gille O, Vital JM (2017) Outcomes of the Bryan cervical disc replacement: fifteen year follow-up. Int Orthop. https://doi.org/10.1007/s00264-017-3745-2

Porchet F, Metcalf NH (2004) Clinical outcomes with the Prestige II cervical disc: preliminary results from a prospective randomized clinical trial. Neurosurg Focus 17:E6

Puttlitz CM, Rousseau MA, Xu Z, Hu S, Tay BK, Lotz JC (2004) Intervertebral disc replacement maintains cervical spine kinetics. Spine (Phila Pa 1976) 29:2809–2814

Quan GM, Vital JM, Hansen S, Pointillart V (2011) Eight-year clinical and radiological follow-up of the Bryan cervical disc arthroplasty. Spine (Phila Pa 1976) 36:639–646. https://doi.org/10.1097/BRS.0b013e3181dc9b51

Radcliff K, Zigler J, Zigler J (2015) Costs of cervical disc replacement versus anterior cervical discectomy and fusion for treatment of single-level cervical disc disease: an analysis of the Blue Health Intelligence database for acute and long-term costs and complications. Spine (Phila Pa 1976) 40:521–529. https://doi.org/10.1097/BRS.0000000000000822

Radcliff K, Coric D, Albert T (2016a) Five-year clinical results of cervical total disc replacement compared with anterior discectomy and fusion for treatment of 2-level symptomatic degenerative disc disease: a prospective, randomized, controlled, multicenter investigational device exemption clinical trial. J Neurosurg Spine 25:213–224. https://doi.org/10.3171/2015.12.SPINE15824

Radcliff K, Lerner J, Yang C, Bernard T, Zigler JE (2016b) Seven-year cost-effectiveness of ProDisc-C total disc replacement: results from investigational device exemption and post-approval studies. J Neurosurg Spine 24:760–768. https://doi.org/10.3171/2015.10.SPINE15505

Reitman CA, Hipp JA, Nguyen L, Esses SI (2004) Changes in segmental intervertebral motion adjacent to cervical arthrodesis: a prospective study. Spine (Phila Pa 1976) 29:E221–E226

Ren X, Wang W, Chu T, Wang J, Li C, Jiang T (2011) The intermediate clinical outcome and its limitations of Bryan cervical arthroplasty for treatment of cervical disc herniation. J Spinal Disord Tech 24:221–229. https://doi.org/10.1097/BSD.0b013e3181e9f309

Riew KD, Buchowski JM, Sasso R, Zdeblick T, Metcalf NH, Anderson PA (2008) Cervical disc arthroplasty compared with arthrodesis for the treatment of myelopathy. J Bone Joint Surg Am 90:2354–2364. https://doi.org/10.2106/JBJS.G.01608

Riina J, Patel A, Dietz JW, Hoskins JS, Trammell TR, Schwartz DD (2008) Comparison of single-level cervical fusion and a metal-on-metal cervical disc replacement device. Am J Orthop (Belle Mead NJ) 37:E71–E77

Robertson JT, Metcalf NH (2004) Long-term outcome after implantation of the Prestige I disc in an end-stage indication: 4-year results from a pilot study. Neurosurg Focus 17:E10

Robertson JT, Papadopoulos SM, Traynelis VC (2005) Assessment of adjacent-segment disease in patients treated with cervical fusion or arthroplasty: a prospective 2-year study. J Neurosurg Spine 3:417–423. https://doi.org/10.3171/spi.2005.3.6.0417

Rozankovic M, Marasanov SM, Vukic M (2017) Cervical disk replacement with Discover versus fusion in a single-level cervical disk disease: a prospective single-center randomized trial with a minimum 2-year follow-up. Clin Spine Surg 30:E515–E522. https://doi.org/10.1097/BSD.0000000000000170

Ryu KS, Park CK, Jun SC, Huh HY (2010) Radiological changes of the operated and adjacent segments following cervical arthroplasty after a minimum 24-month follow-up: comparison between the Bryan and Prodisc-C devices. J Neurosurg Spine 13:299–307. https://doi.org/10.3171/2010.3.SPINE09445

Sasso RC, Smucker JD, Hacker RJ, Heller JG (2007a) Artificial disc versus fusion: a prospective, randomized study with 2-year follow-up on 99 patients. Spine (Phila Pa 1976) 32:2933–2940; discussion 2941–2932. https://doi.org/10.1097/BRS.0b013e31815d0034

Sasso RC, Smucker JD, Hacker RJ, Heller JG (2007b) Clinical outcomes of BRYAN cervical disc arthroplasty: a prospective, randomized, controlled, multicenter trial with 24-month follow-up. J Spinal Disord Tech 20:481–491. https://doi.org/10.1097/BSD.0b013e3180310534

Sasso RC, Anderson PA, Riew KD, Heller JG (2011) Results of cervical arthroplasty compared with anterior discectomy and fusion: four-year clinical outcomes in a prospective, randomized controlled trial. J Bone Joint Surg Am 93:1684–1692. https://doi.org/10.2106/JBJS.J.00476

Sasso WR, Smucker JD, Sasso MP, Sasso RC (2017) Long-term clinical outcomes of cervical disc arthroplasty: a prospective, randomized, controlled trial. Spine (Phila Pa 1976) 42:209–216. https://doi.org/10.1097/BRS.0000000000001746

Sekhon LH (2003) Cervical arthroplasty in the management of spondylotic myelopathy. J Spinal Disord Tech 16:307–313

Sekhon LH, Sears W, Duggal N (2005) Cervical arthroplasty after previous surgery: results of treating 24 discs in 15 patients. J Neurosurg Spine 3:335–341. https://doi.org/10.3171/spi.2005.3.5.0335

Shang Z, Zhang Y, Zhang D, Ding W, Shen Y (2017) Clinical and radiological analysis of Bryan cervical artificial disc replacement for "skip" multi-segment cervical spondylosis: long-term follow-up results. Med Sci Monit 23:5254–5263

Shi S, Zheng S, Li XF, Yang LL, Liu ZD, Yuan W (2016) Comparison of 2 zero-profile implants in the treatment of single-level cervical spondylotic myelopathy: a preliminary clinical study of Cervical Disc Arthroplasty versus fusion. PLoS One 11:e0159761. https://doi.org/10.1371/journal.pone.0159761

Shim CS, Lee SH, Park HJ, Kang HS, Hwang JH (2006) Early clinical and radiologic outcomes of cervical arthroplasty with Bryan Cervical Disc prosthesis. J Spinal Disord Tech 19:465–470. https://doi.org/10.1097/01.bsd.0000211235.76093.6b

Skeppholm M, Lindgren L, Henriques T, Vavruch L, Lofgren H, Olerud C (2015) The Discover artificial disc replacement versus fusion in cervical radiculopathy – a randomized controlled outcome trial with 2-year follow-up. Spine J 15:1284–1294. https://doi.org/10.1016/j.spinee.2015.02.039

Smith GW, Robinson RA (1958) The treatment of certain cervical-spine disorders by anterior removal of the intervertebral disc and interbody fusion. J Bone Joint Surg Am 40-A:607–624

Steinmetz MP, Patel R, Traynelis V, Resnick DK, Anderson PA (2008) Cervical disc arthroplasty compared with fusion in a workers' compensation population. Neurosurgery 63:741–747; discussion 747. https://doi.org/10.1227/01.NEU.0000325495.79104.DB

Suchomel P, Jurak L, Benes V 3rd, Brabec R, Bradac O, Elgawhary S (2010) Clinical results and development of heterotopic ossification in total cervical disc replacement during a 4-year follow-up. Eur Spine J 19:307–315. https://doi.org/10.1007/s00586-009-1259-3

Traynelis VC (2004) The Prestige cervical disc replacement. Spine J 4:310S–314S. https://doi.org/10.1016/j.spinee.2004.07.025

Tu TH, Wu JC, Huang WC, Guo WY, Wu CL, Shih YH, Cheng H (2011) Heterotopic ossification after cervical total disc replacement: determination by CT and effects on clinical outcomes. J Neurosurg Spine 14:457–465. https://doi.org/10.3171/2010.11.SPINE10444

Turel MK, Kerolus MG, Adogwa O, Traynelis VC (2017) Cervical arthroplasty: what does the labeling say? Neurosurg Focus 42:E2. https://doi.org/10.3171/2016.11.FOCUS16414

Upadhyaya CD et al (2012) Analysis of the three United States Food and Drug Administration investigational device exemption cervical arthroplasty trials. J Neurosurg Spine 16:216–228. https://doi.org/10.3171/2011.6.SPINE10623

Vaccaro A et al (2013) Clinical outcomes with selectively constrained SECURE-C cervical disc arthroplasty: two-year results from a prospective, randomized, controlled, multicenter investigational device exemption study. Spine (Phila Pa 1976) 38:2227–2239. https://doi.org/10.1097/BRS.0000000000000031

Walraevens J, Demaerel P, Suetens P, Van Calenbergh F, van Loon J, Vander Sloten J, Goffin J (2010) Longitudinal prospective long-term radiographic follow-up after treatment of single-level cervical disk disease with the Bryan Cervical Disc. Neurosurgery 67:679–687; discussion 687. https://doi.org/10.1227/01.NEU.0000377039.89725.F3

Wang Y et al (2008) Clinical outcomes of single level Bryan cervical disc arthroplasty: a prospective controlled study. Zhonghua Wai Ke Za Zhi 46:328–332

Wigfield C, Gill S, Nelson R, Langdon I, Metcalf N, Robertson J (2002a) Influence of an artificial cervical joint compared with fusion on adjacent-level motion in the treatment of degenerative cervical disc disease. J Neurosurg 96:17–21

Wigfield CC, Gill SS, Nelson RJ, Metcalf NH, Robertson JT (2002b) The new Frenchay artificial cervical joint: results from a two-year pilot study. Spine (Phila Pa 1976) 27:2446–2452. https://doi.org/10.1097/01.BRS.0000032365.21711.5E

Xing D, Ma XL, Ma JX, Wang J, Ma T, Chen Y (2013) A meta-analysis of cervical arthroplasty compared to anterior cervical discectomy and fusion for single-level cervical disc disease. J Clin Neurosci 20:970–978. https://doi.org/10.1016/j.jocn.2012.03.046

Yang S et al (2008) Early and intermediate follow-up results after treatment of degenerative disc disease with the Bryan cervical disc prosthesis: single- and multiple-level. Spine (Phila Pa 1976) 33:E371–E377. https://doi.org/10.1097/BRS.0b013e31817343a6

Yi S, Lee DY, Ahn PG, Kim KN, Yoon DH, Shin HC (2009) Radiologically documented adjacent-segment djegeneration after cervical arthroplasty: characteristics and review of cases. Surg Neurol 72:325–329; discussion 329. https://doi.org/10.1016/j.surneu.2009.02.013

Yi S et al (2010) Difference in occurrence of heterotopic ossification according to prosthesis type in the cervical artificial disc replacement. Spine (Phila Pa 1976) 35:1556–1561. https://doi.org/10.1097/BRS.0b013e3181c6526b

Yin S, Yu X, Zhou S, Yin Z, Qiu Y (2013) Is cervical disc arthroplasty superior to fusion for treatment of symptomatic cervical disc disease? A meta-analysis. Clin Orthop Relat Res 471:1904–1919. https://doi.org/10.1007/s11999-013-2830-0

Yoon DH, Yi S, Shin HC, Kim KN, Kim SH (2006) Clinical and radiological results following cervical arthroplasty. Acta Neurochir 148:943–950. https://doi.org/10.1007/s00701-006-0805-6

Zhang X et al (2012) Randomized, controlled, multicenter, clinical trial comparing BRYAN cervical disc arthroplasty with anterior cervical decompression and fusion in China. Spine (Phila Pa 1976) 37:433–438. https://doi.org/10.1097/BRS.0b013e31822699fa

Zigler JE, Delamarter R, Murrey D, Spivak J, Janssen M (2013) ProDisc-C and anterior cervical discectomy and fusion as surgical treatment for single-level cervical symptomatic degenerative disc disease: five-year results of a Food and Drug Administration study. Spine (Phila Pa 1976) 38:203–209. https://doi.org/10.1097/BRS.0b013e318278eb38

Adjacent-Level Disease: Fact and Fiction

49

Jonathan Parish and Domagoj Coric

Contents

Introduction 885

Historical Perspective 886

Motion Preservation Devices 887

Motion Preservation Effect 887

Conclusions 889

References 890

Keywords

Adjacent disease · Arthroplasty · Cervical fusion · Disc replacement · Motion preservation

J. Parish (✉)
Department of Neurological Surgery, Carolinas Medical Center, Charlotte, NC, USA

Carolina Neurosurgery and Spine Associates, Charlotte, NC, USA
e-mail: john.parrish@cnsa.com

D. Coric
Department of Neurological Surgery, Carolinas Medical Center and Carolina Neurosurgery and Spine Associates, Charlotte, NC, USA
e-mail:
Domagoj.Coric@CNSA.com

© Springer Nature Switzerland AG 2021
B. C. Cheng (ed.), *Handbook of Spine Technology*,
https://doi.org/10.1007/978-3-319-44424-6_82

Introduction

The topics of adjacent segment (AS) degeneration and disease have been increasingly discussed with the development and adoption of motion preserving devices. AS degeneration is defined as new degenerative radiographic changes at a spinal level immediately above or below surgically treated levels. When this degeneration is associated with clinical symptoms, including radiculopathy, myelopathy, or mechanical instability, then the appropriate terminology is AS disease. Controversy exists as to whether AS disease is primarily due to the natural progression of an underlying degenerative process or an accelerated process due to increased forces placed on adjacent segments following fusion surgery. In theory, motion preserving devices would eliminate or significantly decrease any accelerated degeneration related to fusion and increased biomechanical stress. Both clinical and laboratory studies have addressed

AS degeneration and disease as well as the factors leading to their development. In this chapter, we will review these studies as well as examine the evidence basis regarding the effect of motion preservation technology on the incidence of AS disease.

Historical Perspective

The etiology of AS disease has been controversial with some studies suggesting that fusion places significantly increased stress on adjacent segments while others arguing that AS disease is primarily due to the natural progression of underlying disease. Furthermore, there is debate over whether motion preservation devices with their ability to eliminate increased forces on the adjacent discs can decrease AS disease.

Historically, the annual incidence of AS disease following fusion is generally reported to range from 1.5% to 4.5% (Bohlman et al. 1993; Cauthen et al. 1998; Gore and Sepic 1998; Hilibrand et al. 1999). Hilibrand et al. (1999) reported on 409 total procedures in 374 patients followed for 10 years. In this series, symptomatic AS disease was defined as a combination of new radicular or myelopathy symptoms referable to an adjacent degenerated level on two consecutive office visits based on chart review and surgical records (a nonvalidated outcome measure). The annual incidence was 2.9% per year over the 10-year study period (range, 0.0–4.8% per year). In this frequently cited study, only 27 patients (6.6%) had adjacent level surgery with an annual adjacent level reoperation rate of 0.7%. A similar study by Goffin et al. (2004) evaluated long term outcomes in 180 patients with a mean follow-up of 30.9 months. 92% of patients had radiographic evidence of increased degeneration at long-term follow-up. Interestingly, age and number of levels fused showed no correlation with degeneration (Spearmen $r_s = -0.033$, $P = 0.660$ and Spearman $r_s = -0.011$, $P = 0.879$, respectively), but the length of time after operation was correlated with degeneration (Spearman $r_s = 0.156$, $P = 0.036$). This suggests a multifactorial etiology to AS

degeneration given such a high incidence after fusion surgery, but the correlation with length of time after operation suggestive of natural progression.

Though these studies addressed the incidence of AS disease, they did not provide a definitive etiology. Biomechanical studies by Eck et al. (2002) were performed to evaluate the intradiscal pressure after cervical fusion. In cadaveric specimens, the authors found that increased intradiscal pressure resulted with normal range of motion after fusion. Increased segmental motion adjacent to fusion segment resulted in increased pressures. They were unable to make conclusions regarding increased intradiscal pressure and effect on normal degenerative changes. Additional biomechanical studies using a finite element model of the cervical spine by Lopez-Espina et al. (2006) showed significant increases in stress of up to 96% on the annulus, nucleus, and endplates of adjacent levels in fused (single and double level) versus normal cervical spines. The authors argued that increased rotation and stress may explain the disc degeneration and osteophyte formation after fusion.

The counter argument for natural progression of spinal degeneration over time is also well supported with radiographic and clinical data. Matsumoto et al. (1998) performed 497 MRI on asymptomatic subjects and found a significant occurrence of degenerative changes and age. In their initial study, 17% of men and 12% of women in their 20s had evidence of degenerative changes compared to 86% of men and 89% of women over 60 years of age. A follow-up of 223 of those patients showed progression of degenerative changes in 81.1% of patients with only 34.1% developing clinical symptoms. These studies suggest a rate of natural progression with age for AS degeneration. Similarly, Gore et al. (2002) followed 159 patients for 10 years with asymptomatic cervical disease. Radiographic degeneration was seen in 72 patients at initial imaging and degeneration progressed in 70 (97.2%) of these patients with15% of patients developing pain over the 10-year study period. These studies identify a clear progression of degeneration over time. In regard to the effect of cervical surgery on the rate of AS degeneration, Lunsford et al. (1980) reported on 253 patients who underwent anterior cervical discectomy with and without fusion (ACD and ACDF). There was no difference in symptomatic relief and recurrence of symptoms. Further, there was no difference in subsequent development of AS degeneration requiring re-operation.

Motion Preservation Devices

Given the rate of AS degeneration and need for further surgery following fusion, motion preserving devices were developed to theoretically reduce effects of AS disease. Initially developed for the lumbar spine, artificial disc replacement has been performed to prevent loss of vertebral interspace height and reduce pain while maintaining motion. Cadaver studies by Wigfield et al. (2003) showed that artificial disc resulted in reduced stresses in the annulus of neighboring cervical segments compared to simulated fusion. These studies supported the theory that motion preservation resulted in less adjacent segment mechanical stress compared to fusion. The earliest clinical reports of disc replacement in the cervical spine were reported by Fernstrom in 1966. His device was used in a series of 32 patients with 74 cervical disc prosthesis reported by Reitz and Joubert (1964) with good results in all patients and preservation of mobility. The earliest reports of AS degeneration after artificial disc replacement were reported by Cummins et al. (1988). In 18 patients with 5-year follow-up, there was no reported adjacent joint degeneration and motion was preserved on flexion and extension x-ray films.

Motion Preservation Effect

Early US Investigational Device Exemption (IDE) trials of artificial disc replacement showed that results were equivalent in regard to neurologic outcome and surgical success, but data regarding AS degeneration was more difficult to assess given the short follow-up. Heller et al.

(2009) reported on 24-month outcome for BRYAN cervical disc (Medtronic Sofamor Danek, Memphis, TN). 242 patients were randomized to the BRYAN cervical disc and 221 were in the control ACDF group. The rate of secondary surgical procedures at the treated level was 2.5% in the total disc replacement (TDR) patients and 3.6% in the fusion group though this was not statistically significant. Interestingly, composite overall success was achieved in 82.6% artificial disc patients and only 72.7% of fusion patients ($p = 0.010$). Another randomized, controlled IDE study by Mummaneni et al. (2007) enrolled 276 patients to arthroplasty with PRESTIGE ST cervical Disc System (Medtronic Sofamor Danek, Memphis, TN) and 265 patients to ACDF with 24-month follow-up. The groups showed similar improvement in validated outcome measures (NDI and VAS arm/neck pain scores), but the composite overall success rate was significantly higher at 24 months in the arthroplasty group than ACDF control group (79.3% vs. 67.8%, $p = 0.0053$). The reoperation rate in the arthroplasty group was lower (1.1% vs. 3.4%, respectively, $p = 0.0492$, log-rank test) for AS disease than the control group. Though these 2-year outcomes showed equivalence in this noninferiority statistical design and the effect on AS degeneration was promising, long-term studies of the effect on AS disease with motion preservation were still needed.

One of the earliest attempts to analyze AS disease following cervical artificial disc replacement was performed by Jawahar et al. (2010). In this study, a total of 93 patients were enrolled in 3 prospective randomized trials of artificial cervical discs. Patients showed equivalence in symptomatic relief (71% in TDA vs. 73.5% in ACDF). At last follow-up (median 36.4 months), 15% of patients with ACDF and 18% of TDA had clinical and radiographic AS disease which was not statistically different. A follow-up study by Nunley et al. (2012) included 170 patients with 3- and 4-year follow-up after treatment for 1 and 2-level cervical disc degeneration with cervical artificial disc or ACDF. AS degeneration and disease was reported in 16.5% of patients during follow-up ranging from 32 to 54 months (median 38 months) though only 4.1% of patients required a second surgery at adjacent level. At 4 years, adjacent level degeneration-free rate was 76.7% in artificial disc group and 78.3% in the ACDF group, suggesting no difference in development of AS disease after arthroplasty.

Another study by Maldonado et al. (2011) prospectively studied 190 patients with a minimum of 3-year follow-up after ACDF or artificial disc to evaluate the incidence of AS degeneration. Radiographic evidence of AS degeneration was defined as new or enlarging anterior osteophytes or new or increased calcification of the anterior longitudinal ligament. AS degeneration was found in 10.5% of patients in the ACDF group and in 8.8% of patients in the arthroplasty group though this did not reach clinical significance ($p = 0.69$). This study did not address AS disease requiring operative intervention.

Another prospective, randomized IDE trial by Davis et al. (2015) followed 291 patients for 48 months after arthroplasty with MOBI-C cervical artificial disc (LDR Medical; Troyes, France) and ACDF. At 4-year follow-up, TDR group had significantly less AS degeneration than the ACDF group (41.5% vs. 89.5%, respectively, $p < 0.0001$). Re-operation at the index level was significantly lower for TDR group (4.0%) versus ACDF group (15.2%, $p < 0.0001$). Indication for TDR group re-operation was stenosis, device migration, poor endplate fixation, and persist neck and/or shoulder pain. The most common indication for re-operation in ACDF group was symptomatic pseudarthrosis. This study also did not address AS disease.

Studies addressing AS re-operation rate provide a more objective assessment of the effect of motion preservation on adjacent levels. In a single institution study by Coric et al. (2010) with 3 separate prospective randomized trials for artificial cervical discs, lower re-operation rates were observed for arthroplasty than fusion. 90 patients were randomized to ACDF (37 patients) or cervical disc arthroplasty (53 patients) with 2-year minimum follow-up (mean 38 months). Clinical success, defined as a composite measure of five separate components, was significantly higher in the arthroplasty group (85%) compared to the ACDF group (70%, $p = 0.035$). Adjacent level disease requiring re-operation occurred at a rate of 1.7% (0.5%/year) in the arthroplasty group which was lower (but not statistically significant) than the rate of 8.1% (2.6%/year) in the ACDF group. A multicenter randomized US FDA IDE trial also by Coric et al. (2011) addressed radiographic adjacent-level changes and re-operation rate. A total of 269 patients were enrolled with 135 patients randomized to TDR with the Kineflex-C disc and 133 to ACDF. There were no preoperative differences in the radiographic changes at adjacent levels. Radiographic deterioration was graded as none, mild, moderate, or severe. At 2-year follow-up, severe adjacent-level deterioration was evident in 24.8% of ACDF patients and only 9% in TDR group ($p < 0.0001$). Index-level re-operation rate was similar (5.0% TDR vs. 6.1% ACDF) and there was no significant difference in AS re-operation rate (7.6% for TDR and 6.1% for ACDF).

Given the low incidence of AS disease requiring re-operation, long term studies and large number of subjects are required to adequately assess the potential positive effect of motion preservation. A single institution study by Coric et al. included two devices (Bryan Disc or Kineflex/C) and enrolled 41 patients in CDR and 33 patients in ACDF control. A total of 63 patients had a minimum of 4-year follow-up. Both arthroplasty and ACDF patients showed a low rate of index level re-operation rate (2.4% vs. 0%, respectively) and adjacent level re-operation (4.9% vs. 3.0%, respectively) without statistically significant differences. Two studies have presented 7-year follow-up on arthroplasty outcomes. Vaccaro et al. (2013) reported a US FDA IDE trial of the SECURE-C device. At 24 months, patients in the arthroplasty group had statistically lower index level re-operations than ACDF (2.5% vs. 9.7%, respectively) and similar AS re-operation rate at 2-years (1.7% vs. 1.4% respectively). Recently, follow-up 7-year data was released that showed very significant differences in index and adjacent level re-operation rates. Index level re-operation rate was significantly lower in TDR group (4.2% vs. 15.3%). For AS re-operation rates, the incidence for cervical TDR was 4.2% compared to 16.0% in the ACDF group. Another long-term

7-year study by Burkus et al. (2014) reported on the efficacy of cervical disc replacement with Prestige Disc (Medtronic, Memphis, TN). 541 patients were randomized at 31 investigational sites to TDR or ACDF. At 84 months, surgery at the index level were lower for TDR than ACDF (4.8% vs. 13.7%, $p < 0.001$) as well as at adjacent levels (4.6% vs. 11.9%, $p = 0.008$).

Long term results have also been observed to be significant for 2 level cervical disc arthroplasty compared to ACDF. Radcliff et al. (2015) reported on 5-year results of TDR and ACDF for 2-level degenerative cervical disease. A total of 225 patients underwent 2-level TDR and 105 patients underwent 2 level ACDF. At 60-month follow-up, there were significantly fewer second surgeries in TDR group than in the ACDF group (71% vs. 21.0%, $p = 0.0006$). In regard to AS degeneration, there also were significantly less AS degeneration in TDR group than in the ACDF group (50.7% vs. 90.5%, $p < 0.0001$). Furthermore, there were significantly fewer AS reoperations in TDR group than in the ACDF group (3.1% vs. 11.4%, $p = 0.0004$). For TDR, the annual rate of AS re-operation was 0.6%/year which is similar to the actual re-operation rate (0.66%/year) reported by Hillebrand.

Radcliff et al. (2015) also reported on a "real-world" application of arthroplasty versus ACDF. A retrospective, matched cohort analysis of patients enrolled in a Blue Cross Plan assessed a "real-world" population with symptomatic cervical disease treated with TDR or ACDF. A total of 6635 patients in the ACDF group and 327 patients in the cervical TDR group. At 36 months, the incidence of reoperation at index level in TDR group was 5.7% compared to 10.5% in ACDF group ($p = 0.0214$). Further, AS re-operation rate was significantly lower for cervical TDR group compared to ACDF (3.1% vs. 11.4%, respectively). This study was performed outside of randomized trials and therefore represents "real world" outcomes supporting a lower incidence of index and adjacent level re-operation after cervical TDR than ACDF. Interestingly, this study also showed a significant reduction in all costs at 2 years of 12% in the TDR group ($34, 979 vs. ACDF $39,820).

Two meta-analysis have also addressed AS disease after cervical arthroplasty and ACDF. Upadhyaya et al. (2012) included 3 randomized, multicenter, US FDA IDE studies. A total of 621 patients received an artificial disc and 592 patients were treated with ACDF. At 24 months, 1098 patients were available for follow up. The rate of secondary surgery at the index level was significantly lower for arthroplasty with an RR of 0.44 (95% CI 0.26–0.77, $p = 0.004$, $I^2 = 0\%$). There was also a significant reduction in the adjacent-level reoperation risk favoring arthroplasty with an RR of 0.460 (95% CI 0.229–0.926, $p = 0.030$, $I^2 = 2.9\%$). McAfee et al. (2012) meta-analysis of the 3 FDA-approved TDR IDE studies above and PCM cervical disc (NuVasive Inc., San Diego, CA). A total of 1226 patients had a with minimum 2-year follow-up. Overall survivorship was defined as the absence of revision, reoperation, supplemental fixation, or device removal within 24-month follow-up period. Survivorship was achieved in 96.6% of arthroplasty patients (804 of 832) and 93.4% of ACDF patients (725 of 776). The difference in proportions was 3.2% (95% CI:1.1–5.3%, $P = 0.004$), suggesting that arthroplasty is superior to ACDF in regard to secondary surgical procedure. Unfortunately, this meta-analysis did not specifically address AS re-operation rate.

Conclusions

AS degeneration leading to re-operation is a multifactorial process. Factors contributing to the etiology of this process include: (a) the natural history of the underlying degenerative disease, (b) surgical technique, e.g., minimally invasive, muscle, and ligament sparing versus open procedures, (c) surgical decision-making, e.g., single versus multilevel surgery, (d) surgical procedure, i.e., fusion versus decompression alone versus arthroplasty, (e) patient specific factors such as overall sagittal balance. Due to inherently low incidence of AS re-operation following cervical spine surgery ($<1\%$), long-term follow-up and/or large patient numbers are needed to demonstrate statistically significant differences between procedures such as arthroplasty and fusion. Studies aim at detecting differences with only 2-year follow-up with less than several thousand patients are simply not powered to show statistically significant differences. Biomechanical studies have indicated cervical arthroplasty puts less stress on adjacent segments compared to fusion. Some prospective, randomized clinical studies indicate that arthroplasty decreases the rate of AS degeneration. Limited studies with long-term follow-up also support that arthroplasty may lead to less subsequent surgical intervention at index and adjacent segments. But continued long term data is required to confirm that this trend remains significant.

References

Bohlman HH, Emery SE, Goodfellow DB, Jones PK (1993) Robinson anterior cervical discectomy and arthrodesis for cervical radiculopathy. Long-term follow-up of one hundred and twenty-two patients. J Bone Joint Surg Am 75:1298–1307

Burkus JK, Traynelis VC, Haid RW, Mummaneni PV (2014) Clinical and radiographic analysis of an artificial cervical disc: 7-year follow-up from the prestige prospective randomized controlled clinical trial. J Neurosurg Spine 21:516–528

Cauthen JC, Kinard RE, Vogler JB, Jackson DE, DePaz OB, Hunter OL, Wasserburger LB, Williams VM (1998) Outcome analysis of noninstrumented anterior cervical discectomy and interbody fusion in 348 patients. Spine 23:188–192

Coric D, Cassis J, Carew JD, Bolets MO (2010) Prospective study of cervical arthroplasty in 98 patients involved in 1 of 3 separate investigation device exemption studies from a single investigation site with a minimum 2-year follow-up. J Neurosurg Spine 13:715–721

Coric D, Nunley PD, Guyer RD, Musante D, Carmody CN, Gordon CR, Lauryssen C, Ohnmeiss DD, Med D, Boltes MO (2011) A prospective, randomized, multicenter study of cervical arthroplasty: 269 patients from the Kineflex/C artificial disc investigational device exemption study with a minimum of 2-year follow up. J Neurosurg Spine 15:348–358

Coric D, Kim PK, Clemente JD, Boltes MO, Nussbaum M, James S (2013) Prospective randomized study of cervical arthroplasty and anterior cerical discectomy and fusion with long-term follow-up: results in 74 patients from a single site. J Neurosurg Spine 18:36–42

Cummins BH, Robertson JT, Gill SS (1988) Surgical experience with an implanted artificial cervical joint. J Neurosurg 88:943–948

Davis RJ, Nunley PD, Kim KD, Hisey MS, Jackson RJ, Bae HW, Hoffman GA, Gaede SE, Danielson GO, Gordon C, Stone MB (2015) Two-level total disc replacement with Mobi-C cervical artificial disc versus anterior discectomy and fusion: a prospective, randomized, controlled multicenter clinical trial with 4-year follow-up results. J Neurosurg Spine 22:15–25

Eck JC, Humhreys SC, Lim TH, Jeong ST, Kim JG, Hodges SD, An HS (2002) Biomechanical study on the effect of cervical spine fusion on adjacent-level intradiscal pressure and segmental motion. Spine 27:2431–2434

Fernstrom U (1966) Arthroplasty with intercorporal endoprosthesis in herniated disc and in painful disc. Acta Chir Scand Suppl 355:154–159

Goffin J, Geusens E, Vantomme N, Quints E, Waerzeggers Y, Depreitere B, Van Calenbergh F, van Loon J (2004) Long-term follow-up after interbody fusion of the cervical spine. J Spinal Disord Tech 17:79–85

Gore DR, Sepic SB (1998) Anterior discectomy and fusion for painful cervical disc disease. A report of 50 patients with an average follow-up of 21 years. Spine 23:2047–2051

Heller JG, Sasso RC, Papadopoulos SM, Anderson PA, Fessler RG, Hacker RJ, Coric D, Cauthen JC, Riew DK (2009) Comparison of BRYAN cervical disc arthroplasty with anterior cervical decompression and fusion. Spine 34:101–107

Hilibrand AS, Carlson GD, Palumbo MA, Jones PK, Bohlman HH (1999) Radiculopathy and myelopathy at segments adjacent to the site of a previous anterior cervical arthrodesis. J Bone Joint Surg 81:519–528

Javedan SP, Dickman CA (1999) Cause of adjacent-segment disease after spinal fusion. Lancet 354:530–531

Lopez-Espina CG, Amirouche F, Havalad V (2006) Multilevel cervical fusion and its effect on disc degeneration and osteophyte formation. Spine 31:972–978

Lunsford LD, Bissonette DJ, Jannetta PJ, Sheptak PE, Zorub DS (1980) Anterior surgery for cervical disease. Part 1: treatment of lateral cervical disc herniation in 253 cases. J Neurosurg 53:1–11

Maldonado CV, Paz RC, Martin CB (2011) Adjacent-level degeneration after cervical arthroplasty versus fusion. Eur Spine J 20:S403–S407

Matsumoto M, Fujimura Y, Suzuki N, Nishi Y, Nakamura M, Yabe Y, Shiga H (1998) MRI of cervical intervertebral discs in asymptomatic subjects. J Bone Joint Surg Br 80:19–24

McAfee PC, Reah C, Gilder K, Eisermann L, Cunningham B (2012) A meta-analysis of comparative outcomes following cervical arthroplasty or anterior cervical fusion. Spine 37:943–952

Mummaneni PV, Burkus JK, Haid RW, Traynelis VC, Zdeblick TA (2007) Clinical and radiographic analysis of cervical disc arthroplasty compared with allograft fusion: a randomized controlled clinical trial. J Neurosurg Spine 6:198–209

Nunley PD, Jawahar A, Kerr EJ, Gordon CJ, Cavanaugh DA, Birdson EM, Stocks M, Danielson G (2012)

Factors affecting the incidence of symptomatic adjacent-level disease in cervical spine after total disc arthroplasty. Spine 37:445–451

Okada E, Matsumoto M, Ichihara D, Chiba K, Toyama Y, Fujiwara H, Momoshima S, Nishiwaki Y, Hashimoto T, Ogawa J, Watanabe M, Takahata T (2009) Aging of the ervical spine in healthy volunteers: a 10-year longitudinal magnetic resonance imaging study. Spine 34:706–712

Radcliff K, Zigler J, Zigler J (2015) Costs of cervical disc replacement versus anterior cervical discectomy and fusion for treatment of single-level cervical disc disease: an analysis of the blue health intelligence database for acute and long-term costs and complications. Spine 40:521–529

Radcliff K, Coric D, Albert T (2016) Five-year clinical results of cervical total disc replacement compared with anterior discectomy and fusion for treatment of 2-level symptomatic degenerative disc disease: a prospective, randomized, controlled, multicenter investigational device exemption clinical trial. J Neurosurg Spine 25:213–224

Reitz H, Joubert MJ (1964) Intractable headache and cervico-brachialgia treated by complete replacement of cervical intervertebral dis with a metal prosthesis. S Afr Med J 38:881–884

Saavedra-Pozo FM, Deusdara RAM, Benzel ED (2015) Adjacent segment disease perspective and review of the literature. Ochsner J 14:78–83

Upadhyaya CD, Wu JC, Haid RW, Traynelis VC, Tay B, Coric D, Trost G, Mummaneni PV (2012) Analysis of three United States Food and Drug Administration investigational device exemption cervical arthroplasty trials. J Neurosurg Spine 16:216–228

Vaccaro A, Beutler W, Peppelman W, Marzluff JM, Highsmith J, Mugglin A, DeMuth G, Gudipally M, Baker KH (2013) Clinical outcomes with selectively constrained SECURE-C cervical disc arthroplasty. Spine 38:2227–2239

Wigfield CC, Skrzypiec D, Jackowski A, Adams MA (2003) Internal stress distribution in cervical intervertebral discs. J Spinal Disord Tech 16:441–449

Posterior Dynamic Stabilization

50

Dorian Kusyk, Chen Xu, and Donald M. Whiting

Contents

Introduction	893
Biomechanics of Dynamic Stabilization	894
PDS Devices	895
Pedicle Based Systems	895
Facet Replacements	896
Interspinous Process Spacers	896
PDS as an Adjunct to Fusion	897
Cross-References	897
References	897

Abstract

Posterior dynamic stabilization (PDS) systems arose with the promise of stability without fixation. In particular, these systems address the two prevailing models of spinal biomechanics – the Panjabi model of the Neutral Zone and Mulholland-Segupta theory of abnormal load transmission. By both limiting the range of motion of the diseased level and off-loading the disc space of some axial stress, PDS systems hope to treat back pain while preserving motion. However, these design constraints post a significant design challenge, as demonstrated by the multiple models that have been visited over the years. Though a successful PDS system has yet to emerge, surgeons have found other ways to use the technology, including as an adjunct to improve fusion rates when paired with interbody devices.

Keywords

Posterior dynamic stabilization · Neutral zone · Fusion biomechanics · Graf ligament · Dynesys

D. Kusyk · C. Xu
Department of Neurosurgery, Neuroscience Institute, Allegheny Health Network, Pittsburgh, PA, USA
e-mail: Dorian.Kusyk@AHN.ORG; Chen.XU@ahn.org

D. M. Whiting (✉)
Neuroscience Institute, Allegheny Health Network, Pittsburgh, PA, USA
e-mail: donald.whiting@ahn.org

© Springer Nature Switzerland AG 2021
B. C. Cheng (ed.), *Handbook of Spine Technology*,
https://doi.org/10.1007/978-3-319-44424-6_53

Introduction

The question of the treatment of axial back pain, along with the concept of spinal instability, is still incompletely understood. The current gold standard is spinal fusion, which internally fixes spinal elements until a patient can undergo a boney fusion across the levels in questions. However, a spinal fusion will abnormally fix two spinal vertebral bodies, which not only reduce a patient's mobility, but increases the risk of developing adjacent level pathology.

To begin addressing these concerns, posterior dynamic stabilization (PDS) was introduced with the development of the Graf ligament in 1989 and the Dynesys in 1994 (Gomleksiz et al. 2012). The promise of these devices and the multiple iterations since was stability without fixation and therefore pain relief without the long-term repercussions of a rigid construct. However, enthusiasm for these devices has waned considerably in recent years given the mechanical failure of the devices and the failure of clinical success has become increasingly documented (Sengupta and Herkowitz 2012).

The following chapter will review some of the conceptual underpinnings of dynamic stabilization and its proposed benefits, briefly describe some of the key devices designed in this space along with lessons learned from them, and finally, discuss the current literature regarding the use of dynamic stabilization devices in hybrid constructs as a potential path forward with the technology.

Biomechanics of Dynamic Stabilization

Two prevailing theories of spinal biomechanics are used to explain low back pain, and PDS theoretically would address both as pain generators. Panjabi's model describes pain in terms of Neutral zone (NZ) and the range of motion (ROM) of vertebral segments (Panjabi 1992). In his 2003 paper, he described the NZ as the range of motion to which there is minimal resistance to vertebral motion. In the nonpathologic spine, the neutral zone encompasses a smaller ROM than the joint's painful zone – that is, the ligaments and other support structures of the spine limit vertebral motion before it causes pain. However, with ligament laxity or other pathology, the neutral zone of the spine can expand and permit positions of flexion and extension which are painful (Panjabi

2003). PDS seeks to reduce a patient's pain by restoring a more physiologic NZ. Where spinal fusion reduces the neutral zone to very limited motion, dynamic stabilization would theoretically restore a more natural NZ.

According to a second hypothesis, spinal instability should not be thought of as unnatural movement of the segment, rather as abnormal load transmission at the level. According to Mulholland and Segupta, disc degeneration makes it nonhomogeneous with areas with increasingly larger loads transmitted through the annulus. Load transmission therefore becomes uneven, which in turn leads to focal in-folding of endplate cartilage and subchondral bony trabeculae, analogous to a stone a shoe (Mulholland and Sengupta 2002; Simpson et al. 2001; Keller et al. 1989). PDS can help unload the disc space by providing a posterior tension band, ultimately reducing the back pain caused by the uneven load distribution. Some researchers argue that by unloading the disc, the patient will actually begin to repair the damage (Beckmann et al. 2019; Cho et al. 2010).

In addition to addressing the biomechanical foundations of back pain, PDS also aims to limit the risk of adjacent level disease. One of the largest limitations of a rigid fusion construct is the stress it places on levels above and below the initial pathology, likely due to an increased lever arm and the development of a nonphysiologic center of motion. This in turn leads to fractional increases in joint ROM, which in time develops to gross spinal instability and low back pain (Park et al. 2004). Put another way, hard fusion constructs force adjacent levels to expand their neutral zones to adjust to limited mobility, until the neutral zone of the adjacent level extends beyond the pain-free ROM of the joint. PDS would theoretically limit this risk by maintain a physiologic ROM at the diseased level and decrease the stress placed on adjacent levels (Bono et al. 2009; Aygun et al. 2017).

Though PDS devices have many theoretical advantages, it should be noted that these pose a significant design challenge. First, in order for a device to be considered a dynamic stabilization device, it must limit joint motion to a physiologic range and unload the pathologic disk space. Additionally, the dynamic stabilization devices are fundamentally different from a fusion construct – whereas a traditional fusion construct needs to last just long enough for the patient's own bone remodeling to fuse across the segment, the PDS systems need to last indefinitely. Dilip Sengupta argues that these devices need to have uniform motion restriction and load-sharing throughout the ROM (Sengupta and Herkowitz 2012). Any asymmetry in load or motion would lead to an increase of stress on the device and lead to its premature failure.

PDS Devices

There are three main categories of PDS systems depending on the location of where the device is implanted. These focus on the pedicles, the facets, or the spinous processes.

Pedicle Based Systems

The Graf Ligament was one of the earliest pedicle based systems first reported in 1992 (Graf 1992). It was developed in Europe and used braided polyester cables looped around pedicle screws (Fig. 1). There have been several studies that have shown inconsistent outcomes of the device. One study reported patients undergoing the Graf ligamentoplasty doing clinically better in comparison to anterior lumbar interbody fusions (ALIFs) (Madan and Boeree 2003), while another showed worse outcome at 1 year and increased revision rate at 2 years with the Graf ligamentoplasty when compared to posterolateral fusions in the management of low back pain (Hadlow et al. 1998). As regards to patient satisfaction, studies have reported anywhere from 96% of patient feeling that the operation was worthwhile to 41% of patients stating they would not have chosen to have the operation again (Grevitt et al. 1995; Rigby et al. 2001).

The Dynesys system manufactured by Zimmer Spine uses nylon cords combined with plastic spacers (Fig. 2). In comparison to traditional posterior lumbar interbody fusions (PLIFs), the Dynesys was shown to offer similar improvement in clinical outcomes for lumbar degenerative disease. In addition, the Dynesys system was reported to have significantly less adjacent segment disease radiographically when compared to PLIFs and also offered more range of motion (ROM) (Zhang et al. 2016).

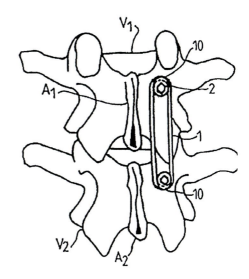

Fig. 1 Graf ligament. (Taken from original 1990 patent filing)

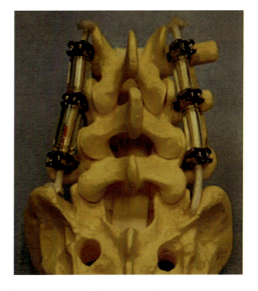

Fig. 2 Dynesys pedicle screws and spacer

The Dynamic Soft Stabilization (DSS) system uses pedicle screws with metal coils connecting the screws to control motion (Figs. 3 and 4). And the Isobar is essentially the tradition rod system but with a mobile joint within the rod.

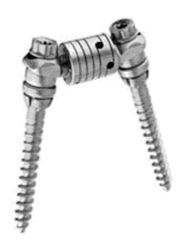

Fig. 3 Dynamic soft stabilization (DSS) (Courtesy of Paradigm Spine, Device not available in the U.S.)

Facet Replacements

The majority of the facet replacement systems have some involvement of the pedicles, although the focus is on motion preservation at the facet joints. The Total Posterior Element Replacement (TOPS) system requires complete removal of the posterior elements for the device to be implanted and is anchored through the pedicles and is the only such device currently in active clinical study.

Interspinous Process Spacers

The Coflex system is the most common interspinous implant currently used in clinical practice. The device is a "U" shape allowing for distraction of the neuroforamina as well as controlled forward and backward bending. Previous studies have shown favorable outcomes with reports of 33% and 66% reductions in back pain and leg pain severities, respectively, with 95% satisfaction form patients (Errico et al. 2009).

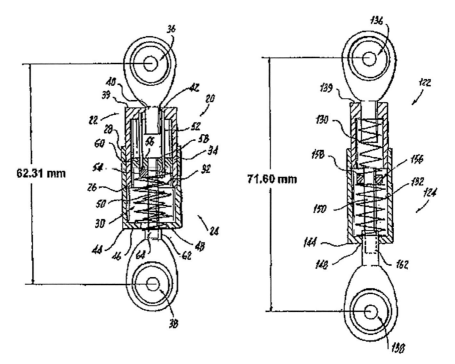

Fig. 4 DSS detail (Hildebrand and Trimm 2005)

This device does require a fair amount of laminotomy to be performed as well as drilling of the spinous processes for implantation with avoidance of posterior canal compression.

PDS as an Adjunct to Fusion

Given the long list of design requirements for a successful device, it is not surprising that in the United States, PDS devices have not been approved as a stand-alone construct. Instead, PDS devices are used as adjuncts to interbody devices to theoretically improve fusion rates.

The foundation of any fusion can be described by Wolff's Law (Wolff 1986), also known as the law of bone remodeling. In brief, Wolff's law describes cellular mechano-transduction – the conversion of mechanical stressors into biochemical signals. In the context of spinal fusion, the law implies that bone remodeling and growth may be enhanced through greater loading on the graft. Though rigid spinal fusion constructs are adequate in this regard, they are still plagued by high rates of pseudoarthrosis, and one reason may be the stress shielding phenomenon. If the pedicle screw-rod complex is too rigid, it can theoretically offload the anterior column. According to Wolff's law, that can only undermine the efficacy of fusion and could potentially play into rates of pseudoarthrosis. In contrast, PDS can maintain a controlled amount of motion when paired with an interbody. As the bone settles and remodels, the microadjustments allowed by a dynamic system can ensure a constant loading force on the anterior column and theoretically a better rate of fusion (Yu et al. 2016). Though this use of the PDS is contrary to the device's initial intent, it is a welcome windfall as the research community continues to search for the ideal motion preservation stabilization device.

Cross-References

▶ Design Rationale for Posterior Dynamic Stabilization Relevant for Spine Surgery

References

Aygun H, Yaray O, Mutlu M (2017) Does the addition of a dynamic pedicle screw to a fusion segment prevent adjacent segment pathology in the lumbar spine? Asian Spine J 11(5):715–721

Beckmann A et al (2019) Biomechanical testing of a polycarbonate-urethane-based dynamic instrumentation system under physiological conditions. Clin Biomech 61:112–119

Bono CM, Kadaba M, Vaccaro AR (2009) Posterior pedicle fixation-based dynamic stabilization devices for the treatment of degenerative diseases of the lumbar spine. J Spinal Disord Tech 22(5):376–383

Cho BY et al (2010) Lumbar disc rehydration postimplantation of a posterior dynamic stabilization system. J Neurosurg Spine 13(5):576–580

Errico TJ, Kamerlink JR, Quirno M, Samani J, Chomiak RJ (2009) Survivorship of coflex interlaminar-interspinous implant. SAS J 3(2):59–67

Gomleksiz C et al (2012) A short history of posterior dynamic stabilization. Adv Orthop 2012:629698

Graf H (1992) Lumbar instability. Surgical treatment without fusion. Rachis 412:123–137

Grevitt MP, Gardner AD, Spilsbury J et al (1995) The Graf stabilisation system: early results in 50 patients. Eur Spine J 4:169–175

Hadlow S, Fagan AB, Glas H et al (1998) The Graf ligamentoplasty procedure: comparison with posterolateral fusion in the management of low back pain. Spine 23:1172–1179

Hildebrand B, Trimm JP (2005) European Patent No EP 1747760B1

Keller TS et al (1989) Regional variations in the compressive properties of lumbar vertebral trabeculae. Effects of disc degeneration. Spine 14(9):1012–1019

Madan S, Boeree NR (2003) Outcome of the Graf ligamentoplasty procedure compared with anterior lumbar interbody fusion with the Hartshill horseshoe cage. Eur Spine J 12:361–368

Mulholland RC, Sengupta DK (2002) Rationale, principles and experimental evaluation of the concept of soft stabilization. Eur Spine J 11(Suppl 2):S198–S205

Panjabi MM (1992) The stabilizing system of the spine. Part II. Neutral zone and instability hypothesis. J Spinal Disord 5(4):390–396; discussion 397

Panjabi MM (2003) Clinical spinal instability and low back pain. J Electromyogr Kinesiol 13(4):371–379

Park P et al (2004) Adjacent segment disease after lumbar or lumbosacral fusion: review of the literature. Spine 29(17):1938–1944

Rigby MC, Selmon GPF, Foy MA, Fogg AJB (2001) Graf ligament stabilisation: mid- to long-term follow-up. Eur Spine J 10:234–236

Sengupta DK, Herkowitz HN (2012) Pedicle screw-based posterior dynamic stabilization: literature review. Adv Orthop 2012:424268

Simpson EK et al (2001) Intervertebral disc disorganization is related to trabecular bone architecture in the lumbar spine. J Bone Miner Res 16(4):681–687

Wolff J (1986) The law of bone remodelling. Springer-Verlag, Berlin, p 126

Yu AK et al (2016) Biomechanics of posterior dynamic fusion Systems in the Lumbar Spine: implications for stabilization with improved arthrodesis. Clin Spine Surg 29(7):E325–E330

Zhang Y, Shan J, Liu X, Guan K, Sun T (2016) Comparison of the dynesys dynamic stabilization system and posterior lumbar interbody fusion for lumbar degenerative disease. PLoS One 11(1): e0148071

Total Disc Arthroplasty

51

Benjamin Ebben and Miranda Bice

Contents

Introduction	900
Surgical Techniques: Cervical Disc Replacement	905
Indications	905
Contraindications	906
Relevant Anatomy	906
Positioning and Approach	907
Implant-Specific Instrumentation	908
Postoperative Protocol	911
Complications	911
Revision Options	912
Outcomes	912
Surgical Techniques: Lumbar Disc Arthroplasty	913
Indications	913
Contraindications	913
Relevant Anatomy	913
Positioning and Approach	914
Implant-Specific Instrumentation	915
Postoperative Protocol	915
Complications	915
Revision Options	916
Outcomes	916
Conclusion	917
Cross-References	918
References	918

B. Ebben
University of Wisconsin, Madison, WI, USA
e-mail: bebben@uwhealth.org; bebben@wisc.edu

M. Bice (✉)
University of Wisconsin School of Medicine and Public Health, Madison, WI, USA
e-mail: bice@ortho.wisc.edu

© Springer Nature Switzerland AG 2021
B. C. Cheng (ed.), *Handbook of Spine Technology*,
https://doi.org/10.1007/978-3-319-44424-6_58

Abstract

The concept of total disc replacement in the spine has been present for decades because of the desire to maintain physiologic motion of spinal segments while treating underlying pain-generating pathology. There has been considerable evolution of this technology, with successes, failures, and the popularity of these procedures waxing and waning over time. Much in vitro and in vivo research has been done on both past and current devices to facilitate understanding of this technology and optimize utilization for clinical success and progress. This chapter describes some of the historical background, current uses and approved devices, surgical techniques, complications, revision options, and outcomes of both lumbar and cervical disc replacement.

Keywords

Lumbar disc replacement · Cervical disc replacement · Disc arthroplasty · Adjacent segment degeneration · Adjacent segment disease · Motion sparing · Spine arthroplasty · Artificial disc

Introduction

Historically, the initial management of painful degenerative spinal disc disease has been conservative and supportive measures. When these efforts fail to provide meaningful relief, decompression and arthrodesis is generally considered the accepted surgical intervention for its effectiveness in maintaining intervertebral height, establishing segmental stability, and improving pain. Overall, arthrodesis has proven quite successful over time. However, the reported reoperation rates cannot be ignored. These reoperations are frequently reported due to persistent or recurrent pain from symptomatic adjacent level degeneration or pseudarthrosis. Although heavily debated, current thought suggests that the complications associated with arthrodesis, namely, adjacent level disease, exist secondary to the alteration of normal spine biomechanics associated with the fusion of a previously mobile segment. There has been a considerable amount of literature dedicated to not only uncovering the presumed association between arthrodesis and adjacent level deterioration but also to investigating the biomechanical and biochemical basis behind this theoretical relationship.

In vitro cadaveric studies have demonstrated increased stresses at mobile segments adjacent to the site of fusion in the cervical spine. Eck et al. found that intradiscal pressure (IDP) increased significantly both cranial and caudal to a cervical fusion during flexion compared to an intact spine by 73% and 45%, respectively (Eck et al. 2002). Similarly, Chang and colleagues reported significantly elevated IDP in the cranial mobile segment during both flexion and extension following cervical fusion. These investigators also demonstrated effects on posterior element stress levels following cervical fusion and found that facet joint forces were significantly greater at both adjacent mobile segments during extension (Chang et al. 2007). A similar group of cadaveric biomechanical studies have been performed in the lumbar spine following instrumented arthrodesis with comparable findings of increased stress within the intervertebral discs and/or facet joints (Cunningham et al. 1997; Lee and Langrana 1984). Examination of intervertebral disc physiology shows that the health of this avascular structure is related to the relative concentrations of specific collagen and proteoglycan subtypes. The maintenance of this extracellular matrix is, in turn, reliant upon adequate diffusion of nutrients through the vertebral body cartilaginous endplate. It can be reasonably inferred that the discs within adjacent mobile segments exposed to chronically elevated intradiscal hydrostatic pressures following spinal arthrodesis may degenerate at an accelerated rate due to the disruption of this intricate metabolic balance (Buckwalter 1995; Hutton et al. 1998).

Long-term radiologic follow-up studies after spinal fusion have reported high incidences of adjacent level degenerative changes. In 2004, Goffin et al. published their radiologic findings for a series of 180 patients an average of 8 years following cervical interbody fusion. They found that 92% of the patients demonstrated an increase in degeneration score at adjacent levels at long-term follow-up. A suggestive trend of correlation, albeit not statistically significant, was appreciated between adjacent level radiologic degeneration and clinical outcomes (Goffin et al. 2004). Other authors have tried correlating these observed radiologic changes with clinical outcomes. In a landmark study, Hilibrand and colleagues studied the development of new radiculopathy or myelopathy referable to mobile segments adjacent to previous anterior cervical arthrodesis in 374 patients available for 10-year follow-up. They reported a nearly 3% annual incidence of symptomatic adjacent segment degeneration and a Kaplan-Meier survival analysis predicted an overall prevalence of 25.6% within the first 10 years after the procedure. Twenty-seven patients underwent a second operation for fusion at the adjacent symptomatic level (Hilibrand et al. 1999). Ghiselli et al. studied adjacent segment disease in the lumbar spine and reported similar clinical outcomes. Fifty-nine of 215 patients, followed for an average of 6.7 years after posterior lumbar arthrodesis, developed symptomatic adjacent segment degeneration that warranted additional surgery. The authors reported a nearly 4% annual incidence of surgical intervention for adjacent segment disease and their survivorship analysis predicted that 36.1% of patients would have new disease requiring reoperation within the first 10 years following the index procedure (Ghiselli et al. 2004). There is a sizeable amount of literature further investigating clinical outcomes following spinal arthrodesis with a focus on defining its contribution to the development of symptomatic adjacent segment degeneration (Park et al. 2004; Gore and Sepic 1998).

Despite the substantial supporting data, no causation has been definitively proven. Randomized controlled trials investigating the relative rates of symptomatic adjacent segment disease with and without arthrodesis do not exist as it would be unethical to deny patients a fusion operation for a situation in which they would otherwise be indicated. Some experts would argue that adjacent segment degeneration is a consequence of natural history and can be expected as an inherent fate in a spine that has already shown signs of degenerative disease. To this end, studies have attempted to decipher the relative contributions of fusion and the natural aging process. Matsumoto et al. evaluated the pre-surgery and 10-year follow-up MRI images of 64 patients who underwent anterior cervical

decompression and fusion (ACDF). They compared the observed radiologic changes to a group of asymptomatic volunteers who, likewise, underwent a baseline and 10-year follow-up MRI. The incidence of progression of degenerative disc disease was significantly higher in the ACDF group (Matsumoto et al. 2010). Nonetheless, this study was limited by differences in group characteristics including both a higher mean age and observed frequency of baseline MRI degenerative findings in the ACDF group. Interestingly, two of the landmark publications referenced earlier found that multilevel fusion is actually protective rather than promotive when it comes to adjacent segment degeneration. Hilibrand et al. discovered that only 12% of patients who underwent multilevel arthrodesis developed symptomatic adjacent segment degeneration, an odds ratio of 0.64 when compared to single level (Hilibrand et al. 1999). In the lumbar spine, mobile segments adjacent to single-level arthrodesis were three times more likely to develop symptomatic adjacent segment degeneration than segments adjacent to a multi-level arthrodesis (Ghiselli et al. 2004).

Another frequently studied complication of spine arthrodesis is the development of a symptomatic pseudarthrosis. There are established but quite variable rates of pseudarthrosis within the cervical and lumbar spine literature. Rates are technique-dependent and vary based on multiple factors including the use of an interbody device, fixation rigidity, whether or not instrumentation was performed, choice of graft, etc. Martin and colleagues used a registry of statewide (Washington) hospital discharges to investigate rates of reoperation following lumbar spinal surgery and found that the cumulative 11-year incidence of reoperation following an index fusion procedure was 20%. Of the 471 reoperations following an index fusion, 23.6% were associated with a coding of pseudarthrosis (Martin et al. 2007). A 47-article meta-analysis conducted to determine success and complication rates for lumbar spinal fusion found pseudarthrosis as the most frequently reported complication (14%). Authors also noted a positive relationship between satisfactory patient outcomes and achievement of solid

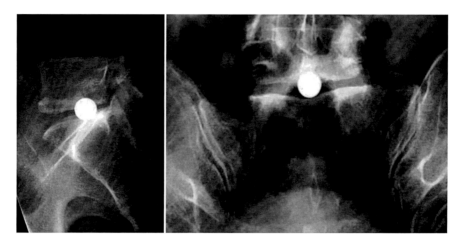

Fig. 1 Fernstrom Ball prosthesis. (Reprinted with permission from Szpalski et al. *Eur Spine J* 2002)

arthrodesis (Turner et al. 1992). A similar meta-analysis investigating the overall incidence of pseudarthrosis following fusion in the cervical spine found a much lower overall rate of 2.6% (Shriver et al. 2015). The true incidence of spine pseudarthrosis is probably underestimated as a percentage are asymptomatic and prompt no further diagnostic workup or additional management.

To combat the pitfalls discussed above that are associated with spinal fusion, the field of spinal arthroplasty and the concept of motion sparing spinal implants evolved. The growth of this field was heavily influenced by the technologic successes of motion-preserving joint prostheses for the treatment of degenerative joint disease in the hip and knee. Motion sparing technology could potentially circumvent the limitations of arthrodesis. In theory, by implanting a motion sparing prosthetic within the intervertebral space, accelerated adjacent segment degeneration could be mitigated. The potential for pseudarthrosis development could be eliminated with no attempt at surgical fusion. In addition, maintaining the mobility of the spinal segment could lead to preservation of normal spine biomechanics and could maximize patient motion, function, and improve clinical outcomes. Along these lines, investigators began to define the characteristics of an ideal spinal arthroplasty system which would include the reproduction of native disc viscoelastic properties, the reproduction of native disc motion characteristics, and the ability to withstand the mechanical and chemical environment of the intervertebral space.

A Swedish surgeon, Ulf Fernström, is historically credited with implantation of the first artificial disc in a human patient, and his experiences were published in the late 1960s and the early 1970s. His prosthesis was quite simple and consisted of a single, corrosion-resistant stainless steel ball bearing implanted into the center of the intervertebral disc space (Fig. 1). It is estimated that he implanted approximately 250 of these devices in total, both in the lumbar and cervical spine (Le et al. 2004; Basho and Hood 2012; Baaj et al. 2009). A duo of South African surgeons, impressed with Fernström's early results, also implanted 75 of these devices in the cervical spine during the same time period, for the treatment of intractable headache and cervico-brachialgia (Reitz and Joubert 1964). Ultimately, with longer-term follow-up, these mobile bearings failed miserably. The unconstrained nature created segmental spinal hypermobility, and the lack of endplate support resulted in a tendency for subsidence and migration into the superior endplate (Le et al. 2004). These early disappointments lead to a temporary abandonment of spinal arthroplasty surgical practice in favor of arthrodesis until the 1980s. Nonetheless, Fernström was ahead of his time in recognizing the potential benefits of motion sparing devices, and other researchers continued to investigate alternative designs.

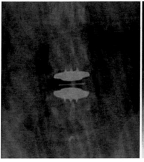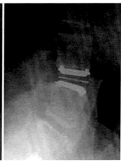

Fig. 2 Charite III prosthesis. (Reprinted with permission from Atkins, et al. *Lumbar Disc Arthroplasty*. In: Essentials of Spinal Stabilization. Holly L., Anderson P. (eds). Springer, Cham. 2017)

Multiple spine arthroplasty models were subsequently developed during the second half of the twentieth century, a majority of which were patented or published but never reached the stage of human implantation (Szpalski et al. 2002).

Spine arthroplasty then garnered renewed interest in 1984 after the maiden implantation of the German-engineered SB Charité I prosthesis, which was the first approved and commercially available lumbar total disc replacement system available in Europe (Link 2002). The SB Charité I was an unconstrained device featuring small, circular, polished steel alloy endplates with anchoring teeth for cementless fixation and a sliding ultrahigh molecular weight polyethylene (UHMWPE) core marked with a radio-opaque circumferential wire (Büttner-Janz et al. 1989). The sliding core allowed for a dynamic instantaneous axis of rotation that could translate during flexion and extension, more closely mimicking normal lumbar spinal motion (Bono and Garfin 2004). Similar to Fernström's ball bearing implants, the earliest SB Charité model lacked sufficient endplate contact surface area secondary to its undersized metal endplates and was noted to subside or migrate axially (Link 2002). This design flaw prompted development of a second version, the SB Charité II, with enlarged metal endplates. Problems with fatigue fractures ultimately lead to the third- and final generation Link SB Charité III (DePuy) device which started production in 1987 in Europe and eventually received FDA approval in the United States in 2004 after 2-year follow-up results from its investigational device exemption (IDE) randomized controlled trial showed noninferiority to lumbar arthrodesis (Fig. 2) (Blumenthal et al. 2005). Subsequent 5-year follow-up data showed a FDA-defined clinical success rate of 58% in the Charité group and 51% in the arthrodesis group (Guyer et al. 2009). Even longer-term follow-up and device retrieval studies have become increasingly available and shed light onto some of the device late failure mechanisms. Punt et al. published a case series analyzing late complications following SB Charité III disc implantation in a group of 75 unsatisfied patients that presented to their institution with persistent leg and back pain. Forty-six of the 75 patients ultimately ended up undergoing a salvage operation, and the authors were directly involved in 37 of these cases. They reported implant subsidence, adjacent disc degeneration, and index-level facet arthrosis as the three most common late complications. Of the 39 cases of observed implant subsidence, they estimated that 24 were secondary to an undersized prosthesis. The authors also reported on 8 cases of anterior-posterior migration and 10 cases of polyethylene core wire breakage (Punt et al. 2007). Van Ooij and colleagues reported very similar findings in their 27 patient case series (van Ooij et al. 2003). In a 2007 international multicenter retrieval study of 21 explanted SB Charité III implants from patients undergoing revision surgery due to persistent pain, Kurtz et al. analyzed polyethylene wear patterns and found the peripheral rim to be susceptible to pinching as evidenced by the observation of plastic deformation, fracture, cracking, and other fatigue damage in most of the specimens (Kurtz et al. 2007). Current long-term clinical outcome data and the results of the most recent FDA IDE randomized controlled trials for the SB Charité III, its contemporaries, and its successors will be covered elsewhere in this chapter. Overall, however, the SB Charité III was quite successful and underwent widespread implantation for many years. It was removed from the US market in 2013 as part of a business decision when DePuy purchased Synthes and elected to sell its lumbar arthroplasty system, the ProDisc-L.

Currently, there are two FDA-approved lumbar arthroplasty systems. The Synthes ProDisc-L was developed concurrently with the SB Charité III in the late 1980s. Like the Charité, it underwent stepwise modifications from its initial design to the release of the current model, which received FDA approval in 2006. Unlike the Charité, the ProDisc-L is a semiconstrained device. There is a single articulating interface between a polyethylene bearing and the superior endplate. The polyethylene bearing is fixed to the inferior endplate and does not slide or translate as in the Charité. The ProDisc-L is secured to the neighboring vertebral bodies via a keel or midline sagittal fin (Bono and Garfin 2004). There is currently a considerable amount of longer-term follow-up studies (>5 years) supporting the use of this device in patients with lumbar degenerative disc disease. The ActivL (Aesculap Implant Systems) prosthesis received FDA approval in 2015 after its 2-year follow-up data showed noninferiority to the other two previously mentioned lumbar arthroplasty prostheses. This implant has been marketed as next generation in that it is designed to be inserted as a single unit, obviating the need for multiple spinal distractions. In addition, its polyethylene inlay is affixed to the inferior endplate in a way that permits a limited amount of translational motion (Garcia et al. 2015).

The technological triumphs in lumbar arthroplasty motivated the pursuit for a counterpart in the cervical spine. The first modern era artificial cervical disc was developed in the United Kingdom and was implanted in 1991. This device came to be known as the Cummins-Bristol and had two distinctive design features when contrasted to the previously discussed lumbar prosthetics: (1) a metal-on-metal articulation with no separate intercalary polyethylene bearing and (2) anterior flanges for the purpose of obtaining immediate anchoring screw fixation into the cranial and caudal vertebral bodies. Early results were quite poor and related to failure of the anterior screw fixation via screw pullout and screw fracture. Following modifications to screw hole positions and the addition of locking screw capabilities, a subsequent group of 20 patients, implanted with the device between 1991 and 1996, fared much better according to Cummins and colleagues. The authors reported that 75% of the patients experienced an improvement in preoperative symptoms and that 88% of the patients available for follow-up in 1996 had radiographic evidence of maintenance of index level motion (Le et al. 2004; Cummins et al. 1998). There were also four patients with persistent dysphagia attributed to the high profile of the anterior flanges. Two years later, a redesigned second-generation version of the Cummins-Bristol artificial disc, known as the Frenchay, was implanted into 15 patients as part of a pilot study (Fig. 3). The Frenchay's superior component "ball" remained hemispherical, while the inferior component "socket" was shallow and ellipsoid making for an incongruent articulation. Theoretically, this permitted the cranial vertebral body to passively align with the dynamic center axis of rotation as dictated by the facet joints. At 2 years, the prosthetic joints remained mobile with an average arc of 6.5° in flexion and extension, there were no cases of joint subluxation or subsidence, and there were 3 reoperations, only one of which involved explanation of the prosthesis for looseness (Wigfield et al. 2002). The Frenchay would eventually become the Prestige (Medtronic), which is one of the commercially available cervical total disc replacement systems on the market today. This device received US FDA approval in 2007, and the latest long-term (7-year) clinical outcome data has been very favorable showing a statistically significant greater overall success rate of 75% in the

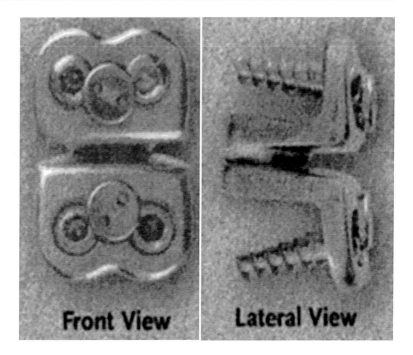

Fig. 3 Frenchay prosthesis. (Reprinted with permission from Buell, et al. Cervical Arthroplasty: Long-Term Outcomes. In: Handbook of Spine Technology. Cheng B. (eds). Springer, Cham. 2019)

arthroplasty group compared to 64% in the control arthrodesis group. These authors also reported maintenance of physiologic segmental angular motion at the index level and an index level secondary surgery 11-year cumulative rate of 4.8% compared to 13.7% in the arthrodesis group (Burkus et al. 2014).

Another unique, albeit unsuccessful, cervical arthroplasty concept is worthy of brief mention. The Pointillart cervical prosthetic entered the scene momentarily between 1998 and 1999, and its concept was influenced by unipolar hip replacement designs (Fig. 4). It featured a single titanium base piece which was anchored via screws into the caudal vertebral body and a carbon sliding cranial surface meant to articulate with the inferior endplate of the cranial vertebral body. The inventing surgeon implanted this device into ten patients and reported "total failure" after 1-year follow-up radiographs showed spontaneous fusion and resultant absence of motion across the index level in eight of the patients (Pointillart 2001).

There are currently six FDA-approved cervical total disc replacement systems: Prestige (Medtronic), Bryan (Medtronic), Mobi-C (Zimmer-Biomet), ProDisc-C (DePuy Synthes), PCM (NuVasive), and Secure-C (Globus Medical). All of these devices have 2–7-year US FDA IDE prospective randomized controlled trial clinical outcome data showing non-inferiority to anterior cervical decompression and fusion (Sasso et al. 2011; Hisey et al. 2016; Janssen et al. 2015; Phillips et al. 2015; Vaccaro et al. 2013). As with any surgical procedure, particularly in the spine, strict adherence to appropriate criteria of both patient selection and surgical indications is paramount for successful outcomes.

Surgical Techniques: Cervical Disc Replacement

Indications

- Subaxial spinal motion segments between C3 and C7
- One or two-level pathology
- Radiculopathy and/or myelopathy secondary to neural element compression by:
  – Soft disc herniation
  – Osteophyte formation

Fig. 4 Pointillart prosthesis. (Reproduced with permission from Pointillart, *Spine* 2001)

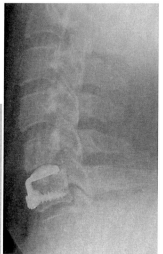

Contraindications

- Spondylolisthesis, instability with translation of greater than 3.5 mm
- Deformity
  - Including kyphosis of greater than 11° at the target level
- Trauma (concern for disruption or irregularity of vertebral endplates)
- Prior cervical laminectomy (concern for disruption of posterior stabilizing elements at the level of interest)
- Prior surgery at the level of interest
- Osteoporosis (T-score less than −2.5)
- Other metabolic bone diseases which may result in abnormal bony architecture and/or stability
  - Rheumatoid arthritis, other inflammatory arthropathies
  - Renal disease
  - Cancer
  - Long-term steroid use
- Infection
- Severe facet arthropathy
- Ankylosing disorders
  - Ankylosing spondylitis
  - Diffuse idiopathic skeletal hyperostosis (DISH)
  - Ossification of the posterior longitudinal ligament (OPLL)
- Metal allergy
- Isolated axial neck pain without radiculopathy or myelopathy

Relevant Anatomy

A standard Smith-Robinson approach to the anterior cervical spine is utilized for cervical disc replacement. While this is generally regarded as a common and safe approach, detailed knowledge and understanding of the local anatomy is necessary to minimize inadvertent injury to several important structures:

Nerves
- **Superior laryngeal nerve** is typically encountered for procedures in the upper cervical spine, at or above C3 and C4. It can be identified traversing from the carotid sheath to the larynx at the thyrohyoid membrane along with the superior laryngeal artery. As this nerve contributes to control of a vocal cords, injury to it may result in difficulty with voice control (dysphonia) and swallowing or aspiration (dysphagia).
- **Recurrent laryngeal nerve** is occasionally visualized on its recurrent path in the tracheoesophageal groove. On the left, once the nerve exits the carotid sheath, it

courses inferiorly under the aortic arch prior to returning cephalad in the tracheoesophageal groove. The recurrent laryngeal nerve on the right is beneath the right subclavian artery and is less constant. For this reason, it is sometimes dogmatically believed to be safer to perform the approach on the left side as this course was previously felt to be more predictable; however this has not been demonstrated clinically, and there are many surgeons that perform this approach on the right side without any increased complication rate related to phonation or swallowing. This nerve also contributes to control of vocal cords well as all of the laryngeal muscles and the esophagus. Similar to injury of the superior laryngeal nerve, injury to this nerve can also result in difficulties with dysphonia and or dysphagia.

- **Sympathetic chain** lies on the ventral surface of the longus coli muscles. Because of this, manipulation in this area is generally avoided, with dissection generally limited to the medial aspect of the longus coli. Injury to the sympathetic chain can result in an ipsilateral Horner's syndrome.

Vessels

- **External jugular vein** lies between the platysma and the caudal mastoid. It often is lateral to the operative field; however occasionally the main external jugular or large branches of it can cross the surgical field. Injury to it may not result in significant functional impairment; however it can bleed quite vigorously, adding difficulty and time to the surgery.
- **Carotid artery** travels within the carotid sheath. It can be easily palpated as a pencil-like structure deep to the sternocleidomastoid muscle belly and used as a landmark for the approach as the entirety of the approach should be medial to this structure along with the other contents of the carotid sheath.
- **Vertebral artery** travels within the foramen transversarium of the cervical vertebrae. It typically enters at C6, although can also enter at C7, and travels proximally to

supply the brainstem and posterior cranial contents. The longus colli muscle lies ventral to the transverse foramen containing these vessels, and so dissection deep to the longus muscle belly is very limited and cautious to avoid injury to the vertebral arteries. However, should a vertebral artery injury occur elsewhere during the procedure, dissection deep to the longus colli can be utilized to gain access to the vessel and control bleeding. Injury to this blood vessel can result in rapid exsanguination. The overall implications of vertebral artery injury varies widely, from asymptomatic to stroke or even death.

Trachea and Esophagus are midline structures medial to the plane of approach. Further mobilization is often necessary for adequate exposure to the targeted disc site(s). Because of its cartilaginous rings, the trachea is more easily identified. The esophagus lies deep to the trachea. As it is composed of smooth muscle of varying degrees of thickness, it is more prone to inadvertent injury during anterior cervical approaches. Injuries to these structures are often occult and not always identified intraoperatively but can lead to profound morbidity and even mortality if not identified and treated appropriately. For these reasons, a high index of suspicion is mandatory during both the index procedure and follow-up if anything is amiss.

Positioning and Approach

The patient is positioned supine on a radiolucent operating table. The authors prefer to have the patient as caudal on the table as patient's height will allow to provide space for the C-arm rostral to the patient when not in use. The neck is positioned in neutral alignment. The arthroplasty devices are not intended to correct or change alignment, and so native alignment is maintained during positioning so as to avoid improper implant placement. If the shoulders preclude adequate visualization of the targeted surgical level, gentle traction can be gained by either taping the shoulders down caudally to the table or placing wraps about the wrists that can then be utilized for intermittent traction. If continuous traction is utilized, the surgeon must ensure that excessive traction is not sustained on the brachial plexus for the entirety of the procedure to decrease the chance of root palsies.

A standard Smith-Robinson approach is performed. This is often a left-sided approach, although can be performed on either side depending on surgeon preference. The location of the incision is planned over the targeted disc space based on manual palpation of landmarks and/or fluoroscopy. If possible, the incision is placed within a natural skin crease for cosmesis. Prior to incision, it can be helpful to mark the sternal notch to facilitate orientation to the midline throughout the procedure, as precise alignment is of utmost importance for accurate placement of arthroplasty implants. A 2–3 cm transverse incision is made, extending approximately from midline to the medial border of the sternocleidomastoid muscle. Subcutaneous fat and platysma are then divided. The superficial layer of the deep cervical fascia is divided in the plane visualized between the sternocleidomastoid laterally and the strap muscles medially. The omohyoid can be sacrificed if needed to gain access to the lower cervical levels. Continued blunt dissection in this plane will then lead to the spine, with the carotid sheath the laterally and the larynx and esophagus medially. When the spine is encountered following this plane, a snap is placed on the annulus of the intended surgical level, and localization is confirmed using lateral cross-table fluoroscopy. Adjacent to the target disc level, the longus colli are gently elevated bilaterally to allow adequate access to the disc space out to the uncovertebral joints, however taking care not to dissect too far laterally as the anterior aspect of the vertebral body slopes down and away from the ventral surface to avoid injury to the vertebral arteries. At this point, self-retaining radiolucent retractors can be placed deep to the elevated longus flaps. The annulotomy is performed followed by the discectomy portion of the procedure.

Implant-Specific Instrumentation

Prestige LP (Medtronic Sofamor Danek) (Prestige LP 2009) (Fig. 5)

- *Device type:*
  - Metal-on-metal (titanium alloy)
  - Ball and socket
- *Procedure*

  Caspar pins are placed in the rostral and caudal vertebral bodies, taking care to ensure that placement is midline, parallel to the endplates and with sufficient distance to prevent violation of the endplates during placement or disc space preparation, and parallel to one another so as not to introduce any kyphosis or lordosis during disc space preparation. Fluoroscopic guidance is highly

Fig. 5 Medtronic Prestige LP prosthesis. (Reproduced with permission from Nasto et al. *Cervical Disc Arthroplasty.* In: Cervical Spine. Menchetti P. (eds) Springer, Cham. 2016)

advised when placing these pins. The remainder of the decompression is completed using Kerrison, curettes, and a bur to facilitate complete osteophyte removal for a wide bilateral foraminal decompression. The posterior longitudinal ligament (PLL) is resected. The endplates are gently burred to provide a flat and parallel disc space; however care is taken to limit amount of cortical bone removed to minimize risk of subsidence. The rasp can facilitate fine-tuning of this step after burring. The anterior vertebral bodies are also flattened with the bur so that to the flanges of the prosthesis will lie flush to the anterior aspect of the vertebral body. Periosteum present on the adjacent vertebral bodies is removed with the monopolar cautery, and all bone dust is copiously irrigated and removed to decrease chance of heterotopic ossification formation. The trial is inserted, and sizing is confirmed using lateral fluoroscopy as well as manual assessment of the resistance encountered for insertion and removal. Ensure the tabs on the trial fit flush with the anterior vertebral body. Compare the trial size and space to adjacent healthy disc spaces and facet joints on fluoroscopy. At this point, the Trial Cutter Guide is placed into the prepared disc space. Confirm that the cutter guide is perfectly midline using fluoroscopy because all steps moving forward will now dictate the final positioning of the implant. The Rail Cutter Bit is then used to prepare the rail tracts; the guide is held in place between rail preps with the Temporary Fixation Pins. When all four rails have been cut, all instruments are removed from the disc space. The Rail Punch is tapped into the disc space to complete the rail preparation. The prosthesis is then implanted into the prepared disc space, with the ball endplate rostral. Bone wax can then be applied over the exposed anterior aspect of the implant and over the exposed vertebral bodies to minimize heterotopic ossification. Ensure that the prosthesis remains parallel and the inserter perpendicular to the prepared disc space. Lateral fluoroscopy is used to guide depth of placement, and AP views confirm accurate coronal positioning.

Mobi-C (Zimmer Biomet) (Mobi-C 2016) (Fig. 6)

- *Device type:*
  - Metal on plastic (ultrahigh molecular weight polyethylene)
  - Semiconstrained
- *Procedure*

Caspar pins are placed in the rostral and caudal vertebral bodies, taking care to ensure that placement is midline, parallel to the endplates and with sufficient distance (5 mm) to prevent violation of the endplates during placement or disc space preparation, and parallel to one another so as not to introduce any kyphosis or lordosis during disc space preparation. The Intervertebral Distractor Device is used to distract the vertebral bodies, and then the distraction is maintained through the Caspar distractor pins. The recommended method of the remainder of the decompression for this device by the manufacture is without

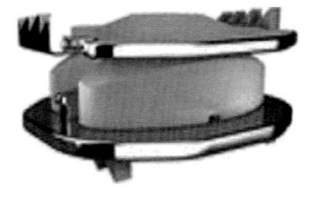

Fig. 6 Mobi-C prosthesis. (Reprinted with permission from Buell, et al. Cervical Arthroplasty: Long-Term Outcomes. In: Handbook of Spine Technology. Cheng B. (eds). Springer, Cham. 2019)

the use of a burr to optimally preserved bony endplate integrity. Bilateral foraminotomies are performed with Kerrison. The PLL is resected to facilitate perpendicular disc space preparation and distraction. The inferior endplate is squared off as wide as possible within the corners of the uncus without complete removal of the uncinates to maximize the width of the footprint of the implant. Next, the Width Gauge is placed into the prepared disc space to determine the width and adequacy of endplate preparation. If this gauge does not lie flat on the endplate, then the uncinates are squared off further using curettes. The Paddle Distractor, Caspar pin, or depth gauge can be used to estimate the depth of the footprint. Do not include anterior osteophytes in this measurement to ensure accuracy of the anterior-posterior footprint measurement. Anterior osteophytes can be removed as needed to create a flat anterior surface; however do not remove the overhang of the superior endplate as this concavity is required to match the shape of the superior endplate of the implant. Place bone wax as needed on exposed or decorticated surfaces of the anterior vertebral body to decrease risk of heterotopic ossification formation. Placed the selected trial with slight distraction on the Caspar pins, and then release the distraction to confirm fit both manually assessing resistances as well as on AP and lateral fluoroscopy. Re-distract the Caspar pins, remove the trial, and place the pre-assembled implant into the prepared disc space, avoiding any rotation during implantation. This can be confirmed using lateral fluoroscopy, ensuring that the Alignment Tabs on the inferior plate remain in line with one another such that only one line is visible without obliquity. The inserter and PEEK cartridge are removed. The implant position can be fine-tuned with the plate impactor and tamp. Prior to removal of the, gently compress through them to seat the prosthesis teeth into the endplates. The Caspar pins are removed and bone wax placed within the defects to control bleeding. Final positioning is confirmed using AP and lateral fluoroscopy.

Bryan Disc (Medtronic Sofamor Danek) (Bryan 2005) (Fig. 7)

- *Device type:*
  - Metal on plastic (soft polyurethane core)
  - Semiconstrained
- *Procedure*

The remainder of the discectomy is performed with hand instruments, taking care not to remove the uncinates to preserve reference anatomy. The overhanging lip of the anterior superior vertebral body is removed, and the anterior vertebral bodies are smoothed to create a flat surface. The Transverse Centering

Fig. 7 Bryan Disc prosthesis. (Reprinted with permission from Duell, et al. Cervical Arthroplasty: Long-Term Outcomes. In: Handbook of Spine Technology. Cheng B. (eds). Springer, Cham. 2019)

Tool and Centering Level are used to identify and mark the center of the superior vertebral body. This can be confirmed with fluoroscopy if needed. Use the Intradiscal Distractor to distract the disc space to 8.5 mm and maintain this for 60 s to stretch the ligaments. Select the appropriate Alignment Guide, attached it to the Milling Guide, and place it into the prepared disc space over a Steinmann pin which has been placed at the reference point previously marked by the Centering Tool. Place the Stabilizer with the Centering Level on the Alignment Guide. Confirm that the alignment Visualization Slots are parallel to and centered between the endplates using fluoroscopy. The drill pilot holes, place Anchor Posts, distract the disc space, and complete a thorough decompression. Prepare the endplates using provided rasps up to 8.5 mm. Mill the superior and inferior endplates with the included Milling Assembly. Fill the implant with sterile saline. Place the implant into the prepared disc space. Irrigate copiously and place bone wax into screw holes and on exposed cortical surfaces to decrease chance of heterotopic ossification formation. Confirm final placement on lateral and AP fluoroscopy.

Postoperative Protocol

Amount of activity as well as the use of a hard or soft collar is at the discretion of the surgeon. A course of nonsteroidal anti-inflammatories is often utilized to decrease heterotopic ossification. The type, amount, and duration are variable, although a 2-week course is common.

Complications

Adverse events related to the approach such as dysphagia, dysphonia, vascular, or tracheoesophageal injury are possible, but reported rates are not significantly different compared to standard anterior cervical discectomy and fusion procedures (Mummaneni et al. 2007). There are, however, complications unique to total disc arthroplasty. While the goal of cervical disc replacement is maintenance of motion to theoretically protect adjacent levels, heterotopic ossification at these levels of preserve motion has been reported. The rates of heterotopic ossification development very widely; however it is felt to infrequently negatively impact range of motion or postoperative outcome (Lee et al. 2010; Chen et al. 2011). Leung reported 17% incidence of heterotopic ossification with the Bryan total disc arthroplasty device as assessed with radiographs. About 11% of these patients had significant loss of motion; however this was not correlated to clinical outcome such as pain or function (Leung et al. 2005). Similarly, Tu assessed the presence of heterotopic ossification using CT. With this more sensitive method, it was detected in 50% of one- and two-level Bryan total disc arthroplasty recipients, but again without adverse effects on clinical outcomes (Tu et al. 2011). Copious irrigation throughout the procedure including endplate preparation as well as postoperative utilization of nonsteroidal anti-inflammatory medications is often recommended to minimize risk for heterotopic ossification formation.

Subsidence is another complication which is often suggested as a possibility; however it is not often demonstrated or reported in the literature (Hacker et al. 2013). Recommendations for avoidance of this complication are relative contraindication in osteoporotic patients, maximizing the footprint of the implant, avoidance of oversizing the disc space, and preserving the endplate integrity during disc space preparation.

Postoperative kyphosis has been observed following total disc arthroplasty. This is also felt to be multifactorial, with contributions such as excessive anterior superior endplate removal during endplate preparation, incorrect angle of insertion, and amount and direction of distraction during endplate preparation (Sears et al. 2007). Again, outcomes have been evaluated in the setting of postoperative kyphosis. Pickett demonstrated preserved range of motion and no significant difference in outcomes despite focal kyphosis, and overall cervical alignment was maintained (Pickett et al. 2004).

Vertebral body fractures are postulated to be a possible complication, particularly with the keeled implant either during insertion or postoperatively. This is potentially more relevant if multilevel keeled implants are placed, but reports are infrequent to date (Shim et al. 2007; Datta et al. 2007).

In an era of heightened awareness to bearing surface wear with resultant particulate debris and metallosis, this is certainly a concern for the majority of cervical disc replacement implant designs. There is, however, a paucity in the literature regarding clinical examples of this problem. In the cervical spine, Cavanaugh presented a case report of metal ion reactivity resulting in hypertrophic tissue formation posterior to the device and subsequent neural compression. This was addressed with removal of the implant, revision decompression, and anterior fusion with resolution of symptoms (Cavanaugh et al. 2009). More instances of bearing wear-related complications have been presented in the lumbar literature, although true incidence remains unknown (Kurtz et al. 2007; van Ooij et al. 2007; Hallab 2009).

Finally, persistent pain is always a concern following any surgical procedure intended to address pain. As related to cervical disc arthroplasty, ongoing radiculopathy is most often due to incomplete decompression, particularly in a motion sparing technique where osteophytes can progress if not completely removed at the time of the index procedure (Goffin et al. 2002).

Revision Options

While interest in cervical disc arthroplasty continues to grow, the extent of need for revision remains to be seen. There is a paucity in the literature regarding this topic at this time. In general, the revision procedure will largely depend on the underlying problem. Replacement of the device may be considered if the issue is positioning or inadequate decompression after the index procedure. If there is particulate reaction, revision may necessitate conversion to fusion. Corpectomy and anterior column reconstruction

may be needed if there is excessive bone loss. Most surgical technique guides recommend simply separating the bone-implant interface with an osteotome or similar device and removing it in a manner similar to which it was placed for implant removal; however in practice this may not always be the case. In the author's experience, some painful cervical arthroplasty devices have been grossly loose and are easily removed during the revision procedure. If radiculopathy is felt to be from recurrent foraminal stenosis secondary to osteophyte formation, some others advocate for posterior foraminotomy to avoid a revision anterior procedure. Likewise, if the pathology dictates, posterior cervical fusion alone is also sometimes a consideration, again to avoid anterior reoperation.

Outcomes

Overall, anterior cervical disc arthroplasty seems to be favorable compared anterior cervical discectomy and fusion in both short- and medium-term studies for both one- and two-level disease (Sasso et al. 2011; Mummaneni et al. 2007; Heller et al. 2009; Murrey et al. 2009; Zou et al. 2017). There is some evidence that two-level cervical arthroplasty procedures may fare better than single-level procedures, perhaps by protection of levels that are already degenerating (Radcliff et al. 2017; Mehren et al. 2018; Sasso et al. 2017). With the technology being available for the better part of two decades at this point, longer-term data are continuing to show favorable outcomes. Some of these longer-term reports are smaller cohorts and without similar rigor as was reported in the original IDE studies that had robust comparisons to traditional anterior cervical fusion, but there is some data suggesting that this option is durable and at least no worse than anterior fusion at these longer intervals. Rates of reoperation for adjacent segment degeneration remain lower than for fusion, although the differences not reach statistical significant (Ghobrial et al. 2018). Sasso and Dejaegher have shown durable outcomes at 10 years, with favorable results and reoperation profiles compared to anterior cervical fusion. Likewise, Pointillart recently reported excellent outcomes in 80% of their patients 15 years out from cervical disc arthroplasty (Sasso et al. 2017; Dejaegher et al. 2017; Pointillart et al. 2018).

Surgical Techniques: Lumbar Disc Arthroplasty

Indications

- Degenerative disc disease
  - Most often single level, although multilevel use has been reported.
  - Demonstrated on MRI, CT, and/or plain radiographs.
  - Utilization of discography for confirmation of degenerative disc disease being causative for low back pain is suggested in some prior studies and technique guides as some have found it helpful for predicting improved outcome after surgery; however subsequent studies have shown increased rates of degenerative disc disease progression with the use of discography (Colhoun et al. 1988; Carragee et al. 2009). At this time, use of discography remains controversial, although anecdotally seems to have largely fallen out of favor.
- L3-S1 levels
- Failure of conservative measures for at least 6 months

Contraindications

- Instability
  - Spondylolisthesis
  - Spondylolysis
- Deformity
- Severe facet degeneration
  - With or without hypertrophy resulting in lateral recess stenosis
- Herniated nucleus pulposus resulting in radiculopathy

- Osteoporosis or osteopenia (T-score less than −1.5)
  - Metabolic disease resulting in compromised integrity of a bone architecture and/or remodeling
- Infection
- Pregnancy
- Prior trauma or fracture at affected level
  - Large Schmorl's nodes involving endplate at the affected levels
- Vascular calcification
- Metal or materials allergy

Relevant Anatomy

For the lumbar total disc replacements discussed in this section, an anterior approach to the spine is utilized. This can be trans- or retroperitoneal, depending on surgeon preference. Some spine surgeons may utilize an access surgeon to perform the approach.

Vessels
- **Aorta** is the largest artery in the body and courses anterior to the spine, left of and ventral to the inferior vena cava. The bifurcation into the common iliac arteries often occurs near the L5 vertebral body. While injury to the aorta itself is rare, if the great vessels need to be mobilized proximal to the bifurcation, segmental lumbar arteries that come directly off the aorta must be identified, isolated, and ligated to prevent significant blood loss, which can be more difficult to control if the vessels retract when avulsed.
- **Inferior vena cava** (IVC) is rarely encountered as it is predominantly a right-sided structure, and most approaches are left sided to (1) avoid injury to the IVC and (2) because there often is a more favorable plane on the left compared to the right of the great vessels leading to the anterior spine. If the IVC or a direct branch going to it is injured, hemorrhage can be massive and swift.

- **Iliac arteries and veins** – Injury to the left common iliac vein is one of the most commonly reported vascular injuries sustained during this approach and can result in massive hemorrhage in a relatively short amount of time. Often, the vessel can be repaired and the remainder of the procedure completed. Anterior lumbar procedures targeted at the L5-S1 level are typically performed caudal to the bifurcation of the aorta and vena cava and between the common iliac arteries and veins. At more proximal levels rostral to the bifurcations, these vessels need to be mobilized to allow adequate access to the targeted disc spaces.
- **Segmental vessels including the iliolumbar vein** can also cause significant bleeding which can be difficult to control unless these vessels are anticipated, identified, and ligated. Particularly the iliolumbar vein, which can be a large but very thin-walled structure traversing from the posterior aspect of the psoas muscle coursing to the left common iliac or IVC at the L4–5 level. This structure can often be identified on preoperative imaging to facilitate planning; however the surgeon must be aware of this vessel to control a prior to avulsion and retraction into the psoas, which can make it particularly difficult to control.

Ureter is a retroperitoneal structure which is identified by its peristalsis and mobilized medially along with the peritoneal contents during a retroperitoneal approach. One must avoid injuring it.

Sympathetic plexus is a latticework of nerve fibers, the superior hypogastric plexus, that runs anterior to the spine and the great vessels and medial to the iliac vessels. Injury to this structure can result in sexual dysfunction, specifically retrograde ejaculation. Patients must be counseled preoperatively on this potential risk, and younger patients may wish to consider further family planning options prior to undergoing an anterior lumbar procedure. A retroperitoneal approach carries a lower risk of injury to the structure compared to a transperitoneal approach. Additionally, blunt or bipolar dissection is recommended at the level and depth of the vessels to minimize risk of injury to these nerve fibers. Although rare, sympathetic dysfunction may occur resulting in ipsilateral lower extremity vasodilation which can mimic deep vein thrombosis. Subjectively the contralateral leg may feel cool relative to the warm ipsilateral lower extremity. This dysfunction typically resolves with observation.

Positioning and Approach

The patient is positioned supine on a radiolucent operating table with the arms out to the sides or crossed over a pillow on the chest. Some surgeons advocate for placement of a bump beneath the sacrum to bring the lumbar spine into a more accessible position. It should be noted, however, that the bump should not be placed beneath the lordotic portion of the lumbar spine so as not to exaggerated lumbar lordosis which may result in improper implant positioning. If possible, the patient position on the operating table should facilitate storage of the fluoroscopy machine when not in use.

There are several options to gain anterior exposure to the lumbar spine such as trans- or retroperitoneal, midline or paramedian, open, mini open, or laparoscopic assisted. For an open, retroperitoneal approach, the incision is localized over the target disc space using lateral fluoroscopy. Subcutaneous dissection is performed down to fascia, which is also incised. The rectus is mobilized either medially or laterally, depending on the approach and the necessary trajectory. The preperitoneal space is identified and entered, and the peritoneum and its contents are mobilized medially to allow access to the retroperitoneum. The ureter should be identified in this plane and mobilized with the peritoneum. The great vessels are identified and gently mobilized as needed for access to the desired disc space. At L4–5, the iliolumbar vein is identified, ligated, and divided to avoid inadvertent avulsion and hemorrhage. At L5-S1, the middle sacral artery is isolated and ligated to allow unimpeded access to this disc space. At the level of the vessels and spine, blunt and bipolar dissection is used to minimize risk of injury to the sympathetic plexus. Fixed retractors can then be placed. The targeted disc is confirmed with lateral fluoroscopy, and the midline is marked using AP fluoroscopy. A standard annulotomy and diskectomy are performed, avoiding violation of the endplates.

Implant-Specific Instrumentation

ProDisc-L II (DePuy Synthes) (Prodisc-L 2017) (Fig. 8)
- *Device Type*
  - Metal on plastic (polyethylene)
  - Ball and socket
- *Procedure*

  After a standard discectomy has been performed, the intervertebral space is distracted with the spreader. A trial is placed to assess the implant height, size, and degree of lordosis. The keel tract is prepared with the chisel. During this step, position and trajectory of the keel must be confirmed as this will establish the implant position. The prosthesis is modular such that there are several options for lordosis of each endplate and insert heights to most accurately reconstruct the native disc space. The selected prosthetic endplates are inserted. Disc space is distracted, and the polyethylene inlay is inserted into the caudal endplate. Final position is confirmed using lateral and AP fluoroscopy.

Postoperative Protocol

Much of the postoperative protocol is at the discretion of the surgeon. In general, avoidance of aggressive bending, twisting, or lifting is recommended for 6 weeks followed by gradual return to full activity thereafter. Postoperative bracing is utilized based on surgeon preference, but not required.

Complications

As can be seen with anterior lumbar interbody fusion, approach-related complications do occur. These include injuries to adjacent vasculature, sympathetic plexus, ureter, and rarely lymphatic ducts. The rates of these complications are similar as to what is seen in anterior lumbar interbody fusion (Blumenthal et al. 2005). Heterotopic ossification has been reported in up to 50% of patients; however this often does not result

Fig. 8 Prodisc-L prosthesis. (Reprinted with permission from Atkins, et al. *Lumbar Disc Arthroplasty*. In: Essentials of Spinal Stabilization. Holly L., Anderson P. (eds). Springer, Cham. 2017)

in inferior clinical outcomes (Park et al. 2018). Jackson et al. did report a case in which heterotopic ossification along with implant malposition resulted in a new radiculopathy (Jackson et al. 2015). Symptoms resolved with revision for implant removal, anterior interbody fusion, posterior decompression, and pedicle screw fixation. Implant-related complications such as subsidence, dislocation, or luxation have been reported (Kurtz et al. 2007; Kostuik 2004). Additionally, bearing surfaces do raise the concern abnormal wear, particulate degeneration, and adjacent inflammatory changes. There are case reports and small series of the instances resulting in inflammation and osteolysis. Authors have postulated that suboptimal local biomechanics such as adjacent level fusion, incorrect implant sizes, and impingement may all be contributing factors. Study of removed implants has demonstrated both abrasive and adhesive wear of the polyethylene (Kurtz et al. 2007; van Ooij et al. 2007). Finally, persistent pain postoperatively has been reported. This is also likely multifactorial. It is well-known that there are multiple possible pain generators in the lumbar spine, and disc replacement does not address all of these. Facet degeneration pre- or postoperatively may be may be a major contributor to ongoing pain. Of 91 patients at a single IDE site, 50% of failures were secondary to facet pathology (Pettine et al. 2017).

Revision Options

As is the case with cervical disc arthroplasty revision, the lumbar revision procedure performed ultimately depends on the underlying pathology to be addressed at the time of surgery. Options include revision for replacement of an arthroplasty device, anterior revision for lumbar interbody fusion with or without posterior instrumentation, or posterior lateral instrumented fusion alone without anterior revision. Repeating an anterior exposure may be needed for situations such as arthroplasty device migration but should otherwise be considered with caution as adhesions can be problematic, and there is higher risk of vascular and visceral injury.

Outcomes

The SB Charité lumbar prosthetic was implanted for a period of nearly 20 years. Despite its eventual withdrawal from the market in 2013, this lumbar device has the longest available follow-up data and permits inquiry into the longevity of lumbar total disc replacement systems. Lemaire and colleagues presented 10-year minimum follow-up results in their retrospective case series of 100 patients implanted with the SB Charité III between 1989 and 1993 for the indication of intractable discogenic back pain. The authors used a modified Stauffer-Coventry scoring system which expresses results as relative gain. A relative gain of $\geq 70\%$ indicates an excellent outcome and is defined as no pain, no medication use, and resumption of activity in the same job after 3 months. Ninety percent of patients in their series had an excellent or good outcome at 10 years, and 92% of eligible patients returned to the work force in some capacity. Radiographic analysis at 10 years showed that the Charité maintained normal range of motion in 95% of patients with a mean flexion/extension arc of 10.3°. Five patients underwent secondary arthrodesis at the index level for poor outcomes and the symptomatic adjacent level disease reoperation rate was 2% (Lemaire et al. 2005). David et al. found very similar positive results (82% with excellent or good outcomes) in their 10-year minimum retrospective case series of 106 patients. These authors reported a 10% index level and a 3% adjacent level reoperation rate (David 2007). The longest prospective data reported is the 5-year results from the US FDA IDE randomized controlled trial comparing the Charité to lumbar fusion. Ninety patients randomized to the Charité group between 2000 and 2002 were available for follow-up 5 years later. Guyer et al. found that Oswestry Disability Index (ODI), SF-36, and Visual Analog Scale (VAS) scores maintained clinically significant improvements over baseline. Overall clinical success, defined by the FDA, was achieved in 58% of the Charité patients and 51% of the arthrodesis patients still after 5 years. Seven of 90 cases were reported as "failures" necessitating index level reoperation, and adjacent level disease

Fig. 9 ActivL prosthesis. (Reprinted with permission from Atkins, et al. *Lumbar Disc Arthroplasty*. In: Essentials of Spinal Stabilization. Holly L., Anderson P. (eds). Springer, Cham. 2017)

reoperation rates were 1.1% and 4.7% for the Charité and arthrodesis, respectively (Guyer et al. 2009).

Outcomes of the ProDisc-L (DePuy Synthes) lumbar artificial disc are perhaps the most relevant at this juncture given that it remains commercially available and has the longest track record. Park et al. followed 35 patients for a mean of 6 years. Subjective outcome surveys were quite encouraging as 31 of 35 patients reported being completely or somewhat satisfied with their results. Similarly, 21 of 35 reported that they would definitely or probably undergo lumbar total disc replacement again if represented the option (Park, Spine 2012). Per the FDA-defined clinical success criteria, 71% of the cases qualified (Park, Spine 2012) (Park et al. 2012). In another retrospective case series of 55 patients with an average follow-up of 8.7 years, 75% had excellent or good results (Tropiano et al. 2005). Prospective data also supports lumbar arthroplasty as a reliable alternative to arthrodesis. Siepe and colleagues prospectively reviewed 181 patients after a mean of 7.4 years and found that both VAS and ODI scores were improved with statistical significance compared to baseline preoperative values (Siepe et al. 2014). Eighty-six percent of their patients were highly satisfied or satisfied. They also reported a low adjacent level disease reoperation rate of 2.2% which was comparable to that of the Charité. The most influential data comes from this device's US FDA IDE randomized controlled trial which showed very comparable results at 5 years between the ProDisc-L and circumferential lumbar arthrodesis. FDA-defined clinical success was met by 54% of the lumbar arthroplasty cases and 50% of the fusion cases. Both groups maintained significant improvements in ODI and SF-36 scores compared with baseline values. Restoration of normal lumbar motion, dictated by level, was achieved in 92% of the ProDisc-L cases with a mean flexion-extension arc of 7.2°. The index level reoperation rate was lower in the arthroplasty group (8%) compared to arthrodesis (12%) (Zigler and Delamarter 2012).

There is no long-term follow-up data for the second FDA-approved lumbar total disc replacement system, the ActivL (Aesculap Implant Systems, Fig. 9). It has only been commercially available since 2015. Nonetheless, its 2-year follow-up data appears to show statistically superiority to its predecessors (Garcia et al. 2015).

Conclusion

The theoretical advantage of motion sparing technology for degenerative spinal pathology is appealing. There has been much research and progress on this topic of intervertebral disc replacement over the last several decades, and the future is promising. Despite the advances, an understanding of the failures remains necessary so as not to repeat them. Currently, cervical disc arthroplasty has outpaced lumbar disc arthroplasty. There are more FDA-approved cervical devices than there are lumbar devices, and anecdotally, cervical disc replacement is more widely favored than lumbar. The greater success of cervical disc replacement may stem from the underlying indications when compared to that of lumbar disc replacement; cervical procedures are indicated for degenerative disc disease resulting in radiculopathy or myelopathy, which are more predictably treatable entities, whereas lumbar disc procedures are often contraindicated in the setting of radiculopathy and predominantly indicated in degenerative disc disease only with axial pain, which is a notoriously difficult entity and patient population to treat successfully and predictably. For both cervical and lumbar disc replacement, early and midrange follow-up are now becoming available up and seemingly favorable, but we will need to continue to follow these technologies for long-term data to show whether it is more definitively a durable alternative to arthrodesis.

Cross-References

► Adjacent-Level Disease: Fact and Fiction
► Biological Treatment Approaches for Degenerative Disc Disease: Injectable Biomaterials and Bioartificial Disc Replacement
► Cervical Arthroplasty: Long-Term Outcomes
► Cervical Spine Anatomy
► Cervical Total Disc Replacement: Biomechanics
► Cervical Total Disc Replacement: Evidence Basis
► Cervical Total Disc Replacement: Expanded Indications
► Cervical Total Disc Replacement: FDA-Approved Devices
► Cervical Total Disc Replacement: Heterotopic Ossification and Complications
► Cervical Total Disc Replacement: Next-Generation Devices
► Cervical Total Disc Replacement: Technique – Pitfalls and Pearls
► Lumbar TDR Revision Strategies
► Posterior Lumbar Facet Replacement and Interspinous Spacers
► Thoracic and Lumbar Spinal Anatomy

References

Baaj AA, Uribe JS, Vale FL, Preul MC, Crawford NR (2009) History of cervical disc arthroplasty. Neurosurg Focus 27(3):E10. https://doi.org/10.3171/2009.6.FOC US09128

Basho R, Hood KA (2012) Cervical total disc arthroplasty. Glob Spine J 2(2):105–108. https://doi.org/10.1055/s-0032-1315453

Blumenthal S, McAfee PC, Guyer RD et al (2005) A prospective, randomized, multicenter Food and Drug Administration investigational device exemptions study of lumbar total disc replacement with the CHARITÉ artificial disc versus lumbar fusion: part I: evaluation of clinical outcomes. Spine 30(14): 1565–1575; discussion E387–E391

Bono CM, Garfin SR (2004) History and evolution of disc replacement. Spine J 4(6):S145–S150. https://doi.org/10.1016/j.spinee.2004.07.005

Bryan cervical disc surgical technique. January 2005:1–26

Buckwalter JA (1995) Aging and degeneration of the human intervertebral disc. Spine 20(11):1307–1314

Burkus JK, Traynelis VC, Haid RW, Mummaneni PV (2014) Clinical and radiographic analysis of an artificial cervical disc: 7-year follow-up from the Prestige prospective randomized controlled clinical trial: clinical article. J Neurosurg Spine 21(4):516–528. https://doi.org/10.3171/2014.6.SPINE13996

Büttner-Janz K, Schellnack K, Zippel H (1989) Biomechanics of the SB Charité lumbar intervertebral disc endoprosthesis. Int Orthop 13(3):173–176

Carragee EJ, Don AS, Hurwitz EL et al (2009) 2009 ISSLS Prize Winner: does discography cause accelerated progression of degeneration changes in the lumbar disc: a ten-year matched cohort study. Spine 34(21): 2338–2345. https://doi.org/10.1097/BRS.0b013e3181 ab5432

Cavanaugh DA, Nunley PD, Kerr EJ, Werner DJ, Jawahar A (2009) Delayed hyper-reactivity to metal ions after cervical disc arthroplasty: a case report and literature review. Spine 34(7):E262–E265. https://doi.org/10.1097/BRS.0b013e318195dd60

Chang U-K, Kim DH, Lee MC, Willenberg R, Kim S-H, Lim J (2007) Changes in adjacent-level disc pressure and facet joint force after cervical arthroplasty compared with cervical discectomy and fusion. J Neurosurg Spine 7(1):33–39. https://doi.org/10.3171/SPI-07/07/033

Chen J, Wang X, Bai W, Shen X, Yuan W (2011) Prevalence of heterotopic ossification after cervical total disc arthroplasty: a meta-analysis. Eur Spine J 21(4):674–680. https://doi.org/10.1007/s00586-011-2094-x

Colhoun E, McCall IW, Williams L, Cassar Pullicino VN (1988) Provocation discography as a guide to planning operations on the spine. J Bone Joint Surg Br 70(2): 267–271. https://doi.org/10.1302/0301-620X.70B2.29 64449

Cummins BH, Robertson JT, Gill SS (1998) Surgical experience with an implanted artificial cervical joint. J Neurosurg 88(6):943–948. https://doi.org/10.3171/jns.1998.88.6.0943

Cunningham BW, Kotani Y, McNulty PS, Cappuccino A, McAfee PC (1997) The effect of spinal destabilization and instrumentation on lumbar intradiscal pressure: an in vitro biomechanical analysis. Spine 22(22):2655–2663

Datta JC, Janssen ME, Beckham R, Ponce C (2007) Sagittal split fractures in multilevel cervical arthroplasty using a keeled prosthesis. J Spinal Disord Tech 20(1):89–92. https://doi.org/10.1097/01.bsd.0000211258.90378.10

David T (2007) Long-term results of one-level lumbar arthroplasty: minimum 10-year follow-up of the CHARITÉ artificial disc in 106 patients. Spine 32(6):661–666. https://doi.org/10.1097/01.brs.0000257554.67505.45

Dejaegher J, Walraevens J, van Loon J, Van Calenbergh F, Demaerel P, Goffin J (2017) 10-year follow-up after implantation of the Bryan cervical disc prosthesis. Eur Spine J 26(4):1191–1198. https://doi.org/10.1007/s00586-016-4897-2

Eck JC, Humphreys SC, Lim T-H et al (2002) Biomechanical study on the effect of cervical spine fusion on adjacent-level intradiscal pressure and segmental motion. Spine 27(22):2431–2434. https://doi.org/10.1097/01.BRS.0000031261.66972.B1

Garcia R Jr, Yue JJ, Blumenthal S et al (2015) Lumbar total disc replacement for discogenic low back pain. Spine 40(24):1873–1881. https://doi.org/10.1097/BRS.0000000000001245

Ghiselli G, Wang JC, Bhatia NN, Hsu WK, Dawson EG (2004) Adjacent segment degeneration in the lumbar spine. J Bone Joint Surg Am 86-A (7):1497–1503

Ghobrial GM, Lavelle WF, Florman JE, Riew KD, Levi AD (2018) Symptomatic adjacent level disease requiring surgery: analysis of 10-year results from a prospective, randomized, clinical trial comparing cervical disc arthroplasty to anterior cervical fusion. Neurosurgery 21(1):34. https://doi.org/10.1093/neuros/nyy118

Goffin J, Casey A, Kehr P et al (2002) Preliminary clinical experience with the Bryan cervical disc prosthesis. Neurosurgery 51(3):840–845; discussion 845–847. https://doi.org/10.1227/01.NEU.0000026100.14273.B3

Goffin J, Geusens E, Vantomme N et al (2004) Long-term follow-up after interbody fusion of the cervical spine. J Spinal Disord Tech 17(2):79–85

Gore DR, Sepic SB (1998) Anterior discectomy and fusion for painful cervical disc disease. A report of 50 patients with an average follow-up of 21 years. Spine 23(19):2047–2051

Guyer RD, McAfee PC, Banco RJ et al (2009) Prospective, randomized, multicenter Food and Drug Administration investigational device exemption study of lumbar total disc replacement with the CHARITÉ artificial disc versus lumbar fusion: five-year follow-up. Spine J 9(5):374–386. https://doi.org/10.1016/j.spinee.2008.08.007

Hacker FM, Babcock RM, Hacker RJ (2013) Very late complications of cervical arthroplasty. Spine 38(26):2223–2226. https://doi.org/10.1097/BRS.0000000000000060

Hallab NJ (2009) A review of the biologic effects of spine implant debris: fact from fiction. Int J Spine Surg 3(4):143–160. https://doi.org/10.1016/j.esas.2009.11.005

Heller JG, Sasso RC, Papadopoulos SM et al (2009) Comparison of BRYAN cervical disc arthroplasty with anterior cervical decompression and fusion: clinical and radiographic results of a randomized, controlled, clinical trial. Spine 34(2):101–107. https://doi.org/10.1097/BRS.0b013e31818ee263

Hilibrand AS, Carlson GD, Palumbo MA, Jones PK, Bohlman HH (1999) Radiculopathy and myelopathy at segments adjacent to the site of a previous anterior cervical arthrodesis. J Bone Joint Surg Am 81(4):519–528

Hisey M, Zigler J, Jackson R et al (2016) Prospective, randomized comparison of one-level Mobi-C cervical total disc replacement vs. anterior cervical discectomy and fusion: results at 5-year follow-up. Int J Spine Surg 10:1–10. https://doi.org/10.14444/3010

Hutton WC, Toribatake Y, Elmer WA, Ganey TM, Tomita K, Whitesides TE (1998) The effect of compressive force applied to the intervertebral disc in vivo. A study of proteoglycans and collagen. Spine 23(23):2524–2537

Jackson KL, Hire JM, Jacobs JM, Key CC, DeVine JG (2015) Heterotopic ossification causing radiculopathy after lumbar total disc arthroplasty. Asian Spine J 9(3):456–460. https://doi.org/10.4184/asj.2015.9.3.456

Janssen ME, Zigler JE, Spivak JM, Delamarter RB, Darden BV II, Kopjar B (2015) ProDisc-C total disc replacement versus anterior cervical discectomy and fusion for single-level symptomatic cervical disc disease. J Bone Joint Surg Am 97(21):1738–1747. https://doi.org/10.2106/JBJS.N.01186

Kostuik JP (2004) Complications and surgical revision for failed disc arthroplasty. Spine J 4(6):S289–S291. https://doi.org/10.1016/j.spinee.2004.07.021

Kurtz SM, van Ooij A, Ross R et al (2007) Polyethylene wear and rim fracture in total disc arthroplasty. Spine J 7(1):12–21. https://doi.org/10.1016/j.spinee.2006.05.012

Le H, Thongtrangan I, Kim DH (2004) Historical review of cervical arthroplasty. Neurosurg Focus 17(3):E1–E9. https://doi.org/10.3171/foc.2004.17.3.1

Lee CK, Langrana NA (1984) Lumbosacral spinal fusion. A biomechanical study. Spine 9(6):574–581

Lee J-H, Jung T-G, Kim H-S, Jang J-S, Lee S-H (2010) Analysis of the incidence and clinical effect of the heterotopic ossification in a single-level cervical artificial disc replacement. Spine J 10(8):676–682. https://doi.org/10.1016/j.spinee.2010.04.017

Lemaire J-P, Carrier H, Sariali E-H, Sari Ali E-H, Skalli W, Lavaste F (2005) Clinical and radiological outcomes with the Charité artificial disc: a 10-year minimum follow-up. J Spinal Disord Tech 18(4):353–359

Leung C, Casey AT, Goffin J et al (2005) Clinical significance of heterotopic ossification in cervical disc replacement: a prospective multicenter clinical trial. Neurosurgery 57(4):759–763; discussion 759–763

Link HD (2002) History, design and biomechanics of the LINK SB Charité artificial disc. Eur Spine J 11(Suppl 2):S98–S105. https://doi.org/10.1007/s00586-002-0475-x

Martin BI, Mirza SK, Comstock BA, Gray DT, Kreuter W, Deyo RA (2007) Reoperation rates following lumbar spine surgery and the influence of spinal fusion procedures. Spine 32(3):382–387. https://doi.org/10.1097/01.brs.0000254104.55716.46

Matsumoto M, Okada E, Ichihara D et al (2010) Anterior cervical decompression and fusion accelerates adjacent segment degeneration: comparison with asymptomatic volunteers in a ten-year magnetic resonance imaging follow-up study. Spine 35(1):36–43. https://doi.org/10.1097/BRS.0b013e3181b8a80d

Mehren C, Heider F, Siepe CJ et al (2018) Clinical and radiological outcome at 10 years of follow-up after total cervical disc replacement. Eur Spine J 26(9):2441–2449. https://doi.org/10.1007/s00586-017-5204-6

Mobi-C cervical disc surgical technique guide. January 2016:1–32

Mummaneni PV, Burkus JK, Haid RW, Traynelis VC, Zdeblick TA (2007) Clinical and radiographic analysis of cervical disc arthroplasty compared with allograft fusion: a randomized controlled clinical trial. J Neurosurg Spine 6(3):198–209. https://doi.org/10.3171/spi.2007.6.3.198

Murrey D, Janssen M, Delamarter R et al (2009) Results of the prospective, randomized, controlled multicenter Food and Drug Administration investigational device exemption study of the ProDisc-C total disc replacement versus anterior discectomy and fusion for the treatment of 1-level symptomatic cervical disc disease. Spine J 9(4):275–286. https://doi.org/10.1016/j.spinee.2008.05.006

Park P, Garton HJ, Gala VC, Hoff JT, McGillicuddy JE (2004) Adjacent segment disease after lumbar or lumbosacral fusion: review of the literature. Spine 29(17):1938–1944

Park C-K, Ryu K-S, Lee K-Y, Lee H-J (2012) Clinical outcome of lumbar total disc replacement using ProDisc-L in degenerative disc disease: minimum 5-year follow-up results at a single institute. Spine 37(8):672–677. https://doi.org/10.1097/BRS.0b013e31822ecd85

Park H-J, Lee C-S, Chung S-S et al (2018) Radiological and clinical long-term results of heterotopic ossification

following lumbar total disc replacement. Spine J 1–7. https://doi.org/10.1016/j.spinee.2017.09.003

Pettine K, Ryu R, Techy F (2017) Why lumbar artificial disk replacements (LADRs) fail. Clin Spine Surg 30(6):E743–E747. https://doi.org/10.1097/BSD.0000000000000310

Phillips FM, Geisler FH, Gilder KM, Reah C, Howell KM, McAfee PC (2015) Long-term outcomes of the US FDA IDE prospective, randomized controlled clinical trial comparing PCM cervical disc arthroplasty with anterior cervical discectomy and fusion. Spine 40(10):674–683. https://doi.org/10.1097/BRS.0000000000000869

Pickett GE, Mitsis DK, Sekhon LH, Sears WR, Duggal N (2004) Effects of a cervical disc prosthesis on segmental and cervical spine alignment. Neurosurg Focus 17(3):E5

Pointillart V (2001) Cervical disc prosthesis in humans: first failure. Spine 26(5):E90–E92

Pointillart V, Castelain J-E, Coudert P, Cawley DT, Gille O, Vital J-M (2018) Outcomes of the Bryan cervical disc replacement: fifteen year follow-up. Int Orthop 42:1–7. https://doi.org/10.1007/s00264-017-3745-2

Prestige LP cervical disc system surgical technique. January 2009:1–24

Prodisc-L surgical technique. January 2017:1–40

Punt IM, Visser VM, van Rhijn LW et al (2007) Complications and reoperations of the SB Charité lumbar disc prosthesis: experience in 75 patients. Eur Spine J 17(1):36–43. https://doi.org/10.1007/s00586-007-0506-8

Radcliff K, Davis R, Hisey M et al (2017) Long-term evaluation of cervical disc arthroplasty with the Mobi-C© cervical disc: a randomized, prospective, multicenter clinical trial with seven-year follow-up. Int J Spine Surg 11(5):244–262. https://doi.org/10.14444/4031

Reitz H, Joubert MJ (1964) Intractable headache and cervico-brachialgia treated by complete replacement of cervical intervertebral discs with a metal prosthesis. S Afr Med J 38:881–884

Sasso RC, Anderson PA, Riew KD, Heller JG (2011) Results of cervical arthroplasty compared with anterior discectomy and fusion: four-year clinical outcomes in a prospective, randomized controlled trial. J Bone Joint Surg Am 93(18):1684–1692. https://doi.org/10.2106/JBJS.J.00476

Sasso WR, Smucker JD, Sasso MP, Sasso RC (2017) Long-term clinical outcomes of cervical disc arthroplasty. Spine 42(4):209–216. https://doi.org/10.1097/BRS.0000000000001746

Sears WR, Duggal N, Sekhon LH, Williamson OD (2007) Segmental malalignment with the Bryan cervical disc prosthesis – contributing factors. J Spinal Disord Tech 20(2):111–117. https://doi.org/10.1097/01.bsd.0000211264.20873.78

Shim CS, Shin H-D, Lee S-H (2007) Posterior avulsion fracture at adjacent vertebral body during cervical disc replacement with ProDisc-C: a case report. J Spinal Disord Tech 20(6):468–472

Shriver MF, Lewis DJ, Kshettry VR, Rosenbaum BP, Benzel EC, Mroz TE (2015) Pseudoarthrosis rates in anterior cervical discectomy and fusion: a meta-analysis. Spine J 15(9):2016–2027. https://doi.org/10.1016/j.spinee.2015.05.010

Siepe CJ, Heider F, Wiechert K, Hitzl W, Ishak B, Mayer MH (2014) Mid- to long-term results of total lumbar disc replacement: a prospective analysis with 5- to 10-year follow-up. Spine J 14(8):1417–1431. https://doi.org/10.1016/j.spinee.2013.08.028

Szpalski M, Gunzburg R, Mayer M (2002) Spine arthroplasty: a historical review. Eur Spine J 11(Suppl 2):S65–S84. https://doi.org/10.1007/s00586-002-0474-y

Tropiano P, Huang RC, Girardi FP, Cammisa FP, Marnay T (2005) Lumbar total disc replacement. Seven to eleven-year follow-up. J Bone Joint Surg Am 87(3):490–496. https://doi.org/10.2106/JBJS.C.01345

Tu T-H, Wu J-C, Huang W-C et al (2011) Heterotopic ossification after cervical total disc replacement: determination by CT and effects on clinical outcomes. J Neurosurg Spine 14(4):457–465. https://doi.org/10.3171/2010.11.SPINE10444

Turner JA, Ersek M, Herron L et al (1992) Patient outcomes after lumbar spinal fusions. JAMA 268(7):907–911

Vaccaro A, Beutler W, Peppelman W et al (2013) Clinical outcomes with selectively constrained SECURE-C cervical disc arthroplasty. Spine 38(26):2227–2239. https://doi.org/10.1097/BRS.0000000000000031

van Ooij A, Oner FC, Verbout AJ (2003) Complications of artificial disc replacement: a report of 27 patients with the SB Charité disc. J Spinal Disord Tech 16(4):369–383

van Ooij A, Kurtz SM, Stessels F, Noten H, van Rhijn L (2007) Polyethylene wear debris and long-term clinical failure of the Charité disc prosthesis: a study of 4 patients. Spine 32(2):223–229. https://doi.org/10.1097/01.brs.0000251370.56327.c6

Wigfield CC, Gill SS, Nelson RJ, Metcalf NH, Robertson JT (2002) The new Frenchay artificial cervical joint: results from a two-year pilot study. Spine 27(22):2446–2452. https://doi.org/10.1097/01.BRS.0000032365.21711.5E

Zigler JE, Delamarter RB (2012) Five-year results of the prospective, randomized, multicenter, Food and Drug Administration investigational device exemption study of the ProDisc-L total disc replacement versus circumferential arthrodesis for the treatment of single-level degenerative disc disease. J Neurosurg Spine 17(6):493–501. https://doi.org/10.3171/2012.9.SPINE11498

Zou S, Gao J, Xu B, Lu X, Han Y, Meng H (2017) Anterior cervical discectomy and fusion (ACDF) versus cervical disc arthroplasty (CDA) for two contiguous levels cervical disc degenerative disease: a meta-analysis of randomized controlled trials. Eur Spine J 26(4):985–997. https://doi.org/10.1007/s00586-016-4655-5

Part VI

International Experience: Surgery

The Diagnostic and the Therapeutic Utility of Radiology in Spinal Care

52

Matthew Lee and Mario G. T. Zotti

Contents

Introduction	926
Comments About Radiation	926
Noninvasive Techniques	927
X-ray	927
Fluoroscopy	929
Computed Tomography	929
MRI	930
Nuclear Medicine	933
Comments About Pain Generators	934
Biomechanical and Chemical Models for Disc and Facet Pain	934
Invasive Interventions	935
Corticosteroid Injections	937
Types of Corticosteroid Injection	938
Sacroiliac Injections	942
Coccyx Injections	942
Other Invasive Forms of Image-Guided Treatment	943
Vertebroplasty and Kyphoplasty	943
Spinal Stimulators	943
Measuring Success of Injections/Radiology Treatments	944
Conclusions	944
References	944

M. Lee (✉)
Western Imaging Group, Blacktown, NSW, Australia
e-mail: info@westernimaginggroup.com.au

M. G. T. Zotti
Orthopaedic Clinics Gold Coast, Robina, QLD, Australia

Gold Coast Spine, Southport, QLD, Australia

© Crown 2021
B. C. Cheng (ed.), *Handbook of Spine Technology*,
https://doi.org/10.1007/978-3-319-44424-6_83

Abstract

Advancements in technology have been a driving force in the development of medical imaging equipment. There are now multiple, varied, and intertwined imaging modalities which visualize the spine in many different formats and positions. This has facilitated an increase in accuracy of diagnosis, and then at the same time allowed medical imaging to be an essential tool, in treatment of spinal conditions. Multiple image-guided therapies are now available to assist physicians/surgeons with treatment regimens and pre- and postoperative surgical planning. This chapter will detail the above.

Keywords

Radiology · Spinal · Neuroradiology · CT · Computed tomography · MRI · Magnetic resonance imaging · Nuclear medicine · EOS · Myelography · Medial branch block · Discography · Intervention · Corticosteroid injection · Aspiration · X-ray · Fluoroscopy

Introduction

Radiology, or medical imaging, has advanced dramatically since the discovery of the X-ray by William Roentgen in 1895. Over the last 120 years, this scientific discovery has evolved from being a novelty nonmedical commercial and social photographic studio tool to a necessity, essential to physicians and surgeons throughout the world (American Society of Radiologic Technologists 2018).

The computer age and advancing technologies allowed the use of X-rays and then other forms of radiation to be progressed into more sophisticated imaging equipment. Thus, medical diagnosis has progressed well beyond the first point of physician-patient contact – history and examination – as the diagnosis or differential diagnoses can be radiologically narrowed or confirmed and the pathology directly viewed within the patient.

The radiology/medical imaging field has taken a dual role of diagnosing and treating. Radiology canbe a primary source of treatment or an adjunctive intervention to both surgical pre- and postoperative care.

Within the field of spine care, radiology has assumed such an important role in the detection, diagnosis, and treatment of spine and spine-related disorders. Thanks to the radiologists, the spine surgeon can deliver a precision diagnosis and with that therapeutic options. When combined with the quantification and prognostication afforded by imaging (e.g., the grade of spondylolisthesis or the amount of sagittal imbalance), this forms an invaluable trinity of diagnosis, quantification, and therapy (see Fig. 1).

The diagnostic and treating armamentarium available to the patient and physician from the radiology specialty is as follows:

- Diagnostic:
  - Noninvasive:
    X-ray
    Fluoroscopy
    CT
    MRI
    Nuclear medicine
  - Invasive:
    Myelography
    Medial branch block
    Discography
- Therapeutic:
  - *Facet joint/medial branch corticosteroid injections*
  - *Epidural/perineural corticosteroid injections*
  - *Synovial cyst puncture and aspiration*
  - *Sacroiliac joint corticosteroid injections*
  - *Coccyx injections*
  - *Facet joint denervations*
  - *Vertebroplasty and kyphoplasty*
  - *Insertion of stimulators*

Comments About Radiation

Plain radiographs, CT, and fluoroscopy all produce ionizing radiation and hence the ability to cause cancer or birth defects, via damage either

Fig. 1 The unity of diagnosis and treatment clearly shown in verification of deformity correction

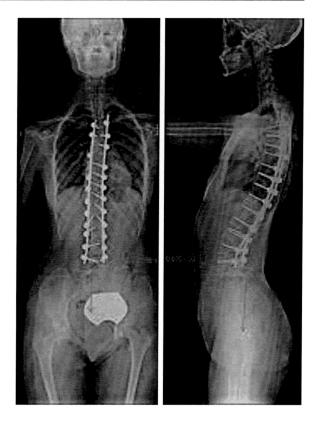

to the reproductive organs or to the developing embryo directly. However, the use of radiation from medical imaging procedures when ordered prudently and for the specific benefit of diagnosis or treatment leads to minimal hypothetical risks especially in relation to cancer deaths and estimated cancers produced and even more so when the principles of ALARA (radiation dose as low as possible), ASARA (medical procedures as safe as reasonably achievable), and AHARA (medical benefits as high as reasonably achievable) are followed (Hendee and O'Connor 2012). With technologies improving all the time, the radiation dose from all forms of imaging is becoming less, and the major equipment suppliers make this a standard in design and development and market accordingly. Despite lessening radiation doses from improving technology, the link of abdominal radiation dose with solid organ malignancy mandates careful assessment of risks and benefits from the ordering of tests involving ionizing radiation.

Noninvasive Techniques

X-ray

The humble radiograph. With all the new modalities for imaging now available, the spine radiograph is of less diagnostic importance as CT and MRI provide far more detail. The fact the plain radiograph cannot show soft tissue details of the spine, only bone, and can only image in limited planes is its major drawback, and the radiograph provides a 2D representation of a 3D structure.

However it still does play an essential role in the investigative role of diagnosing spine and spine disorders and as such should not be dismissed as an irrelevant investigation but a useful investigation in the first line of the diagnostic pathway. Despite government detractors that criticize the plain radiograph from the point of view of ionizing radiation and the lack of benefit, the radiograph provides a positive yield in many situations, clinical and diagnostic, when applied specifically to the clinical situation.

Traditional images are AP and lateral views with oblique or functional views being added depending upon the request of the referrer or individual protocols of the radiology practice. The lateral view plain film will show alignment of the spine – confirming the normal or abnormal lordosis or kyphosis of the cervical, lumbar, and thoracic spines, respectively, and the AP film curvature or more scoliosis. Also disc space narrowing, i.e., degeneration and possibly foraminal stenosis, may also be revealed and of course a bony lesion. Oblique lumbar radiographs may be ordered in the case of spondylolysis to detect pars defects.

For a lumbar spine radiograph, the question of radiation to the reproductive organs is always of some concern. However, like with any radiograph, it must be balanced against its benefit, particularly in the younger person.

An exciting new technology, which has only become available in the last few years, is EOS™. It takes plain film spinal radiography to a new level. Firstly, the radiation dose is about 50% less than for digital radiography; hence the dose is almost negligible. Secondly, the entire skeleton – the chest, upper limbs, entire spine, pelvis, hips, and lower limbs – can be viewed in the weight-bearing position. Both frontal and lateral images are obtained, and from the images, 3D modelling is performed. This allows detailed analysis of the kyphotic and lordotic state of the spine and, of course, scoliosis (see Fig. 2). Hence, the normal distribution of weight, stresses, and angles throughout the axial skeleton can be assessed. Many parameters are measured including the C7 plumb line, kyphosis and lordosis, thoracic and lumbar vertebral and intervertebral rotations, spino-sacral angle, pelvic incidence/version and sacral slope, pelvic obliquity and rotation/tilt, Cobb angle, and scoliosis. Further measurements in relation to lower limb leg lengths and hip and knee angle and alignment parameters can be carried out. With all this additional

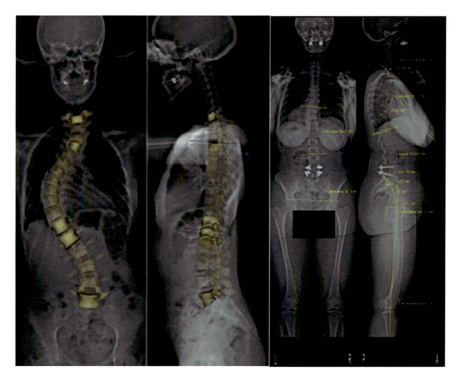

Fig. 2 EOS™ showing images of the whole spine – AP, lateral, and 3D reconstructions – with measurements for assessment of surgical balance as a forerunner to surgical treatment and planning

information, surgical procedures can now be planned (including types and requirements of reconstructions) to take into account the entire axial skeleton rather than solely the symptomatic area in question (Amzallag-Bellenger et al. 2014).

Fluoroscopy

This is using X-rays to allow real-time (i.e., dynamic) imaging. The machinery and technology have developed over time like that of the X-ray machine. It works on a similar principle to the traditional X-ray machine; however it is of low intensity (and hence low radiation) and therefore is coupled with an image intensifier which allows the image to be seen (without the need for a darkened room) (Amzallag-Bellenger et al. 2014).

The fluoroscopy unit has its main use as an adjunct to spinal procedures particularly aiding in needle placement for injections, both for diagnostic, e.g., discography and myelography, and treatment regimens, e.g., corticosteroid – *see below*. It is also used in theater for spinal level checks and aids in planning and confirming spinal surgical hardware placement and position. Such advancements have allowed reduction in malpositioned screws and cages that, if unrecognized, could present problems in the perioperative period for the patient (Amzallag-Bellenger et al. 2014; Goodbody et al. 2017; Deschenes et al. 2010; Laredo et al. 2010).

Computed Tomography

CT was part of the spinal imaging evolution, both diagnostically and therapeutically. The spine could now be imaged in much greater detail than was possible with the plain radiograph and fluoroscopy. The soft tissues, i.e., disc, ligaments, muscle, nerve roots, and CSF, could now be seen as could the size and state of the spinal canal (see Fig. 3). Tumors and fractures were depicted far more clearly. Further benefits were found as CT could image in multiple planes – coronal, sagittal, and axial – as well as oblique planes with rotation and 3D images. Hence, the name changed from computed axial tomography when it originated as a single-slice machine to now just computed tomography with the current cohort being multi-slice/multidetector up to 640. It remained the most accurate method of neural and soft tissue assessment until advancements in the mid-late 1980s made MRI feasible for routine use. CT remains superior to MRI, however, for assessment of bony structures and is still the gold standard for assessing fusion.

Continued refinements in CT have allowed faster, higher resolution and more accurate scans as well as significant reductions in radiation. CT now has the added benefit of CT fluoroscopy

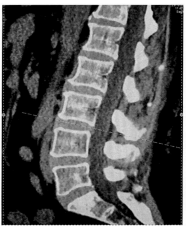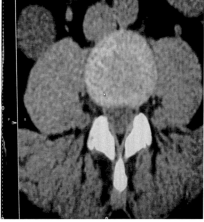

Fig. 3 Sagittal and axial CT of lumbar spine, showing bones, disc, canal, foramina, and nerve roots

– real-time imaging via the CT scanner in procedures.

Diagnostically, CT allows the disc (+/− disc osteophyte complex) to be analyzed, whether or not there is herniation or stenosis and to what degree – both foraminal and central canal. The origin and descending nerve roots can also be seen; hence nerve root compression and displacement becomes available allowing the clinician to diagnose and treat the symptoms with far greater accuracy. This is particularly important in cervical spine surgery where disc osteophytes and uncovertebral complexes need to be cleared to enable unimpeded passage of the nerves. Further diagnostic value is found with discography and myelography, both of which require specific needle tip placement, and this may be done with the use of CT alone or in conjunction with fluoroscopy. All spinal levels – cervical, thoracic, or lumbar – can be analyzed.

Therapeutically, CT (or fluoroscopy alone or CT fluoroscopy) allows the interventionist to perform numerous procedures to treat the patient with spinal pain. Biopsy of perispinal lesions in the case of suspected infection or tumor is an example of a procedure with diagnostic value but also important in planning therapy, e.g., identification of organism or tumor subtype. CT angiogram is particularly useful if an anterior lumbar or high cervical approach is planned and there is concern about vascular anatomy or if there is a thoracic lesion where the spinal cord blood supply is of particular importance when considering embolization of a high vascularity lesion.

MRI

Magnetic resonance imaging is the gold standard for spinal imaging. Like all diagnostic modalities in spinal imaging, it has advanced over time with technology. In particular the availability of high magnetic fields strength systems, increase gradient performance, the use of RF coiler rays and parallel imaging, and increase pulse sequence efficiency allowed for better acquisition speed and improved low signal-to-noise ratio. It provides detailed and conspicuous imaging of the spinal structures, showing greater detail than other modalities (see Figs. 4, 5, 6 and 7). There are categories of MRI available. First is the traditional tunnel lie down 3T MRI (Tesla, the magnetic field strength) which is the most widely used global static imaging tool. The alternative or adjunct to this is the open/upright MRI. The latter provides positional imaging – sitting, standing, flexing, and extending. Different positions can reveal dynamic pathologies that the supine tunnel MRI cannot demonstrate, e.g., instability, herniated discs, and annular tears that may not be detectable when in the unloaded, nonfunctional position (see Fig. 5a–b).

MRI imaging has the advantage of no ionizing radiation and clearly displays the type and extent of spinal pathology. Additional information can be realized with MR imaging. In particular:

(i) Degenerative state of the disc: It can be clearly characterized by MRI, unlike X-ray or CT where, unless there is a decrease in the disc height or a distinctive disc bulge,

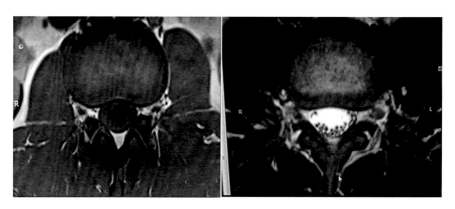

Fig. 4 MRI: note the far clearer delineation of all structures compared to Fig. 3; disc, canal, and nerve roots

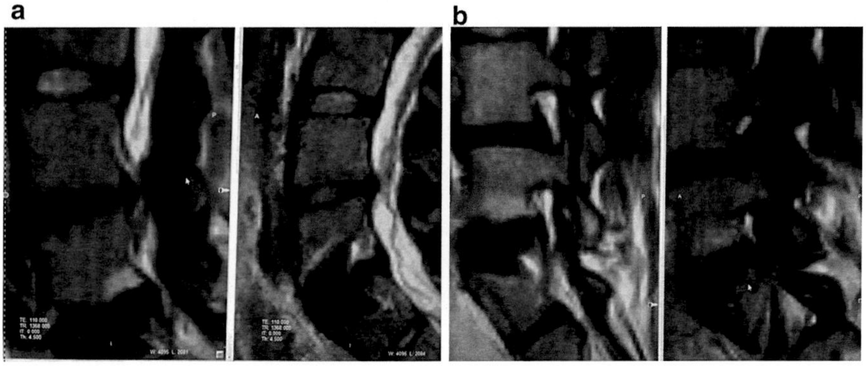

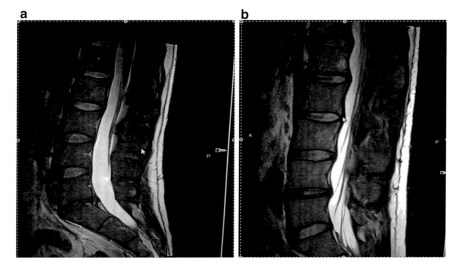

Fig. 5 Note in (**a**) the difference in the degree of herniation and in (**b**) the foraminal stenosis in the weight-bearing position comparing the static lie down images

Fig. 6 (**a**) and (**b**) Note also the state of the discs: L1-4/5, all normal; L5/S1, degenerate grade 4, i.e., nucleus no longer white, loss of height, and with the high signal intensity zone/annular tear. a Note in b the degenerate L4/5 disc, grades 3–4

the morphology of the disc is not ascertained. An MRI classification of the disc degeneration has ensued – Pfirrmann grades I–V. The grading is based upon T2-weighted imaging with the low-signal changes to the nucleus pulposus becoming more pronounced and diffuse within the disc as well as loss of disc height as the degenerative process progresses.

(ii) Further markers of intervertebral disc degeneration shown on MRI are:
 (a) High-signal-intensity zone (HIZ) located in the posterior annulus fibrosis, separated from the nucleus pulposus – a relationship between the HIZ and pain has been observed.
 (b) Modic changes (Modic et al. 1988) – signal changes to the vertebral end plate and bone deep to the cartilage; these are graded I–III combining both T1W and T2W images. Type I, also known as the inflammatory phase, is denoted by inflammation of fibrous tissue, low signal intensity on T1W, and high signal intensity on T2W imaging. Type II, known as

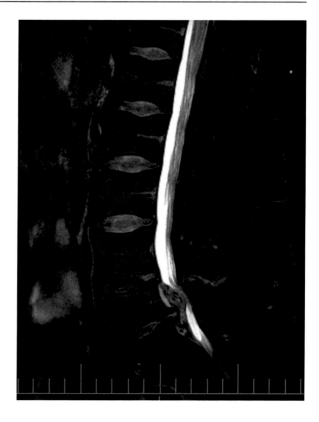

Fig. 7 MRI lumbar spine sagittal slices. Note the detail of the study which enables differentiation of extruded and sequestered disc material in the canal contacting the thecal sac from the broad-based herniation present at L5/S1 and degenerate disc at L4/L5

the fat phase, is marked by a large deposition of fat cells in the end plate and the area underneath it, as well as a high signal intensity on T1W and an equivalent or mildly high signal on T2W imaging. Type III, also known as the bone sclerosis period because the bone becomes hardened in the end plate and the area underneath it, is also characterized by low signal intensity in T1W and T2W imaging (Rahme and Moussa 2008). It has also provided prognostic value for interventions for diagnosed discogenic back pain (Furunes 2018) and has been associated with increased vascular adhesions during anterior lumbar surgery (Malham 2018).

Other value in ordering an MRI includes:

(iii) The exact relation of the herniated disc to the nerve roots and whether or not direct compression is present.

(iv) The status and size of the paraspinal muscles. For example, severe multifidus wasting may suggest radiculopathy and be associated with poorer outcomes for decompression (Zotti et al. 2017) and disc replacement surgery (Le Huec et al. 2005; Storheim et al. 2017).

(v) Vascular pattern: particularly if anterior or oblique or lateral surgery is being considered, then vascular pattern including any anomalies should be studied to anticipate problems.

(vi) Assessment post-surgery for recurrent herniation, stenosis, and/or presence of fusion. This modality can be useful if a patient's leg symptoms recur to the point where intervention would be considered; then MRI with contrast can be of use in assessment to differentiate scar tissue from recurrent disc herniation. Recent studies suggest that MRI is comparable to CT for assessing lumbar spine fusion (Kitchen et al. 2018) (Fig. 7).

MR spectroscopy is an emerging technology whereby differential water and protein contents within the region of interest can be measured and correlated to the patient's symptoms allowing differentiation of painful from non-painful discs (Zuo 2012). This may in time and with maturity enable diagnosis of discogenic pain without invasive provocative discography. Intraoperative MRI can be performed (more in the setting of craniocervical or spinal cord tumor surgery) but is not routine or widespread.

Nuclear Medicine

Radionuclide bone scanning is a well-accepted and sensitive method for uncovering a variety of bony lesions including abnormalities of vertebral bodies or facet joints that may be contributing to spinal pain. It has a more functional basis than the other imaging modalities as it has the ability to detect the most avid area of "inflammation," seen as increased regional blood flow, as determined by the degree of tracer uptake. Single photon emission computed tomography (SPECT) is especially useful in such an evaluation because it allows for precise localization of a lesion to the vertebral body, disc space, or facet joint. Greater diagnostic accuracy is achieved with this dual technique – using both radionuclide tracer, e.g., technetium 99, and integrated CT – allowing the level and anatomical location of pain generation to be imaged. This anatomic distinction is necessary in order to accurately diagnose the underlying condition detected by the bone scan. Most bony abnormalities result in focal areas of abnormal tracer activity but do not affect all components of a vertebra with equal frequency nor have a random pattern of involvement. Vertebral diseases tend to conform to predictable patterns that can be more readily identified by SPECT scan compared to planar imaging (Gates 1988, 1998). In some applications, such as in symptomatic pars defects, SPECT has sensitivity at least equivalent if not superior to MRI (Fig. 8).

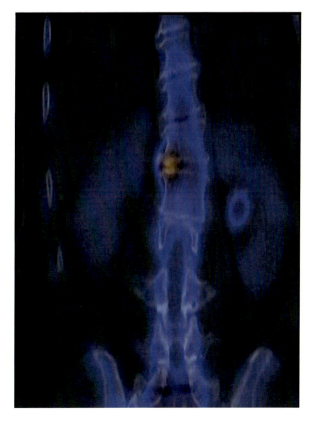

Fig. 8 SPECT scan of the thoracolumbar spine visualized in the coronal plane. Note tracer uptake most pronounced at the T12/L1 end plates asymmetrically which correlated with the patient's pain

Comments About Pain Generators

As with anywhere in the body, the causes and origins of pain are vast and extensive and include referral from extra-spinal regions. In the spine itself, the main pain generators are:

(A) The joints – facet and sacroiliac
(B) The intervertebral disc
(C) The nerve roots
(D) The bones
(E) The muscles

Biomechanical and Chemical Models for Disc and Facet Pain

Intervertebral disc degeneration has been reported to be a source of low back pain in adults. The intervertebral disc consists of the nucleus pulposus, surrounding annulus fibrosus, and the superior and inferior cartilage end plates. Collagen and elastin fibers are present in different orientations lying within a proteoglycan (most prominently aggrecan) and non-cartilaginous protein mixture, forming a complex matrix. Disc degeneration occurs with the breakdown of this matrix with replacement of fibroblasts with chondrocyte-like cells and alteration in the lamellar structure of the annulus and when the nucleus gel becomes fibrous. Annular tears have been strongly associated with the development of degenerative disc disease. In other words as the nucleus can no longer support the load, the annulus can buckle and tear promoting radial and circumferential tears. Neurovascular structures can migrate into these tears. Numerous biomechanical-biochemical studies have shown that following annular tears, the axial load that is normally carried through the center of the disc can shift posteriorly over the nerve concentrated posterior and posterolateral annular fibrosis. Therefore in addition to the painful inflammatory reaction, one can get mechanical irritation of these already inflamed and irritated *nociceptive* fibers in the peripheral annulus. The fundamental basis of this breakdown at the molecular level is the production of an abnormal matrix or an increase in the constituents which cause matrix degradation, e.g., IL-1 and TNF and matrix metalloproteinases (MMPs), and a reduction in the amount of tissue inhibitors of metalloproteinases. The normal disc posteriorly is innervated by branches of the sinuvertebral nerve (from meningeal branches) and sympathetic fibers. Only the outer aspect of the annulus is innervated, and the sensory fibers are primarily nociceptive and proprioceptive (although less so). In a degenerate disc, the number of nerve fibers increases, and nerve nociceptive fibers grow into the normally aneural part of the annulus and nucleus. Many factors may contribute to the degenerative process – genetics, mechanical load, trauma, and nutrition; however, the exact etiology and relationships still require further research.

Studies have linked pathological changes in facet joints with preceding disc degeneration. The intervertebral discs support most of the weight during flexed postures, but the facet joints bear an increasingly greater burden as the lumbar spine is ranged into extension. In addition to stabilizing the spine and guiding segmental motion, facet joints function as weight-bearing structures that support axial loading along with the intervertebral discs. Studies have shown that the facet joints can carry up to 33% of the dynamic axial load. Disc degeneration with associated narrowing of the disc space alters the mechanical load distribution and may result in a degenerative cascade with increased mechanical stress on the facet joint and joint capsule. Within the active range of the lumbar spine, the paraspinal muscles act as the principal contributors to vertebral stability. However, both cyclic and sustained flexion movements decrease the reflexive muscle activity of the paraspinal muscles such as the multifidus muscle. In theory, this may result in increased laxity across the facet joint leading to both decreased stability and increased stress on the facet joint capsule.

The role of the facet joint capsule in stabilizing the motion characteristics of these joints cannot be understated. Studies have suggested that disc degeneration results in increased range of axial rotation. It has been postulated that the increase in axial rotation and subsequent instability place additional stressors upon the facet joint capsules leading to a molecular response, which results in fibrocartilaginous metaplasia in the capsules of facet joints. Boszczyk et al. (2003) reported hypertrophic and fibrocartilaginous changes in the facet joint capsules of patients who had undergone lumbar fusion for degenerative instability.

The facet joint (or zygapophyseal joint) is innervated by the medial branch of the dorsal ramus of the nerve exiting at the same level and also the medial branch of the nerve one level above. The joint has a strong capsule, and hyaline articular cartilage is present.

Changes in load distributions (from a degenerative disc or from spinal malalignment or pelvic tilting or rotation) can lead to osteoarthrosis, osteophyte formation, and inflammation. The cartilage and synovium of facet joints are sources of inflammatory cytokines. It has been proposed that painful symptoms may arise not only from mechanical stress discussed previously but also from the associated inflammatory response involving cytokines such as tumor necrosis factor alpha, interleukin-6, and interleukin-1 beta, oxygen-free radicals such as nitric oxide and inflammatory mediators such as prostaglandins. Interestingly, some have suggested that inflammatory cytokines originating from inflamed synovium may spread to adjacent nerve roots and produce radicular lower extremity symptoms.

The sacroiliac joint (SIJ) is a true diarthrodial joint with unique characteristics not typically found in other diarthrodial joints. The joint differs with others in that it has fibrocartilage in addition to hyaline cartilage, there is discontinuity of the posterior capsule, and articular surfaces have many ridges and depressions. The sacroiliac joint is well innervated. Histological analysis of the sacroiliac joint has verified the presence of nerve fibers within the joint capsule and adjoining ligaments. It has been variously described that the sacroiliac joint receives its innervation from the ventral rami of L4 and L5, the superior gluteal nerve, and the dorsal rami of L5, S1, and S2. Abnormalities with joint function and mobility – hypo- or hypermobility – are the primary cause of the irritation. Inflammatory systemic disease, e.g., ankylosing spondylitis, is of course another reason for pain generation.

As with other diarthrodial joints, the cartilage of facet joints may also be sex-hormone sensitive. Estrogen has been associated with chondrodestruction, although controversy exists as to its actual role in the development of osteoarthritis. However, Ha and Petscavage-Thomas (2014) have found a statistically significant association between the increased expression of estrogen receptors on the articular cartilage of facet joints and the severity of facet arthritis (Binder and Nampiaparampil 2009).

Invasive Interventions

Myelography – an invasive procedure with contrast media (iodinated) being injected into the subarachnoid space, penetrating the thecal sac, to analyze the spinal canal, including the cord, nerve roots, and foramina. With the introduction of MRI, myelography has diminished in importance as a diagnostic tool. Yet it still can play an important role in diagnosis for those for whom MRI is contraindicated, e.g., those with a pacemaker in situ.

Discography – an invasive provocative procedure to determine whether or not the disc is the cause of the pain. One or a number of needles are placed in the nucleus pulposus (i) of the disc(s) at varying levels and then contrast media injected to attempt to reproduce the patients symptoms. Positive discography is defined as follows: (1) abnormal morphology of the examined disc; (2) consistency of pain by provocation; (3) no pain experienced by provocation of the nearest disc; and (4) less than 3 mL of injected contrast agent.

Discography has been the subject of vigorous debate and controversy with strong advocates for and against this functional test. Many studies have shown it to be valid with high correlation to the person's pain (Walsh et al. 1990; Peng et al.

2006). Other studies have questioned the usefulness of the technique. One of the main points of concern was that pain provocation is a subjective measure dependent on the patient, which despite quantification by the VAS, inevitably yields a high rate of false positives in patients with a psychological fear of pain or hyperesthesia from chronic pain or personality trait scores. Also it can be operator dependent with pressure and flow rates of injection leading to reduced stimulation of pain receptors (Derby et al. 2005; Ohnmeiss et al. 1995).

If used it must be critically examined in association with the patients profile, pain diagnosis, and other image-guided treatments performed, e.g., facet joint injections or nerve root blocks.

The Dallas discogram description grade is the mainstay of reporting (Saboeiro 2009) and is a combination of the interventional procedure followed by a diagnostic CT scan.

The Dallas discogram protocol for performance and reporting (or now more appropriately the modified Dallas classification system) is a widely used and accepted method for describing the CT findings of the test in association with the patient's intra-procedural symptoms (Sachs et al. 1987; Resnick et al. 2005; Carragee and Alamin 2001; Cohen and Hurley 2007; Cohen et al. 2005; Madan et al. 2002). When properly performed, low false-positive rates in the order of 6–10% can be anticipated (Bogduk et al. 2013).

There are six possible categories that describe the severity of the radial annular tear.

The grade 0 is a normal disc, where no contract material leaks from the nucleus.

The grade 1 tear will leak contrast material only into the inner 1/3 of the annulus.

The grade 2 tear will leak contrast through the inner 1/3 and into the middle 1/3 of the disc.

The grade 3 tear will leak contrast through the inner and middle annulus. The contrast spills into the outer 1/3 of the annulus.

The grade 4 tear further describes a grade 3 tear. Not only does the contrast extend into the outer 1/3 of the annulus, but it is seen spreading concentrically around the disc. To qualify as a grade 4 tear, the concentric spread must be greater than 30°. Pathologically, this represents the merging of a full-thickness radial tear with a concentric annular tear.

The grade 5 tear describes either a grade 3 or grade 4 radial tear that has completely ruptured that outer layers of the disc and is leaking contract material out of the disc. This type of tear, which one is most likely to suffer from, can cause a chemical radiculopathy in one or both of the extremities and result in persistent leg pain (Fig. 9).

Irrespective of the controversy, it is currently the only test which can directly link symptoms felt to be significant to the patient to the presumed pathology, and studies have shown that patients selected for intervention in this way have improved outcomes compared to those without precision diagnosis (Colhoun et al 1988; Margetic et al. 2013; Xi et al. 2016).

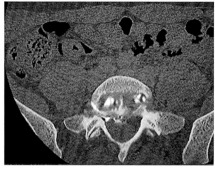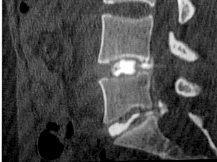

Fig. 9 Note the contrast passing from the nucleus through the outer annulus into the epidural space

Corticosteroid Injections

Which corticosteroid?

- There is great variability in the use of the injected corticosteroid.

Commonly used steroids are:

- Dexamethasone sodium phosphate
- Betamethasone acetate
- Methylprednisolone acetate
- Triamcinolone acetonide

The amount used may also vary considerably and below are examples:

- Dexamethasone sodium phosphate 4–8 mg
- Betamethasone acetate 0.25–1.0 ml
- Methylprednisolone acetate 4–10 mg
- Triamcinolone acetonide 2.5–5 mg

Commonly used local anasthetics and doses:

- Lidocaine hydrochloride (0.25–2 mls)
- Bupivacaine hydrochloride (0.25–2 mls)
- Procaine hydrochloride

Safety

A comprehensive review of the use of injected corticosteroids was undertaken by MacMahon et al. (2009), and this had particular relevance to spinal pain therapy. A number of factors were revealed which previously were not taken into account in terms of safety and protocol. In particular, this related to the particulate composition of steroids. Most corticosteroid preparations contain corticosteroid esters (apart from dexamethasone), which are highly insoluble in water and thus form microcrystalline suspensions. This property cannot only cause adhesions (problematic at subsequent open decompression procedures) but also cause particulate steroid emboli; thus they are likely the primary cause of the reported CNS complications, e.g., paraplegia or stroke. Non-particulate steroid is not known to cause this complication (MacMahon et al. 2009).

Other general complications range from common but minor risks of skin changes or transient hyperglycemia to rare but more significant complications including durotomy causing CSF meningocele and/or arachnoiditis and infection causing osteomyelitis or epidural abscess. Such material risks may be mentioned in discourse if relevant as part of informed consent prior to the injection being performed (Zotti et al. 2012). Cervical injections, particularly, carry the unique risk of vascular injury – particularly radicular artery injury – which can impair spinal cord and brain stem perfusion.

Specific contraindications should be sought and include bleeding diatheses or active use of anticoagulant (for epidural or perineural injections), infection at targeted site (unless for purpose of obtaining a biopsy), immunosuppression, poorly controlled diabetes, and noted contrast or injectable allergy.

Given the above, the alternatives and expected benefits need to be considered for any intervention. In common neurointerventional and spinal surgical practice, corticosteroids when combined with appropriate education and rehabilitation strategies can cure and assist patients with conditions of favorable natural history and who are either unsuitable for or do not wish to undergo formal surgical intervention or, alternatively, palliate patients' conditions.

Mechanism of Action

Corticosteroids predominantly affect the action of cytokines and inflammatory mediators (e.g., substance P, PLA2, arachidonic acid, IL-1, and prostaglandin E2) involved in inflammation. They lead to increased blood flow and down-regulation of immune function, inhibiting cell-mediated immunity, reducing cellular accumulation at inflammatory sites, and decreasing vascular responses. Corticosteroids cause these effects through a mechanism that ultimately involves its active moiety entering cells and combining with receptors to alter messenger RNA production, mainly altering the protein annexin-1 (previously called lipocortin-1) (Barnes 1998; Eymontt et al. 1982; Buckingham et al. 2006; D'Acquisto et al. 2008).

Types of Corticosteroid Injection

Facet joint injection (intra-articular) – the spinal needle is placed into the facet joint cavity and steroid injected along with local anesthetic. Indications include presumed facetogenic lumbar and thoracic or cervical pain. This may include facet-related pain resulting from posterior load-bearing transfer from patients with degenerative disc disease and anterior column pathology where treatment of anterior spinal structures (e.g., intervertebral disc) is thought to be high risk or undesirable. It is important that these patients are counselled that only a portion of their pain will be treated (appropriated to pain relief that may have been experienced from the medial branch block).

There is dispute over the efficacy of these injections, and some of it likely stems from only a limited proportion, perhaps 10–20% of patients having "pure" facetogenic pain. For some, common practice/convention may prevail over scientific evidence as to their efficacy and validity. For greatest accuracy the injection needs to be image controlled. An alternative, which also covers nociceptors from the facet joint but does not violate it, is the medial branch block of the dorsal ramus (Boswell et al. 2007; Sehgal et al. 2007a; Manchikanti et al. 2010; Cohen and Raja 2007; Jackson et al. 1988; Schwarzer et al. 1994, 1997; Sehgal et al. 2007b) (Fig. 10).

Medial branch block – a minimally invasive procedure whereby local anesthetic is injected along the pathway of the medial branch of the dorsal ramus of the spinal nerve, which supplies the facet joints, to determine if the origin of the pain is from the facet joints. It is important to note the innervation of the joint, recognizing that it is not single. The facet joint receives branches from the level above and below. The innervating branch lies with the depression/junction of the transverse process with the body of the vertebra.

Blockade of the medial branch of the posterior ramus nerve is generally preferred over intra-articular facet blocks as it is easy, less traumatic, and less risky than intra-articular injections (including no risk of joint infection) (Dreyfuss et al. 1997). Generally, when facet joint denervation is being considered, it is preferable to assess the patient's response to medial branch blocks given that it allows assessment of analgesic response due to blockade of the anatomic structure to be ablated. A response of 50% or more reduction of pain is an indication for RFD. However, in the presence of inflammation, intra-articular injections may be superior to medial nerve blocks.

Lumbar facet injections and medial branch blocks are both valuable in terms of diagnosis of the patient's pain generator and suitability for other interventions, e.g., radio-frequency neurotomy. However, in themselves there is limited relief of "facetogenic" low back pain. Marks, Houston, and Thulbourne reported limited relief after 3 months with relief of pain diminishing between 1 and 3 months (Marks et al. 1992). Manchikanti and colleagues reported the

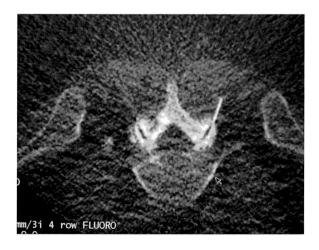

Fig. 10 The needle within the facet joint during an intra-articular injection of L5/S1

majority of patients having improvement in their facet pain at 1 year, however, irrespective of whether treated with local anesthetic alone or with steroid (Manchikanti 2001, 2010).

Most studies report that cervical medial branch facet blocks tend to have longer duration compared to lumbar facets with effect lasting between 3 and 5 months for each injection (Manchikanti 2008). The mean duration of effect for cervical facet block can be up to 8–12 months (Kim 2005), and repeated injections can provide sustained relief at a year and beyond (Manchikanti et al. 2015a). Thoracic facet interventions have not been well studied, and, as such, fair evidence is only available for medial branch blocks in the thoracic spine.

In the lumbar spine, for long-term effectiveness, there is Level II evidence for radio-frequency neurotomy and lumbar facet joint nerve blocks, whereas the evidence is Level III for lumbosacral intra-articular injections. In the cervical spine, for long-term improvement, there is Level II evidence for cervical radio-frequency neurotomy and cervical facet joint nerve blocks and Level IV evidence for cervical intra-articular injections. In the thoracic spine, there is Level II evidence for thoracic facet joint nerve blocks and Level IV evidence for radio-frequency neurotomy for long-term improvement (Manchikanti et al. 2015a). Evidence for diagnosis of cervical facet joint pain with controlled comparative local anesthetic blocks is Level I or II-1. The indicated evidence for therapeutic facet joint interventions is Level II-1 for medial branch blocks and Level II-1 or Level II-2 for radio-frequency neurotomy (Manchikanti et al. 2015a).

Facet joint denervation – this is the "follow on" from a positive medial branch block, which has confirmed that the pain generator is the facet joint. The next step is to denervate the facet joint via radio-frequency ablation or with 90% alcohol. Radio-frequency denervation, involving heating of the targeted nerve typically at 90 °C for 90 s, can provide longer-term relief than the standard facet joint corticosteroid injection. Alcohol denervation also provides significant relief, and some studies show a longer benefit than radio-frequency ablation (Joo et al. 2013).

As a day procedure usually under light sedation and performed with specialized radio-frequency equipment (an addition to standard radiology machine), the medial branches of the dorsal rami are ablated. The technique is very important, and good understanding of anatomy and physical properties of the equipment is paramount. Risks are minimal, but there have been case reports of transient radiculopathy, neural injury, and thermal burns which relate to inappropriate technique and preparation (Barr et al. 2000).

In the cervical spine, the main indication for injections or radio-frequency neurotomy remains facetogenic pain, but facet-pain targeted injections have also been used with varying success for facet pain resulting from herniated nucleus pulposus (load transfer to posterior elements from disc compromise), whiplash, and myofascial pain (Kim et al. 2005).

Like with all forms of thoracolumbar spinal treatment, radio-frequency denervation has been shown in some studies to provide significant pain reduction in patients with chronic low back pain selected with a positive medial branch block for between 6 and 18 months. In addition, this low-morbidity procedure is found to be efficacious on case series when repeated in patients who had a successful prior procedure (Zotti and Osti 2010; Schofferman and Kine 2004; Son et al. 2010) effective in around ~70% of patients for 8–9 months (Zotti and Osti 2010).

Patient selection (i.e., use and quantitative response to intra-articular compared to medial branch blocks) and mode and location of lesioning have been cited for potential inconsistencies in the results of these studies. The majority of patients, in the order of 60–80%, obtain at least 90% relief of pain when selected correctly with a mean effect typically lasting 9–12 months for both cervical and lumbar facet denervation. The evidence for radio-frequency neurotomy for sacroiliac pain is mixed in terms of quality, but sham surgery placebo-controlled trials overall were supportive of this technique (Rupert et al. 2009). However, other studies have shown little benefit to this procedure (Evans et al. 2003; Blasco et al. 2012; Zotti and Osti 2010; Bogduk et al. 2011).

A recent study by Van Tilburg and associates (2016) which was a randomized sham-controlled double-blinded study design was unable to reject the null hypothesis of efficacy for this intervention. However, several studies support the efficacy for this procedure compared to comparative controls (Gallagher et al. 1994; Van Kleef et al. 1999, 2005; Tekin et al. 2007; Kroll 2008).

Synovial cyst puncture and aspiration – a symptomatic synovial cyst from a degenerate facet joint can cause compression upon a descending nerve root and inflammation. The aim of the radiologist to achieve therapeutic relief is to puncture and if possible aspirate the cyst or rupture it, followed by an injection of steroid and local anesthetic. There are two mechanisms for the above:

(i) A direct puncture (which is not always possible due to its position in the canal as access may not be not possible due to the lamina or facet joint covering the anticipated needle pathway).
(ii) An indirect rupture via the facet joint – filling the latter with injectate – steroid and local anesthetic and saline until the cyst ruptures. This technique can be very painful.

Percutaneous treatment for facet cysts has been reported to only fair long-term success, approximately 50–80% of patients in literature reviews, and relief for up to 1 year has been reported (Vad et al. 2002; Carmel et al. 2007). Many of the cysts targeted are gelatinous and not amenable to aspiration, leaving the large residual cyst capsules to continue compressing the neural/dural structures, and cause ongoing neurological dysfunction. Along with the 37.5–50% risk of recurrence is a 45–50% chance of success with repeated cyst rupture attempts (Imai et al. 1998; Rauchwerger et al. 2011; Sabers et al. 2005; Schulz et al. 2011; Shah and Lutz 2003). A further trial can be attempted in refractory cases or recurrence, but a high proportion of these patients (50–60%) will require open spinal surgery. The uncertain efficacy of this intervention has led some authors to advocate for surgical intervention rather than repeated attempts (Epstein and Baisden 2012).

Selective nerve root injections/perineural injections and epidural injections – the spinal needle is placed next to the suspected pain-generating nerve, and a mixture of local anesthetic and steroid (e.g., dexamethasone and bupivacaine being injected) is injected. Again the steroid used varies as does the utilization of the radiology modality and the amount. The technique is most commonly done with fluoroscopy or under CT guidance. There are a number of different techniques/approaches which include transforaminal, interlaminar, and caudal. The most widely used and accepted is the transforaminal approach. The consensus from the literature (and certainly anecdotally) is that epidural steroid injections are effective and of value particularly for limb and girdle pain. However, the degree of efficacy is much and varied. In saying this, the degree of efficacy of the injection is based upon many factors which include the spinal pathology, the severity of the pathology, the expertise and skill of the operator, the exact position of the needle, the patient's mental state, and other systemic or local pathologies. As aforementioned, the mechanism by which the steroid works is manyfold including reducing inflammation/swelling via neutralizing inflammatory mediators, e.g., substance P, PLA2, arachidonic acid, IL-1, and prostaglandin E2. The steroid also increases blood flow and reduces the activity of the immune system (Akuthota et al. 2013; De Smet et al. 2005; Salahadin et al. 2007; Vad et al. 2002; Carmel et al. 2007; Lutz et al. 1998) (Fig. 11).

Many studies report the effectiveness of this intervention, including randomized trials, but large level 1 double-blinded studies with a placebo comparator are lacking. This is particularly so for contained herniated pulposus lesions (MacVicar et al. 2013) with mild neural compression, whereas injections into segments affected by extruded or sequestered disc fragments are thought to be less effective. While the addition of CSI is generally favorable, some studies have suggested that they alter the natural history of the patient and reduce the number of patients who undergo surgery of continued symptoms.

Interestingly, some trials have reported benefit of injection but no additional benefit to

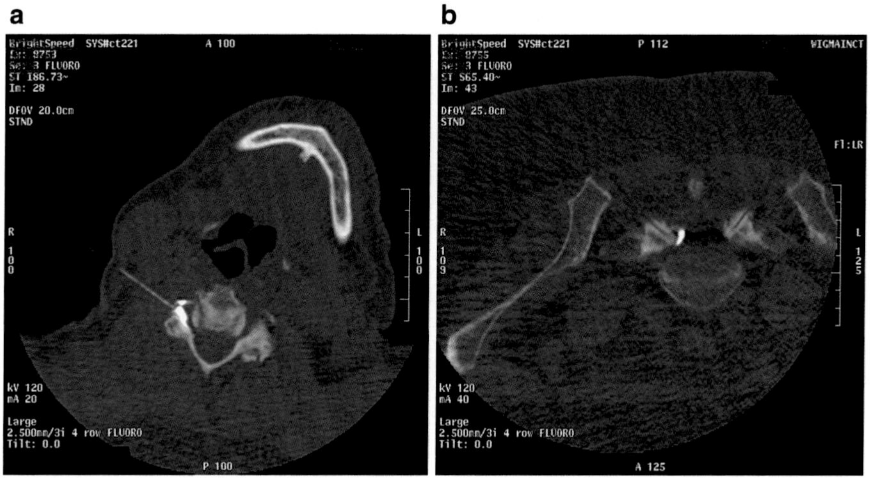

Fig. 11 (**a**) and (**b**) Note the transforaminal approach with contrast (prior to steroid injection) to confirm position around the exiting nerve root and also passing into the epidural space. Note in the next picture the paramedian interlaminar approach noting contrast between the ligamentum flavum and thecal sac in the epidural space

corticosteroids added to local anesthetic (Ng 2005). The majority of patients, in the order of 70–75%, will have significant reduction of their symptoms when they have been presented for less than 3 months. However, patients with symptoms longer than 3 months tend to have more variable success. Furthermore, patients with shorter duration of symptoms can be expected to experience more sustained relief than those with chronic symptoms. When effective, a reduction of at least 50% for 1–2 months can be expected in around 70% of patients and complete resolution in around 30% of patients (Ackerman and Ahmad 2007).

Cervical transforaminal epidural injections are effective for around 70–80% of patients with radiculopathy and have been shown to prevent the need for surgery in around 70% of patients (Costandi 2015; Vallee 2001). While at 3 and 6 months, around 30% of patients have complete resolution of symptoms, this reduces to around 20% at 1 year (Vallee 2001). To achieve sustained and effective relief, repeated injections may be required. For example, Slipman et al. (2000) reported pain reduction, return to full-time work status, reduction or elimination in analgesic use, and satisfaction with treatment in 60% of patients at 12–45 months' follow-up, but treatment on average consisted of 2.2 injections.

Interlaminar injections have good evidence for usage in the setting of herniated discs and radiculitis and fair evidence for axial/discogenic pain without facet joint pain and are technically simpler in the hands of experienced operators. They have been shown to have superior effect for chronic lumbar disc herniation at 2 years compared to caudal and transforaminal injections (Manchikanti 2015b). They have also been shown to be superior to caudal injections for lumbar central spinal stenosis (Manchikanti et al. 2014). In addition, there is Level II and Level II/III evidence for long-term management of cervical disc herniations or stenosis and thoracic disc herniations, respectively. Caudal injections, on the other hand, have good evidence for herniated disc and radiculitis with only fair evidence for axial/discogenic back pain, spinal stenosis, and post-surgery syndrome.

Both interlaminar and caudal injections for axial or discogenic pain are shown to be effective, but interlaminar injections have marginal superiority over caudal injections for this indication (Manchikanti 2015b). Interlaminar and caudal techniques have been reported to be effective for lumbar disc herniation or radiculitis (Kaye et al. 2015); however, some studies have reported them to be less effective than transforaminal injections for radiculopathy due to herniated nucleus pulposus (Kamble et al. 2016; Ackerman and Ahmad 2007; Lee et al. 2009; Thomas et al. 2003).

Sacroiliac Injections

Sacroiliac joint pain – the great mimicker. One of the greatest challenges in diagnosing the pain from this joint is that the symptoms can imitate other pain-generating conditions, e.g., facet joint arthropathy and discogenic or radicular pain from herniated discs with the malady being both around the sacroiliac joint but also radiating down the lower limb or into the groin. As always, imaging can provide both diagnosis and treatment. The issue the clinician faces is that in many cases, the imaging does not directly confirm the provisional diagnosis. Arthropathy may be present; however the joint may show no signs of pathology on plain X-ray, CT, and MRI. The physical examination is therefore paramount to test the suspicion of SI pain with the location of the patient's symptoms and any worsening with provocative tests. Like with other joint-related conditions, steroid and local anesthetic blocks can aid in both diagnosis and treatment.

Patients are prone and the needle advanced before a sensation of entering the joint which is confirmed on multiple planes to be in the joint. It is performed by imaging guidance due to the highly variable morphometry of pelvises between patients for diagnostic and therapeutic purposes. Smaller doses of LA/steroid focusing on the posterior-inferior hyaline portion of the joint tend to be diagnostic, while larger doses that aim to bathe the entire joint are therapeutic. Some clinicians favor the addition of separating more superior injection into the fibrous component of the joint. The controlled diagnostic blocks utilizing the International Association for the Study of Pain (IASP) criteria demonstrated the prevalence of pain of sacroiliac joint origin in 19–30% of the patients suspected to have sacroiliac joint pain (Forst et al. 2006).

Evidence from meta-analyses (Hansen et al. 2007; McKenzie-Brown et al. 2005), albeit based on low quality data, supports the role of SI injections in treating painful sacroiliac dysfunction and spondyloarthropathy. Maugars et al. (1996) performed a double-blinded placebo assessment of CSI versus placebo and found a statistically and clinically important difference. Eighty-six percent had positive effect at 1 month, while the majority continued to have efficacy of the injection with 58% reporting relief at 6 months (Fig. 12).

Coccyx Injections

Diagnostic and therapeutic injections into the coccygeal region are performed for coccydynia. Ideally, the local infiltration blocks the ganglion

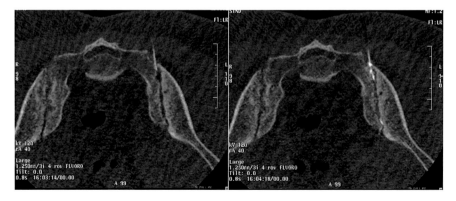

Fig. 12 CT-guided left sacroiliac joint injection. Note the needle confirmed to be within the sacroiliac joint

impar, which is a relay station for nociceptive pain emanating from the sacrococcygeal joint. Indications include coccydynia due to post-traumatic pain/hypermobility or pain from the sacrococcygeal disc. Unique complications to this procedure include rectal laceration and bowel content contamination of the injected field.

The patient is generally prone with sterile preparation and draping and sometimes sedation. Direct percutaneous placement of needle through and proceeding just anterior to the margin of the sacrococcygeal disc with confirmation on lateral and anteroposterior views with dye (if the procedure is done under Xray control rather than CT) followed by injection of LA and CSI to ganglion impar. Occasionally combined with per rectal manipulation in the setting of hyperflexed posture due to trauma or laxity.

Literature for effectiveness is generally limited to smaller cohort studies and case series making it hard to recommend treatments (Howard et al. 2013). Injection alone is effective in around 60–85% of patients with long-term success in around 45–50% with median relief at 6 months (Maigne 2011; Gunduz et al. 2015). Repeated injections were effective in the majority of those presenting with recurrent symptoms (Hodges 2004). Injection combined with manipulation results in around 85% successful outcomes with long-term success in around 60% with the theory for additional manipulation being that abnormally flexed posture of the coccyx leads to increased dural tension.

Other Invasive Forms of Image-Guided Treatment

Vertebroplasty and Kyphoplasty

These are procedures done under image guidance for the treatment of pain due to vertebral compression fractures, usually from osteoporosis. The combination of orthopedic bone cement and direct image guidance of a needle into the vertebra has allowed the past treatments for compression fractures – typically weeks to months of bed rest, analgesia, and sometimes bracing to be replaced or at least supplemented. Kyphoplasty involves partial reduction of fractures by use of an image-guided transpedicular balloon implant prior to cement insertion into the void. These techniques can also be applied to fractures from primary or secondary neoplasia affecting the vertebral body. Success rates vary, but overall significant pain reduction and improvement in the ability to perform ADL have been shown to be statistically significant (Barr et al. 2000; Evans et al. 2003; Blasco et al. 2012). Although felt to be a successful intervention, vertebroplasty for osteoporotic fractures (as distinct from metastases) has been removed from payer coverage in several countries because of equivocal results in sham-controlled procedures.

Spinal Stimulators

Spinal dorsal column stimulators can be inserted either under image guidance or by open techniques in theater with a formal approach and laminectomy. The principle is neuromodulation via electrodes placed onto the spinal cord through interference of emitted frequencies upon pain transmission in the spinal cord. It is believed to take effect through either blockage of pain transmission pathways or upregulation of inhibitory pathways. The patient generally has to meet strict criteria and has a trial period before definitive implantation occurs. The apparatus includes a battery, wires, and an electrode paddle that is applied to the targeted area (depending on pathology).

While the indications are evolving, they are generally indicated for refractory neuropathic pain despite other treatments in patients not amenable to or suitable for any further surgical intervention (low prospect of surgery being able to correct any neuroanatomic abnormality). A classic indication would be arachnoiditis after multiple posterior surgeries but may also include true "failed back surgery syndrome" and complex regional pain syndrome.

Measuring Success of Injections/ Radiology Treatments

There are numerous indicators for pain analysis and benchmarking the premorbid severity and therapeutic impact of interventions and thus providing a means of objectively measuring outcomes and success of treatments. Both statistical and clinical significance of outcomes are both important and measured. More so than other forms of medicine, interventional treatments involving needle injections into joints have undergone extensive analysis against placebo (sham) controls in multiple studies.

Below are listed some of the many available unidimensional assessments relating to pain in such trials but also commonly used in clinical practice (e.g., post-discography or diagnostic injection):

(a) VAS (visual analogue scale)
(b) NRS (numerical rating scale)
   Multidimensional scales (looking at both dimensions of pain and effects on life quality):
(c) Brief Pain Inventory Short form
(d) McGill Pain Questionnaire
(e) West Haven Multidimensional Pain Inventory
(f) SF-36 and Oswestry Disability Questionnaire

All of the above have been combined into the Treatment Outcomes of Pain Survey which is a comprehensive and detailed instrument for measuring pain and outcomes (Younger et al. 2009). Future analytic tools should elaborate upon existing ones by assessing indirect and direct effects upon the patient and the economy including changes in need for aids, opioid usage, employment capability, use of healthcare resources (visits/hospitalizations), and need for care in daily living.

Conclusions

Radiology provides a harmonious and encompassing trinity of diagnosis, quantification, and therapy in relation to spinal pathology. Technology has allowed radiology to become an integral part of diagnosis and treatment, both pre- and postoperatively in those with spinal pain. Radiology provides the clinician with numerous adjuncts to the clinical history and examination by allowing direct analysis of the suspected spinal pain generator and the additional means of providing accurate treatment via targeted imaging. Although the success of image-guided therapeutic techniques is open to some contention, two points should always be kept in mind. First, the skill and subspecialization of the operator are paramount, with them having an interest and formal training and education in the field of spinal pathology. This allows safety for the patient and provides the best chance of obtaining a positive result. Secondly, the majority of patients with back and neck pain will be amenable to several minimally invasive therapeutic technique to obtain relief and return to more "normal" lives, hence, the importance of the first point. With all of the above considered, the usefulness of radiology is self-evident in its ability to provide benefits and alter the natural history of painful conditions with a limited risk profile in selected patients.

References

Ackerman WE 3rd, Ahmad M (2007) The efficacy of lumbar epidural steroid injections in patients with lumbar disc herniations. Anesth Analg 104(5):1217–1222. tables of contents

Akuthota V, Bogduk N, Patel A, Prather H, et al (2013) Lumbar transforaminal epidural steroid injection review & recommendation statement. Review and Recommendation Statement Work Group. North American Spine Society, Burr Ridge, IL

American society of Radiologic Technologists (2018) History of the American Society of Radiologic Technologists. Retrieved https://www.asrt.org/main/about-asrt/asrt-history. Accessed 7 June 2018

Amzallag-Bellenger E, Uyttenhove F, Nectoux E et al (2014) Idiopathic scoliosis in children and adolescents: assessment with a biplanar X-ray device. Insights Imaging 5(5):571–583. https://doi.org/10.1007/s13244-014-0354-0

Barnes PJ (1998) Anti-inflammatory actions of glucocorticoids: molecular mechanisms. Clin Sci (Lond) 94:557–572

Barr J, Lemley T, McCann R (2000) Percutaneous vertebroplasty for pain relief and spinal stabilization. Spine 25(8):923–928

Binder D, Nampiaparampil D (2009) The provocative lumbar facet joint. Spaulding rehabilitation hospital,

Harvard Medical School, Boston, MA USA. Curr Rev Musculoskelet Med 2(1):15–24. Published online 2009 Mar 31

Blasco J, Martinez-Ferrer A, Macho J et al (2012) Effect of vertebroplasty on pain relief, quality of life, and the incidence of new vertebral fractures: a 12-month randomized follow-up, controlled trial. J Bone Miner Res 27(5):1159–1166

Bogduk N, Macintosh J, Marsland A (2011) Technical limitations to the efficacy of radiofrequency neurotomy for spinal pain. Neurosurgery 20(12):2160–2165

Bogduk N, Aprill C, Derby R (2013) Lumbar discogenic pain: state-of-the-art review. Pain Med 14:813–836

Boswell MV, Colson JD, Sehgal N et al (2007) A Systematic review of therapeutic facet joint interventions in chronic spinal pain. Pain Physician 10:229–253

Boszczyk BM, Boszczyk AA, Korge A et al (2003) Immunohistochemical analysis of the extracellular matrix in the posterior capsule of the zygapophysial joints in patients with degenerative L4-5 motion segment instability. J Neurosurg 99(1 Suppl):27–33

Buckingham JC, John CD, Solito E et al (2006) Annexin 1, glucocorticoids, and the neuroendocrine-immune interface. Ann N Y Acad Sci 1088:396–409

Carmel A, Charles E, Samuels J, Backonja M (2007) Assessment: use of epidural steroid injections to treat radicular lumbosacral pain: report of the therapeutics and technology assessment subcommittee of the American Academy of Neurology. Neurology 68(10):723–729

Carragee EJ, Alamin TF (2001) Discography a review. Spine J 5:364–372

Cohen SP, Hurley RW (2007) The ability of diagnostic spinal injections to predict surgical outcomes. Anesth Analg 105:1756–1775

Cohen SP, Raja SN (2007) Pathogenesis, diagnosis and treatment of lumbar zygapophysial (facet) joint pain. Anesthesiology 106:591–614

Cohen SP, Barna SA, Larkin TM et al (2005) Lumbar discography: a comprehensive review of outcome studies, diagnostic accuracy and principles. Reg Anesth Pain Med 30:163–183

Colhoun E, McCall IW, Williams L, Cassar Pullicino VN (1988) Provocation discography as a guide to planning operations on the spine. J Bone Joint Surg 70-B(2):267–271

Costandi SJ, Azer G et al (2015) Cervical transforaminal epidural steroid injections: diagnostic and therapeutic value. Reg Anesth Pain Med 40(6):674–680

D'Acquisto F, Paschalidis N, Raza K, Buckley CD, Flower RJ, Perretti M (2008) Glucocorticoid treatment inhibits annexin-1 expression in rheumatoid arthritis CD4+ T cells. Rheumatology (Oxford) 47:636–639

De Smet A, Jeffrey D, Stanczak J, Fine J et al (2005) Lumbar radiculopathy: treatment with selective lumbar nerve blocks – comparison of effectiveness of triamcinolone and betamethasone injectable suspensions. Radiology 237(2):738

Derby R, Kim BJ, Chen Y, Seo KS, Lee SH (2005) The relation between annular disruption on computed tomography scan and pressure-controlled diskography. Arch Phys Med Rehabil 86:1534–1538

Deschenes S, Charron G, Beaudoin G et al (2010) Diagnostic imaging of spinal deformities. Spine 35(9):989–994

Dreyfuss P, Schwarzer A, Lau P et al (1997) Specificity of lumbar medial branch and L5 dorsal ramus blocks: a computed tomography study. Spine 22(8):p895–p902

Epstein NE, Baisden J (2012) The diagnosis and management of synovial cysts: efficacy of surgery versus cyst aspiration. Surg Neurol Int. https://doi.org/10.4103/2152-7806.98576

Evans AJ, Jensen ME, Kip KE, DeNardo AJ et al (2003) Vertebral compression fractures: pain reduction and improvement in functional mobility after percutaneous polymethylmethacrylate vertebroplasty retrospective report of 245 cases. Radiology 226(2):366–372

Eymontt MJ, Gordon GV, Schumacher HR, Hansell JR (1982) The effects on synovial permeability and synovial fluid leukocyte counts in symptomatic osteoarthritis after intraarticular corticosteroid administration. Medline 9:198–203

Forst S, Wheeler MT, Fortin JD, Vilensky JA (2006) The sacroiliac joint: anatomy, physiology and clinical significance. Pain Physician 9(1):61–67

Furunes H, Hellum C, Brox JI et al (2018) Lumbar total disc replacement: predictors for long-term outcome. Eur Spine J 27(3):709–718. https://doi.org/10.1007/s00586-017-5375-1

Gallagher J, Petriccione di Vadi PL, Wedley JR et al (1994) Radiofrequency facet joint denervation in the treatment of low back pain: a prospective controlled double-blind study to assess its efficacy. Pain Clin 7:193–199

Gates GF (1988) SPECT imaging of the lumbosacral spine and pelvis. Clin Nucl Med 13(12):907–914

Gates GF (1998) SPECT bone scanning of the spine. Semin Nucl Med 28(1):78–94

Goodbody C, Kedem P, Thompson M et al (2017) Reliability and reproducibility of subject positioning with EOS low-dose biplanar X-ray. HSS J 13(3):263–266. https://doi.org/10.1007/s11420-017-9548-6

Gunduz OH, Sencan S, Kenis-Coskun O (2015) Pain relief due to transsacrococcygeal ganglion impar block in chronic coccygodynia: a pilot study. Pain Med 16(7):1278–1281. https://doi.org/10.1111/pme.12752

Ha AS, Petscavage-Thomas JM (2014) Imaging of current spinal hardware: lumbar spine. Am J Roentgenol 203 (3):573–581. https://doi.org/10.2214/AJR.13.12217

Hansen HC, McKenzie-Brown AM, Cohen SP et al (2007) Sacroiliac joint interventions: a systematic review. Pain Physician 10(1):165–184

Hendee WR, O'Connor MK (2012) Radiation risks of medical imaging: separating fact from fantasy radiology. Radiology 264(2). https://doi.org/10.1148/radiol.12112678

Hodges SD, Eck JC, Humphreys SC (2004) A treatment and outcomes analysis of patients with coccydynia. Spine J 4(2):138–140

Howard P, Behrns W, Di Martino M et al (2013) Manual examination in the diagnosis of cervicogenic headache:

a systematic literature review. J Man Manip Ther 23(4):210–218

Imai K, Nakamura K, Inokuchi K, Oda H (1998) Aspiration of intraspinal synovial cyst: recurrence after temporal improvement. Arch Orthop Trauma Surg 118:103–105

Jackson RP, Jacobs RR, Montesano PX (1988) Facet joint injection in low-back pain. A prospective statistical study. Spine 13:966–971

Joo YC et al (2013) Comparison of alcohol ablation with repeated thermal radiofrequency ablation in medial branch neurotomy for the treatment of recurrent thoracolumbar facet joint pain. J Anesth 3:390–395. https://doi.org/10.1007/s00540-012-1525-0

Kamble PC, Sharma A, Singh V et al (2016) Outcome of single level disc prolapse treated with transforaminal steroid versus epidural steroid versus caudal steroids. Eur Spine J 25(1):217–221

Kaye AD, Manchikanti L, Abdi S et al (2015) Efficacy of epidural injections in managing chronic spinal pain: a best evidence synthesis. Pain Physician 18(6): E939–E1004

Kim T, Kim K, Kim C, Shin S et al (2005) Percutaneous vertebroplasty and facet joint block. J Korean Med Sci 20(6):1023–1028

Kitchen D, Rao PJ, Zotti M et al (2018) Fusion assessment by MRI in comparison with CT in anterior lumbar interbody fusion: a prospective study. Glob Spine J 8:586

Kroll H, Duszak R, Nsiah E et al (2008) Trends in lumbar puncture over 2 decades: a dramatic shift to radiology. Am J Roentgenol 204(1):15–19

Laredo J, Wybier M, Bellaiche L et al (2010) Know-how in osteoarticular radiology. Sauramps Med 12:1–9

Le Huec JC, Mathews H, Basso Y et al (2005) Clinical results of Maverick lumbar total disc replacement: two-year prospective follow-up. Orthop Clin North Am 36(3):315–322

Lee JH, Moon J, Lee SH (2009) Comparison of effectiveness according to different approaches of epidural steroid injection in lumbosacral herniated disk and spinal stenosis. J Musculoskelet Rehab 22(2):83–89

Lutz G et al (1998) Fluoroscopic transforaminal lumbar epidural steroids: an outcome study. Arch Phys Med Rehabil 79(11):1362–1366

MacMahon P, Eustace S, Kavanagh E (2009) Injectable corticosteroid and local anesthetic preparations: a review for radiologists. Radiology 252(3):647–661. https://doi.org/10.1148/radiol.2523081929

MacVicar J, King W, Landers MH, Bogduk N (2013) The effectiveness of lumbar transforaminal injection of steroids: a comprehensive review with systematic analysis of the published sata. Pain Med 14(1):14–28. https://doi.org/10.1111/j.1526-4637.2012.01508.x

Madan S, Gundanna M, Harley JM et al (2002) Does provocative discography screening of discogenic back pain improve surgical outcomes? J Spinal Disord Tech 15:245–251

Maigne JY, Pigeau I, Aguer N et al (2011) Chronic coccydynia in adolescents. A series of 53 patients. Eur J Phys Rehabil Med 47(2):245–251

Malham G (2018) Modic 2 changes and smoking predict vascular adherence during anterior lumbar exposure. Presented Sunday 28th April at Spine Society of Australia annual scientific meeting, Adelaide

Manchikanti L (2008) Evidence-based medicine, systematic reviews, and guidelines in interventional pain management, part I: introduction and general considerations. Pain Physician 11(2):161–186

Manchikanti L, Pampati V, Bakhit CE et al (2001) Effectiveness of lumbar facet joint nerve blocks in chronic low back pain: a randomized clinical trial. Pain Physician 4(1):101–117

Manchikanti L, Singh V, Falco FJE, Cash KA, Pampati V (2010) Evaluation of lumbar facet joint nerve blocks in managing chronic low back pain: a randomized, double-blind, controlled trial with a 2-year follow-up. Int J Med Sci 7(3):124–135

Manchikanti L, Cash KA, Pampati V, Falco FJ (2014) Transforaminal epidural injections in chronic lumbar disc herniation: a randomized, double-blind, active-control trial. Pain Physician 17(4):E489–E501

Manchikanti L, Kaye AD, Boswell MV et al (2015a) A systematic review and best evidence synthesis of the effectiveness of therapeutic facet joint interventions in managing chronic spinal pain. Pain Physician 18(4): E535–E582

Manchikanti L, Pampati V, Benjamin RM, Boswell MV (2015b) Analysis of efficacy differences between caudal and lumbar interlaminar epidural injections in chronic lumbar axial discogenic pain: local anesthetic alone vs. local combined with steroids. Int J Med Sci 12(3):214–222. https://doi.org/10.7150/ijms.10870

Margetic P, Pavic R, Stancic MF (2013) Provocative discography screening improves surgical outcome. Wien Klin Wochenschr 125(19–20):600–610

Marks RC, Houston T, Thulbourne T (1992) Facet joint injection and facet nerve block: a randomised comparison in 86 patients with chronic low back pain. Pain 49(3):325–328

Maugars Y, Mathis C, Berthelot JM et al (1996) Assessment of the efficacy of sacroiliac corticosteroid injections in spondylarthropathies: a double-blind study. Br J Rheumatol 35(8):767–770

McKenzie-Brown AM, Shah RV, Sehgal N, Everett CR (2005) A systematic review of sacroiliac joint interventions. Pain Physician 8(1):115–125

Modic MT, Steinberg PM, Ross JS, Masaryk TJ, Carter JR (1988) Degenerative disk disease: assessment of changes in vertebral body marrow with MR imaging. Radiology 166:193–199

Ng L, Chaudhary N, Sell P (2005) The efficacy of corticosteroids in periradicular infiltration for chronic radicular pain: a randomized, double-blind, controlled trial. Spine (Phila Pa 1976) 30(8):857–862

Ohnmeiss DD, Vanharanta H, Guyer RD (1995) The association between pain drawings and computed tomographic/discographic pain responses. Spine (Phila Pa 1976) 20:729–733

Peng B, Hou S, Wu W, Zhang C, Yang Y (2006) The pathogenesis and clinical significance of a high-intensity zone (HIZ) of lumbar intervertebral disc on MR imaging in the patient with discogenic low back pain. Eur Spine J 15:583–587

Rahme R and Moussa R (2008) The Modic Vertebral Endplate and Marrow Changes: Pathologic Significance and Relation to Low Back Pain and Segmental Instability of the Lumbar Spine Am J Neuroradiol 29 (5):838–842. https://doi.org/10.3174/ajnr.A0925

Rauchwerger JJ, Candido KD, Zoarski GH (2011) Technical and imaging report: fluoroscopic guidance for diagnosis and treatment of lumbar synovial cyst. Pain Pract 11:180–184

Resnick D, Choudhri TF et al (2005) Guidelines for the performance fusion procedures for degenerative disease of the lumbar spine. Spine 2:662–669

Rupert MP, Lee M, Manchikanti L et al (2009) Evaluation of sacroiliac joint interventions: a systematic appraisal of the literature. Pain Physician 12(2):399–418

Sabers SR, Ross SR, Grogg BE, Lauder TD (2005) Procedure-based nonsurgical management of lumbar zygapophyseal joint cyst-induced radicular pain. Arch Phys Med Rehabil 86:1767–1771

Saboeiro GR (2009) Lumbar discography. Radiol Clin N Am 47(3):421–433

Sachs BL, Vanharanta H et al (1987) Dallas discogram description. A new classification of CT/discography in low-back disorders. Spine 12:288–294. (Table 2, page 288)

Salahadin A, Sukdeb D, Schultz D, Rajive A et al (2007) Epidural steroids in the management of chronic spinal pain: a systematic review. Pain Physician 10:185–212

Schofferman J, Kine G (2004) Effectiveness of repeated radiofrequency neurotomy for lumbar facet pain. Spine (Phila Pa 1976) 29(21):2471–2473

Schulz C, Danz B, Waldeck S et al (2011) Percutaneous CT-guided destruction versus microsurgical resection of lumbar juxtafacet cysts. Orthopade 40:600–606

Schwarzer AC, Aprill CN, Derby R et al (1994) Clinical features of patients with pain stemming from the lumbar zygapophysial joints. Is the lumbar facet syndrome a clinical entity? Spine 19:1132–1137

Schwarzer AC, Aprill CN, Derby R et al (1997) International Spinal Injection Society guidelines for the performance of spinal injection procedures: I. Zygapophysial joint blocks. Clin J Pain 13:285–302

Sehgal N, Dunbar EE, Shah RV, Colson J (2007a) Systematic review of the diagnostic utility of facet (zygoapophysial) joint injections in chronic spinal pain: an update. Pain Physician 10:213–228

Sehgal N, Rinoo V, Colson J. (2007b) Chronic pain treatment with opioid analgesics: benefits versus harms of long-term therapy. Expert Rev Neurother 13(11): 213–228

Shah RV, Lutz GE (2003) Lumbar intraspinal synovial cysts: conservative management and review of the world's literature. Spine J 3:479–488

Slipman CW, Lipetz JS, Jackson HB et al (2000) Therapeutic selective nerve root block in the non-surgical treatment of atraumatic cervical spondylotic radicular pain: a retrospective analysis with independent clinical review. Arch Phys Med Rehabil 81:741–746

Son JH, Kim SD, Kim SH, Lim DJ, Park JY (2010) The efficacy of repeated radiofrequency medial branch neurotomy for lumbar facet syndrome. J Korean Neurosurg Soc 48(3):240–243

Storheim K, Berg L, Hellum C et al (2017) Fat in the lumbar multifidus muscles – predictive value and change following disc prosthesis surgery and multi-disciplinary rehabilitation in patients with chronic low back pain and degenerative disc: 2-year follow-up of a randomized trial. BMC Musculoskelet Disord 18:145

Tekin I, Mirzai H, Ok G et al (2007) A comparison of conventional and pulsed radiofrequency denervation in the treatment of chronic facet joint pain. Clin J Pain 23(6):524–529

Thomas E, Cyteval C, Abiad L et al (2003) Efficacy of transforaminal versus interspinous corticosteroid injection discal radiculalgia – a prospective, randomised, double-blind study. Clin Rheumatol 22(4–5):299–304

Vad VB, Bhat AL, Lutz GE, Cammisa F (2002) Transforaminal epidural steroid injections in lumbosacral radiculopathy a prospective randomized study. Spine 27(1):11–16

Vallee JN, Feydy A, Carlier RY et al (2001) Chronic cervical radiculopathy: lateral approach periradicular corticosteroid injection. Radiology 218:886–892

van Kleef M, Barendse GA, Kessels A et al (1999) Randomized trial of radiofrequency lumbar facet denervation for chronic low back pain. Spine (Phila Pa 1976) 24(18):1937–1942

van Kleef M, Barendse GA, Kessels A et al (2005) Radiofrequency denervation of lumbar facet joints in the treatment of chronic low back pain: a randomized, double-blind, sham lesion-controlled trial. Clin J Pain 21:335–344

Van Tilburg CWJ, Stronks DL, Groeneweg JG et al (2016) Randomised sham-controlled double-blind multicentre clinical trial to ascertain the effect of percutaneous radiofrequency treatment for lumbar facet joint pain. Bone Joint J 98-B(11):1526–1533

Walsh TR, Weinstein JN, Spratt KF et al (1990) Lumbar discography in normal subjects. A controlled, prospective study. J Bone Joint Surg Am 72:1081–1088

Xi MA, Tong HC, Fahim DK, Perez-Cruet M (2016) Using provocative discography and computed tomography to select patients with refractory discogenic low back pain for lumbar fusion surgery. Cureus 8(2):e514

Younger J, McCue R, Mackey S (2009) Pain outcomes: a brief review of instruments and techniques. Curr Pain Headache Rep 13(1):39–43

Zotti MGT, Osti OL (2010) Repeat percutaneous radio-frequency facet joint denervation for chronic back pain: a prospective study. J Musculoskelet Pain 18(2):153–158

Zotti VA, Zotti MGT, Worswick D (2012) Obtaining informed medical consent: a legal perspective. Bulletin (Law Soc South Australia) 34(1):22–24

Zotti MGT, Boas FV, Clifton T et al (2017) Does pre-operative magnetic resonance imaging of the lumbar multifidus muscle predict clinical outcomes following lumbar spinal decompression for symptomatic spinal stenosis? Eur Spine J 26 (1):2589–2597

Zuo J, Joseph GB, Li X et al (2012) In-vivo intervertebral disc characterization using magnetic resonance spectroscopy and $T_{1\rho}$ imaging: association with discography and Oswestry disability index and SF–36. Spine 37(3):214–221

Surgical Site Infections in Spine Surgery: Prevention, Diagnosis, and Treatment Using a Multidisciplinary Approach

53

Matthew N. Scott-Young, Mario G. T. Zotti, and Robert G. Fassett

Contents

Introduction	950
Epidemiology	950
Etiology	951
Risk Factors	953
Prevention of Spinal SSIs	954
Preoperative	954
Prehospital	954
In Hospital	955
Intraoperative, in Hospital	955
Postoperative, in Hospital	955
Postoperative, Following Hospital Admission	956
Diagnosis of SSI	956
Clinical Presentation	956

M. N. Scott-Young (✉)
Faculty of Health Sciences and Medicine, Bond University, Gold Coast, QLD, Australia

Gold Coast Spine, Southport, QLD, Australia
e-mail: mscott-young@goldcoastspine.com.au;
info@goldcoastspine.com.au

M. G. T. Zotti
Orthopaedic Clinics Gold Coast, Robina, QLD, Australia

Gold Coast Spine, Southport, QLD, Australia
e-mail: mzotti@goldcoastspine.com.au

R. G. Fassett
Faculty of Health Sciences and Medicine, Bond University, Gold Coast, QLD, Australia

Schools of Medicine and Human Movement and Nutrition Sciences, The University of Queensland, St Lucia, QLD, Australia
e-mail: rfassett@me.com

© Springer Nature Switzerland AG 2021
B. C. Cheng (ed.), *Handbook of Spine Technology*,
https://doi.org/10.1007/978-3-319-44424-6_84

Management 958

Conclusions 959

References 959

Abstract

Surgical site infections (SSIs) after spinal surgery are an important cause of postoperative morbidity and a multidisciplinary approach should focus on prevention. Screening for preoperative and recognition of intraoperative risk factors that increase the incidence of SSIs should be routine practice. Factors shown to influence SSIs include diabetes, obesity, previous SSI, complex multilevel procedures, and excessive surgical time and blood loss.

Multidisciplinary awareness and monitoring for SSIs is required with a high index of suspicion based on a combination of clinical findings including pain in the surgical area, swelling, fever, and wound discharge and diagnostic tests including WCC, CRP, ESR, wound microbiology, and blood cultures. Imaging the area with ultrasound, CT, or MRI and guided needle sampling of any detected collections may be required. Prompt treatment with antibiotics reflecting regional bacterial isolates and their sensitivity patterns should be implemented. Surgical wound wash outs, often performed repeatedly, may be necessary in selected cases.

Keywords

Surgical site infection · Prevention · Risk factors · Diagnosis · Management · Staphylococcus aureus · Implant multidisciplinary · Inflammatory markers · Spinal surgery

Introduction

Surgical site infections (SSIs) represent a significant morbidity after spinal surgery and a multidisciplinary approach is required to minimize their risk and manage them when they occur. They are defined as infections as demonstrated by clinical features with support of ancillary tests and microbial testing in the perioperative area of a spinal intervention. These are subclassified into superficial SSIs, which are localized to the skin and subcutaneous tissue, and deep SSIs, which are deep to the fascia and include either deep incision SSI or organ/space SSI (see Fig. 1). Although rates of SSI in spine surgery are low, recognition of high-risk situations for SSI and knowledge of general and specific measures for prevention is critical for the practicing spinal surgical team. Prompt and appropriate diagnosis and intervention, which may be aggressive, is required to prevent a further complication cascade for the patient. This is particularly so as SSIs are not only associated with potential failure of the intended treatment but are also associated with significant increases in hospital inpatient length of stay, readmission, prolonged antibiotics administration, and, hence, increased health costs (Pull Ter Gunne and Cohen 2009; Van Middendorp et al. 2012) (Table 1).

Epidemiology

In most reports analyzing modern spinal surgery in large cohorts with robustly collected data, the incidence of SSI is typically 1–7% (Van Middendorp et al. 2012). However, this figure has been reported to be as low as 0.5% and as high as 25% (Mistovich et al. 2017). It is difficult to generalize and compare the incidence of SSI between study populations given the large variation in patient factors (including pathology to be treated, comorbidities, risk factors, and regional differences in microbial carriage), surgical factors (including approach and complexity), classification (e.g., Superficial SSI which often presents earlier compared to deep), and methods

Table 1. Criteria for Defining a Surgical Site Infection (SSI)*

Superficial Incisional SSI
Infection occurs within 30 days after the operation *and* infection involves only skin or subcutaneous tissue of the incision *and* at least *one* of the following:
1. Purulent drainage, with or without laboratory confirmation, from the superficial incision.
2. Organisms isolated from an aseptically obtained culture of fluid or tissue from the superficial incision.
3. At least one of the following signs or symptoms of infection: pain or tenderness, localized swelling, redness, or heat *and* superficial incision is deliberately opened by surgeon, *unless* incision is culture-negative.
4. Diagnosis of superficial incisional SSI by the surgeon or attending physician.

Do *not* report the following conditions as SSI:
1. Stitch abscess (minimal inflammation and discharge confined to the points of suture penetration).
2. Infection of an episiotomy or newborn circumcision site.
3. Infected burn wound.
4. Incisional SSI that extends into the fascial and muscle layers (see deep incisional SSI).

Note: Specific criteria are used for identifying infected episiotomy and circumcision sites and burn wounds.[433]

Deep incisional SSI
Infection occurs within 30 days after the operation if no implant† is left in place or within 1 year if implant is in place and the infection appears to be related to the operation *and* infection involves deep soft tissues (e.g., fascial and muscle layers) of the incision *and* at least *one* of the following:
1. Purulent drainage from the deep incision but not from the organ/space component of the surgical site.
2. A deep incision spontaneously dehisces or is deliberately opened by a surgeon when the patient has at least one of the following signs or symptoms: fever (>38°C), localized pain, or tenderness, unless site is culture-negative.
3. An abscess or other evidence of infection involving the deep incision is found on direct examination, during reoperation, or by histopathologic or radiologic examination.
4. Diagnosis of a deep incisional SSI by a surgeon or attending physician.

Notes:
1. Report infection that involves both superficial and deep incision sites as deep incisional SSI.
2. Report an organ/space SSI that drains through the incision as a deep incisional SSI.

Organ/space SSI
Infection occurs within 30 days after the operation if no implant† is left in place or within 1 year if implant is in place and the infection appears to be related to the operation *and* infection involves any part of the anatomy (e.g., organs or spaces), other than the incision, which was opened or manipulated during an operation *and* at least *one* of the following:
1. Purulent drainage from a drain that is placed through a stab wound‡ into the organ/space.
2. Organisms isolated from an aseptically obtained culture of fluid or tissue in the organ/space.
3. An abscess or other evidence of infection involving the organ/space that is found on direct examination, during reoperation, or by histopathologic or radiologic examination.
4. Diagnosis of an organ/space SSI by a surgeon or attending physician.

Fig. 1 Classification and criteria for spinal SSI according to the National Noscomial Infections Surveillance System Criteria. (Reprinted with permission from Mangram et al. 1999). (Copyright 1999 by Elsevier)

of diagnosis with variable identification of less virulent organisms (Collins et al. 2008; Schoenfeld et al. 2011). A good example of the variation in incidence and the role of pathology and complexity is within the pediatric scoliosis population – the incidence of SSI for routine AIS correction is in the order of 0.5%, while correction of complex syndromic and neuromuscular scoliosis has a far higher incidence in the order of 20–25% (Mistovich et al. 2017).

Etiology

The understanding of SSI in the spine continues to evolve and the differences in pathophysiology between organisms and processes likely accounts for some of the different presentations observed in the patient clinically. It must be stressed that the etiology is different to that observed in primary spinal infection. Surgical site infections can be sub grouped into temporal (i.e., acute < 3 weeks, subacute 3 weeks–3 months, and

chronic > 3 months) and physical characteristics (superficial and deep). The most common organism isolated is *staph. aureus* followed by *coagulase negative staphylococci* with *staph. Epidermidis, Pseudomonas aeruginosa, E. Coli, Proteus,* and *P. Acnes* all relatively common (Abdul-Jabbar et al. 2013).

Superficial infection (superficial to fascia) as in other areas of the body relates to the favorability of the healing environment at the level of the superficial integument: the genetics and immune system of the host, physical tension, wound oxygen tension, vascularity, dead space, apposition of layers, amount of foreign material, and local bacteria all have a role here. Assuming a truly no fascial breach, either local wound care and antibiotics or simple drainage/aspiration of any abscess with antibiotics may resolve the situation.

Deep infection, however, is a different entity not dissimilar to that seen in deep infection of extremity orthopedic implants. Unlike primary spinal infections, SSI is more commonly a result

Table 1 Criteria for Defining a Surgical Site Infection (SSI)[*]

Superficial incisional SSI
Infection occurs within 30 days after the operation *and* infection involves only skin or subcutaneous tissue of the incision *and* at least *one* of the following:
1. Purulent drainage, with or without laboratory confirmation, from the superficial incision
2. Organisms isolated from an aseptically obtained culture of fluid or tissue from the superficial incision
3. At least one of the following signs or symptoms of infection: Pain or tenderness, localized swelling, redness, or heat *and* superficial incision is deliberately opened by surgeon, *unless* incision is culture-negative
4. Diagnosis of superficial incisional SSI by the surgeon or attending physician
Do *not* report the following conditions as SSI:
1. Stitch abscess (minimal inflammation and discharge confined to the points of suture penetration)
2. Infection of an episiotomy or newborn circumcision site
3. Infected burn wound
4. Incisional SSI that extends into the fascial and muscle layers (see deep incisional SSI)
Note: Specific criteria are used for identifying infected episiotomy and circumcision sites and burn wounds[433]
Deep Incisional SSI
Infection occurs within 30 days after the operation if no implant[†] is left in place or within 1 year if implant is in place and the infection appears to be related to the operation *and* infection involves deep soft tissues (e.g., fascial and muscle layers) of the incision *and* at least *one* of the following:
1. Purulent drainage from the deep incision but not from the organ/space component of the surgical site
2. A deep incision spontaneously dehisces or is deliberately opened by a surgeon when the patient has at least one of the following signs or symptoms: Fever ($>38\ °C$), localized pain, or tenderness, unless site is culture-negative
3. An abscess or other evidence of infection involving the deep incision is found on direct examination, during reoperation, or by histopathologic or radiologic examination
4. Diagnosis of a deep incisional SSI by a surgeon or attending physician
Notes:
1. Report infection that involves both superficial and deep incision sites as deep incisional SSI
2. Report an organ/space SSI that drains through the incision as a deep incisional SSI
Organ/Space SSI
Infection occurs within 30 days after the operation if no implant[†] is left in place or within 1 year if implant is in place and the infection appears to be related to the operation *and* infection involves any part of the anatomy (e.g., organs or spaces), other than the incision, which was opened or manipulated during an operation *and* at least *one* of the following:
1. Purulent drainage from a drain that is placed through a stab wound[‡] into the organ/space
2. Organisms isolated from an aseptically obtained culture of fluid or tissue in the organ/space
3. An abscess or other evidence of infection involving the organ/space that is found on direct examination, during reoperation, or by histopathologic or radiologic examination
4. Diagnosis of an organ/space SSI by a surgeon or attending physician

[*] Horan TC et al.

[†] National Nosocomial Infection Surveillance definition: a nonhuman-derived implantable foreign body (e.g., prosthetic heart valve, nonhuman vascular graft, mechanical heart, or hip prosthesis) that is permanently placed in a patient during surgery.

[‡] If the area around a stab wound becomes infected, it is not an SSI. It is considered a skin or soft tissue infection, depending on its depth.

of direct inoculation during the surgical procedure, although early postoperative contamination (e.g., getting early postoperative wound wet) and hematogenous spread remains a possibility; this is particularly so if gram-negative organisms are isolated (Chahoud et al. 2014). Whatever the cause, the concern in deep infection is biofilm, which is difficult to control without metal removal. There is then the difficult balance and so-called race between fusion and progression of infection where the natural history is unpredictable. Bacteria embedded in biofilms have been shown to develop and colonize on inert surfaces of many spinal implants and electron microscopy of samples taken from implant surfaces have shown biofilms to be the foci of device-related infection for "typical" biofilm forming bacteria such as staphylococci and streptococci (Tofuku et al. 2012). The bacteria within biofilms are protected against host defense mechanisms as there is no native blood supply and antibiotic therapy alone is ineffective because activated phagocytes cannot kill bacteria in biofilms. This is as antibodies released from sessile bacterial cells and antibiotics fail to penetrate biofilms and phagocytosis cannot be achieved (Tofuku et al. 2012). Interestingly, stainless steel implants are particularly vulnerable to biofilm compared to titanium implants. Further development of the use of materials or coatings that release antibiotics in concentrations that kill planktonic bacterial cells around the implant may have a role here. Until then, the only reliable way to disrupt the biofilm is surgical (Chahoud et al. 2014).

There is emerging interest in the atypical presentation of spinal infection, particularly in association with *propionibacterium acnes.* This organism, previously thought to be a contaminant, often causes no fever and a low-grade response with indolent failure of the implant construct. It is notoriously difficult to isolate and culture and, as such, should be sought for with PCR and extended cultures when implant constructs fail without a clear pyogenic presentation (Chahoud et al. 2014; Collins et al. 2008).

Risk Factors

Several papers, including those from large databases, have assessed the risk factors for spinal SSI but are limited in their identification of risk factors by methodology (Van Middendorp et al. 2012). From this viewpoint, several risk factors have been repeatedly identified and should be regarded as established while others have associated conflicting data and should be regarded as relative. Established risk factors include the comorbidities of type II diabetes mellitus,

obesity, multiple spinal operations, and previous SSI with surgical factors being length of open operation $\geq$ 5 h, $\geq$ one liter blood loss and multilevel constructs. Subpopulations identified as high risk of spinal SSIs compared to population norms include oncology patients, syndromic or neuromuscular deformity patients, patients with inflammatory arthritis, and immunosuppressed patients, e.g., HIV/AIDS (Chahoud et al. 2014; Pull Ter Gunne and Cohen 2009).

Relative risk factors which have been variably reported or not robustly studied include several patient and surgical factors. Patient factors include female gender, advanced age, fecal or bladder incontinence, atherosclerotic vascular disease, hypertension, smoking, alcohol, malnutrition, corticosteroids, and multiple operations. Comorbidities such as the above are thought to be compounding and additional to the risk inherent in any spinal procedure (Chan et al. 2014; Oichi et al. 2015; Quan et al. 2011; Walid and Robinson 2011). MRSA carriage is a controversial risk factor in its implications for screening, prevention, and treatment (Catanzano et al. 2014; Mehta et al. 2013; Molinari et al. 2012).

Relative surgical risk factors that have limited or conflicting data include the approach (higher prevalence with posterior and non-same day staged 360° fusions), "invasiveness" of the approach (higher in open compared to minimally invasive for multilevel procedures) (McGirt et al. 2011; Smith et al. 2011), visceral injury (e.g., bowel for lumbar; esophageal for cervical) (Kang et al. 2009; Pichelmann and Dekutoski 2011), thoracic procedures (Smith et al. 2011), instrumented procedures (Smith et al. 2011) particularly stainless steel, transfusion use intraoperatively and use of surgical drains (Kawabata et al. 2017).

The implication of identifying situations of high SSI risk is to optimize the situation for the patient so as to minimize the risk of infection and also to counsel the patient regarding their relative risk for planned surgery. It is self-evident that the relative risk discussion with a patient with minimal comorbidities undergoing a single-level procedure would be different to an elderly, obese, diabetic patient planned for a multilevel reconstruction. Optimizing the patient may then also have benefits with regard to minimizing anesthesia-related comorbidity and reaching the therapeutic goal (e.g., increasing likelihood of fusion of a painful motion segment). Several comorbidity-derived calculators may be useful for this purpose, including screening with the Charlson comorbidity index calculator (Walid and Robinson 2011). At present, while infection risk calculators such as the standardized infected ratio exist, no validated long-term data is available for any spinal specific scoring system (Fukuda et al. 2013).

Prevention of Spinal SSIs

Prevention and treatment of spinal SSIs must be viewed through a truly multifaceted manner. To regard prevention of infections with a narrow focus such as concentrating on patient selection, on which prophylactic antibiotic to administer or on which surgical skin preparation to use misses the very broad range of factors that can be successfully addressed with a holistic approach and a multidisciplinary team. In identifying SSIs postoperatively, a search for factors that have led to the SSI and how they can be prevented in future should be prompted, ideally, in a collaborative audit setting. Involvement of physicians, infection control personnel, and specialty nurses may uncover previously unrecognized factors in the patient care that, if unabated, could lead to continued high incidence of SSIs. These include changes which may seem trivial or banal in patient preoperative skin care, theatre equipment sterilizsation, theatre environment such as laminar flow and cleaning, and in ward care such as showering and wound dressing protocols. However, these are changes which the surgeon may not immediately consider or be aware of in their busy routine of daily practice. With the insight and knowledge of other key allied health personnel involved in the hospital facility and the patient's perioperative journey, such factors are imminently amenable to prevention of further infection episodes. From this viewpoint, below are points considered important and best practice to avoid SSI in our practice.

Preoperative

Prehospital

Patient

- Assess and manage patient-related risk factors as detailed above including diabetic, blood pressure and weight control, cessation of smoking and alcoholism, and optimizing nutrition.
- Educate the patient regarding optimal nutrition and hygiene in planned surgical sites. For example, topical benzoyl peroxide may be useful in decolonizing adolescent patients prior to posterior spinal surgery who would otherwise be at risk of *p. acnes* infection.
- Ensure the patient is free of any intercurrent treatable infections. Although controversial, screening the urine for asymptomatic infection may be appropriate in selected populations.
- Depending on regional prevalence of MRSA and VRE coverage, consider adoption of a screening and eradication program, e.g., Nasal screening, washes, and intranasal mupirocin.
- Referral to or involvement of a dermatologist preoperatively should be considered in selected cases. For example, adolescent patients with widespread dorsal acne planned for scoliosis correction or patients with eczema/dermatitis that may require treatment.
- Involvement of an infectious diseases physician in selected cases. For example in patients with immunodeficiency due to HIV/AIDS or tuberculosis.
- For oncology patients, close collaboration with oncology colleagues regarding the patient's immune status and any radiotherapy or chemotherapy interventions perioperatively.
- Liaison with rheumatologist regarding antirheumatic and immune modulating drugs for patients with inflammatory arthritis and whether they should be discontinued perioperatively.

In Hospital

Theatre

- Robust infection control protocols and procedures including audit of theatre and ward cleanliness and sterilization effectiveness.
- Meticulous hand hygiene of all those in contact with the patient.
- Ensure access to large and clean theatres with laminar flow or clean filters (Gruenberg et al. 2004).
- Properly sterilized equipment available.
- Trays, staff, personnel, and theatre traffic managed by team in a way that minimizes potential for contamination and particulate agitation. Signage that open surgery is in progress to minimize traffic.
- Insist on mask donning and appropriate hair coverage at time of tray opening for all personnel and minimize unnecessary equipment movement, e.g., II.
- Meticulous draping technique and patient warming.
- There are mixed reports regarding efficacy of surgical isolation hoods and ultraviolet light (Cheng et al. 2005).

Patient

- Preincision antibiotics, most commonly cephalosporins in the low-risk population, with appropriate cover of skin organisms (Petignat et al. 2008)
- Clipping of excess hair at the surgical site
- Pre-incision cleaning and scrubbing
- Sterile insertion of any drains or invasive monitoring

Intraoperative, in Hospital

- Minimize soft tissue trauma and retraction times where possible. Where surgery is prolonged, consider release of retraction momentarily to prevent prolonged ischemia. Consider use of minimally invasive muscle sparing techniques if appropriate for pathology and adequate training/skillset.

- Utilization of a "no touch" technique where possible and regular glove changes after heavy soiling.
- Adequate hemostasis given that residual hematoma likely to be nidus for infection.
- Minimizing the repetition of steps and adhering to efficient completion of the operative case to minimize any unnecessary delays in achieving sterile wound closure.
- Intermittent saline bathing of tissues to prevent desiccation and lavage of wounds prior to closure (Brown et al. 2004). Diluted betadine has also been shown to be effective for prevention of infection but its effects on tissue fibroblasts need to also be considered (Cheng et al. 2005).
- Consider use of topical vancomycin powder. Some authors advocate for routine usage in spinal surgery as has been shown in some literature reviews to be effective for prevention of deep SSI (Devin et al. 2015; Godil et al. 2013; Schroeder et al. 2016).
- Apposition of skin edges without excessive tension. Dressings that allow removal of excess moisture on the wound and vapor exchange.

Postoperative, in Hospital

- Prophylactic antibiotics according to local guidelines
- Meticulous hand hygiene and environmental cleaning of the patient's surrounds. Adequate signage and control of visitors if the patient has additional resistant organisms or visitors are found to be unwell.
- Mobilization of the patient with physiotherapy and regular positioning and appropriate mattress selection to offload pressure from skin. Regular pulmonary toilet to prevent respiratory collapse and infection.
- Optimal nutrition with diet on ward having adequate protein, vitamins, and minerals to support healing.
- Regular aperients and encouraging elimination to minimize risk of bowel stasis and urinary tract infection.

- Maintaining adequate perfusion with good hydration and consideration for transfusion if levels critically low, e.g., ≤ 9 g/dL. Perioperative physician care to support the patient in the postoperative period is valuable.
- Preference for mechanical deep vein thrombosis prophylaxis over chemical, where possible, to reduce risk of wound ooze
- Wound changes to be kept to a minimum and performed only where necessary (contaminated wound or excessive fluid).
- Removal of any drains and indwelling devices as soon as practicable to prevent device-related infections

Postoperative, Following Hospital Admission

Patient
- Education and provision of appropriate wound care and hygiene instructions, including washing.
- Maintenance of optimal health and avoiding toxins, e.g., excessive alcohol and tobacco.
- Early review of wound as outpatient and early and aggressive management if infection diagnosed.

Hospital and Surgical Unit
- Hospital infection control and regional infectious disease monitoring to assess local organisms and antibiotic sensitivities.
- Regular auditing and surveillance to correct any unexpected infections or unusual organisms and change practice accordingly. For example, contaminated theatres or malfunctioning sterilization machines.
- Participation in morbidity and mortality meetings and self-reflection on own practice SSI incidence.

Diagnosis of SSI

The diagnosis of SSI can be challenging and a low index of suspicion for infection must be held. With this mentality, neither the "obvious"

pyogenic infections heralded by febrile patients with painful and purulent wounds nor the patients who present in an insidious way with persistent low-grade pain and malaise should be missed. This is particularly important with emerging evidence of previously difficult to diagnose organisms affecting instrumented spinal cases (Collins et al. 2008). Figure 1 shown previously provides objective criteria for superficial, deep, and organ space SSIs.

Clinical Presentation

A presentation of a wound issue, such as a discharging, swollen, purulent, erythematous, or discharging wound, should always be a cause for concern for surgeons and staff alike of a SSI. Systemic symptoms such as lethargy, malaise, loss of appetite, and, in more severe cases, fevers, rigors, and sweats should alarm staff that there is not only a local issue in the spine and its surroundings but potentially a systemic response that, if unabated, could lead to a septic syndrome. The sequelae of this can be heralded by hemodynamic dysfunction and metabolic dysfunction before multiorgan failure ensues.

In fact, the most common presenting complaint of patients later diagnosed with infection is less dramatic in the form of back pain. This presents a quandary for the clinician, where postoperative back pain may have a myriad of causes including residual pathology and "normal" postoperative tissue response in the first month. The onset tends to be insidious and persistent (Collins et al. 2008). The usual duration between procedure and discernible features of infection ranges from 2–30 days and varies depending on the behavior of the organism (Chahoud et al. 2014).

The more dramatic presentation of the unwell patient who is septic is becoming increasingly concerning with different resistant strains of staphylococci identified that have the potential to incite a violent systemic response. However, as an increasingly aging and comorbid population come to spinal surgery, there must be an index of suspicion for hematogenous seeding of implants with the increased incidence of intercurrent infections that can mask an implant infection.

Ancillary investigations that support this diagnosis include blood serology testing, microbiology, and imaging. This also includes evaluation of other potential sites of infection such as the urine, heart, abdomen, and chest with urinalysis, echocardiogram, and radiographs, if the history and physical examination indicates it. Elevated CRP is nonspecific postoperatively but very high levels and levels that do not come down after the early postoperative period is concerning. Likewise, WCC has fair sensitivity for spinal infection only but lacks specificity. ESR can be useful when adjusted for age and sex in the subacute and chronic settings and is concerning if raised but is, again, nonspecific. Novel technologies that are emerging include serology such as serum amyloid and synovial procalcitonin and alpha defensin but these require further validation in the spinal setting (Chen et al. 2017).

Microbiological identification of the causative organism with accurate and careful handling and preparation of sample tissues is critical. It is important, if possible, to avoid the temptation to treat the patient with antibiotics prior to microbiological identification of an organism empirically (unless they are septic) in the cases where a specimen can be obtained. Correct organism identification is paramount to appropriately targeting and tailoring antibiotics to the organism and understanding its behavior. In the case of the septic patient who is progressing to extremis or the patient with probable epidural abscess-related neurological deterioration, blood cultures should at least be obtained prior to commencing empirical antibiotics. Should the patient have inadvertently been given antibiotics then delaying an aspiration for 1 week may be advisable to increase yield. Wound swabs are of mixed value, particularly as they are hard to interpret in the presence of skin colonized by commensals and their often polymicrobial growths. Histopathological examination as an adjunct is useful in assessing for typical pyogenic changes as opposed to granulomatous change or unexpected neoplasm. DNA microarrays and PCR molecular amplification techniques are likely to become the standard of care and offer high sensitivity and specificity (Chahoud et al. 2014).

Imaging can help define and localize the infection and helps to exclude other causes for the patient's presentation. Nuclear medicine, unlike in primary spine infections, have a limited role in SSI given the extent of uptake expected from surgical intervention albeit that continued improvements in technology may make this a more specific and reliable tool in identifying SSI accurately (Cornett et al. 2016). Specimens obtained under sterile conditions either operatively or under image guidance (ultrasound, fluoroscopy or CT) will likely yield the causative organism. In the setting of a spinal SSI, it is important to assess whether there is a collection or any clearly pathological tissue which to target for culture before proceeding with the intervention and it is also important to use a wide-bore cutting needle to maximize tissue yield through a core biopsy (Garg et al. 2016). Imaging here either with MRI, CT (with metal suppression if implants present), or a combination can allow evaluation of the bony and soft tissues in the surgical site and evaluate for evidence of infection as opposed to postoperative edema, hematoma, or pseudomeningocele. For example, if the MRI diagnoses pyomositis and deep soft tissue abscesses around the surgical site, then these present a clear target for diagnostic yield sonographically and likewise suspected vertebral osteomyelitis can be identified readily with CT guided biopsy. While this technique may be useful in trying to avoid the morbidity of addressing a deep SSI, image guided biopsy does not have such a strong indication for superficial infections and when compared to open sampling their diagnostic yield is inferior in the order of 40–60% (Chahoud et al. 2014).

However, there are situations where suspected infected areas are not safely or easily amenable to percutaneous techniques for a diagnostic sample. An example of this is an infection following a decompression procedure that involves an epidural abscess. While some situations allow for safe percutaneous sampling of the abscess, a surgeon must be prepared to perform open sampling with or without drainage of the abscess, depending on the presence of phlegmon and dural adhesions. Infection around pedicle screws and cross links in the absence of a discernible abscess may have limited yield with percutaneous techniques due to difficulty sampling around the metal and fusion mass. In this setting, where the diagnosis of infection is strongly suspected from the patient history, clinical presentation, and supportive serology, the surgeon must consider the role of an open operative biopsy with or without metal exchange. The advantage of metal exchange, other than to reduce the potential for a nidus and debride the involved spinal tissue is to allow sonification of the spinal implants which can greatly improve yield. Again, one should consider PCR and extended cultures for *p. acnes* or staining for fungi or *mycobacterium* if there is an atypical presentation.

Management

Once the diagnosis of a spinal SSI is made, the approach should generally be aggressive. The exception to this is a cellulitis or suture abscess where a more conservative approach with dressings and antibiotics may be advisable. Where the potential for clinical deterioration, instability, and neurological compromise exists that will likely be difficult to control with medical therapy, early and judicious intervention is recommended (Cornett et al. 2016).

Superficial infections should be managed surgically with either wound excision or incision and drainage depending on its size and localization, followed by debridement and lavage of the area. An assessment should then be made on suitability for primary versus delayed primary closure depending on the patient and the extent of wound infection. This can then reverse a potentially catastrophic complication of deep-space involvement with the relatively simple measure of rapid superficial wound treatment. The difficulty in this scenario is often knowing when to intervene when the patient is systemically well, and markers of infection are equivocal but there is a slow-to-settle or oozing wound. Judgment is required here to ensure that wounds progress towards healing and, if this is not the case, then a return to theatre to achieve a clean and sealed wound is advisable (Cornett et al. 2016).

Deep infections present difficulty not just diagnostically but in deciding on the amount of treatment that should be offered to the individual patient. In an ideal world, infection should be treated in accordance with oncological principles achieving wide local clearance and adequate systemic therapy; however, this is not always possible due to the frailty of the patient or the locale of the infection, e.g., on the spinal cord. The intervention also varies depending on which approach the previous operation used and whether it involved instrumentation. In certain circumstances, where there has been very complex surgery, neurological injury, the need for reuse of potentially dangerous approach (e.g., anterior lumbar) or in a physiologically vulnerable patient then nonoperative treatment with suppression may be reasonable if the patient is stable systemically and neurologically. If there is clear infection and the patient is suitable for an intervention, then the surgeon must decide if it is a situation where the natural history will allow for likely resolution of the infection. The un-instrumented patient with a deep infection may be suitable for antibiotics only or in the instrumented complex multilevel posterior case for temporary suppression and later removal of implants (e.g., when arthrodesis or fracture united).

We favor a more aggressive approach for management of SSI when feasible and safe:

- For posterior instrumentation, this would typically involve exchange of any posterior metalwork, washout, grafting, and vancomycin powder application.
- For posterior abscesses in uninstrumented cases, then drainage and washout is recommended provided there is no direct infection with phlegmon on the neural elements, where careful partial de-bulking may be more appropriate.
- For anterior cervical cases (rare), utmost care must be taken given difficulty in reestablishing planes in the revision setting. Assessment for any esophageal breach should be undertaken as a deep cervical infection anteriorly is unusual.

Again, removal of implants and replacement with graft and vancomycin is advisable.

- For posterior cervical and thoracic decompression cases, care must be taken with washout and debridement due to risk to the spinal cord. Gentle lavage and partial de-bulking only along with vancomycin powder is advised for an epidural collection.
- Anterior column infection complicating interbody cage or disc prosthesis insertion requires a considered approach as to whether an anterior or lateral approach is feasible and, generally, either a vertebrectomy or revision with graft and a titanium cage would be preferable depending on the extent of osteomyelitis and post-debridement bony defect.
- In the rare case of exposed metal or large areas of devitalized soft tissue (usually after posterior wound debridement), then a plastic surgeon may be consulted to provide vascularized coverage to the implants.
- In the rare case of an "open space" SSI related to anterior cervical or lumbar fusion, it is advisable to provisionally diagnose the organ(s) which the infection involves and enlist the assistance of a relevant surgical colleague, for example, otorhinolaryngologist for organ involvement in the anterior neck and a vascular or general surgeon for assistance with organ or vessel infection in the abdomen.

Once a sample is taken or the organism is identified, we commence empirical antibiotics transitioning to tailored antibiotics as soon as practicable. Again, the patient should be optimized and cared for in the standard postoperative manner as detailed above. However, the length and course of the antibiotics will be different and should be discussed with a physician so that consensus over the most effective course for eradicating the infection can be achieved. Close follow-up must then be instituted to confirm successful remission and eradication of infection, including in the medium term. This should be undertaken clinically as well as with serology and imaging, such as CT scans, to verify the absence of any implant loosening or compromise.

Conclusions

SSIs represent a significant morbidity associated with spinal surgery and warrant a multidisciplinary approach to management. Early and aggressive treatment can lead to macroscopic eradication and salvage of the situation. A holistic and considered approach to prevention, diagnosis, management, and follow-up is likely to yield a lower incidence of SSI and improve patient outcomes.

References

Abdul-Jabbar A, Berven SH, Hu SS, Chou D, Mummaneni PV, Takemoto S et al (2013) Surgical site infections in spine surgery: identification of microbiologic and surgical characteristics in 239 cases. Spine 38(22):1425–1431. https://doi.org/10.1097/BRS.0b013e3182a42a68

Brown EM, Pople IK, de Louvois J, Hedges A, Bayston R, Eisenstein SM, Lees P (2004) Spine Update. Spine 29 (8):938–945. https://doi.org/10.1097/00007632-200404150-00023

Catanzano A, Phillips M, Dubrovskaya Y, Hutzler L, Bosco J (2014) The standard one gram dose of vancomycin is not adequate prophylaxis for MRSA. Iowa Orthop J 34:111–117. Retrieved from http://www.pubmedcentral.nih.gov/articlerender.fcgi?artid=4127722&tool=pmcentrez&rendertype=abstract

Chahoud J, Kanafani Z, Kanj SS (2014) Surgical site infections following spine surgery: eliminating the controversies in the diagnosis. Front Med 1(March 2015). https://doi.org/10.3389/fmed.2014.00007

Chan T-C, Luk JK-H, Chu L-W, Chan FH-W (2014) Validation study of Charlson comorbidity index in predicting mortality in Chinese older adults. Geriatr Gerontol Int 14(2):452–457. https://doi.org/10.1111/ggi.12129

Chen AF, Nana AD, Nelson SB, McLaren A (2017) What's new in musculoskeletal infection: update across orthopaedic subspecialties. JBJS [Am] 99(14):1232–1243. https://doi.org/10.2106/JBJS.17.00421

Collins I, Wilson-MacDonald J, Chami G, Burgoyne W, Vineyakam P, Berendt T, Fairbank J (2008) The diagnosis and management of infection following instrumented spinal fusion. Eur Spine J 17(3):445–450. https://doi.org/10.1007/s00586-007-0559-8

Cornett CA, Vincent SA, Crow J, Hewlett A (2016) Bacterial spine infections in adults: evaluation and management. J Am Acad Orthop Surg 24(1):11–18. https://doi.org/10.5435/JAAOS-D-13-00102

Devin CJ, Chotai S, McGirt MJ, Vaccaro AR, Youssef JA, Orndorff DG et al (2015) Intrawound vancomycin decreases the risk of surgical site infection after posterior spine surgery–a multicenter analysis. Spine 1. https://doi.org/10.1097/BRS.0000000000001371

Fukuda H, Morikane K, Kuroki M, Taniguchi S, Shinzato T, Sakamoto F et al (2013) Toward the rational use of standardized infection ratios to benchmark surgical site infections. Am J Infect Control 41(9):810–814. https://doi.org/10.1016/j.ajic.2012.10.004

Garg V, Kosmas C, Josan ES, Partovi S, Bhojwani N, Fergus N, . . ., Robbin MR (2016) Hybrid biopsy technique to improve yield 41(August):1–9. https://doi.org/10.3171/2016.4.FOCUS1614

Godil SS, Parker SL, O'Neill KR, Devin CJ, McGirt MJ (2013) Comparative effectiveness and cost-benefit analysis of local application of vancomycin powder in posterior spinal fusion for spine trauma: clinical article. J Neurosurg Spine 19(3):331–335. https://doi.org/10.3171/2013.6.SPINE121105

Gruenberg MF, Campaner GL, Sola CA, Ortolan EG (2004) Ultraclean air for prevention of postoperative infection after posterior spinal fusion with instrumentation: a comparison between surgeries performed with and without a vertical exponential filtered air-flow system. Spine 29(20):2330–2334. https://doi.org/10.1097/01.brs.0000142436.14735.53

Kang B-U, Choi W-C, Lee S-H, Jeon SH, Park JD, Maeng DH, Choi Y-G (2009) An analysis of general surgery–related complications in a series of 412 mini-laparotomic anterior lumbosacral procedures. J Neurosurg Spine 10(1):60–65. https://doi.org/10.3171/2008.10.SPI08215

Kawabata A, Sakai K, Sato H, Sasaki S, Torigoe I, Tomori M et al (2017) Methicillin-resistant Staphylococcus aureus nasal swab and suction drain tip cultures in 4573 spinal surgeries: efficacy in Management of Surgical Site Infections. Spine 1. https://doi.org/10.1097/BRS.0000000000002360

Mangram AJ, Horan TC, Pearson L et al (1999) Guideline for prevention of surgical site infection, 1999. Am J Infect Control 20(4):250–278

McGirt MJ, Parker SL, Lerner J, Engelhart L, Knight T, Wang MY (2011) Comparative analysis of perioperative surgical site infection after minimally invasive versus open posterior/transforaminal lumbar interbody fusion: analysis of hospital billing and discharge data from 5170 patients. J Neurosurg Spine 14(6):771–778. https://doi.org/10.3171/2011.1.SPINE10571

Mehta S, Hadley S, Hutzler L, Slover J, Phillips M, Bosco JA (2013) Impact of preoperative MRSA screening and decolonization on hospital-acquired MRSA burden infection. Clin Orthop Relat Res 471(7):2367–2371. https://doi.org/10.1007/s11999-013-2848-3

Mistovich RJ, Jacobs LJ, Campbell RM, Spiegel DA, Flynn JM, Baldwin KD (2017) Infection control in pediatric spinal deformity surgery. JBJS Reviews 5 (5):e3. https://doi.org/10.2106/JBJS.RVW.16.00071

Molinari RW, Khera OA, Molinari WJ III (2012) Prophylactic intraoperative powdered vancomycin and postoperative deep spinal wound infection: 1,512 consecutive surgical cases over a 6-year period. Eur Spine J 21(S4):476–482. https://doi.org/10.1007/s00586-011-2104-z

Oichi T, Oshima Y, Takeshita K, Chikuda H, Tanaka S (2015) Evaluation of comorbidity indices for a study of patient outcomes following cervical decompression surgery: a retrospective cohort study. Spine 40(24):1941–1947. https://doi.org/10.1097/BRS.0000000000001153

Petignat C, Francioli P, Harbarth S, Regli L, Porchet F, Reverdin A et al (2008) Cefuroxime prophylaxis is effective in noninstrumented spine surgery: a double-blind, placebo-controlled study. Spine 33(18):1919–1924. https://doi.org/10.1097/BRS.0b013e31817d97cf

Pichelmann MA, Dekutoski MB (2011) Complications related to anterior and lateral lumbar surgery. Semin Spine Surg 23(2):91–100. https://doi.org/10.1053/j.semss.2010.12.012

Pull Ter Gunne AF, Cohen DB (2009) Incidence, prevalence, and analysis of risk factors for surgical site infection following adult spinal surgery. Spine 34(13):1422–1428. https://doi.org/10.1097/BRS.0b013e3181a03013

Quan H, Li B, Couris CM, Fushimi K, Graham P, Hider P et al (2011) Updating and validating the charlson comorbidity index and score for risk adjustment in hospital discharge abstracts using data from 6 countries. Am J Epidemiol 173(6):676–682. https://doi.org/10.1093/aje/kwq433

Schoenfeld AJ, Ochoa LM, Bader JO, Belmont PJ Jr (2011) Risk factors for immediate postoperative complications and mortality following spine surgery: a study of 3475 patients from the National Surgical Quality Improvement Program. J Bone Joint Surg Am 93(17):1577–1582. https://doi.org/10.2106/JBJS.J.01048[doi]

Schroeder JE, Girardi FP, Sandhu H, Weinstein J, Cammisa FP, Sama A (2016) The use of local vancomycin powder in degenerative spine surgery. Eur Spine J 25(4):1029–1033. https://doi.org/10.1007/s00586-015-4119-3

Smith JS, Shaffrey CI, Sansur CA, Berven SH, Fu KMG, Broadstone PA et al (2011) Rates of infection after spine surgery based on 108,419 procedures: a report from the Scoliosis Research Society morbidity and mortality committee. Spine 36(7):556–563. https://doi.org/10.1097/BRS.0b013e3181eadd41

Te Cheng M, Chang MC, Wang ST, Yu WK, Liu CL, Chen TH (2005) Efficacy of dilute betadine solution irrigation in the prevention of postoperative infection of spinal surgery. Spine 30(15):1689–1693. https://doi.org/10.1097/01.brs.0000171907.60775.85

Tofuku K, Koga H, Yanase M, Komiya S (2012) The use of antibiotic-impregnated fibrin sealant for the prevention of surgical site infection associated with spinal instrumentation. Eur Spine J 21(10):2027–2033. https://doi.org/10.1007/s00586-012-2435-4

Van Middendorp JJ, Pull Ter Gunne AF, Schuetz M, Habil D, Cohen DB, Hosman AJF, Van Laarhoven CJHM (2012) A methodological systematic review on surgical site infections following spinal surgery: part 2: prophylactic treatments. Spine 37(24):2034–2045. https://doi.org/10.1097/BRS.0b013e31825f6652

Walid MS, Robinson JS (2011) Economic impact of comorbidities in spine surgery. J Neurosurg Spine 14(3):318–321. https://doi.org/10.3171/2010.11.SPINE10139

Lumbar Interbody Fusion Devices and Approaches: When to Use What

54

Laurence P. McEntee and Mario G. T. Zotti

Contents

Introduction	962
Historical Perspective	963
Biomechanics of Interbody Fusion	963
ALIF	965
OLIF and ATP	966
LLIF	966
PLIF and TLIF	968
AxiaLIF	969
Clinical Rationale	970
Anterior Lumbar Interbody Fusion (ALIF)	970
Oblique Lumbar Interbody Fusion (OLIF)	970
Lateral Lumbar Interbody Fusion (LLIF)	971
Transforaminal Lumbar Interbody Fusion (TLIF)	971
Posterior Lumbar Interbody Fusion (PLIF)	972
Axial Lumbar Interbody Fusion (AxiaLIF)	972
Clinical Results	973
ALIF	973
OLIF	977
LLIF	978
TLIF	980
PLIF	981
AxiaLIF	983
Comparative Studies	984

L. P. McEntee (✉)
Gold Coast Spine, Southport, QLD, Australia

Bond University, Varsity Lakes, QLD, Australia
e-mail: lpmcentee@goldcoastspine.com.au

M. G. T. Zotti
Orthopaedic Clinics Gold Coast, Robina, QLD, Australia

Gold Coast Spine, Southport, QLD, Australia
e-mail: mzotti@goldcoastspine.com.au

© Springer Nature Switzerland AG 2021
B. C. Cheng (ed.), *Handbook of Spine Technology*,
https://doi.org/10.1007/978-3-319-44424-6_85

Lumbar Interbody Fusion: When to Use What? 986
Approach by Indication and Level(s) Requiring Interbody Fusion 989

Conclusion 991

References 991

Abstract

Lumbar interbody fusion is an established surgical technique for a variety of conditions affecting the lumbar spine. A large number of interbody fusion devices made of differing materials are now available for use. Approaches for interbody fusion include anterior lumbar interbody fusion, oblique lumbar interbody fusion, lateral lumbar interbody fusion, axial lumbar interbody fusion, transforaminal lumbar interbody fusion, and posterior lumbar interbody fusion. This chapter discusses the biomechanics of lumbar interbody fusion devices and approaches and the clinical rationale and the clinical results of each approach. The advantages and disadvantages of each approach are compared and contrasted. The importance of an appropriate preoperative assessment to determine the best approach for interbody fusion is emphasized, taking into account the condition being treated, sagittal balance, bone quality, and contraindications to a specific approach. The best approach to lumbar interbody fusion by indication and surgical level(s) is discussed.

Keywords

Lumbar interbody fusion · ALIF · OLIF · LLIF · AxiaLIF · TLIF · PLIF · Sagittal balance

Introduction

As with any operation in spinal surgery, the ultimate goal of interbody fusion is to decrease pain and increase function in the patients we treat. The specific technical goals of interbody fusion are to achieve a solid, stable arthrodesis of spinal segments that is able to sustain physiological loads while maintaining disc height and maintaining or restoring sagittal alignment. Any associated neural compression should be addressed as part of the fusion procedure. Maintenance of disc height and lordosis is necessary to preserve the natural alignment of the spine and the dimensions of the neural foramen, thus avoiding compression of the exiting nerve roots.

Spinal fusion has been used for many decades to treat a variety of spinal disorders. Instrumented spinal fusion was introduced to allow surgeons to alter the position of the spine, increase the rate of successful fusion, and allow earlier patient mobilization and recovery. To decrease failure rates of posterior instrumentation, the concept of anterior column interbody support was introduced.

Interbody fusion was originally performed using autograft iliac crest or allograft bone alone (Cloward 1953). The donor site morbidity associated with harvesting large amounts of iliac crest autograft and the inferior mechanical and biological properties of allograft bone were of concern. A significant incidence of collapse and pseudarthrosis was observed. Therefore interbody fusion devices were developed to provide mechanical support, while fusion takes place, thus increasing the rate of successful fusion and maintaining disc height and sagittal alignment while the fusion process occurs.

There are now many devices available for interbody fusion that vary in their material properties and route of implantation. In the lumbar spine, options include anterior lumbar interbody fusion (ALIF), oblique lumbar interbody fusion (OLIF), lateral lumbar interbody fusion (LLIF), posterior lumbar interbody fusion (PLIF), transforaminal lumbar interbody fusion (TLIF), and axial interbody fusion (AxiaLIF).

In determining which approach and device to use for interbody fusion, the surgeon must take into account the specific aims of the surgery in each individual patient. The interbody fusion device itself is only one of many factors in achieving clinical success for the patient. Appropriate patient selection for surgery and technical expertise in performing the surgery are of vital importance. This chapter will discuss the biomechanics of interbody fusion, the various approaches and devices available, the clinical results of interbody fusion, and "when to use what" in various clinical scenarios.

Historical Perspective

Early techniques of interbody fusion using autograft or allograft without instrumentation were associated with a high rate of clinical failure and pseudarthrosis. Stauffer and Coventry (1972) reported a 44% rate of poor clinical outcomes and a 44% rate of pseudarthrosis in 83 patients who underwent anterior interbody arthrodesis. A number of other studies reported similar results. The need for interbody devices to provide mechanical support while fusion occurs was thus established.

Bagby (1988) developed the first interbody fusion cage, "the Bagby basket," a stainless steel basket that was packed with local autograft bone and used in horses undergoing ACDF for wobbler syndrome, a type of spondylytic myelopathy. Bagby and Kuslich developed the first standalone threaded intervertebral cages which were a modified version of the Bagby basket made of titanium alloy and FDA approved in 1996 (Kuslich et al. 1998) (see Figs. 1 and 2). The BAK cage (Sulzer Spine-Tech, Minneapolis, Minnesota) was closely followed by the Ray threaded fusion cage (US Surgical, Norwalk, Connecticut), the threaded interbody fusion device (TIBFD, Medtronic Sofamor-Danek Group, Memphis, Tennessee), the Harms titanium mesh cage (DePuy-Acromed, Cleveland, Ohio), and the Brantigan rectangular and rounded cages (DePuy-Acromed). The cages were designed to stabilize a segment through distraction and tensioning of the annular and ligamentous structures; by partially reaming the end plates, cancellous bone would be exposed for arthrodesis.

Second-generation lumbar cages such as the LT cage (Medtronic, Minneapolis, Minnesota) were designed to be end plate sparing and able to obtain lordosis by threading a wedge-shaped device into the disc space. Since the design of these original devices, the field of interbody fusion technology has advanced significantly with multiple designs now available. The majority of interbody fusion devices are currently made from titanium alloy or polyetheretherketone (PEEK). Other materials include tantalum, carbon fiber, carbon fiber reinforced PEEK (CFRP), and more recently hybrid cages such as those made of titanium-coated PEEK. Emerging technologies include expandable cage technology and 3-D printed cages.

Biomechanics of Interbody Fusion

Interbody fusion offers several biomechanical advantages over posterolateral fusion. The interbody space offers a relatively large area for grafting with excellent vascularity, and the graft is placed under compression further enhancing

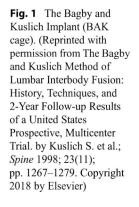

Fig. 1 The Bagby and Kuslich Implant (BAK cage). (Reprinted with permission from The Bagby and Kuslich Method of Lumbar Interbody Fusion: History, Techniques, and 2-Year Follow-up Results of a United States Prospective, Multicenter Trial. by Kuslich S. et al.; *Spine* 1998; 23(11); pp. 1267–1279. Copyright 2018 by Elsevier)

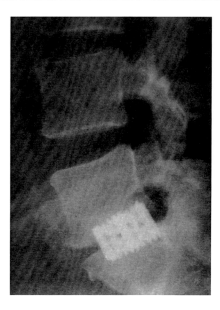

Fig. 2 BAK cage employed for L5–S1 ALIF. (Reprinted with permission from The Bagby and Kuslich Method of Lumbar Interbody Fusion: History, Techniques, and 2-Year Follow-up Results of a United States Prospective, Multicenter Trial. by Kuslich S. et al.; *Spine* 1998; 23(11); pp. 1267–1279. Copyright 2018 by Elsevier)

fusion. As 80% of axial compression forces are transferred through the anterior spinal column, the resulting arthrodesis is biomechanically superior to posterolateral fusion (Cunningham and Polly 2002). Interbody fusion also allows better maintenance or restoration of disc height and segmental lordosis and overall sagittal balance when compared to posterolateral fusion. This is associated with improved outcomes and a reduced rate of adjacent segment degeneration and disease (Rothenfluh et al. 2015).

When choosing a fusion device, the surgeon must consider the biomechanical properties of the device, in particular the implant stiffness which is determined by the material modulus of elasticity (Young's modulus) and the implant design (see Fig. 3). The implant design is in turn determined by the strength of the material used; a stronger material can have a more open architecture for bone graft and vascularization. An implant with a stiffness close to that of bone will allow load sharing between the device and the graft material, thus optimizing the biomechanics for fusion. An implant with a stiffness much higher than bone will load bear, therefore, stress shielding the graft material which can lead to resorption and pseudarthrosis.

When performing interbody fusion, the surgeon must consider the biomechanics of the vertebral end plates. Grant et al. (2002) conducted a biomechanical study assessing regional differences in end plate rigidity and found the posterior part was stronger than the anterior and that the periphery was stronger than the center. The strongest part was the posterolateral area, just in front of the pedicles. The superior end plate was much weaker than the inferior end plate. These findings have implications for implant design and placement when performing interbody fusion. A large footprint cage sitting on the peripheral end plate will be less likely to subside.

Early biomechanical studies suggested that threaded lumbar intervertebral cages were potentially stable enough to be used as stand-alone devices. Brodke et al. (1997) showed that two threaded BAK cages placed from a posterior approach resulted in greater stability in a bovine spine motion segment than a PLIF bone graft alone and equivalent stability to a PLIF bone graft and pedicle screw construct. Kettler et al. (2000) found that posteriorly placed intervertebral cages stabilized the spine in flexion and lateral bending but not extension and rotation and reduced stability under cyclical loading conditions was observed. Similarly, Oxland et al. (2000) found that anteriorly placed threaded cylindrical cages enhanced motion segment stability in all directions except extension and that supplementary translaminar facet screw fixation provided additional stability in extension. Rathonyi et al. (1998) also found that translaminar facet screw fixation greatly improved the stability of anterior threaded cylindrical cages in extension and rotation.

Kanayama et al. (2000) investigated the stability and stress-shielding effect of various lumbar interbody fusion devices; 11 different cages were tested, either threaded cages, non-threaded cages or allograft. No statistical differences were observed in construct stiffness among threaded and non-threaded devices in most of the testing modalities. Threaded cages demonstrated significantly lower intra-cage pressures compared with non-threaded cages and structural

Material	Young Modulus (GPa)	Poisson Ratio
Cortical bone	13.7[51]	0.30[51]
Cancellous bone	1.37[51]	0.30[51]
Titanium implant	110[51]	0.35[51]
Titanium base abutment	110[51]	0.35[51]
Zirconia customized abutment	210[52]	0.30[52]
PEEK customized abutment	3.5*	0.36*
Polymer infiltrated hybrid ceramic	30*	0.23[53]
Translucent zirconia	210*	0.26[54]
Lithium disilicate glass ceramic	95*	0.20[54]
Dual polymerized resin cement	18.6[55]	0.28[55]

PEEK, polyetheretherketone. *Values provided by manufacturer.

Fig. 3 Elastic modulus of materials used in interbody fusion devices. Note near identical elastic modulus of PEEK and cancellous bone. (Reprinted with permission from Effect of different restorative crown and customized abutment materials on stress distribution in single implants and peripheral bone: A three-dimensional finite element analysis study, by Kaleli N et al.; *The Journal of Prosthetic Dentistry* 2018; 119(3); pp. 437–445. Copyright 2018 by Elsevier)

allograft, suggesting a stress shielding effect with threaded cages.

Initial threaded cage stability is dependent on achieving adequate disc space distraction with resultant annular tensioning and is also dependent on compressive preload generated by muscle forces across the disc space. Phillips et al. (2004) found that under low preload conditions such as supine posture, BAK cages were less effective at stabilizing the motion segment in extension, whereas higher compressive preloads such as in sitting or standing led to stability of the motion segment in all motion planes. Supplementary translaminar facet screws provided a significant stabilizing advantage during low preload conditions.

ALIF

Further biomechanical testing of ALIF constructs has defined the relative stability of various stand-alone configurations (usually with integrated fixation) and anterior cage-plate constructs compared to the "gold standard" of ALIF with supplementary bilateral pedicle screw fixation.

Tsantrizos et al. (2000) compared the biomechanical stability of five stand-alone ALIF cages and found that all cage constructs reduced range of motion (ROM) but no cage construct managed to reduce the neutral zone (NZ), in fact the NZ was seen to increase with all constructs suggesting an initial segmental instability. Anteroposterior and mediolateral cage dimensions, cage height, and cage angle all influenced initial stability. Cages with teeth had higher pullout strength. They concluded that stand-alone cages reduced ROM effectively, but the residual ROM present indicated micromotion at the cage-end plate interface which may influence fusion.

Gerber et al. (2006) conducted a human cadaveric biomechanical assessment of ALIF with an anterior plate and screws compared to ALIF with supplementary pedicle screw fixation. ALIF was performed with two INTER FIX cages (Medtronic Sofamor Danek, Memphis Tennessee). Compared to stand-alone cages, supplementary anterior plate and screws reduced ROM by a mean of 41%, and compared to stand-alone cages, supplementary pedicle screw fixation reduced ROM by 61%. Similarly, Beaubien et al. (2005) in another human cadaveric study found that, although not as rigid as pedicle screws or translaminar screws, anterior lumbar plating does add significant stability to an ALIF construct.

A biomechanical evaluation of a stand-alone PEEK ALIF cage (Brigade NuVasive, Inc., San Diego, California) found that the cage alone was significantly more rigid than the intact state in flexion-extension and lateral bending. The addition of three integrated screws was significantly more rigid than the cage alone in all loading directions. The addition of a fourth screw did not significantly increase stability over three screws. The cage with integrated screws allowed more flexion-extension motion than the cage with anterior plate fixation or pedicle screw fixation. Adding a spinous process plate to the 3-screw cage provided the most rigid construct in flexion-extension, providing more stability than the anterior plate and equivalent stability to the pedicle screw construct. Pedicle screw fixation provided the most rigidity in lateral bending and axial rotation although the later was not significant (Kornblum et al. 2013).

Similarly, a comparative biomechanical analysis of a PEEK ALIF cage with integrated fixation anchors (Solus, Alphatec Spine, Carlsbad, California) compared to a standard PEEK cage (ALS, Alphatec Spine), combined with various posterior fixation constructs found that the Solus cage in combination with all posterior constructs provided significant fixation compared to the intact spine. The ALS cage combined with screw-based posterior constructs also provided significant fixation, but the ALS cage combined with an interspinous process clamp showed a significant reduction in stability for lateral bending and axial torsion (Yeager et al. 2015).

Chen et al. (2013) conducted a biomechanical comparison of three different designs of stand-alone ALIF cages using three-dimensional finite element analysis. The cages differed in their method of integrated fixation. All three designs were compared to the "gold standard" of ALIF cage (SynCageOpen, Synthes Spine, Inc., Pennsylvania) with bilateral pedicle screw fixation. The three cages tested were the Latero system (Latero, A-Spine Asia, Taipei, Taiwan) (trapezoidal cage with integrated lateral plate), the Synfix system (Synfix, Synthes Spine Inc., Pennsylvania) (four integrated screws), and the Stabilis system (Stabilis, Stryker, Michigan) (a central threaded cylinder). At the surgical level, the SynCageOpen with bilateral pedicle screws decreased ROM (>76%) in all directions. The Synfix and Latero systems also decreased ROM in all motions compared to the intact model. However, the Stabilis model only decreased ROM slightly in extension, lateral bending, and axial rotation. At the adjacent levels, there was no obvious differences in ROM or annulus stress among all instrumented models. The authors concluded that the Synfix and Latero systems provided adequate stability for clinical use without additional posterior fixation, but the Stabilis cage would require additional fixation. This highlights the importance of biomechanical testing of different cage designs to determine relative stability.

In summary, biomechanical testing suggests that the use of stand-alone ALIF devices without integrated fixation may not offer enough stability to allow fusion to reliably occur. In patients with good bone quality, if an appropriately sized ALIF cage can be placed, then a stand-alone construct with integrated fixation or a cage-plate construct appears to provide adequate stability (see Fig. 4). In a biomechanically challenging environment such as patients with poor bone quality and multilevel surgery or in the setting of sagittal imbalance, the addition of supplementary screw-based posterior fixation provides the most robust biomechanical construct.

OLIF and ATP

First described by Mayer in 1997, there is a paucity of published biomechanical data on OLIF and ATP constructs although in clinical studies the vast majority of OLIF procedures are supplemented with posterior pedicle screw-rod fixation (Sorian-Baron et al. 2017).

The biomechanics of OLIF can be considered similar to LLIF between L1–L2 and L4–L5. At L5–S1 with release of the ALL, the biomechanics are likely more similar to ALIF.

LLIF

The lateral transpsoas approach was developed as a minimally invasive alternative to access the anterior column and places a large surface area interbody device without mobilization of the great

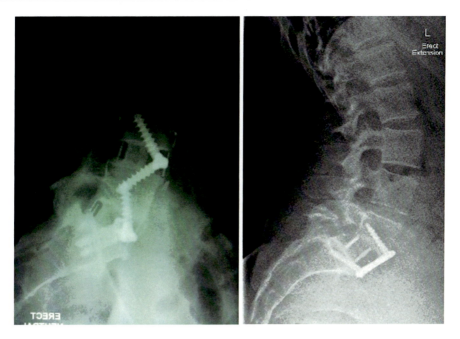

Fig. 4 Examples of stand-alone ALIF cage with integrated fixation, left, and a stand-alone ALIF cage-plate construct, right

vessels. Early biomechanical studies of LLIF have investigated the stability of LLIF compared to ALIF as well as the stability of stand-alone LLIF constructs to the stability of constructs with supplementary lateral plate or posterior fixation.

Laws et al. (2012) performed a human cadaveric biomechanical study to determine the biomechanical differences between ALIF and LLIF with and without supplementary instrumentation. Testing was performed in the intact state, with ALIF or LLIF cage, with cage plus stabilizing plate, with cage plus unilateral pedicle screw fixation, and with cage plus bilateral pedicle screw fixation. The cages used were PEEK cages (Cougar, De Puy Spine, Raynham, Massachusetts). Compared to the intact state, stand-alone LLIF significantly reduced ROM in flexion, extension, and lateral bending, which was not seen with stand-alone ALIF. Addition of a plate increased ALIF group stiffness by 211% in extension and 256% in axial rotation. Compared with stand-alone cages, supplementing with bilateral pedicle screws increased ALIF motion segmental stiffness significantly in flexion (455%) and lateral bending (317%) and LLIF stiffness significantly in flexion (350%) and extension (222%). When bilateral pedicle screws supplemented fusion, ALIF and LLIF were biomechanically equivalent.

Fogel et al. (2014) investigated the biomechanics of lateral lumbar interbody fusion constructs with lateral and interspinous plate fixation and compared these constructs with supplementary bilateral pedicle screw fixation. They found that a stand-alone lateral cage significantly reduced ROM with respect to the intact state in flexion-extension, lateral bending, and axial rotation. Addition of a lateral plate did not alter flexion-extension ROM but significantly reduced lateral bending and axial rotation. Cage with lateral plate was not statistically different from bilateral pedicle screws in lateral bending. Supplementary fixation with a spinous process plate was not statistically different from bilateral pedicle screws in flexion-extension. A combination of lateral plate and spinous process plate was not statistically different from cage with bilateral pedicle screws in all loading modes.

Similarly Reis et al. (2016) found that a stand-alone lateral cage significantly decreased ROM in all directions, with the addition of a lateral plate improving stability in lateral bending, and the addition of an interspinous plate improved stability in flexion-extension; a combination of lateral cage, lateral plate, and interspinous plate was biomechanically equivalent to a lateral cage with bilateral pedicle screws.

One potential disadvantage of LLIF is that, without release of the anterior longitudinal ligament (ALL), there is limited ability to restore segmental lordosis. Thus the LLIF technique has evolved to sometimes include release of the ALL if correction of lordosis is required; this has been termed anterior column realignment (ACR) (Saigal et al. 2016; Berjano et al. 2015).

Melikian et al. (2016) examined the effect of cage angle and surgical technique on segmental lordosis achieved during lateral interbody fusion. They found that insertion of a parallel or 10 ° cage had little effect on lordosis and even insertion of a 30 ° cage with ALL release only led to a modest increase in lordosis (10.5 °s). The addition of spinous process resection and facetectomy was needed to obtain a larger amount of correction (26 °s). None of the cages, including the hyperlordotic cage, caused a decrease in posterior disc height, suggesting hyperlordotic cages do not cause foraminal stenosis.

A cadaveric biomechanical study examining the effect of anterior longitudinal ligament resection on lordosis correction during LLIF found that an 8 mm parallel spacer with an intact ALL provided the greatest stability relative to the intact state but did little to restore lordosis (1.44 ° increase). Conversely, ALL release led to significant improvement in lordosis correction (6.4 ° increase with 8 mm cage and 11 ° increase with 13 mm cage) but significantly destabilized the spine relative to the intact state. Addition of integrated screws to the fusion cage following ALL resection improved stability back to the level of the intact spine (Kim et al. 2017).

There does not appear to be any biomechanical advantage of expandable lateral cages over static lateral cages. Human cadaveric biomechanical testing showed comparable stability between static and expandable lateral stand-alone cages with the most stable construct being a static cage with bilateral pedicle screws. The authors cautioned that there was minimal feedback when expanding the expandable cage which may lead to over-distraction of the disc space and end plate failure (Gonzalez-Blohm et al. 2014).

In summary, a purely stand-alone LLIF cage appears to have more inherent stability than a stand-alone ALIF cage due to preservation of the ALL; however this is at the expense of the ability to effectively restore lordosis. Addition of supplementary lateral or posterior fixation adds further to stability. Release of the ALL significantly destabilizes the motion segment, and much like ALIF, supplementary fixation with integrated screws, lateral plate, or posterior constructs is required to restore stability.

PLIF and TLIF

The biomechanical considerations of posteriorly placed interbody fusion cages differ significantly from ALIF and LLIF. As the ALL is not resected, a ° of inherent stability is retained; however restoration of anterior disc height and lordosis may be somewhat limited. Lordosis can be achieved with compression across a pedicle screw-rod construct with or without osteotomy of the facet joints. This is feasible with an open PLIF or TLIF procedure but difficult to achieve with a MIS-TLIF procedure due to the limited exposure. The interbody fusion devices placed with PLIF or TLIF are typically a lot smaller than those placed for ALIF or LLIF as exposure of the disc space is limited by the neural elements. End plate coverage by the interbody cages is therefore limited and subsidence much more likely if used in a stand-alone fashion. Therefore, in general, PLIF and TLIF cages are supplemented with posterior screw-rod constructs to increase stability and reduce the chances of subsidence.

Brodke et al. (1997) used a calf lumbar spine model to compare PLIF using structural autograft to BAK cages with or without supplementary pedicle screw fixation. PLIF with bone graft alone was the least stiff construct, less stiff than the normal spine. BAK cages alone were similar in stiffness to bone graft with pedicle screw fixation which were both significantly stiffer than the normal spine. BAK cages with pedicle screw fixation were significantly stiffer than all other constructs.

In regard to device material, Xiao et al. (2012) used finite element analysis to compare the biomechanics of PLIF with autogenous iliac bone, PEEK cages, and titanium cages. The lowest stresses on the bone graft and the highest stresses on the end plates were seen with titanium cages, whereas the PEEK cages showed significant stresses on the bone graft and less stresses on surrounding ligaments. The authors concluded that the titanium cage was inferior to the other two models with potentially an increased risk of subsidence due to high end plate stresses and increased risk of pseudarthrosis due to stress shielding of the bone graft.

Vadapalli et al. (2006) also investigated the biomechanics of PEEK and titanium PLIF cages with supplementary pedicle screw fixation. Stresses through the bone graft increased by at least ninefold with the PEEK spacers compared to the titanium spacers. Conversely, end plate stresses increased by at least 2.4-fold with titanium spacers compared to PEEK spacers. There was no difference in stability between the two constructs. Again they concluded that PEEK was superior to titanium in a PLIF construct with similar stability but more graft loading and less chance of subsidence into the end plates.

Slucky et al. (2006) investigated the biomechanics of TLIF with either unilateral or bilateral pedicle screw fixation or unilateral pedicle screws with a contralateral facet screw construct. After TLIF, the unilateral pedicle screw construct provided only half of the improvement in stiffness compared with the other two constructs and allowed for significant off-axis rotational motions.

Similarly, Chen et al. (2012) performed a finite element analysis of unilateral and bilateral pedicle screw fixation for TLIF after decompressive surgery. Finite element analysis was performed for TLIF with a single moon-shaped PEEK cage in the anterior or middle portion of the vertebral bodies and TLIF with a left diagonally placed oval-shaped PEEK cage, all with both unilateral and bilateral pedicle screw fixation. All TLIFs with

bilateral pedicle screws appeared biomechanically stable; however TLIF cages with unilateral pedicle screws on the same side showed increased ROM and annular stress in extension, contralateral lateral bending, and contralateral axial rotation. This was particularly pronounced with the diagonal TLIF cage. The authors cautioned against performing TLIF with unilateral pedicle screws, especially if a diagonal cage is used.

Ames et al. (2005) performed a biomechanical comparison of PLIF and TLIF performed at one and two levels. There was no statistically significant difference in flexion-extension, axial rotation, or lateral bending after either PLIF or TLIF at one level compared to the intact condition. The addition of pedicle screws significantly increased the rigidity for both PLIF and TLIF. Similar findings were seen in the two-level constructs. They concluded that posterior fixation with a pedicle screw-rod construct is suggested for single-level PLIF or TLIF and is necessary to achieve stability with a two-level PLIF or TLIF.

In summary, there is biomechanical evidence supporting the use of PEEK over titanium cages when performing PLIF/TLIF surgery. There is also good evidence suggesting that all PLIF/TLIF constructs should be supplemented with a posterior bilateral screw-rod construct to enhance stability.

AxiaLIF

Biomechanical studies with single-level and two-level AxiaLIF constructs have been performed. In a cadaveric study looking at single-level constructs, ROM was reduced by 40% with a stand-alone trans-sacral rod. Augmentation with facet or pedicle screws reduced ROM between 70% and 90% (Akesen et al. 2008). In two-level stand-alone constructs, ROM decreased by greater than 42% at L4–L5 and 66% at L5–S1. Supplementary pedicle or facet screws further reduced motion (Erkan et al. 2009). Both studies recommended supplementary posterior fixation to provide greater construct stability. In ROM studies, single-level AxiaLIF was shown to be comparable to other fusion types (Ledet et al. 2005).

Clinical Rationale

Anterior Lumbar Interbody Fusion (ALIF)

ALIF is indicated for the treatment of degenerative disc disease (DDD), spondylolisthesis, recurrent disc herniation, and pseudarthrosis, for deformity correction including long fusions to the sacrum/pelvis, and for treatment of adjacent segment degeneration above a previous posterior fusion. It can also be used for debridement and stabilization in cases of infective discitis.

ALIF can be performed via either a retroperitoneal or transperitoneal approach with subsequent mobilization of the great vessels to expose the anterior aspect of the disc. The approach is predominantly used to approach L4–L5 and L5–S1. Access to higher lumbar levels is possible, often involving a transperitoneal approach and the assistance of a vascular surgeon. Renal vein anatomy may preclude accessing L1–L2 and L2–L3 and should be assessed preoperatively if access to the higher lumbar levels is contemplated.

The approach provides extensive exposure of the disc space allowing release of the ALL, thorough discectomy and release of the posterior annulus, excellent end plate preparation, appropriate restoration of disc height and lordosis, and insertion of a large footprint implant. ALIF avoids injury to the paraspinal muscles and direct mobilization of the neural elements.

ALIF is a preferred approach for the treatment of discogenic pain as it allows thorough removal of the pain generator and excellent stabilization of the motion segment. Several studies have shown ALIF to be superior to PLF for the treatment of discogenic pain (Derby et al. 1999; Weatherley et al. 1986).

ALIF is particularly advantageous in patients with severe disc space collapse and spondylolisthesis as it allows a direct and thorough release of the disc space, restoration of disc height and lordosis, and reduction of spondylolisthesis. Restoration of disc height and spondylolisthesis reduction indirectly decompresses the neural foramen.

The biggest driver of disability in adult spinal deformity is loss of lumbar lordosis and sagittal balance. In these patients ALIF is an effective way to release the disc space and correct disc height and lordosis. ALIF is also useful in correcting large coronal plane deformities, especially in rigid curves. In cases where two or more levels are involved, ALIF may be a more efficient strategy than multilevel PLIF or TLIF. In long fusions to the sacrum (L1 or above), ALIF at L4–L5 and L5–S1 has been shown to lower pseudarthrosis rates (Farcy et al. 1992; Kostuik and Hall 1983).

ALIF is the technique of choice in patients who are at high risk of pseudarthrosis such as smokers and in patients with established pseudarthrosis or ongoing discogenic pain after previous posterolateral fusion.

ALIF is also useful in the management of adjacent segment disease after previous posterior fusion as it avoids reoperating through previous scarring, reinjury to the paraspinal muscles, as well as the need for removal and extension of previously placed posterior instrumentation.

ALIF has several potential disadvantages. Many surgeons are unfamiliar with the anterior approach to the lumbar spine, and a volume performance threshold likely exists. Direct decompression and visualization of the neural elements are not possible, and stenosis due to facet joint and ligamentum flavum hypertrophy is not directly addressed.

Anterior approach-related complications can occur. These include vascular and visceral injury, ileus, sympathetic dysfunction, and, in male patients, retrograde ejaculation. Relative contraindications to ALIF include significant obesity, multiple previous abdominal surgeries, significant vascular calcification, and vascular anatomy precluding safe exposure of the disc space.

Oblique Lumbar Interbody Fusion (OLIF)

The OLIF or anterior to psoas approach (ATP) is designed to access the disc space anterior to the psoas muscle (thus avoiding potential injury to the lumbar plexus) without the requirement for mobilization of the great vessels. The indications for OLIF are the same as for ALIF, excepting cases of high-grade spondylolisthesis.

The surgery is performed with the patient in the lateral position either left or right side up, depending on surgeon preference and the pathology being treated. The spine can be accessed from L1–S1 using this technique. Like ALIF, OLIF allows good restoration of disc height and lordosis, although, similar to lateral interbody fusion, adequate release of the ALL is required to achieve significant lordosis correction.

Advantages of OLIF include the minimally invasive approach, facilitating faster patient recovery, and the ability to perform extensive disc space clearance and insert a large interbody implant, thus promoting high fusion rates. The approach is useful in obese patients as the lateral positioning allows the abdomen to fall forward out of the operative field.

As with any minimally invasive retroperitoneal approach, the potential disadvantage is major vascular injury that cannot be easily controlled due to lack of a wide exposure. Sympathetic dysfunction is also a potential complication.

Lateral Lumbar Interbody Fusion (LLIF)

The LLIF technique was pioneered by Pimenta who subsequently first published the technique in the literature with Ozgur et al. (2006). The disc space is accessed via a retroperitoneal transpsoas corridor where no direct mobilization of the vessels is undertaken. The technique allows access from T12–L1 to L4–L5. The L5–S1 level cannot be accessed due to the iliac crest obstructing access.

The patient is positioned laterally either left or right side up depending on surgeon preference and the pathology being treated. A small lateral incision is made followed by placement of a guide wire through the psoas under image intensifier guidance. Serial dilation through the muscle is then undertaken before retractor blades are placed flush on the lateral annulus. Neuromonitoring is essential to avoid injury to the lumbar plexus which courses through the psoas muscle. The plexus is more at risk at the lower lumbar levels where it courses more anteriorly in the psoas muscle.

LLIF allows good restoration of disc height via a minimally invasive approach. If significant lordosis restoration is required, ALL release can be performed before insertion of a hyperlordotic cage. LLIF also allows excellent correction of coronal plane deformity (Arnold et al. 2012) and is a useful approach in obese patents as the abdomen falls forward "out of the way."

LLIF has limited ability to decompress severe central and lateral recess stenosis, and the approach is difficult in high-grade spondylolisthesis (Malham et al. 2015). In general supplementary posterior instrumentation is used, especially in cases with instability and deformity, adjacent to a previous fusion, multiple-level LLIF, and in patients with osteoporosis.

LLIF is contraindicated in patients with prior retroperitoneal surgery and abnormal vascular anatomy. "Flying" or "Mickey Mouse ears" psoas muscles are also a relative contraindication as the lumbar plexus is put at significant risk due to anterior position. The axial MRI should be reviewed preoperatively to assess this.

Complications of LLIF include bowel injury, lumbar plexus injury (particularly at L4–L5), and postoperative lower limb dysesthesia on the side of the approach which is quite common and may last for many months. Quadriceps palsy is a rarer complication but devastating functionally for the patient. Like OLIF, if vascular or bowel injury occurs, it may be difficult to control due to lack of a wide exposure for repair.

Transforaminal Lumbar Interbody Fusion (TLIF)

TLIF allows access to the disc space unilaterally via the neural foramen and can be performed via an open procedure or MIS technique. Indications for TLIF include degenerative disc disease, spondylolisthesis, recurrent disc herniation, spinal stenosis, and degenerative scoliosis. It is useful in patients requiring an interbody fusion who have had previous anterior surgery or who have contraindications to an anterior or lateral approach. In young male patients, it avoids the potential complication of retrograde ejaculation which can occur with an anterior approach. The approach requires less neural retraction than a PLIF procedure.

The patient is positioned prone and a midline or bilateral paramedian incisions used. The disc space is accessed via a unilateral laminectomy and facetectomy. The approach can be used for all lumbar levels.

Advantages of TLIF include direct access and decompression of the neural elements unilaterally and the ability to perform a "circumferential" fusion via a posterior only approach. The approach-related complications of anterior and lateral approaches are avoided. The midline ligamentous structures are preserved due to use of muscle-splitting approaches, thus aiding postoperative stability (Park et al. 2005).

Disadvantages include iatrogenic injury to the paraspinal muscles and limited ability to restore disc height, lordosis, and coronal balance (especially with MIS procedures) (Sakeb and Ahsan 2013; McAfee et al. 2005). If significant lordosis correction is required, then a bilateral open TLIF with complete facetectomies and posterior compression across a pedicle screw-rod construct is an option.

Complications include dural tear/CSF leak, nerve injury, epidural fibrosis, end plate damage with cage subsidence, pseudarthrosis, and iatrogenic injury to the supradjacent facet joint.

Contraindications include severe segmental kyphosis, epidural scarring, conjoined nerve roots, and osteoporosis where a fusion device with a larger footprint (ALIF, OLIF, LLIF) is preferred.

Posterior Lumbar Interbody Fusion (PLIF)

In the PLIF technique, the disc space is accessed posteriorly after laminectomy, facetectomy, and contralateral retraction of the neural elements; traditionally it is performed as a bilateral procedure with supplementary pedicle screw stabilization. The patient is positioned prone, and the procedure performed via an open midline approach or in an MIS fashion via paramedian muscle-splitting incisions. The indications for PLIF are the same as for TLIF.

PLIF has several advantages including direct visualization and decompression of the neural elements and restoration of disc height with preservation of the facet joints, thus aiding in postoperative stability (Lestini et al. 1994). PLIF allows bilateral circumferential fusion through a single incision. Like TLIF, PLIF is useful in patients with contraindications to an anterior or lateral approach. In young males, the risk of retrograde ejaculation is eliminated.

Disadvantages of PLIF include iatrogenic injury to the paraspinal muscles from prolonged retraction (Fan et al. 2010). It may be difficultly to correct coronal imbalance and restore lordosis. The procedure requires significant retraction of the nerve roots which may lead to fibrosis and chronic radiculopathy (Zhang et al. 2014). Other potential complications include dural tear or nerve root injury risking arachnoiditis.

Contraindications include previous posterior surgery, extensive epidural scarring, and arachnoiditis. PLIF is not recommended above L2–L3 as retraction of the conus medullaris can lead to paralysis.

Axial Lumbar Interbody Fusion (AxiaLIF)

Introduced in 2004 and described by Marotta et al. (2006), the AxiaLIF approach was designed to allow distraction and fusion of the L5–S1 disc space through a minimally invasive trans-sacral approach that preserves the anterior and posterior longitudinal ligaments. It is also possible to extend the fusion construct up to include the L4–L5 disc space. AxiaLIF is indicated for degenerative disc disease, low-grade spondylolisthesis, and pseudarthrosis at L5–S1 or L4–L5 and L5–S1.

The patient is positioned prone with care taken to maintain lumbar lordosis. A 2 cm incision is made midline or just left of the paracoccygeal notch, and blunt finger dissection is used to displace the rectum away from the sacrum. The presacral space is developed using blunt dissection followed by docking of a guide pin at the S1–S2 level. The guide pin is advanced into the L5–S1 disc space, followed by over-reaming to create a 10 mm channel and preparation of the disc space using rotating shavers. The disc space is then packed with local autograft +/− other grafting materials. A channel is then drilled into L5, and an interbody screw with a differential thread is then inserted that allows distraction of the disc space as it is inserted.

The advantages of AxiaLIF include its minimally invasive nature using an avascular and aneural corridor, with preservation of the ALL and PLL, thus aiding in initial stability of the construct. AxiaLIF avoids the approach-related complications of anterior, lateral, and posterior interbody fusion techniques.

Although AxiaLIF has some ability to restore disc height, there is little ability to restore lordosis if required, as the ALL is not released. As the disc space is not directly visualized, the quality and quantity of discectomy may be suboptimal. No direct decompression of neural elements is possible. Only the L5–S1 +/− the L4–L5 levels can be accessed. In some patients the procedure is not possible due to the sacrococcygeal morphology. Complications of AxiaLIF include infection, rectal injury, pseudarthrosis, retroperitoneal hematoma, and subsidence/loss of distraction across the disc space.

Clinical Results

ALIF

Early Studies

In one of the first studies comparing fusion to non-operative treatment for DDD, Fritzell et al. (2002) conducted a prospective multicenter randomized trial comparing three lumbar fusion techniques and non-operative treatment in patients with DDD. Two hundred ninety-four patients were randomized; in the interbody fusion group, 56 patients received ALIF using autologous

tricortical bone blocks (19 patients received a PLIF) with additional PLF using pedicle screws and plates. The mean age in this group was 42 years. VAS back pain improved from 65.6 pre-op to 45.7 at 2 years, and ODI improved from 47.3 pre-op to 38.5 at 2 years. The fusion rate was 91% at 2 years. As the biomechanical advantages of graft packed interbody cages over graft alone became apparent, multiple early studies confirmed the clinical effectiveness of the ALIF technique utilizing these devices.

Burkus et al. (2003) reported on 254 patients treated with ALIF using the LT cage device with rhBMP-2 (Medtronic Sofamor Danek, Memphis, Tennessee) and noted a mean 29.3-point improvement in ODI and a 13.5-point improvement in SF-36 PCS. Further follow-up of 146 patients at 6 years (Burkus et al. 2009) was undertaken. The fusion rate was 98%. The ODI improved from 52 pre-op to 20.7 at 6 years. Back pain (using a 20-point scale) improved from 15.4 to 6.9, leg pain improved from 11.6 to 4.8, and SF-36 PCS improved from 27.9 to 43.1 at 6-year follow-up.

Madan and Boeree (2003) conducted a prospective study comparing ALIF, with circumferential fusion through a posterior approach (PLIF and PLF) in patients with degenerative disc disease (DDD). There were 39 patients (47 fusion levels) in the ALIF group. The ALIF procedure was performed using the Hartshill horseshoe cage along with tricortical and cancellous iliac crest autograft. Minimum follow-up was 2 years. Using the subjective score assessment, there was a satisfactory outcome in 71.8% of patients in the ALIF group, and assessment of ODI showed a satisfactory outcome in 79.5% of patients. 64% of patients saw an improvement in their working ability.

Sasso et al. (2004) conducted a prospective randomized controlled trial comparing a cylindrical threaded titanium cage (INTER FIX device, Medtronic Sofamor Danek) to a control group using femoral ring allograft for ALIF. There were 78 patients in the cage group; all had autogenous iliac crest inserted in the cage and had a single-level stand-alone ALIF performed at L4–L5 or L5–S1 for DDD. The fusion rate in the cage group was 97% at 12 months (40% in the control group). ODI improved from 51.1 pre-op to 23.7 at 48 months. SF-36 PCS improved from 28.3 pre-op to 39.8 at 24 months.

Glassman et al. (2006) conducted a multicenter retrospective review of prospectively collected data analyzing clinical outcomes after single-level and two-level lumbar fusion. A total of 497 patients were included in the study; 125 patients underwent ALIF. The ALIF group had a mean age of 42 years. At 2 years post-op, SF-36 improved 13.8 points. Mean improvement in ODI was 27 points at 2 years.

The results of these four relatively early studies of ALIF fusion devices are very similar, with high fusion rates and clinically and statistically significant improvements in ODI and SF-36 scores.

ALIF Using Femoral Ring Allograft

With the advent of rhBMP-2 and other biologics to enhance fusion, a number of studies revisited the use of femoral ring allograft for use in ALIF. Freudenberger et al. (2009) conducted a retrospective review of 59 patients with single-level or two-level lumbar DDD who underwent ALIF with anterior tension band plating or PLIF with pedicle screw instrumentation. ALIF (29 patients) was performed with an allograft bone spacer, BMP, and an anterior tension band plate. In the ALIF group, median estimated blood loss was 112.5 ml and median surgical time 104 min. At 6–9 months post-op, partial to solid fusion was seen in 92% of patients. ODI improved from 24.3 pre-op to 16.0 at 12–18 months post-op. Complications occurred in four patients in the ALIF group (two intra-op common iliac vein injuries, one post-op thrombosis, and one ileus). Compared to the PLIF group, the ALIF group had similar clinical outcomes but with significantly shorter surgical time and decreased blood loss.

Anderson et al. (2011) retrospectively reviewed 50 patients who underwent ALIF using femoral ring allograft and rhBMP-2 with supplementary percutaneous pedicle screws for degenerative lumbar pathology. Twenty-four patients had a single-level fusion and 26 a two-level fusion. Operating time was 131 min anteriorly and 102 min posteriorly. Mean EBL was 288 ml. Follow-up was 12 months. 61% were "definitely fused" and 31% "probably fused." VAS back pain improved from 8 to 3, VAS leg pain improved from 6 to 2. ODI improved from 47 to 28. The overall complication rate was 12%. No intra-operative complications occurred.

ALIF can also be combined with open posterolateral fusion in a true circumferential fusion construct. Zigler and Delamarter (2013) investigated 75 patients treated with 360° lumbar fusion as the control group in a prospective randomized FDA IDE trial of the pro-disc lumbar disc arthroplasty. The fusion patients were treated with femoral ring allograft and DBM for the interbody fusion and open posterolateral fusion. The follow-up rate was 75% at 5 years. VAS improved from 74.9 preoperatively to 40 at 5 years. ODI improved from a mean of 62.7 preoperatively to 36.2, and SF-36 improved from 30.9 preoperatively to 40.1. The fusion rate was 95.6%.

Degenerative Disc Disease

Further studies have examined the role of stand-alone ALIF cages with rhBMP-2 or other biologics in the management of DDD. Gornet et al. (2011) investigated 172 patients undergoing stand-alone ALIF as the control group in a randomized controlled multicenter IDE study of lumbar disc arthroplasty (Maverick). Follow-up was to 24 months. ALIF was performed with tapered fusion cages and rhBMP-2. Mean patient age was 40.2 years. Operative time was 1.4 h and blood loss 95.2 ml. ODI improved from 54.5 pre-op to 24.8 at 24 months. Mean improvement in low back pain was 49 points and mean improvement in leg pain was 23.1 points. The fusion rate was 100%.

Lammli et al. (2014) conducted a retrospective chart review of a consecutive series of patients with DDD treated with single-level or two-level stand-alone ALIF using either a cage and anterior plate or an integrated cage/plate device packed with autograft and rhBMP-2. One hundred eighteen patients were included in the study. The average patient age was 43 years; follow-up was 2 years. VAS score improved from 6.35 to 3.02. Average improvement in ODI was 17%.

Allain et al. (2014) conducted a prospective study involving 65 patients who underwent ALIF using a PEEK cage with integrated intra-corporeal anchoring plates (ROI-A, LDR Medical, Troyes, France) in the treatment of lumbar DDD. The average age was 57 years. In 91% of patients, autologous bone and rhBMP-2 were used within the cage. The mean duration of surgery was 133 min and the mean blood loss was 205.8 ml. At 12-month follow-up, the fusion rate was 96.3%. Statistically significant improvements in back and leg pain were seen by 6 weeks and maintained at 12 months. ODI improved by 26.6 points at 12 months post-op. 88.7% of patients were very satisfied or satisfied with their outcome.

Siepe et al. (2015) reported on 71 patients who underwent stand-alone ALIF at L5–S1 for DDD. ALIF was performed using the Synfix-LR cage (DePuy Synthes, Wet Chester, PA) and rhBMP-2 in the vast majority. The mean follow-up was 35.1 months. Statistically significant and clinically relevant improvements in VAS and ODI were seen at all time points from 3 months post-operatively. The fusion rate was 97.3% at a mean of 27.7 months. Segmental lordosis increased from 16.1 °s to 26.7 °s.

Mobbs et al. (2014) reported on 110 patients who underwent single-level or multilevel ALIF for DDD. Surgery was performed using a stand-alone PEEK cage (Synfix, Synthes) packed with i-FACTOR. Mean follow-up was 24 months. The fusion rate was 93.6% for the whole cohort (98% for single-level fusion and 82% for two-level fusion). SF-12 improved from 68.57 to 92.99. ODI improved from 61.02 to 28.42, and VAS score improved from 7.38 to 2.65. 85.3% of patients reported good to excellent outcomes. They concluded that stand-alone ALIF with i-FACTOR was a viable treatment option for DDD with clinical and radiographic results comparable to ALIF with autograft or rhBMP-2.

Giang et al. (2017) conducted a systematic review of outcomes of stand-alone ALIF. Seventeen studies were included. Mean age of all included patients was 48.5 years. ODI improved by a mean of 26.7 points, VAS back pain improved by 4.1, VAS leg pain improved by 3.3, and SF-36 PCS improved by 12.7. The pooled fusion rate was 79.8%. These studies highlight that in the treatment of DDD, stand-alone ALIF can be expected to give clinically significant improvements in both pain and function with a high fusion rate and patient satisfaction.

Ohtori et al. (2011) also provided level I evidence for ALIF in the treatment of confirmed discogenic low back pain patients without leg pain in a small randomized study of 41 patients. Compared to the non-operative treatment patients, patients who underwent ALIF had reduced ODI and VAS scores.

Spondylolisthesis

ALIF can also be used successfully in the treatment of single-level or multilevel spondylolisthesis.

Kim et al. (2010) compared ALIF and percutaneous pedicle screws with circumferential fusion (ALIF with open instrumented PSF) for adult low-grade isthmic spondylolisthesis. Forty-three patients underwent ALIF and 32 circumferential fusion. ALIF was performed with a stand-alone cage with integrated screws packed with allograft chips. Percutaneous pedicle screws were then inserted. Operative time averaged 189.9 min and blood loss averaged 300 ml. The mean follow-up in the ALIF group was 41.1 months. Significant improvements in segmental lordosis, whole lumbar lordosis, and percentage listhesis were seen. Fusion was seen in 42/43 patients. VAS back pain improved from 7.6 to 2.1, VAS leg pain improved from 7.5 to 2.0, and ODI improved from 49.3 to 13.7.

Hsieh et al. (2017) reported on 23 consecutive patients who underwent ALIF with supplementary percutaneous pedicle screws for multilevel isthmic spondylolisthesis. Twenty-one patients had two-level spondylolisthesis and two patients had three-level slips. The mean follow-up was 22.26 months. Mean operating time was 251.1 min and mean estimated blood loss 346.8 ml. ODI improved from 56.2 to 14.9, VAS back from 8 to 1.7, and VAS leg 7.6 to 1.1. Segmental lordosis improved from 22.7 to 32.7, and total lumbar lordosis improved from 45.8 to 53.1. Successful fusion was seen in all patients.

Adult Spinal Deformity

ALIF has also been used successfully in the treatment of spinal deformity and sagittal imbalance and is an effective way to restore lumbar lordosis (see Fig. 5). Saville et al. (2016) assessed the segmental correction obtained using 20 ° and 30 ° hyperlordotic cages for ALIF in staged anterior-posterior fusions in adults with degenerative pathology and spinal deformity. The authors assessed 69 levels in 41 patients with a mean age 55 years. The average follow-up was 10 months. The cages used were made of either PEEK or carbon fiber reinforced polymer. For 30 ° cages, the mean segmental lordosis achieved was 29 °s; in the presence of spondylolisthesis, this reduced to 19 °s. For 20 ° cages, the mean segmental lordosis achieved was 19 °s. The mean lumbar lordosis increased from 39 °s to 59 °s. The mean SVA reduced from 113 mm to 43 mm. Six cages (9%) displayed a loss of segmental lordosis during follow-up, on average 4.5 °s.

Rao et al. (2015) in a prospective study of 125 patients compared the clinical and radiological outcomes of ALIF based on surgical indication. The mean follow-up was 20 months. Patients with DDD (with or without radiculopathy), spondylolisthesis, and scoliosis had the best clinical response to ALIF with statistically and clinically significant improvement in SF-12, ODI, and VAS scores. The favorable results can be partly explained by the powerful ability of ALIF to obtain lordosis which is vital to at least maintain in the aforementioned indications, where mild deformities can coexist. Failed posterior fusion and adjacent segment disease also showed significant improvement although the mean changes were lower. The overall radiological fusion rate was 94.4%.

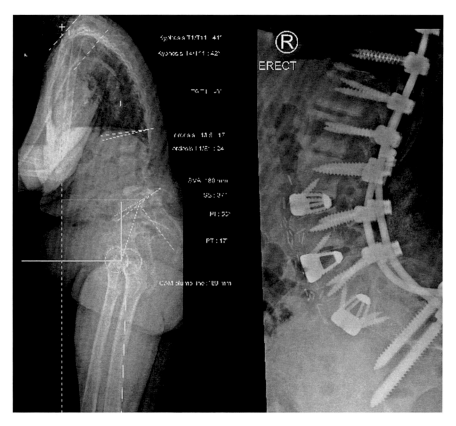

Fig. 5 Preoperative (EOS™) and postoperative standing lateral radiographs of a combined anterior-posterior correction for severe unbalanced kyphoscoliosis utilizing multiple ALIF for anterior column support

In summary, ALIF is an established approach for interbody fusion that can be used successfully to treat DDD, spondylolisthesis, and deformity. ALIF leads to clinically significant improvements in pain and function in appropriately selected patients. The technique is particularly effective at restoring disc height and lordosis, thus maintaining or restoring sagittal balance. Good clinical outcomes are reported for femoral allograft, stand-alone devices with integrated fixation, cage-plate constructs, and circumferential instrumentation.

OLIF

Mayer (1997) were the first to describe the modern OLIF approach in 20/25 patients with DDD, degenerative spondylolisthesis, isthmic spondylolisthesis, or failed back surgery syndrome.

Segments between L2–L3 and L4–L5 were accessed through a retroperitoneal "OLIF" approach and L5–S1 through a traditional transperitoneal ALIF approach. Patients had undergone posterior decompression and pedicle screw fixation 1–2 weeks prior. OLIF was performed using autologous iliac bone graft. The mean follow-up was 10.6 months. For the OLIF approach, mean operating time was 111 min; mean blood loss was 67.8 ml. The fusion rate was 100%.

Lin et al. (2010) presented the results of a prospective clinical study of single-level or two-level OLIF in 46 patients with DDD, low-grade spondylolisthesis, or pseudarthrosis. The mean follow-up was 15 months. OLIF was performed using a titanium cage and plate and autograft bone. The fusion rate was 94.2%. VAS improved from 9.13 to 2.33 and RMDQ improved from 18.58 to 5.43. Good to excellent outcomes were reported in 87% of patients.

Mehren et al. (2016) conducted a chart review of patients who had undergone OLIF over a 12-year period at a single center, specifically focussing on complications. Eight hundred twelve patients with a median age of 63 years were investigated. Patients predominantly suffered from DDD or spondylolisthesis. OLIF was performed using tricortical iliac crest bone graft early in the series, and titanium or PEEK cages filled with autograft, tricalcium phosphate, or rhBMP-2 later in the series. 98% of cases were combined with posterior instrumentation. 62% of patients underwent single-level surgery. The average operating time for OLIF was 110 min. An inhospital complication occurred in 3.7% of patients. The superficial infection rate was 0.24% and deep infection rate 0.37%. There were three intraoperative vascular injuries (0.37%). Nine patients experienced a neurologic injury (1.1%), six of these were meralgia paresthetica due to iliac crest bone harvesting.

Molloy et al. (2016) conducted a prospective cohort study of 64 patients who underwent OLIF using porous tantalum cages for degenerative spine pathology. All patients had additional pedicle screw fixation. The mean follow-up was 1.8 years. VAS back pain improved from 7.5 preoperatively to 1.4, and ODI improved from 64.3 preoperatively to 6.7 at 12 months postoperatively. Radiographic analysis confirmed improvement in multiple lumbopelvic parameters (including pelvic tilt, sacral slope, lumbar lordosis) and the SVA.

Li et al. (2017) conducted a systemic search of the literature to assess operative outcomes and complications of the OLIF procedure. Sixteen studies were included for review, representing 2364 operated levels in 1571 patients. The average follow-up was 22.3 months. The mean blood loss was 109.9 ml, operating time 95.2 min, and postoperative hospital stay 6.3 days. Fusion was achieved in 93% of operated levels. Intraoperative complications occurred in 1.5% and postoperative complications in 9.9% of patients.

In summary, OLIF is a relatively new approach to lumbar interbody fusion that is currently gaining popularity. Initial clinical studies suggest that OLIF is a relatively safe technique that can be used effectively in the treatment of DDD and spondylolisthesis. High fusion rates and the ability to restore lordosis and sagittal balance have been reported in the literature.

LLIF

Ozgur et al. (2006) reported on the first described cases of LLIF, termed XLIF in their series. All procedures were supplemented with percutaneous pedicle screw fixation. Thirteen patients were reported on at short-term follow-up with no early complications noted.

Youssef et al. (2010) conducted a retrospective review of patients treated with LLIF for multiple clinical indications. Single-level, two-level, and three-level procedures were performed using the CoRoent implant (NuVasive, Inc., San Diego, CA). A total of 84 patients were included in the study (15 stand-alone LLIF, 31 with addition of a lateral plate, 38 with supplemental PSF). The mean operating time was 199 min, mean estimated blood loss was 155 ml, and length of hospital stay averaged 2.6 days. The perioperative complication rate was 2.4% and the postoperative complication rate 6.1%. The mean follow-up was 15.7 months. Solid fusion was seen in 68/84 patients (81%). At 1 year postoperatively, VAS improved from 58.9 to 13.7 and ODI improved from 39.7 to 17.3.

Ahmadian et al. (2015) also reported their experience with stand-alone LLIF for the treatment of DDD, spondylolisthesis, and adult degenerative scoliosis. Follow-up was up to 12 months. Fifty-nine patients were included in the study: average age 60 years. Diagnoses were DDD in 63%, degenerative spondylolisthesis in 7%, and ADS in 30%. Surgery was performed using a 10 ° lordotic, 8–12 mm height PEEK cage packed with allograft (and BMP in 19 patients). The fusion rate was 93% at 12 months. VAS improved from 69.1 to 37.8. ODI improved from 51.8 to 31.8. Grade I–II subsidence was seen in 30% of patients. The authors advised that to consider a stand-alone construct, patients should be sagittally balanced, have a coronal Cobb angle less than 30 °s, and be evaluated for osteoporosis preoperatively.

Spondylolisthesis

Ahmadian et al. (2013) reported on the results of LLIF for treatment of L4–L5 spondylolisthesis. Thirty-one patients were included in the study. All patients were treated with LLIF and supplementary percutaneous pedicle screw fixation. The mean follow-up was 18.2 months. Average age of the patients was 61.5 years. Mean blood loss was 94 ml. ODI improved from 50.4 to 30.9, VAS improved from 69.9 to 38.7, and SF-36 improved from 38.1 to 59.5 at latest follow-up. No motor weakness or permanent deficits were noted. Transient anterior thigh numbness was reported in 22.5% of patients. All patients had improvement in spondylolisthesis classification.

Adult Spinal Deformity

Phillips et al. (2013) conducted a prospective multicenter study to evaluate the clinical and radiographic results of patients undergoing LLIF for the treatment of degenerative scoliosis. A total of 107 patients were included in the study. Follow-up was at 24 months. The mean patient age was 68 years. An average of three levels were treated with LLIF at surgery (range: 1–6). ODI, VAS back, VAS leg, and SF-36 PCS scores all improved significantly. Patient satisfaction was 85%. In hypolordotic patients, lumbar lordosis was corrected from a mean of 27.7 °s pre-op to 33.6 °s at 24 months.

Anand et al. (2013) conducted a retrospective review of 2- to 5-year clinical and functional outcomes of minimally invasive surgery for adult scoliosis. Seventy-one patients were included in the study. All underwent a combination of LLIF, AxiaLIF, and posterior instrumentation. LLIF was performed using a PEEK cage filled with rhBMP-2 and Grafton putty. The mean patient age was 64 years and average follow-up was 39 months. On average, patients were sagittally balanced preoperatively and remained that way at latest follow-up. VAS improved from 6.43 to 2.35 and ODI improved from 50.3 to 41 at latest follow-up.

Tempel et al. (2014) reported on 26 patients who underwent combined LLIF and open PSF for the treatment of adult degenerative scoliosis. Patients were aged between 40 and 77 years. LLIF was performed with a 10 ° lordotic PEEK cage 10–12 mm in height packed with DBM. One-year follow-up results were reported. Statistically significant improvement in regional coronal angles and segmental coronal angulation at all operative levels was seen. Coronal Cobb angle improved from a mean of 41.1 °s to 12 °s at latest follow-up. Mean PI-LL mismatch improved from 15 °s to 6.92 °s, and SVA improved from 59.5 mm preoperatively to 34.2 mm at latest follow-up. Significant improvements in clinical outcomes were also observed. VAS back pain improved from 7.5 to 4.3, VAS leg pain improved from 5.8 to 3.1, ODI from 48 to 38, and SF-36 PCS from 27.5 to 35. Three major and 10 minor complications were reported.

Costanzo et al. (2014) conducted a literature review focussing on sagittal balance restoration in adult degenerative scoliosis with the LLIF approach. Fourteen studies were identified representing 476 patients and 1266 operated levels. Only two studies measured global sagittal alignment. They concluded that LLIF was particularly effective when the lumbar lordosis correction goal was less than 10 °s and the sagittal balance correction goal was less than 5 cm.

Phan et al. (2015b) conducted a systematic review to assess the safety and clinical and radiological outcomes of LLIF for the treatment of degenerative spinal deformity. Twenty-one studies were included for review (948 patients, 1920 levels). The median follow-up was 14 months. Mean VAS improved from 6.8 to 2.9 and mean ODI improved from 44.5 to 20.5. Regional lumbar lordosis significantly improved from 35.8 °s to 43.3 °s. Sagittal alignment was unchanged (SVA 34 mm vs. 35.1 mm).

Saigal et al. (2016) performed a literature review of the anterior column realignment (ACR) procedure for sagittal deformity correction. ACR generally involves a LLIF with sectioning of the anterior longitudinal ligament (ALL) and placement of a hyperlordotic cage. ACR usually also involves a second stage posterior column osteotomy. Twelve papers met the inclusion criteria. Segmental lordosis between 10 and 27 °s was reported with use of hyperlordotic cages. A 19 ° increase in mean intradiscal angle was reported when ACR was combined with posterior column osteotomy, 13 °s more than is reported for LLIF alone without a hyperlordotic implant. Complication rates ranged between 18% and 47%. Transient hip flexion weakness was reported in 9.3% and transient paresthesia/dysesthesia in 12%. Motor deficit was reported in 11/75 cases, lower than typically reported for three-column osteotomy procedures.

Berjano et al. (2015) assessed the use of the LLIF-based anterior column reconstruction technique for correcting major sagittal deformity in 11 patients. A mean value of 27 °s lordosis was restored at a single ACR level but two major complications occurred, being a bowel perforation and a postoperative infection requiring posterior debridement. Nevertheless, the authors stated that the ACR technique could provide similar correction to a pedicle subtraction osteotomy with comparable complication profile.°

Complications

Rodgers et al. (2010) reported a prospective analysis of 600 LLIF cases; the paper focussed on intraoperative and perioperative complications. Seven hundred forty-one levels were treated; 99.2% included supplementary internal fixation. The overall incidence of perioperative complications (intraoperative out to 6 weeks postoperative) was 6.2%. Complications were statistically more common at the L4–L5 level. VAS improved from 8.82 preoperatively to 3.12 at 1 year post-op. The patient satisfaction rate was 86.7%.

Lykissas et al. (2014) performed a retrospective analysis of 919 LLIF procedures to identify risk factors for lumbosacral plexus injuries. Four hundred fifty-one patients were included in the study (919 levels). Immediately postoperatively 38.5% of patients reported anterior thigh/groin pain. Sensory deficits were recorded in 38% and motor deficits in 23.9%. At last follow-up, 4.8% of patients reported anterior thigh/groin pain, whereas sensory and motor deficits were recorded in 24.1% and 17.3% of patients, respectively. When patients with preoperative neural deficits were excluded, persistent surgery-related sensory and motor deficits were noted in 9.3% and 3.2% of the patients, respectively. Increased risk of neurological deficit was associated with surgery at the more caudal lumbar levels. The use of rhBMP-2 was associated with persistent motor deficits.

In summary, LLIF is now an established interbody fusion technique with good results reported in the treatment of DDD, spondylolisthesis, and deformity. Supplementary pedicle screw instrumentation is recommended in the majority of cases. To achieve significant lordosis correction, additional release of the ALL is required as part of the procedure. A high incidence of temporary postoperative lower limb neurological symptoms is expected due to the transpsoas approach, particularly at the L4–L5 level. The rate of permanent neurological deficit is low.

TLIF

Despite the popularity of the TLIF procedure, there is a relative paucity of robust published literature regarding the clinical outcomes of the technique. Lowe and Tahernia (2002) conducted a prospective analysis of 40 patients treated with TLIF for degenerative disease of the lumbar spine. TLIF was performed using autograft bone and titanium mesh cages. Bilateral posterolateral fusion and contralateral facet fusion were also performed. The mean follow-up was 3.4 years. The fusion rate was 90% and segmental lordosis was increased in all patients. VAS improved from a mean of 8.3 pre-op to 3.2 at latest follow-up. 85% of patients reported a good to excellent result.

Spondylolisthesis

Rosenberg and Mummaneni (2001) described their early experience with TLIF for the treatment of grade I or II spondylolisthesis. They retrospectively reviewed 22 patients, all presenting with low back pain, the majority with associated radiculopathy. TLIF was performed using Pyramesh titanium cages and iliac crest autograft with supplementary pedicle screw fixation. At a mean follow-up of 5.3 months, low back pain was completely resolved in 16 patients, moderate relief was achieved in 5 patients, and pain was unchanged in 1 patient. One intraoperative durotomy occurred, two postoperative wound infections developed, and one patient had a mild

postoperative L5 motor palsy that resolved quickly.

Hackenberg et al. (2005) reported their minimum 3-year follow-up of 52 consecutive patients treated with TLIF for isthmic spondylolisthesis (22 patients) and lumbar degenerative disorders (30 patients). Thirty-nine cases were single-level, 11 cases were two-level, and 2 cases were three-level fusions. Operating time averaged 173 min for single-level cases and 238 min for multiple-level fusions. TLIF was performed using a curved titanium cage and autograft iliac crest. Four major complications occurred: one deep wound infection, one persistent radiculopathy, one contralateral disc herniation, and one pseudarthrosis. The fusion rate was 89%. VAS and ODI both improved significantly by 3 months postoperatively and remained significantly better than preoperative scores through follow-up.

Lauber et al. (2006) conducted a prospective clinical study evaluating the 2–4-year clinical and radiographic results of TLIF for the treatment of grade I–II degenerative and isthmic spondylolisthesis. Nineteen degenerative, 19 isthmic, and 1 dysplastic spondylolisthesis were treated. Fusion was performed using a curved titanium cage (Micomed Ortho AG, Switzerland) and autograft bone. Minimum follow-up was 24 months, with a mean clinical follow-up of 50 months and mean radiographic follow-up of 35 months. Mean ODI suggested mild to moderate disability preoperatively, but significant improvements were seen postoperatively (23.5 pre-op to 13.5 postoperatively). Significant improvements in VAS were also observed postoperatively. Both VAS and ODI scores began to deteriorate at the 4-year postoperative follow-up. Better results for both VAS and ODI were seen in the isthmic patients compared to degenerative patients, possibly due to the patients being significantly younger. The radiographic fusion rate was 94.8%. There were three serious postoperative complications requiring a return to theatre.

Open Versus Minimally Invasive TLIF

Since these early reports of the open TLIF procedure (O-TLIF), the technique has been modified into the now widely used minimally invasive

TLIF procedure (MI-TLIF). A number of systemic reviews and meta-analyses have been performed comparing the open and minimally invasive procedures.

Khan et al. (2015) performed a systemic review and meta-analysis to investigate early and late outcomes of MI-TLIF in comparison with O-TLIF. Thirty studies were included in the meta-analysis. MI-TLIF was associated with reduced blood loss, length of stay, and complications but increased radiation exposure. Fusion rates and operative times were comparable between the two groups. There were no differences in early and late ODI or early VAS back pain scores, but a (statistically significant but clinically insignificant) decrease in late VAS back pain scores was noted in the MI-TLIF group.

Goldstein et al. (2016) conducted a systematic review and meta-analysis to compare the clinical effectiveness and adverse event rates of MI-TLIF versus O-TLIF. Twenty-six studies met the inclusion criteria; of note, all were of low to very low quality. Overall, there were 856 patients in the MI-TLIF cohort and 806 patients in the O-TLIF cohort. Estimated blood loss, time to ambulation, and length of stay were all in favor of MI-TLIF. Operative times did not differ significantly. There was no difference found in surgical adverse events, but medical adverse events were significantly less likely in the MI-TLIF group. No differences in non-union or reoperation rates were observed. Mean ODI scores were slightly better in the MI-TLIF group at a median follow-up time of 24 months (3.32-point difference). There was no difference observed in VAS back and leg pain scores at 24-month follow-up.

Phan et al. (2015a) conducted a systemic review and meta-analysis of the relative benefits and risks of MI-TLIF and O-TLIF. Twenty-one studies were included in the analysis, representing 966 patients undergoing MI-TLIF and 863 patients undergoing O-TLIF. No significant difference in operating time was noted between the two groups. The median intraoperative blood loss was significantly less in the MI-TLIF group (177 vs. 461 ml). Length of hospital stay was shorter in the MI-TLIF group (4.7 days vs. 8 days). Infection rates were significantly lower in the

MI-TLIF cohort (1.2% vs. 4.6%). There was no difference between the groups in overall complication and reoperation rates. VAS back pain scores and ODI were also slightly lower in the MI-TLIF group, reaching statistical significance but likely not clinically relevant.

Bevevino et al. (2014) conducted a systematic review and meta-analysis to investigate fusion rates of MI-TLIF performed without posterolateral bone grafting. Seven studies with a total of 408 patients were assessed. Average radiographic follow-up was 15.6 months. TLIF was performed using PEEK cages or allograft interbody cages with local autograft bone. In four studies, rhBMP-2 was also used. The overall fusion rate was 94.7% on CT scan, suggesting MI-TLIF without posterolateral bone grafting has similar fusion rates to O-TLIF or MI-TLIF with posterolateral grafting.

In summary, the available literature reports good results for open TLIF in the treatment of degenerative disease of the lumbar spine including spondylolisthesis. Clinically significant improvements in pain and function and a high fusion rate can be expected. In comparison to O-TLIF, MI-TLIF reduces blood loss and hospital stay but does not appear to improve the fusion rate or the clinical outcomes beyond the immediate postoperative period. Restoration of lordosis is possible with O-TLIF but limited with MI-TLIF.

PLIF

In one of the earliest published studies of posterior lumbar interbody fusion, Steffee and Sitkowski (1988) described their experience with PLIF in conjunction with pedicle screws and segmental spine plates. Allograft was used for the PLIF procedure. Thirty-six patients were included in the study with follow-up of 6–12 months. 33/36 had significant improvement in their pain. The fusion rate was reported as 100%.

Degenerative Disc Disease
Barnes et al. (2001) reported their experience with the PLIF procedure for DDD using allograft threaded cortical bone dowels for interbody

fusion. Thirty-five patients with a mean age of 46 years were included in the study. Twenty-three patients underwent a PLIF procedure and 12 patients underwent ALIF. All PLIFs except one were backed up with pedicle screws and rods without posterolateral grafting. ALIFs were performed as a stand-alone procedure. Twenty-eight patients were followed up at a mean of 12.3 months. Satisfactory outcomes were reported in 70% of the PLIF patients. The fusion rate in the PLIF group was 95%. Barnes et al. (2002) also conducted a study comparing the use of allograft cylindrical threaded dowels and allograft impacted wedges for PLIF. There was a 13.6% rate of permanent nerve injury in the threaded dowel group versus 0% in the impacted wedge group. There was no significant difference in fusion rate (95.4% vs. 88.9%). There was a significantly higher rate of satisfactory outcomes in the impacted wedge group.

Chitnavis et al. (2001) reported their experience of PLIF using carbon fiber cages for revision of previous disc surgery. Surgery was performed on patients who had undergone previous discectomy surgery and had ongoing or recurrent low back pain and sciatica with compatible MRI findings. Fifty patients were included in the study; in 40 patients (80%), supplementary pedicle screws were not used. PLIF was performed using paired carbon fiber Brantigan cages packed with autograft bone (iliac crest or spinous process). Symptoms improved in 46 patients (92%) after surgery. Two thirds of patients experienced good to excellent outcomes at early and late follow-up. There was no difference in clinical outcome between those in whom pedicle screws were used or were not used. The fusion rate at 2 years postoperatively was 95%.

Molinari and Gerlinger (2001) reported the functional outcomes of instrumented PLIF in active duty US serviceman and compared these with non-operative management in a non-randomized study. Twenty-nine consecutive patients with single-level lumbar disc degeneration were treated; 15 were treated with instrumented PLIF, and 14 refused surgery and were treated with spinal extensor muscle strengthening, medications, and restricted duty. PLIF

was performed using Brantigan or Harms cages, autogenous iliac crest, and pedicle screw-rod fixation. Average follow-up time was 14 months. Only 5/14 soldiers in the non-operative group returned to full unrestricted military duty. In the PLIF group, 12/15 soldiers were able to return to full duty. Outcomes with respect to posttreatment pain, function, and satisfaction were higher in patients treated with instrumented PLIF.

Haid et al. (2004) published their results of PLIF using rhBMP-2 with cylindrical interbody cages in patients with single-level lumbar DDD. The study was a prospective randomized multicenter study comparing iliac crest autograft and rhBMP-2 in paired cylindrical threaded titanium fusion cages (INTER FIX cages). No supplementary pedicle screw fixation was used. Sixty-seven patients were assigned to one of the two groups. There was no significant difference in operative time or blood loss. The fusion rate at 24 months was 92.3% in the rhBMP-2 group and 77.8% in the autograft group. At all postoperative intervals, mean ODI, VAS back and leg pain, and SF-36 PCS improved in both groups compared to preoperative scores.

Hioki et al. (2005) conducted a retrospective review of 19 patients who underwent two-level PLIF for DDD. PLIF was performed using various interbody cage devices and local autograft bone, with supplementary pedicle screw instrumentation. The mean follow-up was 3.6 years. Lumbar lordosis improved from 25.2 °s preoperatively to 36.6 °s at 6 months postoperatively. Segmental lordosis at the fused levels increased from 12.5 °s preoperatively to 18.7 °s at final follow-up. The fusion rate was 100%. Mean JOA score increased from 12.9 preoperatively to 21.3 at final follow-up. There was a positive correlation between increase in lordotic angle and increase in JOA score. Dural tear occurred intraoperatively in two cases. Postoperatively there were one case of displacement of an interbody spacer, two cases of L3 radiculopathy, and one case of pulmonary embolism.

Spondylolisthesis

Wang et al. (2017) conducted a retrospective analysis of the outcomes of autograft alone versus

PEEK + autograft PLIF in the treatment of adult isthmic spondylolisthesis. Eighty-four patients were included in the study: 44 patients had interbody fusion performed with local autograft alone and 40 patients with autograft + 2 PEEK cages. All cases had supplementary pedicle screw instrumentation. The minimum follow-up was 24 months. At last follow-up, there was no difference between the two groups in clinical outcomes (VAS back and leg pain, ODI, and patient satisfaction). The PEEK + autograft group showed better maintenance of disc height, but this did not reach statistical significance. The fusion rate was 90.9% in the autograft alone group and 92.5% in the autograft + PEEK group.

Sears (2005a) published results of PLIF for lytic spondylolisthesis using insert-and-rotate interbody spacers. The study was a prospective observational study of 18 consecutive patients with lytic spondylolisthesis grades I to IV. The mean age of patients was 50.2 years; the majority of patients presented with predominant radicular symptoms. The mean follow-up was 17.3 months. Intervertebral disc space spreaders and pedicle screw instrumentation were used to reduce the spondylolisthesis. PLIF was then performed using carbon fiber, titanium mesh or PEEK fusion cages, and local and iliac crest autograft bone. VAS improved from 5.0 preoperatively to 2.9 postoperatively. Good to excellent results were reported in 83.3% of patients. Mean preoperative slip reduced from 30.2% to 6.2%. Mean focal lordosis improved from 10.6 °s to 18.1 °s. Total lumbar lordosis did not change, but the lordosis over the lumbar segments above the fusion reduced from 46.8 °s to 34.9 °s.

Sears (2005b) also published results of the same PLIF technique for degenerative spondylolisthesis. Thirty-four patients, mean age 65.1 years, were assessed with the majority presenting with predominantly radicular pain. The mean follow-up was 21.2 months. VAS improved from 5.3 pre-op to 2.2 at last follow-up. 31/34 patients rated their outcome as good to excellent. The mean preoperative slip reduced from 20.2% to 1.7%. The mean focal lordosis increased from 13.1 °s to 16.1 °s.

In summary, PLIF is also a well-established technique for interbody fusion, with good clinical results reported for the treatment of DDD and spondylolisthesis. With appropriate surgical technique, appropriate restoration of segmental lordosis is achievable. A high fusion rate is expected with the procedure.

AxiaLIF

Bohinski et al. (2010) reported on the clinical outcomes, complications, and fusion rate of the AxiaLIF procedure at L5–S1 and L4–S1 in 50 patients presenting with low back pain and radiculopathy due to DDD (37 patients) and spondylolisthesis (6 patients) and in 7 patients with ongoing or recurrent symptoms after previous discectomy. Follow-up was out to 1 year. The procedure was performed at L5–S1 in 35 patients and L4–S1 in 15 patients. Supplementary pedicle screw fixation was used in 45 patients. At 1-year follow-up, VAS score improved by 49% (77 to 39) and ODI by 50% (56 to 28). Fusion was assessed by CT scan and was seen in 44 (88%) patients. One case of bowel perforation occurred.

Bradley et al. (2012) conducted a retrospective review of 41 patients who underwent AxiaLIF at L5–S1 for DDD, the majority combined with posterior fusion. Mean follow-up was 22.2 months. VAS back pain improved from 7.1 to 4.2 and VAS leg pain improved from 6.0 to 3.0. ODI improved from 45.5 to 32.6. Fusion was seen in 26/41 (63.4%) patients. There were four (9.7%) reoperations directly related to the AxiaLIF procedure.

Zeilstra et al. (2017) evaluated the mid- and long-term results of AxiaLIF performed for DDD in 164 patients. Additional facet screw fixation was used in 95 patients. Average follow-up was 54 months with longest follow-up out to 10 years. No intraoperative or perioperative complications were reported. VAS back pain improved from 80 to 34, and VAS leg pain improved from 43 to 24. ODI decreased from 46 to 19. The fusion rate was 89.4%. Female sex, work status (still working), lower BMI, and absence of Modic type II changes were correlated with a good result.

Melgar et al. (2014) measured changes in segmental and global lordosis in patients treated with L4–S1 AxiaLIF and posterior instrumentation. A retrospective multicenter review of 58 patients was performed. The majority of patients suffered from DDD. Mean follow-up was 29 months. VAS back pain improved from 7.8 to 3.3. ODI improved from 60 to 34. Fusion was seen in 96% of treated levels. Maintenance of lordosis was identified in 84% of patients at L4–S1 and 81% of patients at L1–S1. Spino-pelvic parameters and sagittal balance were not assessed.

Schroeder et al. (2015) conducted a systemic review investigating the fusion rate and safety profile of L5–S1 AxiaLIF. Fifteen studies were included in the review. The overall pseudarthrosis rate was 6.9% and the rate of all other complications was 12.9%. The reoperation rate was 14.4% and the infection rate was 5.4%. Deformity studies reported a significantly higher rate of complications (46.3%).

In summary, the published literature suggests that AxiaLIF is a relatively safe procedure with good clinical outcomes in the treatment of DDD (predominantly at L5–S1). The fusion rate of AxiaLIF is comparable with other interbody fusion approaches. However it has a limited role in the treatment of spondylolisthesis and deformity, as fusion is performed "in situ" with little in the way of restoration of disc height or lordosis.

Comparative Studies

Humphreys et al. (2001) conducted a prospective study comparing their early experiences with TLIF to the already established PLIF procedure. Forty TLIFs were compared to 34 PLIFs. TLIF was performed using a Harm's titanium mesh cage packed with iliac crest autograft. Of the 40 TLIF procedures, 17 were single-level, 23 were two-level, and 1 was a three-level fusion. For the single-level fusions, no significant difference in blood loss, operative time, and hospital stay was seen between the TLIF and PLIF groups. Significantly less blood loss occurred in the two-level TLIF group compared to the two-level PLIF group. No complications were seen with the TLIF approach, whereas multiple complications were seen in the PLIF group (four cases of radiculitis, one case of broken hardware, one screw loosening, two cases of screw removal, one superficial infection, and one pseudarthrosis).

Lee et al. (2017b) compared the outcomes of ALIF, PLIF, and TLIF at L5–S1 for the treatment of lumbar degenerative spinal disease in 77 patients. Thirty-four patients were diagnosed with isthmic spondylolisthesis at L5–S1 and the rest with degenerative lumbar spinal stenosis at L5–S1. Specific indications for surgery differed between groups. ALIF was associated with better restoration of segmental lordosis. The fusion rate based on X-ray and CT scan did not differ between the three groups. TLIF was associated with a better postoperative VAS back pain score. PLIF showed the lowest cage subsidence rate.

Lee et al. (2017a) investigated which approach (ALIF, LLIF, or PLIF) is advantageous in preventing development of adjacent segment disease after fusion for L4–L5 spondylolisthesis. Eighty-two patients were included in the study. The mean follow-up was 25 months. ASD was seen in 37% of the ALIF group, 41.7% of the LLIF group, and 64.5% of the PLIF group. The ALIF and LLIF group had significantly increased disc height and foraminal height compared to the PLIF group. The ALIF group had significantly improved lordosis compared to the PLIF and LLIF groups. There was no difference in clinical outcomes (VAS and ODI). This suggests that sagittal profile and avoidance of damage to the posterior spinal structures may help prevent adjacent segment disease.

Phan et al. (2015c) conducted a systematic review and meta-analysis of ALIF versus TLIF. Twelve articles were included in the meta-analysis with a total of 609 ALIF and 631 TLIF patients. Fusion rates and clinical outcomes were comparable between ALIF and TLIF. ALIF was associated with better restoration of disc height, segmental lordosis, and total lumbar lordosis. ALIF was associated with longer hospitalization and, as expected, a lower rate of dural injury and a higher rate of vascular injury.

Dorward et al. (2013) conducted a matched cohort analysis of ALIF versus TLIF in long deformity constructs. There were 42 patients in each group. The average age was 54 years and the number of instrumented vertebrae averaged 13.6 levels. TLIF was associated with less operative time but greater blood loss. Overall complications and neurological complications did not differ. The ALIF group had greater improvement in SRS scores. ODI scores improved similarly in both groups. Segmental lordosis at L4–L5 and L5–S1 as well as regional lordosis (L3–S1) was greater in the ALIF group. TLIF allowed greater correction of coronal plane deformity.

Jiang et al. (2012) conducted a systematic review comparing ALIF and TLIF in the treatment of lumbar spondylosis. Nine studies were included in the review, all retrospective comparative studies. Blood loss and operative time were greater in the ALIF cohort. There was no significant difference in complication rates between ALIF and TLIF. The restoration of disc height, segmental lordosis, and total lumbar lordosis in ALIF was superior to TLIF. Clinical outcomes and fusion rate were not significantly different, but radiological alignment and adjacent segment disease were not outcomes uniformly assessed.

Similarly, Hsieh et al. (2007) compared ALIF and TLIF in regard to restoration of foraminal height, disc angle, lumbar lordosis, and sagittal balance. A retrospective radiographic and clinical analysis was completed, with 32 patients in the ALIF group and 25 patients in the TLIF group. There was no difference in improvement in VAS scores. ALIF was superior to TLIF in its capacity to restore foraminal height, local disc angle, and lumbar lordosis. ALIF increased foraminal height by 18.5%, whereas TLIF decreased it by 0.4%. ALIF increased the local disc angle by 8.3 °s and lumbar lordosis by 6.2 °s, whereas TLIF decreased local disc angle by 0.1 °s and lumbar lordosis by 2.1 °s. They concluded that ALIF may lead to better long-term outcomes compared to TLIF due to better restoration of sagittal balance.

Hoff et al. (2016) conducted a prospective randomized trial comparing the lumbar hybrid procedure (stand-alone ALIF L5–S1 and total disc replacement L4–L5) with two-level TLIF for the treatment of two-level DDD. Sixty-two patients were enrolled, 31 in each arm. TLIF was performed using a PEEK cage, local autograft bone, and pedicle screws. ALIF was performed using a stand-alone PEEK cage with integrated screws. The TDR used was the Maverick™. The mean follow-up was 37 months. Hybrid patients had significantly lower VAS scores immediately postoperatively and at final follow-up compared to fusion patients. There was also a trend for lower ODI scores in the hybrid group although this did not reach statistical significance. Complication rates were low and similar between groups. Lumbar lordosis increased at the operative levels in the hybrid group but not in the TLIF group with a compensatory increase in lordosis at the supradjacent levels. ROM at L3–L4 was significantly higher in fusion patients compared to hybrid patients at final follow-up.

Joseph et al. (2015) conducted a systematic review comparing and contrasting the complication rates of MI-TLIF and LLIF. Fifty-four studies were included for analysis of MI-TLIF, and 42 studies were included for analysis of LLIF. In total 9714 (5454 MI-TLIF and 4260 LLIF) patients were assessed with 13,230 levels fused (6040 MI-TLIF and 7190 LLIF). The total complication rate per patient was 19.2% in the MI-TLIF group and 31.4% in the LLIF group. The rate of sensory deficits, temporary neurological deficits, and permanent neurological deficits in the MI-TLIF group was 20.16%, 2.22%, and 1.01%, respectively. In the LLIF group, the rates were 27.08%, 9.4%, and 2.46%, respectively. Rates of intraoperative and wound complications were 3.57% and 1.63% in the MI-TLIF group compared to 1.93% and 0.8% for the LLIF group, respectively. No significant differences were noted for medical complications or reoperations.

In summary, comparing posterior interbody fusion procedures, TLIF is associated with a lower complication rate than PLIF. ALIF achieves better restoration of disc height and lordosis than TLIF and LLIF (without ALL release). Anterior interbody fusion procedures (ALIF and LLIF) are associated with a lower rate of adjacent segment disease than posterior interbody fusion procedures (PLIF and TLIF). The short-term to midterm clinical outcomes and overall complication rates are similar between anterior and posterior interbody fusion approaches.

Lumbar Interbody Fusion: When to Use What?

When deciding which interbody fusion approach and device to use in each individual patient, the specific goals of lumbar interbody fusion must be kept in mind. These are:

1. Decompression of neural elements, if required
2. Appropriate maintenance or restoration of segmental disc height and lordosis
3. Placement of an interbody fusion device that is able to stabilize the motion segment in the correct position while fusion occurs
4. The use of bone graft and/or other biologic agents to enhance and achieve fusion
5. Avoidance of complications and "collateral damage" from the approach

Decompression of the neural elements can often be achieved indirectly. Restoration of posterior disc height increases foraminal volume allowing decompression of the exiting nerve roots. Restoration of posterior disc height also achieves a ° of central and lateral recess decompression by uncoupling the facet joints and reversing buckling of the ligamentum flavum and posterior disc bulging. In cases of severe lateral recess and central stenosis due to facet arthropathy and ligamentum flavum hypertrophy or a disc protrusion/extrusion, a direct decompressive procedure may still be required. Direct decompression occurs as part of the procedure during PLIF and TLIF procedures. It is also possible to remove disc protrusions/extrusions during an ALIF or LLIF procedure by utilizing rents in the posterior longitudinal ligament to remove the disc fragments. ALIF and LLIF procedures can be combined with a laminectomy if direct decompression is required in addition to the indirect decompression achieved with the procedure.

Increasingly, the importance of maintaining or restoring disc height and lordosis at the time of interbody fusion is being recognized. Fusion of the lumbar spine in appropriate segmental lordosis and sagittal balance leads to a reduced rate of adjacent segment degeneration and disease and therefore a lower incidence of reoperation and superior long-term clinical outcomes (Rothenfluh et al. 2015). With the advent of modern spinal instrumentation and biologics such as rhBMP-2, achieving fusion is no longer a significant issue. Fusion rates for all the approaches to lumbar interbody fusion are high, and as a result the short-term clinical outcomes are fairly similar between the approaches. A high fusion rate and good short-term clinical outcomes should not necessarily be considered a "success" in modern spine surgery, if appropriate segmental lordosis and sagittal balance are not achieved.

As discussed earlier in this chapter, modern interbody device constructs (combined with supplementary pedicle screw-rod fixation when appropriate) are able to achieve a biomechanically stable environment for fusion to occur. In general, a device with a larger footprint is preferred as it provides a larger surface area for fusion to occur and avoids "point loading" and "fish mouthing" through the vertebral end plates, thus reducing the risk of end plate failure and device subsidence. This is of particular importance in osteoporotic bone. The appropriate interbody fusion device combined with bone graft and/or a biologic agent such as rhBMP-2 can be expected to give a high fusion rate, of >90% in most recent published studies.

Avoiding complications and reducing approach-related "collateral damage" rely on appropriate preoperative planning and meticulous surgical technique. Anticipation of approach-related complications preoperatively is vital as it allows modification of the approach or the selection of a different approach for interbody fusion. For example, significant obesity or aortoiliac vascular calcification may preclude performing ALIF. A "flying" or "Mickey Mouse" psoas pattern or large iliolumbar vein at L4–L5 may prevent LLIF at that level. Multiple previous posterior decompressive procedures with epidural scarring may prevent PLIF or TLIF being performed safely. Intraoperatively, each approach for lumbar interbody fusion is associated with its own unique set of approach-related complications which have been discussed earlier in this chapter. Meticulous technique at the time of surgery is required to minimize "collateral damage."

With the above goals of lumbar interbody fusion kept in mind, a number of important factors must be considered preoperatively when planning an interbody fusion.

1. **Are there any absolute or relative contraindications to interbody fusion via a particular approach?**

    An appropriate history, physical examination, and review of imaging studies will determine which approaches are feasible in each individual patient. Contraindications to the various approaches for lumbar interbody fusion have been discussed earlier in the chapter.

2. **What are the patient's lumbopelvic parameters? Is there loss of lumbar lordosis segmentally and globally? Is there sagittal imbalance?**

    In the authors' opinion, every patient who is being worked up for a lumbar interbody fusion should have a preoperative standing full spine X-ray including the pelvis and hips or an EOS scan to assess pelvic parameters (pelvic incidence, sacral slope, pelvic tilt), lumbar lordosis (including where in the lumbar spine it is occurring), and sagittal balance. This is not only appropriate for cases of spondylolisthesis or adult spinal deformity but also in all cases of degenerative disc disease where subtle loss of lumbar lordosis and sagittal balance can occur. When assessing preoperative lumbar lordosis, it is important to assess the relative contribution of each segment to total lumbar lordosis.

    Two thirds of lumbar lordosis occurs between L4 and S1, and it is these levels that are most commonly affected by DDD and spondylolisthesis. A lack of lordosis through the lower lumbar segments can be compensated for by increased lordosis through the upper lumbar segments. Therefore although total lumbar lordosis and sagittal balance may be normal in a patient about to undergo an interbody fusion, it should not be assumed that no increase in lordosis is required as part of the interbody fusion procedure. It is well accepted that fusion of a segment of the lumbar spine in kyphosis increases stresses on the adjacent levels and leads to an accelerated rate of adjacent segment degeneration and disease, even if total lumbar lordosis is normal.

    In general, three patterns are seen when assessing these preoperative images.

(A) Total lumbar lordosis, segmental lordosis, and sagittal balance are all normal.

    This pattern is usually seen in patients with internal disc disruption or early-stage DDD where disc height and lordosis at the involved segment(s) are still relatively well preserved. Interbody fusion must maintain the lordosis of the operative segment(s) to minimize the rate of adjacent segment disease and thus improve long-term outcomes.

(B) Total lumbar lordosis and sagittal balance are normal, but there is loss of segmental lordosis at the operative level(s).

    This pattern is usually seen in the more advanced stages of DDD where loss of disc height and lordosis occurs. It is also commonly seen in low-grade spondylolisthesis. A compensatory increase in lordosis of the unaffected levels normalizes total lumbar lordosis. In these patients the lack of segmental lordosis should be corrected at the time of interbody fusion to minimize the rate of later adjacent segment disease. While the patent may have a good short-term clinical outcome from successful fusion of a segment in kyphosis, the long-term clinical outcome is dubious.

(C) There is loss of both segmental and total lumbar lordosis. Sagittal balance may be maintained by compensatory pelvic retroversion or knee flexion, or there may be loss of sagittal balance.

This pattern is seen in the advanced stages of DDD, spondylolisthesis, and adult spinal deformity. Surgery should plan to correct segmental and global lumbar lordosis and restore sagittal balance. Failure to restore lumbar lordosis and sagittal balance is associated with poor clinical outcomes in both the short and long term.

3. **Does the patient have osteoporosis?**

As osteoporosis affects cancellous bone more than cortical bone, the vertebral bodies are more affected by this disease than the posterior elements. Therefore the presence of osteoporosis is important to consider when planning interbody fusion. Assessment of osteoporosis in the spine can be difficult. DEXA scans through the lumbar spine are performed in cross section and therefore give an "average" of the often osteoporotic vertebral body cancellous bone and the more sclerotic bone of the posterior elements which can give false "normal" readings. The bone density of the femoral neck is a more accurate predictor of vertebral body bone density. Other factors predictive of spinal osteoporosis include a history of osteoporotic fracture elsewhere in the body (femoral neck or distal radius), a family history of osteoporosis and vitamin D deficiency, and a prolonged period of relative inactivity due to the spinal condition or other conditions. Postmenopausal women should be assumed to have spinal osteoporosis until proven otherwise.

Especially in the setting of adult spinal deformity/sagittal imbalance, confirmation of osteoporosis should prompt referral to an endocrinologist for assessment and treatment preoperatively to optimize bone quality which often takes months.

At the time of surgery, patients with osteoporosis are at increased risk of vertebral body end plate failure and subsidence of the interbody device. This can occur either intraoperatively or postoperatively as the patient begins to mobilize and weight bear through the spine. The consequences of subsidence include pseudarthrosis, recurrent foraminal nerve compression, and loss of correction of segmental lordosis. The combination of significant osteoporosis and a very collapsed disc space is especially problematic at the time of surgery as an attempt to increase disc height and place interbody fusion devices without appropriate release of the anterior and posterior longitudinal ligaments will likely result in acute end plate failure.

Strategies to combat osteoporosis at the time of interbody fusion include appropriate (usually circumferential) release of the disc space as noted above and placement of an interbody fusion device with a large surface area that sits on the harder bone of the peripheral end plate ring apophyses. In this regard, ALIF and LLIF are favored over PLIF and TLIF constructs. A device with a relatively low modulus of elasticity (see Fig. 3) is also less likely to subside; therefore PEEK cages are theoretically more appropriate than titanium cages in the setting of osteoporosis. Supplementary pedicle screw-rod fixation is advisable in the setting of osteoporosis. If osteoporosis is severe, cement augmentation of the pedicle screws or vertebroplasty (either delivered directly from an anterior approach into the body or transpedicular) can be used.

4. **Can neural decompression be achieved indirectly or is direct decompression required?**

As discussed above, neural decompression can often be achieved indirectly via anterior interbody fusion approaches (ALIF/OLIF/LLIF). If a direct decompression is required, a posterior interbody fusion approach may be preferred (TLIF/PLIF); however if segmental lordosis needs to be restored also, an anterior approach with supplementary laminectomy will likely achieve this more effectively.

5. **Will supplementary fixation be required?**

Supplementary fixation is often required in biomechanically challenging environments such as in patients with osteoporotic bone, multilevel fusions, large corrections of sagittal imbalance, and spondylolisthesis. In general, due to the smaller footprint of the interbody device used, PLIF and TLIF are supplemented with pedicle screw fixation. Depending on the patient's bone quality and the condition being treated, anterior interbody fusions (ALIF, OLIF, LLIF) may also require supplementary fixation.

Approach by Indication and Level(s) Requiring Interbody Fusion

The decision about which approach to use for interbody fusion is multifactorial and depends on the patient and the condition being treated, the relative advantages and disadvantages of each approach, and the training and experience of the treating surgeon. However based on the biomechanical and clinical literature available, some general recommendations can be made.

Degenerative Disc Disease

ALIF is the preferred approach for management of DDD. The approach avoids iatrogenic injury to the paraspinal muscles and psoas muscles and retraction of the neural elements. With ALIF the pain generator can be almost entirely removed, and the approach allows appropriate restoration of disc height and lordosis with a high fusion rate achieved. DDD typically affects L5–S1 and L4–L5. LLIF cannot be performed at L5–S1 and has a higher complication rate at L4–L5. TLIF and PLIF are options for treatment of DDD with good results published in the literature; however some ° of iatrogenic injury to the paraspinal muscles is expected, and it may be hard to restore disc height and lordosis appropriately. Both these factors likely lead to a higher rate of adjacent segment disease.

Isthmic Spondylolisthesis

Both ALIF and TLIF/PLIF are reasonable treatment options for isthmic spondylolisthesis. Again, this condition is usually seen in the caudal two lumbar levels, limiting the use of LLIF in this condition. ALIF achieves better restoration of disc height and lordosis than TLIF/PLIF.

Supplementary percutaneous pedicle screws of posterior spinal fusion are often required. Care must be taken when performing ALIF for a high-grade L5–S1 spondylolisthesis as L5 nerve injury can occur. Posterior decompression and a pedicle screw-based reduction can be performed prior to ALIF in this setting.

Degenerative Spondylolisthesis

Degenerative spondylolisthesis is often accompanied by some ° of lateral recess stenosis due to facet joint and ligamentum flavum hypertrophy; therefore a posterior approach (TLIF/PLIF) is attractive as it allows both direct and indirect neural decompression and fusion through the one approach. However anterior approaches (ALIF/LLIF) are able to better achieve restoration of disc height and lordosis and as a result achieve significant indirect neural decompression. Each patient should be assessed individually regarding the benefits and risks of an anterior versus posterior approach. If both significant correction of disc height/lordosis and direct neural decompression are required, then ALIF/LLIF can be combined with laminectomy.

Adult Spinal Deformity

ALIF, LLIF/ACR, and TLIF/PLIF are all reasonable options for interbody fusion in the setting of adult spinal deformity, and different approaches can be combined in the same operation if required. ASD is almost always characterized by some ° of sagittal imbalance, and ALIF/ACR are favored over LLIF without ALL release or TLIF/PLIF for correction of sagittal balance. Effective anterior interbody fusion may avoid the need for posterior three-column osteotomy to restore sagittal balance, a procedure associated with a high complication rate, often in elderly comorbid patients. Surgical treatment of ASD generally involves interbody and posterolateral fusion, often up into the thoracic spine. TLIF/PLIF avoids the need for a staged anterior-posterior procedure and allows direct decompression of any associated nerve compression. Each patient should be assessed individually regarding the relative merits of each approach.

Adjacent Segment Disease

There is a paucity of literature to define the best approach for treatment of adjacent segment disease after lumbar fusion. Intuitively, choosing a different interbody approach for the treatment of adjacent segment disease will avoid scarring from the initial surgery and lower the rate of approach-related complications. For example, after a previous L5–S1 ALIF, adjacent segment disease at L4–L5 may be approached using LLIF or TLIF to avoid mobilization of the great vessels in an area of scarring. Conversely, after a previous L5–S1 TLIF, L4–L5 ALIF is a good choice as operating in an area of potential epidural scarring is avoided. Sagittal balance and segment lordosis should be assessed carefully preoperatively, as, if the original fusion was performed in kyphosis, more lordosis will be required when treating the adjacent level.

Pseudarthrosis

Pseudarthrosis after PSF or PLIF/TLIF is best managed with an ALIF as this avoids reinjury to the paraspinal muscles and operating through epidural scarring. Interbody cages previously placed posteriorly can be removed during the ALIF procedure. Previously placed pedicle screws may have to be removed prior to the ALIF if further restoration of disc height/lordosis is required. Pseudarthrosis after a previous ALIF or LLIF procedure is best managed with an open PSF. Re-exposure anteriorly in an attempt to remove the interbody device carries a high risk of vascular, bowel or ureteric injury.

L5–S1

ALIF is the preferred method of interbody fusion for DDD at L5–S1. It allows excellent release of the disc space, restoration of disc height and lordosis, and placement of a large footprint interbody device for fusion (Lee et al. 2016). Male patients who still plan to have children should be warned of the small risk of retrograde ejaculation, and sperm banking can be performed preoperatively or an alternate approach used if deemed an unacceptable risk by the patient. OLIF is also a reasonable option at this level. LLIF is not possible at L5–S1. Open PLIF and TLIF can be considered especially in the setting of associated lateral recess stenosis or recurrent disc herniation; however iatrogenic injury to the paraspinal muscles is of

concern, and restoration of segmental lordosis may be limited. MI-TLIF has no ability to effectively restore segmental lordosis and may not even maintain preoperative lordosis and, therefore, is not a good option at L5–S1 where at least maintaining segmental lordosis is critical for overall lumbar lordosis. AxiaLIF similarly has limited to no ability to restore segmental lordosis.

ALIF can also be used effectively in the treatment of degenerative and low-grade lytic spondylolisthesis, usually combined with supplementary pedicle screw fixation. It can also be used in high-grade lytic spondylolisthesis although a posterior decompression may be required first to avoid L5 nerve injury as the slip is reduced. In these cases, open PLIF/TLIF also is a reasonable treatment option as direct decompression, reduction, and stabilization can all be performed through the one approach. Placement of the interbody cages anteriorly within the disc space, and compression across a posterior pedicle screw-rod construct can help achieve segmental lordosis.

L4–L5

ALIF or OLIF are excellent treatment options for DDD, as disc height and lordosis can be restored effectively and iatrogenic injury to the paraspinal muscles is avoided. The favorability of the vascular anatomy around the bifurcation will dictate which of these two approaches is preferable. LLIF is also a reasonable treatment option for DDD; however a higher rate of lumbar plexus injury is seen with LLIF at L4–L5 compared to higher lumbar levels. PLIF/TLIF can also be considered at L4–L5; however if significant restoration of segmental lordosis is required, ALIF, OLIF, and LLIF with or without ALL release are preferred.

Both anterior interbody approaches (ALIF, OLIF, and LLIF) and posterior interbody approaches (TLIF/LLIF) can also be used effectively to treat L4–L5 spondylolisthesis. If associated spinal stenosis requires direct decompression, then PLIF/TLIF may be preferred. If lordosis restoration is required, then anterior approaches are preferred.

L3–L4 and L2–L3

Exposure for ALIF becomes difficult above L4–L5, and in general, requirement for restoration of segmental lordosis is not as great in the higher lumbar levels. OLIF or LLIF and TLIF/PLIF are therefore good options for DDD and spondylolisthesis of the higher lumbar levels.

L1–L2 and T12–L1

ALIF is not feasible and TLIF/PLIF risk retraction injury to the conus, while OLIF/ATP risks injury to renal vessels and higher retroperitoneal structures. LLIF is, therefore, a good option at these levels.

Multilevel/Deformity Correction

A combination of anterior/lateral and posterior techniques can be used depending on the specific goals of the surgery. Anterior/lateral approaches are preferred if significant correction of sagittal alignment is required, but usually supplementary posterior instrumentation is required necessitating a staged procedure (either same day or delayed).

Conclusion

Interbody fusion is a well-established technique in the treatment of various lumbar pathologies including degenerative disc disease, spondylolisthesis, adult spinal deformity, adjacent segment disease, and pseudarthrosis. Multiple approaches to lumbar interbody fusion exist including ALIF, OLIF, LLIF, TLIF, PLIF, and AxiaLIF with biomechanical and clinical data supporting their use in various contexts. In deciding which approach for interbody fusion is most appropriate in each individual patient, the surgeon must take into account their own skill and experience, the pathology being treated, sagittal balance parameters, bone quality, associated neural compression, and any specific contraindications that may preclude a particular approach. With comprehensive preoperative assessment, appropriate surgical decision-making, and strict surgical technique, lumbar interbody fusion can be expected to yield good to excellent clinical outcomes in both the short-term and long-term with a low complication rate and a high rate of fusion.

References

Ahmadian A, Verma S, Mundis G et al (2013) Minimally invasive lateral retroperitoneal transpsoas interbody fusion for L4-5 spondylolisthesis: clinical outcomes. J Neurosurg Spine 19:314–320

Ahmadian A, Bach K, Bolinger B et al (2015) Stand-alone minimally invasive lateral lumbar interbody fusion: multicenter clinical outcomes. J Clin Neurosci 22: 740–746

Akesen B, Wu C, Mehbod A, Transfeldt E (2008) Biomechanical evaluation of paracoccygeal transsacral fixation. J Spinal Disord Tech 21:39–44

Allain J, Delecrin J, Beaurain J et al (2014) Stand-alone ALIF with integrated intracorporeal anchoring plates in the treatment of degenerative lumbar disc disease: a prospective study of 65 cases. Eur Spine J 23: 2136–2143

Ames C, Acosta F, Chi J et al (2005) Biomechanical comparison of posterior lumbar interbody fusion and transforaminal lumbar interbody fusion performed at 1 and 2 levels. Spine 30:E562–E566

Anand N, Baron E, Khandehroo B, Kahwaty S (2013) Long-term 2- to 5-year clinical and functional outcomes of minimally invasive surgery for adult scoliosis. Spine 38:1566–1575

Anderson D, Sayadipour A, Shelby K et al (2011) Anterior interbody arthrodesis with percutaneous posteror pedicle fixation for degenerative conditions of the lumbar spine. Eur Spine J 20:1323–1330

Arnold P, Anderson K, McGuire R (2012) The lateral transpsoas approach to the lumbar and thoracic spine: a review. Surg Neurol Int 3:S198–S215

Bagby G (1988) Arthrodesis by the distraction-compression method using a stainless steel implant. Orthopaedics 11:931–934

Barnes B, Rodts G, Mclaughlin M, Haid R (2001) Threaded cortical bone dowels for lumbar interbody fusion: over 1-year mean follow-up in 28 patients. J Neurosurg Spine 95:1–4

Barnes B, Rodts G, Haid R et al (2002) Allograft implants for posterior lumbar interbody fusion: results comparing cylindrical dowels and impacted wedges. Neurosurgery 51:1191–1198

Beaubien B, Derincek A, Lew W et al (2005) In vitro, biomechanical comparison of an anterior lumbar interbody fusion with an anteriorly placed, low-profile lumbar plate and posteriorly placed pedicle screws or translaminar screws. Spine 30:1846–1851

Berjano P, Cecchinato R, Sinigaglia A et al (2015) Anterior column realignment from a lateral approach for the treatment of severe sagittal imbalance: a retrospective radiographic study. Eur Spine J 24(3):433–438

Bevevino A, Kang D, Lehman R (2014) Systematic review and meta-analysis of minimally invasive transforaminal lumbar interbody fusion rates performed without posterolateral fusion. J Clin Neurosci 21: 1686–1690

Bohinski R, Jain V, Tobler W (2010) Presacral retroperitoneal approach to axial lumbar interbody fusion: a new, minimally invasive technique at L5-S1: clinical outcomes, complications, and fusion rates in 50 patients at 1-year follow-up. SAS J 4:54–62

Bradley D, Hisey M, Verma-Kurvari S, Ohnmeiss D (2012) Minimally invasive trans-sacral approach to L5-S1 interbody fusion: preliminary results from 1 center and review of the literature. Int J Spine Surg 6:110–114

Brodke D, Dick J, Kunz D et al (1997) Posterior lumbar interbody fusion. A biomechanical comparison, including a new threaded cage. Spine 22:26–31

Burkus J, Heim S, Gornet M, Zdeblick T (2003) Is INFUSE bone graft superior to autograft bone? An integrated analysis of clinical trials using the LT-CAGE lumbar tapered fusion device. J Spinal Disord Tech 16:113–122

Burkus J, Gornet M, Schuler T et al (2009) Six-year outcomes of anterior lumbar interbody arthrodesis with use of interbody fusion cages and recombinant human bone morphogenetic protein-2. J Bone Joint Surg 91:1181–1189

Chen S, Lin S, Tsai W et al (2012) Biomechanical comparison of unilateral and bilateral pedicle screws fixation for transforaminal lumbar interbody fusion after decompressive surgery – a finite element analysis. BMC Musculoskelet Disord 13:72

Chen S, Chiang M, Lin J et al (2013) Biomechanical comparison of three stand-alone lumbar cages – a three-dimensional finite element analysis. BMC Musculoskelet Disord 14:281

Chitnavis B, Barbagallo G, Selway R et al (2001) Posterior lumbar interbody fusion for revision disc surgery: review of 50 cases in which carbon fibre cages were implanted. J Neurosurg Spine 95:190–195

Cloward R (1953) The treatment of ruptured lumbar intervertebral discs by vertebral body fusion. J Neurosurg 10:154–168

Costanzo G, Zoccali C, Maykowski P et al (2014) The role of minimally invasive lateral lumbar interbody fusion in sagittal balance correction and spinal deformity. Eur Spine J 23:S699–S704

Cunningham BW, Polly DW Jr (2002) The use of interbody cage devices for spinal deformity: a biomechanical perspective. Clin Orthop Relat Res 394:73–83

Derby R, Howard M, Grant J et al (1999) The ability of pressure-controlled discography to predict surgical and nonsurgical outcomes. Spine 24:364–371

Dorward I, Lenke L, Bridwell K et al (2013) Transforaminal versus anterior lumbar interbody fusion in long deformity constructs. Spine 38:E755–E762

Erkan S, Wu C, Mehbod A et al (2009) Biomechanical evaluation of a new AxiaLIF technique for two-level lumbar fusion. Eur Spine J 18:807–814

Fan S, Hu Z, Fang X et al (2010) Comparison of paraspinal muscle injury in one-level lumbar posterior inter-body fusion: modified minimally invasive and traditional open approaches. Orthop Surg 2:194–200

Farcy J, Rawlins B, Glassman S (1992) Technique and results of fixation to the sacrum with iliosacral screws. Spine 17:S190–S195

Fogel G, Parikh R, Ryu S, Turner A (2014) Biomechanics of lateral lumbar interbody fusion constructs with lateral and posterior plate fixation. J Neurosurg Spine 20:291–297

Freudenberger C, Lindley E, Beard D et al (2009) Posterior versus anterior lumbar interbody fusion with anterior tension band plating: retrospective analysis. Orthopaedics 32:492–496

Fritzell P, Hagg O, Wessberg P et al (2002) Chronic low back pain and fusion: a comparison of three surgical techniques. Spine 27:1131–1141

Gerber M, Crawford N, Chamberlain R et al (2006) Biomechanical assessment of anterior lumbar interbody fusion with an anterior lumbosacral fixation screw-plate: comparison to stand-along anterior lumbar interbody fusion and anterior lumbar interbody fusion with pedicle screws in an unstable human cadaver model. Spine 31:762–768

Giang G, Mobbs R, Phan S et al (2017) Evaluating outcomes of stand-alone anterior lumbar interbody fusion: a systematic review. World Neurosurg 104:259–271

Glassman S, Gornet M, Branch C et al (2006) MOS short form 36 and Oswestry Disability Index outcomes in lumbar fusion: a multicenter experience. Spine J 6:21–26

Goldstein C, Macwan K, Sundarajan K, Rampersaud R (2016) Perioperative outcomes and adverse events of minimally invasive versus open posterior lumbar fusion: meta-analysis and systematic review. J Neurosurg Spine 24:416–427

Gonzalez-Blohm S, Doulgeris J, Aghayev K et al (2014) In vitro evaluation of a lateral expandable cage and its comparison with a static device for lumbar interbody fusion: a biomechanical investigation. J Neurosurg Spine 20:387–395

Gornet M, Burkus J, Dryer R, Peloza J (2011) Lumbar disc arthroplasty with Maverick disc versus stand-alone interbody fusion. Spine 36:E1600–E1611

Grant J, Oxland T, Dvorak M, Fisher C (2002) The effects of bone density and disc degeneration on the structural property distributions in the lower lumbar vertebral endplates. J Orthop Res 20:1115–1120

Hackenberg L, Halm H, Bullmann V et al (2005) Transforaminal lumbar interbody fusion: a safe technique with satisfactory three to five year results. Eur Spine J 14:551–558

Haid R, Branch C, Alexander J, Burkus J (2004) Posterior lumbar interbody fusion using recombinant human bone morphogenetic protein type 2 with cylindrical interbody cages. Spine J 4:527–539

Hioki A, Miyamoto K, Kodama H et al (2005) Two-level posterior lumbar interbody fusion for degenerative disc disease: improved clinical outcome with restoration of lumbar lordosis. Spine J 5:600–607

Hoff E, Strube P, Pumberger M et al (2016) ALIF and total disc replacement versus 2-level circumferential fusion with TLIF: a prospective, randomized, clinical and radiological trial. Eur Spine J 25:1558–1566

Hsieh P, Koski T, O'Shaughnessy B et al (2007) Anterior lumbar interbody fusion in comparison with transforaminal lumbar interbody fusion: implications for the restoration of foraminal height, local disc angle, lumbar lordosis, and sagittal balance. J Neurosurg Spine 7:379–386

Hsieh C, Lee H, Oh H et al (2017) Anterior lumbar interbody fusion with percutaneous pedicle screw fixation for multiple-level isthmic spondylolisthesis. Clin Neurol Neurosurg 158:49–52

Humphreys C, Hodges S, Patwardhan A et al (2001) Comparison of posterior and transforaminal approaches to lumbar interbody fusion. Spine 26: 567–571

Jiang S, Chen J, Jiang L (2012) Which procedure is better for lumbar interbody fusion: anterior lumbar interbody fusion or transforaminal lumbar interbody fusion. Arch Orthop Trauma Surg 132:1259–1266

Joseph J, Smith B, La Marca F, Park P et al (2015) Comparison of complication rates of minimally invasive transforaminal lumbar interbody fusion and lateral lumbar interbody fusion: a systematic review of the literature. Neurosurg Focus 39:E4

Kaleli N, Sarac D, Külünk S et al (2018) Effect of different restorative crown and customized abutment materials on stress distribution in single implants and peripheral bone: a three-dimensional finite element analysis study. J Prosthet Dent 119(3):437–445. https://doi.org/10.10 16/j.prosdent.2017.03.008

Kanayama M, Cunningham B, Haggerty C et al (2000) In vitro biomechanical investigation of the stability and stress-shielding effect of lumbar interned fusion devices. J Neurosurg Spine 93:259–265

Kettler A, Wilke H, Diets R et al (2000) Stabilizing effect of posterior lumbar interbody fusion cages before and after cyclic loading. J Neurosurg 92:87–92

Khan N, Clark A, Lee S et al (2015) Surgical outcomes for minimally invasive vs open transforaminal lumbar interbody fusion: an updated systematic review and meta-analysis. Neurosurgery 77:847–874

Kim J, Kim D, Lee S et al (2010) Comparison study of the instrumented circumferential fusion with instrumented anterior lumbar interbody fusion as a surgical procedure for adult low-grade isthmic spondylolisthesis. World Neurosurg 73:565–571

Kim C, Harris J, Muzumdar A et al (2017) The effect of anterior longitudinal ligament resection on lordosis correction during minimally invasive lateral lumbar interbody fusion: biomechanical and radiographic feasibility of an integrated spacer/plate interbody reconstruction device. Clin Biomech 43:102–108

Kornblum M, Turner A, Cornwall G et al (2013) Biomechanical evaluation of stand-alone lumbar polyether-ether-ketone interbody cage with integrated screws. Spine J 13:77–84

Kostuik J, Hall B (1983) Spinal fusions to the sacrum in adults with scoliosis. Spine 8:489–500

Kuslich S, Ulstrom C, Griffith S, Ahern J, Dowdle J (1998) The Baby and Kuslich method of lumbar interbody fusion. History, techniques, and 2-year follow-up results of a United States prospective, multicentre trial. Spine 23:1267–1279

Lammli J, Whitaker C, Moskowitz A et al (2014) Standalone anterior lumbar interbody fusion for degenerative disc disease of the lumbar spine. Spine 15:E894–E901

Lauber S, Schulte T, Liljenqvist U et al (2006) Clinical and radiologic 2–4 year results of transforaminal lumbar interbody fusion in degenerative and isthmic spondylolisthesis grades 1 and 2. Spine 15:1693–1698

Laws C, Coughlin D, Lotz J et al (2012) Direct lateral approach to lumbar fusion is a biomechanically equivalent alternative to the anterior approach. Spine 37:819–825

Ledet E, Tymson M, Salerno S et al (2005) A biomechanical evaluation of a novel lumbosacral axial fixation device. J Biomech Eng 127:929–933

Lee YC, Zotti MGT, Osti OL (2016) Operative management of lumbar degenerative disc disease. Asian Spine J 10(4):801–819

Lee C, Yoon K, Ha S (2017a) Which approach is advantageous to preventing development of adjacent segment disease? Comparative analysis of 3 different lumbar interbody fusion techniques (ALIF, LLIF, PLIF) in L4-5 spondylolisthesis. World Neurosurg 105:612–622

Lee N, Kim K, Yi S et al (2017b) Comparison of outcomes of anterior, posterior, and transforaminal lumbar interbody fusion surgery at a single level with degenerative spinal disease. World Neurosurg 101:216–226

Lestini W, Fulghum F, Whitehurst L (1994) Lumbar spinal fusion: advantages of posterior lumbar interbody fusion. Surg Technol Int 3:577–590

Li J, Phan K, Mobbs R (2017) Oblique lumbar interbody fusion: technical aspects, operative outcomes, and complications. World Neurosurg 98:113–123

Lin J, Iundusi R, Tarantino U, Moon M (2010) Intravertebral plate and cage system via lateral trajectory for lumbar interbody fusion – a novel fixation device. Spine J 10:86S

Lowe T, Tahernia D (2002) Unilateral transforaminal posterior lumbar interbody fusion. Clin Orthop Relat Res 394:64–72

Lykissas M, Aichmair A, Hughes A et al (2014) Nerve injury after lateral lumbar interbody fusion: a review of 919 treated levels with identification of risk factors. Spine J 14:749–758

Madan S, Boeree N (2003) Comparison of instrumented anterior interbody fusion with instrumented circumferential fusion. Eur Spine J 12:567–575

Malham G, Parker R, Goss B et al (2015) Clinical results and limitations of indirect decompression in spinal stenosis with laterally implanted interbody cages: results from a prospective cohort study. Eur Spine J 24(Suppl 3):339–345

Marotta N, Cosar M, Pimenta L, Khoo LT (2006) A novel minimally invasive presacral approach and instrumentation technique for anterior L5-S1 intervertebral into account patient needs, expectations, comorbidities, and the surgeon's skills and training.

Keywords

Degenerative disc · Interbody fusion · Disc replacement · Disc height · Static interbody · Dynamic interbody

Introduction

Degeneration of the lumbar intervertebral disc is multifactorial and includes genetic and environmental factors. Disc degeneration is associated with changes in the concentration and fragmentation of the matrix molecules (Singh et al. 2009). Also loss of water content within the nucleus leads to progressive changes in the viscoelastic behavior of the disc (Panagiotacopulos et al. 1987a, b). Similarly, changes in the ratio of collagen contents affect the biomechanical properties of the disc (Melrose and Ghosh 1988; Roberts et al. 1989). Loss of the structural properties of the disc leads to a change in the center of rotation of the disc and instantaneous centers of rotations change from a tightly clustered zone to a long random shape (Gertzbein et al. 1985). Progressive loss of disc height and lordosis leads to a cascade of changes of segmental and global spine biomechanics and alignment. The cascade starts with an internal disc disruption (IDD) and may progress to degenerative kyphoscoliosis affecting the global coronal and sagittal balance.

IDD is the first stage of disc degeneration, and patients usually present with discogenic back pain with a "normal" magnetic resonance imaging (MRI) scan. IDD is diagnosed by performing a provocative discogram, followed by a computed tomography (CT) discogram to confirm the diagnosis. Progression of disc degeneration leads to progressive loss of lumbar disc height resulting in neuroforaminal stenosis in addition to central canal stenosis. At this stage, patients usually present with back and/or leg pain. An MRI scan confirms the degeneration and neural compression. Further disc degeneration results in facet joint arthropathy and further neural compression. Fritzell et al. (2001) showed that surgical treatment for DDD is superior to nonsurgical treatment. The surgical group had a 33% reduction in back pain score and a 25% decrease in disability, measured using the Oswestry Disability Index (ODI), whereas the nonsurgical group had 7% and 6% reductions, respectively.

The art of treating DDD involves obtaining a precision diagnosis, then removal of the pain generator (degenerative disc and the sinuvertebral nerve), and restoration of the disc height, by replacing the disc with a device which either prohibits movement (fusion or static) or maintains it (dynamic or total disc replacement, TDR). Restoring the disc height indirectly decompresses the neuroforamen and the spinal canal. Complete removal of the disc is best achieved by an anterior approach to the lumbar or cervical spine as this allows complete removal of the nucleus pulposus and the cartilaginous end plate. The anterior approach to the spine allows a stand-alone device to be used to replace the disc without the necessary requirement of supplementation with posterior pedicle screws, which would be required if removal of the disc was performed using posterior (or transforaminal) lumbar interbody fusion, for example.

The indications and contraindications for TDR and fusion are discussed in this chapter, as well as the advantages and disadvantages.

Anterior Lumbar Interbody Fusion

Stand-alone anterior lumbar interbody fusion (ALIF) has been performed for many decades in the treatment of DDD with good outcomes (Greenough et al. 1994). It is performed through an anterior approach to the lumbar spine at L2-S1 levels. Anterior approach to the lumbar spine is most commonly performed through a retroperitoneal approach (left or right) to the lumbar spine or transperitoneal approach. Using an appropriate retractor that the surgeon is familiar with is crucial. An example of an anterior retractor is shown in Fig. 1.

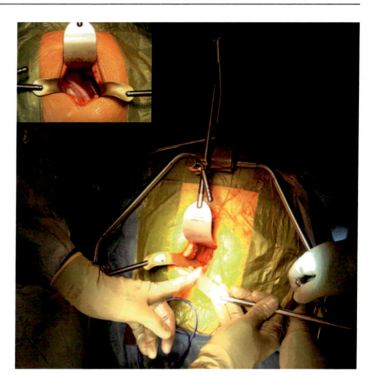

Fig. 1 The Integra Omni-Tract® retractor used during an anterior approach to the lumbar spine

Fig. 2 Modified Steinman pins (OZ pins) are used to hold the vessels while performing the ALIF

Additional retractors are useful during ALIF procedures. In our units (Wansbeck General Hospital and Nuffield Hospital, Newcastle-upon-Tyne, UK), we have adopted the same technique used by Associate Professor Matthew Scott-Young in the Gold Coast Spine practice in Queensland, Australia. Steinman pin retractors are used to hold the vessels after mobilization while performing the ALIF (Fig. 2). Maintaining the anterior annulus and using it as a retractor allows a safe corridor to the disc space and is useful in protecting the surrounding soft tissues and the vessels. This is achieved by performing an H-shaped incision in the annulus and using a stay suture for each arm of the H-shape. This also allows closure of the annulus by suturing the two stay sutures together (Fig. 3).

Anterior approach allows a radical discectomy, removal of the cartilaginous end plate, and the insertion of the interbody cage which contains a bone graft. This allows restoration of the disc height and also restoration of the sagittal and coronal balance (Siepe et al. 2015). Fusion is performed by inserting a cage with or without a bone graft. Cages of various footprint sizes, height, and lordotic angles are available to permit restoration of the appropriate disc height and alignment. Sound interbody fusion is obtained by achieving stability, viability, and proximity of the bone graft to both end plates. This is achieved through meticulous surgical preparation of the end plates, internal fixation, and a good quality

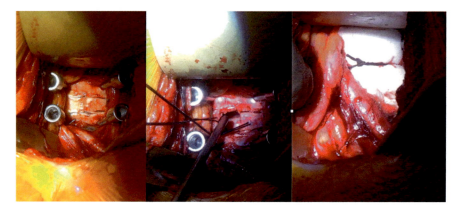

Fig. 3 H-shaped incision in the annulus and suturing the stay sutures to close the annulus on a Spongostan™ as a barrier between the plate and vessels at the end of the procedure

Fig. 4 Comparison between anterior cages (black) of different sizes (small, medium, and large) and a large cage inserted through a posterior approach (white)

bone graft. Inserting the interbody device through an anterior approach places the graft under compression with a large surface area for bony union and with a uniform load transfer. The footprint of an anterior cage is much larger than that of a cage inserted through a posterior approach and allows cross-sectional support on the peripheral ring apophysis (Fig. 4).

Bone graft choice is vital in achieving fusion. The choices for bone graft include autograft, allograft, and synthetic bone graft (Lechner et al. 2017). Combining bone grafts is also an option to improve the fusion rate. Bone morphogenetic protein (BMP-2) has become more popular in recent years (Burkus et al. 2002). Combining allograft (cancellous bone obtained from a femoral head) with BMP-2 INFUSE® is a technique used in Gold Coast Spine, which allows containment of an osteo-inductive bone graft (BMP-2) inside an osteo-conductive allograft-cage construct by drilling holes inside the allograft to host the BMP-2 (Fig. 5). We modified the technique in our unit by inserting a core of autograft (obtained through a vertebral biopsy 10 gauge needle from the iliac crest) inside the allograft. This was used due to the shortage of BMP-2 in Europe in 2016 (Fig. 6).

In our unit, we retrospectively reviewed and compared 50 consecutive patients who underwent ALIF using the bone graft technique described in Fig. 5 with 50 consecutive patient using the technique in Fig. 6. All patients had a CT scan at 5–6 months postoperatively, and the CT scans were reviewed by an independent consultant radiologist. The fusion rate in both groups was identical at 98%.

Fixation of a stand-alone cage is performed either through screws integrated into the cage or through the application of a plate with screws in both vertebral bodies. The aim of the fixation is to stabilize the cage especially in extension which might cause anterior kickout of the cage. Also, fixation limits rotation and flexion. When axial loading is applied through the cage/plate construct, the construct resists axial loading by virtue of its intrinsic cantilever beam with fixed moment arm characteristics.

Gerber et al. (2006) showed that a stand-alone anterior interbody device supplemented by anterior fixation using a plate carries similar stability to a stand-alone cage supplemented by posterior

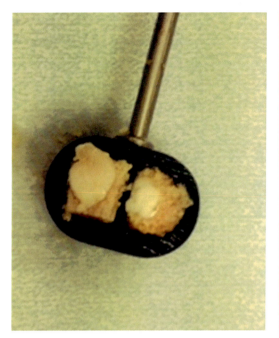

Fig. 5 Combining BMP-2 and femoral head allograft

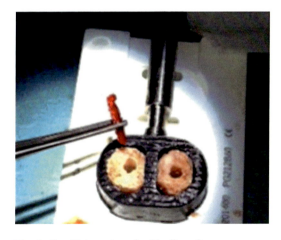

Fig. 6 Combining autograft with allograft

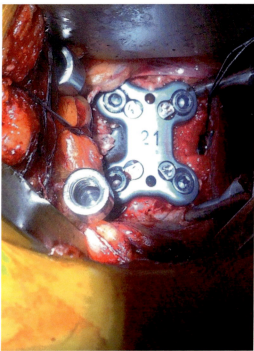

Fig. 7 AEGIS® (DePuy Synthes, Raynham, MA, USA) Anterior Lumbar Plate System. Example of anterior fixation using a plate and four screws with a cam locking mechanism

screw fixation. Both were more stable in all ranges of movement compared to a stand-alone cage with no fixation (Fig. 7).

Treating loss of disc height using the ALIF procedure obtains indirect decompression by increasing the neuroforaminal height and indirectly decompressing the neural elements (Fig. 8). This was proven by an MRI study by Choi et al. (2014).

The indications for ALIF include symptomatic disc pathology which varies from IDD (Blumenthal et al. 1988), DDD, spondylolisthesis, failed prior posterior spine surgery (recurrence/residual disc prolapse (Choi et al. 2005; Vishteh and Dickman 2001) and non-union of posterior fusion (Lee et al. 2006)), and spine deformity. In the absence of spinal instability and deformity, disc degeneration could be treated with stand-alone ALIF. We recommend obtaining a standing lateral and flexion/extension radiograph of the lumbar spine to rule out instability before considering an ALIF procedure (Fig. 9).

Specific complications following an ALIF procedure could be approach-related complications (vascular injury (Fantini et al. 2007; Rajaraman et al. 1999), bowel and ureteric injury, and retrograde ejaculation) or long-term complications such as non-union, metalware complications, and adjacent segment degeneration. Venous thromboembolic (VTE) events are a recognized

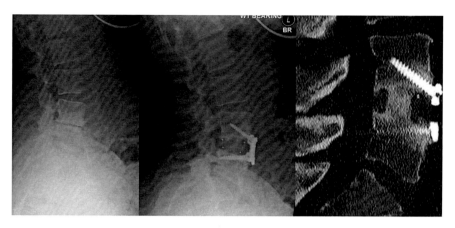

Fig. 8 An example of indirect neuroforaminal decompression obtained by restoring the disc height and a CT scan showing solid fusion at 4 months

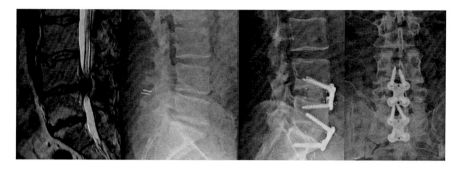

Fig. 9 An example of treating failed posterior surgery using ALIF. A 45-year-old patient who had two-level discectomy at L4/5 and L5/S1 with the insertion of an interspinous spacer. Patient had a two-level ALIF to remove the recurrent disc at L4/5, indirectly decompressing the foramen of L5/S1 and restoring the lordosis

complication after abdominal surgery, and ALIF is no exception since it is done through a mini-laparotomy approach with vessel mobilization. Our unit has adopted a thromboprophylactic regime utilizing physical and chemical prophylactic techniques. This entails TED stockings and compressive calf pumps during surgery and the 24 hours following. The calf pumps are then removed and the TED stockings maintained on the patient for 3 weeks until outpatient review. Postoperatively we use low molecular weight heparin (tinzaparin) 4500 units subcutaneously on the evening before surgery and then daily for 3 to 5 days (while an inpatient) and then Aspirin orally 150 mg daily for 4 weeks after surgery with proton pump inhibitor (PPI) cover. Our unit conducted a study on 160 consecutive patients who underwent ALIF by reviewing their records and also by contacting the patients. There were no symptomatic VTE events in any of the 160 patients. There was no incidence of wound hematomas or bleeding and no symptomatic retroperitoneal hematomas requiring intervention.

Graft migration is often related to the surgical technique and is a difficult management problem in that it involves revision anterior surgery. This is challenging in inexperienced hands and associated with high complication rates. Ideally a CT angiogram and insertion of a ureteric stent should be performed before revision of an anterior approach. Non-union due to patient factors, poor bone quality, or lack of meticulous surgical preparation is also challenging and could be treated through a revision anterior approach or through a posterior approach.

Failure of an ALIF procedure is often due to failure of indication and/or failure of technique. Volume performance threshold and fellowship training is of great importance in performing anterior approach to the lumbar spine in order to reduce the complication rate and improve patient outcomes (Regan et al. 2006).

The aim of ALIF in addition to restoring the disc height is to fuse the spinal segment and eliminate any movement between the two vertebrae. This may change the biomechanics of the adjacent segments resulting in a theoretical increase in the load of the adjacent segment and accelerated degeneration. The effects of post-fusion biomechanics upon adjacent segment disease (ASD) and the role of different approaches remain controversial.

Tang and Meng (2011) showed that restoring disc space height and spinal segmental lordosis is important for preventing ASD by using an anterior cage with appropriate height and lordotic angle. Horsting et al. (2012) followed up 25 patients (minimum 10 year follow-up) who underwent ALIF and posterior fixation for chronic low back pain. The incidence of ASD was 12% above and 20% below the index level. There was significant improvement in patient pain and function up to 2 years with some deterioration at 4 years which was stable at 10 years, but this was not related to the ASD. Kanamori et al. (2012) followed up 20 patients for a minimum of 10 years who had ALIF procedure for degenerative spondylolisthesis and showed progressive ASD. Choi et al. (2014) followed up 49 patients (minimum 10 years) who underwent ALIF for isthmic spondylolisthesis with CT and MRI at 5 and 10 years. The incidence of radiographic ASD was 38.8%, symptomatic ASD was 12.2%, and 4.1% of the patients underwent revision surgery. Also, patients with ASD had more advanced pre-existing facet degeneration compared to those without ASD ($p = 0.01$). They concluded that radiographic ASD is common, while revision for symptomatic ASD is rare, with a risk factor being pre-existing facet joint arthropathy. However, despite the reported incidence of ASD after ALIF procedures, it is reported to be comparatively much less than the incidence of ASD following posterior surgery (Tsuji et al. 2016).

Lumbar TDR

Preventing (or reducing) the incidence of ASD is challenging and has been a subject of debate between surgeons aiming at neural decompression and surgeons aiming at spinal reconstruction. This debate goes back to ASD following knee and hip fusion, which led to the innovation of total hip and total knee replacements and later, ankle and other joint replacements. Similarly in spine, the concept of movement preservation was first introduced in 1966 by Fernström (1966); however the long-term results were disappointing. It was not until 1984 that Schellnack and Buttner-Janz updated this concept by introducing the SB Charité total disc replacement (Buttner-Janz and Schellnack 1990; Buttner-Janz et al. 1987, 1989). Since then the TDR has undergone various modifications in its biomechanics and design.

Based on its biomechanics, the lumbar TDR is classified either as an unconstrained device (Charité III® and In-motion®), as a semi-constrained device (Prodisc®, Maverick®, and Flexicore®), or as a constrained device (the viscoelastic lumbar disc prosthesis-elastic spine pad, LP-ESP®). A mobile core prosthesis (In-motion®) allows uncoupled rotation/translation, while a fixed core prosthesis allows coupled rotation/translation (Fig. 10).

The learning curve of performing lumbar TDR is a steeper curve than that of ALIF (Regan et al. 2006). Inserting the TDR in the midline in the coronal plane is essential, especially when using an unconstrained device. In the sagittal plane, the center of rotation should be at the middle of the end plate or posterior to it (1–3 mm). Correct patient selection for TDR is vital. The condition of the facet joints should be investigated before considering TDR and the type of TDR used. Advanced facet joint arthropathy is a contraindication for performing TDR. Similarly, instability is a contraindication for using unconstrained

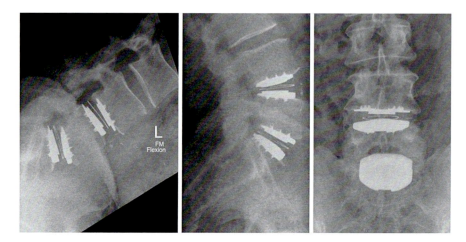

Fig. 10 An example of two-level In-motion TDR for a patient with two-level DDD. (Courtesy of Associate Professor Scott-Young, Gold Coast Spine, Queensland)

Fig. 11 Using the Charité retractor in the intervertebral space to clear the posterior part of the nucleus and the disc allowing assessment of the disc space

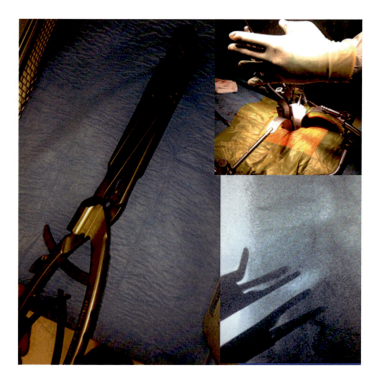

TDR. One key factor when performing lumbar TDR is the way the posterior annulus is handled. Ensuring that the posterior annulus is mobile and assessing the symmetry of disc movement is important during preparation of the disc space. This can be achieved by using a Charité or David retractor, for example, which allows symmetrical distraction of both end plates, and the surgeon can then assess the movement of the disc and its suitability for a TDR (Fig. 11).

Great care should be taken when performing multilevel lumbar TDR as it is essential to obtain near perfect alignment in the coronal and sagittal planes. Eccentric positioning of the TDR leads to suboptimal biomechanical restoration and increased incidence of subsidence and may increase the wear rate. The surgeon should strive to insert the prosthesis with the largest footprint possible to reduce the risk of subsidence. Scott-Young et al. (2012) presented their results, reporting good outcomes of two-level TDR with a minimum follow-up of 4 years.

Lumbar TDR maintains movements of the spinal segment allowing controlled dynamic stability while maintaining near normal disc biomechanics. Maintaining near physiological range of movement reduces the stress on the adjacent segments, and this in turn reduces (and may prevent) ASD and the reoperation rate. Scott-Young et al. (2016) showed a significant reduction of ASD following TDR. In a systematic review, Hiratzka et al. (2015) demonstrated a relative risk of reoperation of 1.7 in the fusion group compared with lumbar TDR, although this risk decreased to 1.1 at 5-year follow-up.

Lumbar TDR is indicated in patients with proven discogenic back pain due to IDD or DDD (with or without disc herniation). Contraindications to lumbar TDR include infection, osteoporosis, scoliosis greater than 20°, spondylolisthesis, failed back surgery syndrome, inflammatory arthropathy, advanced facet joint degeneration, and pregnancy. Adherence to the indication is essential for a good outcome after TDR. Bertagnoli and Kumar (2002) defined criteria for TDR enabling classification of patients as prime, good, borderline, and poor candidates for the procedure. Holt et al. (2007) showed that the incidence of perioperative and postoperative complications for lumbar TDR was similar to that of ALIF, and they recommended that vigilance is necessary with respect to patient indications, training, and correct surgical technique to maintain TDR complications at the levels experienced in the investigational device exemption (IDE) study.

There is robust evidence from class one Food and Drug Administration (FDA) studies of good results achieved using TDR to treat DDD in improvement of both leg and back pain (Delamarter et al. 2011; Gornet et al. 2011; Zigler et al. 2007; Blumenthal et al. 2005). Jacobs et al. (2012) in a systematic literature review found that the TDR had a statistically significant clinical improvement over other methods of treating DDD including fusion; however, they suggested that the differences in clinical improvement were not beyond generally accepted boundaries for clinical relevance.

Combining TDR and Fusion (the Lumbar Hybrid Procedure)

Surgical management of patients presenting with multilevel DDD in various stages of degeneration is challenging. Fusion of multilevel segments increases the risk of ASD (Hiratzka et al. 2015). A good outcome is achieved by matching the technology with the pathology. ALIF could be performed at the symptomatic level if there is significant loss of disc height, facet joint arthropathy, or instability. This allows stabilization of the level, restoring the disc height and treating the facet joint arthropathy. In the level with early stages of DDD, TDR could be performed to maintain movement and reduce the stress on the adjacent segment (Fig. 12).

Aunoble et al. (2010) reported on 42 patients who underwent hybrid lumbar reconstruction with a median follow-up of 26.3 months. The mean improvement of ODI was 53% at 2 years. The visual analogue score improvement for the back pain was 64.6%. Scott-Young et al. (2017) reported that improvements in both back and leg pain and function can be achieved using the hybrid lumbar reconstructive technique and the improvements were maintained at 96 months postoperatively. They reported the largest series in the literature (617 patients) and showed that both statistically and clinically significant ($p < 0.005$) reductions were seen in back and leg pain, which were sustained for at least 8 years post-surgery. In addition, significant improvements ($p < 0.001$) in self-rated disability and function were also maintained for at least 8 years. Patient satisfaction was rated as good or excellent in >90% of cases. Hoff et al. (2016) showed good results of the hybrid procedure in 23 patients in a prospective randomized trial

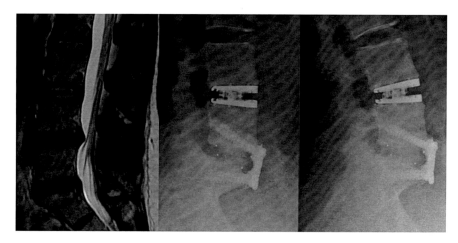

Fig. 12 A 47-year-old patient with two-level DDD, disc prolapse at L4/5 (mobile level) and L5/S1 DDD with facet joint arthropathy who underwent TDR at L4/5 and ALIF at L5/S1

compared with two-level transforaminal lumbar interbody fusion. Yue and Bertagnoli (2006) also showed good results using the hybrid procedure. Chen et al. (2016) showed that hybrid fusion is a valid and viable alternative to ALIF, with at least equal if not better clinical outcomes in terms of survivorship, back pain, and disability scores.

Conclusions

Obtaining a precision diagnosis is paramount to good clinical outcomes in spine surgery. Both ALIF and TDR are valid options in treating DDD; however strict criteria should be met before considering TDR. These include patient factors, surgical training, and also the degree of degeneration. Matching the technology with the pathology is where the art of spine surgery lies.

References

Aunoble S, Meyrat R, Al Sawad Y, Tournier C, Leijssen P, Le Huec JC (2010) Hybrid construct for two levels disc disease in lumbar spine. Eur Spine J 19(2):290–296. https://doi.org/10.1007/s00586-009-1182-7

Bertagnoli R, Kumar S (2002) Indications for full prosthetic disc arthroplasty: a correlation of clinical outcome against a variety of indications. Eur Spine J 11 (Suppl 2):S131–S136. https://doi.org/10.1007/s00586-002-0428-4

Blumenthal SL, Baker J, Dossett A, Selby DK (1988) The role of anterior lumbar fusion for internal disc disruption. Spine (Phila Pa 1976) 13(5):566–569. http://www.ncbi.nlm.nih.gov/pubmed/3187702

Blumenthal S, McAfee PC, Guyer RD et al (2005) A prospective, randomized, multicenter Food and Drug Administration investigational device exemptions study of lumbar total disc replacement with the CHARITÉ™ artificial disc versus lumbar fusion – part I: evaluation of clinical outcomes. Spine (Phila Pa 1976) 30(14):1565–1575. https://doi.org/10.1097/01.brs.0000170587.32676.0e

Burkus JK, Transfeldt EE, Kitchel SH, Watkins RG, Balderston RA (2002) Clinical and radiographic outcomes of anterior lumbar interbody fusion using recombinant human bone morphogenetic protein-2. Spine (Phila Pa 1976) 27(21):2396–2408. https://doi.org/10.1097/00007632-200211010-00015

Büttner-Janz K, Schellnack K, Zippel H (1989) Biomechanics of the SB Charité lumbar intervertebral disc endoprosthesis. Int Orthop 13(3):173–176. https://doi.org/10.1007/BF00268042

Buttner-Janz K, Schellnack K (1990) Intervertebral disk endoprosthesis – development and current status. Beitr Orthop Traumatol 37(3):137–147

Buttner-Janz K, Schellnack K, Zippel H (1987) An alternative treatment strategy in lumbar intervertebral disk damage using an SB Charité modular type intervertebral disk endoprosthesis. Z Orthop Ihre Grenzgeb 125(1):1–6. https://doi.org/10.1055/s-2008-1039666

Chen B, Akpolat YT, Williams P, Bergey D, Cheng WK (2016) Survivorship and clinical outcomes after multilevel anterior lumbar reconstruction with stand-alone anterior lumbar interbody fusion or hybrid construct. J Clin Neurosci 28:7–11. https://doi.org/10.1016/j.jocn.2015.10.033

Choi JY, Choi YW, Sung KH (2005) Anterior lumbar interbody fusion in patients with a previous discectomy: minimum 2-year follow-up. J Spinal Disord Tech 18(4):347–352

Choi K-C, Kim J-S, Shim H-K, Ahn Y, Lee S-H (2014) Changes in the adjacent segment 10 years after anterior lumbar interbody fusion for low-grade isthmic spondylolisthesis. Clin Orthop Relat Res 472(6):1845–1854. https://doi.org/10.1007/s11999-013-3256-4

Delamarter R, Zigler JE, Balderston RA, Cammisa FP, Goldstein JA, Spivak JM (2011) Prospective, randomized, multicenter food and drug administration investigational device exemption study of the ProDisc-L total disc replacement compared with circumferential arthrodesis for the treatment of two-level lumbar degenerative disc disease: res. J Bone Jt Surg Ser A 93(8):705–715. https://doi.org/10.2106/JBJS.I.00680

Fantini GA, Pappou IP, Girardi FP, Sandhu HS, Cammisa FP (2007) Major vascular injury during anterior lumbar spinal surgery: incidence, risk factors, and management. Spine (Phila Pa 1976) 32(24):2751–2758. https://doi.org/10.1097/BRS.0b013e31815a996e

Fernström U (1966) Arthroplasty with intercorporal endoprosthesis in herniated disc and in painful disc. Orthop Scan Suppl 10:287–289

Fritzell P, Hagg O, Wessberg P, Nordwall A (2001) Volvo award winner in clinical studies: lumbar fusion versus nonsurgical treatment for chronic low back pain: a multicenter randomized controlled trial from the Swedish lumbar spine study group. Spine (Phila Pa 1976) 26 (23):2521–2524. https://doi.org/10.1097/00007632-200112010-00002

Gerber M, Crawford NR, Chamberlain RH, Fifield MS, LeHuec JC, Dickman CA (2006) Biomechanical assessment of anterior lumbar interbody fusion with an anterior lumbosacral fixation screw-plate: comparison to standalone anterior lumbar interbody fusion and anterior lumbar interbody fusion with pedicle screws in an unstable human cadaver. Spine (Phila Pa 1976) 31(7):762–768. https://doi.org/10.1097/01.brs.0000206360.83728.d2

Gertzbein SD, Seligman J, Holtby R et al (1985) Centrode patterns and segmental instability in degenerative disc disease. Spine (Phila Pa 1976) 10(3):257–261. http://www.ncbi.nlm.nih.gov/pubmed/3992346. Accessed 19 Nov 2017

Gornet MF, Burkus JK, Dryer RF, Peloza JH (2011) Lumbar disc arthroplasty with MAVERICK disc versus stand-alone interbody fusion: a prospective, randomized, controlled, multicenter investigational device exemption trial. Spine (Phila Pa 1976) 36(25). https://doi.org/10.1097/BRS.0b013e318217668f

Greenough CG, Taylor LJ, Fraser RD (1994) Anterior lumbar fusion: results, assessment techniques and prognostic factors. Eur Spine J 3(4):225–230. http://www.ncbi.nlm.nih.gov/pubmed/7866841. Accessed 19 Nov 2017

Hiratzka J, Rastegar F, Contag AG, Norvell DC, Anderson PA, Hart RA (2015) Adverse event recording and reporting in clinical trials comparing lumbar disk

replacement with lumbar fusion: a systematic review. Glob Spine J 5(6):486–495. https://doi.org/10.1055/s-0035-1567835

Hoff EK, Strube P, Pumberger M, Zahn RK, Putzier M (2016) ALIF and total disc replacement versus 2-level circumferential fusion with TLIF: a prospective, randomized, clinical and radiological trial. Eur Spine J 25 (5):1558–1566. https://doi.org/10.1007/s00586-015-3852-y

Holt RT, Majd ME, Isaza JE et al (2007) Complications of lumbar artificial disc replacement compared to fusion: results from the prospective, randomized, multicenter US Food and Drug Administration investigational device exemption study of the Charité artificial disc. SAS J 1(1):20–27. https://doi.org/10.1016/S1935-9810(07)70043-9

Horsting PP, Pavlov PW, Jacobs WCH, Obradov-Rajic M, de Kleuver M (2012) Good functional outcome and adjacent segment disc quality 10 years after single-level anterior lumbar interbody fusion with posterior fixation. Glob Spine J 2(1):21–26. https://doi.org/10.1055/s-0032-1307264

Jacobs W, Van der Gaag NA, Tuschel A et al (2012) Total disc replacement for chronic back pain in the presence of disc degeneration. Cochrane Database Syst Rev (9): CD008326. https://doi.org/10.1002/14651858.CD008326.pub2

Kanamori M, Yasuda T, Hori T, Suzuki K, Kawaguchi Y (2012) Minimum 10-year follow-up study of anterior lumbar interbody fusion for degenerative spondylolisthesis: progressive pattern of the adjacent disc degeneration. Asian Spine J 6(2):105–114. https://doi.org/10.4184/asj.2012.6.2.105

Lechner R, Putzer D, Liebensteiner M, Bach C, Thaler M (2017) Fusion rate and clinical outcome in anterior lumbar interbody fusion with beta-tricalcium phosphate and bone marrow aspirate as a bone graft substitute. A prospective clinical study in fifty patients. Int Orthop 41(2):333–339. https://doi.org/10.1007/s00264-

Lee S-H, Kang B-U, Jeon SH et al (2006) Revision surgery of the lumbar spine: anterior lumbar interbody fusion followed by percutaneous pedicle screw fixation. J Neurosurg Spine 5(3):228–233. https://doi.org/10.3171/spi.2006.5.3.228

Melrose J, Ghosh P (1988) The quantitative discrimination of corneal type I, but not skeletal type II, keratan sulfate in glycosaminoglycan mixtures by using a combination of dimethylmethylene blue and endo-beta-D-galactosidase digestion. Anal Biochem 170(2):293–300. http://www.ncbi.nlm.nih.gov/pubmed/2969201. Accessed 19 Nov 2017

Panagiotacopulos ND, Pope MH, Bloch R, Krag MH (1987a) Water content in human intervertebral discs. Part II. Viscoelastic behavior. Spine (Phila Pa 1976) 12 (9):918–924. http://www.ncbi.nlm.nih.gov/pubmed/3441838. Accessed 19 Nov 2017

Panagiotacopulos ND, Pope MH, Krag MH, Block R (1987b) Water content in human intervertebral discs.

Part I. Measurement by magnetic resonance imaging. Spine (Phila Pa 1976) 12(9):912–917. http://www.ncbi.nlm.nih.gov/pubmed/3441837. Accessed 19 Nov 2017

Rajaraman V, Vingan R, Roth P, Heary RF, Conklin L, Jacobs GB (1999) Visceral and vascular complications resulting from anterior lumbar interbody fusion. J Neurosurg 91(September 1996):60–64. https://doi.org/10.3171/spi.1999.91.1.0060

Regan JJ, McAfee PC, Blumenthal SL et al (2006) Evaluation of surgical volume and the early experience with lumbar total disc replacement as part of the investigational device exemption study of the Charité artificial disc. Spine (Phila Pa 1976) 31(19):2270–2276. https://doi.org/10.1097/01.brs.0000234726.55383.0c

Roberts S, Menage J, Urban JP (1989) Biochemical and structural properties of the cartilage end-plate and its relation to the intervertebral disc. Spine (Phila Pa 1976) 14(2):166–174. http://www.ncbi.nlm.nih.gov/pubmed/2922637. Accessed 19 Nov 2017

Scott-Young M, Kasis A, Magno C, Nielsen C, Mitchell E, Blanch N (2012) Clinical results of two-level lumbar disc replacement vs. combined arthroplasty & fusion (hybrid procedure) in 200 patients with a minimum of 4 years follow-up. In: ISASS12 Barcelon

Scott-Young M, McEntee L, Cho J, Luukkonen I (2016) The incidence of adjacent segment disease and index level revision and outcomes of secondary surgical intervention following single level lumbar total disc replacement. In: ISASS 16, Las Vegas

Scott-Young M, McEntee L, Schram B, Rathbone E, Hing W, Nielsen D (2017) The concurrent use of lumbar total disc arthroplasty and anterior lumbar interbody fusion. Spine (Phila Pa 1976) 1. https://doi.org/10.1097/BRS.0000000000002263

Siepe CJ, Stosch-Wiechert K, Heider F et al (2015) Anterior stand-alone fusion revisited: a prospective clinical, X-ray and CT investigation. Eur Spine J 24(4):838–851. https://doi.org/10.1007/s00586-014-3642-y

Singh K, Masuda K, Thonar EJ-MA, An HS, Cs-Szabo G (2009) Age-related changes in the extracellular matrix of nucleus pulposus and anulus fibrosus of human intervertebral disc. Spine (Phila Pa 1976) 34(1):10–16. https://doi.org/10.1097/BRS.0b013e31818e5ddd

Tang S, Meng X (2011) Does disc space height of fused segment affect adjacent degeneration in ALIF? A finite element study. Turk Neurosurg 21(3):296–303. https://doi.org/10.5137/1019-5149.JTN.4018-10.0

Tsuji T, Watanabe K, Hosogane N et al (2016) Risk factors of radiological adjacent disc degeneration with lumbar interbody fusion for degenerative spondylolisthesis. J Orthop Sci 21(2):133–137. https://doi.org/10.1016/j.jos.2015.12.007

Vishteh AG, Dickman CA (2001) Anterior lumbar micro-discectomy and interbody fusion for the treatment of recurrent disc herniation. Neurosurgery 48(2):334–337. Discussion 338. http://www.ncbi.nlm.nih.gov/entrez/query.fcgi?cmd=Retrieve&db=PubMed&dopt=Citation&list_uids=11220376

Yue J, Bertagnoli RF-MA (2006) The concurrent use of lumbar total disc arthroplasty and adjacent level lumbar fusion: hybrid lumbar disc arthroplasty: a prospective study. Spine J 6(6):152S–152S

Zigler J, Delamarter R, Spivak JM et al (2007) Results of the prospective, randomized, multicenter food and drug administration investigational device exemption study of the ProDisc®-L total disc replacement versus circumferential fusion for the treatment of 1-level degenerative disc disease. Spine (Phila Pa 1976) 32(11):1155–1162. https://doi.org/10.1097/BRS.0b013e318054e377

Allograft Use in Modern Spinal Surgery

56

Matthew N. Scott-Young and Mario G. T. Zotti

Contents

Introduction 1010

Basic Process of Bone Formation and Union 1011
Direct Bony Healing 1011
Indirect Fracture Healing 1013

Bone Graft Classification System 1015
Allograft-Based Grafts 1015
Factor-Based Grafts 1016
Cell-Based Grafts 1016
Ceramic-Based Grafts 1017
Polymer-Based Grafts 1017

Allografts 1017
Introduction 1017
Allograft Safety 1018
Bone Banking Overview 1018
Types of Bone Allograft 1019

Allograft Use in Spine Surgery 1020
Allograft in the Anterior Column 1021
Posterior Elements 1025

Conclusions 1025

References 1026

M. N. Scott-Young (✉)
Faculty of Health Sciences and Medicine, Bond
University, Gold Coast, QLD, Australia

Gold Coast Spine, Southport, QLD, Australia
e-mail: info@goldcoastspine.com.au; mscott-young@goldcoastspine.com.au

M. G. T. Zotti
Orthopaedic Clinics Gold Coast, Robina, QLD, Australia

Gold Coast Spine, Southport, QLD, Australia
e-mail: mzotti@goldcoastspine.com.au

© Springer Nature Switzerland AG 2021
B. C. Cheng (ed.), *Handbook of Spine Technology*,
https://doi.org/10.1007/978-3-319-44424-6_88

Abstract

Allograft use continues to be important in modern spinal surgery due to its abundant supply, ability to customize to shape, and avoidance of donor site morbidity. However, surgeons must be aware of the limitations of the grafts when used in isolation and how to obtain bony healing. These limitations include subsidence from altered mechanical properties, a lack of osteoinduction and risk of immunogenicity. Optimal healing can be achieved through optimizing the host, selecting the correct graft for the bony environment where the healing is required, and optimizing local graft site biology and stability. Tissue engineering in arthrodesis through obtaining a stable mechanical construct, use of an appropriate structural allograft, and placement of a biologic component (e.g., BMP-2) has shown to be a reliable means to obtain union and achieved satisfactory outcomes. Novel biological agents show promise and will continue to mature in their clinical application.

Keywords

Allograft · Bone banking · Corticocancellous · Femoral ring · Demineralized bone matrix · Bone morphogenetic protein (BMP) · Union · Arthrodesis · Outcomes

Introduction

It is estimated that the number of people aged 65 and older globally will grow from an estimated 524 million in 2010 to nearly 1.5 billion in 2050 with most of the increase occurring in developed countries. Many individuals will suffer from chronic conditions such as hypertension, hypercholesterolemia, osteoarthritis, diabetes, heart disease, cancer, dementia, and congestive cardiac failure. It is thought that heart disease, stroke, and cancer will be the leading chronic conditions that have the greatest impact on the ageing population. However, also to be considered is that a significant number of these individuals will require oncologic surgery, corrective surgery for trauma, revision arthroplasty surgery, and spinal reconstructive surgery which all require bone grafting. At present, it is estimated that there are over two million bone graft procedures performed in the world in a given year. Bone grafting is indeed the second most frequent tissue transplantation worldwide, after blood transfusion. It is estimated that the global bone graft and substitute market accounts for $3.02 billion in 2014 and, it is expected to rise in excess of $4 billion by 2022, growing at a rate of approximately 5% per year.

The use of bone graft and substitutes in the spinal market has increased significantly in conjunction with the aging population and the demand for better standards of health care outcomes. North America is sharing the largest market revenue but Asia-Pacific accounts for the highest growth rate led by the vast ageing population. As a result of the change in demographics, there has been a surge in bone- and joint-related disorders and diseases. Because of this demand, a wide range of products have become available in the market place. These include allograft-based bone graft substitutes, factor-based bone graft substitutes, cell-based bone graft substitutes, ceramic-based bone graft substitutes, and polymer bone graft substitutes. Combining this with the fact that there are limitations associated with autograft, the use of bone graft substitutes has dramatically increased over the last decade. This is due to their ease of use and handling, improved safety profiles, intraoperative cost and time advantages, and adaptability to a variety of clinical challenges.

There are currently over 200 different bone grafts available for surgeons to choose from. This number continues to grow year on year; representing an ever-changing overabundance of options with respect to bone grafting options. Currently, over 90% of reconstructive spinal procedures still utilize autograft and allograft tissue. The current *gold standard* is autogenously sourced bone harvested for reconstructive surgery from the patient. This usually requires an additional surgery from the donor site (usually the pelvis) and has complications such as inflammation, blood loss, infection, and chronic site pain. However, the autogenous bone graft possesses all the necessary characteristics for new bone growth: osteoconductivity, osteogenicity, and osteoinductivity. The ideal synthetic bone graft substitute material would be osteoinductive, osteoconductive, able to bear weight, be resorbable and biologically acceptable, and have a proven safety profile with no adverse local or systemic effects. As yet, the perfect material does not exist, although many materials address one or more of these features.

Basic Process of Bone Formation and Union

In order to achieve solid arthrodesis in the spine, it is a prerequisite to understand the process of bone formation and healing.

Bone is a regenerative organ and maintains this capability in adult life. There is periodic remodeling of the skeleton throughout life and this unique quality allows for fractures to heal and bone grafts to incorporate. The complex and coordinated pathway of bone healing requires an understanding of biomechanical principles, physiological mechanisms, molecular factors, and genetic expression that involves spatial and temporal events. This provides solutions to improve fracture healing or bone grafting and subsequent regeneration. Much of the basic science related to bony union comes from the study of extremity fractures and, as such, the following section talks of union in a spinal (e.g., vertebral body fracture or interbody arthrodesis) as well as an extremity fracture context.

During bone repair, the osteogenic process (under the influence of bone-derived bioactive factors) commences after the inflammatory phase and is initiated by precursor cells from the periosteum adjacent to the fracture site. This generates hard callus by intramembranous bone formation. An autologous bone graft or bone substitute is often required to assist in the healing of an extensive traumatic or postsurgical bone defect and of osseous congenital deformities. The majority of bone formation, however, is by endochondral ossification of the soft callus that appears after infiltrated mesenchyme cells are induced to chondrogenesis. This improved understanding of repair and regeneration has helped with the development of orthopedic tissue engineering.

Understanding the complicated process of bone healing is essential knowledge for a spine surgeon. The primary goal of achieving a successful fusion requires extensive knowledge of bone generation and union and implies a thorough understanding of molecular, physiological, and biomechanical principles. Selection of graft material depends on its properties, the biological status of the patient, comorbidities, mechanical environment, supplemental fixation, availability, cost, efficacy, and the patient's expectations. Inserting materials expecting union or arthrodesis without appreciation of what the role of that material is in bony healing in that particular patient will likely lead to suboptimal results.

Selection of a graft is complex and the three important biological prerequisites need to be considered: osteogenicity, osteoinduction, and osteoconduction. This triangular shaped complex has been extensively studied; however, a fourth element (see Fig. 1 below) should be given the same recognition in terms of significance: mechanical stability. Mechanical stability is a critical factor for bone healing. Progressive maturation of the callus from woven to lamellar bone requires stability. The AO group popularized open reduction and internal fixation techniques to improve union rates. They recognized the role parameters such as fracture rigidity, fracture contact and gap healing, inter-fragmentary strain, and significant role of the soft tissue envelope and vascular environment at the fracture site and coined "The Diamond Concept" (see Fig. 1) (Giannoudis et al. 2007).

To understand new concepts and strategies to enhance the healing of a spinal arthrodesis, a basic summary of the current knowledge on the repair process is required. There are several pathways in which bone can repair, and these are discussed below. One needs to consider that achieving a spinal arthrodesis is a form of tissue regeneration and if fibrous scars form instead then a pseudoarthrosis has developed.

Direct Bony Healing

Direct healing does not commonly occur in the natural process of bone healing. It refers to a direct attempt of the cortical cells to reestablish structural continuity, which requires anatomical reduction of the fragment ends without any gap formation and a stable fixation. This is the primary goal of open reduction and internal fixation surgery. It is also an important type of healing process that occurs when structural allografts are placed under compression in the anterior spine.

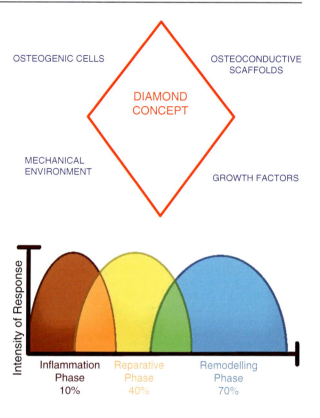

Fig. 1 The diamond concept of bony healing. Reprinted with permission from Fracture healing: The diamond concept, by P. V. Giannoudis; *Injury* 2007 (supp 4; S3–6). (Copyright 2007 by Elsevier)

Fig. 2 The overlapping phases of bone healing and relative intensities of response over different time points with display of the relative contribution of the three phases to the overall time taken to heal

Table 1 Phases of bone formation

The acute inflammatory response	Necrotic debris stimulates release of signaling molecules
MSC recruitment	Multipotent mesenchymal stem cells are recruited and transformation to osteoblasts occurs
Vasculogenesis and neoangiogenesis	Stimulation of osteoblast and osteoclast function
Cartilage reabsorption and mineralization	Bone forms around the new scaffold
Remodeling	Continual process occurs for years

Direct bone healing can now occur by direct remodeling of lamellar bone, the Haversian canals, and blood vessels. Complete healing takes several months to a year and progresses through overlapping phases (see Fig. 2 and Table 1). Although time honored, the old dogmatic mnemonic of "coapt coplanar large cross-sectional cancellous surfaces under compression" still holds wisdom.

Contact Healing

To directly reestablish an anatomically correct and biomechanically competent lamellar bone structure, primary healing can occur through contact or gap healing. If bone on one side of the interface is to unite with bone on the other side of the interface there must be anatomic restoration of the bone fragment surfaces and rigid fixation. This will result in a favorable biomechanical environment that will reduce the inter-fragmentary strain (Shapiro 1988). With gaps less than 0.01 mm, the inter-fragmentary strain is less than 2%, thus the vertebral bodies will unite by contact healing (Rahn 2002). Cutting cones consisting of osteoclasts cross the fracture at a rate of 50–100 μm/day. Osteoblasts then occupy the cavity at the end of the cutting cones (Einhorn 2005). This results in simultaneous generation of a bony union and the restoration of Haversian systems formed in an axial direction (Bais et al. 2009). The reestablished Haversian systems allow for penetration of blood vessels carrying osteoblastic precursors and, with that, complete remodeling ensues in an axial direction (Einhorn 2005).

Gap Healing

If conditions are stable, an anatomical reduction is achieved and the gap is less than 800 μm to 1 mm, gap healing can occur. This differs from contact healing in that the bony union and Haversian remodeling do not occur simultaneously. In this process, the fracture site is primarily filled by lamellar bone oriented perpendicular to the long axis (Dimitriou et al. 2005). This is then replaced by longitudinal revascularized osteons carrying undifferentiated multipotent mesenchyme stem cells (MSC). These MSC differentiate into osteoblasts and produce lamellar bone on each surface of the gap (Shapiro 1988). This lamellar bone, however, is laid down perpendicular to the long axis and is mechanically weak. A secondary remodeling resembling the contact-healing cascade with cutting cones now takes place.

Indirect Fracture Healing

Secondary or indirect fracture healing is the most common form of fracture healing. It involves intramembranous and endochondral ossification leading to callus formation and bone healing (Gerstenfeld et al. 2006). Anatomical reduction or rigid stability is not required. Micro-motion and load bearing enhance the healing process. Excessive load and motion can result in delayed healing or even non-union (Green et al. 2005). Secondary bone healing occurs in the fracture site in non-operative fracture treatment and in operative treatments such as intramedullary nailing, external fixation, or internal fixation of comminuted fractures.

The Acute Inflammatory Response

When a fracture occurs, there is formation of a hematoma and an associated inflammatory response. This consists of cells from both peripheral and intramedullary blood as well as bone marrow cells. The injury initiates an inflammatory response, causing the hematoma to coagulate in between and around the fracture ends as well as within the medulla, forming a template for callus formation (Gerstenfeld et al. 2003). Many local and systemic regulatory factors, hormones, and cytokines work in conjunction with the extracellular osteoconductive matrix producing various cell types and are the first step in the repair process (Gerstenfeld et al. 2003). The acute inflammatory response peaks within the first 24 h and is complete after 7 days. This fracture hematoma produces signaling molecules such as tumor necrosis factor-α (TNF-α); interleukin-1 (IL-1), IL-6, IL-11, and IL-18; fibroblastic growth factor (FGF); insulin-like growth factor(IGF); vascular endothelial growth factor (VEGF); and transforming growth factor-beta (TGFB) that induce a cascade of cellular events that initiate healing (Gerstenfeld et al. 2003). These factors recruit inflammatory cells and promote angiogenesis (Sfeir et al. 2005). The TNF-α concentration has been shown to peak at 24 h and to return to baseline within 72 h post trauma. These factors are secreted by macrophages, MSCs, platelets, chondrocytes, osteoblasts, and other inflammatory cells, and it is believed to mediate an effect by inducing secondary inflammatory signals and act as a chemotactic agent to recruit the necessary cells (Lee and Lorenzo 2006). Differentiation of MSCs down the osteoblastic line is induced by TNF-α. These effects are mediated by activation of the two receptors TNFR1 and TNFR2, which are expressed on both osteoblasts and osteoclasts (Balga et al. 2006).

Interleukins, IL-1 and IL-6, are believed to be most important for fracture healing. There is an overlapping biphasic mode of expression of IL-6 and TNF-α. IL-1 is produced by macrophages in the acute phase of inflammation and induces production of IL-6 in osteoblasts, facilitating the production of the primary cartilaginous callus and angiogenesis at the injured site. IL-6 is produced during the acute phase and stimulates angiogenesis, VEGF production, and the differentiation of osteoblasts and osteoclasts (Lee et al. 2006).

MSC Recruitment

Bone regeneration requires mesenchyme stem cells (MSCs) to be recruited, proliferate, and differentiate into osteogenic cells. MSCs are derived from surrounding soft tissues and bone marrow; there is evidence that a systemic recruitment of circulating MSCs to the injured site might be of great importance for an optimal healing response (Granero-Molto et al. 2009). Which molecular events mediate this recruitment is still under debate. BMP-2 has an important role in this recruitment, but other BMPs such as BMP-7 may play a more important role in the recruitment of progenitor cells (Rahn 2002).

Vasculogenesis and Angiogenesis

There is a greater understanding of the molecular mechanisms controlling callus vascularization. Bony healing requires a blood supply to the healing site, and revascularization is essential. New blood vessels form most commonly, by angiogenesis as well as by vasculogenesis (endothelial progenitor cells, EGC). The vascularization process is mainly regulated by two molecular pathways, an angiopoietin-dependent pathway, and a vascular endothelial growth factor (VEGF)-dependent pathway (Tsiridis et al. 2007). The primarily angiopoetin-1 and 2 are vascular morphogenetic proteins. They are induced early in the healing cascade, promoting vascular ingrowth from existing vessels in the periosteal tissues. The VEGF pathway is considered to be the key regulator of vascular regeneration. High levels of VEGF are expressed by chondrocytes and osteoblasts, promoting the penetration of blood vessels and transforming the avascular cartilaginous matrix into a vascularized osseous tissue. VEGF plays a critical role in the neo-angiogenesis and revascularization by promoting vasculogenesis (aggregation and proliferation of EGC) and angiogenesis (growth of new vessels from already existing ones). It has been observed that blocking VEGF signaling with antibodies demonstrates that intramembranous bone formation during distraction osteogenesis is dependent on VEGF signaling. The blocking of VEGF-receptors inhibits vascular in-growth and delays or disrupts the regenerative process (Keramaris et al. 2008). Other factors promote neoangiogenesis and vasculogenesis such as the synergistic interactions of the BMPs with VEGF.

Cartilage Resorption and Mineralization

Bony healing is a combination of cellular proliferation and differentiation, increasing cellular volume, and increasing matrix deposition. As fracture callus chondrocytes proliferate, they become hypertrophic and the extracellular matrix becomes calcified (Breur et al. 1991). The primary soft cartilaginous callus needs to be resorbed and replaced by a hard bony callus. Resorption of this mineralized cartilage is initiated by osteoprotegerin (OPG) and TNF-α, macrophage colony stimulating factor (M-CSF), and receptor activator of nuclear kappa factor ligand (RANKL) also known as the RANKL-OPG pathway (Gerstenfeld et al. 2003). These molecules mediate and recruit bone cells and osteoclasts to form woven bone. TNF-α further promotes the recruitment of MSCs and has an important role in initiating chondrocyte apoptosis. The mitochondria accumulate calcium-containing granules, which are transported into the extracellular matrix where they precipitate with phosphate. This becomes a nidus for the formation of apatite crystals. The hard callus formation progresses and the calcified cartilage is replaced with woven bone, the callus becomes more solid and mechanically rigid.

Remodeling of Bone

Hard callus provides biomechanical stability but does not fully restore the biomechanical properties of normal bone. The fracture-healing cascade initiates a second resorptive phase, remodeling hard callus into a lamellar bone structure with a central medullary cavity. This phase is characterized by temporal changes in signaling molecules. IL-1 and TNF-α show high expression levels during this stage, as opposed to most members of the TGF-β family, which has diminished in expression (Ai-Aql et al. 2008). BMP2 is also involved in this phase with reasonably high expression levels (Marsell and Einhorn 2009).

The remodeling process occurs through a combination of resorption by osteoclasts, and lamellar bone deposition by osteoblasts. The remodeling may take years to be completed to achieve a fully regenerated bone structure. There are many factors which affect the speed and efficiency of the healing process, such as the production of electrical polarity created when pressure is applied in a crystalline environment. This in turn affects the bone modeling. The creation of one electropositive convex surface, and one electronegative concave surface, caused by axial loading, activates osteoclastic and osteoblastic activity. The external callus is gradually replaced by a lamellar bone structure, whereas the internal callus remodeling reestablishes the medullary cavity.

The process of bone healing involves osteogenesis, osteoinduction, and osteoconduction. For bone remodeling to be successful, an adequate blood supply and a gradual increase in mechanical stability is crucial (Gerstenfeld et al. 2003). This is clearly demonstrated in cases where neither is achieved, resulting in the development of an atrophic fibrous non-union. However, in cases where there is good vascularity but unstable fixation, the healing process progresses to form a cartilaginous callus resulting in a hypertrophic non-union or a pseudoarthrosis.

Bone Graft Classification System

Autologous bone represents the *gold standard* source whenever a skeletal deficiency needs to be grafted. It is osteoconductive, osteoinductive, and osteogenic. It has the passive ability of a scaffold to be progressively substituted by viable bone; the capacity to stimulate the osteoblastic differentiation of local and systemic mesenchymal stem cells through specific growth factors, such as bone morphogenetic proteins (BMPs); and the ability to form new bone from the living osteoblasts and MSCs present within the graft material.

Autograft is non-immunogenic and cannot transmit infectious agents unless contaminated during donor site harvesting. It represents the first choice in several procedures such as fracture non-union surgery, spinal fusion, and orthopedic reconstructive surgery (Kannan et al. 2015).

Autografts provide the best replacement tissue to a defect site. However, autografts from the donor site require an additional surgery. This can result in complications such as inflammation, lengthening of surgical time, infection, blood loss, hematoma formation, and chronic local pain, especially in older patients who are common candidates for primary and revision spine surgery. This donor-site morbidity occurs in approximately 20% of all cases (Perry 1999). In addition, the volumetric supply limitations reduce its desirability. The quality of the bone in the elderly as a source for grafting is questionable in many cases and the osteoinductive potential may be variable in patients.

As a consequence of the donor issues, market forces have developed scaffolds as bone graft **extenders** (or expanders), which are mixed with the bone graft to augment its volume; bone graft **substitutes**, that may be used alone in place of the bone graft; and bone graft **enhancers**, that are adjuvant therapies aimed at improving the biological performance of the bone grafts by adding cells or growth factors.

There are several categories of bone graft substitutes encompassing varied materials, material sources, and origin (natural vs. synthetic). A bone graft classification system has been developed that describes these groups based on their material makeup (Laurencin and Khan 2013) (see Table 2).

A brief discussion on each group is appropriate on the basis it will provide a basic understanding of each group, advantages and disadvantages, as well as knowledge of when, where, and which product to apply to achieve stability and, ultimately, a solid fusion mass.

Allograft-Based Grafts

Allograft bone refers to bone that is harvested from cadavers and donors. Initially, it was primarily used as a substitute for autografts in large defect sites. Its use has expanded as a result of the absence of autograft donor morbidity, the expansion of bone grafting procedures, and the evolution of bone banks. The coordination of bone bank regulations resulted in donor

Table 2 Different classes of bone graft substitutes

Class	Description
Allograft-based	Allograft bone used alone or with other materials
Factor-based	Natural and recombinant growth factors used alone or in combination with other materials
Cell-based	Cells used to generate new tissue alone or seeded into a support matrix
Ceramic-based	Includes calcium phosphate, calcium sulfate, and bioactive glasses used alone or in combination
Polymer-based	Degradable and nondegradable polymers used alone or combined

screening, tissue processing techniques, and reduction in the risk of disease transmission. This resulted in the acceptance of allographic tissues as a source of grafting and an emergence of allographic materials. A variety of allograft forms are available, including osteochondral, cortical, cancellous, and processed bone derivatives such as demineralized bone matrix (DBM). The processing of allograft ensures there are no viable cells, and consequently, these grafts themselves are not osteogenic. Allograft is osteoconductive and, sometimes, mildly osteoinductive.

Factor-Based Grafts

Growth factors have been extensively researched and have been proven to regulate cellular activity. Proteins and/or growth factors bind to cell receptor sites and stimulate the transcription of messenger RNA that leads to formation of proteins that regulate intra- and extracellular homeostasis. These factors include platelet-derived growth factors (PDGFs), IL-1, IL-3, IL-6, macrophage colony-stimulating factors, the transforming growth factor-β family (TGF-β), BMPs, insulin-like growth factors (IGFs), and vascular endothelial growth factor (VEGF). These factors act in a coordinated manner and influence inflammation, cellular migration and differentiation, angiogenesis, and cellular proliferation. They have both paracrine and autocrine capabilities. Recent approaches have focused on using a combination of factors so as to evoke a synergistic response in

the healing of non-union fractures (Kempen et al. 2010). These combinations are delivered in a manner such that they artificially recreate the native microenvironment of healing bone. The use of single factors requires supraphysiological doses to obtain desirable effects. Such high dosages have led to complications such as ectopic bone, osteolysis, and immunological reactions. Multiple growth factor delivery is advantageous because of its ability to promote two or more diverse functions such as mineralization and angiogenesis. This approach requires study on growth factor combinations dosage, temporal events, and release kinetics. Few studies have been performed in this direction, and the delivery or scaffolds to house and release these factors at the appropriate time have yet to be proven.

Cell-Based Grafts

In significant breakthrough in 2006, Takahashi and Yamanaka discovered how mature cells treated with the right factors could be engineered back to a pluripotent stem cell state capable of producing any cell in the body (Takahashi and Yamanaka 2006). Various stem cells are now available for use in conjunction with bone graft substitutes including mesenchymal stem cells (MSCs), adipose-derived stem cells (ADSCs), induced pluripotent stem cells (iPSCs), and embryonic stem cells (ESC).

MSCs are more primitive than ESCs, consequently in the presence of TGF-b and BMP-2, −4, and −7 to culture media can be differentiated down the osteogenic line. ESCs' somatic cell differentiation requires more steps. ESCs are characterized by their unlimited proliferation and ability to differentiate to any somatic cell type, which makes them a great cell source for tissue regeneration.

ADSCs are an attractive source of stem cells because supply limitations and ease of harvesting is less of a problem given the ready access of adipose tissue deposits found under the dermal layers. These cells induced back to an earlier lineage became known as iPSCs. Stem cell technological advances lead to a greater understanding about the interaction between stem cells and their potential use in bone graft substitutes for clinically relevant applications.

Ceramic-Based Grafts

Ceramics are highly crystalline structures formed by heating non-metallic mineral salts to high temperatures in a process known as sintering. Many ceramics are used in various orthopedic applications (White and Shors 1986). There are resorbable ceramics such as tricalcium phosphate (TCP) and ceramics with highly reactive surfaces such as bioactive glasses and calcium phosphates. The least reactive ceramics are in use in hip arthroplasty (zirconia).

Currently available bone graft substitutes contain ceramics, including calcium sulfate, bioactive glass, and calcium phosphates. Ceramics contain calcium hydroxyapatite (HA), a subset of the calcium phosphate group, which is the primary inorganic component of bone. Calcium phosphates can come close to mimicking the natural matrix of bones, depending on the porosity and structure hence the widespread use of bone graft substitutes that contain HA-based biomaterials.

Calcium phosphates are also osteoconductive, osteointegrative, and, in some instances, can be osteoinductive by the addition of MSCs (Urist and Strates 1970). The structure and crystallinity can influence osteoblastic proliferation and differentiation when in contact with calcium phosphate. The crystallinity of the HA can vary the spatiotemporal proliferation of osteoblasts, and therefore the biological repair activity. The manufacturing of ceramics requires exposure to high temperatures. This complicates the addition of biological molecules. They also tend to be brittle, making them challenging in certain bone graft applications. They are frequently combined with other materials to form a composite.

Polymer-Based Grafts

Polymers are chemical compounds or mixtures of compounds formed by polymerization and consist of repeating structural units. They are classified into natural polymers and synthetic polymers, which can be divided further into degradable and nondegradable. Natural polymers, such collagen, are derived from living sources, whereas synthetic polymers are manufactured.

Synthetic polymers can be used in a wide variety of medical applications. The polymerization process can achieve an extraordinary range of physical and chemical properties. They can be utilized because of their structural and mechanical properties that allow for complex shapes. Poly-lactideco-glycolide (PLGA) is an example of a synthetic, degradable polymer for bone graft applications. With the addition of water, it can break down to lactic acid and glycolic acid, which are natural human metabolites.

Natural polymers, such as collagen hydrogels, are used in scaffolds in tissue engineering as cells adhere and grow on the collagen fibers within the hydrogel.

Due to their carbon-based chemistry, polymers are closer to biological tissue than inorganic materials. This can be used for targeted interaction between the material and the body. As with ceramics, the functionality of polymers can be enhanced if used in combination with other materials, such as ceramics, to form composites.

Allografts

Introduction

Bone grafting is an essential component of spinal surgery. The need commonly arises in spinal fusion (Takaso et al. 2011) and is utilized in trauma, infection, and tumor resection (Finkemeier 2002), where bony defects can arise. These defects can be filled with autograft, which is generally preferred as it is osteoconductive and osteoinductive. However, harvesting for autograft creates donor morbidity, extended operating time, and may be contaminated by systemic processes (e.g., sepsis, metastases) (Aro and Aho 1993). In addition, autografts are limited in number, shape, size, and volume to which allografts are not constrained.

Consequently, the need for bone banks arose. Currently bone is a commonly transplanted tissue, second only to blood.

Therefore allografts supplied by a bone bank are commonly used instead of autografts because of their immediate availability and unlimited volume. The use of allograft bone eliminates the operative time required for harvesting and associated donor morbidity. Virtually any shape or size can be fashioned and are often combined with other enhancers, extenders, and substitutes to achieve union. Allogenic bone has osteoconductive activity; it serves as an acellular mineralized frame against which newly formed bone gets deposited. It is, however, variably and weakly osteoinductive (depending on source and processing) and is not osteogenic.

The obvious disadvantages of allografts are the potential for disease transmission, host incompatibility, and infection from contaminated tissue.

The Food and Drug Administration (FDA) has established guidelines that all bone and tissue banks must adhere to. The American Association of Tissue Banks (AATB) established guidelines that accredited banks must follow, which have set the industry standards. These guidelines include training and certification of employees as well as regular inspections of the facility through to assessing documentation and auditing.

The AATB is a not for profit organization that regulates and monitors the safety, consistency and availability of allografts across the United States. The AATB ensures and accreditates tissue banks to ensure their compliance with the FDA guidelines. Governments have imposed the standardization of bone bank operations to ensure the safety of transplanted tissues. The performance of a bone bank depends on its organization, donor selection and procurement, documentation, storage and processing, and implementation.

There are two types of donor of homologous tissues: live donors, consisting mainly of donations of femoral heads after fractures and total hip arthroplasty and cadaver donors, from which much greater quantities of tissues can be harvested, from any segment of the skeleton. Cadaveric donors are usually younger and have better bone quality.

Allograft Safety

The aggregate risk of disease transmission with allograft is a reflection of the rigor of the screening and testing of donors and the type of allogenic tissue that is transplanted. The risk of transmission of viral diseases (human immunodeficiency virus (HIV), hepatitis B (HBV), hepatitis C (HCV)) depends on the product type and its preparation. Regulatory guidelines in Europe, the United States, and Australia have developed critical pathways for manufacturers of biological products to follow. Simply stated, these involve screening tissues for infectious agents (viral/bacterial), processing techniques to ensure sterilization (ethylene oxide/radiation), and follow up mechanisms to track the patient and product. By AATB standards, serologic testing is used including for hepatitis B surface antigen, hepatitis B core antibody, hepatitis C antibody, syphilis, human T-lymphotrophic virus-1 antibody, HIV 1 and 11 antibodies, and HIV P24 antigen. All tissue and blood samples are tested for infectious diseases, including for AIDS with Nucleic Acid Testing (NAT by TMA), HCV, HIV-1, HIV-2, HTLV-1, HTLV-11, HB Core, RPR, HCV-Ab, HBs Ag, HIV-1 NAT, and HCV-NAT.

Mroz et al. (2009) performed a retrospective review of data analyzing 54,476 allograft specimens recalled by the FDA. They found that despite the large number of allograft recalls in the USA, there was only one documented case of disease transmission (HIV) in spine surgery. The review found no reports of bacterial transmission from the use of allograft. They admitted that the precise incidence of disease transmission linked to tissue allografts was unknown. They concluded that there appears to be no overt risk associated with the use of allograft bone in spine surgery.

Bone Banking Overview

The presence of microorganisms on processed tissues is inevitable and unavoidable. In order to prevent disease transmission the appropriate donor selection process, proper tissue processing, and adequate sterilization is of paramount importance.

The most valuable method for determination of suitability is the donor's medical history. In general, there should no history of infectious, malignant, neurological, and autoimmune diseases. In addition, there should be no metabolic bone disease, drug abuse, or exposures to radiation and toxic substances present in the history. Essentially, any condition in which there is the possibility of disease transmission or where the quality of the bone may be compromised could warrant exclusion from donation.

The procurement and processing of allograft has to be performed with a sterile and hygienic technique. Grafts may be harvested from two sources. Bone is procured aseptically or cleanly. The first is from live donors where femoral heads are typically collected aseptically in the operating room by a surgeon in the course of a hip replacement. The second source is cadaver bone which can also be procured under aseptic conditions in the operating room. Removal of bone should take place within 12 h of death or within 24 h if the body has been refrigerated at 4 °C to reduce bacterial growth and bone autolysis. Standard sterile draping of the cadaver and sterile gowning of all trained procurement personnel are essential.

A sterilization process that has a high inactivation process prevents the transmission of diseases from donor to recipient. Various techniques are utilized such as ethylene oxide (Dziedzic-Goclawska 2005), irradiation (Nguyen et al. 2007), thermo disinfection (Folsch et al. 2015), and other techniques such as antibiotic soaks. Ethylene oxide sterilization is of limited use because of limited tissue penetration and can cause an inflammatory reaction to ethylene chlorohydrin, a by-product formed from ethylene oxide. Gamma irradiation has bactericidal and viricidal properties and has been proven to be successful in sterilizing medical products. The concept of sterility assurance level (SAL) is derived from studies on bacteria, fungi, and spores where a sterilization dose is high enough that the probability of an organism surviving is no greater than one in one million units tested. The dosage of 25 kGy has been the recommended for terminal sterilization of allograft tissues. This is 40% above the minimum dose required to kill resistant microorganisms. Research has shown dose-dependent reductions in biomechanical properties of allografts at high levels of gamma irradiation (>30 kGy). This prompted bone banks to employ lower doses that are efficient at deactivating microorganisms, while protecting the biomechanical properties of bone allograft. Several bone and tissue banks around the world utilize minimum doses to achieve sterility with dosages as low as 11 kGy. The radiation process is a cold sterilization process and therefore preserves the properties and characteristics of tissues (Singh et al. 2016).

Gamma irradiation of allografts is a safe and highly effective sterilization method and offers a clear advantage in terms of safety compared with other sterilization techniques. There is no substitute for donor screening and rigorous tissue processing procurement techniques and, when combined with gamma irradiation, the sterile product is safe for clinical use.

Types of Bone Allograft

There are three commonly used forms: fresh frozen allograft, freeze dried allograft and demineralized bone matrix (DBM).

Allografts are either fresh or processed. Fresh allografts are sometimes utilized because they are alive. However the immunologic reaction and the risk of disease transmission has led to most allografts being processed.

Fresh frozen allograft has improved osteoinductive and biomechanical properties relative to irradiated materials. After harvest, the allograft is cleaned by a high-pressure lavage and antibiotic solutions. The removal of marrow elements removes a significant antigenic cell population. After cleaning, the tissue is cultured and if sterile the graft is then frozen and released for implantation in due course. The grafts are packaged without solution and frozen to −70 °C to −80 °C. The stored life of sterile fresh frozen allograft is 3–5 years. Fresh frozen graft has the greatest strength of any type of structural allograft, however carries the risk of disease transmission, which can be reduced by donor screening and sterile harvesting techniques. Any grafts that have positive cultures after processing are secondarily sterilized. This secondary sterilization process occurs with either gamma irradiation or ethylene oxide.

Freeze-dried allograft is prepared by tissue water being replaced with alcohol to a moisture level of 5% with the alcohol, then being removed under vacuum. Freeze-dried grafts can be stored at room temperature for 3–5 years and are therefore easier to maintain. The graft requires a 30 min period of rehydration prior to implantation. The risk of disease transmission is also reduced with no cases of HIV transmission being reported. It is estimated the risk is 1 in 2.8 billion compared to fresh frozen allograft risk of 1 in 1.6 million (Costain et al. 2000). Freeze drying does not eliminate the HIV virus, however it does reduce the immunogenicity. The effect of freeze-drying on the mechanical characteristics of the graft is dependent on the method and rate of rehydration. Compared with fresh frozen bone, it has been found that the process results in a small but significant reduction in stress (18%) and stiffness (20.2%) (Cornu et al. 2000). A study comparing the compressive strengths of fresh frozen, freeze dried and ethylene oxide treated allograft showed no significant differences (Brantigan et al. 1993). The effect of irradiation on freeze-dried allograft has been shown to reduce graft strength further (Hamer et al. 1996). As well as the mechanical stability reduction, irradiation results in the denaturing of endogenous BMPs eliminating much of its osteoinductive capacity.

The discovery of osteoinductive proteins within demineralized bone matrix has resulted in the widespread use of DBM in grafting procedures (Urist et al. 1967). The production of DBM is a multistep process commencing with cortical bone being cleaned followed by machining into small particles. Acid is then applied to reduce the calcium content while maintaining the organic matrix and growth factors. The demineralization process releases these cytokines that participate in the complex cascade of events leading to bone repair. This renders DBM weakly osteoinductive. Approximately 93% is collagen and 5% are growth factors, a portion of which is BMPs. There is variability in the osteoinductivity depending on the donor, the site of the harvest and the method of processing. This variability resulted in different osteoinductive capabilities between different manufacturers and within a single manufacturer's product (Wang et al. 2007). The end product is a powder and therefore the handling characteristics were problematic. If placed in a cavity the product could be easily displaced from its desired location by blood and other fluids. Hence, most DBMs are placed in carriers such as putty or glycerol. DBMs have been found to be useful as graft extenders for both local bone and iliac crest bone graft (Schizas et al. 2008). The literature supports the utility of DBMs as enhancers and extenders but not as bone substitutes in isolation.

It is important to select an allograft product that is applicable to the clinical situation. In addition, surgeons should be familiar with the bone bank and its regulations, the products and their preparation, the biologic activity and biomechanical characteristics, as well as cost, safety, and efficacy of the product.

Allograft Use in Spine Surgery

It is clear that there are four essential elements of bone grafts for successful bone regeneration. Osteoconductivity, osteoinductivity, osteogenicity, and the mechanical environment is the diamond concept and all aspects need to be considered when selecting an allograft in a particular operation to fulfill a particular role.

Osteoconductivity is the ability of a material to provide a three-dimensional structure for the ingrowth of host capillaries, perivascular tissue, and osteoprogenitor cells.

Osteoinductivity is defined as the ability of a material to stimulate primitive, undifferentiated, and pluripotent cells to develop into the bone-forming cell lineage with the capacity to form new bone.

Osteogenicity implies that a bone grafting material has the intrinsic capacity to stimulate bone healing by the presence of mesenchymal stem cells (MSCs) or osteoprogenitors cells.

The mechanical environment in which the allograft is placed is critical to the success: the construct needs to be rigid, the graft under compression and the recipient vascular bed viable.

The choice of bone graft for achieving an arthrodesis or reconstruction in the spine should be made on the evidence available from the literature and not from a salesman in a suit or a glossy brochure. The primary goal is to achieve an interbody arthrodesis, therefore, many factors need to be taken into account. Less commonly, graft is used to recreate and restore anatomy in the case of spinal deformities. For example, remodeled endplates can be recreated in cases of high grade dysplastic spondylolisthesis to support an interbody cage/graft and anterior fixation. The recipient's biological status, age, comorbidities, graft harvest site, the vascularity and local tissue viability, the mechanical environment and the use of supplemental fixation influences the environment for bone healing to occur. In selecting a graft, one needs to take into account the diamond concept: osteoinductivity, osteoconductivity, osteogenicity and the mechanical environment.

Although iliac crest bone graft is the gold standard, there are numerous reports on the success of alternate approaches, especially in combinations. For example, with large or segmental bone defects, the healing process is impaired; thus the use of tissue engineering techniques becomes a necessity. To mimic the natural bone healing process, three major components are required: a mechanically stable graft, a suitable cell source and the presence of chemical and biological factors. Extensive research has been done in all three components mentioned above; however, we are still at the laboratory bench stage, in which biomaterials, growth factors, and cell sources are being examined and optimized for the regeneration of bone. In addition new efforts by several major orthopedic companies have expanded the role of allografts by tailoring them for specific surgical procedures. For example, dowels and wedges can be utilized in spinal fusions and DBMs with carriers to provide better handling and performance characteristics.

Allografts are now manufactured in a variety of forms and consequently offer versatility to meet the requirements of an ideal graft. They can be processed to offer mechanical support in load bearing environments, provide the ability to incorporate and remodel, to be biocompatible, to be osteoinductive when demineralized and to provide a high level of safety. As mentioned earlier factor, cellular, ceramic and polymer based bone enhancers, extenders and substitutes are used singularly or in combinations to enhance the role of allografts in spinal fusion. As new technologies are developed, tissue engineering and gene therapies are likely to add to the biological characteristics already available with allografts.

Allograft in the Anterior Column

Multiple studies on the radiographic success of allograft in the anterior column of the cervical spine have been reported. Few studies have studied the clinical efficacy. Generally, the use of allografts is supported for use in anterior column support (level I-IV evidence). The utilization of supplemental fixation, such as cages and plates, has resulted in a substantial increase in fusion rate as well as maintenance of lordosis, which in turn appears to reduce adjacent motion segment degeneration (AMSD) and improves the clinical efficacy. Many options exist for the usage of allograft in the anterior column: these include cortical or cortico-cancellous allograft, with or without supplemental cage or plate fixation, and DBM, which always requires supplemental fixation. It should be noted that allograft alone with BMP-2 leads to high union rates but is prone to subsidence (Vaidya et al. 2007) and, as such, additional measures to improve the biomechanical environment such as cage support and/or plate/posterior fixation is important (Slosar et al. 2007).

The Cervical Spine

Allograft has been shown to be at least equivalent to autograft when used for anterior cervical procedures with the exception of multilevel constructs. Tuchman et al. (2017), in a systemic review compared the effectiveness and safety between iliac crest bone graft (ICBG), non-ICBG autologous bone, and allograft in cervical spine fusion. The review identified 13 comparative studies: 2 prospective cohort and 11 retrospective cohort studies. Twelve cohort studies compared allograft with ICBG autograft during anterior cervical fusion and demonstrated with a low evidence level of support that there are no differences in fusion percentages, pain scores, or functional results. There was insufficient evidence comparing patients receiving allograft with non-ICBG autograft for fusion, pain, revision, and functional and safety outcomes.

The FDA IDE studies on disc replacement have provided information on allograft fusion in the anterior cervical spine (Coric et al. 2018; Gornet et al. 2017). These studies have provided valuable information on the control group in regards to the incidence of fusion, reoperation, non-union and data on the patient-reported outcome measures (PROM) and have generally found high fusion rates and acceptable outcomes. These, along with other studies with single level allograft constructs supported with internal fixation show that meticulous surgical technique could result in fusion and improvement in PROMs irrespective of graft choice (Yeh et al. 2017; Fraser and Härtl 2007).

ACDF using allografts have in general shown clinical and radiological success. Muzević et al. (2018) investigated clinical parameters of ACDF treatment and outcomes using osseous allografts in different age groups, studying the postoperative results of restoration of lordosis and evaluating the utility of bone allografts for ACDF, including graft subsidence. Fifty-two patients had disc herniation and 102 had spondylosis. Surgery was performed on a total of 313 levels. The median duration of follow-up was 24 months, and no patients were lost to follow-up. Human cortical allografts were used in 51 segments (16.3%), and corticocancellous allografts were used in 262 segments (83.7%). Solid fusion was achieved in 97.92% of patients and 98.37% of levels at a mean follow-up of 5.97 ± 2.86 months. Graft sizes ranged from 8 mm to 15 mm. The most frequently used graft size for fusion was 11 mm (119 levels; 38%), followed by 10 mm (72 levels;

23%) and 12 mm (70 levels; 22.4%). Anterior cervical plates and screws were used in all patients. The importance of a plate in load sharing is recognized, especially in patients with a kyphotic cervical spine. Treatment outcomes achieved excellent or good outcomes in more than 80% of patients, regardless of age. Yeh et al. (2017) retrospectively collected preoperative and postoperative radiographic and clinical data of 50 patients from 2005 to 2009, with a diagnosis of multilevel cervical spondolytic myelopathy (MCSM), who received 2-level anterior cervical corpectomy and fusion (ACCF) with a fresh frozen cortical strut allograft (FFCSA) fibular shaft and an anterior dynamic plate (see Fig. 3). The cervical curvature lordosis improved and the neurogenic function recovered well postoperatively. The VAS-neck and NDI scores both decreased after 12 and 48 months following surgery. The Japanese Orthopaedic Association score recovery rate at postoperative 4 years was 87.5%. Fusion rates achieved were 100% at 12 months. They stated the results were satisfying and the complication rate was low. The authors emphasized meticulous graft preparation on both the donor bone ends and the recipient endplates so the graft could be inserted with a press fit technique.

Evidence for the efficacy regarding the use of allograft in multi-level anterior cases is mixed. A study by Park et al. (2017) demonstrated similar clinical and radiologic outcomes between patients treated with corticocancellous composite allograft or autograft for ACDF, with a decreased subsidence rate in the corticocancellous composite allograft group. They utilized freeze dried, fully machined corticocancellous composite allograft cages. APS was utilized for primary fixation. Corticocancellous composite allograft is composed of cortical lateral walls with a cancellous centre. The cortical portion provides structural support for the disc space, while the cancellous portion provides a scaffold for bone in-growth that can minimize graft subsidence with an enhanced fusion rate. The authors stated the allograft group took longer to fuse and in multi level cases, may result in hardware failure. Peppers et al. (2017) reported on the results of a prospective multicentre clinical trial assessing the safety

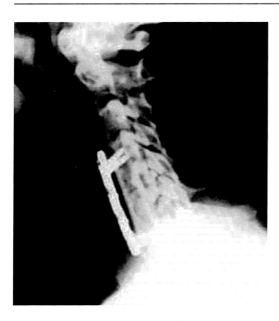

Fig. 3 Example of corpectomy and fibular strut allograft with plate/screw stabilization for reconstruction of the anterior column

and effectiveness of the viable cellular bone allograft in combination with a polyetheretherketone (PEEK) interbody spacer in two-level ACDF using patient reported and radiological outcome measures. The per subject fusion rate increased over time and was determined to be 65.7% of subjects fused at 6 months and 89.4% at 12 months. This study did not have a control group and thus treatment was not directly compared to autograft or non-cellular allograft treatments. Samartzis et al. (2003) however found equivalent and high rates of fusion (in the order of 97–98%) with rigid plating emphasizing that mechanical stability may be a factor influencing the difference between autograft and allograft in older studies without rigid fixation.

A near 100% union rate is achievable with the addition of BMP to allograft but the side effects related to dosage require caution. Burkus et al. (2017) reported on a prospective study evaluating the safety and efficacy of BMP-2 with allograft for ACDF in single level degenerative disc disease (DDD). The investigational group had 0.6 mg of BMP-2 inside a PEEK cage/APS with a fusion rate of 99.4% versus a control group treated with allograft spacer/APS with a fusion rate of 87.2%.

A higher rate of adverse events such as swelling, dysphagia and oropharyngeal pain was noted. Butterman (2008) similarly found high rates of union but increased incidence of neck swelling and higher cost of allograft with BMP-2 compared to iliac crest autograft for ACDF. Higher dosage has also been linked with increased osteoclastic activity via the RANKL-OPG pathway and osteolysis (Gerstenfeld et al. 2003) that can lead to subsidence of implants.

While allograft and autograft showed near equivalence in the literature for the cervical spine, the same cannot be said for synthetic bone graft. Buser et al. (2016) reviewed the efficacy and safety of synthetic bone graft substitutes versus autograft or allograft for the treatment of cervical degenerative disc disease. Data from 8 comparative studies were included: 4 RCTs and 4 cohort studies (1 prospective and 3 retrospective studies). Synthetic grafts included HA, β-TCP/HA, PMMA, and biocompatible osteoconductive polymer (BOP). The PMMA and BOP grafts led to lower fusion rates, and PMMA, HA, and BOP had greater risks of graft fragmentation, settling, and instrumentation problems compared with iliac crest bone graft. The authors stated that conclusions regarding the efficacy and effectiveness of these products are low and insufficient. Most of these studies were sponsored and the sample size inadequate, therefore with the potential for bias. The use of bone substitutes, extenders and enhancers has escalated rapidly and of course buyers are being charged a premium without clinical evidence to support the use of such products. In this review, synthetic grafts performed similarly or worse than autologous grafts in achieving fusion. A detailed review of the level of evidence, safety, and efficacy is required.

It should be discussed briefly also that allografts play a role in providing structural integrity or bridging significant bone loss in certain circumstances in the anterior cervical spine. This includes structural allograft for vertebrectomy cases anteriorly in the setting of metastatic tumor or systemic infection or significant traumatic bony comminution in a burst fracture where the patient's autograft may be less desirable for reasons of contamination, healing problems or increased morbidity.

In summary, the evidence suggests that ICBG and allograft demonstrated clinical equipoise in terms of fusion rates, pain scores, and functional outcomes following anterior cervical fusion. Recognition about patient factors and surgical techniques play a significant role in the outcome of anterior cervical fusion surgery. Age, osteoporosis, the number of levels and tobacco use can affect the fusion rate. In addition, surgeon factors such as graft doweling, endplate preparation and the use of supplemental fixation can also affect fusion rates.

Allograft utilization removes donor site complications. While the preparation and sterilization reduce or eliminate transmission of disease, it also reduces or eliminates osteoinductive capabilities and can mechanically alter the graft and the ability to withstand compressive and torsional loads. This can result in longer fusion times, subsidence, and loss of correction.

The Lumbar Spine

A large variety of devices have been developed and used for structural support of the spine in the anterior and middle columns. These structural deficits arise from anterior discectomies, trauma, tumor resections and following osteotomies for correction of deformity. The anterior lumbar spine is similar to the cervical spine in that, following disc resection or vertebrectomy, the graft is placed under compression. An additional advantage is the larger cross sectional area of the lumbar spine vertebral body on which to seat the allograft. The loads are mainly axial compression, although some rotational stresses are also applied. The two most commonly used allografts in the anterior spine are femoral ring allografts (FRA) and tricortical iliac crest. Mechanical testing of the vertebral body and FRAs has revealed compressive strength of 8000n and 25000n, respectively (Voor et al. 1998). These biomechanical tests revealed the importance of placement, surface area coverage, and the importance of endplate preservation.

Clinically, studies of allograft use in the lumbar spine for degenerative disc disease and deformity have reported favorable outcomes in use with an anterior approach (ALIF). Burkus et al., in a randomized controlled trial of 131 patients used ALIF threaded cortical allograft dowels and BMP equivalent rates to autograft without additional morbidity (Burkus et al. 2005). In fact, some other multicenter studies have even reported superior results compared with allograft (Burkus et al. 2003).

Regarding tumor and deformity, allograft has been a useful tool in the surgeon's armamentarium. Bridwell et al. (1995) reported in a prospective study of 24 patients with kyphosis or anterior column defects treated with fresh frozen allograft and posterior instrumentation and autogenous grafting. Only two patients showed some subsidence and the other 22 maintained the correction. Bridwell (Bridwell et al. 1995) also found that deformity correction using anterior allograft support was effective on the proviso it was combined with rigid posterior fixation and autograft. Janssen et al. (2005) reported on the clinical and radiographic outcomes of 137 patients who were treated with a FRA allograft packed with ICBG ($n = 117$) and DBM ($n = 13$) and supplemental posterior fixation. They were able to achieve a 94% fusion rate.

Systemic review and consensus of expert opinion in the setting of reconstructions following en bloc tumor resection have recommended cages packed with morcelized allograft and suitable autograft for single level vertebrectomies with strut bone grafting used in the thoracic spine (Glennie et al. 2016).

Other interbody approaches have had variable success. Generally, PLIF studies have reported inferior outcomes from the use of allograft compared with autograft alone (Jorgenson et al. 1994), be it with or without instrumentation (Brantigan 1994). Anand et al. (2006) in a study of allograft laden TLIF cages have been more favorable with regard to fusion at 99% and satisfaction 96%. The literature on interbody fusion via lateral and oblique approaches utilizing allograft is limited so an inference as to their efficacy cannot currently be drawn.

It is important that one is aware of the potential for allograft resorption when used in isolation with BMP-2 (Pradhan et al. 2006). An anterior cage-allograft-BMP-2 combination may provide the best synergy in terms of initial support,

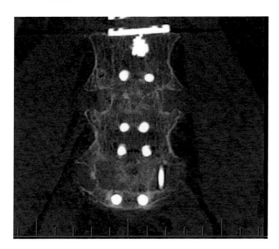

Fig. 4 Solid arthrodesis achieved in the lumbar anterior column as demonstrated across a coronal CT plane through a tissue engineering construct of plate-PEEK cage-femoral allograft and BMP-2

scaffolding and growth factor stimulus combining to support effective arthrodesis. The cage (usually with supplemental screw fixation within or extraneous to the cage) provides the mechanical stability, the cancellous structural graft the osteoconductive matrix and the BMP-2 the osteoinductive protein (see Fig. 4). This is effectively tissue engineering. There is good evidence for the utility of spinal allografts for structural reconstruction in the anterior spine. They have the advantage of immediate strength, are under compression and have comparable fusion rates with autograft if combined with anterior and/or posterior instrumentation.

Posterior Elements

Cervical

There is limited literature on use of allograft for fusion of the posterior elements of the cervical spine as most studies relate to anterior cervical reconstructions. However, studies do highlight their utility with regard to availability of shape and size in more complex anatomy. For example, posterior interventions including occipitocervical and C1–2 posterior based fusions can benefit from access to spanning shaped segmental support (Nockers et al. 2007; Aryan et al. 2008).

Thoracolumbar

Posterior correction of scoliosis is a particularly important clinical situation requiring large amounts and surface area of bone graft to achieve a solid posterior fusion. In the most common cases of either adolescent idiopathic or pediatric neuromuscular scoliosis, adequate quantity and quality of autograft bone may not result from simple local bone grafting alone. In this instance, the choice of graft(s) is important in minimizing further blood loss and morbidity from an already invasive procedure for the child. To this end, freeze-dried allograft chips have been employed with high success in some literature with the advantage of reduced blood loss and morbidity (Blanco et al. 1997; Montgomery et al. 1990) but some studies have reported inferior outcomes when compared to autograft only or composite autograft-demineralized bone matrix combinations (Price et al. 2003).

Regarding fusion of the posterior elements in posterolateral fusion (PLF), studies are mixed on the efficacy of allograft bone making it difficult to recommend for or against for posterior fusion, e.g., in a 360° fusion strategy. An and colleagues reported inferior clinical and radiological results of allografts compared to autografts in PLF (An et al. 1995). However, in a randomized controlled trial instrumented allograft use in PLF without BMP has been found to have equivalent fusion rates and outcomes scores to autograft (Gibson et al. 2002). In general, there is little or no evidence for any utilization of structural allograft in the posterior spine except at the occipital-cervical junction. There is some evidence for cancellous allograft chips being used as an extender in PLF. There is evidence for the use of DBM as an extender and possibly as an enhancer in the posterior spine.

Conclusions

There is a wide range of osseous allografts available for use in spinal surgery. Allograft has gained popularity because of its abundant supply and absence of donor site morbidity. The goal is to produce comparable or superior outcomes when used as a substitute for autograft. The surgeon needs to take into account the type of procedure and therefore the graft best suited for that environment.

The ideal bone replacement material should be osteoinductive or conductive, non-pathogenic, minimally antigenic and mechanically stable. Compared with autografts, allografts show delayed vascularization and remodeling of the fusion mass. Allogenous bone has limited osteoinductive properties and carries the risk of subsidence due to delayed union or non-union. Currently, several modified allograft cages have been introduced to enhance union rate and structural stability, including corticocancellous composite allograft. Despite these facts, allografts are in plentiful supply, have a proven track record, and are an effective adjunct when used in the correct clinical situation.

There are many patients who may not have enough available bone for the prescribed procedure and any additional surgery may result in added blood loss, pain, infection, contamination and an increased hospital stay. Therefore, allogeneic bone from cadaver donors or live donors has been used successfully for a number of procedures and has several advantages including long-term storage, large available quantities, and specific types and sizes of bone.

Regenerative engineering is emerging at a rapid rate. More is now understood about material science, stem cells, signaling molecules, growth factors, and the strategies available to integrate these components to produce the functional biological system we regard as bone. One can envisage a time when structural allografts will be composites of minerals and signaling molecules with growth factors in their structure that will facilitate bone union.

References

Ai-Aql ZS, Alagl AS, Graves DT et al (2008) Molecular mechanisms controlling bone formation during fracture healing and distraction osteogenesis. J Dent Res 87(2):107–118

An HS et al (1995) Prospective comparison of autograft vs. allograft for adult posterolateral lumbar spine fusion:

differences among freeze-drier, frozen and mixed grafts. J Spinal Disord 8(2):131–135

Anand N et al (2006) Cantilever TLIF with structural allograft and RhBMP2 for correction and maintenance of segmental sagittal lordosis: long-term clinical, radiographic, and functional outcome. Spine (Phila Pa 1976) 31(20):E748–E753

Aro HT, Aho AJ (1993) Clinical use of bone allografts. Ann Med 25(4):403–412

Aryan HE et al (2008) Stabilization of the atlantoaxial complex via C-1 lateral mass and C-2 pedicle screw fixation in a multicenter clinical experience in 102 patients: modification of the Harms and Goel techniques. J Neurosurg Spine 8(3):222–229. https://doi.org/10.3171/SPI/2008/8/3/222

Bais MV, Wigner N, Young M et al (2009) BMP2 is essential for post natal osteogenesis but not for recruitment of osteogenic stem cells. Bone 45(2):254–266

Balga R et al (2006) Tumor necrosis factor-alpha: alternative role as an inhibitor of osteoclast formation in vitro. Bone 39(2):325–335

Blanco JS et al (1997) Allograft bone use during instrumentation and fusion in the treatment of adolescent idiopathic scoliosis. Spine (Phila Pa 1976) 22(12):1338–1342

Brantigan JW (1994) Pseudarthrosis rate after allograft posterior lumbar interbody fusion with pedicle screw and plate fixation. Spine (Phila Pa 1976) 19(11):1271–1279; discussion 1280

Brantigan J et al (1993) Compression strength of donor bone for posterior interbody fusion. Spine 18(9):1213–1221

Breur GJ, VanEnkevort BA, Farnum CE et al (1991) Linear relationship between the volume of hypertrophic chondrocytes and the rate of longitudinal bone growth in growth plates. J Orthop Res 9(3):348–359

Bridwell KH et al (1995) Anterior fresh frozen structural allografts in the thoracic and lumbar spine. Do they work if combined with posterior fusion and instrumentation in adult patients with kyphosis or anterior column defects? Spine (Phila Pa 1976) 20(12):1410–1418

Burkus JK et al (2003) Is INFUSE bone graft superior to autograft bone? An integrated analysis of clinical trials using the LT-CAGE lumbar tapered fusion device. J Spinal Disord Tech 16(2):113–122

Burkus JK et al (2005) Use of rhBMP-2 in combination with structural cortical allografts: clinical and radiographic outcomes in anterior lumbar spinal surgery. J Bone Joint Surg Am 87(6):1205–1212

Burkus JK et al (2017) Clinical and radiographic outcomes in patients undergoing single-level anterior cervical arthrodesis: a prospective trial comparing allograft to a reduced dose of rhBMP-2. Clin Spine Surg 30(9):E1321–E1332

Buser Z et al (2016) Synthetic bone graft versus autograft or allograft for spinal fusion: a systematic review. J Neurosurg Spine 25(4):509–516

Butterman GR (2008) Prospective nonrandomized comparison of an allograft with bone morphogenic protein versus an iliac-crest autograft in anterior cervical discectomy and fusion. Spine J 8(3):426–435. Epub 2007 Mar 7

Coric D et al (2018) Prospective, randomized multicenter study of cervical arthroplasty versus anterior cervical discectomy and fusion: 5-year results with a metal-on-metal artificial disc. J Neurosurg Spine 28(3):252–261. https://doi.org/10.3171/2017.5.SPINE16824. Epub 2018 Jan 5.

Cornu O et al (2000) Effect of freeze drying and gamma irradiation on the mechanical properties of human cancellous bone. J Orthop Res 18(3):426–431

Costain DJ et al (2000) Fresh frozen vs irradiated allograft bone in orthopaedic reconstructive surgery. Clin Orthop Relat Res 371:38–45

Dimitriou R, Tsiridis E, Giannoudis PV (2005) Current concepts of molecular aspects of bone healing. Injury 36(12):1392–1404

Dziedzic-Goclawska A (2005) Irradiation as a safety procedure in tissue banking. Cell Tissue Bank 6:201–219

Einhorn TA (2005) The science of fracture healing. J Orthop Trauma 19(Suppl 10):S4–S6

Finkemeier CG (2002) Bone-grafting and bone-graft substitutes. J Bone Joint Surg Am 84-A(3):454–464

Folsch C et al (2015) Influence of thermal disinfection and duration of cryopreservation at different temperatures on pull out strength of cancellous bone. Cell Tissue Bank 16:73–81

Fraser JF, Härtl R (2007) Anterior approaches to fusion of the cervical spine: a metaanalysis of fusion rates. J Neurosurg Spine 6(4):298–303

Gerstenfeld LC et al (2003) Fracture healing as a post-natal developmental process: molecular, spatial, and temporal aspects of its regulation. J Cell Biochem 88(5): 873–884

Gerstenfeld LC, Alkhiary YM, Krall EA et al (2006) Three-dimensional reconstruction of fracture callus morphogenesis. J Histochem Cytochem 54(11):1215–1228

Giannoudis PV, Einhorn T, Marsh D (2007) Fracture healing: the diamond concept. Injury 38s4:s3–s6

Gibson S et al (2002) Allograft versus autograft in instrumented posterolateral lumbar spinal fusion: a randomized control trial. Spine 27(15):1599–1603

Glennie RA et al (2016) A systematic review with consensus expert opinion of best reconstructive techniques after osseous en bloc spinal column tumor resection. Spine (Phila Pa 1976) 41(Suppl 20):S205–S211

Gornet M et al (2017) Cervical disc arthroplasty with the prestige LP disc versus anterior cervical discectomy and fusion, at 2 levels: results of a prospective, multicenter randomized controlled clinical trial at 24 months. J Neurosurg Spine 26(6):p653–p667

Granero-Molto F, Weis JA, Miga MI et al (2009) Regenerative effects of transplanted mesenchymal stem cells in fracture healing. Stem Cells 27(8):1887–1898

Green E, Lubahn JD, Evans J (2005) Risk factors, treatment, and outcomes associated with nonunion of the midshaft humerus fracture. J Surg Orthop Adv 14(2): 64–72

Hamer AJ et al (1996) Biomechanical properties of cortical bone allograft using a new method of bone strength measurement. A comparison of, fresh, fresh-frozen and irradiated bone. J Bone Joint Surg Br 78(3):363–368

Janssen ME et al (2005) Anterior lumbar interbody fusion using femoral ring allograft for treatment of degenerative disc disease. Semin Spine Surg 17:251–258

Jorgenson SS et al (1994) A prospective analysis of autograft versus allograft in posterolateral lumbar fusion in the same patient. A minimum 1-year follow-up in 144 patients. Spine (Phila Pa 1976) 19(18):2048–2053

Kannan A, Dodwad SN, Hsu WK (2015) Biologics in spine arthrodesis. J Spinal Disord Tech 28:163–170

Kempen DHR, Creemers LB, Alblas J, Lu L, Verbout AJ, Yaszemski MJ, Dhert WJA (2010) Growth factor interactions in bone regeneration. Tissue Eng B Rev 16:551–566

Keramaris NC, Calori GM, Nikolaou VS et al (2008) Fracture vascularity and bone healing: a systematic review of the role of VEGF. Injury 39(Suppl 2): S45–S57

Laurencin CT, Khan YM (2013) Regenerative engineering. CRC Press, Boca Raton

Lee SK, Lorenzo J (2006) Cytokines regulating osteoclast formation and function. Curr Opin Rheumatol 18(4): 411–418

Lee SK et al (2006) Cytokines regulating osteoclast formation and function. Curr Opin Rheumatol 18(4): 411–418

Marsell R, Einhorn TA (2009) The role of endogenous bone morphogenetic proteins in normal skeletal repair. Injury 40(Suppl 3):S4–S7

Montgomery DM et al (1990) Posterior spinal fusion: allograft versus autograft bone. J Spinal Disord 3(4):370–375

Mroz T et al (2009) The use of allograft bone in spine surgery:is it safe? Spine J 9:303–308

Muzevic D et al (2018) Anterior cervical discectomy with instrumented allograft fusion: lordosis restoration and comparison of functional outcomes among patients of different age groups. World Neurosurg 109:e233–e243

Nguyen H et al (2007) Sterilization of allograft bone: effects of gamma irradiation on allograft biology and biomechanics. Cell Tissue Bank 8:93–105

Nockers RP et al (2007) Occipitocervical fusion with rigid internal fixation: long-term follow-up data in 69 patients. J Neurosurg Spine 7(2):117–123

Park JH et al (2017) Efficacy of cortico/cancellous composite allograft in treatment of cervical spondylosis. Medicine (Baltimore) 96(33):e7803

Peppers TA et al (2017) Prospective clinical and radiographic evaluation of an allogeneic bone matrix containing stem cells (Trinity Evolution® Viable Cellular Bone Matrix) in patients undergoing two-level anterior cervicaldiscectomy and fusion. J Orthop Surg Res 12(1):67. https://doi.org/10.1186/s13018-017-0564-5

Perry CR (1999) Bone repair techniques, bone graft, and bone graft substitutes. Clin Orthop Relat Res 360:71–86

Pradhan BB et al (2006) Graft resorption with the use of bone morphogenetic protein: lessons from anterior lumbar interbody fusion using femoral ring allografts and recombinant human bone morphogenetic protein-2. Spine (Phila Pa 1976) 31(10):E277–E284

Price CT et al (2003) Comparison of bone grafts for posterior spinal fusion in adolescent idiopathic scoliosis. Spine (Phila Pa 1976) 28(8):793–798

Rahn BA (2002) Bone healing: histologic and physiologic concepts. In: Fackelman GE (ed) Bone in clinical orthopedics. Thieme, Stuttgart, pp 287–326

Samartzis D et al (2003) Comparison of allograft to autograft in multilevel anterior cervical discectomy and fusion with rigid plate fixation. Spine J 3(6):451–459

Schizas C et al (2008) Posterolateral lumbar spine fusion using a novel demineralized bone matrix: a controlled pilot study. Arch Orthop Trauma Surg 128:621–625

Sfeir C et al (2005) Fracture repair. In: Leiberman JR, Freidlander GE (eds) Bone regeneration and repair. Humana Press, Totowa, pp 21–44

Shapiro F (1988) Cortical bone repair. The relationship of the lacunar-canalicular system and intercellular gap junctions to the repair process. J Bone Joint Surg Am 70(7):1067–1081

Singh R et al (2016) Radiation sterilization of tissue allografts: a review. World J Radiol 8(4):355–336

Slosar PJ et al (2007) Accelerating lumbar fusions by combining rhBMP-2 with allograft bone: a prospective analysis of interbody fusion rates and clinical outcomes. Spine J 7(3):301–307

Takahashi K, Yamanaka S (2006) Induction of pluripotent stem cells from mouse embryonic and adult fibroblast cultures by defined factors. Cell 126:663–676

Takaso M, Nakazawa T, Imura T, Ueno M, Saito W, Shintani R, Fukushima K, Toyama M, Sukegawa K, Okada T, Fukuda M (2011) Surgical treatment of scoliosis using allograft bone from a regional bone bank. Arch Orthop Trauma Surg 131(2):149–155

Tsiridis E, Upadhyay N, Giannoudis P (2007) Molecular aspects of fracture healing: which are the important molecules? Injury 38(Suppl 1):S11–S25

Tuchman A et al (2017) Autograft versus allograft for cervical spinal fusion: a systematic review. Global Spine J 7(1):59–70

Urist MR, Strates BS (1970) Bone formation in implants of partially and wholly demineralized bone matrix. Including observations on acetone-fixed intra and extracellular proteins. Clin Orthop Relat Res 71:271–278

Urist M et al (1967) The bone induction principle. Clin Orthop Relat Res 53:243–283

Vaidya R et al (2007) Interbody fusion with allograft and rhBMP-2 leads to consistent fusion but early subsidence. J Bone Joint Surg Br 89(3):342–345

Voor MJ et al (1998) Biomechanical evaluation of posterior and anterior lumbar interbody fusion techniques. J Spinal Disord 11(4):328–334

Wang JC et al (2007) A comparison of commercially available demineralized bone matrix for spinal fusion. Eur Spine J 16:1223–1240

White E, Shors EC (1986) Biomaterial aspects of Interpore-200 porous hydroxyapatite. Dent Clin N Am 30:49–67

Yeh KT et al (2017) Fresh frozen cortical strut allograft in two-level anterior cervical corpectomy and fusion. PLoS One 12(8):e0183112. https://doi.org/10.1371/journal.pone.0183112. eCollection 2017

Posterior Approaches to the Thoracolumbar Spine: Open Versus MISS

57

Yingda Li and Andrew Kam

Contents

Introduction	1030
Brief History of Open and Minimally Invasive Spinal Fusion	1031
Regional and Global Trends in Spinal Fusion	1031
Selected Indications and Evidence for Spinal Fusion	1032
Spondylolisthesis	1032
Axial Back Pain	1033
Thoracolumbar Burst Fractures	1033
MISS Fusion	1033
Open Lumbar Fusion	1034
Positioning	1034
Laminectomy	1034
Facetectomy	1035
Interbody	1035
Pedicle Screw Placement	1036
Selected Variations in Open Lumbar Fusion	1038
Posterolateral Fusion	1038
Pedicle Screws Via a Wiltse Approach	1038
Cortical Bone Trajectory Screws	1039
Hybrid Percutaneous Screws with Miniopen Interbody	1040
Minimally Invasive Lumbar Fusion	1040
Fluoroscopy Nuances	1040
Jamshidi Needle Advancement	1041

Y. Li (✉)
Department of Neurosurgery, Westmead Hospital, Sydney, NSW, Australia

Department of Neurological Surgery, University of Miami, Miami, FL, USA

A. Kam (✉)
Department of Neurosurgery, Westmead Hospital, Sydney, NSW, Australia

© Springer Nature Switzerland AG 2021
B. C. Cheng (ed.), *Handbook of Spine Technology*,
https://doi.org/10.1007/978-3-319-44424-6_89

Kirshner Wire Management and Screw Placement 1042
Interbody 1042
Rod Passage 1043

Selected Variations in Minimally Invasive Lumbar Fusion 1043
Tubular Retractors 1043
Cross over the Top Decompression for Bilateral Stenosis 1043
Endoscopy 1044

Thoracic Instrumentation and Selected Variations 1044

Conclusion 1046

Cross-References 1047

References 1047

Abstract

The traditional open approach to the thoracolumbar spine remains one of the most powerful and widely practiced approaches in all of spine surgery. Over the past 2 decades or so, minimally invasive options have gained increasing traction and have been associated with reduced blood loss, paraspinal musculature disruption, infection rates, and length of stay, as well as hospitalization costs, without compromising clinical outcomes or radiographic fusion rates. The minimally invasive approach is not necessarily appropriate for all patients and pathologies, and the two approaches are not mutually exclusive. Currently an array of open and minimally invasive options exist for posterior thoracolumbar fusion, including midline and paramedian approaches, conventional and tubular retractors, posterior and transforaminal interbody as well as posterolateral fusion options, static and expandable cages, and various fixation systems, including pedicle (both open and percutaneous) and cortical bone trajectory screws. More recently, endoscopic spine surgery has garnered growing attention as an ultra minimally invasive alternative and may yet play a significant role in neural decompression and spinal fusion. Furthermore, advances in navigation, robotics, osteobiologics, and perioperative protocols will hopefully translate into increased safety, efficacy, and reproducibility for posterior thoracolumbar fusion procedures.

Keywords

Thoracolumbar fusion · Open · Minimally invasive · Posterior lumbar interbody fusion · Transforaminal lumbar interbody fusion · Tubular retractor · Percutaneous pedicle screw · Cortical bone trajectory screw · Endoscopy

Introduction

The posterior approach to the thoracolumbar spine is one of the most powerful tools in the spine surgeon's armamentarium. This approach is the oldest, and most widely practiced and accepted technique in all spinal surgery (Knoeller and Seifried 2000). It affords the surgeon access to all three columns of the spine through a single stand-alone approach, obviating the need for patient repositioning and staged procedures. It enables direct decompression of the common thecal sac and nerve roots and provides an avenue for fixation and fusion and therefore correction of instability and deformity.

Despite these advantages, the traditional open approach to the thoracolumbar spine is associated with significant iatrogenic disruption of normal surrounding tissue, in particular collateral damage to the paraspinal musculature, leading to devascularization, pain, atrophy, and disability (Fan et al. 2010; Kim et al. 2005). Minimally invasive spine surgery (MISS) has gained much popularity in recent years owing to the reductions in patient morbidity, length of hospital stay, and costs. This has been supported by advances in technology, including access, instrumentation, neuromonitoring, biologics, navigation, and robotics (Yoon and Wang 2019).

In this chapter, we will address the history of lumbar instrumentation and fusion, and the increasing adoption of minimally invasive techniques. We will also address current trends in spinal procedures performed in Australia and outline the common indications for lumbar fusion, including reviewing the most contemporaneous literature on the subject. Rather than exhaustively detailing each step involved in common thoracolumbar fusion operations, we will endeavor to share with our reader specific nuances accumulated through our surgical experience.

Brief History of Open and Minimally Invasive Spinal Fusion

Harrington in the 1950s is credited with the birth of spinal instrumentation (Harrington 1962). He revolutionized the treatment of pediatric scoliosis with his stainless-steel rod construct. While these were effective in correcting coronal deformities, it created a generation of patients with flat back deformities. The next major revolution in instrumentation came in the form of segmental transpedicular screw fixation, and while described a few decades earlier (Knoeller and Seifried 2000), Roy Camille is often credited with their popularization in the 1970s (Roy-Camille et al. 1976). While Hibbs had harvested iliac crest bone graft in the 1910s for posterolateral graft (Hibbs 1911), Cloward in the 1940s was the first to describe its placement in the interbody space (Cloward 1952), now considered the first iteration of the posterior lumbar interbody fusion (PLIF). To mitigate the forceful retraction applied to the traversing nerve root and thecal sac, Harms (Harms and Rolinger 1982) modified this technique in the 1980s to a more lateral approach, now termed transforaminal lumbar interbody fusion (TLIF) involving total facetectomy and entrance through a corridor referred to as Kambin's triangle (Kambin and Zhou 1996), formed by the obliquely oriented exiting nerve as its hypotenuse, the longitudinally oriented traversing nerve root medially, and the transversely oriented disc space and vertebral endplates inferiorly. Following the lead of our general surgical colleagues and their widespread adoption of laparoscopic techniques over traditional open laparotomies, the search for less invasive approaches to the spine had started to gain momentum. Magerl's percutaneous adaptation of the pedicle screw in the 1980s (Magerl 1982) and Foley's introduction of the tubular retractor[7] a decade later are often considered two of the most significant landmarks in MISS. Kambin, in addition to his eponymous anatomical triangle, is also credited with the development of percutaneous and later endoscopic approaches to the intervertebral space (Kambin and Zhou 1996), and thus spinal endoscopy was born.

Regional and Global Trends in Spinal Fusion

The World Health Organization estimated that low back pain (LBP) affects approximately two-thirds of people in industrialized countries at some point in their lives (Duthey 2013). Epidemiological studies have ranked LBP as the second commonest cause of disability in adults (Prevalence and most common causes 2009), and number one in Years Lived with Disability (Hoy et al. 2014). In parallel to the growing disability incurred by spinal pathology, the number of spinal surgeries performed has also increased, particularly fusion procedures. In Australia, where we practice, the number of simple spinal fusion procedures doubled between 2003 and 2013, while complex fusion procedures quadrupled (Machado et al. 2017). Similar trends have been demonstrated in the United States, with the fastest increases seen in the over 65 age group (Martin et al. 2019). Over a similar epoch, MISS has also gained increasing traction. According to a recent global survey of nearly 300 spinal surgeons, most respondents (71%) regarded MISS as mainstream, while the majority (86%) practiced some form of

MISS (Lewandrowski et al. 2020). In parallel with this trend, based on patient surveys, most patients (80%) prefer MIS over open surgery, provided that long-term outcomes and complication risk are comparable (Narain et al. 2018).

Selected Indications and Evidence for Spinal Fusion

Most spine surgeons would support the addition of fixation and fusion in patients with evidence of instability, classically manifesting as spondylolisthesis with abnormal movement on dynamic radiographs, although indirect signs such as sagittally oriented facets, intra-articular effusions, and synovial cysts may sway a surgeon toward fusion out of concern for creating iatrogenic instability following decompression (Blumenthal et al. 2013). Furthermore, the predominance of mechanical LBP in patients with neurogenic claudication or radiculopathy significantly reduces probability of improvement following decompression alone and may provide further impetus to fusion (Pearson et al. 2011). More recently, our growing understanding of spinal deformity and the negative impact of sagittal imbalance and spinopelvic mismatch on outcomes following spine surgery (Glassman et al. 2005; Schwab et al. 2013) has contemporized our understanding of the longitudinal impact of segmental fusion upon regional and global spinal alignment, as well as the potential benefits and pitfalls of long segment fusion, strategic placement of interbody devices and osteotomies, and deformity correction.

Spondylolisthesis

The rate of *fusions* around the world has more than doubled from the start of the twenty-first century and is only continuing to *increase* from *year* to *year* (Makanji et al. 2018). Despite this, evidence from large randomized controlled trials remains either lacking or conflicting. Certainly, the as-treated results from the spondylolisthesis arm of the Spine Patient Outcomes Research Trial (SPORT) supported surgery over conservative management for patients with degenerative spondylolisthesis (Abdu et al. 2018). However, the significant crossover rate mitigated the benefits of randomization, and the heterogeneity in surgical methods prevented any firm conclusions regarding whether fusion afforded additional benefit to decompression alone.

The two recent randomized controlled trials published in the New England Journal of Medicine addressing whether the addition of fusion to decompression in patients with low-grade degenerative spondylolisthesis raised more questions than they answered. The Swedish study (SSSS) randomized more patients (Försth et al. 2016), around 250, but only half had spondylolisthesis, and important patient characteristics such as dynamic instability and relative contributions of mechanical LBP versus leg pain were not addressed. They concluded that fusion was no better than laminectomy alone in all outcome measures and resulted in longer length of stay and higher costs. The North American study (SLIP) compared the addition of fusion to laminectomy alone in approximately 60 patients (Ghogawala et al. 2016). Patients with mechanical LBP and dynamic instability, generally considered relative indications for fusion, were excluded, potentially reducing the applicability of their patient population to real-world practice. Their results suggested a small but statistically significant improvement in the physical component of the 36-item Short Form Health Survey (SF-36). Neither trial was able to explore the nuances in decision-making spine surgeons face every day in this diverse patient population, and both largely used a surgical strategy, instrumented posterolateral fusion with autologous iliac crest bone graft without interbody that some would consider outdated today. Certainly, no minimally invasive techniques were utilized. Some evidence does also exist supporting the use of interbody over posterolateral fusion with respect to fusion and reoperation rates (Liu et al. 2014). Furthermore, interbody graft provides additional potential benefits of anterior column support and load sharing, fusion under compression and over a shorter distance, as well as indirect foraminal decompression and restoration of segmental lordosis.

Axial Back Pain

Fusion specifically for LBP has remained a subject of contention for many years. The reduced efficacy of surgery in patients with back-pain predominant symptomatology (Pearson et al. 2011), coupled with difficulties in localizing a specific pain generator in these patients (Brusko et al. 2019), who often possess significant psychological overlay and covert secondary gain, has made this field one of the most controversial in all of spine surgery. The initially positive Swedish trial (Fritzell et al. 2001) on fusion for intractable LBP was later rebutted by the Norwegian trial (Brox et al. 2003), which showed no benefit for fusion over rehabilitation with a cognitive behavioral component. True structured rehabilitation is, however, a scarce commodity in a lot of countries, including Australia, often with lengthy wait times. The latest American Association of Neurological Surgeons (AANS) and Congress of Neurological Surgeons (CNS) guidelines support at least consideration for fusion surgery in the setting of persistent mechanical LBP once all reasonable conservative alternatives have been exhausted (Eck et al. 2014).

Thoracolumbar Burst Fractures

Trials on surgery versus nonoperative management for thoracolumbar burst fractures in neurologically intact patients have shown similarly conflicting results (Abudou et al. 2013), although contemporary minimally invasive methods have not yet been rigorously studied. Certainly, patients with unstable thoracolumbar fractures without need for direct decompression may serve as an ideal cohort for percutaneous fixation to facilitate pain control, mobilization, and fracture union, with minimal collateral soft tissue disruption (Court and Vincent 2012). The instrumentation can often be removed following fracture union to remobilize the involved segment of the spine and prevent long-term adjacent segment issues (Court

and Vincent 2012). Similarly, percutaneous instrumentation has an established role in providing supplemental fixation in the context of interbody fusion approached via a lateral route (Alvi et al. 2018) and holds promise in the realm of spinal infection (Deininger et al. 2009), with minimization of communication with infected tissue, and preservation of paraspinal musculo-vasculature and viability.

MISS Fusion

With an aging population and associated frailty, coupled with increasing emphasis on healthcare economics, there is growing demand for less invasive surgical options. The benefits of MISS have been clearly demonstrated in other subspecialties, such as laparoscopic abdominal surgery and endovascular neurosurgery. There is now a growing body of evidence that MISS fusion provides similar outcomes and fusion rates as traditional open methods. Our meta-analysis on MISS TLIF versus open TLIF showed less blood loss and lower incidence of infection with at least comparable clinical outcomes with regard to axial pain and disability (Phan et al. 2015a). Other studies have consistently shown shorter length of stay (Goldstein et al. 2014), reduced complications (Khan et al. 2015), less disruption of paraspinal musculature (Fan et al. 2010; Kim et al. 2005), less postoperative narcotic use, and earlier return to work (Adogwa et al. 2011), as well as decreased overall costs (Wang et al. 2012). Concerns around increased fluoroscopic exposure to the surgical team (Khan et al. 2015) have been counteracted by advances in navigation and robotic technology, which have also resulted in improved fixation accuracy (Kosmopoulos and Schizas 2007). The initial steep learning curve has been overcome to some extent by widespread dissemination of techniques, and opportunities to learn and practice at cadaveric workshops. The unique challenges raised by patients at risk for nonunion, including osteoporosis (Benglis et al. 2008), have led to strategies such as augmenting pedicle screws with cement to increase pull-out strength, and bone morphogenetic protein (BMP) to improve fusion (Mccoy et al. 2019). Understanding the dose-dependent properties of BMP, and risks of radiculitis, heterotopic ossification, and osteolysis (Fu et al. 2013), has led to more controlled application of smaller doses in carefully selected patients without malignancy to areas without exposed dura or nerve root, or endplate violation.

The classic tenets of MISS involving small incisions and tubular retractors have shifted toward an overarching paradigm of minimizing collateral tissue disruption to reduce disability, and a greater appreciation for the importance of multidisciplinary teams in enhancing recovery after surgery (ERAS). (Dietz et al. 2019) Patient selection remains key, and while indications for minimally invasive approaches have expanded, there remain pathologies, including but not limited to severe adult spinal deformity, especially if concomitantly rigid, which may be better suited to an open approach (Mummaneni et al. 2019).

Open Lumbar Fusion

There are several variations on the traditional open PLIF technique. We prefer to decompress then instrument to allow us to palpate and visualize the pedicular walls, although the opposite sequence is equally valid. This guides our pedicle screw trajectory both in the craniocaudal as well as medio-lateral planes, thereby minimizing risk of breaching. We also remove most if not the entire facet, comparable to a traditional Ponte osteotomy or Schwab grade 2 osteotomy (Schwab et al. 2014) and affording a similar lateral trajectory as TLIF. Not only does this minimize the amount of nerve root retraction necessary, it also increases the amount of autologous bone available for fusion, and mobilizes the spine to facilitate interbody insertion, foraminal height restoration, spondylolisthesis reduction, and deformity correction. Topical hemostatic agents such as thrombin and gelatin are essential to minimize blood loss, and cell saver technology should be considered if available. Retractors are intermittently released throughout the case to minimize muscle ischemic time. In closing, the muscle is approximated to obliterate dead space, but not so tightly

as to risk ischemia. The fascia is closed tightly, particularly if there has been incidental durotomy. We prefer to do this is in an interrupted fashion so that suture line integrity is not reliant on a single knot at each end. We often place an epidural catheter for narcotic infusion (Klatt et al. 2013) postoperatively in addition to a wound drain. There is also some evidence to suggest that topical vancomycin placed in the wound may reduce the incidence of postoperative infection, particularly following instrumentation (Khan et al. 2014). Loupe magnification with headlight illumination is used to enhance visualization.

Positioning

Following appropriate timeout, intravenous antibiosis, and application of mechanical lower limb antithrombotic devices, the patient is positioned prone on the operating table. Particular emphasis is paid to the position of the arms to avoid undue traction on the brachial plexus, padding of all potential pressure areas, sufficient room for the abdomen so as to not impede venous return, and slight reverse Trendelenburg position and avoidance of any direct pressure on the globes to prevent ischemic optic neuropathy.

Laminectomy

In exposing the spine, it is critical to avoid, if possible, violating the capsules of the facet joints of uninvolved levels, particularly at the upper-instrumented vertebra, to minimize acceleration of adjacent segment disease (ASD). Furthermore, clear delineation of bone and bony edges is paramount and facilitates surgeon orientation, particularly in revision cases where the anatomy may be distorted. Laminectomy is performed with a combination of Leksell bone nibblers, high-speed drill, and Kerrison punches. There is usually a deficiency in the midline where ligamentum flavum attaches to the undersurface of the lamina, where epidural fat is encountered, heralding entrance into the spinal canal. The thinner the bone is egg-shelled, the easier it is to enter the canal with rongeurs. Significant dural adhesions may be encountered, especially in revision cases, which require careful separation with blunt dissectors such as curettes. Not all epidural adhesions or scar tissue require excision, provided the necessary neural elements have been detethered and decompressed.

Facetectomy

Following laminectomy, attention is turned to the facetectomy. The inferior articular process (IAP) is disarticulated by drilling or osteotomizing across the pars interarticularis, allowing it to be removed en bloc and saved as graft. Care must be taken to avoid violating the superior pedicle. The naked articular surface of the superior articular process (SAP) is then exposed. The SAP can be similarly removed en bloc by first palpating the superior border of the inferior pedicle with a blunt dissecting instrument such as the Woodson elevator. This defines the inferior limit of drilling or osteotomy (Fig. 1). The pars artery (Macnab and Dall 1971) is often encountered during these maneuvers and must be secured for hemostasis. In excising both the IAP and SAP en bloc, it is important that bony leverage occurs in the upward direction to avoid neural injury. Alternatively, Kerrison punches can be used to skeletonize the medial and superior borders of the inferior pedicle until sufficient space is created for interbody insertion. Care is taken superiorly and laterally in the foramen to avoid injury to the exiting nerve root. Foraminal ligament can be preserved as a protective barrier over the exiting nerve root if satisfactory direct and indirect decompression has otherwise been achieved.

Interbody

The epidural veins are cauterized with the bipolar tips parallel to the traversing nerve root to avoid inadvertent thermal injury, and divided to avoid neural traction. In cases where the disc is severely collapsed, it may be difficult to gain entrance into the disc space with traditional interbody instruments. It may be effective in these situations to enter the space with a smaller blunt tipped instrument, such as a pedicle probe, under lateral fluoroscopic guidance. Gradual distraction can then be achieved by sequentially upsizing spacers placed contralateral to the side that discectomy and endplate preparation is occurring if bilateral interbody devices are planned. Alternatively, laminar spreaders or ones anchored to pedicle screw heads can be used. Aggressive distraction must be avoided in the latter instance to avoid pedicular fracture, particularly in patients with osteoporosis.

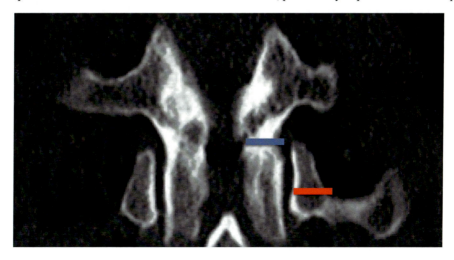

Fig. 1 Coronal lumbar spine computed tomography (CT) demonstrating the relationship of the IAP and SAP. The osteotomies performed are indicated by the blue (IAP) and red (SAP) lines, taking care not to violate the cranial and caudal pedicles

Similarly, care must be taken to avoid violating the bony endplate with forceful use of oversized shavers. The final implant is then inserted and impacted as ventrally as possible to take advantage of the strong apophyseal ring as well as maximize segmental lordosis. However, care must be taken to avoid breaching the anterior longitudinal ligament, as ventrally displaced cages are notoriously difficult to retrieve (Murase et al. 2017). Autogenous bone, supplemental allograft and BMP, if necessary, is packed into the disc space to enhance fusion (ventral to the implant in the case of BMP to prevent predural seroma and radiculitis), as implants themselves often contain very little space to accommodate graft. Traditionally, polyetheretherketone (PEEK) cages have been used, although titanium technologies are gaining popularity due to their osteo-integrative potential (Rao et al. 2014), at the cost of possibly increased risk of subsidence due to higher modulus of elasticity (Seaman et al. 2017), radio-opacity, and difficulties visualizing fusion mass. Insert and rotate devices (Sears 2005), as well as expandable cages (Fig. 2), offer further options in disc height and segmental lordosis restoration (Boktor et al. 2018). Autologous iliac crest bone graft, the gold standard to which other interbody devices and biologics have historically been compared, has been used with decreasing frequency due to the morbidity associated with its procurement (Banwart et al. 1995).

Pedicle Screw Placement

There are several methods for placing pedicle screws, including free hand and fluoroscopic techniques. Advances in navigation and robotics have improved placement accuracy (Kosmopoulos and Schizas 2007). We do not routinely use neuromonitoring due to its expense, lack of availability at our institution, and lack of substantive evidence demonstrating efficacy in preventing neurological harm outside of deformity, lateral transpsoas, and intramedullary tumor surgery (Fehlings et al. 2010). The safety of the freehand method is enhanced by intimate understanding of anatomy, visualization and palpation of the pedicular walls, tactile feedback, and subtle adjustments made based on detailed study of

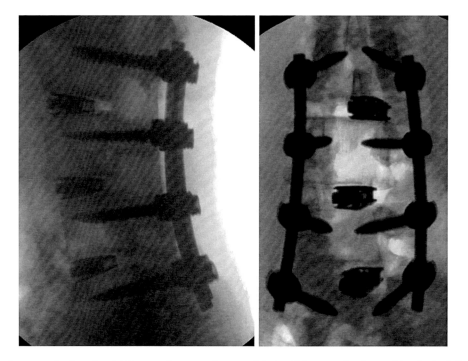

Fig. 2 Intraoperative lateral and AP x-rays demonstrating open L2–5 pedicle screw fixation and interbody fusion with expandable cages to restore foraminal height as well as segmental lordosis

preoperative imaging. Aiming perpendicularly toward the floor in the craniocaudal plane at L4 and adding approximately 5° of medialization per level to a baseline of 10° at L1 serve as useful additional guides.

The entry point is at the junction between the SAP and the bisected transverse process, where the mammillary process may be visualized (Fig. 3). To identify the entry point, it is often necessary to remove the lateral overhang of hypertrophic facets. This also serves to create sufficient room to house the head of the screw. Entry points can be customized to facilitate easier rod passage, particularly if multiple levels are instrumented. Furthermore, the trajectory of open pedicle screws is usually less medialized than their percutaneous counterparts due to the significantly increased amount of tissue dissection necessary in order to achieve a sufficiently lateral starting point, and the hindrance of both paraspinal musculature and retractors to medialization.

In probing the pedicle, tactile feedback is provided by the crunchiness of cancellous bone (in contrast to the hardness of cortical bone), and visual feedback by the marrow blush of the cancellous bone. It is critical that the screw goes down the same tapped hole, and can be aided by marking the trajectory on the skin edge, and to avoid forcefully tightening the screw against the facet, losing its poly-axiality and potentially stripping the screw. Screw symmetry can be achieved by leaving the handle on the contralateral screw as a guide or using fluoroscopic control. Pull-out strength is improved by using the longest screw possible with the widest diameter and augmenting with cement in osteoporotic patients. Given the largely cancellous nature of S1, it may be desirable to achieve bicortical purchase through the sacral promontory (the most corticated part of the vertebra) at this level. Compression and reduction are achieved against a final tightened screw if necessary, aided by extension tabs on the screw head, lordotically contoured rods, and cantilever maneuvers, although a significant degree of reduction is often already accomplished through the interbody work.

One must also be adept at managing breaches of the pedicular wall. While medial and inferior breaches have classically been associated with injury to the traversing and exiting nerve roots, respectively, lateral breaches can be equally undesirable, with potential injury to the adjacent intrapsoas lumbar plexus, as well as lumbosacral trunk at the caudalmost levels. While existing pilot holes can sometimes be rescued by redirecting the pedicle probe, including using ones with curved tips, it is often easier to fashion

Fig. 3 Axial CT demonstrating the typical latero-medial trajectory of a lumbar pedicle screw (asterisk represents the mammillary process, an ideal entry point)

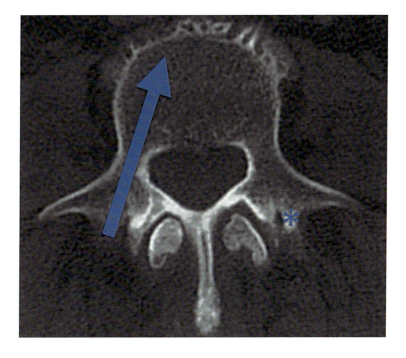

new entry points in order to avoid existing tracts. Careful examination of preoperative imaging can aide in preventing pedicular breach, including accounting for rotational deformities, as well as accounting for narrow, dysmorphic, or sclerotic pedicles, particularly on the concavity of a scoliotic curve in the latter.

Selected Variations in Open Lumbar Fusion

Posterolateral Fusion

We place interbody grafts routinely due to the aforementioned benefits. However, there may be clinical scenarios such as significant disc space collapse, weakened osteoporotic endplates, or minimal neuro-foraminal stenosis, in which interbody fusion may be difficult, inappropriate, or unnecessary. In these cases, posterolateral fusion serves as a reasonable alternative. Equally, posterolateral fusion may serve as a useful adjunct to interbody fusion in patients at risk for nonunion and in revision cases for pseudoarthrosis. It is critical that meticulous decortication of the transverse processes down to bleeding cancellous bone is performed to create an ideal fusion environment, a process that is often neglected. The remaining facet joint may also be decorticated. A cottonoid may be temporarily placed over the thecal sac as a barrier against bone graft inadvertently placed epidurally, preventing iatrogenic stenosis.

Pedicle Screws Via a Wiltse Approach

One of the criticisms of open pedicle screws is the difficulty in achieving the desired medialization due to hindrance by paraspinal muscles and retractors. Idealized exposures often require extensive lateral dissection and lengthy incisions. To mitigate this, bilateral incisions can be made in the lumbodorsal fascia through a single midline skin incision. Dissection is then carried down between the multifidus and longissimus muscles, often through a natural avascular cleavage plane, landing directly onto the junction between the facet joint and transverse process (Wiltse et al. 1968). This plane between the two muscles is measurable from the midline on preoperative imaging (Fig. 4), and often palpable and visible

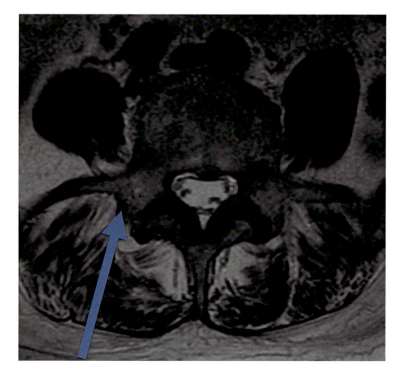

Fig. 4 Axial T2-weighted magnetic resonance imaging (MRI) illustrating the Wiltse paraspinal plane between medial multifidus and lateral longissimus, with a muscle-sparing approach (arrow) landing directly onto the facet-transverse process junction

intraoperatively. Pedicle screw insertion then proceeds in the aforementioned fashion. However, the extensive suprafascial undermining required creates significant dead space, which must be obliterated to prevent postoperative seroma and potential infection.

Cortical Bone Trajectory Screws

Some of the other criticisms of the open approach to pedicle screw placement are the amount of lateral muscular dissection required and the propensity to violate the facet capsule at the upper-instrumented level, thus potentially accelerating adjacent segment degeneration (Sakaura et al. 2019). Furthermore, pedicle screws reside mostly in cancellous bone, which is significantly weaker than cortical bone, an issue accentuated in osteoporotic patients. Within the last decade, a medial to lateral and inferior to superior screw trajectory has been proposed to address these issues, including maximizing purchase into cortical bone (Santoni et al. 2009). The entry point is in the pars, and the upward and outward trajectory is analogous to lateral mass screws in the cervical spine (Fig. 5). The poorer definition on fluoroscopy of the pars on fluoroscopy and the lack of tactile feedback due to the cortical nature of the traversed bone can be mitigated by use of intraoperative navigation. The spinous process navigation clamp, if used, should be placed at the cranial end of the exposure (rather than caudal end in navigated pedicle screws) to ensure that it remains between the surgeon and navigation camera, as well as maximizing the amount of

Fig. 5 Lateral and AP radiographs contrasting the latero-medial trajectory of traditional pedicle screws (blue arrows) versus the infero-superior and medio-lateral trajectories of CBT screws (red arrows)

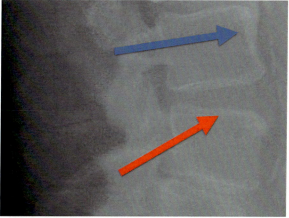
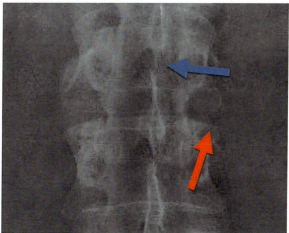

working space given the caudo-cranial trajectory of these screws. The diameter and length of cortical bone trajectory (CBT) screws are typically narrower and shorter. While laboratory studies have demonstrated comparable biomechanical strength and some evidence exists to support similar short-term clinical and radiographic outcomes compared to traditional pedicle screws, long-term follow-up data remains pending (Phan et al. 2015b). The CBT screw certainly represents a less invasive open alternative to traditional pedicle screws, with a potential specific role in osteoporotic patients, although its efficacy in multilevel constructs, high-grade spondylolistheses, and deformity remains unknown.

Hybrid Percutaneous Screws with Miniopen Interbody

A minimally invasive variation on the traditional open PLIF combines percutaneous pedicle screw fixation, described later, with a miniopen midline incision for laminectomy and interbody work (Mobbs et al. 2012). This reduces the amount of lateral muscular dissection required and shortens the midline incision. In these hybrid cases, we prefer transversely oriented stab incisions for pedicle screw placement to longitudinal ones to minimize devascularization of overlying skin and soft tissues. A further variation involves paramedian stab incisions in the fascia through a single midline incision to avoid multiple unsightly skin incisions. The same percutaneous instrumentation can then be used through the fascial incisions. However, this often necessitates a longer incision, such as a traditional open approach, as well as extensive undermining of the skin alluded to previously.

Minimally Invasive Lumbar Fusion

An MISS TLIF is the archetypal MISS fusion procedure. It is often synonymous with tubular retractors and percutaneous pedicle screws, (Foley et al. 2003) although several variations

exist. We prefer the miniopen paramedian Wiltse approach on the side of interbody, dissecting between the multifidus and longissimus muscles as this represents a natural cleavage plane, landing the surgeon directly onto the junction between the SAP and transverse process. Critics of the unilateral transforaminal approach cite poor disc clearance and endplate preparation for fusion, comparative biomechanical weakness in lateral bending compared to bilateral PLIF constructs (Sim et al. 2010), and inability to induce significant segmental lordosis (Carlson et al. 2019) as justification against minimally invasive TLIF. However, in cases of immobile facets, or where significant segmental lordosis induction (Jagannathan et al. 2009) or spondylolisthesis reduction is desirable, we often perform bilateral facetectomies for complete segmental mobilization through short bilateral paramedian incisions and muscle splitting Wiltse approaches. Expandable cages can further facilitate induction of segmental lordosis without compromising disc and foraminal height. Percutaneous pedicle screws are inserted through the Wiltse incision on the side of the interbody and small contralateral stab incisions. The MISS transforaminal approach also naturally lends itself to revision cases where florid epidural scar makes reapproaching through the midline technically challenging and potentially hazardous, with heightened risks of durotomy and cerebrospinal fluid leak.

Fluoroscopy Nuances

Once the patient is positioned, prepped, and draped, the C-arm is positioned in the anteroposterior (AP) plane. Kirschner wires are used to identify the desired level, as well as mark out the lateral border of the pedicle in the vertical plane and the bisected pedicle in the transverse plane. It is imperative that a true AP image of the desired vertebra is obtained, with a clearly defined superior endplate without any elliptical shadow, and midline spinous processes (Fig. 6). The C-arm should be locked in this position and any adjustments from the orthogonal plane recorded to ensure ease of return to the same desired position.

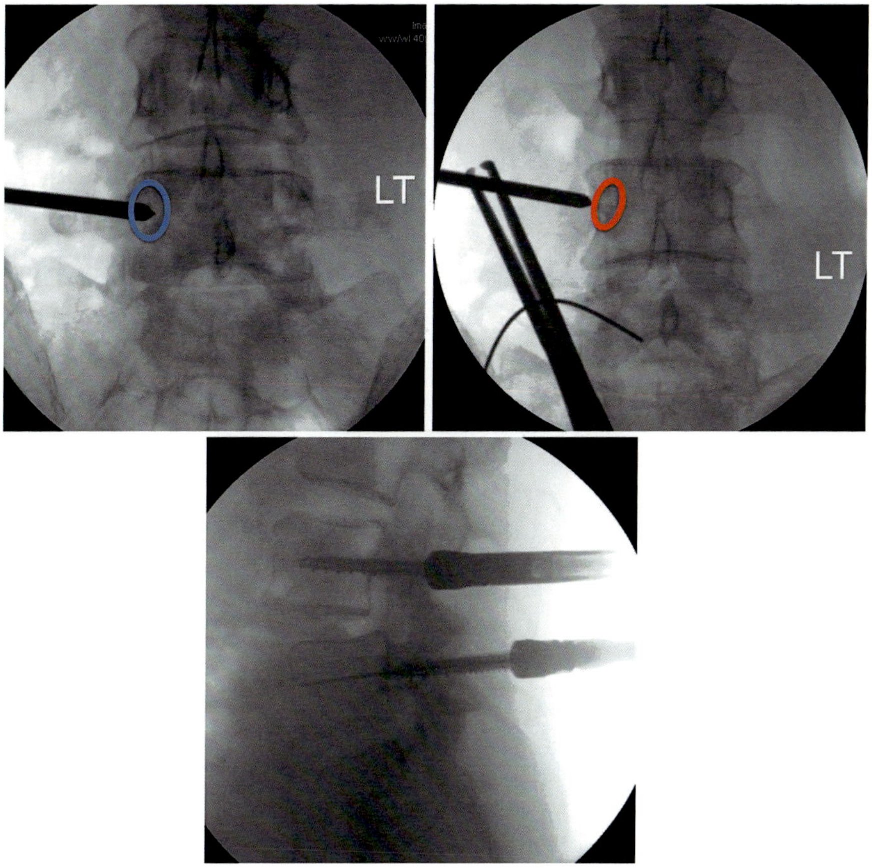

Fig. 6 True AP fluoroscopy with crisp L5 superior endplate (top image) and midline spinous process, demonstrating passage of Jamshidi needle and K-wire through the right L5 pedicle (blue circle, top image), followed by L4 (red circle, middle image), starting at 9 o'clock. At an approximate depth of 2 cm (usually heralding the junction between pedicle and vertebral body), the tip of the needle should not transgress the medial border of the pedicle. Screws are subsequently placed under lateral fluoroscopy (bottom images)

The importance of having a skilled radiographer experienced in the percutaneous workflow cannot be overemphasized. Draping of the C-arm and absolute attention to sterility are also of paramount importance, as any adjustments to the C-arm, particularly switching between AP and lateral views, can desterilize the drape and endanger the operative field.

Jamshidi Needle Advancement

Stab incisions are made approximately 1–2 cm lateral to the outer border of the pedicle, depending on the body habitus of the patient and the depth of intervening soft tissue. The bull's eye technique is used for pedicle cannulation (Fig. 6). The Jamshidi needle is docked at the junction of the SAP and the transverse process. It is often useful to walk the tip of the needle along the superior and inferior borders of the transverse process and the lateral wall of the facet joint for secondary anatomical confirmation. Close examination of preoperative imaging is crucial, as a severely hypertrophied facet joint can significantly alter the desired entry point as well as increase the depth the Jamshidi needle needs to be advanced in order to traverse the pedicle, traditionally considered to be 2 cm in patients without distorted anatomy. Failure to account for this can lead to complications, including medial pedicular breach, and injury to the traversing nerve root and common thecal sac. The craniocaudal trajectory of the Jamshidi needle should match the degree of tilt or Ferguson on the C-arm.

While fluoroscopic control is critical, a degree of both tactile and aural feedback, similar to traditional open pedicle screw probing, remains possible and serves as secondary confirmation. Tactilely, advancement through crunchy cancellous bone should be relatively unhindered. Resistance often heralds proximity to cortical bone and forewarns against imminent pedicular breach. Similarly, the sound the Jamshidi needle makes against cortical bone when using the mallet is usually lower in frequency and duller in quality. There is often a small amount of toggle within the cancellous part of the pedicle to allow subtle redirections of the Jamshidi needle. Excessive force should, however, be avoided as the needle may bend, making passage of the Kirschner wire and subsequent needle removal from the vertebra difficult. It is critical to be constantly cognizant and wary of the length of the needle that has been advanced. Sclerotic pedicles pose a specific challenge to Jamshidi needle advancement and may necessitate gentle coring out of the pedicle with a high-speed drill to facilitate Kirschner wire passage, both carefully performed under fluoroscopic control but nonetheless often still achievable percutaneously.

Kirshner Wire Management and Screw Placement

The Kirschner wire can often be manually advanced up to 1 cm further into the cancellous bone through the Jamshidi needle without need for the mallet. Tip position within the vertebral body is confirmed if a bottom is palpable, analogous to using the ball tip feeler in open cases. At all stages, including Jamshidi needle removal, tissue dilation, tapping, and screw insertion, care must be taken to avoid inadvertent loss of wire position, including pullout or advancement, and undue twisting and bending. This is achieved both manually with judicious control with the noninstrumenting hand as well as constant attention to fluoroscopy.

Pedicle screws are inserted down the Kirschner wire through the tapped hole under lateral fluoroscopy (Fig. 6). When advancing the pedicle screw, resistance is met once the screw head meets the facet capsule. Further forceful advancement may cause stripping of the screw and loss of poly-axiality of its head. Systems incorporating a sharp-tipped stylet into a self-tapping screw now exist and further streamline the percutaneous workflow (Huang et al. 2020), although possibly at the cost of reduced tactile feedback. Navigation and robotic technologies that marry percutaneous pedicle screw systems also exist, reducing radiation exposure for the surgeon and other operating room staff, while maintaining high rates of placement accuracy (Kochanski et al. 2019).

Interbody

Once the contralateral pedicle screws and ipsilateral Kirschner wires have been placed, the ipsilateral skin and fascial incision are connected, and the facet landed upon by dissecting down through the natural cleavage plane between multifidus and longissimus. This is often accomplishable by spreading the tips of the bipolar forceps, coagulation and division of any small bridging fibers, and gradual retractor advancement. Blunt finger dissection is also often effective. Upon landing on the facet joint, we use a bladed retractor system such as the McCullough, with the short blade medial and long lateral, to maintain exposure. Kirschner wires can often be engaged into the teeth of the retractor blades and kept out of instruments' way. Further medial dissection with electrocautery is carried out, partially exposing the lamina. The steps that follow are like the interbody portion of the open approach detailed earlier, performed either under loupe magnification and headlight illumination, or microscopic visualization. A laminotomy is performed, followed by facetectomy, discectomy, and endplate preparation. Disc removal and fusion bed preparation can be optimized by gradual medialization of interbody instruments and deployment of forward angled rongeurs. Distraction on the contralateral screws can be performed if necessary, to facilitate entrance into the disc space and maintenance of working corridor. Several interbody options exist, including banana-shaped devices, initially inserted vertically then gradually horizontalized to optimize ventral and medial positioning, maximizing cortical apophyseal ring contact, and potentially inducing lordosis, as well as bulleted and the expandable technologies previously described.

Rod Passage

After interbody and once the pedicle screws have been inserted bilaterally, attention is turned to rod placement. The incision through which the rod is placed may need to be extended to facilitate passage to avoid excessive skin tension. The tip of the rod is inserted initially vertically to engage the screw head and then advanced through each successive tower. This not only ensures subfascial placement, but also minimizes the amount of paraspinal muscle captured, preventing possible compartment syndrome. The rod is maneuvered with subtle movements to engage each tower, including medially or laterally rotating the rod holder. Screw engagement is confirmed if the overlying tower no longer rotates, by dropping a specialized measuring tool down the tower, by direct visualization, or by fluoroscopy. Placing the set screw into the tulip closest to the rod holder first brings the rod beyond the screw head, ensuring sufficient rod proximally. Reduction can be achieved through a variety of means, including rod contouring, extension, and cantilever maneuvers, as well as specialized reduction tools.

Selected Variations in Minimally Invasive Lumbar Fusion

Tubular Retractors

Traditionally, MIS lumbar fusions have been associated with the tubular retractor (Foley and Smith 1997). This requires gradual dilation through the paraspinal musculature and docking of the final tube on the facet joint prior to securement onto a table-mounted arm. Despite gradual dilation, a small amount of muscle is invariably encountered at the depth of the retractor, which then requires excision for exposure. If the tubular system is used, we advise against the use of the initial Kirschner wire due to the risk of inadvertent dural puncture and neural injury. The retractor should be docked onto the facet joint with sufficient exposure of the adjacent lamina, and ideally orthogonal to both the desired disc space as well as the floor to optimize disc access and surgical ergonomics. Given the narrow working corridor, specialized angled and bayonetted instruments are necessary. Similarly, the protected portion of the conventionally straight monopolar tip can be manually bent to facilitate use. Various other retractor systems, including bladed and screw-based assemblies, are also available.

Cross over the Top Decompression for Bilateral Stenosis

If decompression of the contralateral subarticular zone is desired, the retractor can be wanded medially (Fig. 7) or the bed rotated to facilitate over-the-top decompression. In this method, also known as unilateral laminotomy for bilateral decompression (ULBD) or ipsilateral-contralateral approach, the ligamentum is left intact while the base of the spinous process and under surface of the contralateral lamina are drilled to protect the underlying dura. Flavum and contralateral medial facet can subsequently be removed till the hump of the thecal sac drops away and the contralateral traversing nerve root visualized. The dura is at greatest risk of injury when rongeuring medially due to the upward slope of the thecal sac, though the risk of overt cerebrospinal fluid leak is low as the paraspinal muscles remain largely intact and reapproximate following retractor removal, obliterating any dead space. Use of upward angled Kerrison punches can also be useful in this approach to achieve contralateral decompression. The results of this approach are comparable to the traditional midline laminectomy, while largely

Fig. 7 Axial MRI simulating wanding (red arrow) of the tubular retractor (blue cylinders) to facilitate decompression of the contralateral lateral recess from a unilateral approach

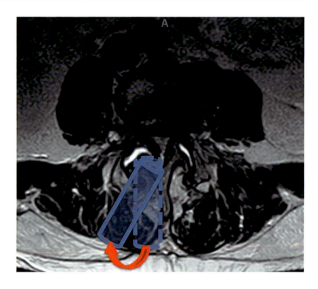

preserving the posterior tension band (Mobbs et al. 2014).

Endoscopy

More recently, endoscopic techniques have been applied to minimally invasive TLIFs, permitting even smaller incisions and less tissue destruction. This has been combined with awake anesthetic techniques, application of long-acting liposomal local anesthetic agents, expandable technologies, biologic materials, and ERAS protocols to treat a range of lumbar spondylotic conditions (Kolcun et al. 2019). The intervertebral disc is accessed via percutaneous transforaminal route through Kambin's triangle using a spinal needle, followed by nitinol wire insertion and sequential dilation and docking of an endoscopic channel, all under constant fluoroscopic control (Fig. 8). Discectomy and endplate preparation are accomplished using specialized endoscopic rongeurs and curettes, and percutaneous reamers, shavers, and stainless-steel brushes, followed by sizing and insertion of an expandable interbody device. The procedure is completed by standard insertion of percutaneous pedicle screws. While long-term and comparative data are eagerly awaited, this technique, representing the least anatomically and physiologically disruptive of all MIS fusion methods, holds promise for elderly and infirm patients who may not otherwise tolerate lengthy prone general anaesthetics (Kolcun et al. 2019).

Thoracic Instrumentation and Selected Variations

A comprehensive description of the multitude of approaches to the thoracic spine is beyond the scope of this chapter. We will, however, endeavor to describe the various options for posterior thoracic instrumentation, both open and minimally invasive. MIS thoracic instrumentation naturally lends itself to scenarios in which direct decompression or fusion is unnecessary, such as burst fractures in patients without neurological compromise, while traditional open methods remain valid, particularly if concomitant direct decompression, fusion, or anterior column reconstruction is required, such as in oncologic pathologies.

The traditional entry point for thoracic pedicle screws is immediately inferior to the intersection between the superior border of the transverse process and the lateral border of the SAP, classically

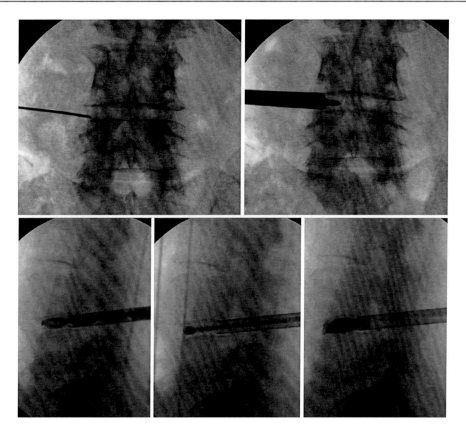

Fig. 8 Intraoperative fluoroscopy demonstrating transforaminal entrance into the L4–5 intervertebral space via Kambin's triangle using a spinal needle (**a**), followed by sequential dilation (**b**), introduction of percutaneous reamer and stainless steel brush (**c** and **d**), and measurement of extent of discectomy and sizing of interbody graft size by inflation of a balloon with radio-opaque contrast (**e**)

described as the junction between the medial two-thirds and lateral one-third of the base of the SAP (Fig. 9) (Chung et al. 2008). The ideal entry point moves slightly laterally and inferiorly as one progresses toward the cranial and caudal ends of the thoracic spine (Kim et al. 2004). Adjustments to the entry point in the axial plane can also be made based on the patient's unique anatomy on preoperative CT. Furthermore, bleeding cancellous pedicular bone can often be exposed by removing the tip of the transverse process, particularly at T12 (Fig. 10). Medialization increases at the superior-most segments of the thoracic spine, while both straight forward and the more caudally directed anatomical trajectories (Fig. 9) in the sagittal plane are acceptable (Puvanesarajah et al. 2014). The in-out-in technique (Fig. 9) with a more lateral entry point along the superior edge of the transverse process to minimize risk of medial breach has also been advocated and may be especially useful in patients with narrow pedicles, particularly in the mid-thoracic spine, enabling the insertion of wider and longer screws with tri-cortical purchase (Jeswani et al. 2014).

Percutaneous thoracic pedicle screw insertion follows the same principles described previously for the lumbar spine. We adopt a strategy of erring on the side of less medialization of the Jamshidi until the pedicle-vertebral body junction is reached to minimize risk of medial breach, followed by subtle toggling of the needle to

Fig. 9 CT comparing the straight forward (blue arrow) trajectory, parallel to the endplates, with anatomical (red arrow), parallel with the superior and inferior pedicular borders, in the sagittal plane (top image), and traditional intrapedicular (blue arrow) and in-out-in (red arrow) techniques in the axial plane (bottom image)

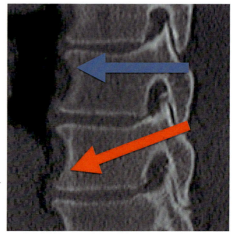

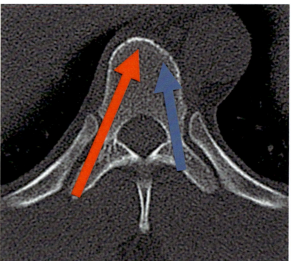

achieve more medialization to ensure that it remains within the vertebral body. Navigation and robotics may also improve accuracy of thoracic pedicle screw insertion, both open and percutaneous, especially in cases with narrow pedicles or significant deformity (Kochanski et al. 2019).

Conclusion

In summary, the posterior approach to the thoracolumbar spine is a versatile workhorse for the spine surgeon, affording access to all three columns of the spine and enabling the trinity of decompression, instrumentation, and interbody through a single approach. Both open and minimally invasive approaches present valid options, and the modern spine surgeon should be adept at both in order to cater to the needs of different patient populations with contrasting pathologies. Advances in navigation and robotics, biologics, access, instrumentation, and expandable technologies have improved the safety and efficacy of minimally invasive thoracolumbar fusions. These advances, coupled with progress in perioperative protocols and multidisciplinary care, will continue to deliver improvements in posterior thoracolumbar surgery.

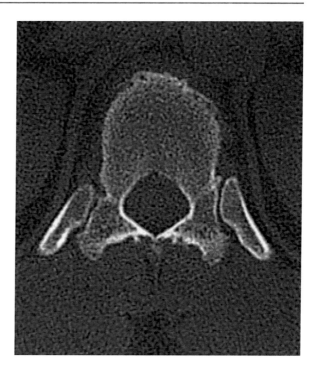

Fig. 10 Removing the tip of the T12 transverse process (asterisk) can aide in the exposure of the underlying cancellous pedicular bone

Cross-References

▶ Lumbar Interbody Fusion Devices and Approaches: When to Use What
▶ Minimally Invasive Spine Surgery
▶ Pedicle Screw Fixation

References

Abdu WA, Sacks OA, Tosteson ANA, Zhao W, Tosteson TD, Morgan TS et al (2018) Long-term results of surgery compared with nonoperative treatment for lumbar degenerative spondylolisthesis in the spine patient outcomes research trial (SPORT). Spine (Phila Pa 1976) 43:1619–1630

Abudou M, Chen X, Kong X, Wu T (2013) Surgical versus non-surgical treatment for thoracolumbar burst fractures without neurological deficit. Cochrane Database Syst Rev 6:CD005079

Adogwa O, Parker SL, Bydon A, Cheng J, Mcgirt MJ (2011) Comparative effectiveness of minimally invasive versus open transforaminal lumbar interbody fusion: 2-year assessment of narcotic use, return to work, disability, and quality of life. J Spinal Disord Tech 24(8):479–484

Alvi MA, Alkhataybeh R, Wahood W et al (2018) The impact of adding posterior instrumentation to transpsoas lateral fusion: a systematic review and meta-analysis. J Neurosurg Spine 30(2):211–221

Banwart JC, Asher MA, Hassanein RS (1995) Iliac crest bone graft harvest donor site morbidity. A statistical evaluation. Spine 20(9):1055–1060

Benglis D, Wang MY, Levi AD (2008) A comprehensive review of the safety profile of bone morphogenetic protein in spine surgery. Neurosurgery 62(5 Suppl 2): ONS423–ONS431

Blumenthal C, Curran J, Benzel EC et al (2013) Radiographic predictors of delayed instability following decompression without fusion for degenerative grade I lumbar spondylolisthesis. J Neurosurg Spine 18 (4):340–346

Boktor JG, Pockett RD, Verghese N (2018) The expandable transforaminal lumbar interbody fusion – two years follow-up. J Craniovertebr Junction Spine 9 (1):50–55

Brox JI, Sørensen R, Friis A, Nygaard Ø, Indahl A, Keller A et al (2003) Randomized clinical trial of lumbar instrumented fusion and cognitive intervention and exercises in patients with chronic low back pain and disc degeneration. Spine (Phila Pa 1976) 28:1913–1921

Brusko GD, Perez-roman RJ, Tapamo H, Burks SS, Serafini AN, Wang MY (2019) Preoperative SPECT imaging as a tool for surgical planning in patients with axial neck and back pain. Neurosurg Focus 47(6):E19

Carlson BB, Saville P, Dowdell J et al (2019) Restoration of lumbar lordosis after minimally invasive transforaminal lumbar interbody fusion: a systematic review. Spine J 19(5):951–958

Chung KJ, Suh SW, Desai S, Song HR (2008) Ideal entry point for the thoracic pedicle screw during the free hand technique. Int Orthop 32(5):657–662

Cloward RB (1952) The treatment of ruptured lumbar intervertebral disc by vertebral body fusion. III. Method of use of banked bone. Ann Surg 136:987–992

Court C, Vincent C (2012) Percutaneous fixation of thoracolumbar fractures: current concepts. Orthop Traumatol Surg Res 98(8):900–909

Deininger MH, Unfried MI, Vougioukas VI, Hubbe U (2009) Minimally invasive dorsal percutaneous spondylodesis for the treatment of adult pyogenic spondylodiscitis. Acta Neurochir 151(11):1451–1457

Dietz N, Sharma M, Adams S et al (2019) Enhanced recovery after surgery (ERAS) for spine surgery: a systematic review. World Neurosurg 130:415–426

Duthey B (2013) Priority Medicines for Europe and the World "A Public Health Approach to Innovation", Background Paper 6.24, Low back pain. WHO, Geneva

Eck JC, Sharan A, Ghogawala Z et al (2014) Guideline update for the performance of fusion procedures for degenerative disease of the lumbar spine. Part 7: lumbar fusion for intractable low-back pain without stenosis or spondylolisthesis. J Neurosurg Spine 21(1):42–47

Fan SW, Hu ZJ, Fang XQ, Zhao FD, Huang Y, Yu HJ (2010) Comparison of paraspinal muscle injury in one-level lumbar posterior inter-body fusion: modified minimally invasive and traditional open approaches. Orthop Surg 2(3):194–200

Fehlings MG, Brodke DS, Norvell DC, Dettori JR (2010) The evidence for intraoperative neurophysiological monitoring in spine surgery: does it make a difference? Spine 35(9 Suppl):S37–S46

Foley KT, Smith MM (1997) Microendoscopic discectomy. Tech Neurosurg 3:301–307

Foley KT, Holly LT, Schwender JD (2003) Minimally invasive lumbar fusion. Spine 28(15 Suppl):S26–S35

Försth P, Ólafsson G, Carlsson T, Frost A, Borgström F, Fritzell P et al (2016) A randomized, controlled trial of fusion surgery for lumbar spinal stenosis. N Engl J Med 374:1413–1423

Fritzell P, Hägg O, Wessberg P, Nordwall A (2001) Lumbar fusion versus nonsurgical treatment for chronic low back pain: a multicenter randomized controlled trial from the Swedish lumbar spine study group. Spine (Phila Pa 1976) 26:2521–2534

Fu R, Selph S, Mcdonagh M et al (2013) Effectiveness and harms of recombinant human bone morphogenetic protein-2 in spine fusion: a systematic review and meta-analysis. Ann Intern Med 158(12):890–902

Ghogawala Z, Dziura J, Butler WE, Dai F, Terrin N, Magge SN et al (2016) Laminectomy plus fusion versus laminectomy alone for lumbar spondylolisthesis. N Engl J Med 374:1424–1434

Glassman SD, Bridwell K, Dimar JR, Horton W, Berven S, Schwab F (2005) The impact of positive sagittal balance in adult spinal deformity. Spine 30(18):2024–2029

Goldstein CL, Macwan K, Sundararajan K, Rampersaud YR (2014) Comparative outcomes of minimally invasive surgery for posterior lumbar fusion: a systematic review. Clin Orthop Relat Res 472(6):1727–1737

Harms J, Rolinger H (1982) A one-stager procedure in operative treatment of spondylolistheses: dorsal traction-reposition and anterior fusion. Z Orthop Ihre Grenzgeb 120:343–347

Harrington PR (1962) Treatment of scoliosis. Correction and internal fixation by spine instrumentation. J Bone Joint Surg Am 44:591–610

Hibbs RA (1911) An operation for progressive spinal deformities. N Y Med 121:1013

Hoy D, March L, Brooks P et al (2014) The global burden of low back pain: estimates from the global burden of disease 2010 study. Ann Rheum Dis 73(6):968–974

Huang M, Brusko GD, Borowsky PA, Kolcun JPG, Heger JA, Epstein RH, Grossman J, Wang MY (2020) The University of Miami spine surgery ERAS protocol: a review of our journey. J Spine Surg 6(Suppl 1):S29–S34. https://doi.org/10.21037/jss.2019.11.10

Jagannathan J, Sansur CA, Oskouian RJ, Fu KM, Shaffrey CI (2009) Radiographic restoration of lumbar alignment after transforaminal lumbar interbody fusion. Neurosurgery 64(5):955–963

Jeswani S, Drazin D, Hsieh JC et al (2014) Instrumenting the small thoracic pedicle: the role of intraoperative computed tomography image-guided surgery. Neurosurg Focus 36(3):E6

Kambin P, Zhou L (1996) History and current status of percutaneous arthroscopic disc surgery. Spine (Phila Pa 1976) 21(Suppl 24):S57–S61

Khan NR, Thompson CJ, Decuypere M et al (2014) A meta-analysis of spinal surgical site infection and vancomycin powder. J Neurosurg Spine 21(6):974–983

Khan NR, Clark AJ, Lee SL, Venable GT, Rossi NB, Foley KT (2015) Surgical outcomes for minimally invasive vs open Transforaminal lumbar interbody fusion: an updated systematic review and meta-analysis. Neurosurgery 77(6):847–874

Kim YJ, Lenke LG, Bridwell KH, Cho YS, Riew KD (2004) Free hand pedicle screw placement in the thoracic spine: is it safe? Spine 29(3):333–342

Kim DY, Lee SH, Chung SK, Lee HY (2005) Comparison of multifidus muscle atrophy and trunk extension muscle strength: percutaneous versus open pedicle screw fixation. Spine 30(1):123–129

Klatt JW, Mickelson J, Hung M, Durcan S, Miller C, Smith JT (2013) A randomized prospective evaluation of 3 techniques of postoperative pain management after posterior spinal instrumentation and fusion. Spine 38(19):1626–1631

Knoeller SM, Seifried C (2000) Historical perspective: history of spinal surgery. Spine 25(21):2838–2843

Kochanski RB, Lombardi JM, Laratta JL, Lehman RA, O'toole JE (2019) Image-guided navigation and robotics in spine surgery. Neurosurgery 84(6):1179–1189

Kolcun JPG, Brusko GD, Basil GW, Epstein R, Wang MY (2019) Endoscopic transforaminal lumbar interbody fusion without general anesthesia: operative and clinical outcomes in 100 consecutive patients with a minimum 1-year follow-up. Neurosurg Focus 46(4):E14

Kosmopoulos V, Schizas C (2007) Pedicle screw placement accuracy: a meta-analysis. Spine 32(3):E111–E120

Lewandrowski KU, Soriano-Sanchez JA, Zhang X et al (2020) Regional variations in acceptance, and utilization of minimally invasive spinal surgery techniques among spine surgeons: results of a global survey. J Spine Surg 6(Suppl 1):S260–S274

Liu XY, Qiu GX, Weng XS, Yu B, Wang YP (2014) What is the optimum fusion technique for adult spondylolisthesis-PLIF or PLF or PLIF plus PLF? A meta-analysis from 17 comparative studies. Spine 39(22):1887–1898

Machado GC, Maher CG, Ferreira PH et al (2017) Trends, complications, and costs for hospital admission and surgery for lumbar spinal stenosis. Spine 42(22):1737–1743

Macnab I, Dall D (1971) The blood supply of the lumbar spine and its application to the technique of intertransverse lumbar fusion. J Bone Joint Surg Br 53(4):628–638

Magerl F (1982) External skeletal fixation of the lower thoracic and the lumbar spine. In: Uhthoff HK, Stahl E (eds) Current concepts of external fixation of fractures. Springer, New York, pp 353–366

Makanji H, Schoenfeld AJ, Bhalla A, Bono CM (2018) Critical analysis of trends in lumbar fusion for degenerative disorders revisited: influence of technique on fusion rate and clinical outcomes. Eur Spine J 27:1868–1876

Martin BI, Mirza SK, Spina N, Spiker WR, Lawrence B, Brodke DS (2019) Trends in lumbar fusion procedure rates and associated hospital costs for degenerative spinal diseases in the United States, 2004 to 2015. Spine 44(5):369–376

Mccoy S, Tundo F, Chidambaram S, Baaj AA (2019) Clinical considerations for spinal surgery in the osteoporotic patient: a comprehensive review. Clin Neurol Neurosurg 180:40–47

Mobbs RJ, Sivabalan P, Li J (2012) Minimally invasive surgery compared to open spinal fusion for the treatment of degenerative lumbar spine pathologies. J Clin Neurosci 19(6):829–835

Mobbs RJ, Li J, Sivabalan P, Raley D, Rao PJ (2014) Outcomes after decompressive laminectomy for lumbar spinal stenosis: comparison between minimally invasive unilateral laminectomy for bilateral decompression and open laminectomy: clinical article. J Neurosurg Spine 21(2):179–186

Mummaneni PV, Park P, Shaffrey CI et al (2019) The MISDEF2 algorithm: an updated algorithm for patient selection in minimally invasive deformity surgery. J Neurosurg Spine 32(2):221–228

Murase S, Oshima Y, Takeshita Y et al (2017) Anterior cage dislodgement in posterior lumbar interbody fusion: a review of 12 patients. J Neurosurg Spine 27(1):48–55

Narain AS, Hijji FY, Duhancioglu G et al (2018) Patient perceptions of minimally invasive versus open spine surgery. Clin Spine Surg 31(3):E184–E192

Pearson A, Blood E, Lurie J et al (2011) Predominant leg pain is associated with better surgical outcomes in degenerative spondylolisthesis and spinal stenosis: results from the spine patient outcomes research trial (SPORT). Spine 36(3):219–229

Phan K, Rao PJ, Kam AC, Mobbs RJ (2015a) Minimally invasive versus open transforaminal lumbar interbody fusion for treatment of degenerative lumbar disease: systematic review and meta-analysis. Eur Spine J 24(5):1017–1030

Phan K, Hogan J, Maharaj M, Mobbs RJ (2015b) Cortical bone trajectory for lumbar pedicle screw placement: a review of published reports. Orthop Surg 7(3):213–221

Prevalence and most common causes of disability among adults – United States, 2005 (2009) MMWR Morb Mortal Wkly Rep 58(16):421–426

Puvanesarajah V, Liauw JA, Lo SF, Lina IA, Witham TF (2014) Techniques and accuracy of thoracolumbar pedicle screw placement. World J Orthop 5(2):112–123

Rao PJ, Pelletier MH, Walsh WR, Mobbs RJ (2014) Spine interbody implants: material selection and modification, functionalization and bioactivation of surfaces to improve osseointegration. Orthop Surg 6(2):81–89

Roy-Camille R, Saillant G, Berteaux D et al (1976) Osteosynthesis of thoracolumbar spine fractures with metal plates screwed through the vertebral pedicles. Reconstr Surg Traumatol 15:2

Sakaura H, Ikegami D, Fujimori T et al (2019) Early cephalad adjacent segment degeneration after posterior lumbar interbody fusion: a comparative study between cortical bone trajectory screw fixation and traditional trajectory screw fixation. J Neurosurg Spine 32(2):155–159

Santoni BG, Hynes RA, Mcgilvray KC et al (2009) Cortical bone trajectory for lumbar pedicle screws. Spine J 9(5):366–373

Schwab FJ, Blondel B, Bess S et al (2013) Radiographical spinopelvic parameters and disability in the setting of adult spinal deformity: a prospective multicenter analysis. Spine 38(13):E803–E812

Schwab F, Blondel B, Chay E et al (2014) The comprehensive anatomical spinal osteotomy classification. Neurosurgery 74(1):112–120

Seaman S, Kerezoudis P, Bydon M, Torner JC, Hitchon PW (2017) Titanium vs. polyetheretherketone (PEEK) interbody fusion: meta-analysis and review of the literature. J Clin Neurosci 44:23–29

Sears W (2005) Posterior lumbar interbody fusion for lytic spondylolisthesis: restoration of sagittal balance using insert-and-rotate interbody spacers. Spine J 5(2):161–169

Sim HB, Murovic JA, Cho BY, Lim TJ, Park J (2010) Biomechanical comparison of single-level posterior versus transforaminal lumbar interbody fusions with bilateral pedicle screw fixation: segmental stability and the effects on adjacent motion segments. J Neurosurg Spine 12(6):700–708

Wang MY, Lerner J, Lesko J, Mcgirt MJ (2012) Acute hospital costs after minimally invasive versus open lumbar interbody fusion: data from a US national database with 6106 patients. J Spinal Disord Tech 25(6):324–328

Wiltse LL, Bateman JG, Hutchinson RH, Nelson WE (1968) The paraspinal sacrospinalis-splitting approach to the lumbar spine. J Bone Joint Surg Am 50:919–926

Yoon JW, Wang MY (2019) The evolution of minimally invasive spine surgery. J Neurosurg Spine 30:149–158

Lateral Approach to the Thoracolumbar Junction: Open and MIS Techniques

58

Mario G. T. Zotti, Laurence P. McEntee, John Ferguson, and Matthew N. Scott-Young

Contents

Introduction	1052
Patient Presentation	1052
Indications and Comparison of Approaches	1052
Relevant Anatomy of Thoracolumbar Junction	1055
Operative Considerations	1056
Preoperative Workup	1056
Intraoperative	1058
Postoperative	1063
References	1065

M. G. T. Zotti
Orthopaedic Clinics Gold Coast, Robina, QLD, Australia

Gold Coast Spine, Southport, QLD, Australia
e-mail: mzotti@goldcoastspine.com.au

L. P. McEntee
Gold Coast Spine, Southport, QLD, Australia

Bond University, Varsity Lakes, QLD, Australia
e-mail: lpmcentee@goldcoastspine.com.au

J. Ferguson
Ascot Hospital, Remuera, Auckland, New Zealand
e-mail: dr.jaiferguson@mac.com

M. N. Scott-Young (✉)
Faculty of Health Sciences and Medicine, Bond University, Gold Coast, QLD, Australia

Gold Coast Spine, Southport, QLD, Australia
e-mail: info@goldcoastspine.com.au;
mscott-young@goldcoastspine.com.au

Abstract

The thoracolumbar junction is a site less commonly affected by degenerative disease but disproportionately affected by unstable spinal pathologies such as fractures, infections, and neoplasms. Varying approaches for the treatment of these pathologies have been historically described, but there has been a shift toward more pathology being treated by minimally invasive approaches. While a working knowledge of the open approach and its advantages and disadvantages is important, the MIS approach is favored where feasible due to being less disruptive with reduced cardiovascular, respiratory, muscular, and cosmetic morbidity. The unique anatomy of the thoracolumbar junction is discussed with respect to the separation of and structures relevant to the abdominal and thoracic cavities. Evaluation of and preparation for surgery

© Springer Nature Switzerland AG 2021
B. C. Cheng (ed.), *Handbook of Spine Technology*,
https://doi.org/10.1007/978-3-319-44424-6_90

of patients with thoracolumbar junction pathologies are also discussed. With considered techniques, an expansion of the safe treatment of spinal pathologies has been made possible.

Keywords

Thoracolumbar junction · High lumbar · Lower thoracic · Lateral approach · Lateral interbody fusion · Minimally invasive (MIS) · Open

Introduction

The thoracolumbar junction (TLJ) is a transitional zone between the rigid thoracic spine and relatively mobile lumbar spine, making it a common site for spinal fractures, as well as being afflicted by other diseases including infectious, neoplastic, and degenerative lesions. With the advent of modern instrumentation, pathology and deformity involving the thoracic spine are being increasingly treated via posterior techniques in isolation. However, the sensitivity of the spinal cord to any manipulation, high rates of durotomy, and the limited access afforded by other posterior-based approaches such as costotransversectomy limit the safe and complete treatment of anterior column-based pathology with posterior approaches (Arnold et al. 2012; Malham and Parker 2015). Concurrently, there has been a significant advancement of surgical approaches and techniques to approach the thoracolumbar and lower thoracic vertebrae safely causing a paradigm shift from open thoracotomy-based techniques to thoracoscopic and lateral minimally invasive surgery (MIS) exposures. This development has led to expanding indications for the lateral and anterolateral approaches to treating pathology of the thoracolumbar region such as disc herniation, vertebral body/pedicle tumor, and infection as well as allowing efficient and safe intervertebral space access for interbody anterior column support.

Patient Presentation

Pathology of the TLJ such as tumor, trauma, and infection of the thoracolumbar anterior column will generally present with significant pain with or without radiculopathy or myelopathy. Trauma or systemic malaise in these cases will cause the patient to present for medical investigation where further imaging may reveal a TLJ lesion. Deformity, where utilization of the anterior column of the TLJ may form part of the surgical strategy, may include coronal or sagittal plane deformities and may be idiopathic or syndromic in nature. While many of these deformities will present in childhood, there is an increasing burden of ageing patients presenting for treatment with de novo deformities in middle age and beyond due to pain or deformity.

The degenerate or herniated thoracic disc presents variably. While many thoracic disc lesions are asymptomatic, significantly symptomatic thoracic herniations account for around 1% of disc-related presentations. Anand and Regan (2002) developed a classification system for different presentations of symptomatic thoracic disc herniations and their outcome. These include presentations of axial pain only (28%), thoracic radicular only (5%), axial pain and thoracic radiculopathy (38%), axial pain and lower leg pain (19%), myelopathy (8%), and paralysis (2%). Unlike symptomatic lumbar disc disruption, symptomatic thoracic disc disruption does not tend to respond favorably to nonsurgical measures, and surgery is generally recommended. Arce and Dohrmann (1985) in their series of 280 thoracic discs found that 75% are in the lower thoracic segments (T8–T12).

Indications and Comparison of Approaches

Indications for thoracolumbar spinal surgery include treatment of anterior- and middle column-based tumors or infections (Kawahara et al. 1997; Uribe et al. 2010; Pimenta 2015), as an adjunct in a deformity correction strategy for short segment fusion of focal coronal malalignment or focal kyphosis (usually with posterior instrumentation) (Good et al. 2010; Min et al. 2012), in the trauma setting for vertebrectomy of burst fractures, and treatment of pseudarthrosis and degenerative lesions. Regarding the latter, disc pathology either herniated nucleus pulposus causing ventral cord compression or significant degenerative disc disease can be treated with microdiscectomy or interbody fusion, respectively. While microdiscectomy has been published with favorable results (Malone and Ogden 2013; Nacar et al. 2013; Oskouian et al. 2002; Otani et al. 1982; Berjano et al. 2014; Roelz et al. 2016), the desire to avoid further approach involving the pleura for the patient in the form of a revision fusion has led some authors to advocate for primary fusion, with satisfactory results (Berjano et al. 2012, 2015; Malham and Parker 2015; Meredith et al. 2013). Posterior-based approaches can be considered for dorsal cord compression, but treatment of anterior-based pathology from a posterior-only approach historically yielded unsatisfactory results, and an example of this is shown in Fig. 1 (Love and Kiefer 1950; Logue 1952; Benson and Byrnes 1975; Perot and Munro 1969).

Complications, regardless of mode of approach, remain significant and include pleural effusion, pneumothorax, intercostal neuralgia, vascular injury, pain associated with chest tube (if required), and diaphragmatic paresis or herniations (if affected) (Boriani et al. 2010). Open procedures, despite being considered "gold standard," were particularly affected by intercostal neuralgia and incisional pain and required a thoracic access surgeon. Open approaches have more classically been either the transthoracic retroperitoneal or lateral retropleural approaches and, more recently, a lateral extensile extracavitary approach. There is significant patient discomfort and pulmonary complications that stem from need for resection of a rib, deflation of ipsilateral lung, and insertion of a chest tube which can all contribute to postoperative pain, atelectasis, and pneumonia (see Fig. 2). Thoracotomy-associated major complications occur in 11% to 11.5% of patients, tend to extend hospitalizations, and augment medical resource use (Faciszewski et al. 1995).

Treatment of thoracolumbar lesions that was once marked by significant morbidity and patient risk that resulted from traditional open techniques has become feasible and technically simpler due

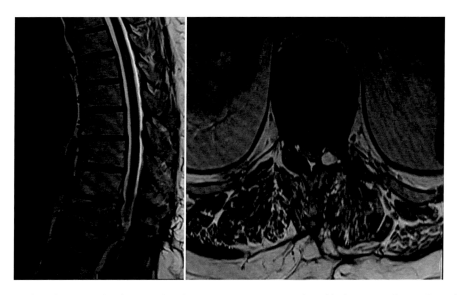

Fig. 1 Ventral cord compression from T11/12 degenerative disc-osteophyte complex causing clinical myelopathy. Posterior decompression failed to improve the patient's symptoms and residual compression and cord signal change is demonstrated

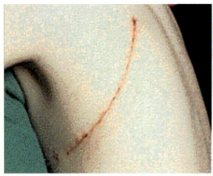
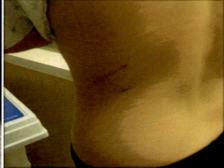

Fig. 2 Significant scar and morbidity from traditional open transthoracic approaches to the thoracic spine (left); MIS incision required to approach limited TLJ levels (right). (Copyright 2009, used with permission from NuVasive. All rights reserved)

to the evolution of surgical techniques stemming from development of MIS approaches to the thoracic spine (Smith and Fessler 2012). The advantage of thoracoscopic and MIS techniques is that they are minimally disruptive (not usually requiring rib resection), deflation of the lung is not routinely required, and there is minimal blood loss, direct visualization, less risk of aortic injury, and typically no requirement for any chest tube. However, circumstances may dictate that such approaches may not be appropriate for the patient despite indicating pathology, e.g., significant thoracic pulmonary injury and hemothorax involving the operative segments for trauma vertebrectomy or a primary bone tumor in the anterior column necessitation open approach to minimize contamination and safely perform spondylectomy (Gandhoke et al. 2015).

Limitations of the thoracoscopic (also known as video-assisted thoracic surgery (VATS) or thoracic endoscopic) technique include a high learning curve, limited visualization compared to direct stereoscopic assessment, and need for lung deflation. VATS had been first reported in 1993 for spinal disease by Mack et al. (1993) VATS allows a significant reduction in chest wall morbidity related to the traditional thoracotomy (Cunningham et al. 1998; Newton et al. 2003). These included a reduction of the postoperative incisional pain and intercostal neuralgia. While thoracoscopy is capable of producing the same exposure as the transthoracic route without the need for a large incision or rib resection, there is, however, still a significant decrease in vital capacity by up to 30% (Faro et al. 2005). This technique provides a greater access to more vertebral levels through smaller incisions, when compared to transthoracic approach, but still presents some complications, such as intercostal neuralgia (7.7%), symptomatic atelectasis (6.4%), excessive (>2000 cc) intraoperative blood loss (2.5–5.5%), pneumonia (1–3%), wound infections (1–3%), chylothorax (1%), tension pneumothorax, long thoracic nerve injury, and pulmonary embolism (Pimenta Journal of Spine 2015). Although this is an effective technique in specialized practices that have sufficient experience and expertise with this skill set, MIS techniques are technically simpler, without a steep learning curve, and arguably safer for spine surgeons not trained in endoscopic surgery, and, as such, we will focus on open and MIS TLJ approaches.

The MIS approach referred to herein is either a direct or extreme lateral approach depending on the proprietary company and is either in the form of a coelomic (transthoracic) or extracoelomic (retropleural) approach, while the lumbar spine requires a retroperitoneal approach (Uribe et al. 2010). Because this approach remains in the retroperitoneal/retropleural space, it can be performed from the right or left side depending on surgeon preference or location of the pathology without interference by the liver, spleen, or other peritoneal structures and theoretically can access from T4 to L5 blocked superiorly by the axilla and inferiorly by the iliac crest (Malham and Parker

2015). Because the approach remains in the extra-coelomic space and the diaphragm is not incised, there is no need for any repair of the diaphragm (Pimenta 2015). The MIS approach also can traverse the diaphragm with care, enabling the surgeon to treat low thoracic and high lumbar levels either side of the TLJ by coming from pleural to retroperitoneal through a diaphragmatic incision. Above T12–L1 there is typically no need for dividing the diaphragm. Diaphragmatic incisions need to be biased toward the chest wall and periphery to minimize diaphragmatic innervation disruption and enable a tendinous repair, and regarding incisions, there is evidence that incision <4 cm of the diaphragm heal without suture.

With regard to pitfalls and contraindications to the MIS approach, it is important to understand that there is a long working distance in a relatively narrow working space, and as such, the operative tools typically used may not be long enough to perform the procedure in some patients. Further, retropleural and retroperitoneal dissection may not be feasible after a previous ipsilateral thoracotomy or retroperitoneal approach such as patients who have had osteomyelitis of the spine and spinal metastases, where marked paraspinal pleural reactions with adhesive thickening of the parietal pleura and infiltration of the pleura by tumor or inflamed fibrous tissue can occur (Uribe et al. 2010).

Relevant Anatomy of Thoracolumbar Junction

The complex relationship between the diaphragm, ribs, pleura, and peritoneum poses a notable challenge when surgically approaching the lower thoracic and upper lumbar vertebrae (Pimenta 2015), and the relevant structures at risk to be aware of must be studied and anticipated.

The basic path of MIS approaches is through the rib cage laterally onto the targeted disc space. After incision through the skin and fat, the palpable targeted rib is sought and a plane developed above the rib. This is as the intercostal neurovascular bundle lies directly inferior and deep to the rib. One can do a rib osteotomy for access and graft material but places the neurovascular bundle at risk unless it is explicitly protected. The parietal pleura which is the next layer is plastered over the inner surface of the rib-intercostal complex. This must be penetrated, ideally with blunt dissection, prior to entering the chest cavity. The visceral pleura, on the other hand, overlies the lung parenchyma, and a pneumothorax can result with its injury.

The relevant vascular anatomy includes on the right, the vena cava; on the left, the aorta (variable course); and traversing the field, the segmentals. If approaching on the left-hand side levels T8–L1, it is imperative to assess on preoperative imaging and angiogram for a dominant vascular cord supply in the form of the artery of Adamkewicz, as ligation of this segmental vessel unknowingly may disturb region flow to the cord. Left-sided lateral approaches will require ligation of the segmentals at the level of planned treatment and possibly adjacent levels for safety as they come off the left-side positioned aorta and transmit segmental branches posteriorly which will be encountered in cases of vertebrectomy and plate positioning.

The diaphragm is a complex but important three-dimensional dome-shaped structure to understand, particularly if planning TL approaches as it will be in the surgical access path when approaching levels from T10 to L1. Uncontrolled injuries to the diaphragm can lead to atelectasis, reduced vital capacity, and hypoxemia. It has multiple attachments and separates the thoracic from the abdominal cavity between T12 and L1 with a convex superior thoracic floor and concave inferior abdominal roof. The crura extend along the anterolateral vertebral body on each side with the left crus extending to L2 while the right extends to L3. It is innervated central to peripheral (important for incision and dissection planning) with the phrenic nerve supplies entering medially toward the periphery, formed by the C3-4-5 cervical nerves, more prevalent in its central portion (Joaquim et al. 2012). The peripheral portions of the diaphragm have sensorial afferents from the intercostal nerves (T5–11) and the subcostal nerve (T12) (Joaquim et al. 2012).

Regarding the relationship of the diaphragm to the peritoneum, the posterior portion of the diaphragm is separated from the peritoneum by a fat layer and by the superior aspect of the kidneys and adjacent structures. Posteriorly, the diaphragm forms two ligamentous bands, on either side: the medial and lateral arcuate ligament. The former arises from the tip of the 12th rib and expands around the quadratus lumborum, while the latter spans across the psoas muscles. These ligamentous bands meet on the transverse processes of L1. Also to be aware of are diaphragmatic openings which transmit structures. These are the aortic hiatus (containing the aorta, azygos vein, and thoracic duct between the left and the right crura), the esophageal hiatus (containing the esophagus and anterior and posterior vagal trunks), and the caval hiatus (inferior vena cava and branches of the right phrenic nerve) (Joaquim et al. 2012). Other small openings are also present, especially near the crus (containing the splanchnic nerves and the hemiazygos veins).

Regional wall muscles include the latissimus dorsi, intercostal, transversus abdominis, and external and internal oblique muscles as well as the quadratus lumborum. The superficial nerves such as the subcoastal (T12), Iliohypogastric (L1), and ilioinguinal (L1) nerve can run in the field of dissection and the abdominal wall planes after and be injured if any sharp dissection or diathermy is undertaken between the muscular layers. The position and morphology of the psoas muscle should also be considered for high lumbar procedures given that trans-psoas approaches to L1 and L2 still have the potential to injure the upper lumbar plexus. They generally consist of two main portions in their origins: the anterior and lower edges of the lumbar transverse processes and the lateral vertebral bodies and annuli of the lumbar vertebrae. Slight flexion at the hip in positioning can help to relax each psoas and reduce the tendency for anterior traction on the lumbar plexus.

As the approach goes further caudally to the L1 and L2 levels, the retroperitoneal structures come into play. The lumbar plexus, unless there are vertebral anomalies, does not tend to be at risk unless approach levels at T12 and lower as it is generally formed from ventral roots of L1–L4 (with variable subcostal T12 contribution) and sits in the posterior third of the psoas. Renal vessels and the position of the ureters and bowel must be considered when approaching L2 and caudally, particularly if the approach strays anteriorly. Ventral exiting branches can still be injured by retractors placed too posteriorly, even above the level of the psoas origin, and for this reason, retractor placement in the ventral anterior three quarters of the body is recommended (Arnold et al. 2012).

Operative Considerations

Preoperative Workup

The patient should be assessed from the point of view of fitness for surgery and anesthesia in the perioperative period. Where possible, all acute illness should be dealt with and the patient physiology and nutrition optimized, in a multidisciplinary fashion, before they undergo surgery. Relevant biochemistry, hematological parameters, as well as blood typing for possible transfusion should be assessed and prepared for surgery. Poor nutritional status (preoperative albumin and ferritin), current or recent smoking, and lung parenchymal radiotherapy may be more subtle to patients at risks of pulmonary complications than the obvious patients with morbid obesity and obstructive airway disease.

Pulmonary function tests are particularly relevant in the elective or semi-urgent setting where poor reserve or functional asymmetry of the pulmonological system may warrant delaying the surgery or modifying the side and type of approach.

AP/lateral X-ray erect films are ideal to assess for vertebral body size and alignment (see Fig. 3). EOS standing films, if available, give detailed study of the spine type and alignment which is important when treating lesions that have likely significantly altered the sagittal profile of the patient. CT or MRI assessment of structures at the planned operative level should be undertaken especially looking for calcification, which may cause

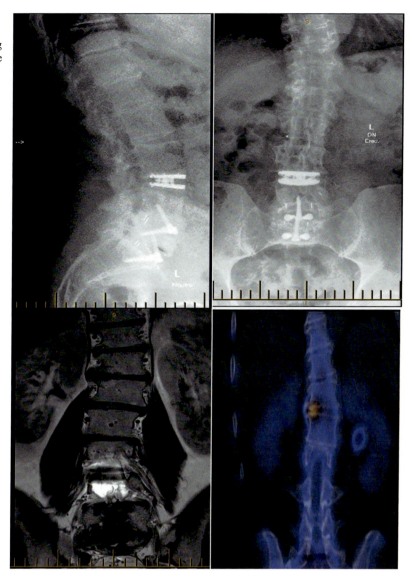

Fig. 3 Erect lateral and anteroposterior films and coronal MRI demonstrating collapsed T12/L1 disc. Note "hot spot" on SPECT scan due to collapse and focal scoliosis

operative difficulty (see Fig. 3). In the case of trauma, it will also assist with non-contiguous injury and evaluate for any posterior element injury or hematoma that may change management. The MRI also allows adequate assessment of the neural elements and position of the conus medullaris.

Neurological studies such as nerve conduction studies and/or electromyography may characterize and reveal any preoperative myelopathy or radiculopathy prior to treatment that may have implications for prognosis and patient expectations.

In the case of suspected neoplastic or infectious lesions, the local and systemic staging followed by biopsy and/or microbiological testing should be instituted to characterize the pathology (Liljenqvist et al. 2008). Diagnosis of thoracic DDD has not been well characterized. Axial ache and stiffness that changes with rest, recumbency, and different positions is suggestive. Extrapolating from lumbar studies, discogram is the gold standard, while combinations of findings on CT/MRI and pain blocks or activity on nuclear medicine scans can support a diagnosis.

Depending on the level(s) to be approached, the position of the aorta, the sympathetic plexus, and renal vessels and their relation to the psoas muscle and the spinal curvature should be studied. Scoliosis may place the aorta in the path of the lateral surgical corridor and may warrant consideration for right- rather than left-sided approach. If there is significant anomalous anatomy, then obtaining an arteriogram and/or venogram may be prudent.

Intraoperative

The procedure is undertaken under general intravenous anesthesia so as not to attenuate motor pathway signals of the neuromonitoring. As to the choice of endotracheal tube, a double lumen tube for selective lung ventilation is utilized above T8 level, whereas below T8 it is not usually necessary for MIS.

Open Lateral Approach to the TLJ

The patient is positioned lateral and commonly left side up with the most common approach to access T10–L2 being the left-sided thoracolumbar junction approach and is described adapted from the retropleural thoracotomy of McCormick (1995). Alternatives include the mini-thoracotomy and transdiaphragmatic, extrapleural (Balasubramaniam et al. 2016; Foreman et al. 2016; Graham et al. 1997; Otani et al. 1988), or left-sided thoracotomy. Pertinently, high thoracic lesions tend to be approached from the right to limit exposure to cardiovascular structures, while more caudal thoracic lesions are best approached from the left to avoid the inferior vena cava and liver.

An oblique incision is centered over the pathological vertebra/disc space as confirmed by image intensifier. The length of the incision will depend upon the number and location of levels to be treated. The incision is made over the rib belonging to the vertebrae 2 levels above the targeted discovertebral space given the caudal sloping of the ribs, e.g., 10th rib for T12. The thoracic muscles over the rib and abdominal muscles distal to the costal cartilage are divided with electrocautery. There is subperiosteal exposure of the rib, and depending whether bone grafting or interbody fusion is planned, it can either be preserved or osteotomized, protecting the caudal neurovascular bundle, for later use. The parietal pleura is opened at the posterior aspect of the wound revealing parenchyma/visceral pleura of the lung and the diaphragmatic attachment to the chest wall anteriorly.

Next, the retroperitoneal space is accessed via splitting the layers of the abdominal wall at the anterior part of the incision, and the peritoneum is peeled off from the inferior surface of the diaphragm down to the crus adjacent to the spine. It is important to leave a small (1–2 cm) cuff for later repair, and marking stay sutures periodically can be useful to aid later approximation. Smooth handheld deep retractors may be used here, but a dedicated table-mounted retraction system allows ease of access and ergonomics in a limited field.

Further steps of the approach depend on the discovertebral level(s) to be accessed as is discussed below (Vialle et al. 2015). If above T12 then the approach converts to a left-sided thoracotomy. If T12/L1 has to be exposed, the diaphragm has to be cut. If L1/2 has to be exposed, the principles of the retroperitoneal lumbar approach should be followed.

For levels T12 and above, a left-sided lung deflation is completed (see Fig. 4) and the deflated lobe retracted anterior to expose the ventral thoracic spine (Shen and Haller 2010). The parietal pleura is incised over the planned operative level (s) at the disc space and then to adjacent levels, being aware that the segmental vessels are in the midline and needing to be isolated and ligated. A subperiosteal exposure from posterior and lateral to anterior is then completed with blunt instruments protecting the great vessels anteriorly. Posterior dissection exposes the rib heads which often have to be osteotomized for adequate disc access.

For access to the TLJ, the diaphragm has to be incised to be able to access either side. As discussed in the anatomy section, the diaphragmatic innervation is from the phrenic nerve near the esophagus. Thus, it is important to incise the diaphragmatic attachments as close as possible to the chest wall as any muscle left in situ after

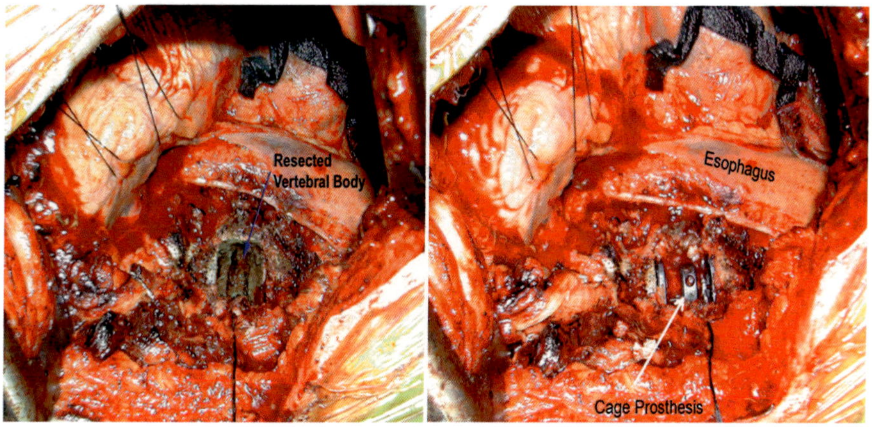

Fig. 4 Visualization via a lateral thoracotomy of low thoracic corpectomy and cage prosthesis insertion for fracture. (Copyright 2008, used with permission from CTSNet (www.ctsnet.org). All rights reserved)

incision in the periphery will undergo denervation. In the case of isolated TLJ approach (T12–L1), the diaphragm can be retracted anteriorly with the lung being retracted with a moist sponge and not requiring selective deflation. The parietal pleura is then incised longitudinally along the lateral aspect of the spine at least one level proximal to the most proximal vertebra to be treated. As in the thoracotomy approach, the segmental vessels are isolated and ligated, and the sympathetic chain is bluntly retracted and the great vessels protected.

For high lumbar levels (L1–2), the peritoneum is retracted along with the sympathetic chain anteriorly after being shifted away from the lateral abdominal wall and the psoas muscle reflected posteriorly after careful dissection on the anterior border of the muscle over the disc space and isolation and ligation of segmental vessels. Care for nearby renal vessels and the nearby great vessels lying anterior to the anterior longitudinal ligament is paramount.

Any approach with pleural breach, as distinct from an extrapleural approach, will require a chest drain insertion. One or two chest tubes are typically inserted in the midaxillary line although some surgeons prefer an anterior drain to extract air from the chest cavity and a posteriorly placed drain for drainage of blood and fluid.

The diaphragm is reattached using interrupted sutures, making sure that the parietal pleura is approximated and preferably closed over the instrumentation. The chest wall muscles that have been divided are then re-approximated.

MIS Approach

The minimally invasive approach to the thoracolumbar spine has been described, publicized, and refined since the first report by Pimenta in 2000 (Pimenta 2015). The patient is positioned in the right lateral decubitus position on a radiolucent table with the hip and knees flexed. Bony prominences and pressure points are padded, arms placed onto table attachments assessing for brachial plexus and ulnar pressure, and the torso secured either with a beanbag or table-mounted bolsters and sports tape (Elastoplast™) is also used. The choice of how to secure the torso and keep the posterior spinal elements exposed to prevent rolling during the procedure will depend on whether supplemental pedicle screws are planned from the same patient position or repositioning or a staged posterior fixation is to take place (see Figs. 5 and 6). They are taped in such a way as to afford good biplanar visualization of the affected levels. The skin can be secured with taping if access to lower lumbar levels is planned and a side bolster or break in the thoracolumbar region can be placed to open up the ipsilateral ribs and operative disc space at the TLJ. Neural monitoring including transcranial motor evoked potentials and somatosensory evoked potentials along with pertinent nerve roots is, we believe, imperative when performing direct lateral approaches.

The acquisition of proper image intensifier (II) images is important for ease of workflow. The II

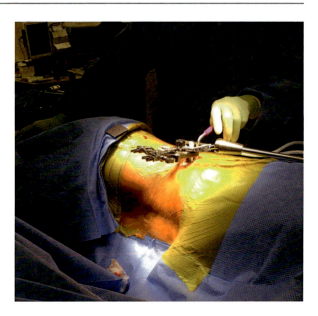

Fig. 5 Setup and retractors in place for direct, left side up lateral approach to the TLJ

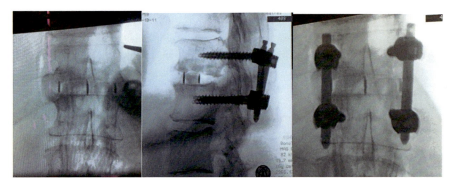

Fig. 6 Interbody insertion at T12/L1 level for same patient as in Fig. 5, followed by posterior instrumentation, all carried out without changing the position of the patient or need for redraping

generally will go over the patient who is in the lateral position and first obtain a true AP of their pedicles, and the table is adjusted until this is achieved. This allows us to correct the rotation of the spine and makes confirmation of endplate alignment in the lateral position easier. Once this correct AP position is confirmed and marked on the II machine, a position 90° orthogonal to this is utilized for the disc space access. Note, it is very important to have a well-trained radiographer for these procedures, and often it is easier to describe the AP view as being a side to side C-arm image and the lateral as being a top to bottom image so as to avoid confusion when conversing with members of the team. Given the relatively limited visual field of standard C-arms, if one cannot see the affected level and the sacrum in a single view, then it is advised to have a metallic marker placed in the affected pedicle under CT guidance prior to surgery.

The patient is then prepped and sterile draped over the shoulder and pelvis. It is important to have a large enough surgical window and imperative to have in mind an extensile approach in the event of significant bleeding occurring through a small incision. Incision planning is undertaken using fluoroscopy for level check and the 12th rib as landmark. One can use the ribs to trace up rib 2 levels above planned thoracic level(s) to choose for resection as a guide. Placement of a Kirschner wire in the spinous process immediately inferior to the pathological thoracic disc/

level has been proposed as a strategy to avoid wrong site surgery (Malham and Parker 2015).

A 4–6-cm-long oblique incision (parallel to and following the trajectory of the rib at the index level) is made at the midaxillary line, 90° lateral to the disc space (Uribe et al. 2010). Sharp dissection is carried to deep fascia, and a small rent is typically made in the deep fascia with the Bovie cautery. Blunt dissection is then carried through the intercostal muscles above the rib. The parietal pleura is then pierced in the same way as one would pass a chest drain in the setting of thoracic trauma. Typically, the right index finger can then be swept down the undersurface of the rib to palpate the transverse process. The blunt dilator can then be passed on the palmar aspect of the finger to the level of the transverse process and walked out onto either the disc or the affected vertebral body (see Figs. 7 and 8). This ensures that no visceral pleura or lung parenchyma is caught by the dilator. The position of the neurovascular bundle under the rib is considered at all times. The dilator is initially directed posteriorly away from the lung parenchyma before being brought anteriorly to the rib head over the target disc space and confirmed by lateral fluoroscopy (see Figs. 8 and 9). It is important to note the reversal of usual retractor blade positions in the thoracic spine, with the posterior blade actually being anterior with the lung retractor blade.

Multiple retractors exist for the purpose of this surgery. When crossing to T12 or L1 or lower, to obtain retroperitoneal access, we prefer a tubular retractor such as a NuVasive Maximum Access™ as this allows a smaller hole to be made in the pleura and this retractor is passed through the proximal diaphragmatic attachment at the chest wall and with the dilator passing below apex of the diaphragm (Fig. 10).

Following treatment of the targeted region, provided that there be no visceral injury observed and it is a low thoracic level performed through a direct lateral approach, a drain is seldom required as the small air leak that may result is usually subclinical. Thus far, in our experience, this T12/L1 approach has never necessitated diaphragmatic repair or the passage of a chest drain. Obviously, when removing the retractor, a positive-pressure ventilation, a "bubble test," should be performed prior to definitive closure and a chest X-ray gained prior to waking the patient to ensure no significant pneumo- or hemothorax has occurred.

If the approach is to be made targeting higher lumbar levels, the transversalis fascia is one of the main components that maintain structural integrity of the retroperitoneal space. A 4 cm transverse incision is made along the lateral flank at the midline level of the index vertebral body. The incision should be made parallel to the direction

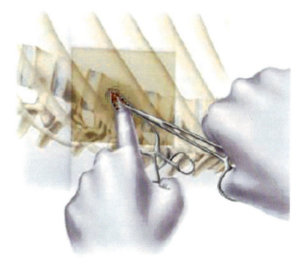

Fig. 7 Blunt dissection carried out above the rib. (Copyright 2009, used with permission from NuVasive. All rights reserved)

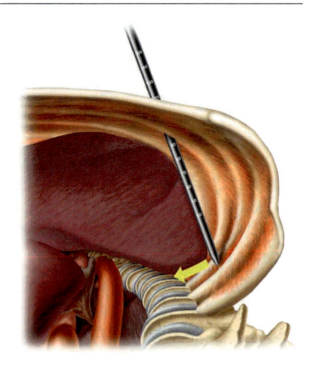

Fig. 8 The blunt dilator is initially brought posteriorly and walked down the rib until reaching the rib head overlying the operative level. (Copyright 2009, used with permission from NuVasive. All rights reserved)

Fig. 9 Retractor placement and confirmation of correct operative level on lateral fluoroscopy. (Copyright 2009, used with permission from NuVasive. All rights reserved)

of the fibers of the external oblique to minimize the possibility of injury to the motor nerves supplying them. This prevents abdominal wall pseudo-hernia formation from loss of tone to these abdominal wall muscles. Blunt dissection with anterior sweeping movements of the retroperitoneal contents is then performed to enable palpation of the psoas muscle and the transverse process of the index vertebra.

Once the operative levels are reached and illuminated retractors are in place, a removal of the rib head via osteotomy over the operative site is typically necessary for adequate access to the disc space (Fig. 11). The operative strategy can then be carried out, whether discectomy, interbody fusion, or corpectomy (Fig. 12). In the case of an interbody fusion, the macroscopic discectomy is carried out first from anterior to posterior disc

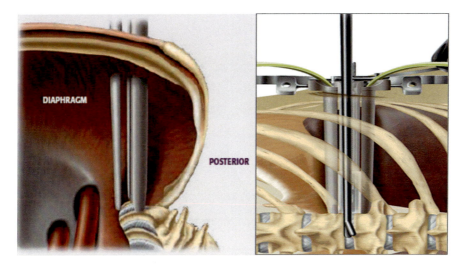

Fig. 10 Crossing of instrumentation through the diaphragm from thoracic to upper lumbar levels through small diaphragmatic incision. With the retractors in place, long instruments, such as pituitary rongeurs, can be safely passed. (Copyright 2009, used with permission from NuVasive. All rights reserved)

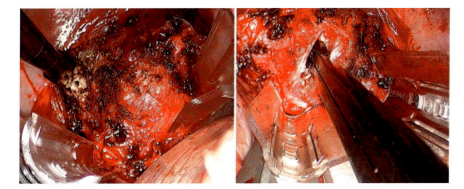

Fig. 11 Rib head osteotomy to enable adequate access to the lateral thoracic intervertebral space. (Copyright 2009, used with permission from NuVasive. All rights reserved)

space before careful resection of the posterior longitudinal ligament.

Postoperative

A chest radiograph in recovery is recommended if no chest drain has been inserted to assess for any interim pneumothorax development (Fig. 13).

Following a decision for insertion and securing of a chest drain, a chest radiograph should be obtained the first day postoperatively. If the pulmonary sacs remain expanded with no evidence of pneumothorax and the drain continues to "bubble," await approximately 24 h and remove at that time. If there is no re-expansion, then either the drain should be checked (if there is no bubbling or swinging) or a respiratory physician opinion should be obtained (Henry et al. 2003).

The ward care postoperatively will depend on the procedure performed. The common path, however, is for early graduated mobilization and aggressive chest exercises to prevent atelectasis and pneumonia. Analgesia for open thoracotomy approaches will typically be considerably more aggressive than for MIS procedures and may

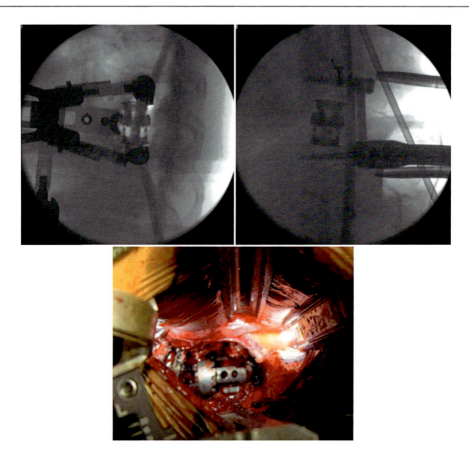

Fig. 12 Example of corpectomy and pedicle screw stabilization carried out through a lateral MIS approach. View of operative site between retractor blades. (Copyright 2009, used with permission from NuVasive. All rights reserved)

Fig. 13 Immediate postoperative chest radiograph in recovery after TLJ approach to exclude clinically significant pneumothorax with a chest drain in situ

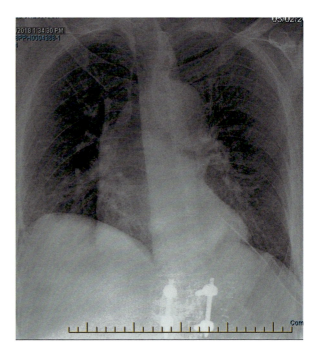

involve a patient-controlled analgesia setup in an attempt to minimize pain-mediated pulmonary dysfunction. Hemoglobin and electrolytes are assessed routinely and corrected as required. Erect radiographs to verify position and alignment of hardware is arranged as an inpatient when the patient is able to stand.

Once safe and comfortable, the patient is discharged, and follow-up is arranged as an outpatient as postoperative week 2 for a wound review. The patient is encouraged to keep active, but avoidance of heavy lifting and explosive or rotational activities during the first 6–12 weeks is advised. The patient is then followed clinically and radiographically from a fusion perspective if interbody fusion or corpectomy or clinically if a microdiscectomy was performed.

References

Anand N, Regan JJ (2002) Video-assisted thoracoscopic surgery for thoracic disc disease: classification and outcome study of 100 consecutive cases with a 2-year minimum follow-up period. Spine 27(8):871–879. https://doi.org/10.1097/00007632-200204150-00018

Arce CA, Dohrmann GJ (1985) Herniated thoracic disks. Neurol Clin 3:383–392

Arnold P, Anderson K, McGuire R (2012) The lateral transpsoas approach to the lumbar and thoracic spine: a review. Surg Neurol Int 3(4):198. https://doi.org/10.4103/2152-7806.98583

Balasubramaniam S, Tyagi D, Zafar S, Savant H (2016) Transthoracic approach for lesions involving the anterior dorsal spine: a multidisciplinary approach with good outcomes. J Craniovertebral Junction Spine 7(4):237. https://doi.org/10.4103/0974-8237.193254

Benson MKD, Byrnes DP (1975) The clinical syndromes and surgical treatment of thoracic intervertebral disc prolapse. J Bone Joint Surg 57(4):471–477

Berjano P, Balsano M, Buric J, Petruzzi M, Lamartina C (2012) Direct lateral access lumbar and thoracolumbar fusion: preliminary results. Eur Spine J 21(1):37–42. https://doi.org/10.1007/s00586-012-2217-z

Berjano P, Garbossa D, Damilano M, Pejrona M, Bassani R, Doria C (2014) Transthoracic lateral retropleural minimally invasive microdiscectomy for T9-T10 disc herniation. Eur Spine J 23(6):1376–1378. https://doi.org/10.1007/s00586-014-3369-9

Berjano P, Langella F, Damilano M, Pejrona M, Buric J, Ismael M, Villafañe JH, Lamartina C (2015) Fusion rate following extreme lateral lumbar interbody fusion. European Spine Journal 24 (3):369–371

Boriani S, Bandiera S, Donthineni R, Amendola L, Cappuccio M, De Iure F, Gasbarrini A (2010) Morbidity of en bloc resections in the spine. Eur Spine J 19(2):231–241. https://doi.org/10.1007/s00586-009-1137-z

Cunningham BW, Kotani Y, McNulty PS, Cappuccino A, Kanayama M, Fedder IL, McAfee PC (1998) Video-assisted thoracoscopic surgery versus open thoracotomy for anterior thoracic spinal fusion: a comparative radiographic, biomechanical, and histologic analysis in a sheep model. Spine. https://doi.org/10.1097/00007632-199806150-00008

Faciszewski T, Winter RB, Lonstein JE, Denis F, Johnson L (1995) The surgical and medical perioperative complications of anterior spinal fusion surgery in the thoracic and lumbar spine in adults: a review of 1223 procedures. Spine (Phila Pa 1976) 20(14):1592–1599

Faro FD, Marks MC, Newton PO, Blanke K, Lenke LG (2005) Perioperative changes in pulmonary function after anterior scoliosis instrumentation: thoracoscopic versus open approaches. Spine (Phila Pa 1976) 30 (9):1058–1063

Foreman PM, Naftel RP, Moore TA, Hadley MN (2016) The lateral extracavitary approach to the thoracolumbar spine: a case series and systematic review. J Neurosurg Spine 24(4):570–579. https://doi.org/10.3171/2015.6.SPINE15169

Gandhoke GS, Tempel ZJ, Bonfield CM, Madhok R, Okonkwo DO, Kanter AS (2015) Technical nuances of the minimally invasive extreme lateral approach to treat thoracolumbar burst fractures. Eur Spine J 24:353–360. https://doi.org/10.1007/s00586-015-3880-7

Good CR, Lenke LG, Bridwell KH, O'Leary PT, Pichelmann MA, Keeler KA, Koester LA (2010) Can posterior-only surgery provide similar radiographic and clinical results as combined anterior (thoracotomy/thoracoabdominal)/posterior approaches for adult scoliosis? Spine 35(2):210–218. https://doi.org/10.1097/BRS.0b013e3181c91163

Graham AW, Mac Millan M, Fessler RG (1997) Lateral extracavitary approach to the thoracic and thoracolumbar spine. Orthopedics 20(7):605–610. Retrieved from http://www.ncbi.nlm.nih.gov/pubmed/9766306

Henry M, Arnold T, Harvey J (2003) BTS guidelines for the management of spontaneous pneumothorax. Thorax 58(Suppl II):ii39–ii52

Joaquim AF, Giacomini L, Ghizoni E, Mudo ML, Tedeschi H (2012) Surgical anatomy and approaches to the anterior thoracolumbar spine region. J Brasileiro de Neurocirurgia 23(4):295–300

Kawahara N, Tomita K, Fujita T et al (1997) Osteosarcoma of the thoracolumbar spine: Total en bloc Spondylectomy a case report. J Bone Joint Surg 79(3):453–458

Liljenqvist U, Lerner T, Halm H, Buerger H, Gosheger G, Winkelmann W (2008) En bloc spondylectomy in malignant tumors of the spine. Eur Spine J 17(4):600–609. https://doi.org/10.1007/s00586-008-0599-8

Logue V (1952) Thoracic intervertebral disc prolapse with spinal cord compression. J Neurol Neurosurg Psychiatry 15:227–241

Love JG, Kiefer EJ (1950) Root pain and paraplegia due to protrusions of thoracic intervertebral disks. J Neurosurg 7:62–69

Mack MJ, Regan JJ, Bobechko WP, Acuff TE (1993) Application of thoracoscopy for diseases of the spine. Ann Thorac Surg 56:736–738

Malham GM, Parker RM (2015) Treatment of symptomatic thoracic disc herniations with lateral interbody fusion. J Spine Surg 1(1):86–93. https://doi.org/10.3978/j.issn.2414-469X.2015.10.02

Malone HR, Ogden AT (2013) Thoracic Microdiskectomy: lateral and posterolateral approaches. Surg Anat Tech Spine 30:282–293

McCormick PC (1995) Retropleural approach to the thoracic and thoracolumbar spine. Neurosurgery 37:908–914

Meredith DS, Kepler CK, Huang RC, Hegde VV (2013) Extreme lateral interbody fusion (XLIF) in the thoracic and thoracolumbar spine: technical report and early outcomes. HSS J 9(1):25–31. https://doi.org/10.1007/s11420-012-9312-x

Min K, Haefeli M, Mueller D, Klammer G, Hahn F (2012) Anterior short correction in thoracic adolescent idiopathic scoliosis with mini-open thoracotomy approach: prospective clinical, radiological and pulmonary function results. Eur Spine J 21(SUPPL. 6):765–773. https://doi.org/10.1007/s00586-012-2156-8

Nacar OA, Ulu MO, Pekmezci M, Deviren V (2013) Surgical treatment of thoracic disc disease via minimally invasive lateral transthoracic trans/retropleural approach: analysis of 33 patients. Neurosurg Rev 36(3):455–465. https://doi.org/10.1007/s10143-013-0461-2

Newton PO, Marks M, Faro F, Betz R, Clements D et al (2003) Use of video-assisted thoracoscopic surgery to reduce perioperative morbidity in scoliosis surgery. Spine 28:S249–S254

Oskouian RJ, Johnson JP, Regan JJ (2002) Thoracoscopic microdiscectomy. Neurosurgery 50(1):103–109. https://doi.org/10.1097/00006123-200201000-00018

Otani K, Nakai S, Fujimura Y, Manzoku K, Shibasaki K (1982) Surgical treatment of thoracic disc herniation using the anterior approach. J Bone Joint Surg Br Vol 64(3):340–343

Otani KI, Yoshida M, Fujii E et al (1988) Thoracic disc herniation: surgical treatment in 23 patients. Spine 13:1262–1267

Perot PL Jr, Munro DD (1969) Transthoracic removal of midline thoracic disc protrusions causing spinal cord compression. J Neurosurg 31:452–458

Pimenta L (2015) Minimally invasive lateral approach to the thoracic spine – case report and literature overview. J Spine 4(4). https://doi.org/10.4172/21657939.1000240

Roelz R, Scholz C, Klingler JH, Scheiwe C, Sircar R, Hubbe U (2016) Giant central thoracic disc herniations: surgical outcome in 17 consecutive patients treated by mini-thoracotomy. Eur Spine J 25(5):1443–1451. https://doi.org/10.1007/s00586-016-4380-0

Shen FH, Haller J (2010) Extracavitary approach to the thoracolumbar spine. Semin Spine Surg 22(2):84–91

Smith ZA, Fessler RG (2012) Paradigm changes in spine surgery-evolution of minimally invasive techniques. Nat Rev Neurol 8(8):443–450. https://doi.org/10.1038/nrneurol.2012.110

Uribe JS, Dakwar E, Le TV, Christian G, Serrano S, Smith WD (2010) Minimally invasive surgery treatment for thoracic spine tumor removal: a mini-open, lateral approach. Spine 35(SUPPL. 26S):347–354. https://doi.org/10.1097/BRS.0b013e3182022d0f

Uribe JS, Dakwar E, Cardona RF, Vale FL (2010) Minimally invasive lateral retropleural thoracolumbar approach: cadaveric feasibility study and report of 4 clinical cases. Operative Neurosurgery 68(SUPPL. 1):32–39. https://doi.org/10.1227/NEU.0b013e318207b6cb

Vialle LR, Bellabarba C, Kandziora F (2015) Thoracolumbar spine trauma, AOSpine masters series, vol 6, 1st edn. Thieme, New York, p 236

Surgical Approaches to the Cervical Spine: Principles and Practicalities

59

Cyrus D. Jensen

Contents

Introduction	1068
Anterior Surgery	1069
Posterior Surgery	1069
Planning	1070
Surgical Strategy	1070
Surgical Approach	1070
Surgical Setup	1071
Anterior Transoral Approach to C1-C2	1071
Indication	1071
Positioning, Preparation, and Practicalities	1071
Approach	1072
Procedure	1072
Closure	1072
Structures at Risk	1072
Complications	1072
Anterior Retropharyngeal Approach (C1-C3)	1074
Indication	1074
Positioning, Preparation, and Practicalities	1074
Approach	1074
Procedure	1075
Closure	1075
Structures at Risk	1075
Complications	1076
Lateral Approach (Verbiest)	1076
Indication	1076
Positioning, Preparation, and Practicalities	1076
Approach	1077
Procedure	1077

C. D. Jensen (✉)
Department of Trauma and Orthopaedic Spine Surgery,
Northumbria Healthcare NHS Foundation Trust,
Newcastle upon Tyne, UK
e-mail: jensen@doctors.org.uk

© Springer Nature Switzerland AG 2021
B. C. Cheng (ed.), *Handbook of Spine Technology*,
https://doi.org/10.1007/978-3-319-44424-6_91

Closure	1077
Structures at Risk	1078
Complications	1078
Anterolateral Approach (Smith-Robinson)	1078
Indication	1078
Positioning, Preparation, and Practicalities	1079
Approach	1079
Procedure	1081
Closure	1083
Structures at Risk	1083
Complications	1083
Posterior Approach (C3-C7)	1084
Indication	1084
Positioning, Preparation, and Practicalities	1084
Approach	1084
Procedure	1084
Closure	1086
Structures at Risk	1086
Complications	1086
Posterior Approach (Occiput-C2)	1086
Indication	1086
Positioning, Preparation, and Practicalities	1086
Approach	1087
Procedure	1087
Closure	1087
Structures at Risk	1087
Complications	1087
Conclusions	1088
Cross-References	1088
References	1088

Abstract

In order for surgeons to be able to achieve the best possible outcomes for their patients, it is important that they tailor their treatments to each individual patient and their condition. To achieve this, spinal surgeons in particular must be familiar with a variety of different implants and the various approaches that can be used to access the pathology. Patient positioning and preoperative planning are important, as well as identifying soft tissue corridors, which avoid damage to the numerous neurovascular structures found in close proximity to the axial skeleton.

Focusing on the cervical spine, this chapter provides discussion regarding indications, patient positioning, preparation, practicalities, as well as illustrations of the commonest approaches and surgical procedures. Approaches include the anterior (retropharyngeal, transoral), anterolateral (Smith-Robinson), lateral (Verbiest), and posterior, and for each approach, there is also a discussion of the structures at risk and a review of the literature regarding complications.

Keywords

Cervical spine · Surgical approach · Transoral · Odontoid · Retropharyngeal · Verbiest · Anterolateral · Smith-Robinson · Posterior approach

Introduction

Surgery in the cervical spine can be incredibly rewarding. The pathology at times may be severe, with a high chance of permanent disability if left untreated. Despite this, the outcomes are often remarkably good. The complications associated with cervical spine surgery are potentially devastating; however, thankfully they are infrequent, and some are avoidable if one takes care to identify anatomy and plan the surgery diligently. This chapter will present approaches commonly, and in some instances less commonly, used to treat pathologies of the cervical spine. This is by no means an exhaustive list; there are several eponymous approaches which are infrequently used and will not be discussed at length as they are excellently described by their creators in the wider literature. This chapter will also explore the practicalities of performing these approaches, including the positioning of the patients, the structures at risk, the procedures themselves including their indications and the complications which may occur.

Anterior Surgery

Up until the 1950s, cervical spine surgery was almost exclusively performed via a posterior approach, including decompression and stabilization. The "workhorse" of cervical spine surgeons was the posterior cervical laminectomy. However, the ability to retrieve the prolapsed intervertebral disc was limited by the unretractable characteristics of the spinal cord through this approach. In 1954, Lahey (Lahey and Warren 1954) described an anterior cervical approach which he used to approach esophageal diverticula. In addition, neurosurgeons such as Dereymaeker from Louvain, Belgium, were discussing the potential of using this approach in spinal surgery. Two years later, Smith and Robinson (Robinson and Smith 1955) were the first to formally describe an anterolateral cervical approach for the purpose of spine surgery. In 1958, Smith and Robinson (1958) went on to publish the results of their first 14 cases of Anterior Cervical Discectomy and Fusion (ACDF) surgery. Popularity for this anterolateral approach grew, as it provided a new safe corridor for surgeons to access the intervertebral disc without having to venture around the spinal cord. New retractor systems were quickly developed to facilitate this approach, and surgeons such as Cloward (1958) began to refine the technique for achieving an interbody fusion. As more surgeons began to favor the anterior cervical approach further developments were seen in the implant materials and designs, including static and dynamic plates, cement and bone block spacers, interbody and corpectomy cages, and arthroplasty devices. Further anterior and lateral approaches to the cervical spine were later developed in order to give better access to the proximal and distal most vertebrae, the neuroforaminal roots and vertebral arteries. These include the transoral approach to the odontoid, the anterior retropharyngeal approach (McAfee et al. 1987) to the C1-3, the lateral (Verbiest) approach (Verbiest 1968), and the transsternal approach to the cervicothoracic junction.

Posterior Surgery

In the late 1970s, led by spinal surgeons in Japan, there was also a drive to further develop the posterior cervical operations. This was still the favored approach for the treatment of certain spinal conditions, such as myelopathy caused by ossification of the Posterior Longitudinal Ligament (OPLL,) which has a higher prevalence in patients of Japanese and Asian heritage. Posterior laminectomy was known to be associated with postoperative segmental instability, kyphosis, perineural adhesions, and late neurological deterioration. Hirabayashi (Hirabayashi et al. 1983) and Kurokawa (Kurokawa et al. 1982) developed the techniques of "open door" and "double door" laminoplasty in an attempt to avoid these problems. Since their inception, there have been several variations on these techniques including the introduction of different "spacer" implants designed to hold open the elevated lamina. Another great advancement was the introduction of the surgical microscope into spine surgery (Yasargil 1977). The superior magnification and light they provided enabled surgical wounds to shrink, thus reducing the damage to surrounding tissues, and led to the introduction of endoscopic surgery. Posterior stabilization of the cervical spine was difficult to achieve as it relied almost entirely upon wiring techniques which were not good at resisting extension, rotation, or lateral bending forces. The introduction of translaminar, lateral mass and pedicle screws insertion techniques by surgeons such as Roy-Camille (Roy-Camille et al. 1989) and Magerl (Magerl and Seemann 1987) has greatly improved the ability of surgeons to stabilize the cervical spine from a posterior approach.

As surgical approaches to the anterior, lateral, and posterior aspects of the cervical spine have evolved, so too has our understanding of the various pathologies, facilitated through basic science research and the advancements in imaging including computerized tomography (CT) and magnetic resonance imaging (MRI). With the help of these imaging modalities, clinical evaluation, electromyography (EMG), and diagnostic injections, surgeons are now able to be more confident about their diagnoses, and in turn, more likely to select a suitable surgical strategy to achieve the goal of treatment – a good outcome for their patients.

Planning

As with any surgical treatment, it is of paramount importance that the surgeon takes care to plan the procedure fully before entering the operating theater.

Surgical Strategy

Planning starts with identifying the surgical strategy which will be used to address the pathology and achieve the surgical goal. Unfortunately, it is still commonplace to find ill-considered operations being performed on patients, which can fail to help with symptom relief and potentially cause them harm. Surgeons must make every effort to identify the underlying cause of a symptom and formulate a surgical strategy to deal with this cause, as opposed to treating just the symptom. For example, it should come as no surprise that stabilizing a degenerative C5/6 disc in a patient will do little to relieve the neck pain that they have referred up from their arthritic shoulder.

In general, the pathologies requiring treatment in the cervical spine tend to be related to tumor, trauma, infection, degenerative, or inflammatory musculoskeletal and disc disease. The sequelae of these pathologies are often either instability or compression of neural structures, which then gives a target for surgical treatment. Consequently, surgical strategies usually involve decompression, stabilization, or both.

Surgical Approach

Next, the surgeon needs to decide which surgical approach will best enable them to carry out the surgical strategy, whilst also minimizing the risk of complications by avoiding structures at risk. Most pathologies can be addressed with either an anterolateral or posterior approach, although some require both. The lateral, transoral, retropharyngeal, or transsternal approaches are more specialist approaches with narrow indications which will be highlighted within the relevant subsection of this chapter.

Intuitively, pathologies which affect the anterior structures – that is structures lying ventral to the spinal cord and roots – are often best approached from the anterior aspect. This approach enables removal of any compressive material (disc, bone, pus, tumor, or hematoma) with little manipulation of the neural structures, followed by stabilization, fusion or motion preservation, depending on the implants used. Kyphotic deformity is also an indication for an anterior approach as it enables restoration of the normal cervical lordosis by lengthening of the anterior column. Most degenerative pathologies affecting the cervical spine benefit from an anterior approach for these reasons. Trauma, tumor, or infection which compromises the integrity of the anterior column is an absolute indication for an anterior approach to enable reconstruction of the anterior column, preventing a later kyphotic deformity.

Posterior cervical pathologies are often best treated with a posterior approach. These pathologies include cord or neural compression caused by trauma, tumor infection, or degenerative disease such as osteophytes or an in-folding of the ligamentum flavum. Dissecting through the posterior muscles and stripping them from their boney attachments is more likely to give the patients postoperative pain compared with the anterior intermuscular approach; however, it does enable a more complete decompression of the cord for the entire length of the spine. It is an extensile approach and is, therefore, ideal for treating pathology which is found at the occipito-cervical and cervico-thoracic junctions, or extensive pathology occurring at several different levels in the cervical spine. If stabilization needs to bridge to the occiput or the thoracic spine, then the posterior approach is indicated. The posterior approach is also used if intradural pathology is encountered, or if there is OPLL which is often adherent to the ventral surface of the cord.

Whichever approach is selected by the surgeon to best treat the pathology, it must be one which they are familiar with and can perform competently. There are many pathologies, such as cervical spondylotic myelopathy, where research is ongoing in trying to identify the approach which delivers the best outcomes. It is, therefore, advisable for surgeons to carefully consider all options when planning their surgical approaches.

Surgical Setup

Finally, once the tactics and approach have been confirmed, consideration should then be given to the surgical setup, including patient positioning, operating room layout, instruments and implants, list management, and additional support which may be required. Patient positioning will be covered in more detail later in the chapter. One common setup strategy is to have the cervical spine patient positioned on the operating table with their feet by the anesthetist, and their head at the other end nearer to the surgeon. This is to provide the least obstruction for the surgeon to be able to operate with the microscope and reduces the risk of contamination of the surgical field. It is important to use an operating table which is radiolucent over the areas where intraoperative X-rays may

need to be taken. The need for instruments for insertion (and removal) of implants should be anticipated in advance and present before commencement of the anesthetic, along with any support staff such as neurophysiologists for intraoperative cord monitoring. With the advent of navigation and robotic-assisted surgery, spinal operating rooms can easily become cluttered with this additional equipment, implant trays, instrument trolleys, microscopes, and image intensifier equipment. It is often worth planning the theater layout with adequate space assigned for all the items with the largest footprints to ensure the operation can go ahead safely. Although we are reluctant to admit it, surgeons do fatigue during the day, and so it is preferable to schedule bigger cases for the morning to reduce the risk of avoidable complications.

Ensuring the key members of the operating room team are fully informed of the preoperative plan is the simplest way of preventing mistakes and negative outcomes when unforeseen events occur in theater.

Anterior Transoral Approach to C1-C2

Indication

The transoral approach (Crockard 1985; Fang and Ong 1962; Menezes and VanGilder 1988) to the cervical spine is ideal for pathology in the anterior midline at the craniocervical junction, such as tumors. Previously, cord compression and instability caused by rheumatoid arthritis were the leading indication; however, this disease is now better treated by medication and hence the overall incidence of transoral surgery to the spine has greatly reduced (Choi and Crockard 2013).

Positioning, Preparation, and Practicalities

Nasopharyngeal bacterial swabs are taken 3 days pre-op along with antiseptic mouth washes (Watkins III 2015b). Intravenous antibiotics are given at induction and for 7 days post-op. The patient is positioned supine with their head on a Mayfield headrest. A reinforced nasotracheal tube is inserted endoscopically and skull traction (3 kg) applied via a HALO ring. Some surgeons feel a tracheostomy is advisable in order to ensure a patent airway post-op. Antiemetics are given and a nasogastric (NG) tube is passed to reduce the risk of any regurgitation of stomach contents affecting the wound. The NG tube is retained for 5 days post-op to enable feeding as the patient will remain nil by mouth for at least 5 days post-op. A rubber loop is passed down from the un-intubated nostril and hooked round the uvula and pulled cephalad to bring it out of the field of surgery. If this is insufficient, the soft palate can be incised with a curvilinear incision around the uvula in a cephalad direction and the two flaps held clear with stay sutures. A Boyles-Davis mouth gag is then inserted to retract the tongue, being released and reapplied every 30 min to avoid tongue necrosis. The nasopharynx and laryngopharynx are packed to catch any secretions. The oropharynx is prepped with Betadine and the posterior pharyngeal tissues are copiously infiltrated with a lidocaine and 1:200,000 adrenaline solution for hemostasis (Fig. 1).

Approach

The anterior tubercle on the atlas is palpated and confirmed on X-ray. A 3 cm longitudinal midline incision is made – 1 cm cephalad and 2 cm caudal to this tubercle. Incision is made through the posterior pharyngeal mucosa, the constrictor muscles, the prevertebral fascia, and the anterior longitudinal ligament (ALL). Remaining soft tissues are then bluntly dissected off the body of C2 (below the odontoid) and the anterior tubercle of C1. At this stage the blunt dissection can be extended further laterally and longitudinally to expose the lateral masses of C1 and C2. However, one should remember that the vertebral arteries lie a minimum of 2 cm from the anterior tubercle of C1 in the foramen transversarium on either side. There is often venous bleeding just lateral to the base of the odontoid, and sharp dissection may be needed

to detach the longus coli muscles from the anterior aspect of C1 and C2.

Procedure

With the C1 and C2 now exposed, the definitive treatment can begin, usually with burring away of the anterior arch of C2 to reveal the odontoid which can also be excised if indicated. The dura should be completely decompressed and if necessary, a durotomy can be performed. Access can be extended proximally to reach the clivus, although this may require further soft palate incision or formal maxillotomy. Depending on stability and the extent of bony excision, a HALO jacket or posterior stabilization may now be required.

Closure

The longus coli muscles, constrictor pharyngeal muscles, and mucosa are all closed as individual layers if possible with absorbable sutures. Any soft plate extensions are closed in one layer with an absorbable suture.

Structures at Risk

The vertebral arteries lie a minimum of 2 cm from the anterior tubercle of C1 in the foramen tranversarium on either side. They are found at the lateral edges of the joints between the C1 and C2 lateral masses, and so care must be taken to use blunt retractors laterally if exposing these joints and stay sutures should not go too deep within the lateral pharangeal wall. Care should be taken to avoid damaging the tongue or the pharyngeal mucosa with the retractors. The spinal cord and dura are located beneath the cruciate ligaments at the back of the odontoid.

Complications

In 2016, Shriver et al. published a meta-analysis of the complications related to 1238 transoral

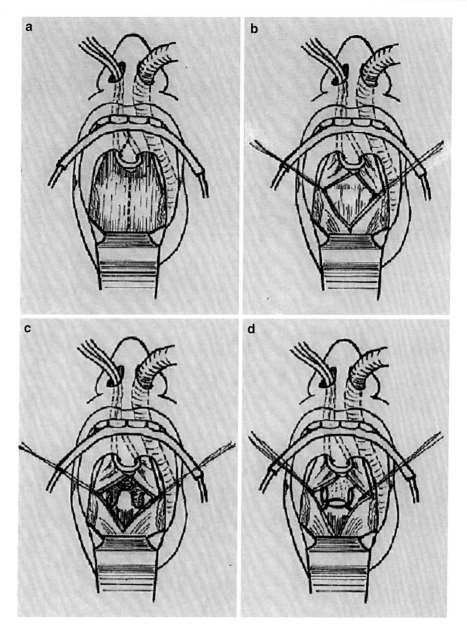

Fig. 1 Anterior transoral approach to C1-C2. (**a**) Transoral incision, including soft palate peri-uvula extension (dotted line), showing vascular loop elevating uvula and McIvor mouth gag exposing the posterior pharyngeal mucosa. (**b**) Pharyngeal mucosa, constrictor muscles, and prevertebral fascia, incised and retracted with stay sutures, exposing the ALL and longus coli muscles over the anterior tubercle of C1 and body of C2. (**c**) Anterior tubercle and arch of C1 removed with high-speed burr to reveal the odontoid process. (**d**) Ondontoidectomy can then be performed carefully with Kerrison rongeurs to decompress the cord

odontoidectomy patients (Shriver et al. 2016). The commonest reported complications were medical complications (13.9%), which were mostly respiratory with some cardiac. The mortality rate at 30 days was 2.9%. Other complications included arterial injury (1.9%), CSF leak (0.8%), meningitis (1%), pharyngeal wound dehiscence (1.7%), dysphagia (3.8%), velopharyngeal insufficiency (VFI) (3.3%), and tracheostomy (10.8%).

Anterior Retropharyngeal Approach (C1-C3)

Indication

The anterior retropharyngeal approach (De Andrade and Macnab 1969; McAfee et al. 1987) is indicated in tumor, trauma, infection, and instability cases affecting the upper cervical spine, especially where there may be a need to insert bone graft and implants. It also enables the surgeon to extend distally past C3, which is not possible from the transoral approach without splitting the mandible.

Positioning, Preparation, and Practicalities

The patient is positioned supine with their head on a Mayfield headrest, and rotated slightly contralaterally. A reinforced nasotracheal tube is inserted endoscopically to allow better elevation of mandible. Depending upon the pathology skull traction (3 kg) can be applied using Gardener-Wells tongs. Due to the slightly awkward approach angle (cephalad under the mandible) early introduction of the microscope is advised as it serves as an excellent light source. The handheld Cloward retractor facilitates the blunt dissection and approach. The skin around the incision can be pre-infiltrated with a lidocaine and 1:200,000 adrenaline solution to reduce the bleeding from the superficial tissues during the procedure.

Approach

A 2–4 cm skin incision is made below the angle of the mandible from midline to the mid-axis of the sternocleidomastoid muscle, curving up toward the mastoid process (Watkins III 2015a). The platysma and superficial fascia are divided in line with the incision. The subplatysmal plane is developed a little proximally and distally. Care is taken to identify and protect the greater auricular nerve and anterior cervical nerve branches of the ansa cervicalis. These course around from the posterior border of sternocleidomastoid, crossing the proximal end of the muscle on their route up to give sensory distribution to the auricular and mandibular skin. The investing fascia is incised along the deep medial edge of the sternocleidomastoid muscle. The pulsation of the carotid sheath is palpated, and blunt dissection preferably with the index fingers is used to develop a plane medial to the carotid sheath and lateral to the trachea, esophagus, and strap muscles.

The neurovascular structures crossing from lateral to medial, across the carotid triangle (Fig. 2), are found in this layer. Before ligating any of these vessels it is important, at this stage, to identify the hypoglossal nerve. After descending between the internal carotid artery and jugular vein, it turns horizontally and becomes superficial over the proximal external carotid artery, traversing medial-lateral over the proximal carotid triangle, before passing deep to the digastric tendon to supply the muscles of the tongue. A nerve stimulator should be used to confirm correct localization of the hypoglossal nerve. It should then be mobilized from the surrounding tissue to enable gentle cephalad retraction along with the posterior belly of the digastric muscle. If necessary, the digastric (and stylohyoid) muscles can be divided at their tendinous junction and tagged for later repair. Access into the retropharyngeal space is prevented by the lateral-medial traversing branches of the external carotid artery, including the superior thyroid, lingual, and facial arteries. These vessels, along with their accompanying veins, are divided. The submandibular duct can be tied off and the gland removed to facilitate proximal exposure.

The carotid sheath is retracted along with the ligated stumps of the superior thyroid, lingual, and facial vessels laterally. The digastric muscle and hypoglossal nerve are retracted proximally, and the musculovisceral column is retracted medially. Blunt finger dissection is used to locate the ALL and longus coli muscles on anterior cervical spine.

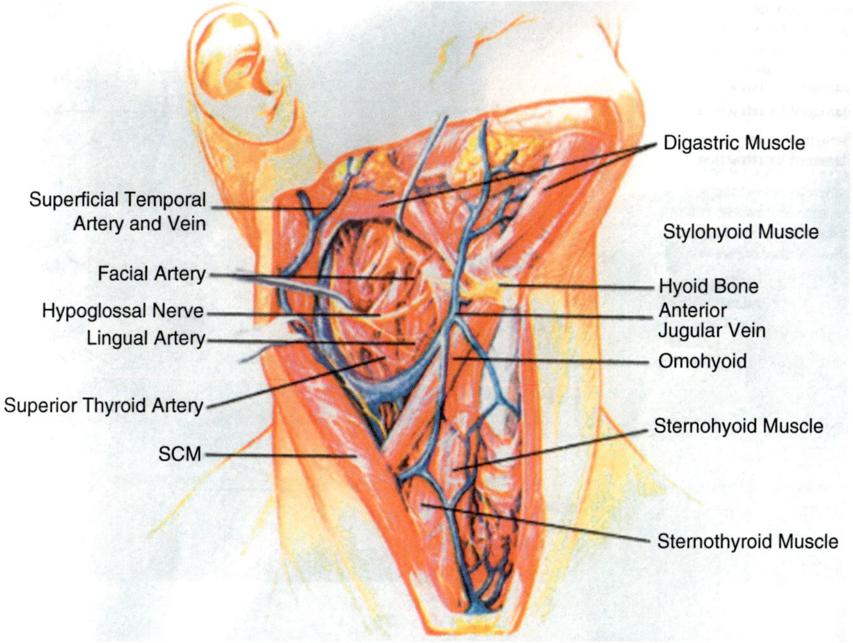

Fig. 2 Anterior retropharyngeal approach with the exposure of the carotid triangle. (From Watkins III 2015a)

The anterior tubercle on the atlas serves as a useful landmark and can be confirmed on X-ray. The prevertebral fascia and longus coli fibers are elevated, and the blunt handheld Cloward retractors are substituted for some sharp, clawed retractor blades under longus coli.

Procedure

Once the desired bony anatomy is identified, a high-speed burr is used to remove the anterior arch of C1 and the odontoid process. The tip of the odontoid process is removed before sectioning of the base to prevent superior retraction of this structure in cases of basilar invagination. The body of the odontoid is drilled, leaving a thin shell of bone posteriorly, which is carefully removed with Kerrison rongeurs. The lateral pillars should be retained as they are the primary load-bearing structures of the atlantoaxial articulation. Following decompression, supplementary stabilization is usually required. This can be accomplished with anterior or posterior instrumentation, or a halo device.

Closure

Deep closure is not required. The digastric muscle tendon is repaired, if it was divided. The platysma muscle is closed over a drain, and then fat and skin are closed in separate layers. Suture/clip removal equipment accompanies the patient during the first 12 h in case an urgent evacuation of hematoma is needed to prevent airway compromise.

Structures at Risk

Take care to identify and protect the greater auricular nerve and anterior cervical nerve branches of the ansa cervicalis crossing over the proximal sternocleidomastoid muscle. Injury to these can cause loss of sensation to the auricular and mandibular skin. When operating in the submandibular triangle, it is important to avoid damaging the marginal mandibular branch of the facial nerve. Ensure the hypoglossal nerve has been correctly identified before ligating any other vessels. Avoid excessive retraction on the stylohyoid muscle as this can injure the facial nerve as it exits the skull

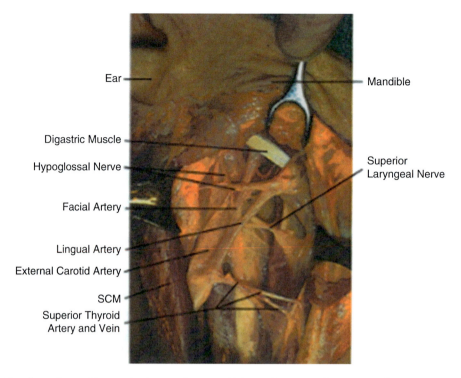

Fig. 3 Note the position of the superior laryngeal nerve, which must be protected. (From Watkins III 2015a)

through the stylomastoid foramen. It is important to identify the superior laryngeal nerves (both internal and external branches) (Fig. 3) as injury to these can lead to increased voice fatiguability, loss of high-pitched notes in singing, or even aspiration pneumonia as a result of reduced laryngeal cough reflex. They are located in the carotid triangle adjacent to the superior thyroid artery and are commonly injured by retractors. The superior laryngeal nerve should be mobilized in order to prevent retraction injury to its internal and external branches.

Complications

The commonest complaint with the retropharyngeal approach is dysphagia. Other complications include facial nerve palsy (9.4%), hypoglossal nerve plasy (4.7%), non-union/graft displacement (6.3%), infection (1.6%), hypopharynx injuries (3.1%), and a mortality rate of 3.1%.

Lateral Approach (Verbiest)

Indication

The lateral approach to the vertebral artery was described in detail by Henry (1970) and later modified by Verbiest (Verbiest 1968) to enable additional access to the lateral aspects of the cervical vertebral bodies, the neuroforamina, and the portions of the anterior rami of the brachial plexus. Bony spurs compressing the vertebral artery or the cervical nerve roots, and neural tumors can be addressed with this extensile approach (Watkins 2015).

Positioning, Preparation, and Practicalities

The patient is positioned supine with their head on a Mayfield headrest, without any rotation. This is to avoid occlusion of the vertebral artery in cases of spondylotic compression. The skin around the incision can be pre-infiltrated with a lidocaine and 1:200,000 adrenaline solution to reduce the bleeding from the superficial tissues during the procedure.

Approach

An oblique incision is made along the medial border of the sterncleidomastoid muscle. Skin and subcutaneous tissues are retracted, and the platysma muscle is divided longitudinally in line with its fibers. The investing layer of the cervical fascia is incised to reveal the medial border of sternocleidomastoid muscle. The plane between the muscle and the medial strap muscles is developed with finger dissection. Care is taken to identify and protect the greater auricular nerve and anterior cervical nerve branches of the ansa cervicalis. These course around from the posterior border of sternocleidomastoid, crossing the proximal end of the muscle on their route up to give sensory distribution to the auricular and mandibular skin. The sternocleidomastoid muscle is then retracted laterally and the carotid sheath is identified by its pulsation against a fingertip. The plane between the carotid sheath and sternocleidomastoid (laterally) and the musculovisceral column (medially) is further developed with blunt dissection with a finger and mounted pledgets, down to the anterior tubercle of the transverse process. The sternocleidomastoid can be released from its proximal attachment on the mastoid process if it is required. However, care must be taken to identify and protect the accessory nerve which enters the muscle 4–6 cm from the mastoid tip. After identifying the anterior tubercle of the transverse process, where longus coli, longus capitis, and anterior scalene muscles attach, it is possible to palpate medially to feel the vertebral body and disc. The longitudinal sulcus between the tubercle (laterally) and the body (medially) is the costotransverse lamellae, which forms the roof of the foramen transversarium covering the vertebral artery. Incise the prevertebral fascia longitudinally to expose the longus coli and longus capitis muscles. The longus coli muscle has

three parts – two oblique parts lying over one longitudinal part. The longus capitis muscle lies longitudinally over the upper oblique longus coli. The upper vertical longus coli arises from the anterior tubercle of the atlas and then inserts on the body of C4, while the lower stretches from C5 to T3. The oblique parts insert (with the longus capitis) on the anterior tubercles on the transverse processes. The sympathetic chain lies directly anterior to the transverse processes, between the prevertebral fascia and the carotid sheath, sometimes being adherent to the latter. The sympathetic chain, including its superior (at C1) and middle (at C6) cervical ganglions, should be moved as one with the longus coli muscles from lateral to medial to access the underlying structures. In order to achieve this retraction, the muscular attachments of the oblique longus coli and the longus capitis are sharply dissected off the anterior tubercles. The dissector should be working in the acute angle the muscle fibers make with the bone (the stripping angle), which is an inferolateral direction when working on the proximal levels, and a superolateral direction when working on the inferior levels (Fig. 4).

Procedure

The bone of the anterior tubercle is removed with a rongeur, and the proximal and distal margins of the costotransverse lamellae are identified. With a small Kerrison rongeur, the bone of the costotransverse lamellae is carefully removed, and the foramen transversarium is de-roofed. In doing so it is inevitable that venous plexus surrounding the artery will bleed, and this should be controlled with light pressure. The nerve root should now be accessible underneath the mobile vertebral artery. Again, further bleeding may be encountered from the main vertebral veins which may require additional hemostatic products.

Closure

Deep closure is not required. The platysma muscle is closed over a deep drain, and then fat and skin

Fig. 4 Lateral approach revealing the vertebral artery running through the foramen transversarium. (From Watkins 2015)

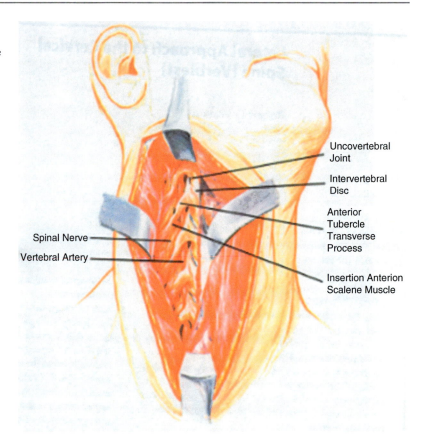

are closed in separate layers. Suture/clip removal equipment accompanies the patient during the first 12 h in case an urgent evacuation of hematoma is needed to prevent airway compromise.

Structures at Risk

Take care to identify and protect the greater auricular nerve and anterior cervical nerve branches of the ansa cervicalis crossing over the proximal sternocleidomastoid muscle. Injury to these can cause loss of sensation to the auricular and mandibular skin. If proximal detachment of the sternocleidomastoid is required, care must be taken to identify and protect the accessory nerve which enters the muscle 4–6 cm from the mastoid tip. The sympathetic chain should be moved as one with the longus coli muscles from lateral to medial to avoid injury and a subsequent Horner's syndrome (partial ptosis, meiosis, and anhidrosis). The vertebral artery is at risk during the longus coli strip and the excision of bony costotransverse lamellae. If the vertebral artery is injured, bleeding may be controlled by a ligature above and below the lesion.

Complications

Due to a lack of published data on lateral-approach cervical surgery, it is not possible to present the complications rates from these procedures.

Anterolateral Approach (Smith-Robinson)

Indication

The anterolateral approach was borne out of the desire of spinal surgeons to treat diseases of cervical discs and the vertebral bodies, without having to negotiate a path around the immoveable cord and nerve roots. It is indicated primarily for treatment of pathologies which lie ventral to the cord such as neural compression caused by disc, osteophyte, tumor, abscess, or bone. It is also useful in correcting instability and deformities caused by degeneration, trauma, infection, or tumor. Cervical disc arthroplasty is only possible through an anterolateral approach. This approach should enable access from C2-T2, although this range may be reduced to C3-C7 in certain patients where the mandible or the clavicle restricts further access.

Positioning, Preparation, and Practicalities

The patient is positioned supine with their head on a Mayfield headrest, without any rotation (Fig. 5). A small rolled towel, or alternatively a 500 ml bag of iv fluid, is placed between the scapulae approximately at the level of T2 spinous process. This allows for slight extension of the neck and permits the shoulders to drop back toward the table, providing more room for the surgeons' instruments. The head and neck are secured in position with soft tape or, if traction is known to be needed, Gardener-Wells skull traction. If the arms and shoulders need to be pulled caudally to enable a lateral X-ray to be taken, then they can be held in this position with soft tape to facilitate intraoperative X-rays. An inserted nasogastric tube can facilitate identification of the esophagus in difficult revision approaches; however, this is not necessary in primary cervical surgery. Although palpable anatomic landmarks such as the hyoid (C3), thyroid cartilage (C4/5), carotid tubercle (C6), or the cricoid cartilage (C6) can provide fairly reliable indication of incision location (Fig. 6), it is preferable to use a radio-opaque marking stick and an X-ray machine to map out the incision. The skin around the incision can be pre-infiltrated with a lidocaine and 1:200,000 adrenaline solution to reduce the bleeding from the superficial tissues during the procedure.

Historically, the side of the approach has generated much research interest regarding the effect of side of approach on the risk of recurrent laryngeal nerve (RLN) injury. In the original description of the procedure (1955) (Robinson and Smith 1955), the authors suggest an apparent increased traction applied on the RLN when approaching from the right side and hence they recommend a left-sided approach. Another of the earliest users of this technique (Cloward 1958) used the right-sided approach. There has yet to be any conclusive published evidence that a right-sided approach significantly increases the RLN injury rate, although a cadaveric study (Rajabian et al. 2020) does seem to show that below C5 the RLN requires more retraction to keep it from the field of surgery. For ergonomic reasons, many surgeons favor use of the same side as their hand dominance and this is likely to lead to other indirect benefits such as reduced length of surgery and retractor time. It is helpful for the theater team and the anesthetist to know of the side of the approach in advance, as the endotracheal tube may be positioned to the contralateral side and theater layout can be tailored to the advantage of the surgeon.

Approach

A transverse skin-crease incision is made from the midline to the medial border of the sternocleidomastoid at the desired level. The subcutaneous tissues are dissected and then retracted with a Mollinson retractor, exposing the platysma. The platysma is incised in line with the wound, and the subplatysmal plane is only developed proximally and distally if more than one cervical level is to be approached. The investing layer of the cervical fascia, which envelopes the sternocleidomastoid muscle, is incised along the medial muscle border. The medial musculovisceral tissues (thyroid, esophagus, trachea, strap muscles) are retracted medially with the handheld Cloward retractor, while the surgeon's index finger is used to dissect through the pretracheal fascia and palpate for the pulsation of the carotid sheath underneath the sternocleidomastoid. The superior and inferior thyroid vessels may be encountered at C3/4 and C6/7 levels, respectively, and they can either be

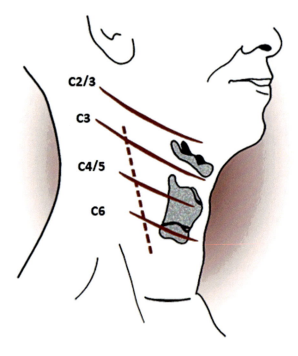

Fig. 5 Incisions related to bony anatomical landmarks including lower border of the mandible (anterior retropharyngeal approach C2/3), hyoid bone (C3), upper aspect of thyroid cartilage (C4/5), cricoid cartilage (C6)

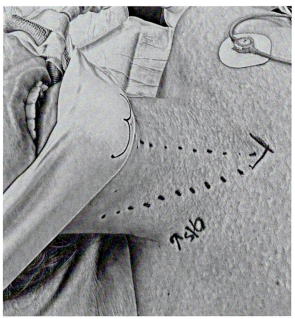

Fig. 6 Anterolateral approach positioning

retracted or ligated if necessary. Omohyoid is often encountered at the level of C6/7, and can also either be retracted or divided laterally. The former option is the author's preferred choice. A plane between the pulsatile carotid sheath and sternocleidomastoid laterally, and the musculovisceral tissues medially is further developed down to the front to the prevertebral fascia (Fig. 7). Through the smooth, white prevertebral fascia it should be easy to palpate the anterior spinal column, made up of alternating ridges and sulci of the intervertebral discs and vertebral

Fig. 7 Cross section of the anterolateral approach at C5/6 showing medial retraction of the thyroid, strap muscles, trachea, and esophagus, and lateral retraction of the sternocleidomastoid and carotid sheath. The longus coli muscle is elevated along its medial border

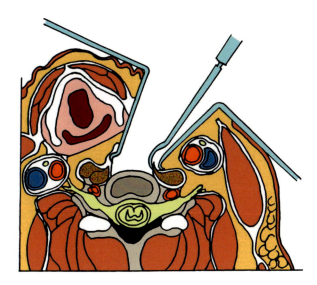

bodies, respectively. Sweeping the index finger medially and laterally should confirm the location of the midline, between the prominent longus coli and capitis muscles running longitudinally. Using toothed forceps, the prevertebral fascia is lifted at this midline point, and incised proximally and distally to reveal the ALL. A small clip is applied to the ALL and X-ray is used to confirm that the correct level has been approached. The medial borders of longus coli are elevated with insulated diathermy, and toothed retractor blades are inserted under each side (Fig. 8).

Procedure

This approach can be used to gain access to the spine for fracture fixation, such as interfragmentary lag screw fixation of the odontoid process. In this scenario dual image-intensifiers are used to simultaneously obtain anteroposterior and lateral fluoroscopy images of the odontoid. Through a C5/6 anterolateral approach, blunt finger dissection is used to dilate the prevertebral space proximally as far as the C2/3 disc ridge. A curved, radiolucent retractor is inserted into this space, and with the fulcrum of the retractor on the distal C2 body, the anterior musculovisceral tissues of the neck are levered off the anterior spine. This enables two guidewires to be driven up from the anterior inferior rim of the C2 vertebral body, across and perpendicular to the oblique fracture and as far as the subcortical bone of the proximal dens.

The anterolateral approach is invariably followed by one or more discectomies. After the correct level is confirmed on X-ray and retractor blades are secured under the medial boarders of longus coli, diathermy is used to expose the distal half of the cranial vertebra and the proximal half of the caudal vertebra, either side of the disc being removed. The midline can be identified using the midpoint between the two uncovertebral (Luschka's) joints. A scalpel blade is used to make a horizontal rectangular annulotomy, removing the ALL and anterior annulus fibrosus of the disc in one piece. A series of fine curettes are then used to scrape the cartilage from the bony endplates, taking care not to violate the end plates as this is likely to lead to spacer implant subsidence. Once some of the disc and cartilage have been removed, it is possible to increase the space inside the disc by inserting a pin distractor. This may not be necessary if the patient has skull traction already fitted. Two pins are placed in the midline, 3 mm from the edge of cranial and cauda vertebra, taking care to insert them parallel to the endplates. This will enable improved visualization of the deeper structures which are to be carefully removed under microscope guidance with 1–2 mm Kerrison rongeurs and a high-speed burr. Complete discectomy is achieved when the posterior annulus, posterior longitudinal ligament, sequestrated disc fragments, or compressive osteophytes are removed and both neuroforamen are probed to confirm lack of any compression on the nerve roots. In the event of a corpectomy, it is useful to perform complete discectomy at two adjacent levels, as detailed here. Corpectomy can then be performed, without the risk of violating the cranial or caudal endplates which are needed to support the subsequent corpectomy cage endplates. For ACDF surgery a spacer containing a bone substitute or autologous graft is inserted into the empty disc space and a plate over the cage is optional.

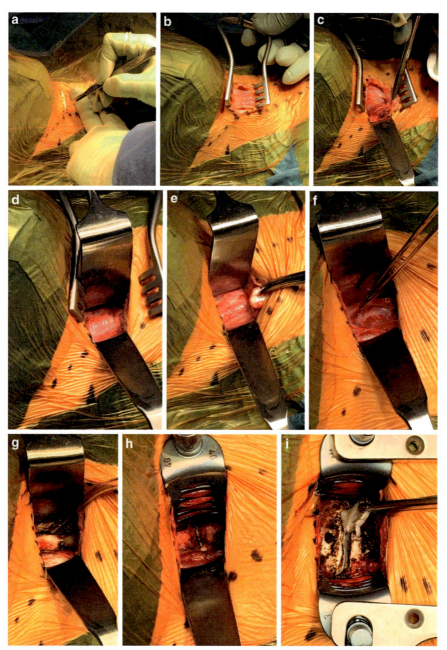

Fig. 8 (continued)

Fig. 8 Anterolateral approach. (**a**) Incision. (**b**) Platysma layer. (**c**) Platysma undermined, revealing the sternocleidomastoid (laterally) and strap muscles (medially). (**d**) Medial retraction of musculovisceral column and lateral retraction of sternocleidomastoid, revealing the carotid sheath (laterally) and pretracheal fascia (medially). (**e**) Medial retraction of the musculovisceral column, and lateral retraction of the carotid sheath, revealing the prevertebral fascia. (**f**) Elevation of the prevertebral fascia and longitudinal incision. (**g**) After the correct level is confirmed on X-ray, the medial edges of longus coli are elevated either side of the disc ridge (covered in ALL, between the longitudinal muscle bellies). (**h**) Toothed retractor blade is carefully tucked in under the longus coli. (**i**) Disc margins are cauterized, annulotomy performed, and disc removed with rongeurs and curettes. (**j**) After complete discectomy and clearance of the posterior longitudinal ligament and osteophytes, an implant is inserted (in this case a cervical disc replacement)

Closure

The platysma muscle is closed over a deep drain, and then fat and skin are closed in separate layers. Suture/clip removal equipment accompanies the patient during the first 12 h in case an urgent evacuation of hematoma is needed to prevent airway compromise.

Structures at Risk

The RLN is at risk during the approach and the time the retractor blades are in situ. It lies in the tracheoesophageal groove and so is not usually encountered during a standard anterolateral approach; it is retracted medially along with the musculovisceral structures and, hence, lies under tension throughout the procedure. The esophagus and pharynx are also at risk, and perforations of either tissue can lead to devastating complications including abscess, tracheoesophageal fistula or mediastinitis. Any abscesses need to be drained and washed out, tears should be repaired, and the patient must remain nil by mouth and receive nasogastric feeds. The sympathetic chain, which lies between the carotid sheath and the longus coli muscle, is at risk from injury if the toothed retractor blades inadvertently slip out from under the longus coli muscles, resulting in Horner's syndrome. The vertebral arteries are at risk as they lie immediately lateral to the uncovertebral joints, which may be opened by either skull traction or segmental pin distractors.

Complications

ACDF is the most commonly performed surgery using the anterolateral approach, with an average of 137,000 ACDF surgeries performed per year in the United States (Epstein 2019). A recent comprehensive review of complications associated with ACDF surgery (Smith and Robinson 1958) has found an overall morbidity rate of 13.2–19.3%. The complications rates are detailed in Table 1. Psueudarthrosis rates rose with the numbers of ACDF levels: 0.4–3% (1 level), 24% (2 level), 42% (3 level), to 56% (4 level).

Table 1 Breakdown of complication rates in ACDF surgery from literature review (1989–2019) (Epstein 2019)

Complication	Rate (%)
Dysphagia	1.7–9.5
Postoperative hematoma	0.4–5.6
Exacerbation of myelopathy	0.2–3.3
Symptomatic RLN palsy	0.9–3.1
Cerebrospinal fluid (CSF) leak	0.5–1.7
Wound infection	0.1–1.6
Worsening of radiculopathy	1.3
Horner's syndrome	0.06–1.1
Respiratory insufficiency	1.1
Esophageal perforation	0.3–0.9
Instrumentation failure	0.1–0.9

Posterior Approach (C3-C7)

Indication

In many respects the posterior approach is more straightforward, and many spine surgeons find it familiar given the large number of posterior lumbar decompressions which are performed each year. This approach is ideally suited to treating any compression of the cord or neural structures caused by posterior structures, such as ligamentum flavum, bone spurs, fracture fragments or hematomas, epidural abscesses, or tumors. Intradural surgery always requires posterior approach and laminectomy. In multilevel stenosis, the posterior approach is often preferred as it enables several levels to be addressed without the morbidity of having prolonged compression on the anterior neurovascular and visceral structures. Posterior approach surgery is also preferred in treating OPLL, large calcified disc prolapse with myelopathy, and as a revision option where anterior approach has already been performed.

Positioning, Preparation, and Practicalities

A neurosurgical head fixation device is fitted to the patient. They are then turned over into a prone position, and the device is fixed to the table (some surgeons prefer to position patients upright). The arms remain by their side on arm supports, and the bed is tilted in slight reverse Trendelenburg position to increase venous drainage and reduce blood loss. The neck is flexed to open the facet joints and to aid the exposure (Fig. 9). Excessive soft tissues around the scapular region can be pulled caudally and taped to the contralateral buttock. The spinous processes of C2 and C7 are usually palpable landmarks. However, it is always advisable to plan the incisions using lateral view X-rays. The skin around the incision should be pre-infiltrated with a lidocaine and 1:200,000 adrenaline solution to reduce the bleeding from the superficial tissues during the procedure.

Approach

A longitudinal midline incision is made in the skin, and sharp dissection down to the ligamentum nuchae. Overzealous stripping of the subcutaneous layer too widely off the ligament should be avoided, as this creates a large dead-space and increases the risk of hematoma and seroma formation. The bifid spinous processes should be palpable through the ligament. The dissection is continued through the ligamentum nuchae in the midline in order to reduce the amount of bleeding. Once the spinous process tip is visible, a small clip is attached to it, and confirmatory lateral fluoroscopic X-rays are obtained. The periosteal dissection is then continued down each spinous process, lamina, and out onto the start of the lateral masses. If posterior instrumentation is to be inserted, the dissection should continue out to the lateral border of the lateral masses. The approach is extensile distally as far as the coccyx, and proximally as far as the occiput (Fig. 10).

Procedure

All or part of the laminae can be removed to enable access to the spinal canal. Care should be taken when inserting the first rongeur or probe through the ligamentum flavum, as at that stage the canal diameter is at its tightest and the cord

Fig. 9 Posterior approach positioning, slight Trendelenburg position with neck flexed

Fig. 10 Posterior cervical approach from external occipital protuberance (inion) to C7 spinous process, out as far as the lateral borders of the lateral masses in preparation for lateral mass screw fixation and fusion

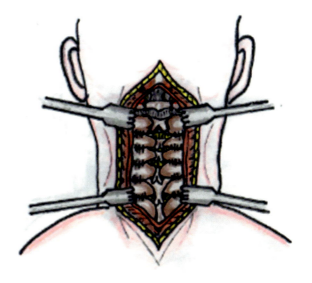

may be vulnerable to injury. The high-speed burr should be used to thin the base of the lamina and only a 1 mm Kerrison rongeur should be used until the canal is mostly decompressed.

If a muscle sparing, lamina-splitting approach was planned, a high-speed burr is used between the bifid spinous process and down through the middle of the lamina. The lateral edges of the laminae are then thinned with the burr until they can be bent out or detach from the base where they meet the lateral masses. This keeps the external muscular attachments while allowing access to the base of the lamina and then the spinal canal. Decompression can be achieved at two levels through this single split spinous process. Various laminoplasty techniques are also possible, either with or without implants.

If stabilization is required due to the extent of the decompression, the lateral masses are cleared of all remaining facet capsule tissue and the boundaries are clearly defined. The screws are then inserted using the surgeon's choice of technique.

Closure

The ligamentum nuchae is repaired over a drain. Copious infiltration with local anesthetic into the deep and superficial tissues around the wound helps to reduce the need for opiates in the immediate post-op phase of recovery.

Structures at Risk

The vertebral artery is largely shielded from the surgeon by the lateral masses, although it is at risk at the proximal end of the approach at C2, or during instrumentation of the lateral masses. It is advisable to have a 14 mm lateral mass screw loaded and ready to immediately insert, in the event of a vertebral artery injury with the lateral mass drill.

Complications

Post-laminectomy kyphosis is a risk associated with posterior approach surgery due to muscle denervation or excessive decompression. The risk increases with age, increasing numbers of levels being decompressed and if C2 is approached, and it can be as high as 47% in these groups (Nishizawa et al. 2012). Other

complications associated with the posterior cervical approach include epidural hematoma (1.3%), wound infection (1.2%), C5 nerve palsy (4.6%), and spinal cord injury (0.18–2.6%) (Cheung and Luk 2016; Nishizawa et al. 2012). Complications relating to lateral mass screws include misplacement (0–7%), vertebral artery injury (1.3–4%), and neural injury (1.3%) (Cheung and Luk 2016; Nishizawa et al. 2012).

Posterior Approach (Occiput-C2)

Indication

The posterior approach to the C1/C2 may be indicated for insertion of C1/2 fixation screws, for reduction and stabilization of a fracture or dislocation, for decompression for degenerative disease, or excision of a tumor. There may be a need to extend the fixation up to the occiput.

Positioning, Preparation, and Practicalities

A neurosurgical head fixation device is fitted to the patient. They are then carefully tuned over into a prone position, and the device is fixed to the table. If the indication is for fracture fixation, a lateral X-ray is obtained to see if adjustments are needed to reduce any displacement which may have occurred during the proning of the patient. The arms remain by their side on arm supports, and the bed is tilted in slight reverse Trendelenburg position to increase venous drainage and reduce blood loss. The neck is flexed under fluoroscopic guidance to allow correct placement of the C1/2 fixation screws. The external occipital protuberance, the first palpable spinous process (C2), and the most prominent spinous process (C7) are usually palpable landmarks. However, it is always advisable to plan the incisions using lateral view X-rays. The skin around the incision should be pre-infiltrated with a lidocaine and 1:200,000 adrenaline solution to reduce the bleeding from the superficial tissues during the procedure.

Approach

A midline incision is made from the external occipital protuberance down to the mid-cervical spine. Overzealous stripping of the subcutaneous layer too widely off the ligament should be avoided, as this creates a large dead-space and increases the risk of hematoma and seroma formation. The bifid spinous processes should be palpable through the ligament, especially those of C2 and C7. The dissection is continued through the ligamentum nuchae in the midline, in order to reduce the amount of bleeding. Once the spinous process tip is visible, a small clip is attached to it, and confirmatory lateral fluoroscopic X-rays are obtained. The periosteal dissection is then continued down the C2 spinous process, lamina, and out onto the lateral masses, and up onto the occiput. The C1 posterior arch is deep between the C2 spinous process and the occiput. The tissues over the posterior arch are carefully dissected outward up to about 1.5 cm from the midline, using either a knife or bipolar diathermy. The vertebral artery runs over the cranial part of the arch and enters the spinal canal behind the lateral mass of atlas (Jeanneret 2015). Only the superior part of the midportion of the posterior C1 arch is prepared. A small sharp elevator is used to identify and subperiosteally clear the cranial surface of the C2 lamina. On sliding along the edge of the lamina, the elevator will meet the isthmus of C2, at which point the soft tissues cranial to the elevator contain the C2 root with its venous plexus. This can be elevated to expose the C1/2 articulation. It can also be ligated leading to occipital numbness but reducing the chance of a painful neuroma caused by constant impingement by a C1 lateral mass screw on a retained C2 nerve root.

Procedure

At this stage either transarticular screws can be passed supplemented with a wired fusion of the spinous process tips. Alternatively, smooth-shank screws can be passed into the lateral masses of C1 and pedicle screws into C2, once again using the fine elevator to feel the trajectory of the C2 pedicles and the C1 lateral masses. The two pairs of polyaxial screws are then connected to the rods to complete the stabilization. If C1 is not to be instrumented, its posterior arch need not be exposed, and rather more exposure of the occiput is required to accommodate the occipital plate. Once fixed to the occiput, the plate can be connected to the cervical screws with two rods bent to 90° to fix the head in a position which will permit a forward gaze.

Closure

The ligamentum nuchae is repaired over a drain. Copious infiltration with local anesthetic into the deep and superficial tissues around the wound helps to reduce the need for opiates in the immediate post-op phase of recovery.

Structures at Risk

The main structure at risk is the vertebral artery. It is most vulnerable to injury during lateral dissection along the posterior arch of the atlas, and one should not attempt to dissect tissues more lateral than the greater occipital nerve (Singh et al. 2011).

Complications

Complications related specifically to upper cervical approaches include adjacent segment disease (7%), pseudoarthrosis (6%) and problems with the occipital plate screws, loosening (4.2–7%), and CSF leak (0–4.2%). As with lower cervical surgery, post-laminectomy kyphosis is a risk associated with posterior approach surgery due to muscle denervation or excessive decompression. The risk increases with age, increasing numbers of levels being decompressed and if C2 is approached, and it can be as high as 47% in these groups (Nishizawa et al. 2012). Other complications associated with the posterior cervical approach include epidural hematoma (1.3%), wound infection (1.2%), C5 nerve palsy (4.6%), and spinal cord injury (0.18–2.6%) (Cheung and Luk 2016; Memtsoudis et al. 2011; Nishizawa et al. 2012). Complications relating to lateral mass screws include misplacement (0–7%), vertebral artery injury (1.3–4%), and neural injury (1.3%) (Nishizawa et al. 2012).

Conclusions

Preparation for cervical spine surgery commences in the outpatient clinic when a condition is diagnosed based upon symptoms, signs, and investigation findings. Surgeons will then make plans on how best to treat the condition and weigh up the advantages and disadvantages of each treatment option. Eventually, they will arrive upon the approach, which is least invasive for the patient, most likely to achieve the surgical goal, and is within their capabilities – in that order of importance. It is clear from reading this chapter that there are often several options for treating the same conditions, which is fortunate as not all patients suit the same approach for a single condition. For example, a music teacher who leads a choir may reasonably have their radiculopathy treated with a posterior cervical foraminotomy to avoid the rare complication risk of an RLN palsy, whereas their pathology could equally have been treated with an ACDF, and may have resulted in less neck pain afterwards.

Once the surgical plan has been made, it is also important to have contingency plans in place to help the surgeon navigate through unexpected challenges that may be encountered in the operating room.

Knowing the anatomy before performing surgery around the cervical spine is of critical importance given the devastating consequences of the complications which can occur. Hopefully, with a better understanding of the various approaches available and their anatomical detail, the surgeon will be better equipped to select the best options for their patients and deliver their management proficiently.

Cross-References

▶ Anterior Spinal Plates: Cervical
▶ Cervical Spine Anatomy
▶ Cervical Total Disc Replacement: Heterotopic Ossification and Complications
▶ Cervical Total Disc Replacement: Technique – Pitfalls and Pearls
▶ Interbody Cages: Cervical

References

Cheung JP, Luk KD (2016) Complications of anterior and posterior cervical spine surgery. Asian Spine J 10 (2):385

Choi D, Crockard HA (2013) Evolution of transoral surgery: three decades of change in patients, pathologies and indications. Neurosurgery 73:296–304

Cloward RB (1958) The anterior approach for removal of ruptured cervical disks. Neurosurgery 16(6):602–617

Crockard HA (1985) The transoral approach to the base of the brain and upper cervical spine. Ann R Col Surg Engl 67(5):321–325

De Andrade JR, Macnab I (1969) Anterior occipito-cervical fusion using an extra-pharyngeal exposure. J Bone Joint Surg 51(8):1621–1626

Epstein NE (2019) A review of complication rates for anterior cervical discectomy and fusion (ACDF). Surg Neurol Int 10:100

Fang HSY, Ong GB (1962) Direct anterior approach to the upper cervical spine. J Bone Joint Surg 44(8): 1588–1604

Henry AK (1970) Extensile exposures, 2nd edn. E and S Livingstone, Edinburgh/London, pp 58–72

Hirabayashi K, Watanabe K, Wakano K, Suzuki N, Satomi K, Ishii Y (1983) Expansive open-door laminoplasty for cervical spinal stenotic myelopathy. Spine 8:693–699

Jeanneret B (2015) Posterior approach to the C1-C2 joints. In: Watkins RG III, Watkins RG IV (eds) Surgical approaches to the spine. Springer, New York, pp 259–267

Kurokawa T, Tsuyama N, Tanaka H, Kobayashi M, Machida H, Nakamura K, Iizuka T, Hoshino Y (1982) Enlargement of spinal canal by the sagittal splitting of spinous processes. Bessatsu Seikeigeka 2:234–240

Lahey FH, Warren KW (1954) Esophageal diverticula. Surg Gynecol Obstet 98:1–28

Magerl F, Seemann PS (1987) Stable posterior fusion of the atlas and axis by transarticular screw fixation. In: Kehr P, Weidner A (eds) Cervical spine, vol 1. Springer, Wien, pp 322–327

McAfee PC, Bohlman HH, Riley JL, Robinson RA, Southwick WO, Nachlas NE (1987) The anterior retropharyngeal approach to the upper part of the cervical spine. J Bone Joint Surg (Am) 69(9):1371–1383

Memtsoudis SG, Hughes A, Ma Y, Chiu YL, Sama AA, Girardi FP (2011) Increased in-hospital complications after primary posterior versus primary anterior cervical fusion. Clin Orthop Relat Res 469(3):649–657

Menezes AH, VanGilder JC (1988) Transoral-transpharyngeal approach to the anterior craniocervical junction: ten-year experience with 72 patients. J Neurosurg 69(6):895–903

Nishizawa K, Mori K, Saruhashi Y, Matsusue Y (2012) Operative outcomes for cervical degenerative disease: a review of the literature. ISRN Orthop 16:2012

Rajabian A, Walsh M, Quraishi NA (2020) Right-versus left-sided exposures of the recurrent laryngeal nerve and considerations of cervical spinal surgical corridor: a fresh-cadaveric surgical anatomy of RLN pertinent to spine. Spine 45(1):10–17

Robinson RA, Smith GW (1955) Anterolateral cervical disc removal and interbody fusion for cervical disc syndrome. Bull Johns Hopkins Hosp 96:223–224

Roy-Camille R, Saillant G, Mazel C (1989) Internal fixation of the unstable cervical spine by a posterior osteosynthesis with plates and screws. In: Sherk HH, Dunn HJ, Eismont FJ, Fielding JW, Long DM, Ono K, Penning L, Raynor R (eds) The cervical spine, 2nd edn. JB Lippincott, Philadelphia, pp 390–403

Shriver MF, Kshettry VR, Sindwani R, Woodard T, Benzel EC, Recinos PF (2016) Transoral and transnasal odontoidectomy complications: a systematic review and meta-analysis. Clin Neurol Neurosurg 148:121–129

Singh K, Munns J, Park DK, Vaccaro AR (2011) Surgical approaches to the cervical spine. In: Devlin VJ (ed) Spine secrets plus. Elsevier Health Sciences, Philadelphia, pp 177–182

Smith GW, Robinson RA (1958) The treatment of certain cervical-spine disorders by anterior removal of the intervertebral disc and interbody fusion. J Bone Joint Surg Am 40:607–624

Verbiest H (1968) A lateral approach to the cervical spine: technique and indications. J Neurosurg 28:191–203

Watkins GW (2015) Lateral approach to the cervical spine (Verbiest). In: Watkins RG III, Watkins RG IV (eds) Surgical approaches to the spine. Springer, New York, pp 45–50

Watkins GW III (2015a) Anterior medial approach to C1, C2 and C3. In: Watkins RG III, Watkins RG IV (eds) Surgical approaches to the spine. Springer, New York, pp 27–32

Watkins GW III (2015b) Transoral approach to C1-C2. In: Watkins RG III, Watkins RG IV (eds) Surgical approaches to the spine. Springer, New York, pp 7–10

Yasargil MG (1977) Microsurgical operation of herniated lumbar disc. Adv Neurosurg 4:81–83

Anti-Inflammatory Agents	1102
Interleukin-6 Receptor Monoclonal Antibody	1102
Glucocorticoids	1103
Celecoxib	1104
Summary of Anti-Inflammatory Agents	1104
Summary	1104
Conclusion	1104
References	1106

Abstract

Chronic low back pain is one of the leading causes of adult disability globally. Currently there are no therapeutic options that target the commonest pathophysiology of back pain which is intervertebral disc degeneration. Intradiscal therapeutics aim to treat some aspects of the pathological disease processes, by regenerating or assisting biophysical characteristics of the intervertebral disc. Intradiscal therapies include biological therapy, cell-free implantation, and anti-inflammatory agents and research conducted into these areas will in time elucidate an effective intradiscal therapeutic for disc degeneration. This chapter reviews the horizon of this exciting area of development.

Keywords

Intradiscal therapeutics · OP-1 · GDF-5 · GDF-6 · HGF · PRP · Link-N · Statins · HSC · Hypertonic dextrose · Chymopapain · Gellified ethanol · Oxygen-ozone · Methylene blue · Hyaluronate hydrogel · Fibrin sealant · Il-6R MAB · Glucocorticoids · Celecoxib

Introduction

Chronic low back pain is one of the most common musculoskeletal diseases worldwide, with 70–85% of adults experiencing back pain at some time in life (Andersson 1999). The accepted leading cause for chronic low back pain is intervertebral disc (IVD) degeneration, with sciatic pain associated with posterior IVD herniation (Luoma et al. 2000). Current treatments such as discectomy or analgesia only target the clinical symptoms of the disease rather than treating the pathological process itself. Intradiscal therapeutics aim to treat aspects of the pathological disease process, by regenerating or assisting (by way of aiding or augmenting) biological processes and thereby physical characteristics of the IVD.

Biological Therapy

Biological therapies function by regenerating an aspect of the IVD. Within early disc degeneration biological therapies involve enhancing anabolic extracellular matrix (ECM) molecules, or to lower the levels of catabolic ECM molecules. Biological therapies focus growth factors to promote anabolic molecule synthesis, either by direct or indirect stimulation (Moriguchi et al. 2016). A presentation of promising and unfavorable results allow insight into which model would be the most appropriate to pursue for bedside therapy.

Bone Morphogenic Protein 7

Bone morphogenic protein 7 (BMP-7), also known as osteogenic protein-1 (OP-1), is part of the bone morphogenic protein (BMP) family which regulates many aspects of embryonic skeletal development, including osteoblast and chondrocyte differentiation, cartilage and bone formation, mesoderm patterning, and craniofacial and limb development (Wan and Cao 2005). OP-1 is used clinically as an adjunct to treatment of fractures and atrophic long bone nonunion. OP-1 has been used within posterolateral lumbar fusion as a substitution for autologous bone. However a large prospective, randomized, multicenter clinical trial of 295 patients suggested in the 4 year follow-up that autograft appeared to have superiority in bridging bone formation (Hustedt and Blizzard 2014). Hence, it was reasonable to investigate a clinically available molecule as a possible intervertebral disc regenerator rather than a spinal fusion enhancer.

The rationale of OP-1 is to increase anabolic ECM proteins. Within the context of IVD regeneration, OP-1 has been experimented in animal trials involving rats, rabbits, and canines, with the disc degeneration models including needle puncture, nucleus pulposus aspiration, compressive loading, chondroitinase induction, and spontaneous degeneration. Delivery of OP-1 into the IVDs has either been direct OP-1 protein injection or cell-based and viral-based OP-1 gene transfer. OP-1 has been demonstrated to increase disc height and MRI T2 intensities compared to the control, indicating its strong ability to regenerate degenerated discs. OP-1 was found to enhance disc mechanical function, maintain spine stability, and alleviate pain-related behaviors. However, the rabbit and mice models used mostly keep the notochordal cells into adulthood, whereas in humans they disappear before the age of 10 and hence the animal IVDs may contain more regenerative capabilities than the human IVD. Moreover, within humans the natural degeneration is a slow process, whereas within the models the applied degeneration is acute and may not mimic the natural human degenerative process (Li et al. 2017). On the other hand, notochordal cells disappear when degeneration is induced prior to OP-1 or other therapeutic injections. Care should be taken when OP-1 delivery is used to treat human disc degeneration due to the differences between human and animal discs.

Research into human IVD regeneration has undergone in vitro experiments. Within a cadaveric model, human nucleus pulposus (NP) cells grown in a medium containing OP-1upregulated proteoglycan synthesis and accumulation, and prevented a decrease in cell numbers (Imai et al. 2007; Wei et al. 2008). In moderately degenerated lumbar discs, capacitively coupled electric stimulation increased BMP-7-induced upregulation of aggrecan and collagen type II (Wang et al. 2017). However, degenerative human NP cells cultured with OP-1 alone, or released from a slowly degrading biomaterial, had no positive effect on the cell growth (van Dijk et al. 2017).

OP-1 has been effective in halting disc degeneration and regenerating degenerative discs within animal models. However, there has been varying results demonstrated within human NP cell regeneration. A phase 2 study was conducted in late 2000's to evaluate the safety and efficacy of OP-1 as a disc regenerator. The results of this trial are not available in the public domain. Further research is mandatory to assess the efficacy and safety of use of OP-1 into human subjects prior to consideration as an intradiscal therapeutic option for disc degeneration.

Growth and Differentiation Factor 5

Growth and differentiation factor 5 (GDF-5) is part of the BMP family and is also known as BMP-14 and cartilage-derived morphogenic protein 1 (CDMP-1). The function of GDF-5 is primarily chondrogenesis during embryonic development. It is present within the joint regions of skeletal precursors to subdivide the precursor into the individual skeletal elements, and is required in all joints for normal development (Settle et al. 2003). Mutation of the GDF-5 gene results in anomalies of development such as Du Pan syndrome, Hunter–Thomson dysplasia, and lumbar disc degeneration.

GDF-5 has been demonstrated to have some reasonable effects on intervertebral disc regeneration. Within live animal models, GDF-5 has regenerated mouse and rabbit intervertebral discs. GDF-5 has also been demonstrated within human cell experiments to increase cell numbers and matrix proteins, establishing it as a protein that can possibly cause IVD regeneration. Delivery methods of GDF-5 into live animals resulting in positive results include protein injection, viral-based gene transfer, and slow releasing biomaterials.

However, there are some disadvantages presented by GDF-5. Repeated injections of GDF-5 results in inflammation of the IVD, which impedes the biological effect of the growth factor (Walsh et al. 2004). GDF-5 has been shown to function in a dose-dependent manner, with a short half-life of 20 min, which means repeated injections would be required in human subjects (Cui et al. 2008). The subsequent inflammation may negate any positive effects of GDF-5. A recent clinical trial NCT00813813 has been conducted assessing injections of GDF-5 into the IVD, with results yet to be published. Our center was involved in the escalating dose arm of the phase 2 multicentric outside the US study. While the average reduction in visual analogue scores for pain was good, the variability in individual patient's response and lack of sponsor support led to cessation of the development of the molecule. Finally, Clarke et al. have demonstrated within an in vitro study that GDF-5 lacked the ability to convert stem cells effectively to disc cells when compared to GDF-6 (Clarke et al. 2014).

Growth and Differentiation Factor 6

Growth and differentiation factor 6 (GDF-6) is part of the BMP family and is also known as BMP-13 and CDMP-2. During embryonic development, GDF-6 is expressed strongly in the notochordal cells of the NP, and key extracellular matrix molecules aggrecan and collagen type II coincide with GDF-6 expression (Wei et al. 2016). GDF-6 mutations are related to several disease states including Klippel–Feil syndrome, and age-related macular degeneration.

GDF-6 has had positive results in regeneration of the IVD. GDF-6 was found within mouse mesenchymal stem cell lines to induce anabolic extracellular matrix (ECM) molecules, without causing ossification, validating its possibility as a regenerative therapy. Conversely with GDF-5, the adenoviral vector transmission did not effectively increase anabolic ECM molecules to a significant level (Zhang et al. 2006, 2007). However, direct culturing of NP cells with GDF-6 elicited positive results (Gulati et al. 2015). An ovine annular puncture model with GDF-6 treatment applied simultaneously with the annular injury showed complete restoration of the IVD height after 4 months (Le Maitre et al. 2015). A rabbit annular puncture model with GDF-6 treatment applied 4 weeks after injury demonstrated faster recovery

with GDF-6, and some recovery of the IVD height (Miyazaki et al. 2018).

GDF-6 has been demonstrated to be a potential future intradiscal biological therapy for IVD degeneration. There are encouraging results from in vitro and in vivo trials providing some evidence that GDF-6 can regenerate the IVD after injury. However, disadvantages to GDF-6 treatment include its cost, time to manufacture, and lack of human data. Future research would include clinical trials to assess the safety and efficacy of the molecule.

Hepatocyte Growth Factor

Hepatocyte growth factor (HGF) was first identified as a potent mitogen of primary cultured hepatocytes. HGF is essential in mammalian development, as disruption of the HGF gene resulted in impaired organogenesis of the liver and placenta and is incompatible with life. HGF is also involved in formation and trophic support of vital organs such as the kidneys, lungs, heart, and brain. The receptor for HGF was identified as c-met proto-oncogene product (c-Met) and the HGF-c-Met signalling pathway leads to multiple biological responses in a variety of cells including mitogenic, morphogenic, and antiapoptotic activities (Funakoshi and Nakamura 2003; Nakamura et al. 2011)

The therapeutic approaches for HGFwithin animal models have been very effective for the treatment of chronic fibrosis in various disease models, such as liver cirrhosis, chronic kidney disease, dilated cardiomyopathy, and lung fibrosis (Nakamura et al. 2011). There have been few human clinical studies with HGF, however a recent randomized, double-blind, placebo-controlled clinical trial of hepatocyte growth factor plasmid for critical limb ischemia demonstrated the safety and efficacy of HGF treatment using naked plasmid (Shigematsu et al. 2010). There are currently several clinical trials occurring in the context of cardiac therapy for conditions such as acute coronary syndrome, myocardial infarction, and critical limb ischaemic.

For the treatment of intervertebral disc degeneration limited research has been conducted. A rat tail disc degeneration model was applied, and a slow release biomaterial was used as the vehicle to deliver HGF. The experiment resulted in an increase in T2-weighted signal intensity on MRI, improved histological score, and stronger immunohistochemical responses demonstrating some prevention of IVD degeneration (Zou et al. 2013). In a rabbit cellular in vitro model HGF promoted NP cell proliferation, inhibited apoptosis and inflammatory cytokine expression, however did not affect matrix protein production (Ishibashi et al. 2016). The results suggest that HGF alters matrix catabolism and hence may be used as an adjunct to prevent degeneration. However as there is no improved anabolic effect, HGF alone may not induce IVD regeneration. Further studies within animal models to elucidate the exact expression of HGF and its receptor c-Met within the IVD is required to research the anabolic response elicited. HGF may be a potential adjunct within future intradiscal IVD therapy.

Platelet-Rich Plasma

Platelet-rich-plasma (PRP) is obtained by concentrating platelets and blood products with the use of a centrifuge. PRP can contain up to eight times the products found in blood and includes not only platelets but a variety of growth factors such as epithelial growth factor (EGF), platelet-derived growth factor (PDGF), transforming growth factors (TGF), and vascular endothelial growth factor (VEGF). Many of these growth factors are stored in granular storage compartments within the platelets, which must be activated to release their contents. The concentration of growth factors parallels the increased concentration of platelets. PRP can increase collagen content, accelerate endothelial regeneration, and promote angiogenesis. Whilst the scientific basis remains to be elucidated, due to ease of availability, and no intellectual property disputes, PRP has been used for various applications such as assisting repairs of the common extensor tendon of the elbow, rotator cuff tendons, and knee articular cartilage (Monfett et al. 2016).

PRP contains growth factors which are claimed to be beneficial in regeneration of IVD degeneration. Within in vitro models of both human and animal cells, PRP increased proteoglycan and NP cell proliferation, as well as upregulated ECM synthesis. PRP also downregulates proinflammatory cytokines and reduces their detrimental effect upon the NP cells (Yang et al. 2016). Within rabbit in vivo models, PRP increases the water content present on T2 MRI within the IVD suggesting a restoration of matrix function, and in some studies increased the disc height of the PRP-injected IVDs. Within rat models, there was an increase of water content present within T2 MRI. The animal models suggested PRP to be appropriate for treatment of IVD degeneration (Monfett et al. 2016; Obata et al. 2012). It remains unclear whether a specific factor helps the PRP achieve its intradiscal effect, or it is the cells and associated media.

There have been clinical trials assessing the safety and efficacy of PRP. Autologous blood cells are used, and are activated by calcium chloride, which is advantageous as there is little side effects using this technique. However, a disadvantage with this technique is that there is variability in the concentration of growth factors within the PRP. Another disadvantage is the vehicle of delivery, as injections themselves can induce inflammatory responses. However, there have been encouraging results presented from a double-blinded randomized control trial with 47 patients, with reduced pain and increased function even up to 2 years later (Tuakli-Wosornu et al. 2016). A feasibility trial observed the effects of stromal vascular fraction and PRP, essentially PRP enhanced with adipose-derived stem cells (ADSC), and within 15 patients after 6 months there was decrease in pain, increase in flexion, and no adverse effects (Comella et al. 2017). The trials suggest PRP may have potential to lower back pain.

PRP has encouraging findings particularly within the clinical setting. Future research would involve phase 2 and 3 trials discovering which group of patients would receive the most benefit from the treatment, finding the optimum concentration for PRP, and combining with other biological and stem cell therapy to find the most effective treatment.

Link-N

Link-N, also known as DHLSDNYTLDHDRAIH is the N-terminal peptide of the link protein. The role of Link-N is to stabilize the proteoglycan aggregates, by binding both to aggrecan and hyaluronate. Link-N is generated by the cleavage of human link protein by stromelysins 1 and 2, gelatinase A and B, and collagenase between His(16) and Ile(17) (Mwale et al. 2003). As Link-N is a synthetic peptide, it has the possibility of being financially beneficial as it is cheap to produce.

Link-N has been researched mainly within the context of IVD repair. Link-N functions by binding to the BMP receptor II, which establishes a classical feed-forward circuit converging on SMAD 1/5 activation. The feed-forward loop is the BMP-RII inducing BMP-4/7 synthesis which activates BMP-RI to produce SMAD 1/5, whilst the BMP-RII also itself induces SMAD 1/5 production (Wang et al. 2013). Link-N is proteolytically cleaved by AF cells into a bioactive short form of the peptide (sLink-N), into a residue spanning amino acid residues 1–8.

Link-N has been demonstrated to increase anabolic ECM proteins. Within both animal and human in vitro models, Link-N has been shown to induce the production of collagen type II, aggrecan, and glycosaminoglycan content. Within an in vivo rabbit annular puncture model, Link-N increased the disc height, increased anabolic proteins, and decreased catabolic cytokine levels (Mwale et al. 2011; Wang et al. 2013). Within a human in vitro model, sLink-N increased the levels of collagen type II and aggrecan. However, Link-N has a limited capacity to overcome catabolic and proinflammatory cytokine expression presented within severely degenerated NP cells (AlGarni et al. 2016; Bach et al. 2017).

Link-N has the capability to upregulate anabolic ECM proteins within human IVD cells, however this may be limited within the late stages of degeneration. It may be able to prevent or regenerate early stages of IVD degeneration. As part of the mechanism of Link-N is activating the BMP-4 and BMP-7 pathways, caution should be taken as these are potentially osteogenic growth factors. The safety and efficacy of Link-N is yet to be established, however the promising results and financial benefit lead Link-N to be an attractive molecule for further research for an intradiscal therapeutic for disc degeneration.

Statins

Statins function by inhibiting 3-hydroxy-3-methylglutaryl-coenzyme A (HMG-CoA) reductase and are the most efficient agent to reduce blood cholesterol with good tolerance. Statins alter the conformation of the enzyme, making the effect very specific. Statins also have anti-atherosclerosis effects, reducing incidents of coronary events (Stancu and Sima 2001).

Simvastatin activates BMP-2 gene expression, a growth factor which stimulates osteoblast proliferation but inhibits osteoblast differentiation, making it an ideal candidate for stimulation of bone formation (Mundy et al. 1999). However within the human IVD cells, BMP-2 facilitated the chondrogenic gene expression of human IVD cells with no evidence of bone nodule formation, and simvastatin was demonstrated within rat IVD cells to upregulate BMP-2 expression and stimulate the production of anabolic ECM molecules (Zhang and Lin 2008). Simvastatin loaded into a hydrogel carrier, injected into rat tails which had undergone annular needle puncture, increased anabolic ECM molecules, collagen type II, and the density on MRI improved (Zhang et al. 2009). Simvastatin was demonstrated within a rat model to function more effectively if injected within a hydrogel carrier, than a saline model, with improvements within MRI and histology and higher gene expression of anabolic ECM molecules (Than et al. 2014). Simvastatin was demonstrated to suppress several catabolic ECM molecules, suggesting that simvastatin may not only be effective in inducing NP cells to express regenerative ECM molecules, but may also help prevent any further degeneration (Tu et al. 2017).

Lovastatin is very similar to simvastatin, with the same mechanism of action, similar hydrophobicity, and same method of metabolization by cytochrome P450 3A. A difference is in reduction of LDL-cholesterol. To reduce cholesterol by 25–30%, 10 mg of simvastatin are required, whereas 20 mg of lovastatin is needed for the same effect (Neuvonen et al. 2008). Simvastatin was shown to stimulate BMP-2 production, and as they have a similar mechanism of action, lovastatin was researched. Within human nucleus pulposus cells in vitro lovastatin upregulated collagen type II genes and chondrogenesis gene SOX9 (Hu et al. 2011). Within a discography-induced rat model, lovastatin was shown to increase anabolic ECM molecules, whilst suppressing collagen type I production which leads to fibrotic scar tissue (Hu et al. 2014).

The early research presented has shown statins to have positive results for intervertebral disc regeneration. However, as the statins mechanism of action for intervertebral disc regeneration rely on the BMP-2 pathway, it is appropriate to consider results of trials investigating BMP-2. BMP-2 has positive results, including inducing mitosis and increasing anabolic ECM molecules, however it has also been demonstrated to reduce proteoglycan content, initiate ossification of the annulus fibrosus, and accelerate osteophyte formation (Belykh et al. 2015). Furthermore, there has been no assessment of regeneraion on disc height and whether catabolic molecules have an effect upon statins function. Whilst early findings are promising, more research is required to assess if statins can be a feasible treatment for intradiscal therapy of disc degeneration.

Haematopoietic Stem Cells

Haematopoietic stem cells (HSC) are progenitor cells which have the capability to differentiate into cellular elements of blood and maintain blood cell production throughout an individual's whole life. HSC transplantation was one of the first proven clinical uses of stem cells, and is being researched to be used as a cure for haematological malignancies (Barriga et al. 2012). HSCs were the first cell therapy used with living humans for treatment of IVD diseases (Wu et al. 2018).

There has been very limited research between HSCs and IVD treatments. According to a prospective analysis of 10 patients found after 1 year, intradiscal HSC injections had 0% reduction in their pain (Haufe and Mork 2006). Moreover, a rabbit model of determining the fate of injected human HSC and MSC cells within coccygeal discs discovered no detection of HSC cells after 21 days, whereas the MSC cells survived and differentiated into a chondrocytic phenotype (Wei et al. 2009). These were the only two studies found assessing HSC in the context of IVD degeneration.

Currently there is little evidence to suggest HSCs have the potential to assist in the treatment of IVD diseases. However, there have been very few studies and hence there is a potential for future research to investigate the relationship between HSC and the IVD, including if within cellular contexts HSC can survive in an avascular tissue such as the IVD, and if they can be differentiated into a chondrocytic phenotype. HSCs are currently unsuitable to be an intradiscal therapy for IVD degeneration.

Hypertonic Dextrose

Prolotherapy is a minimally invasive technique where percutaneous delivery of a therapeutic results in influx of macrophages, fibroblasts, and other molecules with the final goal of new collagen formation strengthening the connective tissue, reducing pain, and disability (Linetsky and Manchikanti 2005). Hypertonic dextrose has been used as a prolotherapy agent for a range of chronic pain conditions with positive results. Prolotherapy within the intervertebral disc may result in improvement of discogenic pain due to modulation of chemoreceptors for pain (Miller et al. 2006).

There have been clinical trials assessing the use of hypertonic dextrose within IVD degeneration. Miller et al. studied advanced lumbar disc pain within 76 patients, with 43.4% of patients having a sustained treatment response with an improvement in pain reduction by 71% after 18 months (Miller et al. 2006). Many other trials have been conducted in treating chronic low back pain, however the hypertonic dextrose has been injected into the surrounding soft tissue, or other areas of the vertebrae including the lamina, spinous process, and vertebral body, with prolotherapy providing significantly greater long-term pain reduction than corticosteroid injection in patients with sacroiliac joint pain (Hauser et al. 2016).

The principle of prolotherapy is introducing a small irritant, to create a low-level inflammatory reaction to induce fibroblasts, growth factors, and other cytokines to induce proliferation and create new connective tissue to strengthen the area (Hauser et al. 2016). Within the intervertebral disc, inflammation induces a degenerative response, and can accelerate degenerative disease (Carragee et al. 2009). Future research into hypertonicdextrose would involve its effect on the IVD, and its long-term safety. Due to the mechanism of prolotherapy theoretically inducing disc degeneration, despite the prospective trial presented having some benefits, it will be with great caution and skepticism that clinicians will assess hypertonic dextrose as a future therapeutic intradiscal treatment for degenerative IVD diseases.

Summary of Biological Therapies

The growth factors BMP-7, GDF-5, and GDF-6 have been demonstrated to be possible sources of future therapy for IVD degeneration, however there is still more research to be conducted. HGF may have a role, but as an adjunct to therapy as an anticatabolic agent. PRP is currently being used in clinical trials, however further results are needed to assess its efficacy. Link-N and statins function through other growth factors which are yet to be established for effective IVD therapy, yet they might bring about possible positive results. HSC and hypertonic dextrose have shown little improvements and probably do not have a role in future intradiscal treatments of IVD degeneration.

Chemonucleolysis

Chemonucleolysis is a percutaneous intradiscal injection to dissolve the nucleus pulposus. This aims to reduce intradiscal pressure, reducing pain

within cervical and lumbar disc herniations. Moreover, destruction of the dermal nerve endings which are present in IVD degeneration within the cartilaginous end plate may also result in pain relief. Interestingly, this technique is used within animal models to induce disc degeneration (Norcross et al. 2003). An exploration of chemonucleolysis will elucidate whether it is an effective treatment for pain relief, or an inducer of degeneration.

Chymopapain

Chymopapain is a proteolytic enzyme derived from the latex of papaya (*Carica papaya*). It is a chemonucleolytic agent, which functions by causing dehydration and degradation of nuclear proteoglycans. Chymopapain was first introduced in 1964 and used in the treatment of lumbar disc herniation until the 1980s, with production discontinued in 2003. Chymopapain treatment involved using a minimally invasive injection with local anesthetic.

There were several clinical trials assessing chymopapain's treatment of symptomatic lumbar disc herniations. Two meta-analyses have been completed investigating chymopapain, which have been summarized by Varshney and Chapman (2012; Couto et al. 2007; Gibson and Waddell 2007). The meta-analyses concluded that chymopapain was superior to placebo in the treatment of symptomatic lumbar disc herniations, with fewer patients having subsequent surgery. In comparison with the gold standard of microdiscectomy, despite gross heterogeneity between 12 studies, surgery was found to be superior to chymopapain for treatment and long-term outcomes. Concerning the safety of chymopapain, the anaphylaxis rate was 0.5%, with a total mortality rate of 0.02%. Compared to microdiscectomy mortality rate of 0.1%, it was considered a safer procedure (Nordby et al. 1993).

The FDA maintains chymopapain on its discontinued product list, however not for reasons of effectiveness. Simmons and Fraser suggest several reasons for the loss of popularity and eventual discontinuation of chymopapain, including poor techniques leading to complications, inappropriate patient selection including patients allergic to papaya, changes in attitude of early rehabilitation after surgery, and use of targeted epidural steroids (Simmons and Fraser 2005). Regarding future research for treatment of degenerated IVDs, chymopapain would not treat the disease state, only causing possible symptomatic relief through chemonucleolysis.

Gelified Ethanol

Pure ethanol is an effective agent to induce chemonucleolysis. However, it is radiopaque and is injected blindly, with a risk of an undetected leak into the epidural space. The leak can damage all components of the IVD, giving rise to severe pain. Gelified ethanol (GE) is pure ethanol enhanced with ethyl cellulose to increase the viscosity of the liquid and regulate its diffusion, thus reducing the risk of epidural leaks, and radiopaque tungsten for visualization (Theron et al. 2007, 2010). GE has advantages over similar techniques as it does not cause the allergic reactions of chymopapain, and it has a more lytic effect than oxygen-ozone.

There have been several trials assessing the safety and efficacy of GE. A preliminary study of 276 patients demonstrated safety of the technique with no pathologic event recorded after 4 years, and patients had an increase in function and reduction in pain (Theron et al. 2007). Thirty-two patients who failed treatment of oxygen-ozone chemonucleolysis therapy were treated with GE, with success within 75% of the patients (Stagni et al. 2012). A further 80 patients with lumbar and cervical disc herniations were treated with GE, and 3 months later 85% of lumbar disc herniation patients and 83% of cervical disc herniations obtained significant improvement in function and reduction of pain (Bellini et al. 2015). Twenty-nine patients with L5-S1 disc herniations with failed conservative treatments were treated with GE, and at the 6–12 month follow-up 66% of patients obtained a 50% relief of pain, with only three patients not experiencing any pain relief from the treatment (Houra et al. 2017).

GE has been demonstrated to be effective at treatment of radicular leg pain, compared with low back pain. The pathophysiological mechanism of radicular pain is most likely a combination of somatic pain from the outer annulus and adjacent ligaments, and neuropathic pain from nerve root compression and chemical inflammatory reaction. GE, by reducing the pressure caused by herniations, serves to reduce this radicular pain. GE's mechanism of action is destruction of the nucleus pulposus, rendering it unsuitable for treatment of degenerative disorders other than disc herniation.

Oxygen-Ozone

Oxygen-ozone (O_2-O_3) is a chemonucleolysis agent which is used in the treatment of lumbar disc herniation. The mechanism of action relies on the chemical properties of both oxygen and ozone. Oxygenation leads to reduced pain due to the oxidization of proinflammatory mediators and improves microcirculation within the compressed areas. Ozone is an unstable allotropic form of oxygen, and reacts with proteoglycan glycosaminoglycans (GAG) to form oxidization products smaller than the original GAGs, reducing the osmotic pressure of the NP, and causing dehydration and shrinking the disc (Murphy et al. 2016; Muto et al. 2004). Empirical studies have demonstrated the optimal concentration of ozone per millimetre of oxygen to be 27 μg (Iliakis et al. 2001).

Several clinical studies have been undertaken assessing O_2-O_3 therapy for treatment of lumbar disc herniation. Intradiscal injections of O_2-O_3 have been demonstrated to reduce pain and improve function after 1 month (Murphy et al. 2015), 6 months (Lehnert et al. 2012), 6 months with an additional paraganglionic injection (Andreula et al. 2003), and after 2 years intradiscal injections alone (Das et al. 2009). Intraformational injections have also been demonstrated to have success after 6 months with about 75% reduction in pain (Bonetti et al. 2005; Perri et al. 2015). And peri radicular with paraganglionic injections have had 80% success after 6 months and 75% at 18 months (Muto et al. 2004).

O_2-O_3 therapy has been demonstrated to be effective in treating the symptoms of disc herniation. However, there are some risks to the therapy, as the dose of ozone must not exceed the capacity of antioxidant enzymes. Excess ozone leads to accumulation of the superoxide anion (O_2-) and hydrogen peroxide (H_2O_2), which can cause cell membrane degradation (Andreula et al. 2003). The injections also present a source of infection, as demonstrated by a case study involving an infection of Achromobacter xylosoxidans following O_2-O_3 therapy (Fort et al. 2014). Similarly, to GE, O_2-O_3 is unsuitable for treatment of IVD disorders other than pain caused by herniation.

Methylene Blue

Methylene blue, also known as methylthioninium chloride, can be used as both medication and a dye. It is part of the WHO Model List of Essential Medicines, particularly as an antidote for poisonings. Methylene blue is an inhibitor of nitric oxide synthase, and guanylate synthase. It has been used in a wide variety of treatments, including improving arterial pressure in septic shock, treatment of methemoglobinemia, neutralization of heparin, and more. Methylene blue has also been used as a dye to locate lesions within parathyroid glands, intestinal lumen, lymph nodes, and as a rapid detection method for Helicobacter pylori on histology. Adverse effects of methylene blue include toxicity in high doses to the renal system, cardiovascular system, and pulmonary system. Contraindications include patients with renal insufficiency, and G6PD-deficient patients (Ginimuge and Jyothi 2010).

Methylene blue has shown to destroy dermal nerve endings. Part of the pathophysiology of pain from intervertebral disc degeneration is ingrowth of new nerve vessels. A pilot study of 72 patients was conducted under the rationale that methylene blue would destroy nerve fibres growing within the annulus, alleviating discogenic pain. The study found after 2 years from treatment, 91.6% of patients were satisfied with their outcome, and 89% of patients had obvious to complete alleviation of pain (Peng et al. 2010). A clinical trial with 33 patients assessed the use of methylene blue injections for low back pain, and had 81% success at 3 months, however after 12 months only 54% of patients had reduced pain and increase in function (van Dijk et al. 2017). A study of 24 patients discovered 87% of patients to have alleviation of back pain, and improvements in physical function after an average of 18 months (Peng et al. 2007). Another study of 20 patients found after 1 year that only 20% of patients had successful alleviation of pain (Kim et al. 2012), and a retrospective case series of 8 patients only identified 1 to have any benefits from methylene blue injections (Gupta et al. 2012).

Methylene blue has been demonstrated within the previous clinical studies to have varying results. There are differences in measurements of pain and assessments of function; however the general trend is that methylene blue is appropriate for reduction of short-term pain, and loses its potency over time. Factors contributing to this may include dissipation of the methylene blue, and regrowth of nerve endings, or a more complex pathophysiological system may be causing the pain. There have been little assessments on the effect of methylene blue on the histology of the intervertebral discs, and future research would identify the exact response of the intervertebral disc to methylene blue. Methylene blue has been shown to have varying degrees of success, however a lack of understanding about its function on disc cells makes its use simply for pain alleviation, and symptom control, but not treating the root cause of the disease state and may be an effective adjunct in finding a cure for disc degeneration.

Summary of Chemonucleolysis

Chemonucleolysis is a destructive tool which can be very effective in the reduction of pain during disc herniation. Chymopapain was found to cause anaphylaxis, due to the papaya content. The other agents have found to provide useful short-term benefits, however long-term outcomes vary. This technique does not treat the cause of the degeneration and is only useful for the treatment of pain caused by herniation of disc content.

Chemonucleolysis may be useful in destruction of the intradiscal nerve endings, however, it will only serve as either a very specific treatment or an adjunct to any future therapy developed.

Cell-Free Intradiscal Implantation

Intradiscal implantations are used either after nucleotomy procedures to mimic the mechanical properties of the IVD, or as a mechanism to repair annular fibrosus defects. They can also become a medium in which the NP cells can migrate into and proliferate within. They are able to be manufactured prior to surgery and then implanted.

Hyaluronate Hydrogel

Injectable hyaluronate hydrogels (HH) are cell-free intradiscal matrix implantations that have been proposed to limit disc degeneration following a nucleotomy procedure. HH have an anti-inflammatory effect within the intervertebral disc, and also provide a growth-permissive environment for NP cells and MSCs (Isa et al. 2015; Priyadarshani et al. 2016).

HH has been investigated within several animal models. Hyaluronic acid alone reduced degeneration on imaging following nucleotomy procedures in nonhuman primate lumbar spines (Pfeiffer et al. 2003). Within a rabbit annular puncture model of IVD degeneration, cross-linked HH and cross-linked chondroitin sulphate hydrogel retained MRI T2 intensity for 3 months, and histologically had increased proteoglycan staining (Nakashima et al. 2009). Within a sheep nucleotomy model, after 6 months HH completely regenerated sheep IVD histologically, biochemically, and radiologically (Benz et al. 2012). However, within a porcine nucleotomy model, HH did not affect the regeneration of the IVD, and caused further annular scarring and localized annular inflammation (Omlor et al. 2012). Within each of these animal models, the exact makeup of the HH is different, with the best results of Benz et al. incorporating autologous serum solution into their injectable HH solution.

HH has been demonstrated to be an effective medium for IVD cells to proliferate. A gel culture with 4% hyaluronan cross-linked with serum albumin was demonstrated to be a viable medium for the culturing of human IVD cells, and chondrogenic MSCs, stimulating the release of anabolic ECM molecules (Benz et al. 2010). Human NP cells cultured for 8 weeks within a HH showed functional matrix accumulation and synthesis, and these results were higher at lower density of NP cells (20 million cells per millilitre) (Kim et al. 2015). A study cross-linking type-II collagen HH with 1-ethyl-3(3-dimethyl aminopropyl) carbodiimide (EDC) was also demonstrated to be a viable growth medium for culturing of NP cells (Priyadarshani et al. 2016). Another hydrogel with chitosan and hyaluronic acid cross-linked with glycerol phosphate promoted ADSC proliferation and nucleus pulposus differentiation (Zhu et al. 2017).

HH is a potential future intradiscal therapy for degenerative IVD disease. Advantages of HH include its versatility and ability to create different cross-linking to find the optimal hydrogel for regeneration of the NP. It also has the capacity to be a growth medium for NP cells and MSCs, potentially improving its regeneration ability. However, as the variability of the animal models demonstrate, not only is there a capacity for HH to cause complete regeneration, there is also the possibility of HH solutions to cause more degeneration. Care must be taken in future research to develop an HH which is both efficacious and safe. HH has the potential to be an effective intradiscal therapy for IVD degeneration.

Fibrin Sealant

Fibrin sealant (FS) can be used within surgery to prevent local haemorrhaging complications. It functions by consisting of two components, factor XII and fibronectin (the sealant) and thrombin and albumin (the catalyst) (Canonico 2003). When the sealant and catalyst are mixed, they create a fibrin monomer, and a stable clot. Fibrin is required for normal wound healing, and factor XII and fibronectin is important for fibroblast proliferation and adhesion. Within the IVD the fibrin sealant would function to seal annular fissures from proinflammatory substances and facilitate disc healing (Buser et al. 2011).

Fibrin has been shown to be an alternative three-dimensional cell carrier to cultivate porcine and rabbit NP cells in vitro (Sha'Ban et al. 2008; Stern et al. 2000). In vitro FS was demonstrated to decrease proinflammatory cytokine levels in both human and porcine cells (Buser et al. 2014). Within a porcine nucleotomy model FS preserved the disc architecture, reduced section of inflammatory cytokines, and recovered the mechanical properties lost from the nucleotomy. FS also increased proteoglycan synthesis and inhibited progressive fibrosis of the NP (Buser et al. 2011). A prospective clinical trial of 15 patients found improvements in pain relief and function at weeks 15, 26, 52, and 104 with low complication rates (Yin et al. 2014).

Guterl et al. added genipin (Fib-Gen), a plant-based chemical cross-linker with low cytotoxicity, alongside cell adhesion molecules fibronectin and collagen and successfully improved the shear properties of FS (Guterl et al. 2014). Fib-Gen was injected into in vitro bovine IVDs which had undergone nucleotomy, and shown to effectively seal any defects, and prevented IVD height loss from the induced compressive force. Fib-Gen maintained AF cells, and NP cells migrated into the gel (Likhitpanichkul et al. 2014). Fib-Gen has also been demonstrated to be an effective drug carrier, maintaining a constant slow release of infliximab, and more effectively than FS alone (Likhitpanichkul et al. 2015). Long et al. elucidated the biomechanical properties of Fib-Gen, by performing annular injuries on bovine coccygeal IVDs in vitro and repairing the defect with Fib-Gen within a scaffold. Long et al. demonstrates Fib-Gen to reduce disc height loss, had little herniation risk, but could only partially restore disc biomechanical behaviors (Long et al. 2016).

FS is a promising therapy due to its potential for multipurpose action. Whist originally intended to seal annular fissures and prevent ingrowth of nerve vessels leading to pain, the possibility of being loaded with pharmaceuticals allows for optimization of the treatment. Moreover, the creation of Fib-Gen and attempts to biomechanically replicate the NP provide an adjunct treatment to perform alongside discectomy. Cellular migration of NP and AF cells within FS and Fib-Gen also may strengthen the material within in vivo trials. However, care must be taken as there may be fibrin instability and solubility over time, particularly when exposed to cells, and currently it does not provide the same mechanical properties as the IVD (Colombini et al. 2014). FS has the potential to be an effective intradiscal therapeutic, either as an adjunct to pharmacological therapy or as an adjunct to discectomy.

Summary of Cell-Free Intradiscal Implantations

Both HH and FS are promising as they can not only repair or restore some mechanical function of the IVD, but they can also house either NP or stem cells for continued treatment. However, large drawbacks include trying to compare the studies, as each proposes a different formula for the creation of the hydrogel. Another disadvantage is that they cannot fully match the durability and mechanical properties of the IVD, leading to degradation and possible replacement after an unknown period of time. Intradiscal implantations have the potential to be used alongside biological treatments to be an effective intradiscal therapy for IVD degeneration.

Anti-Inflammatory Agents

Anti-inflammatory agents are used to reduce the catabolic ECM molecules within IVD degeneration. They also reduce proinflammatory cytokines which are a source of pain. They achieve this through preventing their synthesis, and directly inhibiting their action upon the cells.

Interleukin-6 Receptor Monoclonal Antibody

Tocilizumab is a recombinant monoclonal IgG1 antihuman interleukin-6 receptor antibody (IL-6R mAB). Also known as Atilizumab, it is an immunosuppressive drug used within the treatments of disorders such as rheumatoid arthritis, and systemic juvenile idiopathic arthritis (Rosman et al. 2013). IL-6 is an inflammatory cytokine which functions by activating the 130 gp signal transducer, inducing angiogenesis, and amplifying activity of adhesion molecules. Tocilizumab was found to be beneficial and safe for treatment of rheumatoid arthritis in cases of nonresponse to anti-TNF-alpha therapy or when anti-TNF-alpha therapy is contraindicated (Rosman et al. 2013).

IL-6 expression is secreted by intervertebral discs, with raised expression present in herniated discs. IL-6 levels are also upregulated in degenerative IVDs, and upregulates other catabolic cytokines that attribute to intervertebral disc degeneration. IL-6 is also a cause of discogenic pain, as it induces apoptosis of neuronal cells in the dorsal root ganglion, which may contribute to allodynia and hyperalgesia (Risbud and Shapiro 2014).

There have been few experiments concerning IL-6 and IVD regeneration. A mouse degeneration model demonstrated IL-6R mAB to reduce IL-6 expression, and decreased pain-related peptide release within the dorsal root ganglions (Sainoh et al. 2015). A prospective comparative cohort study revealed tocilizumab to provide short-term, 2 weeks, relief of back pain (Sainoh et al. 2016). An in vitro trial elucidated human degenerative annular fibrosus cells induces the expression of IL-6 through the JAK/STAT pathway, and therefore may be causally linked to IVD degeneration (Suzuki et al. 2017).

IL-6 is a catabolic factor which contributes towards IVD degeneration. It is unclear if IL-6 is the principal cytokine involved in the development of IVD degeneration, or functions as an ancillary to other cytokines. Future research is required to clarify this relationship, however IL-6 treatment would only slow the degenerative process, and would not be able to induce regenerative capabilities.

Glucocorticoids

Glucocorticoids are a class of corticosteroids and are created endogenously within the adrenal cortex. The function of glucocorticoids is to reduce the synthesis and release of a variety of inflammatory mediators (Becker 2013). Vertebral degenerative changes at the lumbar spine can be classified into a three-stage system using Modic classification (Modic et al. 1988). Within Modic type 1, the cartilaginous end plates have a high amount of proinflammatory cytokines, and increased vascularity, indicating an inflammatory reaction and Modic type 1 changes are closely associated with chronic lower back pain (Beaudreuil et al. 2012). The rationale of glucocorticoids is to stop the inflammatory process, and provide symptomatic relief and prevent further degeneration.

Clinical trials have illustrated glucocorticoids to have a mixed effect in the treatment of chronic low back pain. Glucocorticoids have a mostly positive effect at reducing pain and improving function within the short term, up to 1 month (Beaudreuil et al. 2012; Benyahya et al. 2004; Buttermann 2004; Fayad et al. 2007; Nguyen et al. 2017) and no effect after 12 months (Khot et al. 2004) with only one paper revealing glucocorticoids to have no effect within early time periods (Simmons et al. 1992). Conversely, Benyahya et al. found after 6 months 43.5% of 67 patients had improvements in pain reduction and function (Benyahya et al. 2004). Long-term results were improved when combined with an alternative therapy such as O_2-O_3, with paraganglionic glucocorticoid injections providing treatment and increased function at 1 month and 6 months (Andreula et al. 2003; Murphy et al. 2015) and combined with a polypeptide, positive results after 3 and 6 months (Cao et al. 2011).

Concerns about glucocorticoids include a risk that it can cause degeneration and primary calcification as demonstrated within a rabbit model (Aoki et al. 1997). Moreover intradiscal procedures can create inflammation (Ulrich et al. 2007) and possibly contribute to IVD degeneration (Carragee et al. 2009), with the potential of glucocorticoids to have a toxic effect on intradiscal cells. Glucocorticoids has been generally very well tolerated within the clinical trials presented. Glucocorticoids may have symptomatic relief for a short time of chronic low back pain, however as its mechanism is only preventing inflammation, future research may reveal it to be a helpful adjunct with other treatments, as opposed to a cure. As an intradiscal treatment glucocorticoid is currently only appropriate for short-term symptomatic relief, and inappropriate for curative treatment.

Celecoxib

Celecoxib is a nonsteroidal anti-inflammatory drug (NSAID), and was the first cyclooxygenase-2 (COX-2) selective inhibitor introduced into clinical practice. Inhibition of COX-2 results in anti-inflammatory and analgesic effects, with fewer gastrointestinal side effects than NSAIDs. Celecoxib is eliminated by hepatic metabolism involving primarily the CYP2C9 protein, with a half-life of 11 to 16 h (McCormack 2011; Shi and Klotz 2008). COX-2 inhibitors can be effective in reducing back pain and reducing degeneration, as they prevent the formation of proinflammatory cytokines such as prostaglandin E2 (PGE2), which shifts the environment to a catabolic state leading to regeneration. However, systemic delivery of COX-2 at levels high enough to alleviate chronic low back pain is associated with comorbidities and side effects, hence a localized delivery system would provide more effective pain relief with fewer side effects (Tellegen et al. 2018).

Celecoxib does not reduce the rate of nerve growth within the nucleus pulposus (Olmarker 2005). Canine models have demonstrated that separate hydrogels are safe and feasible when loaded with celecoxib, with one study having significant improvement in 9 out of 10 canines, and at 3 months only 3 having recurring back pain, however no changes in %DHI or MRI T2 brightness (Tellegen et al. 2018; van Dijk et al. 2015). Within a bovine NP cellular model replicating herniation, celecoxib reduced PGE2 proving within the intervertebral disc celecoxib is effective in reducing catabolic enzymes (van Dijk et al. 2015). Within a cox-2 knockout mouse experiment (Cox-2 $-/-$), the deficiency causes delay in the ossification of lumbar vertebral endplates, and plays a role in IVD degeneration by affecting the sonic hedgehog and BMP signaling pathways (Ding et al. 2018).

Celecoxib can reduce the levels of the catabolic molecule PGE2. However, there has been no evidence so far that celecoxib has other effects than in reducing pain, and potentially reducing the rate of degeneration. Future research would involve perfecting the hydrogel to load with celecoxib, and then observing if within humans it can reduce pain levels as demonstrated within canines. Other future research into the relationship between COX-2 and intervertebral disc development may lead to a novel solution to prevent early degenerative disease. However with the current evidence, Celecoxib may be an effective adjunct in preventing pain alongside other intradiscal therapies for disc degeneration.

Summary of Anti-Inflammatory Agents

Intradiscal anti-inflammatory agents have mixed results within the treatment of IVD degeneration. Tocilizumab has promising results, but has not been trialled in clinical practice. Glucocorticoids have been demonstrated to have a positive early response, albeit with negative long-term results, and celecoxib is unknown if it can reduce pain. More evidence is required for the effectiveness of intradiscal applications of anti-inflammatory agents, and at most they might be a useful adjunct with treatments for intradiscal therapies of IVD degeneration.

Summary

The following Table 1 is a summary of the current research of intradiscal therapies.

Conclusion

Biological therapies, intradiscal implantations, and anti-inflammatory agents have been explored as possible future intradiscal therapies for IVD degeneration. The most promising research is within the biological and intradiscal therapies, as they have the possibilities of working in conjunction to halt degeneration, induce regeneration, and

Table 1 A summary of the studies of intradiscal injections and implants. BMP-7, bone morphogenic protein 7; OP-1, osteogenic protein 1; GDF-5, growth and differentiation factor 5; GDF-6, growth and differentiation factor 6; HGF, hepatocyte growth factor; IL-6R mAB, interleukin-6 receptor monoclonal antibody; HSC, haematopoietic stem cell

Studies of intradiscal injection/implants

Category	Drug/Material	Product name	Development stage		Available on market	Reference
			Preclinical study (animal)	Clinical trial		
Biological	BMP-7	OP-1	X (rabbit)	X (Phase 1)	X	Imai Y et al. 2007
	GDF-5	rhGDF5		X (Phase 1 and 2a completed) n = 40		NCT01158924
	GDF-6		X (sheep)			Wei A et al. 2009
	HGF		X (rat)			Zou F et al. 2013
	Platelet-rich plasma		X (rabbit)	X (Phase 2) n = 112	Autologous	NCT02983747 Monfett M et al. 2016
	Link-N		X (rabbit)			Mwale F et al. 2011
	Simvastatin	Zocor	X (rat)		X	Than KD et al. 2014
	Lovastatin	Mevacor	X (rat)		X	Hu MH et al. 2014
	HSC			X (Pilot study) n = 10	Autologous	Haufe SMW and Mork AR 2006
	Hypertonic dextrose			X (Pilot study)		Miller MR et al. 2006
Chemonucleolysis	Chymopapain			X (Pilot study) n = 17	X	Jenner JR et al. 1986
	Gelified ethanol	Discogel	X	X (Phase 1) n = 40	X	NCT02343484 Stagni S et al. 2012
	Oxygen-ozone			X		Muto M et al. 2004
	Methylene blue		X	X (Phase 1) n = 40	X	NTR2547 (NL) Geurts JW et al. 2015
Cell-free intradiscal implantation	Hyaluronate hydrogel		X (rabbit)			Nakashima S et al. 2009
	Chondroitin sulfate hydrogel		X (rabbit)			Nakashima S et al. 2009
	Fibrin sealant	BIOSTAT BIOLOGX		X (Phase 1) terminated	X	NCT01011816 Yin W et al. 2014
Anti-inflammatory	IL-6R mAb	Tocilizumab;Actemra; RoActemra		X (Phase 1) n = 31	X	Sainoh T et al. 2016
	Glucocorticoid	Hydrocortancyl (Prednisolone)		X (Phase 4) n = 137	X	NCT00804531 Nguyen C et al. 2017
	Celecoxib	Celebrex	X (dog)		X	Tellegen AR et al. 2016 OARSI Abstract

provide support for the recovering IVD. Anti-inflammatory agents may be used as an useful adjunct for therapies, however their efficacy is not yet conclusive. There are currently no intradiscal therapies available, however with the development of so many different avenues of research the promise for treatment being soon available is great.

References

AlGarni N, Grant MP, Epure LM, Salem O, Bokhari R, Antoniou J, Mwale F (2016) Short Link N stimulates intervertebral disc repair in a novel long-term organ culture model that includes the bony vertebrae. Tissue Eng A 22:1252–1257

Andersson GBJ (1999) Epidemiological features of chronic low-back pain. Lancet 354:581–585. https://doi.org/10.1016/s0140-6736(99)01312-4

Andreula CF, Simonetti L, De Santis F, Agati R, Ricci R, Leonardi M (2003) Minimally invasive oxygen-ozone therapy for lumbar disk herniation. Am J Neuroradiol 24:996–1000

Aoki M, Kato F, Mimatsu K, Iwata H (1997) Histologic changes in the intervertebral disc after intradiscal injections of methylprednisolone acetate in rabbits. Spine 22:127–131

Bach FC et al (2017) Link-N: the missing link towards intervertebral disc repair is species-specific. PLoS One 12:e0187831

Barriga F, Ramírez P, Wietstruck A, Rojas N (2012) Hematopoietic stem cell transplantation: clinical use and perspectives. Biol Res 45:307–316

Beaudreuil J, Dieude P, Poiraudeau S, Revel M (2012) Disabling chronic low back pain with Modic type 1 MRI signal: acute reduction in pain with intradiscal corticotherapy. Ann Phys Rehabil Med 55:139–147

Becker DE (2013) Basic and clinical pharmacology of glucocorticosteroids. Anesth Prog 60:25–32

Bellini M et al (2015) Percutaneous injection of radiopaque gelified ethanol for the treatment of lumbar and cervical intervertebral disk herniations: experience and clinical outcome in 80 patients. Am J Neuroradiol 36:600–605

Belykh E, Giers M, Bardonova L, Theodore N, Preul M, Byvaltsev V (2015) The role of bone morphogenetic proteins 2, 7, and 14 in approaches for intervertebral disk restoration. World Neurosurg 84:871–873

Benyahya R et al (2004) Intradiscal injection of acetate of prednisolone in severe low back pain: complications and patients' assessment of effectiveness. Ann Readapt Med Phys 9:621–626

Benz K et al (2010) A polyethylene glycol-crosslinked serum albumin/hyaluronan hydrogel for the cultivation of chondrogenic cell types. Adv Eng Mater 12:B539

Benz K et al (2012) Intervertebral disc cell-and hydrogel-supported and spontaneous intervertebral disc repair in nucleotomized sheep. Eur Spine J 21:1758–1768

Bonetti M, Fontana A, Cotticelli B, Dalla Volta G, Guindani M, Leonardi M (2005) Intraforaminal O_2-O_3 versus periradicular steroidal infiltrations in lower back pain: randomized controlled study. Am J Neuroradiol 26:996–1000

Buser Z, Kuelling F, Liu J, Liebenberg E, Thorne KJ, Coughlin D, Lotz JC (2011) Biological and biomechanical effects of fibrin injection into porcine intervertebral discs. Spine 36:E1201–E1209

Buser Z, Liu J, Thorne KJ, Coughlin D, Lotz JC (2014) Inflammatory response of intervertebral disc cells is reduced by fibrin sealant scaffold in vitro. J Tissue Eng Regen Med 8:77–84

Buttermann GR (2004) The effect of spinal steroid injections for degenerative disc disease. Spine J 4:495–505

Canonico S (2003) The use of human fibrin glue in the surgical operations. Acta Biomed 74:21–25

Cao P, Jiang L, Zhuang C, Yang Y, Zhang Z, Chen W, Zheng T (2011) Intradiscal injection therapy for degenerative chronic discogenic low back pain with end plate Modic changes. Spine J 11:100–106

Carragee EJ, Don AS, Hurwitz EL, Cuellar JM, Carrino J, Herzog R (2009) 2009 ISSLS prize winner: does discography cause accelerated progression of degeneration changes in the lumbar disc: a ten-year matched cohort study. Spine 34:2338–2345

Clarke LE, McConnell JC, Sherratt MJ, Derby B, Richardson SM, Hoyland JA (2014) Growth differentiation factor 6 and transforming growth factor-beta differentially mediate mesenchymal stem cell differentiation, composition, and micromechanical properties of nucleus pulposus constructs. Arthritis Res Ther 16:12

Colombini A, Ceriani C, Banfi G, Brayda-Bruno M, Moretti M (2014) Fibrin in intervertebral disc tissue engineering. Tissue Eng Part B Rev 20:713–721

Comella K, Silbert R, Parlo M (2017) Effects of the intradiscal implantation of stromal vascular fraction plus platelet rich plasma in patients with degenerative disc disease. J Transl Med 15:12

Couto JMC, Castilho EA, Menezes PR (2007) Chemonucleolysis in lumbar disc herniation: a meta-analysis. Clinics 62:175–180

Cui M et al (2008) Mouse growth and differentiation factor-5 protein and DNA therapy potentiates intervertebral disc cell aggregation and chondrogenic gene expression. Spine J 8:287–295. https://doi.org/10.1016/j.spinee.2007.05.012

Das G, Ray S, Ishwarari S, Roy M, Ghosh P (2009) Ozone nucleolysis for management of pain and disability in prolapsed lumber intervertebral disc: a prospective cohort study. Interv Neuroradiol 15:330–334

Ding Q et al (2018) Cyclooxygenase-2 deficiency causes delayed ossification of lumbar vertebral endplates. Am J Transl Res 10:718

Fayad F et al (2007) Relation of inflammatory modic changes to intradiscal steroid injection outcome in chronic low back pain. Eur Spine J 16:925–931

Fort NM, Aichmair A, Miller AO, Girardi FP (2014) L5–S1 Achromobacter xylosoxidans infection secondary to oxygen-ozone therapy for the treatment of lumbosacral disc herniation: a case report and review of the literature. Spine 39:E413–E416

Funakoshi H, Nakamura T (2003) Hepatocyte growth factor: from diagnosis to clinical applications. Clin Chim Acta 327:1–23

Geurts JW, Kallewaard J-W, Kessels A, Willems PC, Van Santbrink H, Dirksen C, Van Kleef M (2015) Efficacy and cost-effectiveness of intradiscal methylen blue injection for chronic discogenic low back pain: study protocol for a randomized controlled trial. Trials 16 (1):532

Gibson JNA, Waddell G (2007) Surgical interventions for lumbar disc prolapse: updated Cochrane review. Spine 32:1735–1747

Ginimuge PR, Jyothi S (2010) Methylene blue: revisited. J Anaesthesiol Clin Pharmacol 26:517

Gulati T, Chung SA, Wei A-Q, Diwan AD (2015) Localization of bone morphogenetic protein 13 in human intervertebral disc and its molecular and functional effects in vitro in 3D culture. J Orthop Res 33: 1769–1775

Gupta G, Radhakrishna M, Chankowsky J, Asenjo JF (2012) Methylene blue in the treatment of discogenic low back pain. Pain Physician 15:333–338

Guterl CC et al (2014) Characterization of mechanics and cytocompatibility of fibrin-genipin annulus fibrosus sealant with the addition of cell adhesion molecules. Tissue Eng A 20:2536–2545

Haufe SM, Mork AR (2006) Intradiscal injection of hematopoietic stem cells in an attempt to rejuvenate the intervertebral discs. Stem Cells Dev 15:136–137

Hauser RA, Lackner JB, Steilen-Matias D, Harris DK (2016) A systematic review of dextrose prolotherapy for chronic musculoskeletal pain. Clin Med Insights: Arthritis Musculoskelet Disord 9:CMAMD. S39160

Houra K, Perovic D, Rados I, Kvesic D (2017) Radiopaque gelified ethanol application in lumbar intervertebral soft disc herniations: Croatian multicentric study. Pain Med 19:1550

Hu MH, Hung LW, Yang SH, Sun YH, Shih TTF, Lin FH (2011) Lovastatin promotes redifferentiation of human nucleus pulposus cells during expansion in monolayer culture. Artif Organs 35:411–416

Hu M-H, Yang K-C, Chen Y-J, Sun Y-H, Yang S-H (2014) Lovastatin prevents discography-associated degeneration and maintains the functional morphology of intervertebral discs. Spine J 14:2459–2466

Hustedt JW, Blizzard DJ (2014) The controversy surrounding bone morphogenetic proteins in the spine: a review of current research. Yale J Biol Med 87:549

Iliakis E, Valadakis V, Vynios D, Tsiganos C, Agapitos E (2001) Rationalization of the activity of medical ozone on intervertebral disc a histological and biochemical study. Riv Neuroradiol 14:23–30

Imai Y, Miyamoto K, An HS, Eugene J-MT, Andersson GB, Masuda K (2007) Recombinant human osteogenic protein-1 upregulates proteoglycan metabolism of human anulus fibrosus and nucleus pulposus cells. Spine 32:1303–1309

Isa ILM, Srivastava A, Tiernan D, Owens P, Rooney P, Dockery P, Pandit A (2015) Hyaluronic acid based hydrogels attenuate inflammatory receptors and neurotrophins in interleukin-1β induced inflammation model of nucleus pulposus cells. Biomacromolecules 16:1714–1725

Ishibashi H et al (2016) Hepatocyte growth factor/c-met promotes proliferation, suppresses apoptosis, and improves matrix metabolism in rabbit nucleus pulposus cells in vitro. J Orthop Res 34:709–716

Jenner J, Buttle DJ, Dixon A (1986) Mechanism of action of intradiscal chymopapain in the treatment of sciatica: a clinical, biochemical, and radiological study. Ann Rheum Dis 45(6):441

Khot A, Bowditch M, Powell J, Sharp D (2004) The use of intradiscal steroid therapy for lumbar spinal discogenic pain: a randomized controlled trial. Spine 29:833–836

Kim S-H, Ahn S-H, Cho Y-W, Lee D-G (2012) Effect of intradiscal methylene blue injection for the chronic discogenic low back pain: one year prospective follow-up study. Ann Rehabil Med 36:657–664

Kim DH, Martin JT, Elliott DM, Smith LJ, Mauck RL (2015) Phenotypic stability, matrix elaboration and functional maturation of nucleus pulposus cells encapsulated in photocrosslinkable hyaluronic acid hydrogels. Acta Biomater 12:21–29

Le Maitre C, Binch A, Thorpe A, Hughes S (2015) Degeneration of the intervertebral disc with new approaches for treating low back pain. J Neurosurg Sci 59:47–62

Lehnert T et al (2012) Analysis of disk volume before and after CT-guided intradiscal and periganglionic ozone–oxygen injection for the treatment of lumbar disk herniation. J Vasc Interv Radiol 23:1430–1436

Li P et al (2017) Effects of osteogenic protein-1 on intervertebral disc regeneration: a systematic review of animal studies. Biomed Pharmacother 88:260–266

Likhitpanichkul M et al (2014) Fibrin-genipin adhesive hydrogel for annulus fibrosus repair: performance evaluation with large animal organ culture, in situ biomechanics, and in vivo degradation tests. Eur Cell Mater 28:25

Likhitpanichkul M et al (2015) Fibrin-genipin annulus fibrosus sealant as a delivery system for anti-TNFα drug. Spine J 15:2045–2054

Linetsky FS, Manchikanti L (2005) Regenerative injection therapy for axial pain. Techn Reg Anesth Pain Manage 9:40–49

Long RG et al (2016) Mechanical restoration and failure analyses of a hydrogel and scaffold composite strategy for annulus fibrosus repair. Acta Biomater 30:116–125

Luoma K, Riihimaki H, Luukkonen R, Raininko R, Viikari-Juntura E, Lamminen A (2000) Low back pain in relation to lumbar disc degeneration. Spine 25:487–492. https://doi.org/10.1097/00007632-200002150-00016

McCormack PL (2011) Celecoxib. Drugs 71:2457–2489

Miller M, Mathews R, Reeves K (2006) Treatment of painful advanced internal lumbar disc derangement with intradiscal injection of hypertonic dextrose. Pain Physician 9:115–121

Miyazaki S et al (2018) ISSLS PRIZE IN BASIC SCIENCE 2018: growth differentiation factor-6 attenuated pro-inflammatory molecular changes in the rabbit anular-puncture model and degenerated disc-induced pain generation in the rat xenograft radiculopathy model. Eur Spine J 27:739–751

Modic M, Steinberg P, Ross J, Masaryk T, Carter J (1988) Degenerative disk disease: assessment of changes in vertebral body marrow with MR imaging. Radiology 166:193–199

Monfett M, Harrison J, Boachie-Adjei K, Lutz G (2016) Intradiscal platelet-rich plasma (PRP) injections for discogenic low back pain: an update. Int Orthop 40:1321–1328

Moriguchi Y, Alimi M, Khair T, Manolarakis G, Berlin C, Bonassar LJ, Härtl R (2016) Biological treatment approaches for degenerative disk disease: a literature review of in vivo animal and clinical data. Global Spine J 6:497–518

Mundy G et al (1999) Stimulation of bone formation in vitro and in rodents by statins. Science 286: 1946–1949

Murphy K, Muto M, Steppan J, Meaders T, Boxley C (2015) Treatment of contained herniated lumbar discs with ozone and corticosteroid: a pilot clinical study. Can Assoc Radiol J 66:377–384

Murphy K et al (2016) Percutaneous treatment of herniated lumbar discs with ozone: investigation of the mechanisms of action. J Vasc Interv Radiol 27:1242–1250. e1243

Muto M, Andreula C, Leonardi M (2004) Treatment of herniated lumbar disc by intradiscal and intraforaminal oxygen-ozone (O_2-O_3) injection. J Neuroradiol 31: 183–189

Mwale F et al (2003) A synthetic peptide of link protein stimulates the biosynthesis of collagens II, IX and proteoglycan by cells of the intervertebral disc. J Cell Biochem 88:1202–1213

Mwale F et al (2011) The efficacy of Link N as a mediator of repair in a rabbit model of intervertebral disc degeneration. Arthritis Res Ther 13:R120

Nakamura T, Sakai K, Nakamura T, Matsumoto K (2011) Hepatocyte growth factor twenty years on: much more than a growth factor. J Gastroenterol Hepatol 26: 188–202

Nakashima S et al (2009) Regeneration of intervertebral disc by the intradiscal application of cross-linked hyaluronate hydrogel and cross-linked chondroitin sulfate hydrogel in a rabbit model of intervertebral disc injury. Biomed Mater Eng 19:421–429

Neuvonen PJ, Backman JT, Niemi M (2008) Pharmacokinetic comparison of the potential over-the-counter statins simvastatin, lovastatin, fluvastatin and pravastatin. Clin Pharmacokinet 47:463–474

Nguyen C et al (2017) Intradiscal glucocorticoid injection for patients with chronic low back pain associated with active discopathy: a randomized trial. Ann Intern Med 166:547–556

Norcross JP, Lester GE, Weinhold P, Dahners LE (2003) An in vivo model of degenerative disc disease. J Orthop Res 21:183–188

Nordby EJ, Wright PH, Schofield SR (1993) Safety of chemonucleolysis. Adverse effects reported in the United States, 1982–1991. Clin Orthop Relat Res 293:122–134

Obata S et al (2012) Effect of autologous platelet-rich plasma-releasate on intervertebral disc degeneration in the rabbit anular puncture model: a preclinical study. Arthritis Res Ther 14:R241

Olmarker K (2005) Neovascularization and neoinnervation of subcutaneously placed nucleus pulposus and the inhibitory effects of certain drugs. Spine 30:1501–1504

Omlor G, Nerlich A, Lorenz H, Bruckner T, Richter W, Pfeiffer M, Gühring T (2012) Injection of a polymerized hyaluronic acid/collagen hydrogel matrix in an in vivo porcine disc degeneration model. Eur Spine J 21:1700–1708

Peng B, Zhang Y, Hou S, Wu W, Fu X (2007) Intradiscal methylene blue injection for the treatment of chronic discogenic low back pain. Eur Spine J 16:33–38

Peng B, Pang X, Wu Y, Zhao C, Song X (2010) A randomized placebo-controlled trial of intradiscal methylene blue injection for the treatment of chronic discogenic low back pain. Pain 149:124–129

Perri M et al (2015) T2 shine-through phenomena in diffusion-weighted MR imaging of lumbar discs after oxygen–ozone discolysis: a randomized, double-blind trial with steroid and O_2-O_3 discolysis versus steroid only. Radiol Med 120:941–950

Pfeiffer M, Boudriot U, Pfeiffer D, Ishaque N, Goetz W, Wilke A (2003) Intradiscal application of hyaluronic acid in the non-human primate lumbar spine: radiological results. Eur Spine J 12:76–83

Priyadarshani P, Li Y, Yang S, Yao L (2016) Injectable hydrogel provides growth-permissive environment for human nucleus pulposus cells. J Biomed Mater Res A 104:419–426

Risbud MV, Shapiro IM (2014) Role of cytokines in intervertebral disc degeneration: pain and disc content. Nat Rev Rheumatol 10:44

Rosman Z, Shoenfeld Y, Zandman-Goddard G (2013) Biologic therapy for autoimmune diseases: an update. BMC Med 11:88

Sainoh T et al (2015) Interleukin-6 and interleukin-6 receptor expression, localization, and involvement in pain-sensing neuron activation in a mouse intervertebral disc injury model. J Orthop Res 33:1508–1514

Sainoh T et al (2016) Single intradiscal injection of the interleukin-6 receptor antibody tocilizumab provides short-term relief of discogenic low back pain; prospective comparative cohort study. J Orthop Sci 21:2–6

Settle SH Jr, Rountree RB, Sinha A, Thacker A, Higgins K, Kingsley DM (2003) Multiple joint and skeletal patterning defects caused by single and double mutations in the mouse Gdf6 and Gdf5 genes. Dev Biol 254: 116–130

Sha'Ban M et al (2008) Fibrin promotes proliferation and matrix production of intervertebral disc cells cultured in three-dimensional poly (lactic-co-glycolic acid) scaffold. J Biomater Sci Polym Ed 19:1219–1237

Shi S, Klotz U (2008) Clinical use and pharmacological properties of selective COX-2 inhibitors. Eur J Clin Pharmacol 64:233–252

Shigematsu H et al (2010) Randomized, double-blind, placebo-controlled clinical trial of hepatocyte growth factor plasmid for critical limb ischemia. Gene Ther 17:1152

Simmons JW, Fraser RD (2005) The rise and fall of chemonucleolysis. In: Arthroscopic and endoscopic spinal surgery. Springer, New York, pp 351–358

Simmons J, McMillin J, Emery S, Kimmich S (1992) Intradiscal steroids. A prospective double-blind clinical trial. Spine 17:S172–S175

Stagni S et al (2012) A minimally invasive treatment for lumbar disc herniation: DiscoGel® chemonucleolysis in patients unresponsive to chemonucleolysis with oxygen-ozone. Interv Neuroradiol 18:97–104

Stancu C, Sima A (2001) Statins: mechanism of action and effects. J Cell Mol Med 5:378–387

Stern S, Lindenhayn K, Schultz O, Perka C (2000) Cultivation of porcine cells from the nucleus pulposus in a fibrin/hyaluronic acid matrix. Acta Orthop Scand 71:496–502

Suzuki S et al (2017) Potential involvement of the IL-6/JAK/STAT3 pathway in the pathogenesis of intervertebral disc degeneration. Spine 42:E817–E824

Tellegen A, Beukers M, Miranda-Bedate A, Willems N, De Leeuw M, Van Dijk M, Creemers L, Tryfonidou M, Meij B (2016) Intradiscal injection of a slow release formulation of celecoxib for the treatment of dogs with low back pain. Global Spine J 24:S480

Tellegen AR et al (2018) Intradiscal application of a PCLA–PEG–PCLA hydrogel loaded with celecoxib for the treatment of back pain in canines: What's in it for humans? J Tissue Eng Regen Med 12:642–652

Than KD et al (2014) Intradiscal injection of simvastatin results in radiologic, histologic, and genetic evidence of disc regeneration in a rat model of degenerative disc disease. Spine J 14:1017–1028

Theron J, Guimaraens L, Casasco A, Sola T, Cuellar H, Courtheoux P (2007) Percutaneous treatment of lumbar intervertebral disk hernias with radiopaque gelified ethanol: a preliminary study. Clin Spine Surg 20:526–532

Theron J, Cuellar H, Sola T, Guimaraens L, Casasco A, Courtheoux P (2010) Percutaneous treatment of cervical disk hernias using gelified ethanol. Am J Neuroradiol 31:1454–1456

Tu J et al (2017) Simvastatin inhibits IL-1β-induced apoptosis and extracellular matrix degradation by suppressing the NF-kB and MAPK pathways in nucleus pulposus cells. Inflammation 40:725–734

Tuakli-Wosornu YA et al (2016) Lumbar intradiskal platelet-rich plasma (PRP) injections: a prospective, double-blind, randomized controlled study. PM&R 8:1–10

Ulrich JA, Liebenberg EC, Thuillier DU, Lotz JC (2007) ISSLS prize winner: repeated disc injury causes persistent inflammation. Spine 32:2812–2819

van Dijk B, Potier E, van Dijk M, Langelaan M, Papen-Botterhuis N, Ito K (2015) Reduced tonicity stimulates an inflammatory response in nucleus pulposus tissue that can be limited by a COX-2-specific inhibitor. J Orthop Res 33:1724–1731

van Dijk BG, Potier E, van Dijk M, Creemers LB, Ito K (2017) Osteogenic protein 1 does not stimulate a regenerative effect in cultured human degenerated nucleus pulposus tissue. J Tissue Eng Regen Med 11:2127–2135

Varshney A, Chapman JR (2012) A review of chymopapain for chemonucleolysis of lumbar disc herniation. Curr Orthop Pract 23:203–208

Walsh AJL, Bradford DS, Lotz JC (2004) In vivo growth factor treatment of degenerated intervertebral discs. Spine 29:156–163. https://doi.org/10.1097/01.brs.0000107231.67854.9f

Wan M, Cao X (2005) BMP signaling in skeletal development. Biochem Biophys Res Commun 328:651–657

Wang Z, Weitzmann MN, Sangadala S, Hutton WC, Yoon ST (2013) Link protein N-terminal peptide binds to bone morphogenetic protein (BMP) type II receptor and drives matrix protein expression in rabbit intervertebral disc cells. J Biol Chem 288:28243–28253

Wang Z, Hutton WC, Yoon ST (2017) The effect of capacitively coupled (CC) electrical stimulation on human disc nucleus pulposus cells and the relationship between CC and BMP-7. Eur Spine J 26:240–247

Wei A, Brisby H, Chung SA, Diwan AD (2008) Bone morphogenetic protein-7 protects human intervertebral disc cells in vitro from apoptosis. Spine J 8:466–474

Wei A, Tao H, Chung SA, Brisby H, Ma DD, Diwan AD (2009) The fate of transplanted xenogeneic bone marrow-derived stem cells in rat intervertebral discs. J Orthop Res 27:374–379

Wei A et al (2016) Expression of growth differentiation factor 6 in the human developing fetal spine retreats from vertebral ossifying regions and is restricted to cartilaginous tissues. J Orthop Res 34:279–289

Wu T, Song H-x, Dong Y, Li J-h (2018) Cell-based therapies for lumbar discogenic low back pain: systematic review and single-arm meta-analysis. Spine 43:49–57

Yang H et al (2016) The role of TGF-β1/Smad2/3 pathway in platelet-rich plasma in retarding intervertebral disc degeneration. J Cell Mol Med 20:1542–1549

Yin W, Pauza K, Olan WJ, Doerzbacher JF, Thorne KJ (2014) Intradiscal injection of fibrin sealant for the treatment of symptomatic lumbar internal disc disruption: results of a prospective multicenter pilot study with 24-month follow-up. Pain Med 15:16–31

Zhang H, Lin C-Y (2008) Simvastatin stimulates chondrogenic phenotype of intervertebral disc cells partially through BMP-2 pathway. Spine 33:E525–E531

Zhang Y, An HS, Eugene J-MT, Chubinskaya S, He T-C, Phillips FM (2006) Comparative effects of bone morphogenetic proteins and sox9 overexpression on extracellular matrix metabolism of bovine nucleus pulposus cells. Spine 31:2173–2179

Zhang Y, Anderson DG, Phillips FM, Thonar EJ-M, He T-C, Pietryla D, An HS (2007) Comparative effects of bone morphogenetic proteins and Sox9 overexpression on matrix accumulation by bovine anulus fibrosus cells: implications for anular repair. Spine 32:2515–2520

Zhang H et al (2009) Intradiscal injection of simvastatin retards progression of intervertebral disc degeneration induced by stab injury. Arthritis Res Ther 11:R172

Zhu Y et al (2017) Development of kartogenin-conjugated chitosan–hyaluronic acid hydrogel for nucleus pulposus regeneration. Biomater Sci 5:784–791

Zou F, Jiang J, Lu F, Ma X, Xia X, Wang L, Wang H (2013) Efficacy of intradiscal hepatocyte growth factor injection for the treatment of intervertebral disc degeneration. Mol Med Rep 8:118–122

Replacing the Nucleus Pulposus for Degenerative Disc Disease and Disc Herniation: Disc Preservation Following Discectomy

61

Uphar Chamoli, Maurice Lam, and Ashish D. Diwan

Contents

Introduction	1112
Structure and Function of a Healthy IVD	1112
Degenerative Disc Disease (DDD)	1114
Biochemical Changes	1114
Morphological Changes	1115
Biomechanical Changes	1115
Nucleus Replacement Implants	1118
Essential Design Criteria	1118
Classification	1120
Clinical Outcomes: A Systematic Review of Literature	1120
Preformed Mechanical	1120
Preformed Elastomer	1120
In Situ Curing	1120
Lessons Learnt	1122

U. Chamoli (✉)
Spine Service, Department of Orthopaedic Surgery,
St. George & Sutherland Clinical School, University of
New South Wales, Kogarah, NSW, Australia

School of Biomedical Engineering, Faculty of Engineering
and Information Technology, University of Technology
Sydney, Sydney, NSW, Australia
e-mail: u.chamoli@unsw.edu.au

M. Lam · A. D. Diwan
Spine Service, Department of Orthopaedic Surgery,
St. George & Sutherland Clinical School, University of
New South Wales, Kogarah, NSW, Australia
e-mail: andox6@hotmail.com; a.diwan@unsw.edu.au;
A.Diwan@spine-service.org

© Springer Nature Switzerland AG 2021
B. C. Cheng (ed.), *Handbook of Spine Technology*,
https://doi.org/10.1007/978-3-319-44424-6_94

Disc Preservation Following Lumbar Discectomy	1123
Lumbar Discectomy: Clinical Outcomes	1123
Nucleus Replacement Implants as an Adjunct to Discectomy	1125
Other Potential Uses of Nucleus Replacement Implants	1126

Conclusions ... 1126

References ... 1127

Abstract

Low back pain is the leading cause of years lived with disability worldwide and thus a significant burden on the economy and healthcare systems. Degenerative changes and/or repetitive abnormal loading in the lumbar spine could lead to structural failures of the intervertebral disc and herniation of the nucleus pulposus, all of which may manifest as chronic back and/or leg pain. Although lumbar discectomy is a clinically beneficial procedure for appropriately selected disc herniation patients, revision discectomy rates range from 2% to 18% within the first decade of the primary discectomy, especially in patients younger than 65 years. Discectomy being a tissue discarding procedure may compromise the biomechanical integrity of the disc and accelerate its degeneration. Nucleus replacement (NR) implants present a promising option to address some of the challenges surrounding lumbar discectomy. An NR implant may be used as an adjunct to discectomy to preserve the biomechanical integrity of the disc and minimize recurrent herniation of the nuclear tissue. Nonetheless, a systematic review of the literature on clinical outcomes for NR implants revealed high rates for endplate remodeling and implant subsidence. A detailed multiscale understanding of the mechanisms of disc herniation and reherniation, closure of the annular defect, and the ability to tailor geometry and material properties for individual patients are needed to develop the next generation of NR implants.

Keywords

Intervertebral disc · Disc degeneration · Disc herniation · Discectomy · Nucleus replacement implants

Introduction

Low back pain (LBP) is the leading cause of years lived with disability (YLD) worldwide, contributing approximately 57.6 million years to the total YLDs in 2016 (followed by migraine contributing 45.1 million years) and a lifetime prevalence that exceeds 80% in the industrial world (Connelly et al. 2006; Vos et al. 2017). In the United States alone, the total costs associated with low back pain exceed US$ 100 billion per year, two-thirds of which are a result of lost wages and reduced productivity (Katz 2006). Although the factors leading to LBP are largely unknown, it is frequently due to the defects or failures of the intervertebral disc (IVD) resulting in the herniation of the inner disc (pulposus) material which causes irritation and/or mechanical compression of the spinal cord or the exiting nerves, often resulting in pain, neurologic deficit, or both.

Structure and Function of a Healthy IVD

The IVD is a fibrocartilaginous structure that has a mechanical role of absorbing and transmitting loads acting on the spinal column. Together with the facet joints, the IVD completes the three-joint complex at each motion segment in the spinal column. There are three primary components in an IVD: an inner jellylike material called the nucleus pulposus (NP), an outer tough fibrocartilaginous structure called the annulus fibrosus (AF), and the vertebral endplates (EP) which serve as a transitional zone joining the IVD to the vertebrae above and below (Fig. 1).

The NP is a hydrated mass of gelatinous tissue in the center of the IVD, primarily composed of large amounts of proteoglycans with sparsely arranged collagen fibrils serving as supporting

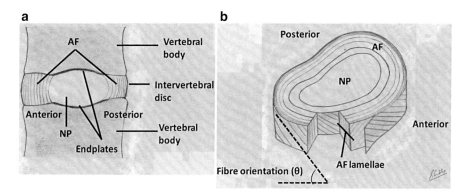

Fig. 1 Schematic representation of an adult intervertebral disc (IVD). (**a**) Midsagittal section of the IVD attached to the vertebral bodies, showing its primary components: nucleus pulposus (NP); annulus fibrosus (AF); and endplates (EP). (**b**) Three-dimensional view of the IVD illustrating the oblique and counter-oblique fiber orientation in adjacent lamellae

matrix. Bottlebrush-shaped proteoglycan molecules contain a protein core and hydrophilic glycosaminoglycan (GAG) chains (Cassinelli and Kang 2000). The high concentration of GAGs increases the osmotic pressure of the NP and allows it to swell and resist large compressive loads (Buckwalter 1995; Kroeber et al. 2002). While proteoglycans make up roughly 50% dry weight of the NP, the NP is also composed of approximately 25% collagen (Cassinelli and Kang 2000; Maroudas et al. 1975; Trout et al. 1982). Collagen type II is highly prevalent in the NP, and its concentration decreases toward the peripheral AF (Cassinelli and Kang 2000). The cells within the NP are sparse and are responsible for the maintenance of the extracellular matrix (Cappello et al. 2006).

The AF is a composite structure comprising ground substance and concentric layers (lamellae) of collagen type I fibers arranged in a regular crisscross pattern and attached circumferentially to the endplates (Hukins 2005). The AF is composed of more than two-thirds collagen, and proteoglycans make up only a small percentage of its composition (Cassinelli and Kang 2000; Maroudas et al. 1975). Within each lamella, the orientation of the fibers varies, alternating at approximately $\pm 20°$ on the ventral side and continuously changing up to $\pm 45°$ toward the dorsal side (Cassidy et al. 1989). The outer lamellae are stiffer and more densely packed than the inner ones (Holzapfel et al. 2005; Mengoni et al. 2015). In addition to axial compression, the AF withstands stresses in the tangential and radial directions. The NP and the inner AF contain only chondrocytes, while the outer AF contains mostly fibrochondrocytes (Melrose et al. 2008).

The top and the bottom of the IVDs are capped with EP (Fig. 1). The EP consists of a bony and a cartilaginous component which serves to balance conflicting biophysical demands. The bony endplates (BEP) provide the strength required to resist mechanical failure, whereas the cartilaginous endplates (CEP) facilitate chemical transport due to their porous nature (Lotz et al. 2013). In the lumbar spine, the cranial BEP is significantly thicker (1.03 ± 0.24 mm) and denser than the caudal BEP (0.78 ± 0.16 mm) (Wang et al. 2011). Within lumbar EPs, regional variation in stiffness exist, with the periphery stronger than the center, posterior stronger than the anterior, and the posterolateral sites in front of the pedicles being the strongest (Grant et al. 2001).

Together, the AF and the EPs serve to contain the NP, and in a healthy IVD, an osmotic pressure gradient provides the flow of nutrients to the NP. The IVD does not contain nerve tissues beyond the outer layers of the AF and the EP (Fagan et al. 2003; Ozawa et al. 2006). With no direct vascular supply, the IVD is the largest avascular component of the human body, relying entirely on diffusion for nutrition as well as the elimination of the waste products.

Degenerative Disc Disease (DDD)

Disc degeneration is one of the many progressive changes in the human body primarily attributable to natural ageing and is not a disease as much as it is a process. Magnetic resonance imaging (MRI) studies have shown that morphologically similar degenerated discs can be symptomatic or asymptomatic, which supports the hypothesis that a painful disc is a result of biochemical rather than morphological changes (Boos et al. 1995; Brinjikji et al. 2015). As early as the first two decades of life, the disc starts undergoing a progressive alteration in biochemical and morphological characteristics, which subsequently alters its biomechanical properties (Haefeli et al. 2006; Vernon-Roberts et al. 2007).

The etiology of disc degeneration is not well understood. One of the primary causes is thought to be the failure of nutrient supply to the disc cells, which may happen due to endplate calcification or other factors that affect the blood supply to the vertebral body such as atherosclerosis (Nachemson et al. 1970). A relationship has been found between loss of cell viability and fall in nutrient transport in scoliotic discs (Urban et al. 2001). Abnormal mechanical loads are also thought to initiate injury that leads to disc degeneration (Lotz et al. 1998; Stokes and Iatridis 2004). Lastly, genetic predisposition has been confirmed in twin studies as well as by reports of an association between disc degeneration and polymorphisms of matrix macromolecules (Paassilta et al. 2001).

Biochemical Changes

The composition and organization of the extracellular matrix in an IVD largely govern its mechanical properties. The balance between synthesis, breakdown, and accumulation of the matrix macromolecules determines the quality and integrity of the matrix and thus the mechanical behavior of the disc itself. The extracellular matrix in an IVD comprises two main macromolecules which have distinct composition, structure, and function. The collagen network, formed mostly of type I and type II collagen fibrils and making up approximately 70% and 20% of the dry weight of the AF and NP, respectively, provides tensile strength to the disc and anchors the tissue to the bone (Eyre and Muir 1977). Aggrecan, the major proteoglycan of the disc, has negatively charged GAG chains which attract water molecules, thereby maintaining tissue hydration and creating an osmotic pressure gradient within the NP (Johnstone and Bayliss 1995).

The most significant change to occur in an IVD with degeneration is the loss of proteoglycan molecules (Lyons et al. 1981). The bigger proteoglycan molecules break down into smaller fragments resulting in a loss of GAG chains, which leads to a gradual loss of hydration in the disc matrix and is primarily responsible for a fall in osmotic pressure within the NP. With degeneration, the collagen population in the disc can alter in type and distribution, but the absolute quantity of collagen does not change significantly. The relatively thin type II collagen in the nucleus is replaced by denser type I collagen with increased cross-linking between the collagen fibrils, making the nucleus more fibrotic, which is thought to further hinder tissue-fluid exchange (Duance et al. 1998). The barriers created by increased cross-linking reduce the rate of turnover and repair of collagen and proteoglycans, altering homeostasis within the IVD and resulting in the retention of damaged macromolecules (Adams and Roughley 2006).

In a normal disc, aggrecans, because of their high concentration and charge, prevent the movement of large uncharged molecules such as serum proteins and cytokines into and through the disc matrix (Maroudas 1975). The fall in concentration of aggrecans could result in an unchecked loss of osmotically active small aggrecan fragments from the disc, and increased penetration of large molecules such as growth factor complexes and cytokines into the disc. The increased vascular and neural ingrowth observed in degenerated discs is likely associated with proteoglycan loss because disc aggrecan has been shown to inhibit neural ingrowth (Melrose et al. 2002).

Hydration of the AF extracellular matrix, which serves to facilitate waste and nutrient exchange, is crucial to imparting viscoelasticity properties to the disc (Gu et al. 1999; Travascio et al. 2009). Both extra- and intrafibrillar fluids are responsible for the AF hydration. The intrafibrillar fluid that closely adheres to collagen fibers provides long-term AF hydration. The earliest known compositional change in proteoglycan loss reduces both the AF extra- and intrafibrillar fluid capacity and swelling pressure while impairing its overall load-bearing capability (Johannessen and Elliott 2005; Yao et al. 2002). On the other hand, extrafibrillar fluid, which is responsible for nutrient and waste (i.e., lactic acid) transport, moves freely across the AF. Drop in the AF osmotic pressure secondary to the lack of extrafibrillar fluid reduces the efficiency of nutrient-waste exchange and leads to a decrease in pH and consequently lactic acid accumulation (Iatridis et al. 2007; McMillan et al. 1996). Both decrease in pH, which leads to acidity increase, and insufficient nutrition across the AF impair cellular metabolism and increase the risk of disc degeneration (Cassinelli et al. 2001). Therefore, any undesirable changes in the AF extra- and intrafibrillar fluids lead to dehydration, alter osmotic and viscoelastic properties, and may result in disc degeneration (Gu et al. 2014; Murakami et al. 2010).

Morphological Changes

Macroscopic analysis of midsagittal slices of human lumbar IVD of individuals ranging from newborn to senile age has revealed the temporospatial variation of age-related morphological changes in the disc (Haefeli et al. 2006). Degenerative processes in the disc start in the first two decades of life with transformation in the NP, mucous degeneration, AF disorganization, alteration of the EP, and osteophyte formation (Fig. 2). After the initial phase of significant alterations, degenerative changes remain constant over the next two decades before increasing again after the fourth decade (Haefeli et al. 2006).

The initial morphological changes are followed by the appearance of nuclear cleft and subsequent radial and concentric tears in the annulus in the fifth decade of life. Radiating annular tears rarely extend to the outer AF and are thought to be a consequence of clefts originating in the NP (Vernon-Roberts et al. 2007). Rim lesions typically occur independent of annular tears and substantially later in life (Haefeli et al. 2006; Vernon-Roberts et al. 2007). With progressive degeneration, an ingrowth of nerve fibers and blood vessels beyond the outer AF is observed which is often associated with discogenic pain (García-Cosamalón et al. 2010; Stefanakis et al. 2011). Although disc height narrowing with DDD has been reported in some studies, Twomey and Taylor (1987) contradicted this opinion by showing that the average disc height is maintained in old age, with the distance between the anterior and posterior corners of the vertebral bodies decreasing and the IVD expanding centrally to become increasingly convex (Butler et al. 1990; Twomey and Taylor 1987). Nonetheless, in some cases, severe tissue destruction (including cleft and tears formation as well as rim lesions) may occur in the first two decades of life, which presents an enormous challenge for any prophylactic tissue engineering repair attempt.

Disc herniation can be considered as one specific feature of disc degeneration that is much more closely related to mechanical loading (exceeding tissue strength) and pain than other features of degeneration such as signal intensity on MR scans and biochemical changes. Examination of autopsy or surgical specimens suggests that some degenerative changes, such as nuclear desiccation and fragmentation and preexisting tears in the AF, are necessary before a disc can herniate (Moore et al. 1996). In extruded disc tissue material, isolated fragments of AF and EP are much less common than the NP (Moore et al. 1996).

Biomechanical Changes

In healthy conditions, the high water content within the NP creates hydrostatic pressure which contributes to sustaining large loads acting on the spinal column. The compressive spinal loads are uniformly distributed to the AF through hydrostatic pressure, which creates hoop stresses within the AF. The fiber orientation of the AF is suitable

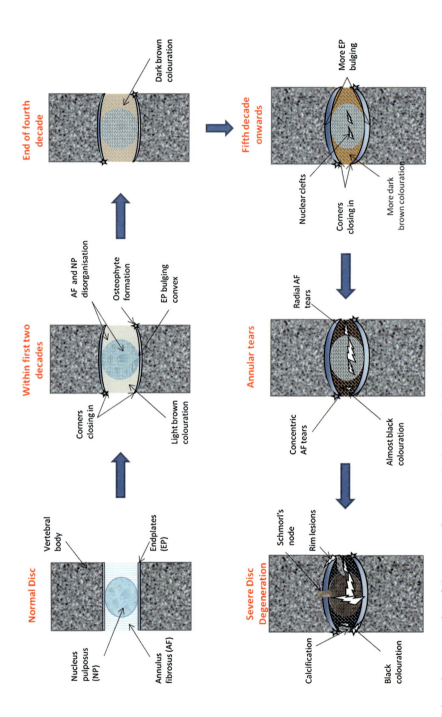

Fig. 2 Schematic representation of the course of macroscopic degeneration in the human lumbar intervertebral discs. Age or degeneration-related structural abnormalities combined with complex repetitive mechanical loading in the disc could result in disc prolapse or herniation (Fig. 3). A disc may, however, herniate without degeneration if loaded severely enough, and the degenerative changes found in the herniated disc material probably occur after herniation has taken place, as a result of tissue swelling, leaching of proteoglycans, and revascularization (Fig. 3d) (Adams and Hutton 1986; Lama et al. 2013)

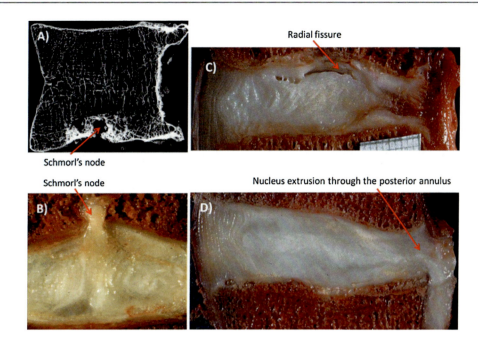

Fig. 3 Structural defects in lumbar intervertebral discs due to degeneration and/or mechanical overloading. (**a**) Microradiograph of a midsagittal slice of a cadaver vertebral body showing Schmorl's node and bone remodeling around it. (**b**) A cadaver disc which has herniated through the endplate in response to compressive loading shows decompressed nucleus and bucking of inner annulus walls. (**c**) A cadaver disc showing a complete radial fissure in the posterior annulus. (**d**) Nucleus extrusion through the posterior annulus due to abnormal bending and compressive loads in an otherwise nondegenerate disc. (Images adapted from Adams and Dolan 2016, with permission)

to resist hoop stresses generated by the hydrostatic pressure.

Degenerative changes in the biomechanical properties of the disc can occur due to changes in either material properties of the individual NP and AF tissues or due to consequent morphological changes in the substructure of the disc. The process is thought to initiate in the NP, with a decrease in its proteoglycan concentration and a gradual change in collagen type, making the nucleus more fibrotic, stiffer, and severely limited in its ability to generate hydrostatic pressure. Degenerated NP tissues have significantly lower swelling stress ($P_{sw} = 0.037 \pm 0.038$ MPa degenerate, $P_{sw} = 0.138 \pm 0.029$ MPa nondegenerate), lower effective aggregate modulus ($H_A^{eff} = 0.44 \pm 0.19$ MPa degenerate, $H_A^{eff} = 1.01 \pm 0.43$ MPa nondegenerate), and a higher permeability ($k_a = 1.4 \pm 0.58 \times 10^{-15}$ m^4/N-s degenerate, $k_a = 0.9 \pm 0.43 \times 10^{-15}$ m^4/N-s nondegenerate) (Johannessen and Elliott 2005). In the degenerate AF, the fiber orientation becomes disorganized, and the nonlinear elastic response also varies consequently (Schollum et al. 2010). The response of the degenerate AF tissue has been shown to be of a twofold increase in the toe-region modulus in tensile testing, which correlated with age, as well as fiber reorientation toward the loading direction (Guerin and Elliott 2006; O'Connell et al. 2009). Although the water content within the AF is not affected by degeneration, material property parameters such as Poisson's ratio, failure stress, and strain energy density are strongly influenced by the level of degeneration (Michalek et al. 2009).

The above changes in biochemical, morphological, and biomechanical properties of the disc combined with complex repetitive loading may result in the structural failure of the disc and present in the form of disc herniation: either protrusion or complete rupture of the AF walls followed by the expulsion of the NP material (extrusion or sequestration). The inflammatory material from the disc (particularly the NP) may cause chemical irritation and mechanical compression of the cord and the exiting nerves and contribute to radicular back and/or leg pain (Goupille et al. 1998; Omarker and Myers 1998). Therefore, a common approach to the clinical treatment for painful disc herniation is surgical removal of the herniated material to unload the nerves (discectomy) (DeLeo and Winkelstein 2002; Loupasis et al. 1999).

Herniation and discectomy may accelerate disc degeneration. The loss of NP material results in decreased pressure within the disc, progressive loss in disc height, inward buckling of the inner annulus, and an increased bulging of the annulus under compression (Brinckmann and Grootenboer 1991; Frei et al. 2001; Meakin and Hukins 2000). In an in vitro study on human lumbar discs, Brinckmann and Grootenboer (1991) demonstrated that, on average, removal of 1 g of disc tissue resulted in a height decrease of 0.8 mm and a radial bulge increase of 0.2 mm under compressive loads. Removal of 3 g of central disc tissue lowered the intradiscal pressure to approximately 40% of its initial value (Brinckmann and Grootenboer 1991).

Although discectomy provides immediate relief from leg/back pain in most cases, the procedure is a tissue discarding one, in which the most frequent adverse events are the recurrent herniation of the residual nuclear tissue, progressive disc height loss, and reoccurring back/leg pain. An additional risk following discectomy is the loss of disc height which has been linked to the amount of nucleus material removed at the time of surgery (Tibrewal and Pearcy 1985; Yorimitsu et al. 2001). In turn, there is an alteration of the entire spinal column kinematics, and as the more nuclear material is removed from within the disc, the less capable it is of supporting the spinal loads.

Filling the nucleotomized cavity with a biologically inert replacement material has the potential to restore biomechanical characteristics of the disc and mitigate the progressive loss in disc height. Our group has conducted preliminary work to assess biomechanical efficacy of a non-hydrogel silicone-based in situ curing nucleus replacement (NR) implant, the Kunovus Disc Device (KDD,

Kunovus Pty Ltd., Australia) in restoring the bending stiffness of a human lumbar motion segment following discectomy. A finite element modelling study to evaluate changes in bending stiffness of a L3–L4 motion segment revealed that compared with the baseline intact state, a complete nucleotomy significantly increases annular bulge (flexion, 0.65 mm; extension, 0.18 mm) and average Von Mises stress in the annulus (flexion, 38%; extension, 6%) (Fig. 4). Although partial filling of the nucleotomized cavity with the KDD was not able to restore the synergistic biomechanical interaction between the AF, EP, remnant NP, and the NR implant; complete filling of the cavity restored the biomechanical characteristics of the motion segment close to the normal intact state levels (Fig. 4).

Nucleus Replacement Implants

Nucleus replacement (NR) implants present a promising option for restoring and preserving the biomechanical integrity of an IVD and address some of the challenges surrounding discectomy procedures. These implants fill the treatment gap between nonsurgical care and invasive surgical procedures such as fusion and total disc replacement (TDR).

Essential Design Criteria

An NR implant must meet five essential criteria to be considered for clinical use:

1. Biocompatible and durable to survive the lifespan of the recipient.
2. Maintain the disc height, stabilize motions across all axes of movement, and restore normal distribution of loads in the motion segment.
3. High conformity in the nucleotomized cavity to avoid device migration and subsidence
4. Optimal stiffness to avoid excessive wear and/or remodeling of the EP. An overly compliant implant will overload the AF and fail to maintain the disc height.

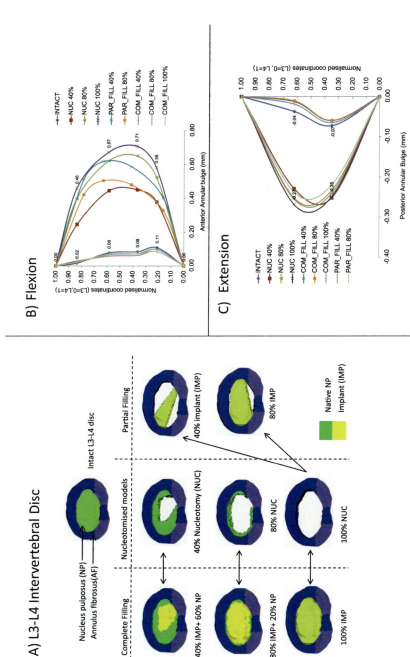

Fig. 4 (**a**) Finite element (FE) models of the L3–L4 intervertebral disc were assembled representing the disc in the intact state, different states of nucleotomy, partial and complete filling with the Kunovus Disc Device (KDD, Kunovus Pty Ltd., Australia). Multisegment FE models of the lumbar spine (L1–L5) were loaded in flexion and extension bending moments (±10 Nm) with the L5 vertebra fixed in space. (**b**) Anterior annulus bulge in the midsagittal plane of the L3–L4 disc in different states during peak flexion loading. (**c**) Posterior annulus bulge in the midsagittal plane of the L3–L4 disc in different states during peak extension loading. Only complete filling of the nucleotomized disc with the KDD implant restored the annular bulge to the intact state levels

5. Easy to implant as an adjunct to discectomy, minimize any additional damage to the AF during implantation. Easy to remove in the instance of any adverse event during or after surgery.

Classification

A number of NR implants have been developed so far, with an unaccounted number of them still under development. Some of these NR implants are at different stages of clinical use, while others have been abandoned. Based on the design principles and materials used, NR implants may be divided into various categories (Fig. 5).

Clinical Outcomes: A Systematic Review of Literature

A systematic literature search into currently available clinical data on NR implants published in various journals and book chapters between January 1988 and March 2017 was conducted using Scopus and Medline online databases. After removing duplicate articles across both the databases, and further screening through reading abstracts, a total of 12 articles were found, which presented short-term ($\leq$1 year), mid-term (1–3 years), and long-term (>3 years) clinical and radiological follow-up data on NR implants. One article reported data on three different NR implants (PDN, NuBac, PNR) (Pimenta et al. 2012). Two articles reported short-term and mid-term clinical results (Ahrens et al. 2009; Balsano et al. 2011). Due to variations in implant designs and materials used, studies were not directly comparable; and therefore, articles were grouped based on the type of implant reported (Table 1).

Preformed Mechanical

NuBac (Pioneer Surgical Technology, Michigan): NuBac is a two-piece prosthesis made from PEEK, with a ball-and-socket-type articulation between the two pieces (Fig. 6). Short-term clinical follow-up data for 49 patients were presented in two separate studies (Alpizar-Aguirre et al. 2008; Balsano et al. 2011). Balsano et al. further presented mid-term clinical data for 166 patients implanted with the NuBac device (Balsano 2014; Balsano et al. 2011). Pimenta et al. (2012) reported long-term follow-up data for 19 patients implanted with the NuBac device. Table 2 presents clinical and radiological follow-up data for the NuBac implant.

Preformed Elastomer

Prosthetic disc nucleus PDN (Raymedica, Minnesota): The PDN implant comprises a special hydrogel pellet core encased in a polyethylene jacket that helps maintain device shape when subjected to heavy spinal loads (Fig. 7). The expanding hydrogel constrained within the jacket is designed to provide the lifting force in the intervertebral disc space to maintain the disc height and remain flexible at the same time. In order to minimize the size of annular opening required for implantation, the device is implanted as two separate units, connected by means of a tethering suture (Klara and Ray 2002).

Among NR implants, clinical data for PDN is most widely reported in the literature. Three separate studies have reported short-term clinical data for a total of 84 patients implanted with PDN (Bertagnoli and Vazquez 2003; Jin et al. 2003; Shim et al. 2003). Four additional studies have reported long-term follow-up data for a total of 199 patients implanted with PDN (Klara and Ray 2002; Pimenta et al. 2012; Selviaridis et al. 2010; Zhang et al. 2009). Table 3 presents clinical and radiological follow-up data for the PDN implant.

In Situ Curing

In situ curing injectable materials have been the recent focus of research in NR implants due to their ability to conform to the shape of the nucleotomized cavity and cure within the disc. These NR implants can be delivered using a minimally invasive surgery and, in principle, are designed to overcome endplate remodeling and

61 Replacing the Nucleus Pulposus for Degenerative Disc Disease and Disc Herniation: Disc... 1121

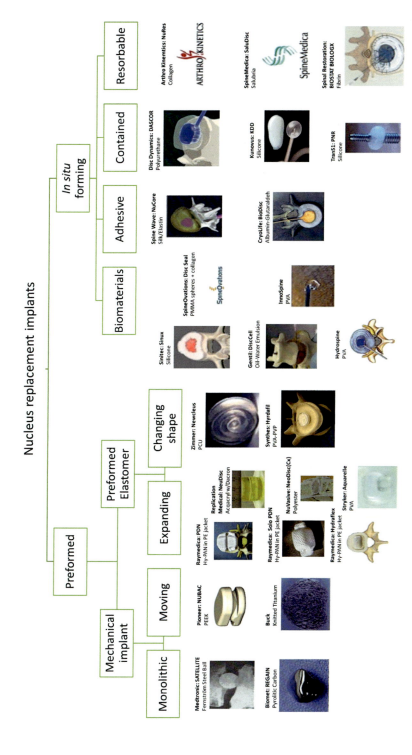

Fig. 5 A broad classification scheme for nucleus replacement implants

Table 1 Medline and Scopus online database search between January 1988 and March 2017 revealed that clinical and radiological follow-up data are available for five nucleus replacement implants

Type of implant	Implant name	Studies with clinical data	Manufacturer	Material
Preformed mechanical	NuBac	2 (short-term), 2(mid-term), 1(long-term)	Pioneer Surgical Technology, Michigan	PEEK
Preformed elastomer	PDN	3(short-term), 4(long-term)	Raymedica, Minnesota	Hydrogel pellet in PE jacket
In situ curing	DASCOR	1(short-term), 1(mid-term)	Disc Dynamics, Minnesota	Polyurethane
	NuCore	1(mid-term)	Spine Wave, Connecticut	Silk and elastin
	PNR	1(long-term)	TranS1, Colorado	Silicone

Fig. 6 NuBac implant comprises two articulating pieces made using PEEK. (Image adapted from Ordway et al. 2013, with permission)

implant migration issues associated with the preformed NR implants. A number of in situ curing NR implants have been developed using a variety of materials (Fig. 5); however, clinical and radiological follow-up data is only available for the following: DASCOR (Disc Dynamics, Minnesota), NuCore (Spine Wave, Connecticut), and Percutaneous Nucleus Replacement (PNR) (TranS1, Colorado) (Fig. 8).

Ahrens et al. (2009) reported short-term (n = 70, 1 year) and mid-term (n = 41, 2 years) clinical data for 85 patients implanted with the DASCOR device (Ahrens et al. 2009). Berlemann and Schwarzenbach (2009) reported mid-term clinical data for 14 patients implanted with the NuCore device. Pimenta et al. (2012) reported long-term clinical data for 26 patients implanted with the PNR device. Table 4 presents clinical and radiological follow-up data for the above three in situ curing implants.

Lessons Learnt

In all of the above studies, the average age of patients receiving an NR implant was 35–45 years, and therefore the implant was expected to function for five to six decades. The premise behind using an NR implant is to restore mobility and salvage structures in a functionally suboptimal disc which would otherwise be sacrificed in more invasive spine surgeries. Although short-term and mid-term clinical results have been promising (pain scores, functional outcomes, disc height preservation, intra-op, and post-op complication rates), reoperation rates in the long-term remain a matter of serious concern, particularly for the mechanical and preformed NR implants. In 199 patients implanted with PDN and followed for a minimum of 4 years, endplate remodeling rate was 32%, subsidence rate was 26%, and reoperation rate was 27%. For in situ curing implants, although conformity with the shape of the nucleotomized cavity has theoretical advantages in distributing loads to the adjoining structures, there is a dearth of long-term clinical follow-up data to show any translational benefits of this design principle.

A stiff implant in the nucleus space could lead to remodeling of the endplates and result in implant subsidence, whereas a compliant implant (or a nucleotomized cavity) may offload the endplates and overload the annulus, consequently increasing the likelihood of annulus degeneration, implant extrusion, or both. Perhaps, the *one-size-fits-all* philosophy for the material properties of NR implants may not be able to address all the design objectives in individual patients, and future

Table 2 Table summarizing short-term (≤1 year), mid-term (1–3 years), and long-term (>3 years) clinical and radiological follow-up data (average values) for patients implanted with NuBac device (Pioneer Surgical Technology, Michigan)

NuBac	VAS score	ODI score	Disc height	References
Short-term (49 patients)	7.3 (pre-op) ➔ 2.9 (3mths) ➔ 1.8 (1yr)	58 (pre-op) ➔ 23.2 (3mths) ➔ 18 (1yr)	9.4 mm (pre-op) ➔ 13 mm (6wks) ➔ 12.5 mm (3mths)	Alpizar-Aguirre et al. (2008) and Balsano et al. (2011)
Mid-term (166 patients)	7.7 (pre-op) ➔ 2.5 (2 yrs)	55.4 (pre-op) ➔ 15.7 (2 yrs)	N/r	Balsano et al. (2011) and Balsano (2014)
Long-term (19 patients)	N/r	N/r	N/r	Pimenta et al. (2012)

NuBac	Endplate changes	Migration rate	Subsidence rate	Reoperation rate	References
Short-term (49 patients)	0%	0%	0%	0%	Alpizar-Aguirre et al. (2008) and Balsano et al. (2011)
Mid-term (166 patients)	N/r	0%	0%	0%	Balsano et al. (2011) and Balsano (2014)
Long-term (19 patients)	31.6%	21.1%	21.1%	52.6%	Pimenta et al. (2012)

N/r not reported, *VAS* visual analog scale, *ODI* oswestry disability index

Fig. 7 Prosthetic disc nucleus (PDN) comprises a hydrogel pellet core encased in a polyethylene jacket. (Image adapted from Schnake and Kandziora 2016, with permission)

NR implants will need to provide clinicians with tools to customize material properties and geometry to suit their patient's needs.

While the implant geometry and material properties are important parameters in meeting the design objectives for an NR implant, the clinical success will also rely on the quality of vertebral subchondral bone, extent of endplate calcification, and structural integrity of the AF; and therefore a careful selection of patients is important.

Disc Preservation Following Lumbar Discectomy

Lumbar Discectomy: Clinical Outcomes

Lumbar discectomy in disc herniation patients results in significantly better clinical outcomes when compared with nonsurgically treated patients (Atlas et al. 2005; Weinstein et al.

Table 3 Table summarizing short-term (≤1 year) and long-term (>3 years) clinical and radiological follow-up data (average values) for patients implanted with prosthetic disc nucleus PDN (Raymedica, Minnesota)

PDN	VAS score	ODI score	Prolo score	Disc height	References
Short-term (84 patients)	8.5 (pre-op) → 3.1 (1 yr)	53.9 (pre-op) → 18 (6 mths) → 16.5 (1 yr)	5 (pre-op) → 7.6 (6mths) → 7.3 (1 yr)	9.4 mm (pre-op) → 10.8 mm (6 mths) → 10.8 mm (1yr)	Bertagnoli and Vazquez (2003), Jin et al. (2003), and Shim et al. (2003)
Long-term (199 patients)	6.6 (pre-op) → 1.6 (8yrs)	52 (pre-op) → 10.3 (4 yrs) → 6.2 (8 yrs)	4.5 (pre-op) → 8.9 (3 yrs)	8.5 mm (pre-op) → 8.7 mm (4 yrs)	Pimenta et al. (2012), Klara and Ray (2002), Selviaridis et al. (2010), and Zhang et al. (2009)

PDN	Endplate changes	Migration rate	Subsidence rate	Reoperation rate	References
Short-term (84 patients)	Scleroses, 33.3% Modic changes, 28.6%	0%	12%	7.1%	Bertagnoli and Vazquez (2003), Jin et al. (2003), and Shim et al. (2003)
Long-term (199 patients)	32.2%	13.6%	26.1%	27.1%	Pimenta et al. (2012), Klara and Ray (2002), Selviaridis et al. (2010), and Zhang et al. (2009)

VAS visual analog scale, *ODI* oswestry disability index

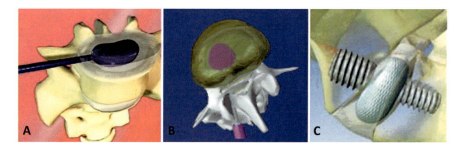

Fig. 8 (**a**) DASCOR comprises two-part curable polyurethane and an expandable balloon. (**b**) NuCore implant is an injectable 100% synthetic recombinant protein hydrogel. (**c**) PNR consists of a titanium screw system anchoring itself onto the superior and inferior vertebrae, with a central membrane that is filled with curable material and acts as the nucleus. (Image adapted from Serhan et al. 2011, with permission)

2008). Although lumbar discectomy is a clinically beneficial procedure for appropriately selected patients, revision discectomy rates range from 2% to 18% within the first decade of the primary discectomy (Virk et al. 2017; Watters and McGirt 2009). The survivorship rate for first lumbar discectomy is particularly lower for patients younger than 65 years when compared with older patients (Virk et al. 2017).

It remains unknown whether limited discectomy (LD) or aggressive discectomy (AD) provides better clinical outcomes for the treatment of lumbar disc herniation patients with radiculopathy. In a systematic review of 44 studies, the reported incidence of short-term (<2 years) recurrent leg or back pain was similar after LD (mean, 14.5%; range, 7–16%) and AD (mean, 14.1%; range, 6–43%) (McGirt et al. 2009a). In the long-term (>2 years), the reported incidence of recurrent back or leg pain was 2.5-fold less after LD (mean, 11.6%; range, 7–16%) compared with the AD (mean, 27.8%; range, 19–37%). However, the reported incidence of recurrent disc herniation after LD (mean, 7%;

Table 4 Table summarizing mid-term (1–3 years) and long-term (>3 years) clinical and radiological follow-up data (average values) for patients implanted with in situ curing nucleus replacement implants (DASCOR, NuCore, and PNR)

	VAS score	ODI score	Disc height	References
DASCOR – mid-term (70 patients)	7.6 (pre-op) ➔ 3.3 (1yr) ➔ 3.3 (2yrs)	57.5 (pre-op) ➔ 25.2 (1yr) ➔ 23.2 (2yrs)	N/r	Ahrens et al. (2009)
NuCore – mid-term (14 patients)	3.6 (pre-op) ➔ 1.1 (2yrs)	43 (pre-op) ➔ 10 (2yrs)	100% (pre-op) ➔ 93% (2yrs)	Berlemann and Schwarzenbach (2009)
PNR – long-term (26 patients)	N/r	N/r	N/r	Pimenta et al. (2012)

	Endplate changes	Migration rate	Subsidence rate	Reoperation rate	References
DASCOR – mid-term (70 patients)	1.4%	1.4%	2.9%	10%	Ahrens et al. (2009)
NuCore – mid-term (14 patients)	64.3%	0%	0%	0%	Berlemann and Schwarzenbach (2009)
PNR – long-term (26 patients)	N/r	N/r	N/r	57.7%	Pimenta et al. (2012)

VAS visual analog scale, *ODI* oswestry disability index

range, 2–18%) was significantly greater than that reported after AD (mean, 3.5%; range, 0–9.5%) (McGirt et al. 2009a).

LD may result in shorter operative time, a quick return to work, and a decreased incidence of long-term recurrent back pain, but at a significantly greater risk of long-term recurrent herniation compared with AD. Aggressive removal of the remanent nucleus is effective at decreasing reherniation but at the cost of significantly poor long-term clinical outcomes and low patient satisfaction (McGirt et al. 2009a).

Various risk factors have been identified for the recurrent herniation of the disc. Smoking and occupational lifting are known to increase the likelihood of recurrent herniation (Miwa et al. 2015). Discectomy patients with preserved disc height postoperatively generally have favorable results, but the risk of recurrent disc herniation is high in this population (Yorimitsu et al. 2001). In a retrospective study of 75 lumbar disc herniation patients, 8 of whom re-herniated after primary microdiscectomy, the authors found that the mean body mass index (BMI) of patients with recurrent herniation (33.6 ± 5.1) was significantly higher than those without recurrence (26.9 ± 3.9) (Meredith et al. 2010). In a prospective study of 108 lumbar disc herniation patients undergoing first-time discectomy and followed for up to 2 years, the authors observed that the mean annular defect area was significantly greater in recurrent herniation patients compared with no-recurrence patients (46 ± 20 vs. 32 ± 14 mm^2) (McGirt et al. 2009b). Mean annular defect was also significantly larger in patients with symptomatic early reherniation (within 4 months after surgery) compared to later herniation (57 vs. 39 mm^2) (McGirt et al. 2009b). Clinically silent recurrent disc herniation is common after lumbar discectomy, but treatment is recommended only when correlating radicular symptoms exist (Lebow et al. 2011).

Nucleus Replacement Implants as an Adjunct to Discectomy

Annulus repair following discectomy may be beneficial for retaining the intradiscal material. While patients with tall and healthy discs preoperatively have the most to gain with annular closure (thus reducing the amount of nucleus that needs to be removed), repair of the annulus is not able to restore the biomechanical characteristics of the disc, and further annular tear adjacent to the repair is possible due to persistent fragmented nuclear material present in the disc cavity. A prospective, multicenter, randomized control trial of 750 patients treated for herniated lumbar discs and randomly assigned in a 2:1 ratio to discectomy with annular closure and discectomy without annular closure found no significant difference in the clinical outcomes (including the recurrent herniation rates) between the two surgical cohorts at 2-year follow-up mark (Bailey et al. 2013).

Discectomy remains one of the rare surgical procedures where the tissues lost to herniation and removed during the surgery are not replaced with any prosthesis. Nucleus replacement implants can be used as void-fillers during a standard discectomy procedure to (1) restore the structural and mechanical integrity of the disc; (2) restore load sharing and synergistic interaction between the implant, remnant NP, AF, and the EP; (3) minimize loss in disc height and accelerated degeneration of the AF; and (4) minimize recurrent disc herniation by acting as an annular-closure-plug.

Furthermore, there has been a growing interest in the development of alternative minimally invasive technologies for the treatment of degenerative disc disease in otherwise healthy patients who suffer from unremitting pain due to damaged and bulging intervertebral discs, and are not responsive to nonoperative care. These patients are candidates for spinal fusion but retain a workable disc height and undamaged facet joints. These patients would benefit from an option which would provide lower risks than spinal fusion and similar improvement in quality of life. The loss of disc function can be mitigated by replacement of the NP with a biologically inert material with a goal of maintaining the disc height and function.

Other Potential Uses of Nucleus Replacement Implants

Pedicle screw-based posterior dynamic stabilizers (PDS) are nonfusion spinal implants that aim to restore normal load sharing and kinematics in a degenerate spinal motion segment (Chamoli et al. 2014). Because these implants are posteriorly placed, the center of rotation of the motion segment is shifted posteriorly upon implantation compared with that of an intact spine; and therefore anterior load sharing cannot be satisfactorily achieved using a PDS implant alone. The cyclic nature of the pedicle screw loading and the micromotions at the bone-screw interface likely increase screw loosening and pullout rates and is one of the major reasons for the implant failure. Replacing the core of the degenerated disc with an NR implant and using PDS as an augmenting device may have the potential to overcome this problem.

An NR implant may be used in hybrid constructs for prophylactic dynamic stabilization of segments adjacent to the fused levels, which could reduce hypermobility and impede the accelerated degeneration of the adjacent segments. The NR implants are advantageous over commonly used TDR implants in hybrid constructs, as they are less invasive, salvage disc structures, and more closely mimic the kinematic signature of an intact motion segment.

Conclusions

Degenerative changes and/or repetitive abnormal loading in the lumbar spine could lead to structural failures of the intervertebral disc and herniation of the nucleus pulposus, all of which may manifest as chronic back and/or leg pain. Although lumbar discectomy is a clinically beneficial procedure for appropriately selected disc herniation patients, revision discectomy rates range from 2% to 18% within the first decade of the primary discectomy, especially in patients younger than 65 years. Nucleus replacement (NR) implants present a promising option to address some of the challenges surrounding standard lumbar discectomy. These implants could be used as void-fillers during a standard discectomy procedure to: restore and preserve the biomechanical integrity of the disc, impede progressive disc degeneration and loss in disc height, and minimize the incidence of recurrent herniation. Nonetheless, long-term follow-up results for the present NR implants reveal high rates for endplate remodeling and implant subsidence. A detailed multiscale understanding of the mechanisms of disc herniation and reherniation, closure of the annular defect, and the ability to tailor geometry and material properties for individual patients are needed to develop the next generation of NR implants.

References

Adams MA, Dolan P (2016) Lumbar intervertebral disk injury, herniation and degeneration. In: Pinheiro-Franco J, Vaccaro AR, Benzel EC, Mayer HM (eds) Advanced concepts in lumbar intervertebral disk disease. Springer, Heidelberg, pp 23–39

Adams MA, Hutton WC (1986) The effect of posture on diffusion into lumbar intervertebral discs. J Anat 147:121–134

Adams MA, Roughley PJ (2006) What is intervertebral disc degeneration, and what causes it? Spine 31:2151–2161

Ahrens M et al (2009) Nucleus replacement with the DASCOR disc arthroplasty device: interim two-year efficacy and safety results from two prospective, non-randomized multicenter European studies. Spine 34:1376–1384

Alpizar-Aguirre A, Mireles-Cano JN, Rosales-Olivares M, Miramontes-Martinez V, Reyes-Sanchez A (2008) Clinical and radiological follow-up of Nubac disc prosthesis. Preliminary report. Cir Cir 76:317–321

Atlas SJ, Keller RB, Wu YA, Deyo RA, Singer DE (2005) Long-term outcomes of surgical and nonsurgical management of sciatica secondary to a lumbar disc herniation: 10 year results from the Maine lumbar spine study. Spine 30:927–935

Bailey A, Araghi A, Blumenthal S, Huffmon GV (2013) Prospective, multicenter, randomized, controlled study of anular repair in lumbar discectomy: two-year follow-up. Spine 38:1161–1169

Balsano M (2014) Lumbar nucleus replacement. In: Menchetti PPM (ed) Minimally invasive surgery of the lumbar spine. Springer, London, pp 229–242

Balsano M, Zachos A, Ruggiu A, Barca F, Tranquilli-Leali P, Doria C (2011) Nucleus disc arthroplasty with the NUBAC device: 2-year clinical experience. Eur Spine J 20(Suppl 1):S36–S40

Berlemann U, Schwarzenbach O (2009) An injectable nucleus replacement as an adjunct to microdiscectomy: 2 year follow-up in a pilot clinical study. Eur Spine J 18:1706–1712

Bertagnoli R, Vazquez RJ (2003) The Anterolateral Trans-Psoatic Approach (ALPA): a new technique for implanting prosthetic disc-nucleus devices. J Spinal Disord Tech 16:398–404

Boos N, Rieder R, Schade V, Spratt KF, Semmer N, Aebi M (1995) 1995 Volvo Award in clinical sciences. The diagnostic accuracy of magnetic resonance imaging, work perception, and psychosocial factors in identifying symptomatic disc herniations. Spine 20:2613–2625

Brinckmann P, Grootenboer H (1991) Change of disc height, radial disc bulge, and intradiscal pressure from discectomy. An in vitro investigation on human lumbar discs. Spine 16:641–646

Brinjikji W et al (2015) Systematic literature review of imaging features of spinal degeneration in asymptomatic populations AJNR. Am J Neuroradiol 36:811–816

Buckwalter JA (1995) Aging and degeneration of the human intervertebral disc. Spine 20:1307–1314

Butler D, Trafimow JH, Andersson GB, McNeill TW, Huckman MS (1990) Discs degenerate before facets. Spine 15:111–113

Cappello R, Bird JL, Pfeiffer D, Bayliss MT, Dudhia J (2006) Notochordal cell produce and assemble extracellular matrix in a distinct manner, which may be responsible for the maintenance of healthy nucleus pulposus. Spine 31:873–882. Discussion 883

Cassidy JJ, Hiltner A, Baer E (1989) Hierarchical structure of the intervertebral disc. Connect Tissue Res 23:75–88

Cassinelli EH, Kang JD (2000) Current understanding of lumbar disc degeneration. Oper Tech Orthop 10:254–262

Cassinelli EH, Hall RA, Kang JD (2001) Biochemistry of intervertebral disc degeneration and the potential for gene therapy applications. Spine J 1:205–214

Chamoli U, Diwan AD, Tsafnat N (2014) Pedicle screw-based posterior dynamic stabilizers for degenerative spine: in vitro biomechanical testing and clinical outcomes. J Biomed Mater Res A 102:3324–3340

Connelly LB, Woolf A, Brooks P (2006) Cost-effectiveness of interventions for musculoskeletal conditions. In: Jamison DT et al (eds) Disease control priorities in developing countries. The International Bank for Reconstruction and Development/The World Bank Group, Washington DC

DeLeo JA, Winkelstein BA (2002) Physiology of chronic spinal pain syndromes: from animal models to biomechanics. Spine 27:2526–2537

Duance VC et al (1998) Changes in collagen cross-linking in degenerative disc disease and scoliosis. Spine 23:2545–2551

Eyre DR, Muir H (1977) Quantitative analysis of types I and II collagens in human intervertebral discs at various ages. Biochim Biophys Acta 492:29–42

Fagan A, Moore R, Vernon Roberts B, Blumbergs P, Fraser R (2003) ISSLS prize winner: the innervation of the intervertebral disc: a quantitative analysis. Spine 28:2570–2576

Frei H, Oxland TR, Rathonyi GC, Nolte LP (2001) The effect of nucleotomy on lumbar spine mechanics in compression and shear loading. Spine 26:2080–2089

García-Cosamalón J, del Valle ME, Calavia MG, García-Suárez O, López-Muñiz A, Otero J, Vega JA (2010) Intervertebral disc, sensory nerves and neurotrophins: who is who in discogenic pain? J Anat 217:1–15

Goupille P, Jayson MI, Valat JP, Freemont AJ (1998) The role of inflammation in disk herniation-associated radiculopathy. Semin Arthritis Rheum 28:60–71

Grant JP, Oxland TR, Dvorak MF (2001) Mapping the structural properties of the lumbosacral vertebral endplates. Spine 26:889–896

Gu WY, Mao XG, Foster RJ, Weidenbaum M, Mow VC, Rawlins BA (1999) The anisotropic hydraulic permeability of human lumbar annulus fibrosus. Influence of age, degeneration, direction, and water content. Spine 24:2449–2455

Gu W, Zhu Q, Gao X, Brown MD (2014) Simulation of the progression of intervertebral disc degeneration due to decreased nutritional supply. Spine 39:E1411–E1417

Guerin HA, Elliott DM (2006) Degeneration affects the fiber reorientation of human annulus fibrosus under tensile load. J Biomech 39:1410–1418

Haefeli M, Kalberer F, Saegesser D, Nerlich AG, Boos N, Paesold G (2006) The course of macroscopic degeneration in the human lumbar intervertebral disc. Spine 31:1522–1531

Holzapfel GA, Schulze-Bauer CA, Feigl G, Regitnig P (2005) Single lamellar mechanics of the human lumbar anulus fibrosus. Biomech Model Mechanobiol 3:125–140

Hukins DWL (2005) Tissue engineering: a live disc. Nat Mater 4:881–882

Iatridis JC, MacLean JJ, O'Brien M, Stokes IA (2007) Measurements of proteoglycan and water content distribution in human lumbar intervertebral discs. Spine 32:1493–1497

Jin D, Qu D, Zhao L, Chen J, Jiang J (2003) Prosthetic disc nucleus (PDN) replacement for lumbar disc herniation: preliminary report with six months' follow-up. J Spinal Disord Tech 16:331–337

Johannessen W, Elliott DM (2005) Effects of degeneration on the biphasic material properties of human nucleus pulposus in confined compression. Spine 30:E724–E729

Johnstone B, Bayliss MT (1995) The large proteoglycans of the human intervertebral disc. Changes in their biosynthesis and structure with age, topography, and pathology. Spine 20:674–684

Katz JN (2006) Lumbar disc disorders and low-back pain: socioeconomic factors and consequences. J Bone Joint Surg Am vol 88(Suppl 2):21–24

Klara PM, Ray CD (2002) Artificial nucleus replacement: clinical experience. Spine 27:1374–1377

Kroeber MW, Unglaub F, Wang H, Schmid C, Thomsen M, Nerlich A, Richter W (2002) New in vivo animal model to create intervertebral disc degeneration and to investigate the effects of therapeutic strategies to stimulate disc regeneration. Spine 27:2684–2690

Lama P, Le Maitre CL, Dolan P, Tarlton JF, Harding IJ, Adams MA (2013) Do intervertebral discs degenerate before they herniate, or after? Bone Joint J 95-B:1127–1133

Lebow RL, Adogwa O, Parker SL, Sharma A, Cheng J, McGirt MJ (2011) Asymptomatic same-site recurrent disc herniation after lumbar discectomy: results of a prospective longitudinal study with 2-year serial imaging. Spine 36:2147–2151

Lotz JC, Colliou OK, Chin JR, Duncan NA, Liebenberg E (1998) Compression-induced degeneration of the intervertebral disc: an in vivo mouse model and finite-element study. Spine 23:2493–2506

Lotz JC, Fields AJ, Liebenberg EC (2013) The role of the vertebral end plate in low back pain. Global Spine J 3:153–164

Loupasis GA, Stamos K, Katonis PG, Sapkas G, Korres DS, Hartofilakidis G (1999) Seven- to 20-year outcome of lumbar discectomy. Spine 24:2313–2317

Lyons G, Eisenstein SM, Sweet MB (1981) Biochemical changes in intervertebral disc degeneration. Biochim Biophys Acta 673:443–453

Maroudas A (1975) Biophysical chemistry of cartilaginous tissues with special reference to solute and fluid transport. Biorheology 12:233–248

Maroudas A, Stockwell RA, Nachemson A, Urban J (1975) Factors involved in the nutrition of the human lumbar intervertebral disc: cellularity and diffusion of glucose in vitro. J Anat 120:113–130

McGirt MJ et al (2009a) Recurrent disc herniation and long-term back pain after primary lumbar discectomy: review of outcomes reported for limited versus aggressive disc removal. Neurosurgery 64:338–344. Discussion 344–335

McGirt MJ et al (2009b) A prospective cohort study of close interval computed tomography and magnetic resonance imaging after primary lumbar discectomy: factors associated with recurrent disc herniation and disc height loss. Spine 34:2044–2051

McMillan DW, Garbutt G, Adams MA (1996) Effect of sustained loading on the water content of intervertebral discs: implications for disc metabolism. Ann Rheum Dis 55:880–887

Meakin JR, Hukins DW (2000) Effect of removing the nucleus pulposus on the deformation of the annulus fibrosus during compression of the intervertebral disc. J Biomech 33:575–580

Melrose J, Roberts S, Smith S, Menage J, Ghosh P (2002) Increased nerve and blood vessel ingrowth associated with proteoglycan depletion in an ovine anular lesion model of experimental disc degeneration. Spine 27:1278–1285

Melrose J, Smith SM, Appleyard RC, Little CB (2008) Aggrecan, versican and type VI collagen are components of annular translamellar crossbridges in the intervertebral disc. Eur Spine J 17:314–324

Mengoni M, Luxmoore BJ, Wijayathunga VN, Jones AC, Broom ND, Wilcox RK (2015) Derivation of interlamellar behaviour of the intervertebral disc annulus. J Mech Behav Biomed Mater 48:164–172

Meredith DS, Huang RC, Nguyen J, Lyman S (2010) Obesity increases the risk of recurrent herniated nucleus pulposus after lumbar microdiscectomy. Spine J 10:575–580

Michalek AJ, Buckley MR, Bonassar LJ, Cohen I, Iatridis JC (2009) Measurement of local strains in intervertebral disc anulus fibrosus tissue under dynamic shear: contributions of matrix fiber orientation and elastin content. J Biomech 42:2279–2285

Miwa S, Yokogawa A, Kobayashi T, Nishimura T, Igarashi K, Inatani H, Tsuchiya H (2015) Risk factors of recurrent lumbar disk herniation: a single center study and review of the literature. J Spinal Disord Tech 28:E265–E269

Moore RJ, Vernon-Roberts B, Fraser RD, Osti OL, Schembri M (1996) The origin and fate of herniated lumbar intervertebral disc tissue. Spine 21:2149–2155

Murakami H, Yoon TS, Attallah-Wasif ES, Kraiwattanapong C, Kikkawa I, Hutton WC (2010) Quantitative differences in intervertebral disc-matrix composition with age-related degeneration. Med Biol Eng Comput 48:469–474

Nachemson A, Lewin T, Maroudas A, Freeman MA (1970) In vitro diffusion of dye through the end-plates and the annulus fibrosus of human lumbar inter-vertebral discs. Acta Orthop Scand 41:589–607

O'Connell GD, Guerin HL, Elliott DM (2009) Theoretical and uniaxial experimental evaluation of human annulus fibrosus degeneration. J Biomech Eng 131:111007

Omarker K, Myers RR (1998) Pathogenesis of sciatic pain: role of herniated nucleus pulposus and deformation of spinal nerve root and dorsal root ganglion. Pain 78:99–105

Ordway NR, Lavelle WF, Brown T, Bao QB (2013) Biomechanical assessment and fatigue characteristics of an articulating nucleus implant. Int J Spine Surg 7:e109–e117

Ozawa T, Ohtori S, Inoue G, Aoki Y, Moriya H, Takahashi K (2006) The degenerated lumbar intervertebral disc is innervated primarily by peptide-containing sensory nerve fibers in humans. Spine 31:2418–2422

Paassilta P et al (2001) Identification of a novel common genetic risk factor for lumbar disk disease. Jama 285:1843–1849

Pimenta L, Marchi L, Coutinho E, Oliveira L (2012) Lessons learned after 9 Years' clinical experience with 3 different nucleus replacement devices. Semin Spine Surg 24:43–47

Schnake KJ, Kandziora F (2016) Disk arthroplasty: a 30-year history. In: Pinheiro-Franco J, Vaccaro A, Benzel E, Mayer H (eds) Advanced concepts in lumbar degenerative disk disease. Springer, Berlin/Heidelberg

Schollum ML, Robertson PA, Broom ND (2010) How age influences unravelling morphology of annular lamellae – a study of interfibre cohesivity in the lumbar disc. J Anat 216:310–319

Selviaridis P, Foroglou N, Tsitlakidis A, Hatzisotiriou A, Magras I, Patsalas I (2010) Long-term outcome after implantation of prosthetic disc nucleus device (PDN) in lumbar disc disease. Hippokratia 14:176–184

Serhan H, Mhatre D, Defossez H, Bono CM (2011) Motion-preserving technologies for degenerative lumbar spine: the past, present, and future horizons. SAS J 5:75–89

Shim CS et al (2003) Partial disc replacement with the PDN prosthetic disc nucleus device: early clinical results. J Spinal Disord Tech 16:324–330

Stefanakis M, Sychev I, Summers BA, Dolan P, Harding I, Adams MA (2011) Ingrowth of nerves and blood vessels into painful intervertebral discs: GP88. Spine J Meet Abstr https://journals.lww.com/spinejournalabstracts/Fulltext/2011/10001/INGROWTH_OF_NERVES_AND_BLOOD_VESSELS_INTO_PAINFUL.85.aspx #print-article-link

Stokes IA, Iatridis JC (2004) Mechanical conditions that accelerate intervertebral disc degeneration: overload versus immobilization. Spine 29:2724–2732

Tibrewal SB, Pearcy MJ (1985) Lumbar intervertebral disc heights in normal subjects and patients with disc herniation. Spine 10:452–454

Travascio F, Jackson AR, Brown MD, Gu WY (2009) Relationship between solute transport properties and tissue morphology in human annulus fibrosus. J Orthop Res 27:1625–1630

Trout JJ, Buckwalter JA, Moore KC (1982) Ultrastructure of the human intervertebral disc: II. Cells of the nucleus pulposus. Anat Rec 204:307–314

Twomey LT, Taylor JR (1987) Age changes in lumbar vertebrae and intervertebral discs. Clin Orthop Relat Res 224:97–104

Urban MR, Fairbank JC, Bibby SR, Urban JP (2001) Intervertebral disc composition in neuromuscular scoliosis: changes in cell density and glycosaminoglycan concentration at the curve apex. Spine 26:610–617

Vernon-Roberts B, Moore RJ, Fraser RD (2007) The natural history of age-related disc degeneration: the pathology and sequelae of tears. Spine 32:2797–2804

Virk SS, Diwan A, Phillips FM, Sandhu H, Khan SN (2017) What is the rate of revision discectomies after primary discectomy on a National Scale? Clin Orthop Relat Res 475:2752–2762

Vos T et al (2017) Global, regional, and national incidence, prevalence, and years lived with disability for 328 diseases and injuries for 195 countries, 1990–2016: a systematic analysis for the Global Burden of Disease Study 2016. Lancet 390:1211–1259

Wang Y, Battie MC, Boyd SK, Videman T (2011) The osseous endplates in lumbar vertebrae: thickness, bone mineral density and their associations with age and disk degeneration. Bone 48:804–809

Watters WC, McGirt MJ (2009) An evidence-based review of the literature on the consequences of conservative versus aggressive discectomy for the treatment of primary disc herniation with radiculopathy. Spine J 9:240–257

Weinstein JN et al (2008) Surgical versus nonoperative treatment for lumbar disc herniation: four-year results for the Spine Patient Outcomes Research Trial (SPORT). Spine 33:2789–2800

Yao H, Justiz MA, Flagler D, Gu WY (2002) Effects of swelling pressure and hydraulic permeability on dynamic compressive behavior of lumbar annulus fibrosus. Ann Biomed Eng 30:1234–1241

Yorimitsu E, Chiba K, Toyama Y, Hirabayashi K (2001) Long-term outcomes of standard discectomy for lumbar disc herniation: a follow-up study of more than 10 years. Spine 26:652–657

Zhang ZM, Zhao L, Qu DB, Jin DD (2009) Artificial nucleus replacement: surgical and clinical experience. Orthop Surg 1:52–57 spine conditions can be addressed with anterior lumbar interbody fusion, direct lateral (transpsoas) lumbar interbody fusion, or oblique lumbar interbody fusion (using an anterior to psoas approach). The cervical spine with degenerative changes can be treated with fusion anteriorly, using anterior cervical discectomy and fusion, or posteriorly, with posterior cervical decompression and fusion. The success or failure of the fusion procedure is determined on the absence or presence of pseudarthrosis based on clinical findings supplemented by diagnostic evidence of bridging bone. Imaging modalities most commonly used for evaluation of fusion include radiographs, magnetic resonance imaging, and computed tomography, with CT as the gold standard for assessment. Recent studies support CT as the imaging modality of choice, with some studies presenting different techniques that may aid in the evaluation of fusion status. In the hope of attaining higher outcomes for fusion along with decrease in morbidity associated with graft harvest, several bone graft substitutes and extenders have been developed. Several studies have been produced supporting their use. None have provided clear-cut evidence or recommendations that would help determine any advantage of one bone graft substitute/extender over the other.

Keywords

Fusion · Pseudarthrosis · Cervical spine · Lumbar spine · Radiographic evaluation · Bone graft · Graft substitutes

---

Introduction

Degenerative spinal conditions can lead to abnormal motion and biomechanical instability of affected spinal segments, which can result in pain, deformity, neural element compromise, or deterioration. The degenerative cascade is believed to have a complex multifactorial etiology. The process itself may be secondary to aging, genetic factors, metabolic disorders, low-grade infection, neurogenic

inflammation, autoimmune response, toxic, or mechanical factors (Hadjipavlou et al. 2008). In order to address spinal segment instability, numerous surgical fusion techniques have been developed over the years.

Several fusion techniques have been described in literature. For the cervical spine, surgeons utilize either anterior cervical discectomy and fusion or posterior cervical decompression and fusion. The most commonly performed fusions are on the lumbar spine. Degenerative conditions of the lumbar spine may be addressed with different techniques such as posterolateral fusion, posterior lumbar interbody fusion, transforaminal interbody fusion, and anterior lumbar interbody fusion. Recently introduced is the so-called lateral interbody fusion [direct lateral interbody fusion (DLIF)/extreme lateral interbody fusion (XLIF™)], which employs a transpsoas muscle approach, and its variation, the oblique lateral interbody fusion (OLIF), a technique that uses the plane anterior to the psoas muscle for its approach. All these procedures do come at a cost, as patients who undergo these operations run the risk of developing complications related to surgery. Fusion strategies have been improved upon with the use of minimally invasive techniques, with goals of limiting blood loss, soft tissue injury, operative time, immobilization, incidence of wound infections, and hospital stay in properly selected patients (Mummaneni et al. 2013; Bach et al. 2014).

Since the inception of various fusion techniques, several studies have been produced that focus on the evaluation and analysis of their clinical and radiologic outcomes. The development of new bone graft substitutes has also prompted investigators to closely examine these products to better understand the possible benefits that they can offer to patients undergoing fusion surgery. With this in mind, the objective of this chapter is to present a comprehensive review and discussion of spinal fusions in human studies.

Clinical Presentation of Pseudarthrosis

Fusion is said to occur once the bone graft within or around a spinal motion segment is deemed absent of any pseudarthrosis or non-union. Pseudarthrosis refers to the failure of spinal fusion diagnosed more than a year after surgery. Patients with pseudarthrosis may or may not present with any symptoms. If symptomatic, they may complain of axial or radicular pain and be diagnosed with "refractory back syndrome." These may also be associated with claudication or myelopathy as well. Clinical findings are not reliable in the diagnosis of pseudarthrosis. Confirmation is achieved preferably through surgical exploration, but this can be assessed noninvasively through a number of imaging modalities. These imaging modalities include X-rays, computed tomography (CT), magnetic resonance imaging (MRI), bone scan, ultrasound, and radiostereometric analysis.

Imaging Techniques

Routine orthogonal radiographs are requested upon follow-up to determine the progression of arthrodesis, which can be observed as increasing opacification and bridging trabecular bone at the bone graft margins. Static orthogonal radiographs may be inadequate for some surgeons due to its inability to identify motion at the spinal fusion segment. Although controversial, lateral flexion-extension radiographs have been used to rule out motion at the spinal fusion segment, hardware failure, and issues with sagittal alignment (Raizman et al. 2009; Gruskay et al. 2014b). Further assessment can be done by making use of different methods such as measuring the Cobb angle and also by making use of Simmons method and Hutter methods (Hutter 1983; Simmons 1985). Although widely used by both spine surgeons and radiologists, these have problems in terms of concordance with findings on CT images (Figs. 1 and 2).

Computed tomography can be very helpful in instances where there is no evidence of progression of fusion on radiographs. Fusion is present once note of bony trabeculation is found across the fusion level with absence of radiolucency at the graft-vertebral body interface. The downside with using CT scans is its decreased sensitivity to fusion due to metallic artifacts secondary to instrumentation implanted within the spine. The incidence of metallic artifacts has lessened with

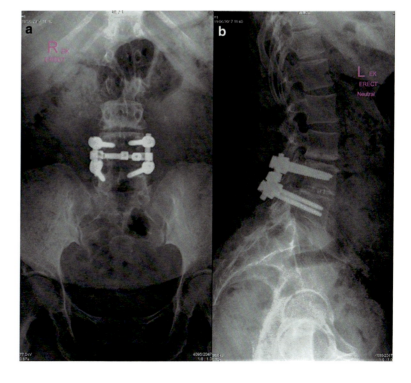

Fig. 1 Standard AP (**a**) and lateral (**b**) X-ray views of the lumbar spine. Posterior decompression and TLIF were performed at L4–5. Note presence of bridging bone at said level

Fig. 2 Flexion (**a**) and extension (**b**) views performed to rule out motion at fused segment

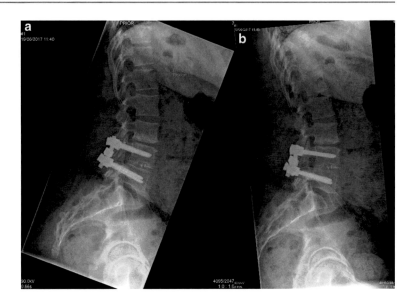

the advent of titanium implants. Despite this disadvantage, computed tomography has been deemed as the gold standard imaging modality for assessing the presence of fusion (Raizman et al. 2009; Gruskay et al. 2014b) (Fig. 3).

Other imaging modalities available for assessing spinal fusion include magnetic resonance imaging, bone scintigraphy, ultrasound, and radiostereometric analysis. MRI is not routinely used for spinal fusion evaluation due to its susceptibility to artifact formation from metallic implants, and its clinical utility, while promising, is not yet established (Kitchen et al. 2018). It is more useful in identifying stenosis and presence of adjacent segment degeneration. Bone scintigraphy is able to ascertain the level of metabolic activity of the spine. Increased uptake is noted in areas of heightened biologic activity and blood supply which may suggest non-union. This is of limited use due to its low sensitivity (50%) and specificity (58%) when compared to surgical exploration. Ultrasound was found to be able to ascertain the presence of fusion if there was note of hyperechoic and shadowing interface across vertebral segments. On the other hand, the presence of scattered and nonbridging echogenic foci indicates possible pseudarthrosis. This imaging modality was found to be more suitable for patients who have undergone posterior instrumentation which may produce artifacts in CT or MRI scans. Radiostereometric analysis allows for three-dimensional imaging of spinal motion in vivo. The use of this imaging modality has been limited to research purposes, despite its high accuracy, and requires bony insertion of tantalum beads (Raizman et al. 2009; Gruskay et al. 2014b).

Multitudes of studies have been performed to evaluate fusion in different regions of the spine. Evaluation of success for a particular technique relies on a combination of clinical and radiologic outcomes. The difficulty lies on properly assessing the quality of evidence and strength of recommendations presented by these studies (Guyatt et al. 2009) (Table 1).

Fusion of the Cervical Spine

Anterior Cervical Discectomy and Fusion (ACDF)

Cervical disk degeneration may present as axial neck pain associated with radiculopathy or myelopathy. Anterior cervical discectomy and fusion is the accepted standard treatment for symptomatic cervical disk degeneration. After removal of the pathologic disc, fusion is carried out via stand-alone cages inserted into the intervertebral disk space or supplemented by an anterior plate. The cage allows for bridging union of vertebral bodies

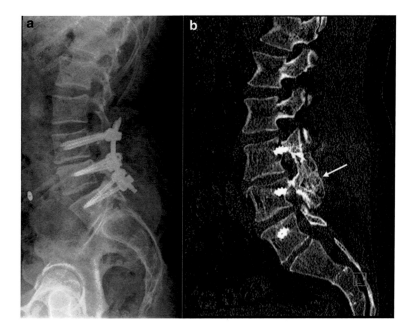

Fig. 3 Lateral radiograph (**a**) of instrumented posterolateral fusion performed from L3 to L5. Sagittal cut of CT scan (**b**) shows abundant fusion mass (arrow) over posterolateral gutter adjoining the levels from L3 to L4. Bridging bone was absent in between L4 and L5

Table 1 Imaging modalities for evaluation of fusion

Imaging modality	Advantages	Disadvantages	Utility
Radiographs	Picks up opacification and bridging bone along graft margins Low cost Easy to perform Dynamic radiographs can be used to assess for translational or angular motion of fusion segment	Occasionally miss presence of bridging bone or bony trabeculations Low radiation exposure	++
Computed tomography	Greater ability to determine presence or absence of bridging bone or bony trabeculations between fusion segments Identification of hardware failure/loosening	Higher radiation exposure compared to radiographs Dynamic CT not common practice Artifact formation from metallic implants may prevent visualization of bone bridges	++++
Magnetic resonance imaging	More useful in identifying stenosis and adjacent segment degeneration	Prone to artifact formation from metallic implants Static imaging modality	+
Bone scintigraphy	Increased uptake due to heightened biologic activity may be secondary to non-union	Low sensitivity (50%) and specificity (58%)	–
Ultrasound	Alternative imaging modality that identifies pseudarthrosis which may present as areas of nonbridging echogenic foci between vertebral segments No radiation	Limited to patients with posterior instrumentation	+
Radiostereometric analysis	Capable of three-dimensional imaging of spinal motion in vivo High accuracy	Limited presently to research purposes Requires tantalum bead insertion around prosthesis	++

to take place once fusion has set in. The addition of a plate is believed to allow for higher fusion rates, decreased pseudarthrosis, decreased graft dislodgement, resistance to segmental kyphosis, and less need for external immobilization.

Several imaging modalities can be used for the assessment of fusion at the cervical level. Besides assessing the quality of fusion present, specific measurements are utilized depending on the specific type of imaging modality used. These measurements are most often taken from dynamic radiographs through the use of lateral flexion and extension views.

The assessment of fusion status through dynamic radiographs is dependent on currently accepted criteria. For example, the US FDA previously defined successful fusion as less than 3 mm of translational motion and less than 5° of angular motion in the lumbar spine. These criteria are also applied in the assessment of successful fusion in the cervical spine (USSDHHS et al. 2000). Traditionally, motion was assessed by measuring the Cobb angle. This has been difficult to evaluate as there is very minimal motion detected with this method. An alternative to the Cobb angle method is the interspinous method (Cannada et al. 2003). In general, there is also low interobserver agreement as well with regard to the assessment of cervical spine fusion on dynamic films (Taylor et al. 2007) (Fig. 4).

A retrospective chart review of 383 patients treated ACDF with a total of 1155 postoperative visits was performed from 2002 to 2007 to determine the utility of radiographs in postoperative monitoring of fusion status. These patients were classified according to normal versus abnormal history and physical examination presentations and also by normal and abnormal radiographs. Patients with normal history and physical examination findings were rarely found to be managed/not left alone [5/879 (0.57%)]. There were 276 visits with abnormal history and physical examination findings. Abnormal radiographic findings were found in 34 out of the 276 visits (12.3%). Revision surgery was advised in 44% of these visits with abnormal radiographic findings (15/34). This study concluded that postoperative radiographs had limited utility in patients with normal history and physical examination findings independent of normal or abnormal radiographic results (Grimm et al. 2013).

A prospective clinical study attempted to evaluate the reliability of detecting pseudarthrosis after anterior cervical fusion using radiography, CT, and MRI as compared to surgical exploration. The investigators found that assessment of fusion status via CT is most closely related with findings upon surgical exploration when compared to the other imaging modalities. This was assessed by evaluating the agreement between findings on surgical exploration and the different imaging modalities and through paired interobserver reliability. Mean Kappa statistics for agreement between intraoperative and radiographic findings were 0.67 (plain radiographs), 0.81 (CT), and 0.48 (MRI), while mean Kappa statistics for paired interobserver reliability were 0.46 (plain

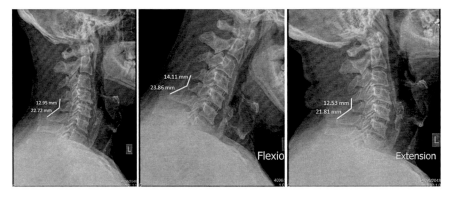

Fig. 4 Lateral X-ray views of the cervical spine showing method of measuring interspinous process distance

radiographs), 0.82 (CT), and 0.32 (MRI) (Buchowski et al. 2008).

On the other hand, Park et al. (2015) found that CT scans may overestimate the rate of fusion after ACDF. They performed a radiographic analysis of patients who had undergone ACDF. This study attempted to compare the fusion rates with CT scans and dynamic radiographs at 3, 6, and 12 months postoperatively. Their results show that the fusion rates post ACDF assessed via radiographs and CT scans were 26% versus 79%, 41% versus 79%, and 65% versus 91% at 3, 6, and 12 months, respectively. This study concluded that overestimation of fusion rates was present upon assessment with CT scans. This is due to the static nature of the fusion levels being assessed. Radiographs, when compared to CT scans, are assessed taking into account dynamic and static factors for fusion.

Ouchida et al. (2015) performed a study comparing the ability of functional computed tomography in determining fusion status versus functional radiography in patients post ACDF. This study was able to determine that functional CT scanning was superior in detecting non-union when compared to functional radiography. Radiographs showed fusion rates of 83.9% and 91.1% at 6 months and 12 months postoperatively, respectively. On the other hand, CT scans revealed considerably lower fusion rates at 6 months (55.3%) and 12 months (78.6%). Patients found to have incomplete union were shown to develop more neck pain postoperatively.

Ghiselli et al. (2011) attempted to determine the gold standard for assessing pseudarthrosis of the cervical spine. This study investigated the ability of CT imaging versus quantitative motion analysis with dynamic radiography to examine for presence of fusion. Findings from imaging were then compared to those found in the patients intraoperatively. The amount of angular motion on radiographs thought to correlate with pseudarthrosis was at more than $4°$. The study found this parameter provided for high specificity with a positive predictive value of 100% but was coupled with a low sensitivity as shown by a negative predictive value of 52%. Fusion on CT was defined as presence of bridging bone. CT scans were shown to have a positive predictive value of 100% and a negative predictive value of 73%. This was comparable to the negative predictive value with the use of quantitative motion analysis on dynamic radiographs once the accepted angular motion was changed to a value of greater than one degree (73%). The researchers were then able to determine that by making use of both CT scans and the modified angular motion ($1°$) on radiographs produces a positive predictive value of 100% and negative predictive value of 85%, thus improving the specificity in detecting for presence of pseudarthrosis in the cervical spine.

Posterior Cervical Decompression and Fusion (PCDF)

Posterior cervical decompression and fusion can be performed through varying techniques. Decompression in itself can be done via different methods. These include laminectomies and laminoplasties, often done in patients with cervical spondylotic myelopathy, which allow for extensive decompression of the cervical canal with potential for postoperative instability of the motion segments. Foraminotomies limit the extent of destabilization and are strictly indicated in patients with stenosis at the foraminal level only. Fusion can be achieved via multiple techniques supplemented with instrumentation through the use of screw fixation. The instrumentation can either be performed with application of lateral mass screws or pedicle screws (Fig. 5).

Lee et al. (2017) analyzed the fusion and graft resorption rates in 56 patients who underwent posterior cervical fusion with pedicle screw instrumentation 1 year postoperatively under CT imaging and dynamic radiography. The patients who participated in this study were classified into three groups according to the type of graft used for posterolateral fusion (autograft, allograft, or mixture of both). Mean resorption rates were as follows: 56.2% (autograft), 75.9% (mixture), and 91.5% (allograft). Despite the resorption rates mentioned, the overall fusion rate was 98.2%.

Fig. 5 Posterior cervical decompression and fusion at C4–5 with lateral mass screws

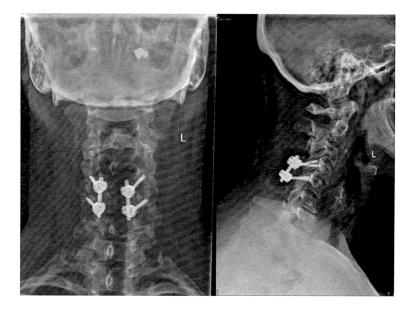

A retrospective case series by Dorward et al. reported on the fusion rates and complications associated with the use of recombinant human bone morphogenetic protein-2 (rhBMP-2) in patients treated with posterior cervical fusion. A majority of patients included in the study were operated on to revise previous surgeries (84.2%). Several patients were also found to have preexisting cervical pseudarthrosis (42.1%). Successful fusion was noted in 89.5% of patients. Radiographic evidence of non-union was found in the remaining 10.5%. Pseudarthrosis occurred in patients with fusions that spanned either occipitocervical or cervicothoracic junctions. Complications were noted in 14 patients (24.6%), with superficial wound infection, pain from fusion levels, and pseudarthrosis being among the most common (Dorward et al. 2016).

Fusion of the Lumbar Spine

Degenerative conditions of the lumbar spine may present as back pain with radiculopathy or other forms of neurologic deficits in patients. These warrant fusion of spinal motion segments with or without decompression of the pathologic intervertebral disk. Techniques utilized for the fusion of the lumbar spine will depend on the type of approach. Several techniques available for the posterior approach include posterolateral fusion, posterior lumbar interbody fusion, and transforaminal interbody fusion, which are then supplemented by posterior pedicle screw instrumentation. Another method for achieving arthrodesis of the spine is by performing anterior lumbar interbody fusion. This technique utilizes a retroperitoneal approach to access the lumbar spine. Recently developed variations of this technique include the direct lateral interbody fusion and oblique lateral interbody fusion. The former gains access to the intervertebral disk via transpsoas technique, while the latter approaches the disk space through a corridor just anterior to the psoas muscle.

Posterior Lumbar Decompression and Fusion

Fogel et al. (2008) determined the accuracy of fusion assessment of posterior lumbar interbody fusions with the use of plain radiographs and computed tomography in comparison to findings on surgical exploration. Successful fusion was found in 168 out of 172 (97%) motion segments fused upon surgical exploration. Interbody fusion had a success rate of 87% (X-ray) and 77% (CT), while posterolateral fusion was 75% (X-ray) and 68% (CT). They were able to determine that plain radiographs and CT imaging were both accurate in the assessment of fusion status when compared to findings on surgical exploration of the lumbar interbody and posterolateral fusion levels. The investigators also found that CT is unlikely to provide any additional significant information if there is strong evidence of fusion or pseudarthrosis on plain radiographs (Fig. 6).

Carreon et al. (2007b) performed a blinded cross-sectional study on plain radiographs and fine-cut CT scans to evaluate intra- and interobserver reliability and agreement for assessing single-level instrumented posterolateral fusions. Plain radiographs with anteroposterior and lateral flexion-extension views and fine-cut CT scans with sagittal and coronal reconstructions were performed 1 year postoperatively. Fine-cut CT scans were found to have greater interobserver and intraobserver agreement as compared to plain radiographs. Agreement on fusion status between plain radiographs and CT scans ranged from 46% to 59% only. Another study by Carreon et al. (2007a) showed that radiographic findings of facet fusion and posterolateral fusion on fine-cut CT scans yielded 96% fusion rate on surgical exploration. The kappa statistic for interobserver reliability for evaluating facet fusions was moderate (0.42) and for posterolateral fusions was found to be substantial (0.62). The probability of solid fusion on surgical exploration was higher when bilateral posterolateral gutter fusion was present (89%) compared to bilateral facet fusion on CT scan (74%). The study found several poor predictors for non-union upon surgical exploration, which include absence of fusion of unilateral or bilateral facets or through one posterolateral gutter.

The utility of magnetic resonance imaging (MRI) for the evaluation of fusion status post PLIF has been explored in the past. A prospective study investigated the use of MRI in the assessment of PLIF using carbon fiber reinforced polymer cages. The most reliable radiographic finding is the presence of bridging bone within the carbon cage on the coronal planes based on MRI. Kroner et al. concluded that MRI is a reliable imaging

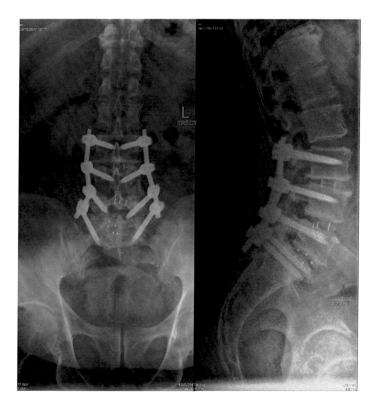

Fig. 6 Multilevel TLIF spanning L3 to S1

method for detecting pseudarthrosis after a period of 2 years post fusion (Kröner et al. 2006).

Nakashima et al. (2011) compared the ability of dynamic radiographs and CT scans with the lumbar spine in flexion and in extension to identify the quality of arthrodesis or fusion in patients who had undergone PLIF. There were 81 patients with a total of 97 fused levels included in the study and were followed for more than 12 months after PLIF. Dynamic radiographs revealed fusion in 90.7% of all operative levels at 10.7 months postoperatively. Patients were then further evaluated, revealing 87.6% fusion on flexion CT and 69.1% fusion on extension CT. Pseudarthrosis detection on extension CT was found to be significantly higher compared to dynamic radiography and flexion CT.

Anterior Lumbar Interbody Fusion (ALIF)

Carreon et al. (2008) once again evaluated the reliability and accuracy of CT imaging versus surgical exploration in determining fusion status, this time around on patients who had undergone anterior lumbar interbody fusion with metallic cages implanted. The study aimed to establish the interobserver reliability with regard to the presence or absence of bridging bone, as well as the anterior and posterior sentinel signs on CT and actual findings of fusion or pseudarthrosis upon surgical exploration of the lumbar spine. On average, 67% of the cases were correctly classified as fused (93% sensitivity, 46% specificity). Interobserver reliability was found to be fair (kappa 0.25). The investigators were able to conclude that their CT scans had a high false-positive rate of determining fusion (Fig. 7).

Lateral Lumbar Interbody Fusion [Direct Lateral LIF (Transpsoas) and Oblique LIF (Anterior to Psoas)]

Clinical and radiologic evaluation of fusion rates with the use of lumbar interbody fusion via direct

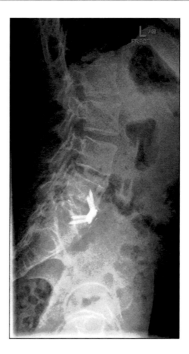

Fig. 7 ALIF performed at L5-S1. Bridging bone well appreciated at fused segment

lateral/transpsoas approach (XLIF) along with varying types of graft material (autograft, calcium triphosphate, and Attrax) was performed by Berjano et al. (2015). Patients who had undergone the procedure from 2009 to 2013 were assessed through clinical evaluation and CT scans with a minimum of 1-year follow-up. CT scans were evaluated with complete fusion defined as presence of bridging bone within the interbody space. Pseudarthrosis was defined as complete absence of graft material within the cage or where there was presence of radiolucency. Complete fusion was found in 68 out of 78 operated interbody levels (87.1%). Stable but incomplete fusion was noted in eight levels (10%), while pseudarthrosis was identified in only two operated levels (2.6%). Comparison of fusion rate by graft material used revealed successful arthrodesis with autograft in 75% in patients, Attrax in 83%, and calcium trisphosphate having the highest fusion rate of 89% (Figs. 8 and 9).

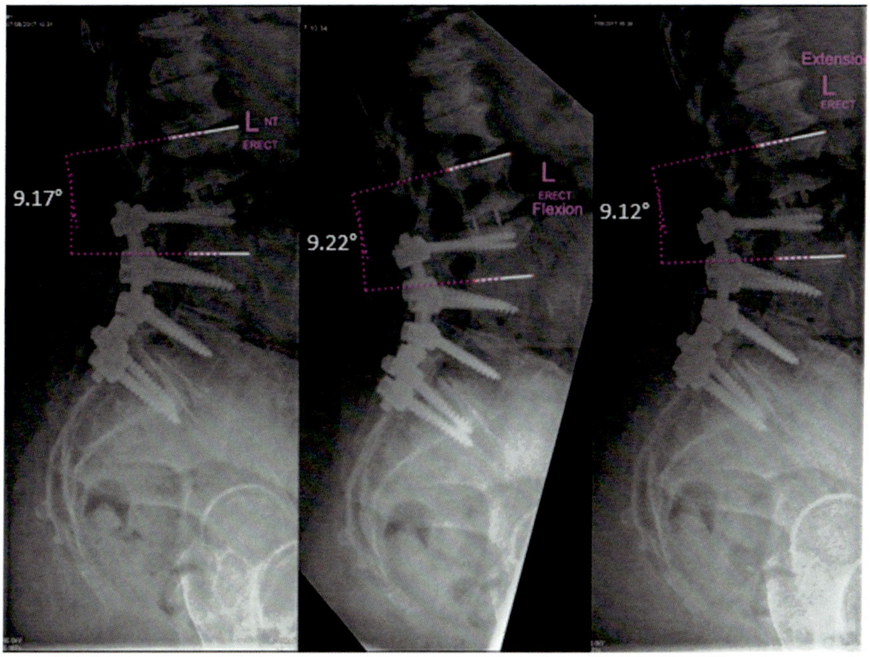

Fig. 8 Lateral views of the lumbar spine post XLIF for adjacent segment degeneration at L2–3. Cobb method was utilized to assess for motion at fusion segment

Fig. 9 Degenerative lumbar scoliosis treated with XLIF from T12 to L4. Preoperative curve was at 44°. Correction of 23.33° was achieved after XLIF

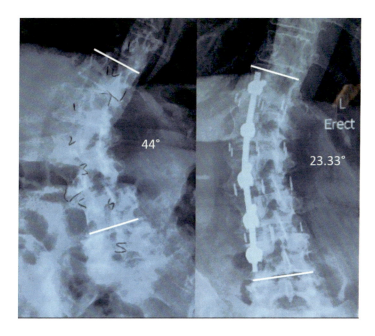

Factors to Consider for Further Imaging Evaluation in Pseudarthrosis

Plain radiographs are routinely requested for initial postoperative evaluation of postoperative spinal fusion. These have been found to be inadequate in detecting pseudarthrosis. A retrospective review by Klineberg et al. (2016) evaluated the reliability of anterior fusion grading systems and the ability of health-related quality of life (HRQOL) outcomes to predict pseudarthrosis. Its results showed that the grading system used to evaluate for the presence of fusion missed many pseudarthroses that were confirmed surgically and with fine-cut CT. Low Scoliosis Research Society (SRS) and Oswestry Disability Index (ODI) scores compared to preoperative baseline were found in patients with apparent pseudarthrosis. Patients with poor HRQOL scores indicate the need for further advanced investigations in order to address the possibility of pseudarthrosis (Klineberg et al. 2016).

In light of clinical findings and despite the presence of additional imaging investigations to validate any suspicions of non-union after fusion, there is still great difficulty in reliable confirmation of pseudarthrosis. Besides the prerequisite identification of de novo bridging bone and absence of translational and angular motion spanning the fusion segment, one must also take into account the resorption of the bone grafts, substitutes, and extenders as well. Once pseudarthrosis is confirmed, revision surgery may then be performed in appropriate cases. Autografts or allografts are used again, with possible augmentation through bone graft substitutes or extenders, if these were not used during the index surgery, to increase the possible chances of successful rate of fusion.

Evaluation of Bone Graft Materials for Fusion

With all the techniques available for fusion of the spine, the question of which bone graft material should be used is still up for debate. Autogenous iliac crest bone graft provided similar effectiveness in terms of fusion rate, pain scores, and functional outcomes when compared to local autogenous bone graft and allograft. The use of iliac crest bone graft in spinal fusion is still considered the gold standard due to its osteogenic, osteoconductive, and osteoinductive properties, despite the possible complications associated with harvesting. These included postoperative donor site pain, hematoma, infection, pelvic fracture, and nerve palsy. One study found that use of autogenous iliac crest bone graft in spinal fusions was associated with increase postoperative blood transfusion, extended operative time, and increased length of stay in hospital (Gruskay et al. 2014a; Tuchman et al. 2016). Bone graft substitutes and extenders have thus been developed, which not only address the possible problems that arise with bone graft harvesting but also increase the potential for solid arthrodesis of the spine (Kaiser et al. 2014). Despite the increased number of bone graft substitutes at hand, evidence to support their superiority over autogenous bone graft and allografts is low and hard to come by. Herein lies the difficulty in selecting the most appropriate bone graft material or substitute to aid in achieving successful arthrodesis of the spine (Buser et al. 2016).

Bone allografts can be differentiated into fresh-frozen and freeze-dried products. Both are said to be incorporated slower and to a lesser degree compared to autografts. In terms of fusion rates, autografts are still superior versus allografts alone or in combination with autografts. Fresh-frozen allografts are also found to be stronger, more immunogenic, and more completely incorporated as opposed to freeze-dried allografts. This has also been found to work well when used in anterior lumbar fusion for reconstruction procedures with resulting good fusion rates (Ehrler and Vaccaro 2000). Strong recommendations have also been made with regard to the use of allografts in ACDF, ALIF, and posterolateral lumbar fusion. At present, there is still need for more investigations on its use in TLIF (Gibson et al. 2002; Thalgott et al. 2009; Miller and Block 2011; Buser et al. 2016).

Demineralized bone matrix (DBM) functions as both an extender and augmenter of bone graft material. Its advantages include safety in terms of having no risk of disease transmission and its respectable storage and shelf life. This bone graft material is not without its disadvantages. Studies have shown varying efficacy and poor predictability, with variable degrees of osteoconductive potential with unknown effect on bone formation due to the different carrier types. The manufacturing process is also unregulated, thus leading to fluctuating amounts of bone morphogenetic proteins found in the different DBM products available in the market (Rihn et al. 2010; Aghdasi et al. 2013; Tavakol et al. 2013). Thalgott et al. (2001) investigated the efficacy of coralline hydroxyapatite with or without DBM as an extender of autogenous bone graft in patients who had undergone instrumented posterolateral fusion. There was an overall fusion rate of 92.5% for this study, but a lower fusion rate was found in patients that had DBM added into the autogenous bone graft (89.3%).

Bone morphogenetic proteins (BMPs) are known for their powerful osteoinductive potential evidenced by their ability to produce ectopic bone, promote spinal fusion, and induce fracture healing. rhBMP-2 and BMP-7 have been found to be safe for use in certain conditions for spine surgery, particularly in ALIF and in revision PLIF (Burkus et al. 2002; Vaccaro et al. 2005). The said products have been involved in off-label use as well, such as in ACDF, TLIF, and PLF. However, judicious use and awareness of their side effects and different dosage effects in different anatomic locations are critical. Concern has risen with regard to the safety and effectiveness of the off label of BMPs. In a systemic and review and meta-analysis by Fu et al. on the effectiveness and harms of rhBMP-2 on spinal fusion, it was found that rhBMP-2 had no clinical advantage over iliac crest bone graft but were not associated with the morbidity of harvesting and difficulties with autograft volume. Several complications have also been found to occur in respect to the type and region of the spine where the fusion was performed. Retrograde ejaculation and other urogenital problems were found to be associated with the use of rhBMP-2 in ALIF although it has been proposed to be approach related, where other authors have found a 0.7% rate with retroperitoneal approaches (Scott-Young 2014). There was an increased risk of wound complications and dysphagia when this particular BMP was used for ACDF which were likely dose related.

Although event rates were found to be low, this isolated study also found an increased risk of developing cancer in association with the use of rhBMP-2. These are contrary to the findings of Simmonds et al., who found increased fusion rates with the use of rhBMP-2 and inconclusive evidence with regard to the increased incidence of cancer (Fu et al. 2013; Simmonds et al. 2013). Currently, there does not appear to be any cause-effect relationship between rhBMP2 and cancer.

Ceramics act as bone graft extenders designed with optimized porosity and pore size to allow for bony ingrowth. The advantages attributed with this type of bone graft material are its availability, cost effectiveness, and no risk of disease transmission. The varying resorption rate across the different types (calcium sulfate, beta-tricalcium phosphate, and hydroxyapatite) of ceramics presents as a limiting factor, particularly calcium sulfate, in spine surgery (Rihn et al. 2010). The material is best used to augment the fusion mass and is not recommended as a stand-alone substitute for actual bone graft material. A prospective study found that hydroxyapatite-bioactive glass ceramic used as a stand-alone bone graft substitute had a high incidence of resorption with poor consolidation noted in 95% (21/22) of patients who had undergone instrumented posterolateral fusion (Acharya et al. 2008). Other studies that claim good results with the use of calcium phosphate ceramics were found to have the ceramic mixed with local autograft harvested from the site of decompression (Fujibayashi et al. 2001; Dai and Jiang 2008). Tricalcium phosphate has been found to provide excellent results in ACDF and posterolateral lumbar fusion. The use of beta-tricalcium phosphate (94%) for ACDF yielded fusion rates comparable to those of hydroxyapatite (90%) after 2 years of follow-up. Early fusion rates at 6 months (46%) and 1 year (69%) post ACDF were higher with beta-tricalcium phosphate versus hydroxyapatite (24% at 6 months and 49% at 1 year). Beta-tricalcium phosphate was found viable in patients treated with instrumented posterolateral fusion. Radiographs taken at 24 and 36 months post fusion showed good fusion status with no signs of motion on flexion or extension (Dai and Jiang 2008; Sugawara et al. 2011) (Table 2).

Table 2 Known spinal fusion evaluation methods for various regions and approaches

Fusion region	Approach		Method(s) of fusion assessment	Comments

Cervical	Anterior (ACDF)		X-rays
			  Flexion-extension views
			    Cobb angle
			    Simmons
			      Uses reference points
			      $>2°$ extension is
			pseudarthrosis
			    Hutter
			      Overlap of images
			    Cannada et al. 2003
			      Measure interspinous
			process distance
			      doi: 10.1097/00007632-200301010-00012
			Findings on CT have higher correlation with findings on surgical exploration when compared to x-ray and MRI (Buchowski et al. 2008)
			  doi: 10.1097/BRS.0b013e318171927c
			Combination of CT and dynamic radiographs improved specificity in determining pseudarthrosis (Ghiselli et al. 2011)
			  Modified angular motion ($>1°$) on radiographs
			  doi: 10.1097/BRS.0b013e3181d7a81a
			Functional (flexion and extension) CT was found superior over functional (dynamic) radiography in determining non-union (Ouchida et al. 2015)
			  doi: 10.1007/s00586-014-3722-z
				Postoperative x-rays of limited value for determining fusion in patients with normal history and physical findings (Grimm et al. 2013)
				  doi: 10.1016/j.spinee.2013.01.018
				CT images may tend to lead to overestimation of the presence of fusion due to their static nature (Park et al. 2015)
				  doi: 10.1097/BSD.0b013e31829a37ac

	Posterior (PCDF)		X-rays
			  Flexion-extension views
			    Cobb angle
			    Simmons
			    Hutter
			CT
				High resorption rates noted for allografts and auto/allograft mixture compared to autograft alone (Lee et al. 2017)
				  Despite resorption rates, average fusion rate across different types of grafts was 98.2%
				  Doi: 10.1016/j.wneu.2016.12.027
				Non-union in 10.5% of patients after posterior cervical fusion with rhBMP2 use (Dorward et al. 2016)
				  Most common in occipitocervical or cervicothoracic junctions
				  doi: 10.1097/BSD.0b013e318286fa7e

Lumbar	Interbody fusion	Anterior (ALIF, XLIF)	X-rays
			  Flexion-extension views
			    Cobb angle
			    Simmons
				Overestimation of fusion status on CT when compared to findings on surgical exploration (Carreon et al. 2008)

(continued)

Table 2 (continued)

Fusion region	Approach		Method(s) of fusion assessment	Comments
			Hutter CT	doi: 10.1016/j. spinee.2007.12.004 XLIF fusion success rate (Berjano et al. 2015) Complete fusion 87.1% Pseudarthrosis in 2.6% of patients Difference in fusion rate between Attrax (83%) and calcium triphosphate (89%) was not statistically significant doi: 10.1007/s00586-015-3929-7
		Posterior (PLIF, TLIF)	MRI was a viable tool for detecting pseudarthrosis in patients 2 years post PLIF (Kroner et al. 2006) doi: 10.1097/01. brs.0000218583.43398.e3 Extension CT had higher ability to detect pseudarthrosis compared to flexion CT and dynamic radiographs (Nakashima et al. 2011) doi: 10.1007/s00586-011-1739-0	Fusion success rate was higher using on x-ray (87%) versus CT (77%) (Fogel et al. 2008) doi: 10.1016/j. spinee.2007.03.013
	Posterolateral fusion		Interobserver and intraobserver agreement greater with CT compared to radiographs (Carreon et al. 2007b) doi: 10.1016/j. spinee.2006.04.005	Solid fusion on surgical exploration noted when there is presence of bilateral posterolateral fusion compared to facet fusion (Carreon et al. 2007b) doi: 10.1097/01. brs.0000259808.47104.dd Fusion success rate was higher using x-ray (87%) than CT (77%) (Fogel et al. 2008) doi: 10.1016/j. spinee.2007.03.013

References

Acharya NK, Kumar RJ, Varma HK, Menon VK (2008) Hydroxyapatite-bioactive glass ceramic composite as stand-alone graft substitute for posterolateral fusion of lumbar spine: a prospective, matched, and controlled study. J Spinal Disord Tech 21:106–111. https://doi.org/10.1097/BSD.0b013e31805fea1f

Aghdasi B, Montgomery SR, Daubs MD, Wang JC (2013) A review of demineralized bone matrices for spinal fusion: the evidence for efficacy. Surgeon 11:39–48. https://doi.org/10.1016/j.surge.2012.08.001

Bach K, Ahmadian A, Deukmedjian A, Uribe JS (2014) Minimally invasive surgical techniques in adult degenerative spinal deformity: a systematic review. Clin Orthop Relat Res 472:1749–1761. https://doi.org/10.1007/s11999-013-3441-5

Berjano P, Langella F, Damilano M et al (2015) Fusion rate following extreme lateral lumbar interbody fusion. Eur Spine J 24:369–371. https://doi.org/10.1007/s00586-015-3929-7

Buchowski JM, Liu G, Bunmaprasert T et al (2008) Anterior cervical fusion assessment surgical exploration versus radiographic evaluation. Spine (Phila Pa 1976) 33:1185–1191

Burkus JK, Gornet MF, Dickman CA, Zdeblick TA (2002) Anterior lumbar interbody fusion using rhBMP-2 with tapered interbody changes anterior lumbar interbody fusion using rhBMP-2 with tapered lnterbody cages. J Spinal Disord Tech 15:337–349

Buser Z, Brodke DS, Youssef JA et al (2016) Synthetic bone graft versus autograft or allograft for spinal fusion: a systematic review. J Neurosurg Spine 25:509–516. https://doi.org/10.3171/2016.1.SPINE151005

Cannada LK, Scherping SC, Yoo JU et al (2003) Pseudoarthrosis of the cervical spine: a comparison of radiographic diagnostic measures. Spine (Phila Pa 1976) 28:46–51. https://doi.org/10.1097/00007632-200301010-00012

Carreon LY, Djurasovic M, Glassman SD, Sailer P (2007a) Diagnostic accuracy and reliability of fine-cut CT scans with reconstructions to determine the status of an instrumented posterolateral fusion with surgical exploration as reference standard. Spine (Phila Pa 1976) 32:892–895. https://doi.org/10.1097/01.brs.0000259808.47104.dd

Carreon LY, Glassman SD, Djurasovic M (2007b) Reliability and agreement between fine-cut CT scans and plain radiography in the evaluation of posterolateral fusions. Spine J 7:39–43. https://doi.org/10.1016/j.spinee.2006.04.005

Carreon LY, Glassman SD, Schwender JD et al (2008) Reliability and accuracy of fine-cut computed tomography scans to determine the status of anterior interbody fusions with metallic cages. Spine J 8:998–1002. https://doi.org/10.1016/j.spinee.2007.12.004

Dai LY, Jiang LS (2008) Single-level instrumented posterolateral fusion of lumbar spine with B-tricalcium phosphate versus autograft: a prospective, randomized study with 3-year follow-up. Spine (Phila Pa 1976) 33:1299–1304. https://doi.org/10.1097/BRS.0b013e3181732a8e

Dorward IG, Buchowski JM, Stoker GE, Zebala LP (2016) Posterior cervical fusion with recombinant human bone. Clin Spine Surg 29:276–281

Ehrler DM, Vaccaro AR (2000) The use of allograft bone in lumbar spine surgery. Clin Orthop Relat Res 371:38–45

Fogel GR, Toohey JS, Neidre A, Brantigan JW (2008) Fusion assessment of posterior lumbar interbody fusion using radiolucent cages: X-ray films and helical computed tomography scans compared with surgical exploration of fusion. Spine J 8:570–577. https://doi.org/10.1016/j.spinee.2007.03.013

Fu R, Selph S, McDonagh M et al (2013) Effectiveness and harms of recombinant human bone morphogenetic protein-2 in spine fusion: a systematic review and meta-analysis. Ann Intern Med 158:890–902

Fujibayashi S, Shikata J, Tanaka C et al (2001) Lumbar posterolateral fusion with biphasic calcium phosphate ceramic. J Spinal Disord 14:214–221

Ghiselli G, Wharton N, Hipp JA et al (2011) Prospective analysis of imaging prediction of Pseudarthrosis after anterior cervical discectomy and fusion: computed tomography versus flexion-extension motion analysis with intraoperative correlation. Spine (Phila Pa 1976) 36:463–468. https://doi.org/10.1097/BRS.0b013e3181d7a81a

Gibson S, McLeod I, Wardlaw D, Urbaniak S (2002) Allograft versus autograft in instrumented posterolateral lumbar spinal fusion: a randomized control trial. Spine (Phila Pa 1976) 27:1599–1603. https://doi.org/10.1097/00007632-200208010-00002

Grimm BD, Leas DP, Glaser JA (2013) The utility of routine postoperative radiographs after cervical spine fusion. Spine J 13:764–769. https://doi.org/10.1016/j.spinee.2013.01.018

Gruskay JA, Basques BA, Bohl DD et al (2014a) Short-term adverse events, length of stay, and readmission after iliac crest bone graft for spinal fusion. Spine (Phila Pa 1976) 39:1718–1724. https://doi.org/10.1097/BRS.0000000000000476

Gruskay JA, Webb ML, Grauer JN (2014b) Methods of evaluating lumbar and cervical fusion. Spine J 14:531–539. https://doi.org/10.1016/j.spinee.2013.07.459

Guyatt GH, Oxman AD, Vist GE et al (2009) GRADE: an emerging consensus on rating quality of evidence and strength of recommendations. Chin J Evid-Based Med 9:8–11. https://doi.org/10.1136/bmj.39489.470347.AD

Hadjipavlou AG, Tzermiadianos MN, Bogduk N, Zindrick MR (2008) The pathophysiology of disc degeneration: a critical review. J Bone Joint Surg Br 90–B:1261–1270. https://doi.org/10.1302/0301-620X.90B10.20910

Hutter CG (1983) Posterior intervertebral body fusion. A 25-year study. Clin Orthop Relat Res 179:86–96

Kaiser MG, Groff MW, Watters WC et al (2014) Guidelines for the performance of fusion procedures for degenerative disease of the lumbar spine. Part 16: bone graft extenders and substitutes. J Neurosurg Spine 21:106–132. https://doi.org/10.3171/spi.2005.2.6.0733

Kitchen D, Rao P, Zotti M et al (2018) Fusion assessment by MRI in comparison With CT in anterior lumbar interbody fusion: a prospective study. Global Spine J. https://doi.org/10.1177/2192568218757483. First Published March 26, 2018

Klineberg E, Gupta M, McCarthy I, Hostin R (2016) Detection of Pseudarthrosis in adult spinal deformity: the use of health-related quality-of-life outcomes. Clin Spine Surg 29:318–322

Kröner AH, Eyb R, Lange A et al (2006) Magnetic resonance imaging evaluation of posterior lumbar interbody fusion. Spine (Phila Pa 1976) 31:1365–1371. https://doi.org/10.1097/01.brs.0000218583.43398.e3

Lee JK, Jung SK, Lee YS et al (2017) Analysis of the fusion and graft resorption rates, as measured by computed tomography, 1 year after posterior cervical fusion using a cervical pedicle screw. World Neurosurg 99:171–178. https://doi.org/10.1016/j.wneu.2016.12.027

Miller LE, Block JE (2011) Safety and effectiveness of bone allografts in anterior cervical discectomy and fusion surgery. Spine (Phila Pa 1976) 36:2045–2050. https://doi.org/10.1097/BRS.0b013e3181ff37eb

Mummaneni PV, Tu T-H, Ziewacz JE et al (2013) The role of minimally invasive techniques in the treatment of adult spinal deformity. Neurosurg Clin N Am 24:231–248. https://doi.org/10.1016/j.nec.2012.12.004

Nakashima H, Yukawa Y, Ito K et al (2011) Extension CT scan: its suitability for assessing fusion after posterior lumbar interbody fusion. Eur Spine J 20:1496–1502. https://doi.org/10.1007/s00586-011-1739-0

Ouchida J, Yukawa Y, Ito K et al (2015) Functional computed tomography scanning for evaluating fusion status after anterior cervical decompression fusion. Eur Spine J 24:2924–2929. https://doi.org/10.1007/s00586-014-3722-z

Park DK, Rhee JM, Kim SS et al (2015) Do CT scans overestimate the fusion rate after anterior cervical discectomy and fusion? J Spinal Disord Tech 28:41–46. https://doi.org/10.1097/BSD.0b013e31829a37ac

Raizman NM, O'Brien JR, Poehling-Monaghan KL, Yu WD (2009) Pseudarthrosis of the spine. J Am Acad Orthop Surg 17:494–503. https://doi.org/10.1097/00003086-199211000-00011

Rihn JA, Kirkpatrick K, Albert TJ (2010) Graft options in posterolateral and posterior interbody lumbar fusion. Spine (Phila Pa 1976) 35:1629–1639. https://doi.org/10.1097/BRS.0b013e3181d25803

Scott-Young M (2014) The incidence of retrograde ejaculation following anterior lumbar interbody fusion using RHBMP2 in 376 male patients. In: Proceedings of the Spine Society of Australia 25th annual scientific meeting combined with the Scoliosis Research Society worldwide course, Apr 11, Brisbane

Simmonds MC, Brown JE, Heirs MK et al (2013) Safety and effectiveness of recombinant human bone morphogenetic protein-2 for spinal fusion: a meta-analysis of individual-participant data. Ann Intern Med 158:877–889

Simmons JW (1985) Posterior lumbar interbody fusion with posterior elements as chip grafts. Clin Orthop Relat Res 85–89

Sugawara T, Itoh Y, Hirano Y et al (2011) B-tricalcium phosphate promotes bony fusion after anterior cervical discectomy and fusion using titanium cages. Spine (Phila Pa 1976) 36:1509–1514. https://doi.org/10.1097/BRS.0b013e31820e60d9

Tavakol S, Khoshzaban A, Azami M et al (2013) The effect of carrier type on bone regeneration of demineralized bone matrix in vivo. J Craniofac Surg 24:2135–2140. https://doi.org/10.1097/SCS.0b013e3182a243d4

Taylor M, Hipp JA, Gertzbein SD et al (2007) Observer agreement in assessing flexion-extension X-rays of the cervical spine, with and without the use of quantitative measurements of intervertebral motion. Spine J 7:654–658. https://doi.org/10.1016/j.spinee.2006.10.017

Thalgott JS, Giuffre JM, Fritts K et al (2001) Instrumented posterolateral lumbar fusion using coralline hydroxy-apatite with or without demineralized bone matrix, as an adjunct to autologous bone. Spine J 1:131–137. https://doi.org/10.1016/S1529-9430(01)00011-0

Thalgott JS, Fogarty ME, Giuffre JM et al (2009) A prospective, randomized, blinded, single-site study to evaluate the clinical and radiographic differences between frozen and freeze-dried allograft when used as part of a circumferential anterior lumbar interbody fusion procedure. Spine (Phila Pa 1976) 34:1251–1256. https://doi.org/10.1097/BRS.0b013e3181a005d7

Tuchman A, Brodke DS, Youssef JA et al (2016) Iliac crest bone graft versus local autograft or allograft for lumbar spinal fusion: a systematic review. Glob Spine J 6:592–606. https://doi.org/10.1055/s-0035-1570749

USSDHHS, FDA, CDRH (2000) Guidance document for the preparation of IDEs for spinal systems. Device Evaluation Office. http://www.fda.gov/cdrh/ode/87.pdf

Vaccaro AR, Patel T, Fischgrund J et al (2005) A 2-year follow-up pilot study evaluating the safety and efficacy of op-1 putty (rhbmp-7) as an adjunct to iliac crest autograft in posterolateral lumbar fusions. Eur Spine J 14:623–629. https://doi.org/10.1007/s00586-004-0845-7

Effects of Reimbursement and Regulation on the Delivery of Spinal Device Innovation and Technology: An Industry Perspective

63

Emma Young

Contents

Introduction 1150

The Regulatory Effect 1153

The Reimbursement Effect 1158

Conclusions 1162

Cross-References 1163

References 1163

Abstract

Since the 1990s regulation has been relatively unchanged in the world's second largest medical technology sector, the European Union. However, recent medical device-related incidents involving breast implants and hip replacements have prompted urgent regulatory and compliance reforms. Against the background of increasing global healthcare costs and an aging population, medical devices including spinal devices are about to undergo one of the industry's most transformational regulatory changes. What does this mean for current and potential innovators of new medical device technologies and their beneficiaries? What will future reimbursement and regulatory frameworks look like, and what will be their impact on medical device technology investment? Navigating the complex requirements of innovative medical device development such as increasing regulatory burden and a multitude of differing payer uncertainties can often be the hurdle to sustained device innovation for many companies.

Keywords

Reimbursement · Regulatory · Innovation · Healthcare expenditure · Medical Device Directive · Medical device regulation · Therapeutic Goods Administration · CE mark · Spinal devices · Medical device

---

NB: this work is not related to the views or opinions of Prism Surgical

---

E. Young (✉)
Prism Surgical Designs Pty Ltd, Brisbane, QLD, Australia
e-mail: emma@prismsurgical.com.au

© Springer Nature Switzerland AG 2021
B. C. Cheng (ed.), *Handbook of Spine Technology*,
https://doi.org/10.1007/978-3-319-44424-6_95

Introduction

Healthcare systems are organized, financed, and regulated in different ways around the world, but most would agree that universal access to innovative technology and quality healthcare at an affordable cost to both the individual and society is an elementary need. Almost all OECD (Textbox 1) countries have universal health coverage for a core set of services (OECD 2015). Statistics on healthcare expenditure and financing are often used to evaluate how a country's healthcare system responds to the challenge of universal access to quality healthcare.

> **Textbox 1 What is an OECD Country?**
> The Organisation for Economic Co-operation and Development (OECD) is an intergovernmental economic organization with 36 member countries, founded in 1961 to stimulate economic progress, stability, and world trade (OECD 2018). It is a forum of countries describing themselves as committed to democracy and the market economy, providing a platform to compare policy experiences, seeking answers to common problems, identify good practices, and coordinate domestic and international policies of its members. Most OECD members are high-income economies and are regarded as developed countries.
>
> Source: Based on information from OECD (2018).

Healthcare expenditure is defined as expenditure on health goods and services, including investment in equipment and facilities. As defined by the Australian Institute of Health and Welfare (2017), this definition closely follows the definition that the OECD System of Health Accounts framework provides. It excludes expenditure that is incurred outside the health sector, personal activities, and where health is not the primary expected benefit.

At the macrolevel, health expenditure in Australia is considered within the context of changes in the economy and population growth. The focus is on total health expenditure. Total health expenditure in 2015–2016 in Australia was reported at $170.4 billion–$6.0 billion higher in real terms than 2014–2015 and $63.2 billion higher than in 2005–2006 (AIHW 2017). Health's share of Australian gross domestic product (GDP) has continued to rise, from 8.68% in 2005–2006 to 10.3% (Table 1) in 2015–2016 (AIHW 2017). The share of the economy (or GDP) represented by health continues to steadily grow as reflected in Table 1. Non-government sources (individuals, private health insurance funds, and other non-government sources) contributed an estimated 32.7% toward total health spending in 2015–2016 (AIHW 2017).

The European Union (EU) healthcare spend, as a percentage of GDP, aligns itself closely with the Australian experience. The European Union (EU) reports total health expenditure in 2014, in terms of GDP, ranging as high as 10.9–11.4% in countries such as Germany, the Netherlands, France, the UK, and Switzerland (Table 2) (Eurostat 2017).

In the USA, the National Center for Health Statistics (2017) recorded the total US national health expenditure in 2015 to be US$3.2 trillion, 17.8% of the national GDP. In 2017, it was reported that health expenditure rose to 18.3% of national GDP.

Health systems across the globe are required to rapidly develop and respond to a multitude of factors such as new medical technologies, new health services, and greater access to them, changing health policies and organizational structures and more complex financing mechanisms. Access to healthcare and greater patient choice are increasingly scrutinized against the background of financial sustainability. New medical technologies are improving diagnoses and treatments, but they are also increasing health spending. Medical technology is often viewed as a primary contributor to increased healthcare expenditure. The prospect that spending pressures will escalate raises questions around the benefits and costs of technology advances and the processes for evaluating them. Life expectancy continues to increase steadily in OECD countries, rising on average by

Table 1 Total Australian health expenditure and GDP, current prices, and annual health to GDP ratios of 2005–2006 to 2015–2016

Year	Total health expenditure ($million)	GDP ($million)	Ratio of health expenditure to GDP (%)
2005–2006	86,685	998,458	8.68
2006–2007	94,938	1,087,440	8.73
2007–2008	103,563	1,178,809	8.79
2008–2009	114,401	1,259,280	9.08
2009–2010	121,710	1,297,508	9.38
2010–2011	131,612	1,410,442	9.33
2011–2012	141,957	1,491,741	9.52
2012–2013	146,953	1,527,529	9.62
2013–2014	154,671	1,589,940	9.73
2014–2015	161,617	1,617,016	9.99
2015–2016	170,386	1,654,928	10.3

Data source: AIHW (2017)

3–4 months each year. In 2013, OECD (2015) life expectancy at birth reached 80.5 years, an increase of over 10 years since 1970. This rise in average life expectancy continues to increase the need for treatment of illness and chronic diseases. Interestingly and despite the highest OECD health spend, the US Center for Disease Control and Prevention (2017) recently reported that the 2016 US life expectancy at birth was 78.6 years for the population as a whole – 0.1 year less than it was in 2015. This recent statistic from the largest healthcare economy in the world has sharpened the focus on the relationship between healthcare spend and its effect on overall health improvement.

The Australian Productivity Commission's work on the Economic Implications of an Ageing Australia (Productivity Commission 2005) found that most of the growth in health expenditure over the last 20 years was due to factors such as a greater demand for health services, in combination with the adoption of new technologies. Modelling estimates prepared as part of this report confirm that technology has played an important role in driving total real healthcare expenditure growth. It is reasonable to expect that technologies such as medical devices will continue to play a key role in influencing healthcare expenditure.

As knowledge increases there is a constant uptake of improved, innovative, and inventive medical technology, such as medical devices. However, the rapid advances in device technology continue to drive medical costs upward. It has become increasingly difficult to balance the dual responsibilities of controlling healthcare expenditure while simultaneously enhancing the welfare of its beneficiaries, particularly when it comes to coverage decisions for costly new medical devices. Patient demographics are shifting toward a greater emphasis on notions of well-being and increased activity levels, including an expected standard of health and technology delivery that will enable an aging population to lead more active lives for longer. To be concise, patients expect a better quality of life for longer. Rapid developments in new technologies and their increasing complexity will place demands on regulatory requirements to adapt, as patients increase their awareness of newly developed technologies through sources such as the media and the Internet. At any one time, there are thousands of new technologies undergoing development, most of which will fail to progress to end commercial availability. As a result it is often difficult to identify effective technological advances, let alone the implications for health expenditure.

The influence of a new medical technology on health expenditure will depend on factors such as but not limited to:

- Whether the medical advance will increase or decrease the cost of a particular procedure or treatment
- How the number of procedures undertaken will change as a result of the new medical device

Table 2 2014 EU healthcare expenditure total, per inhabitant and as a percentage of GDP

EU Member	Total health expenditure (EUR million)	EUR per inhabitant	Ratio of health expenditure to GDP (%)
Belgium	41,711	3722	10.4
Bulgaria	3640	504	8.5
Czech Republic	11,841	1125	7.6
Denmark	27,517	4876	10.4
Germany	321,720	3973	11.0
Estonia	1223	931	6.1
Ireland	19,148	4147	9.9
Greece	14,712	1351	8.3
Spain	94,534	2034	9.1
France	236,948	3582	11.1
Croatia	2886	681	6.7
Italy	145,938	2401	9.0
Cyprus	1184	1389	6.8
Latvia	1297	650	5.5
Lithuania	2265	772	6.2
Luxembourg	3091	5556	6.3
Hungary	7473	757	7.2
Malta	–	–	–
Netherlands	72,475	4297	10.9
Austria	33,795	3957	10.3
Poland	25,987	684	6.3
Portugal	15,583	1498	9.0
Romania	7727	388	5.1
Slovenia	3189	1546	8.5
Slovakia	5256	970	7.0
Finland	19,523	3575	9.5
Sweden	48,154	4966	11.1
United Kingdom	222,609	3448	9.9
Iceland	1138	3476	8.8
Liechtenstein	294	7906	–
Norway	35,132	6389	9.4
Switzerland	60,276	7361	11.4

Data Source: Eurostat (2017)

- Whether the advance will change the place of treatment, for example, an inpatient to an outpatient basis

Moreover, it is inappropriate to consider only the potential expenditure effects of technology advances in isolation of their expected efficacy and clinical benefits. Spinal pathology and related back pain are among the most prevalent contributors to health expenditure in the world. Within the Australian adult population, the results of a "cost-of-illness" study (Walker et al. 2003) of low back pain (LBP) estimated the direct cost of LBP in 2001 to be AU\$1.02 billion, with indirect costs at AU\$8.15 billion producing a total cost of AU\$9.17 billion. LBP in Australian adults represents an exponential health problem with significant economic burden.

Care for individuals with a spinal pathology, related symptoms, and operative care have long been associated with rising healthcare costs and productivity losses, costing the world's largest

Table 3 EU medical device classification

Class 1	Low risk Non-sterile. Self-certified. Registered within each member state they are sold within
Class 1 (measuring/ sterile)	Low-medium risk Provided sterile and/or have a measuring function.
Class IIa	Medium risk Special controls. NB assessment. Used to diagnose and monitor. Limited invasiveness
Class IIb	Medium risk Full QMS and targeted review of design or technical files by NB. Surgically invasive/implantable
Class III	High risk Similar to IIB + full design review by NB Implants support or sustain human life. Present a potential, high risk of illness/injury

healthcare market, the USA, US$86 billion in 2005 (Bisschop and Tulder 2016). Bisschop and Tulder (2016) state that spinal pathology and related back and neck symptoms are among the most common health problems and are ranked number one with respect to years lived with disability. The global burden is so great that it has compelling and urgent ramifications for health policy, planning, and research in all jurisdictions.

Growth of the global spinal device market is estimated to reach US$17.27 billion by 2021 as reported in a 2017 market research study (Markets and Markets 2017). A number of contributing factors include the rising incidence and prevalence of spinal disorders, development of technologically advanced medical devices, and the global rise in an aging population. The challenge for the payers and regulators in any jurisdiction is to craft a standard for policy, reimbursement, and regulatory provision that both protect patient safety and efficacy while preserving some incentive to spur authentic device research and innovation. Reimbursement policies and supporting frameworks for innovative medical devices differ significantly from the regulatory requirements. With this in mind, what are the effects of reimbursement and regulation frameworks on the delivery of spinal device innovation?

The Regulatory Effect

Medical devices cannot be placed on the European market without conforming to the strict safety requirements of European legislation. In the European Union or EU, three directives cover the medical device sector. Collectively known as the Medical Device Directive (MDD), this legal framework regulates the safety and marketing of medical devices in Europe and is largely mirrored throughout the Australian medical device approval process under the authority of the Australian Therapeutic Goods Administration (TGA). Of the three directives, it is the Medical Device Directive (93/42/EEC) that regulates implantable spinal devices such as screws, rods, cages, plates, and total disc replacements.

Medical device regulations around the world begin by assessing the risk of a device and, more specifically, the risk of the intended use of the device. The regulatory requirements or "burden" are proportional to the estimated risk. The lowest risk or "class" of device (Table 3), such as a reusable surgical instrument, is self-certified. These low-risk devices are required to meet general controls and be registered in their respective jurisdiction. The highest class of device, such as hip replacements and heart valves, must undergo stringent premarket assessment by a qualified body before market placement. Within the European Union, all medical devices are placed into one of the four graduated categories, using the classification rules listed in the Directive 93/42/EEC Annex IX.

Under the Therapeutic Goods Act 1989, medical devices must be included in the Australian Register of Therapeutic Goods (ARTG) prior to supply in Australia, unless exempt. The TGA's current regulatory framework is based on the model recommended by the Global Harmonization Task Force (GHTF) (Textbox 2). A conformity assessment is the key mechanism for assuring a medical device is safe and performs as intended. Conformity assessment certification is issued by a conformity assessment body, and the degree of assessment rigor is determined by the risk classification of the device.

> **Textbox 2 What is the GHTF?**
>
> *Founded in 1992, the Global Harmonization Task Force or GHTF was created in an effort to respond to the growing need for international harmonization in the regulation of medical devices (IMDRF 2018b). The GHTF was a voluntary group of representatives from regulatory authorities and members of the medical device industry. The representatives from its five founding members (the EU, the USA, Canada, Japan, and Australia) were divided into three geographical areas, Europe, Asia-Pacific, and North America, each of which highly regulates medical devices using their own unique regulatory framework.* GHTF principles are similar to and largely based on the member countries framework in that device classification is risk based and assessed by third-party bodies. There are a number of jurisdictions such as Japan who currently operate their regulatory frameworks on the GHTF principles.
>
> *The GHTF organization had been a mainstay among the regulatory harmonization movements, and the initiative was arguably the most successful effort to harmonize medical device standards around the globe. The GHTF was discontinued with its mission taken over by the International Medical Device Regulators Forum (IMDRF) in late 2011, a successor organization comprised of officials from regulatory agencies around the world (IMDRF 2018b).*

The TGA issues conformity assessment certification under the Australian regulatory framework, while European-notified bodies issue conformity assessment certification under the European regulatory framework. There is great similarity between the Australian and European processes with both frameworks based on the GHTF principles.

Signed in 1998 and effective since 1999, the mutual recognition agreement (MRA) between Australia and the European Community (EC), known as the EU-AU MRA, officially recognizes the competence of conformity assessment bodies located in the EU to assess compliance of certain types of medical devices with Australia's regulatory requirements. Conversely, the MRA recognizes the competence of Australia's Therapeutic Goods Administration (TGA) to assess medical devices for compliance with EU requirements. According to the TGA (2013), this practice recognizes that conformity assessment is an intensive and potentially expensive process and that unnecessary duplication would increase the costs of many medical devices for consumers and create disincentives to supply products in Australia's small medical devices market.

However, changes to the provisions of the EU-AU MRA came into force on the 1 January 2013 to exclude particular medical devices from the scope of the agreement. Medical devices excluded from the MRA include high-risk devices such as active implantable devices (AIMDs) and class III (high risk) medical devices. Exclusion of these medical devices will continue until confidence-building activities have been undertaken by Australia and the European Union. Certain additional medical devices incorporating materials of biological origin are also principally excluded, with no confidence-building phase planned.

Medical devices covered under the terms of the amended MRA that have undergone a conformity assessment procedure by an EU-recognized notified body and are in compliance with Australian medical device regulations are included by the TGA on the Australian Register of Therapeutic Goods (ARTG). In practice, 97% of applications for inclusions in the ARTG are certified medical devices by EU notified bodies (TGA 2013).

> **Textbox 3 What is a CE Mark?**
>
> CE marking is the medical device manufacturer's claim that a product meets the essential requirements of all relevant European Medical Device Directives. The Directives outline the safety and performance

(continued)

> requirements for medical devices in the European Union (EU). The CE mark is a legal requirement to place a device on the market in the EU.

Although a device may be granted an EU approval (CE mark) (Textbox 3), for companies whose devices do not qualify under the amended MRA between Australia and the EC, they must apply for a conformity assessment procedure to be conducted by the TGA. Although the TGA's regulations and approval processes are similar to those applicable in the EU, the full TGA registration process can take anywhere from 2 to 18 months to complete, depending on the risk classification of the device, prior conformity assessment sources, and whether the TGA determines that an audit is required.

A key feature of the European medical device legislation is that it defines what is commonly known as the Essential Requirements or the "ERs." Medical devices can only be placed in the European market if they satisfy the Essential Requirements criteria, specified in Annex I of the Medical Device Directive 93/42/EEC. All medical devices must comply with these requirements. Manufacturers are required to verify each device type or model against each of the requirements, determine whether the requirement is applicable, and demonstrate documented evidence of compliance. In this respect, it is reasonable to assert that Annex I is the foundation of the Medical Device Directive (MDD).

The MDD outlines the core elements that companies need to have in place. It sets out and defines the conformity assessment processes required to assess whether a device is in conformity with the directives, and it lays down precise obligations on the part of the device's legal manufacturer (IOM 2010). Competent authorities (CA), notified bodies (NB), and authorized representatives (AR) are all involved in the CE marking process. Competent authorities exist in each member state and designate, control, and monitor the notified bodies, govern clinical trials, and monitor post market vigilance activities. Notified bodies ensure and certify the safety and compliance of devices and their manufacturers in accordance with the relevant criteria.

The legislation itself is underpinned by "standards." European standards or harmonized standards adopted by a recognized European Standards Organization such as CEN or CENLEC allow manufacturers, other economic operators, or conformity assessment bodies to use these designated harmonized standards to demonstrate that products, services, or processes comply with relevant EU legislation. Other commonly utilized standards are written by the International Organization for Standardization (ISO), while others are in the form of "EU guidelines" called MEDDEVs and GHTF guidance documents.

The European and Australian regulatory frameworks are similar to the US system in that they are all constructed on a risk-based classification system (Table 3). These risk-based frameworks are more similar than different. However, despite the similarities there remains enough difference between them that complying with one does not guarantee compliance with the other. It has been long reported that devices cleared by the FDA have been declined by EU notified bodies and vice versa.

Within Australia, the USA, and the European Union, the highest risk devices have development paths that are heavily regulated and expensive to both commercialize and maintain on the market. In the USA, class III or novel devices require a Premarket Approval (PMA) (see ▶ Chap. 18, "FDA Premarket Review of Orthopedic Spinal Devices"). Sorenson and Drummond (2014) report over the past 10 years approximately 2% of medical devices have undergone the PMA process, and unlike PMA, direct evidence of safety and effectiveness is usually not required for a 510 (k) application with only 10–15% of those applications containing any clinical data. Medium-risk or class II devices in the USA are usually required to undergo the 510(k) review process, which determines principally whether the new device is "substantially equivalent" to previously marketed or "predicate" devices. Class III (high risk) devices or novel technologies deemed to not possess substantial equivalence undergo the more stringent PMA process. These technologies must demonstrate safety and efficacy through clinical studies.

Similarly in Europe the evidence requirement increases with the risk of the device. The majority of spinal devices is approved through demonstrating safety and performance in relation to the intended purpose. In most cases the evidence submitted for premarket approval of spinal devices is from non-clinical origins, extensive literature reviews to similar products or small clinical trials. This pathway presents opportunity to market approval with the least regulatory burden.

The large disparity in regulatory burden between class II and class III medical devices allows the development of smaller companies and more rapid innovation cycles for the low-medium-class devices. While small companies have completed US PMAs, the majority of high-risk devices are developed by the larger device organizations (IOM 2010).

The two factors that have the biggest impact on regulated device development are review cycle time and the level of evidence required for approval. The influence and impact of a short device review cycle can be illustrated when comparing US PMA products to European class III devices. The Institute of Medicine (2010) explains for the very reason that the US review cycle time for these products is twice as long as the current European cycle; not only are inventive products introduced later to the US markets, but the US markets often forego revised models of product as the innovation and approval cycles outside the USA are generally much swifter to market.

However, sweeping reforms of the rules that govern the regulation of the medical device sector in Europe represents one of the most disruptive changes to affect the industry in recent times. When the European medical device regulation commonly referred to as the MDR replaces the current set of Medical Device Directives (MDD), companies will have 3 years to comply with a broad scope of new rules for almost every kind of product in the medical device spectrum. Under the new regulations, medical device companies will have to provide substantially more clinical evidence to gain market access or even maintain existing products on the market. Companies will need to conduct audits to determine the new rules' impact on maintaining and upgrading device portfolios. Companies can expect a significantly more costly path to compliance in the European Union. The costs associated with compliance may force a number of companies to take strong steps, such as discontinuing existing device lines or considering acquisition proposals (De Busscher et al. 2016).

The result of this transformation anticipates a stronger, more accountable device industry that may look considerably different from todays (De Busscher et al. 2016). The EU MDD has been in effect since the 1990s. Incremental changes to the text have occurred along the way due to new and emerging technologies which have both challenged the framework and identified gaps. However, it was a series of well-known device events (Textbox 4) that emphasized to both policymakers and the industry an urgent need for regulatory reform to ensure patient safety concerns were adequately attended. On the 26 September 2012, the EU announced a package of reforms to provide a more stringent regulatory framework for medical devices to ensure a higher level of protection of human health and safety (TGA 2013).

Textbox 4 Why Regulatory Reform Was Required

2011 The US FDA warned of serious complications associated with the use of urogynecologic surgical repair mesh after nearly 2874 medical device reports (MDRs) in a 3-year period including injury, death, and malfunction[1]. One thousand five hundred three were associated with pelvic organ prolapse (POP) and 1371 for stress urinary incontinence (SUI).[1]

2012 Poly Implant Prothèse (PIP), a French company, was revealed to have knowingly sold breast implants made with industrial grade silicone rather than medical grade. The reported probability of PIP rupture at 10 years is 25–30% versus 2–15%

(continued)

reported in standard silicone implants[2]. It is reported that approximately 300,000 women were affected.[2]

2010 DePuy (Johnson & Johnson) voluntarily recalled the ASR (metal on metal) hip replacement system after an Australian National Joint Replacement Register (NJRR) reported a failure rate of 13% at 5 years for the device – unacceptably higher than the average. Ninety-three thousand people worldwide were implanted with an ASR device[3]. The ASR hip replacement was removed from the Australian market in 2009 after intervention by the Australian TGA.[3]

[1]Source: SCENIHR (2013)
[2]Source: CDRH (2011)
[3]Source: TGA (2011)

The introduction of the new EU MDR which entered into force on 25 May 2017 marks the start of a transition period for manufacturers selling medical devices into Europe and other jurisdictions that recognize and approve a CE-marked device under the EU regulations. The MDR, which replaces the Medical Device Directive (93/42/EEC), has a transition period of 3 years. The extensive and long-awaited regulation will come into force in May 2020 and more closely reflects the current FDA scrutiny of product safety by placing stricter requirements on subjects such as clinical evaluation, post market clinical follow-up, and increased traceability of devices through the supply chain. The implications of the new EU MDR on the global medical devices sector are enormous. The European medical technology market is significant and important for the industry, estimated to contribute 31% of the global total (Eucomed 2013).

Traditionally, small and medium companies who did not have the revenue to conduct expensive PMAs in the US viewed the European market as an opportunity to innovate and collect valuable clinical performance data. As more detail has emerged about the composition of the EU MDR, it has become apparent that the full impact of the changes will extend beyond the regulatory framework. Among a plethora of new requirements, a CE mark will require pre- and post market clinical studies, data transparency, tightening of vigilance reporting, and the introduction of a unique device identification (UDI) system similar to that recently introduced into the USA. Additionally, a number of medical devices will be reclassified such as spinal total disc replacement which will increase in its defined risk classification from class IIb (medium risk) to class III (high risk).

According to a recent report by Makower et al. (2010), current FDA and EU regulatory frameworks are creating almost impossible barriers to authentic device innovation. Due to their usually limited financial resources, the future regulatory environment will be particularly challenging for start-up companies who have historically played a key role in driving innovation. Compliance with regulatory legislation is generally viewed as a driver of complexity and cost for medical device businesses, whose regulatory people are tasked with ensuring that the companies who employ them are compliant while curtailing the risks and costs associated with it. However, it is becoming increasingly apparent that tackling frameworks such as the future EU MDR goes beyond the remit of even the most resourced regulatory teams (De Busscher et al. 2016).

Providing data on both new and existing medical devices may require revision and the need to conduct new clinical studies. De Busscher et al. (2016) estimate that some devices will need to have their safety and efficacy validated clinically or be at risk of being removed from the global market. The proven technology concept commonly referred to as "grandfathering" of legacy devices is one of the key changes of the new MDR determined by the commission to now require supporting clinical evidence in compliance with the current standards and regulations.

De Busscher et al. (2016) propose that the additional clinical evidence requirements likely to be stipulated by the EU MDR will mean that products in development may take longer to obtain commercialization, which is likely to impact further on revenues and the raising and allocation of capital. Additionally, the notified bodies that regulate and certify manufacturers of medical devices and technology will undergo significant changes. Under the EU MDR, the notified bodies' requirements and responsibilities will intensify. Many in the industry fear this will lead to further delays in product certification and audit assessments.

These are transformational shifts for medical device companies. Accessing the EU market or gaining CE mark by complying with current and future regulations, restructuring operations, and planning future business pathways will become an increasingly burdensome activity. For at least a decade, Europe has remained a favorable market entrance for many innovative medical devices joining the market. After obtaining a CE mark (EU approval), a manufacturer could further validate their device clinically and thereby build a complete dossier of evidence that would be both required and beneficial to secure future FDA approval. Leading up to and beyond the EU MDR becoming an industry reality in 2020, this familiar path will need to be reviewed with Europe possibly becoming a less attractive first market destination (De Busscher et al. 2016). It is difficult to estimate how the cost of compliance with this future regulatory framework will financially burden a medical device company. In 2013 Eucomed, an industry body, conducted an industry-wide survey on the financial impact of the updated EU MDR (Table 4).

A reflection of increasing burden and its impact is the reported declining numbers of regulatory submissions for new medical devices in the USA over the past several years. The annual Emergo (2017) study examined approximately 15,000 510(k) applications cleared by the FDA

Table 4 Forecast MDR compliance costs (Eucomed 2013)

€17.5b
(US$18.9 billion) cost to industry if a centralized premarket authorization system is implemented
€7.5b
(US$8.1 billion) cost to industry of compliance with a UDI (unique device identification) system, improvements in labelling, and clinical performance data
€17.5 m
(US$18.9 million) cost to small-medium-sized enterprises (SMEs) to bring a new class III product to market under a clinical premarket approval system

between 2012 and 2016, finding not only that numbers of US applicants have steadily declined but also that the overall number of 510(k) applications cleared by the agency hits a 4-year low in 2016. In an era of greater scientific knowledge and more technology advancements than any other time in history, industry experts question what forces are driving genuine medical technology innovation and invention in a negative direction.

Device manufacturers continue to argue that the device industry is fundamentally different from other industries such as the pharmaceutical industry taking into consideration that the engineering and quality frameworks required to support development, market approval, and continuous device innovation are increasingly onerous (Reed et al. 2008). In general, devices tend to have faster commercial cycle times and tend to be characterized by incremental improvements to existing technologies. This difference is increasingly identified by major bodies such as the FDA, evidenced by The FDA Modernization Act defining the "least burdensome approach" (CDRH 2002). The least burdensome concept is defined as a successful means of addressing a premarket issue that involves the smallest investment of time, effort, and resources (e.g., money) on the part of the submitter and FDA, to help ensure scientific integrity while affording a high degree of public health protection and expediting the availability of new device technologies (CDRH 2002).

However, an issue continues to arise when that incremental evolution of technology occurs and the suitability between that technology and existing regulatory pathways does not exist. The prospect of piloting through the increasing requirements of product development and regulatory approvals to then confront a multitude of differing payer uncertainties is often a hurdle to sustained device innovation for many companies.

The Reimbursement Effect

Achieving approval by a regulatory body in any jurisdiction does not imply that a product will be reimbursed by payers or private insurers. Whereas regulatory approval focuses on safety and efficacy, the payers tend to strongly consider the cost savings and cost-effectiveness in determining whether to cover a new technology. Suitable costings for reimbursement must be obtained while ensuring reimbursement rates are appropriate and balanced accordingly with the cost and benefits of new technologies and expected yet increasing standards of care.

For example, in Australia a new technology pending the classification may gain regulatory approval with the Therapeutic Goods Administration (TGA) without delay. However, in order to commercialize the innovation in the Australian private health system, the device must gain payer reimbursement through approval by the Minister for Health via the government-administered Prostheses List (PL). The Prostheses List Advisory Committee or PLAC's primary role is to make recommendations and provide advice to the Minister for Health. The PLAC advises on the listing of medical devices and their reimbursement benefits. Using evidence provided by the manufacturer, the PLAC makes recommendations on the most clinically efficacious and cost-effective devices. This ensures that privately insured Australians have access to a range of medical devices that are both clinically and cost-effective. Similar to many jurisdictions around the world, approval by the Australian regulatory body (TGA) does not guarantee reimbursement by the government's Prostheses List arrangements. Reimbursement and/or payer coverage can be the factor which decides whether a new medical device will succeed or fail, and the common goal of reimbursement or payer policies around the world is maintaining that balance between cost- and clinical effectiveness of a new technology.

Payment issues are of increasing concern to the healthcare industry. Most who are connected with healthcare acknowledge the rising costs of healthcare and the need to find ways of managing costs more effectively. In fact, some continue to point to technology as a major cause of these increasing costs. The emphasis on cost and the role of technology in costs have placed a large part of the onus on the device industry. It is a prominent inclusion of the company consciousness that

innovation is not just creating a better device but includes creating a more cost-effective or financially sustainable alternative to the current offerings (IOM 2001). Of course, it is not that simplistic, and it is challenging for a medical device company to fulfil the criteria for novelty and superiority with cost-effectiveness while also fulfilling the required regulatory principles of clinical safety and efficacy.

Cost is guaranteed to be in the innovation and invention equation, but where do the benefits to patient outcomes place in that equation? In the current global device environment, providers are increasingly pressured to use cost-saving technology as opposed to cost-effective technology. Manufacturers are forced to consolidate, to fashion technologies into commodities, and often to compete on the basis of price alone (IOM 2001). The Institute of Medicine (2001) assert too often pricing contracts and rationing, which are often employed in a number of cost frameworks, fail to balance with the patient benefit and effectiveness side of the equation. There is a multitude of different parties that have an effect on payments for technologies such as multiple insurers and/or government structures who all vary in their approval frameworks for that payment.

Definitive reimbursement is an essential part of the overall concept, design, development, and marketing of the medical device life cycle. For many decades, new innovations have been developed and launched into a market in which prospective payment systems and unanticipated changes have forced key stakeholders to operate in an increasingly efficient and lean manner. The key message is that new innovations must fit within a diminishing yet dynamic operating margin imposed by cost-constrained third parties.

Within the Australian framework, manufacturers of innovative device technology must have an associated procedural code included on the Australian Medicare Benefits Schedule (MBS) for commercial release within the private health system. This is the list of procedures, tests, and consultations for which the Australian Medicare System will subsidize. The MBS is fundamental to the Australian health system and is a prerequisite for private payer device reimbursement.

Devices cannot be included on the Australian Prostheses List (PL) if a related MBS item number does not exist. This requirement can be challenging for the suppliers of authentic innovative or inventive technologies if solid clinical efficacy has not been established. In any market an understanding of these key relationships is essential. Innovative technology that offers substantial clinical and economic improvements may struggle to obtain reimbursement if payer frameworks are not well understood.

Early stage reimbursement strategies are vital for the acceptance and success of an innovative or inventive device. It is absolutely critical for a device company, large or small, to obtain coverage and attract investors. In venture capital there are two important aspects for consideration: feasibility and market acceptance (IOM 2001). The Institute of Medicine (2001) maintains the regulatory and healthcare payment environment which introduces additional levels of risk and uncertainty. In those cases where uncertainty of reimbursement for a particular device under development exists, securing funding for that innovative or inventive technology becomes problematic. While US firms have the dominant position in critical markets, the global industry argues that a large number of innovative devices and clinical breakthroughs are often the product of the smaller businesses. According to the Institute of Medicine (2001), 72% of medical device companies employ fewer than 50 people. Eucomed (2013) reported small-medium-sized companies accounted for 95% of the 25,000 companies in the European medical device industry, while in the USA Stirling and Shehata (2016) report that 80% of the 6500 companies in the medical device sector are small (less than 50 employees).

Smaller companies are nimble and responsive, with a tremendous tolerance for uncertainty and are therefore well suited to be the source of innovation for medical devices. It is the smaller companies that drive a substantial portion of true industry innovation, yet the Institute of Medicine (2001) states that historically out of approximately 6 ideas, only one device will make it to commercial success. An organic advantage by any company is attained by leveraging their knowledge or IP. The Institute of Medicine (2001) asserts that in a global business environment, the small incubators of technology are particularly challenged with respect to increasing financial pressures.

Many small companies lack the infrastructure required to meet worldwide regulatory and marketing activities. The larger companies have an ability to assist with the delivery of innovative technology created by the smaller entities through exchanges such as acquisition, joint ventures, strategic partnerships, contract research, licensing, and royalty agreements (IOM 2001). Large medical device companies have the ability to bring scale to the challenge of globalization and successful product development. The key factors that determine product investment, demand, and eventual commercial success are whether the device is reimbursable or not and the quantum of that reimbursement.

The success of a medical device requires both strategic and coordinated planning. It must be followed by a timely execution of stakeholder-specific promotion, information, and development plans. The most innovative and/or inventive medical devices may never establish commercial success if there is a failure to constrain costs or convince stakeholders of that new technology's increase in value. This was illustrated by the familiar case of the Charité Artificial Disc Replacement (Textbox 5).

The Charité Artificial Disc was originally developed at the Charité University Hospital in Berlin, Germany, in the mid-1980s by leading spine specialist Professor Karin Büttner-Janz and Professor Kurt Schellnack. In 2003 DePuy Spine acquired the Link Spine Group, Inc. for $US325 million gaining exclusive worldwide rights to its principal product, the SB Charité Artificial Disc (Arida et al. 2006). The Charité was FDA approved in October 2004 as an alternative to spinal fusion surgery. While pre-launch, physician and patient dynamics were strongly in DePuy's favor, it did not suppress the negative response from hospitals and payers alike. With strong payer resistance and little hospital and provider enthusiasm, Charité sales waned. In 2006, DePuy announced plans to release 5- and 10-year data in the future for the Charité (Johnson and Johnson 2006) in the hope of reviving the prospects of what was a revolutionary advance in the spinal device sector.

Textbox 5 Innovation and Reimbursement: A Case Study

The Charité Artificial Spinal Disc was once viewed as revolutionary surgical technology. In October 2004 Charité was the first total disc replacement (TDR/ADR) on the market to receive US regulatory approval and be commercialised for the anterior replacement of diseased lumbar discs.

At the time of a 2006 report by Arida et al. (2006) "The Charité: Lessons in the Launch of a New Medical Device," Charité's commercial success was in doubt. Charité had failed to convince the third-party payers that its use should be covered and reimbursed. On the 14 August 2007, the Centers for Medicare and Medicaid Services (CMS) determined that lumbar artificial disc replacement was not reasonable and necessary for the Medicare population over 60 years of age. CMS further determined that for those Medicare beneficiaries less than 60 years of age, there was no national coverage determination, leaving coverage decisions to be made on a local basis (CMS 2007). Payers cited a lack of evidence that the Charité was as effective as promoted. Follow-up clinical evaluations of the device and the complexity of the surgical procedure cast further doubts about its safety and effectiveness (Sparks et al. 2011).

According to Sparks et al. (2011), the Charité device did not become a physician preference item, and its adoption was actively opposed by purchasing departments and other administrative decision-makers. Summarizing the reports of Arida et al. (2006) and Sparks et al. (2011), DePuy failed to support Charité's launch and subsequent commercialization in the following critical areas:

1. *A disconnect occurred between Text the devices positioning as an alternative to a modern spinal fusion method where the clinical trial supporting this claim was a non-inferiority claim comparing Charité to the BAK cage, a somewhat controversial procedure that had been largely discontinued due to poor outcomes.*
2. *The Charité was priced around US$11,500, approximately 2.5 times that of the BAK cage and BMP (US$4500 each), without securing new procedural codes at or before launch. At that time the Charité was being sold in Australia and Europe for US$4500–$5000. DePuy did not possess a proactive strategy or economic data to prove this but relied on patients, physicians, and advocacy associations to push for reimbursement.*
3. *DePuy largely ignored the role of worker compensation insurance carriers in influencing other payers by not studying the long-term effects of Charité on return to work and productivity gains. A clinical paper was also released in 2004 as a result of the 2003 landmark study "Total Disc Replacement for Chronic Low Back Pain....," which concluded that there was no definitive clinical evidence that disc replacement surgery was efficacious or resulted in fewer adjacent segment problems. The net effect created scepticism in the surgical community over the true clinical value of arthroplasty.*

Sparks et al. (2011) summarize that DePuy did not fully appreciate that the conventional clinical endpoints used to secure regulatory approvals are not necessarily the outcomes that the payer uses making coverage decisions nor the outcomes hospitals use for purchasing purposes. DePuy did not build the body of evidence necessary to establish comparative safety and effectiveness of Charité with payers and

(continued)

> physician groups alike (Sparks et al. 2011). Without coverage and clinician support, hospitals were not able to justify the purchase of the Charité device for both cost and coverage reasons, and this very inventive and innovative device failed commercially.

Conclusions

Society places a premium on the delivery of innovative technologies to ensure optimal healthcare. Patients are educated, informed, and interested in their health treatment choices and the outcomes associated with that choice. However, to deliver high-quality innovative technologies, it takes an increasingly substantial investment of time and cost to meet stringent regulatory and payer requirements.

End reimbursement and regulatory obligations are the factors which decide whether a new medical device will succeed or fail. There is no doubt, despite greater knowledge, innovative devices are under increasing regulatory scrutiny, escalating development costs and declining reimbursement. Authentic innovation and invention in the spinal devices sector have slowed down substantially. The spinal device industry is saturated with creative differences or "me too" products that fall within the least burdensome "predicate," "equivalent," or "grandfathering" application processes.

The upcoming European texts defining what constitutes the regulatory requirements of a new or existing medical device will have a profound impact on the innovation pipeline. The challenge for a medical device company to fulfil the criteria for novelty, superiority, and cost-effectiveness while also fulfilling the increased regulatory principles of safety and efficacy will become increasingly difficult. The ultimate impact of regulation on innovation will be viewed empirically. The balance between innovation-inducing factors and the compliance costs generated may differ on a case-by-case basis. Additionally the amount of time required to satisfy regulatory requirements will be essential to enable future innovations.

The investment required to obtain approval will no doubt change, with a profound emphasis on implantable medical devices and small entrepreneurial businesses. The work required to obtain a CE mark under the new EU MDR will place a heavier emphasis on solid clinical dossiers, premarket, and post market, both for new products and iterations of existing products. The assessment process inevitably will take longer than it does today.

This constitutes a concern for device companies who invest huge resources to release a device to market and clinicians who strive to deliver superior technology to their patients. Even after successful clinical data is obtained, the ultimate financial result of investment may be questionable, since payments for products and services through reimbursement mechanisms are not guaranteed.

Accompanying the ratification of the EU MDR is an expectation that a number of intended positive outcomes will transpire. With prodigious access to information, the clinician and the patient are well informed of negative medical device reports such as the incidences presented in Textbox 4. The increased regulatory requirements of the MDR may restore patient and clinician confidence in the medical device industry, ensuring trust in both the quality and the safety of the product. Compliance with these new regulations is predicted to be onerous, costly, and distracting; however all involved with the delivery of healthcare should be reminded of the ultimate goal – better patient safety and quality of product. Investment in transparent clinical processes, better traceability, and the ability to better contain adverse events involving medical devices can only be a positive step toward industry endurance.

The challenge for all stakeholders in the global device market is to craft a standard for policy, reimbursement, and regulatory provision that protect both patient safety and efficacy while carefully preserving incentive to induce the delivery of innovative and inventive technologies.

Cross-References

▶ FDA Premarket Review of Orthopedic Spinal Devices

References

Arida D, Kabra A, Lowe C, Szafranski AM and Milestone D (2006) The Charité; Lessons in the launch of a new medical device. Bio-medical Marketing

Australian Institute of Health and Welfare (AIHW) (2017) Health expenditure Australia 2015–2016. Health and welfare expenditure series no. 58. Cat. No. JWE68. Based on Australian Institute of Health and Welfare material, Canberra

Bisschop A, van Tulder MW (2016) Market approval processes for new types of spinal devices: challenges and recommendations for improvement. Eur Spine J 25:2993–3003

Center for Devices and Radiological Health (CDRH) (2002) The least burdensome provisions of the FDA modernization act of 1997: concept and principles. Final Guidance for FDA and Industry

Center for Devices and Radiological Health (CDRH) Urogynecologic surgical mesh: update on the safety and effectiveness of transvaginal placement for pelvic organ prolapse, July 2011

Centers for Medicare and Medicaid Services (CMS) (2007) Decision memo for lumbar artificial disc replacement (LADR) (CAG-00292R). https://www.cms.gov/medicare-coverage-database/details/nca-decisionmemo.aspx?NCAId=197&NcaName=Lumbar+Artificial+Disc+Replacement+(LADR)&DocID=CAG-00292R. Accessed 01 Aug 2018

De Busscher L, Flockhart A, Baylor-Henry M, Brennan J, Giovannetti G and Kumli F (2016) How the new EU Medical Device Regulation will disrupt and transform the industry. EYGM Limited

Emergo (2017) How long does it take the US FDA to clear medical devices via the 510(k) process? https://www.emergogroup.com/sites/default/files/emergo-fda-510(k)-data-analysis-2017.pdf. Accessed 15 Jan 2018

Eucomed Medical Technology (2013) Factsheet; Financial impact of the revision of the EU Medical Devices Directives on European SME's and industry. www.medtecheurope.org/sites/default/files/resource_items/files/20130910_MTE_Financial%20impact%20of%20the%20Revision%20of%20the%20EU%20Medical%20Devices%20Directives%20on%20European%20SMEs%20and%20industry_factsheet.pdf. Accessed 28 Jan 2018

Eurostat (2017) Healthcare expenditure statistics; Statistics Explained. http://ec.europa.eu/eurostat/statisticsexplained/index.php/Healthcare_expenditure_statistics#Healthcare_expenditure. Accessed 01 Aug 2017

From Recall of DePuy Orthopaedics ASR Hip Replacement Device (2011) Therapeutic Goods Administration, used with permission of the Australian Government. https://www.tga.gov.au/behind-news/recall-depuy-orthopaedics-asr-hip-replacement-device. Accessed 10 Nov 2017

From Regulation Impact Statement: Changes to premarket assessment requirements for medical devices (2013) Therapeutic Goods Administration, used with permission of the Australian Government https://www.tga.gov.au/sites/default/files/devices-reforms-premarket-scrutiny-ris-130626.pdf. Accessed 01 Jan 2018

IMDRF (International Medical Device Regulators Forum) (2018a). Information based on material contained in the IMDRF website – About Us. http://www.imdrf.org/about/about.asp. Accessed 01 Aug 2018

IMDRF (International Medical Device Regulators Forum) (2018b). Information based on material contained in the IMDRF website – GHTF Archives. http://www.imdrf.org/ghtf/ghtf-archives.asp. Accessed 01 Aug 2018

IOM (Institute of Medicine) (2001) (US) Roundtable on research and development of drugs, biologics, and medical devices (ed: Hanna KE, Manning FJ, Bouxsein P, Pope A). The National Academies Press, Washington, DC

IOM (Institute of Medicine) 2010 (US) Public health effectiveness of the FDA 510(k) clearance process: balancing patient safety and innovation: workshop report, The National Academies Press, Washington, DC

Johnson & Johnson (2006) J&J presentation at 26th annual Cowen healthcare conference, March 7, 2006

Makower J, Meer A and Denend L (2010) FDA impact on US Medical Technology Innovation; A survey of over 200 medical technology companies. Stanford University, Stanford

Markets and Markets (2017) Spinal implants and surgical devices market. https://www.marketsandmarkets.com/PressReleases/spine-surgery-devices.asp. Accessed 10 Jan 2018

National Center for Health Statistics (2017) Health, United States, 2016 with Chartbook on long-term trends in health. National Center for Health Statistics, Hyattsville

National Center for Health Statistics (2018) Health, United States, 2017; With special feature on mortality. Hyattsville, MD

OECD (2015) Health at a glance 2015: OECD indicators. OECD Publishing, Paris. https://doi.org/10.1787/health_glance-2015-en

OECD (The Organisation for Economic Co-operation Development) (2018). Information based on material contained within the OECD website. http://www.oecd.org/about/membersandpartners. Accessed 01 Aug 2018

Productivity Commission 2005, Economic implications of an ageing Australia, Research report, Canberra. Copyright Commonwealth of Australia, Productivity Commission reproduced by permission

Reed SD, Shea AM, Schulman KA (2008) Economic implications of potential changes to regulatory and reimbursement policies for medical devices. J Gen Intern Med 23(Suppl 1):50–56. https://doi.org/10.1007/s11606-007-0246-9

Scientific Committee on Emerging and Newly Identified Health Risks (SCENIHR) Scientific opinion on the safety of Poly Implant Prothèse (PIP) silicone breast implants (2013 update) 25 September 2013

Sorenson C, Drummond M (2014) Improving medical device regulation: the United States and Europe in perspective. Milbank Q 92:114–150. https://doi.org/10.1111/1468-0009.12043

Sparks S, Freeman R and Davidson N (2011) Reimbursement and coverage strategies matter, here's why. https://www.mddionline.com/reimbursement-and-coverage-strategies-matter-here's-why, December 8, 2011. Accessed 15 Jan 2018

Stirling C, Shehata A (2016) Collaboration – the future of innovation for the medical device industry. KPMG International Cooperative. https://assets.kpmg/content/dam/kpmg/pdf/2016/05/the-future-of-innovation-for-the-medical.pdf. Accessed 01 Jan 2018

US Center for Disease Control and Prevention (2017) National Center For Health Statistics. Health. United States. With special feature on mortality. Hyattsville, MD. 2018

Walker BF, Muller R, Grant WD (2003) Low back pain in Australian adults; the economic burden. Asia Pac J Public Health 15(2):79–87

Anterior Lumbar Spinal Reconstruction

64

Matthew N. Scott-Young, David M. Grosser, and Mario G. T. Zotti

Contents

Introduction	1166
Indications	1167
Benefits	1167
Contraindications	1169
Biomechanical Rationale	1170
Preoperative Assessment and Preparation	1171
Diagnosis	1171
Preoperative Evaluation	1171
Surgical Planning	1171
Patient Optimization	1173
Procedure	1173
Pre-incision Setup: Equipment and Personnel	1173
Preoperative Setup: Preparation of the Patient in the Perioperative Setting	1175
Intraoperative: Approach for Anterior Reconstruction	1176
Approaches with a Focus on Vascular Access	1176
Retroperitoneal Approach	1176
Transperitoneal Approach	1185

M. N. Scott-Young (✉)
Faculty of Health Sciences and Medicine, Bond
University, Gold Coast, QLD, Australia

Gold Coast Spine, Southport, QLD, Australia
e-mail: info@goldcoastspine.com.au; mscott-young@goldcoastspine.com.au

D. M. Grosser
Southern Queensland Cardiovascular Centre, Southport,
QLD, Australia
e-mail: dgrosser@arteries-veins.com

M. G. T. Zotti
Orthopaedic Clinics Gold Coast, Robina, QLD, Australia

Gold Coast Spine, Southport, QLD, Australia
e-mail: mariozotti@gmail.com

© Springer Nature Switzerland AG 2021
B. C. Cheng (ed.), *Handbook of Spine Technology*,
https://doi.org/10.1007/978-3-319-44424-6_124

Revision and Previous Abdominal Surgery	1187
Ability to Transfer Between Retro- and Transperitoneal Approaches	1188
Vascular Pathology, Injuries, and How to Manage Them	1189
Venous Pathology and Handling Techniques	1189
Arterial Pathology	1193
The Rationale for Synchronous Treatment of Concurrent Spinal and Vascular Lesions with the Anterior Approach	1197
Vascular and Renal Anomalies and Variations	1201
Postoperative Care	1205
Conclusion	1205
References	1206

Abstract

The anterior lumbar approach for spinal pathology is a powerful method to achieve reconstruction. It enables wide discectomy, restores disc and neuroforaminal height, optimizes sagittal alignment, and affords large cross-sectional area of endplates for spinal implants while avoiding neural retraction and posterior muscular damage. It is arguably underutilized given the need for a considerable volume of training and need for comfort with abdominal anatomy and vascular handling that is required in its safe application. The preparation of a patient and surgical setup as well as common techniques for anterior lumbar surgery are described in this chapter. Knowledge of assessment techniques and handling of vascular structures in their normal, anomalous, and pathological states is critical for safe and efficient performance of the procedure. Equally, recognition of vascular problems and measures to deal with them enables spinal surgeons to safely expand their indications for this technique which is aided through a multidisciplinary approach and collaboration with a vascular surgery service.

Keywords

Anterior lumbar interbody fusion · Total disc replacement · Hybrid · Retroperitoneal · Transperitoneal · Vascular anatomy · Atherosclerosis · Vascular injury · Vascular repair · Ureteric injury · Combined vascular and spinal reconstruction

Introduction

The anterior approach offers a number of advantages over the other interbody access techniques currently being utilized. The exposure from L1 to S1 vertebral bodies is possible from the direct anterior transperitoneal approach, the midline rectus splitting approach, the para-rectal splitting retroperitoneal approach, and the oblique and lateral approaches. This provides the surgeon access to the disc and vertebral bodies for multiple indications. The decision about which approach is applicable is dependent on many variables that will be discussed in the review. As such, the anterior approaches that are available provide considerable versatility to treat a variety of pathologies with proven therapeutic success.

Spine surgeons around the world are utilizing the anterior approach to treat pathologies in the anterior column. The incidence of anterior procedures has increased substantially since the publication of multiple Food and Drug Administration class I randomized clinical trials on the results of total disc replacement (TDR) and anterior lumbar interbody fusion (ALIF) that recognized that patient-related outcome measures were improved dramatically and show no signs of decay (Gao et al. 2011). This advance in fusion techniques and the maturation of the therapeutic utility of total disc replacement has led to the increased need for the anterior exposure of the lumbar spine. Most spine surgeons perform their own approaches to the cervical spine but few perform their own approach to the lumbar spine, instead utilizing "access surgeons." There are two main reasons: The first is the fear of the complications, particularly vascular injury, and the second is the variable and low quality of training provided in the orthopedic and neurosurgical training programs with respect to anterior access, techniques, and utility.

Anterior access requires the surgeon to be familiar with intra-abdominal, retroperitoneal, and vascular anatomy and its variances. Thorough preoperative review of the vascular anatomy is essential, as is the ability to provide a safe mobilization or dissection of the vasculature and ability to repair vascular injury should it arise.

One author has been performing their own access to the spine since 1996 with over 5000 approaches, the majority via the midline rectus splitting retroperitoneal approach (Holt et al. 2003). The indications and pathologies that he has treated have expanded over time such that in complex cases, he utilizes the knowledge and skills of a senior vascular surgeon who has had over 35 years' experience in vascular surgery. The involvement of a vascular surgeon in spinal surgery for the inexperienced surgeon contributes to the efficient and immediate repair of vascular injuries. Hamdan et al. (2008) stated that while exposure to the lumbar spine can be readily accomplished via a retroperitoneal approach, minor vascular injuries during exposure, mostly venous, are not uncommon. Most are easily repaired but occur more frequently when L4-5 is part of the exposure and less commonly when L5-S1 alone is exposed. Major injuries occur in less than 2% of patients.

As the surgeon's volume performance threshold increases (Regan et al. 2006), there is a natural expansion of indications to treat complex deformity, tumor, infection, revision anterior procedures at index, and adjacent levels ranging from patients with pristine vessels through to patients with combined multilevel degenerative disc disease and complex vascular pathologies (calcific atherosclerosis and aortoiliac disease). The vascular surgeon can provide expertise with endovascular techniques such as angioplasty and stenting and open techniques such as repair, endarterectomy, and bypass.

Indications

Anterior reconstructions provide biomechanical support by stabilizing the anterior column. This stabilization can now be performed by static and/or dynamic technologies (Scott-Young et al. 2018a). The common indication for anterior reconstructions is for degenerative conditions that have been recalcitrant to conservative therapeutic modalities (Fritzell et al. 2001). These include discogenic pain secondary to degenerative disc disease (DDD) and internal disc disruption (IDD) (Sehgal 2000). Other degenerative conditions include degenerative spondylolisthesis, isthmic and dysplastic spondylolisthesis, DDD with large herniated discs with verified radiculopathy, post discectomy recurrence and instability, iatrogenic instabilities following posterior decompressive procedures, degenerative de novo scoliosis, and anterior reconstruction for pseudoarthrosis following failed posterolateral, transforaminal, and lateral fusions.

Rare and "red flag" conditions that are commonly treated include vertebrectomy for tumor, trauma, and infection. The treatment of flat back syndromes, deformity, and congenital abnormalities to restore sagittal and coronal balance are often best treated with anterior reconstructions whether it be at single or multiple levels (Kim et al. 2017). The spino-pelvic revolution (Labelle et al. 2005) has enlightened the spine community on its relevance and importance. The sagittal plane can be considered as an open chain of interconnected segments from head to pelvis. The shape and orientation of each segment are closely related and influence the adjacent segment biomechanics (Roussouly et al. 2011). Hence, the enormous influence that consideration of sagittal balance is having on spinal surgery (Labelle et al. 2005).

Benefits

There are many benefits in having anterior surgery skills as part of a surgeon's armamentarium. The approach allows access and treatment of disc pathologies, vertebral pathologies, and multiple levels. It is a powerful tool because it enables a complete or near-complete excision of the disc and therefore access to the anterior cauda equina. This thorough discectomy allows for removal of herniated discs, extruded disc material, and sometimes sequestered fragments and minimizes any residual pain generators at the posterior disc-annular complex. Complete vertebral body removal for trauma, deformity, tumor, and infection is also possible. This can then be followed by reconstruction where the cage, allograft, or autograft is placed under compression. The endplate provides a large and strong cross-sectional area for supporting the reconstruction that is not afforded by the other approaches.

The anterior approach historically has been performed throughout the twentieth century and was originally used to treat Pott's disease (Ito et al. 1934). Originally, it involved a large transperitoneal approach to access the spine and would require longer recovery times. Modern anterior approaches recognize the importance of minimizing collateral damage (Fraser 1982) and as such more minimally invasive (Ito et al. 1934) or minimally destructive approaches (Ruey-Mo et al. 2008) have been developed. These incisions are limited to between 5 and 8 centimeters, generally subumbilical and utilize the retroperitoneal space to gain access to the anterior spine (Watkins and Watkins 2015). This allows a direct view into the disc space, thus facilitating proper and complete discectomy and annular release. The anterior approach is the optimal approach to allow correct positioning of the device in the midline. The goal is to maximize the size of the device in the medial lateral and anterior posterior plane. This generally minimizes eccentricity and subsidence while maximizing bone integration interface to the implant or the graft. It also allows direct visualization of the posterior annulus and posterior longitudinal ligament and one can release these tissues and thus decrease the incidence of "fish mouthing." While correct release, balancing, and positioning are important for ALIF interbody devices, it is especially important for disc arthroplasty where point loading and abnormal force transmission can result from incorrect sizing and placement.

It provides the surgeon and patient opportunities for a dynamic or static stabilization reconstruction for most symptomatic disc pathologies. DDD has been shown to cause significant morbidity and after identification of the symptomatic disc(s) they can be reconstructed with anterior lumbar interbody fusion (ALIF) or total disc replacement (TDR) (see Fig. 1). The approach can be performed via a muscle splitting incision (rectus interval) and avoids the muscular collateral damage that is associated with the other approaches. There is no devascularization or denervation of the erector spinae muscles that results from posterior approaches or damage to the psoas muscle that results from the direct lateral approach. This allows for significant reductions in patients' back and leg pain which, in turn, improves patients' function and quality of life. There are generally a shorter hospital stay, less morbidity, and faster recovery. Pradhan et al. (2002) compared ALIF and posterior fusion and found there were less blood loss and reduced transfusion rate, operative time, and hospital stay for patients with anterior fusion procedures.

The access to the anterior structures facilitates complete disc clearance and, with that, the ability to correct deformity and decompress the neuroforamen through disc height restoration. Other benefits include direct or indirect decompression of the neural elements centrally, laterally, and in the neuroforaminal region. In addition, powerful restoration of sagittal balance can be achieved anteriorly.

From a mechanical perspective, addressing the anterior column is a sound strategy in that it transmits a relatively high proportion of force and body weight and has reduced risk of subsidence from the large surface contact afforded in graft placement (see Fig. 2). When used as a stand-alone strategy, there are the additional benefits to the patient of avoidance of posterior soft tissue envelope violation or of neural retraction.

It cannot be emphasized enough that the majority of pathologies occur from diseases in the anterior column. Therefore, it is imperative that every certified spinal surgeon be able to use this access to effectively treat the patient's pathologies. By preserving the posterior and lateral paravertebral muscle and avoidance of facet joint dissection or violation as well as by restoration of disc height

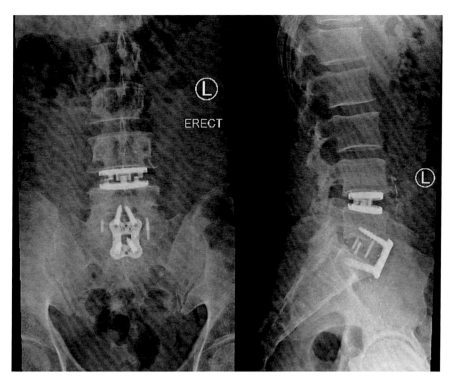

Fig. 1 The anterior approach allows efficient performance of either ALIF, TDR, or a combination of both (hybrid) procedure. Above is an example of a lumbar hybrid procedure with a constrained TDR at L4-5 and an ALIF construct at L5-S1

and lordosis, the incidence of later adjacent segment disease is also reduced. Posterior and lateral procedures do not regularly or predictably restore the disc height and lordosis. Thus, the patient is fused in a kyphotic position and their dynamic stabilizers (erector spinae) that pull them into extension are damaged.

The modern anterior approach has matured and is justified as an alternative for decompressing neural elements, alleviating instability, relieving intractable mechanical back and leg pain, and restoring functionality to the patient or addressing "failed back syndrome."

Contraindications

Contraindications can be classified as absolute and relative. Every patient should undergo a thorough medical and surgical assessment to assess any surgical and/or medical risks in the operative and perioperative period. The history and examination, focusing on cardiac and pulmonary risks, and determination of the patient's functional capacity are essential. The administration of an anesthetic and a surgical procedure is associated with a complex stress response that may present problems in the postoperative period. Assessment of the patient's overall health status is required. This incorporates referrals to specialists to investigate and treat any relevant conditions prior to surgery.

The history should include past and current medical and surgical history, current medications, allergies to drugs, and use of recreational drugs and tobacco use. Height and weight, particularly girth, is important to assess and treat prior to anterior surgery. Patients with a body mass index (BMI) greater than 25 are considered overweight, those greater than 30 are considered obese, and a BMI greater than 35 is considered morbidly obese. Patients who are obese can provide access difficulty and have higher risks and complication rates. These include hernias, wound infection and dehiscence,

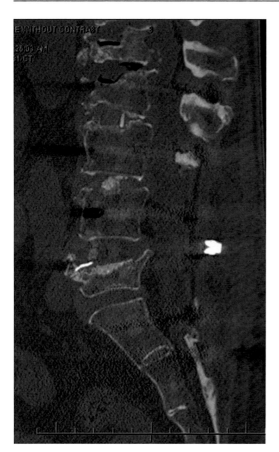

Fig. 2 Example of limitations of laterally placed interbody cages not only from limited endplate visualization and disc space balancing but through limited surface area. Failure of the construct is demonstrated here through multiple endplate/vertebral body failures in a susceptible patient

longer operative times, longer hospital stays, higher incidence of deep venous thrombosis (DVT), and possibly less successful outcomes secondary to these issues. Puvanesarajah et al. (2017) reviewed the obese and morbidly obese patients undergoing spine fusion and found that they are at significant risk of major medical complications, wound infections, and 30-day readmissions. Both groups had a longer length of stay and hospital costs. The obese have to be counselled about these risks and be active participants in weight reduction programmes.

Pregnancy, infection, severe osteoporosis, metabolic bone disease, chronic steroid use, multiple prior abdominal surgeries, implant or metal allergies, and severe psychological disorders, to name a few, are probable absolute contraindications to anterior reconstructions. Relative contraindications include any issue that may impede access or mobilization of the vessels. Therefore, conditions such as obesity, intra-abdominal scarring, retroperitoneal scarring, calcific atherosclerosis, abdominal aortic aneurysm, and prior vascular surgery need to be reviewed with respect to risk-benefit analysis. Complex vascular pathologies can be managed but require a multidisciplinary medical and surgical assessment prior to performing these complex surgeries.

Biomechanical Rationale

The basic mechanics of interbody cage fixation have been well investigated (Panjabi 1988). The strength and stability of a construct are important concepts to understand when reconstructing the anterior spine. The strength relates to the absence of structural failure of either the implant or endplate bone. The stability relates to the lack of motion between adjacent vertebrae, which is important to facilitate bone ingrowth and eventual fusion (Jost et al. 1998).

Gerber et al. (2006) found in a comparative cadaver biomechanical study that an anterior screw-plate and pedicle screws-rod constructs both substantially reduced range of motion and increased stiffness compared to stand-alone interbody cages. There was no significant difference in the amount by which the supplementary fixation devices limited flexion, extension, axial rotation, or anteroposterior shear; pedicle screws-rods better restricted lateral bending. This stability can be enhanced by insertion of ALIF cages with larger anterior/posterior dimensions and broader widths. This has the added benefit of less eccentricity and subsidence. The bone density probably plays an important role in reducing subsidence, as does the location of the cage on the endplate. Steffen et al. (1998) recommended more peripheral ring apophyseal contact was best from a biomechanical and biological perspective.

Cunningham et al. (2002) found that in the treatment of spinal deformities, structural interbody support probably is the best method to minimize longitudinal rod and screw-bone interface strain.

Moreover, anterior load-bearing structural grafts and interbody devices have been shown to increase construct stiffness, decrease the incidence of posterior implant failure, permit the use of smaller diameter longitudinal rods, and enhance the rate of successful spinal arthrodesis. The study reinforces the principles of load sharing between the anterior and posterior spinal columns and affirms the biomechanical dominance of anterior column support in circumferential spinal arthrodesis.

Watkins et al. (2014) compared anterior, transforaminal, and lateral interbody fusion techniques to determine which is most effective for restoring lordosis, increasing disc height, and reducing spondylolisthesis. They found improvement of the lordosis was significant for both the anterior and lateral groups, but not the transforaminal group. Intergroup analysis showed the anterior group had significantly improved lordosis compared to both the other groups. The anterior and lateral groups had significantly increased disk height compared to the transforaminal group. Hsieh et al. (2007) similarly found that ALIF is superior to TLIF in its capacity to restore foraminal height, local disc angle, and lumbar lordosis.

Preoperative Assessment and Preparation

Preoperative assessment and preparation of a patient for anterior spinal reconstruction should reliably and accurately identify those patients in whom operative intervention is indicated and any features in their presentation that need to be addressed or considered to make surgical intervention safe and feasible. We shall focus mainly upon the indication of treating DDD as it is by far the most common disorder.

Diagnosis

A diagnosis of symptomatic degenerative disc disease with or without radiculopathy (the most common indication for operative intervention) is usually made with history and clinical examination. A classical history of axial pain that worsens through the day and with prolonged sitting, bending, standing, or lifting that is relieved with recumbence is suggestive of IDD (Lee et al. 2016). There may be associated leg symptoms that suggest radiculopathy from a nerve root involvement due to disc herniation or multifactorial loss of neuroforaminal height (Lee et al. 2016). The clinical picture is then supported by investigations that may include typical imaging findings (degenerative bony changes on radiographs and disc degeneration with or without Modic (Modic et al. 1988) changes observed on MRI), MR spectroscopy, nerve conduction studies/electromyography, and provocative discography.

Preoperative Evaluation

Findings on the history and clinical examination that should be sought in anticipation of any contraindications to surgery or difficulty in the perioperative period include the patient's height and weight (obesity), any smoking or systemic illness that could affect healing or bone quality (e.g., osteoporosis, poorly controlled diabetes, steroid use), any coagulopathy/anticoagulants or prothrombotic tendencies, and any previous intra-abdominal surgery or conditions. Osteoporosis can be difficult to reliably quantify but a focused history and examination with investigations such as DEXA scan or fine-slice CT assessing the architecture and Hounsfield units can be suggestive of a patient that may require delay in spinal reconstruction until bone quality is sufficiently improved with medical treatment.

Screening clinically for any peripheral vascular disease (i.e., pulses), poor nutrition with sarcopenia, or any systemic illnesses (e.g., features of rheumatoid arthritis, vasculitis, or connective tissue diseases such as Ehlers-Danlos and Marfan's syndrome) is imperative. A physician to screen and optimize high-risk patients and provide a perioperative management plan is particularly useful in this regard.

Surgical Planning

In our practice, plain erect radiographs of the entire spine have been largely superseded by the use of low-dose radiation EOS™ scanning. This enables determination of all relevant sagittal, coronal, and rotational parameters that need to be considered when planning a reconstructive strategy. A quick assessment of the pelvic parameters (pelvic incidence, pelvic tilt, sacral slope), the type of lordosis (I-IV) (Roussouly et al. 2005), and the positioning of C7 plumb line can be done to assist in planning a reconstruction strategy. EOS™ also allows for rapid calculation of the current L5-S1 disc segmental lordosis by subtracting the L1-S1 lordosis from the L1-L5 lordosis.

Restoration of the sagittal balance is the best way to obtain a good result no matter the technique (dynamic, static) or the pathological situation. The only morphological parameter that is constant throughout life is the pelvic incidence (PI). How much the present position of the spino-pelvic complex is in its anatomical position and how much is in its pathologic or functional adaption is difficult to assess (Barrey et al. 2011). When treating patients with a low-grade PI (type I or II) no unnecessary lordosis is required for reconstruction. In patients with higher PI (type 3 or 4) more lordotic augmentation should be considered. Essentially, every spinal reconstruction requires preoperative classification (Roussouly et al. 2005). Patients with hyperlordotic lumbosacral junctions and a high PI can cause difficulty in accessing the disc space given the trajectory and position in relation to the pubic symphysis (see Fig. 3). Flexion and extension films may be useful in revealing the extent of dynamic sagittal instability in the case of spondylolisthesis.

Computed tomography (CT) scans are obtained in the case of any abnormal bony morphology or defects and can help assess the bone quality (relevant to the feasibility of a stand-alone strategy and/or need for augmentation) and status of the facet joints when considering arthroplasty versus arthrodesis for a particular motion segment. For segments that appear close to ankylosed, gas in the disc space indicates movement that can aid in decision-making regarding which levels to include in the reconstruction. CT angiogram has become the standard of care for any multilevel or complex reconstruction to better define the vascular anatomy preoperatively so that any variations or difficulty can be anticipated. This will be further discussed in special vascular cases.

MRI reveals the status of the intervertebral space (quantifies amount of collapse and Modic changes) and surrounding neural structures but

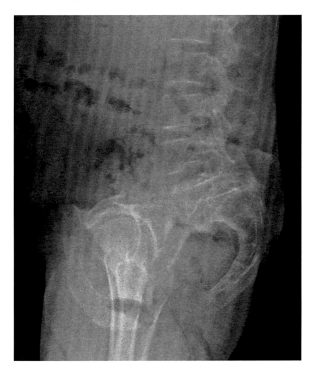

Fig. 3 Example of potential difficulty in anterior access to a L5-S1 disc in a case of dysplastic spondylolisthesis due to obliquity and relation of the disc space to the superior pubic symphysis

also the quality of the posterior elements and paraspinal muscles. This is relevant in deciding the need for retrieval of disc fragments in the canal from anteriorly or for a posterior decompression in addition to an anterior reconstruction in the case of tight multifactorial stenosis. MRI in the axial plane also reveals the presence of a fat plane behind vascular structures or, alternatively, any adhesions to an inflammatory disc.

Provocative discography gives information regarding the contribution of discs to the patient's symptomatology and should generally be performed by an independent third party assessor to exclude or include a questionable motion segment level in a reconstructive strategy (i.e., that particular level is responsible for minor or no symptoms).

Patient Optimization

Infection or bony failure in these cases with complex prostheses would be a disaster and prevention through host optimization is mandatory. Patients need to maximize fitness with maintenance of excellent nutrition including sparse amounts of sugar and simple carbohydrates but adequate protein and vitamin intake (especially vitamin C and complex B). Having optimized "protoplasm" will also favor efficient bony and soft tissue healing. In many cases dietary weight loss is mandatory not only for facilitating exposure but minimizing soft tissue and wound complications. Equally, a frail patient with sarcopenia may benefit from an appropriate strengthening and gentle aerobic program to combat osteopenia as well as higher caloric and protein intake.

Procedure

Pre-incision Setup: Equipment and Personnel

- A large well-illuminated theater and operative headlights are recommended given difficulties in adequate deep abdominal visualization that are associated with standard lights.

- We recommend the use of a Jackson Pro Table Mizuho OSI™ with capability for anterior and posterior approaches (see Fig. 4). Side brackets are placed to enable mount of the abdominal retractor system of choice. The bed is placed in a Trendelenburg position so that gravity allows the intra-abdominal contents to migrate cephalad and pillows are placed under the patient's knees to relax psoas and minimize popliteal compression syndrome.
- A large abdominal retractor is mandatory. We use a modified Thompson retraction system with multiple radiolucent blade variations whose design facilitates attachment and securing at different depths and angles. Alternatives are Omni-Tract™ and Bookwalter systems. A pin insertion (e.g., AUS) or related system for keeping the vessels positioned away from the working corridor is required (see Fig. 5a, b).

Sutures and Needle Holders for Vascular Repair

- 5/0 prolene sutures are on hand to close any cuts or tears as they occur. Smaller than this, the needles are impractical for the control and maneuver in this area of tissue.
- Make sure your vascular needle holders are strong and the teeth and ratchet not worn. In the setting of a large hemorrhage to discover faulty equipment is unsatisfactory.

Vascular Clips and Vascular/Urological Disposables

- Hemoclips also can be used after appropriate clearance of areolar tissue but the surgeon should be aware that they can be displaced by any traction required to mobilize large vessels (see Fig. 6a, b). These are best used prophylactically after isolating and skeletonizing the vessel. They can also be used as an immediate response to small vessel injury with suture as a later step. There are also some well-suited angled clamps which will control the vein edge until a suture is placed and we have these available on all the spinal trays (see Fig. 6).

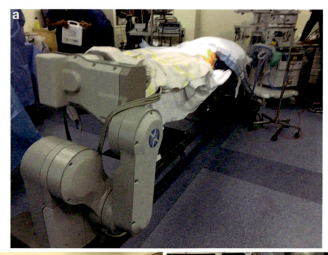

Fig. 4 Examples of Jackson-type table head down in picture above (**a**) and trays of retractors and instrumentation required in pictures below (**b–c**)

- Vascular stents, balloons, and equipment for vascular repair or reconstruction should be available and may be deployed prophylactically, depending on the complexity of the case.
- Ureteric stenting equipment for performance by a urologist is in place for revision cases.

Personnel

- A well-trained and organized theater team is conducive to optimal results. This includes cooperative theater staff, instrument nurses, operating theater technicians, spinal and vascular trained anesthetists, prosthetic device representatives, and office and hospital surgical planners who ensure that a full range of required equipment is available and that the team is well prepared for any unexpected events.
- A Cell Saver™ autologous reinfusion provider is present for all cases.
- An assistant or surgeon skilled at patient positioning, control, and respect for vascular structures and assistance with preparation of interbody grafts facilitates theater safety and efficiency.
- Recovery nurses adept at neurological and vascular observations eases detection of

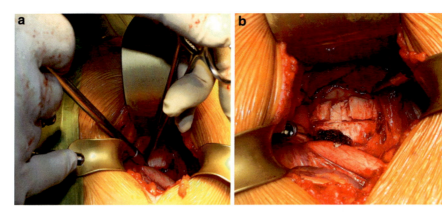

Fig. 5 Exposure of a disc space achieved with use of removable vertebral body pins, in this case AUS pins inserted by mallet and a threaded T-handle, to achieve continued retraction adjacent to a disc space. Insertion under visualization and protection from a peanut retractor (**a**) and sustained retraction for disc access (**b**)

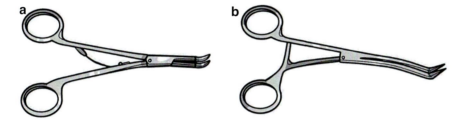

Fig. 6 (**a**) Example of hemoclip applier and (**b**) angled clamp for hemorrhage control

any unexpected immediate postoperative events.
- High dependency or intensive care ward care is recommended for complex procedures or procedures in high-risk patients.
- An inventory and equipment tray layout that not only allows anterior reconstruction but also allows for posterior procedures and vascular repair is mandatory.

Preoperative Setup: Preparation of the Patient in the Perioperative Setting

Skin Care and Preparation
- Preoperative skin care by the patient is the starting point for a successful approach by using appropriate washing and antibacterial soap in the time leading up to the operative dates. Hair should be clipped but not shaved to prevent integument compromise and, in theater, once upon the operating table the patient's skin should be scrubbed and dried. We use Betadine™ antiseptic liquid with the hand sponge from the scrub bay. Then alcoholic Betadine™ solution is used covering the whole area likely required for access including the inguinal areas. The rationale is that rarely one may require access to the vessels for stenting purposes. The skin must then dry before draping and a final cover of the wound is completed with iodine-impregnated adhesive plastic drapes (Parvizi et al. 2017).

Prophylactic Antibiotics
- Usually 2 grams of Kefzol (or an equivalent first-generation cephalosporin) is given on call of the patient to theater and coordinated so that it is in effect at the time of incision. Vancomycin or teicoplanin are considered for high-risk patients.

Anticoagulation
- The strategy is both mechanical and pharmaceutical. Enoxaparin 40 mg subcutaneously is given the evening before the procedure as a

routine. Other anticoagulants are stopped as is reasonably safe in conjunction with the advice of a physician. Positioning of pillows under the knees reduces popliteal obstruction syndrome. The addition of intraoperative calf compression pumps is continued till the next day. TEDS stockings are also worn in the perioperative period until the two-week follow-up.

Elimination
- An indwelling catheter is placed to decompress the bladder, enabling more efficient access, but also enabling monitoring of fluid output and patient comfort while having limited mobility. For retroperitoneal approaches the addition of bowel preparations is not routine, but is recommended to reduce fecal loading for transperitoneal approaches.

Autologous Reinfusion
- Cell Savers™ are set up to be used for all spinal and major vascular procedures so that there is a capability to infuse the patient's own blood in the event of unexpectedly large loss.

Patient Anesthesia and Monitoring
- An array of monitoring of general anesthesia and hemodynamic status including arterial lines is typically instituted.
- For complex deformities, intraoperative neurological monitoring is employed if osteotomies or manipulations are planned posteriorly at the same anesthetic.
- Some surgeons like to utilize an additional Doppler oxygen saturation probe on the left toe to monitor any ischemia caused by arterial retraction.

Intraoperative: Approach for Anterior Reconstruction

Approaches with a Focus on Vascular Access

Once the decision to proceed to surgery is made then the access method to be employed, whether left or right retroperitoneal (LR or RR) or transperitoneal access (TP), needs to be carefully considered. This is really a choice relating to minimizing trauma, safety in mobilizing vessels, and achieving satisfactory exposure to enable completion of the intended reconstruction. Deciding on which approach to utilize in the revision or previous abdominal approach setting has further challenges and presents a unique situation to consider. Generally, an approach to L5-S1 should be a right retroperitoneal. It should be via a rectus midline split and be kept low relative to the disc trajectory. This facilitates future left-sided approaches for adjacent segment disease should it arise. Approaches to L4-5 can be performed via a LR, RR, and TP. Generally, we would recommend a LR. The tip is to not dissect up or down as to minimize vessel or tissue trauma at adjacent levels. Approaches to L3-4, L4-5, and L5-S1 are via a LR if there are no vascular issues and if there is, then a TP approach is appropriate. Any 4 or 5 level approaches are via a TP approach. In every anterior case one should plan for a revision at the index or adjacent level in the preoperative workup. Therefore the approach needs to be atraumatic, direct, and possibly revisable.

A preoperative angiogram is only required if there is vascular pathology or variants in the anatomy that may present issues. Most vascular anatomical patterns and variants are obvious on CT scans and MRI. Always look for the presence of a fat plane under the vessels that will likely need to be mobilized, particularly the venous structures on the axial T2 images (see Fig. 7).

Retroperitoneal Approach

Nonvascular Anatomy
The primary nonvascular area of concern for the team is the genitourinary tract. On the left side, the ureter courses with its blood supply and the gonadal vessels until the ureter tracks medially over the iliac vessels at approximately the level of the common iliac artery bifurcation. More superiorly, the left kidney and its surrounding perirenal fat and fascia are encountered. Care must be taken when mobilizing the ureter to avoid devascularizing it with excessive

Fig. 7 Adequate fat plane visualized behind the great vessels (white arrow) prior to bifurcation as visualized at the inferior L4 vertebral body

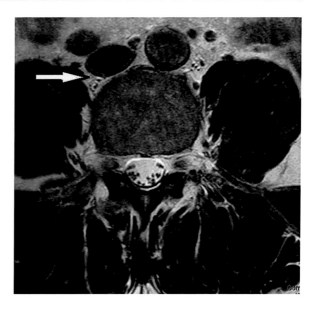

dissection; this can be accomplished by maintaining the ureter with the peritoneum, to which it is adherent. Furthermore, knowledge of the peritoneal folds and space, basic colonic anatomy, and knowledge of the sympathetic nervous anatomy and how they relate to mobilizing the gastric sac en masse are mandatory (see Fig. 8). Details of vascular anatomy and anomalies are further discussed below.

Selection Philosophy

In general, where there is standard anatomy, non-atheromatous arteries, no major venous anomalies, and no previous retroperitoneal surgery, then the retroperitoneal approach is advantageous, with minimal gastrointestinal disturbance and satisfactory access to L5-S1, L4-L5, and L3-L4 levels.

The vast majority of routine cases require an approach to reconstruct three levels or less. The more major cases we have been involved with that required up to four and five levels of disc reconstruction with some cases needing 50 degrees of correction of lordosis are the exception rather than the norm, and we would typically favor a transperitoneal approach in this setting.

The usual approach for the most common problems at L5-S1, L4-5, and occasionally L3-4 is through an infraumbilical midline incision progressing extraperitoneal in the preperitoneum, behind the rectus, and progressing around and then back medially in front of psoas, in the retroperitoneum. This approach is typically limited more proximally by the renal vessels which appear at the L2 vertebral body. The usual approach involves the inferior mesenteric vessels being displaced forward and translated to the opposite side of the approach with the bowel sac. We start the dissection bluntly external to the peritoneum and progress laterally separating the peritoneum below the lower border of the rectus sheath. With experience division of the posterior rectus sheath is rare. The main action initially is using your left index and middle finger to gently peel away the peritoneal sac away from iliacus and then over the psoas muscles. In a single level L5-S1 approach we favor a right retroperitoneal approach as it preserves the option of a left-sided approach for any anticipated future proximal level disease surgery; conversely, in the case of pathology of two or more levels we favor a left retroperitoneal approach. Lift and shift the sac together with the ureter with your hand to the opposite side of the approach away from the iliac blood vessels and eventually the major arteries. The peritoneum is gently dissected bluntly off the vessels so they can be clearly visualized. The hypogastric plexus will move away with the peritoneum. Retraction using the table mounted retractors is then modified to display the vascular

Fig. 8 Schematic of the path of the hand and dissectors around the retroperitoneum in the LR approach to the L4-5 disc. By exploiting the space and mobilizing the peritoneal sac with its contents and the ureter, safe exposure of the vessels and, ultimately, the disc can be achieved

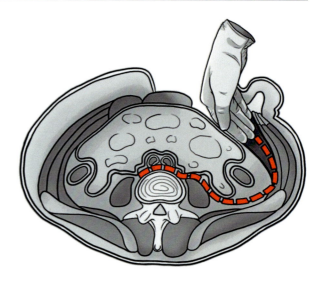

structures and the underlying disc level that is going to be operated on.

Once the peritoneum and the hypogastric plexus have been reflected, one will encounter areolar tissue over the disc space and the anterior spinal artery and venous vessels. Tease with a peanut on a Roberts artery the areolar tissue just to the right of the sacral vessels to the lower right. Insert an AUS-style pin inferior to L5-S1 disc at its lateral margin aiming slightly toward the midline. Next insert a pin superior to the disc on its lateral margin of L5-S1 again aiming slightly toward the midline. These measures keep the right-sided vessels safely away from the surgical area. Ligate the midline sacral vessels and then get under the fat plane and gently mobilize the left common iliac vein up and to the left. Next insert the pin on the superior left of L5-S1. One can also insert another retractor pin on the lower left if needed. We favor the use of AUS™ pins to retract the vessels rather than handheld retractors. Care is needed at this level to avoid injury to the sacral venous plexus which can cause significant bleeding.

The approach to L4-5 is more complicated as the bifurcation of the aorta and the veins varies considerably and is usually closely associated. How the dissection is planned particularly depends on the venous anatomy as this is more compressible and at risk of avulsion and with careful wider dissection these can be mobilized well out of the way. In revision surgery, the fragile vein walls are liable to tears and penetration more easily as dissection is developed through the surrounding adhesive scar tissue.

In the younger patients below 55 years of age with minimal known risk factors, the vessel anatomy and positioning related to the planned operative discs and spinal levels can be deduced from the MRI studies. Often the plan of approach to L4-5 is defined as the dissection above L5-S1 progresses. It is necessary then to understand the options for exposure so that alterations to the exposure can be made – "on the hop" – as the need arises. According to Chiriano and colleagues (Parvizi et al. 2017), in a series of 405 anterior spine procedures, exposure of L4-L5 was able to be accomplished above the left CIA in 44% of cases, between the left CIA and CIV in 45% of cases, and below the left CIV in 11% of cases. An approach between the bifurcations is facilitated by a high location either at the L3 vertebral body or L3-L4 disc space, situations which occur in less than half of patients.

For the older patients it is not only the venous anatomy that is significant but there is increased risk in dealing with aging arteries affected by peripheral vascular disease. The surgeon needs to assess the possibility of problems associated with atheromatous vessels with unstable plaques, ulcers, lumps of medial calcification, displacement and tortuosity, stenosis, occlusion, or aneurysm disease; all of these may be encountered in the access to the operative level(s). The surgeons planning the operation have to assess potential risks of arterial vessel damage, the likelihood of the type of damage, what measures to have in place to resolve problems at the time of surgery, and what to monitor postoperatively and for how long. Many vascular injuries are only detected very late because of limited understanding of the spinal surgeon, and a multidisciplinary approach to planning and assessment has a role here in reducing injury. For example, there are case reports of a major arterial occlusion such as the common iliac being overlooked for 2 weeks or an occlusion of the left common iliac vein for 2 years with the young patient presenting with severe stasis changes in the affected leg. Sachinder et al. (2011) reported on a series of 560 anterior spine exposures: five patients had arterial injuries; four were diagnosed more than 24 h after the operation with one on postoperative day 13.

The risks of manipulation of a calcified artery include fracture of plaque and dissection which can rapidly progress to vessel occlusion with limb- or life-threatening effects. With such dense calcification, safe compression and flexion of the vessel is not possible. It also should be recognized that aneurysms may be unstable and present the risk of rupture or they may be large and stiff enough to prevent access. On occasions the aortae and iliacs were occluded at the time of presentation. Patients presented with severe symptoms related to the disc degeneration and had proven radiculopathy in conjunction with their pain from vascular claudication. Before proceeding with anterior spinal surgery these patients need vascular surgical assessment.

Left-Sided Retroperitoneal Approach to L4-5 and L5-S1

A standard rectus splitting left-sided retroperitoneal approach to L4-5 and L5-S1 would be the most common and is explained here in detail. Other variants on this approach to access other levels described later use the same principles and techniques detailed below in other subsections.

It is best to palpate the abdomen once the patient is asleep on the table and be aware where the disc spaces are relative to the umbilicus and ASIS so that an appropriately placed incision is performed. The sacral promontory can often be felt by simple palpation. If there are any issues in regard to obesity or inability to palpate, then use of the image intensifier to verify the level and to optimize the trajectory of the incision and the approach to the disc is warranted. The usual incision is midline below the umbilicus. It is recommended that a larger incision be used for the obese who are sometimes brought in for surgery prior to any loss of abdominal girth that we can improve the access rather than encountering limited visualization and increased force of retraction through smaller wounds. With increasing fatty abdominal protuberance the wound may have to be extended to above the umbilicus which is typically unnecessary in patients with a BMI < 30. Given that we aim to improve the safety of the approach for patients, referral to dieticians and physiotherapists for weight loss advice is routinely performed. In some cases, where obesity is refractory over long periods to simple weight loss measures, patients are referred for weight loss surgery as a preliminary step to formal spine reconstruction.

MSY prefers to stand on the right side of the patient. The skin is incised and a diathermy is then used to dissect through the subcutaneous tissues to the rectus fascia trying to maintain full-thickness flaps to the level of the fascia (see Fig. 9a). A one centimeter incision in the rectus fascia is performed; then it is undermined with the left index finger developing the plane between the fascia and rectus muscle. The fascia is then incised with the diathermy distally and proximally extending beyond the skin incision so that you are funneling out rather than funneling in. The interval between the recti is usually identifiable with a fat plane that allows entry into the preperitoneum and by careful dissection, you can enter the preperitoneum on the left- or right-hand side depending on your approach (see Fig. 9b, c).

Next, a Langenbeck retractor is placed under the left rectus and one can see the inferior epigastric vessels travelling up to pass between the posterior border of the rectus and the superior surface arcuate ligament or posterior rectus sheath (see Fig. 9d). It is important not to disrupt these vessels, so use your right index finger to lift them gently off the preperitoneum and take the

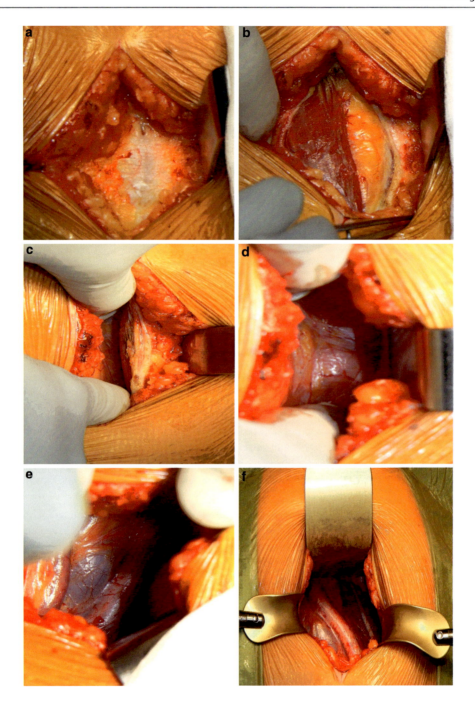

Fig. 9 (a) Development of full-thickness skin flaps to rectus sheath; (b) following incision of linea alba and posterior fascial incision, preperitoneal fat is revealed; (c) the window is expanded with use of fingers; (d) identification of superficial vascular leash (inferior epigastric vessels) for protection; (d) left Iliac vessels revealed after maneuver to mobilize peritoneum and peritoneal sac; (e) exposure of vessels overlying L4-5 following placement of malleable externally mounted retractors and mobilization of peritoneal sac/ureter

peritoneum down with your left index finger. Then, work below the inferior arcuate ligament or the posterior rectus sheath, often palpated as a "soft spot," and move toward the iliac crest. There is usually a small fat pad here that one can peel off the lateral and posterior wall essentially heading toward the anterior superior iliac spine. Take care not to begin the dissection too deep and distal to the anterior superior iliac spine lest the iliac veins are placed at risk of an avulsion injury.

Slim, fit people can have little adipose tissue between the fascial planes, and the trick is to just persevere with use of the index and ring finger, carefully blunt dissecting the peritoneum off the wall. Once in the iliac fossa you can then peel the peritoneum off quite easily over the iliacus followed by the psoas muscle. Once the peritoneum is retracted to the medial border of the psoas, then place your right hand in the gap and then just peel the peritoneum further off the back wall on the psoas and gently roll the peritoneum off the posterior rectus sheath. This will prevent peritoneal tears. Perform this step gently and just roll your cupped or closed hand in the plane and you will feel it gently separate. The next step is to peel the peritoneum off the vessels. Usually, start down on the lumbar sacral junction. Take the peritoneum, the ureter, and the hypogastric plexus from the left side to the right. Once on the sacral promontory move your left index finger down over the promontory and spread your left middle finger right and ventrally. This maneuver helps lift the hypogastric plexus to the right and to safety. Following this, keep your left hand in the wound and then sweep with your right hand, taking the peritoneum off the vessels up at L4-5 and higher if needed. By taking the peritoneum off the vessels it allows you to assess the vascular anatomy (see Fig. 9e, f). The real benefit is that if there is a vessel injury, access above and below allows you to control the breach and repair it safely. Anterior surgery is much easier when you can visualize the vessels and manipulate them appropriately.

Leaving the peritoneum attached to the vessels is hazardous. In this setting, you cannot adequately see the anatomy and it makes a tear difficult to control; thus, working hard on making sure you can get the peritoneum reflected to give clear access is critical. When reflection of the peritoneum and its contents is performed in the manner detailed previously, the superior and inferior hypogastric plexus is reflected safely. As a consequence, the incidence of retrograde ejaculation reduces and any sympathetic side effects also reduce. There are sympathetics in the plexus and there are also sympathetics that run on the medial psoas gutters and it is important to make sure, when you do use dissection, that you avoid these areas.

Study and be aware of the vascular anatomy from the preoperative MRIs, CTs, or angiograms. Place them on radiographic lightbox in your theater and refer to them before and after reflecting the peritoneum and visualizing the vascular anatomy. This mentality and preparedness will reduce vascular injuries. The aorta, lying to the left and anterior to the vena cava, and the vena cava, lying to the right and posterior of the aorta, bifurcate at L4-5 and the venous structures normally duplicate the arterial system. This is the level where most vascular injuries occur. One can take the vessels left to right in most cases. If the bifurcation is above the L4-5 level and the common vein runs at or above the disc level, one can mobilize the vein up and to the left. At the L4-5 level, the common iliac vein usually runs under and inferior to the left common iliac artery. The trick is to obtain blunt dissect initially with Metzenbaum scissors and then with a peanut dissector and just gently tease and strip the tissue off the lateral aspect of the artery at the level of the disc to expose the disc (see Fig. 10a, b). There may be one or two iliac perforators that come off the common iliac vein that come under the artery and these just need to be clipped and ligated.

If performing an approach at L4-5 and the anatomy dictates a right to left exposure, gently tease the tissue off the artery and have a look for the iliolumbar vein. Some routinely ligate it but I do not think you must do this unless it presents a problem. So, in less than 15% of my cases I will ligate the iliolumbar vein. If you do ligate the iliolumbar vein, tie off first at the origin then reinforce it with clips. Move proximally over the disc and tease the tissues up as this helps release the vessels to mobilize them to the right and down.

If there is any vascular vessel calcification or slightly abnormal vasculature where you need to get better visualization, then you can skeletonize the common iliac artery and put a sling around it and take it over to the left. This gives complete exposure to the common iliac vein and allows mobilization down to the right. The vein has to be mobilized enough to be below the level of the

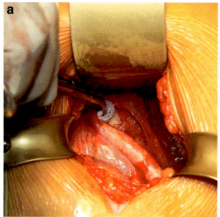 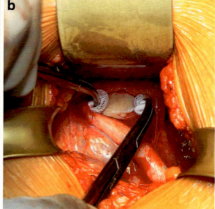

Fig. 10 (**a**) Mobilization and traction provided by periosteal plane dissection with a peanut retractor held with the left hand, (**b**) teasing of areolar tissue across the disc space by a peanut dissector held with the right hand while the left hand maintains traction

disc so that a retractor pin can be placed for retraction and access far enough to the right to allow correct positioning of the prosthesis to be obtained (see Figs. 11 and 12). Next move to the superior right side and insert the retractor pin. Then make your incision in your annulus, reflect the annulus, and insert stay sutures in the annulus to protect the adjacent areas. Normally we just use a 1 or a 2 Vicryl™ (see Fig. 12b).

Then perform a macro-discectomy (see Fig. 12b). This is best done with long pituitaries, Kerrison rongeurs, controlled movements within the intervertebral space with a long sharp Cobb, curettes, and rotational distractors as well as interbody distractors. After the bulk of the anterior disc has been removed by the pituitaries, the Cobb is used to remove the cartilaginous endplate with small, careful, controlled movements. Do not make any lateral movements or rotations as lateral movements will result in the Cobb skiving and sometimes lacerating the adjacent vascular structures and uncontrolled movements can result in penetration into the spinal canal or divots in the endplate which can compromise implantation. One needs to have the appropriate Cobb and the appropriate technique.

The cartilaginous endplate is then removed and the disc space after that is distracted so that we can pare down the posterior annular structures. From there one will usually go to the paddle distractor to increase the disc height and therefore gain access to the posterior annulus and the posterior longitudinal ligament. When one distracts, it is important to balance the lateral annuli to make sure that there is maximum exposure to the endplate and that there is parallel distraction and no evidence of any "fish mouthing."

Initially use a Charité central distractor, then insert a David distractor, and use the Cobb to get the annulus either partly off the posterior endplate superiorly and inferiorly. If it needs to be resected, then Kerrison rongeurs are utilized. After the annulus is released one can perform an internal intradiscal electrothermal therapy with bipolar diathermy, ablating the course of the sinuvertebral nerve around the posterior annulus.

From there, one should reuse the distractor paddle including a T handle paddle to ensure maximal disc height restoration and appropriate posterior release (see Fig. 13a, b). It is important to do this probably three times and to go back and check the disc space and make sure it is cleared and balanced front to back, side to side, and that you have addressed any disc herniations or releases required. If there is a disc herniation, extruded fragment, or a sequestered fragment, they can be retrieved after careful preparation of the disc space and rongeurs. Once you are back at the posterior annulus,

Fig. 11 With vessels held by a peanut retractor and under direct vision, AUS pins are inserted sequentially on a T handle to achieve a working corridor for disc preparation and device implantation

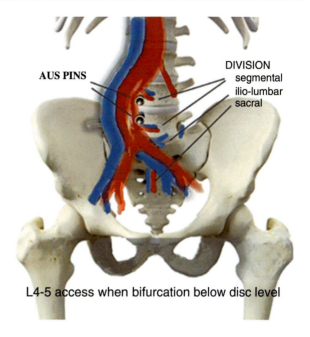

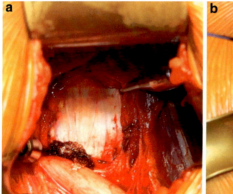
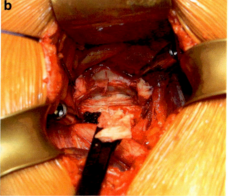

Fig. 12 (**a**) Pin retractor placement to maintain safe exposure of the disc. (**b**) Following an H-shaped annular incision, flaps are mobilized and retracted with stay sutures and a macro-discectomy can then be safely completed with careful use of a Cobb retractor and pituitary rongeurs

insert the David distractor in and rotate the panels out to the left hand side. This gives you clear access into the distracted disc space. From there you can use long Kerrison rongeurs to create a rent in the annulus. If there is a herniation, the rent will already be there. You can then follow that rent and retrieve the herniation. It is essentially just like doing and anterior cervical discectomy except that you have got much more room and you easily evacuate nuclear and annular material.

Once you have got the disc prepared, you can then make a decision about whether the device preoperatively planned for implantation is still appropriate. Whether it be a static or a motion preserving device, the important issue is to achieve maximal cross-sectional area from medial to lateral and anterior to posterior endplate as possible as this aids in supporting the chosen implant (see Fig. 14a, b). This will prevent eccentricity, reduce subsidence, and maximize the graft-bone interface opportunities. Should the endplate

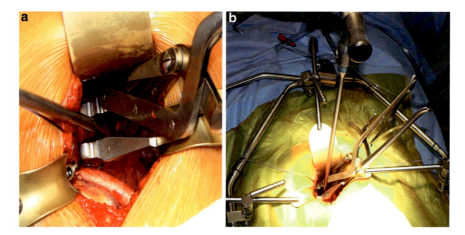

Fig. 13 (**a**) A David distractor is inserted followed by a paddle-style distractor panel to achieve disc distraction for periannular release and full access for discectomy and implantation. (**b**) Example of rotation of a distractor into the disc space splinted by the David retractor to achieve safe distraction while minimizing the risk of endplate breach

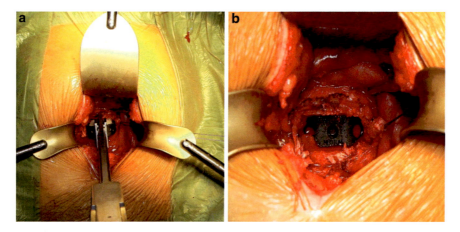

Fig. 14 (**a**) Implantation of an interbody cage loaded with graft. (**b**) Implantation and recession of interbody device prior to annular closure

or vertebral bony cancellous quality be found to be poor (often sensed via limited bony resistance with insertion of AUS pins), then vertebral augmentation should be strongly considered after implantation. This is achieved via insertion of vertebral cement into a Jamshidi needle under fluoroscopy control.

Following implantation, we harvest a fat graft and place it over the implant and then repair the annular flaps and then remove the retracting pins. Should poor quality or sparse fat be available, then use a square of Gelfoam™. The rationale for this is to close the dead space for hemorrhage and reduce the theoretical nidus for adhesions. I think that adhesion barriers should be used generally in young people, who may later require revision or treatment for adjacent segment disease. There is a preference to use Surgiflo™ with thrombin judiciously into the pin holes so that there is no hematoma, which will increase fibrosis in the postoperative period.

Next, carefully release the retractors and massage the peritoneum to the contralateral side of the approach, allowing it to fall back into its natural position. One should generally perform this repositioning because if this is not performed, the ureter can end up in a more central position and develop a kink in its course. On the rare occasion this can result in a hydroureter above the stricture leading possibly to a hydronephrosis.

Once the peritoneum is back in position, verify the position of the implants radiographically. At this point, reconfirm hemostasis is achieved and secure the fascia with a double loop nylon, making sure the sutures pick up the fascia and not the muscle, aiming for normal tension. Close with an absorbable monofilament in the subcutaneous and subcuticular layers.

Tips on Revision Retroperitoneal Surgery

Revision surgery can be very complicated if the approach is attempted through a previous ipsilateral retroperitoneal approach. There are other options available and these include right retroperitoneal and transperitoneal (preferred). If a revision retroperitoneal approach is performed, then a ureteric stent needs to be inserted preoperatively. In revision, anterior surgery it is expected that the approach to the spine will be complicated by adhesions. Peritoneal tears are common and this may predispose to bowel injuries, ileus, and bowel obstructions in the postoperative period. At the index disc and often at adjacent discs there will be adhesions and fibrosis. This can result in distortion and tethering of vascular structures. Both of these situations greatly increase the risk of injury. As stated above our experience has found transperitoneal approach to be safer and provide a greater chance of success. On some occasions, it is not possible to access the revision level safely. This commonly occurs at the L4-L5 level where the vascular anatomy is often challenging. The vessels may be completely immobile and one must be prepared to spend time performing meticulous dissection. Sometimes the decision to abort has to be made. Consider subperiosteal dissection of the tissues which may provide surprising access on some occasions. In regards to minimizing vascular difficulties during index revision procedures, one should reconstruct over the disc replacement with fats graft or anti-adhesion barriers. This will maximize access to the level and help minimize vascular complications during revision index surgery. A team approach is required in regard to anterior revision cases and will be discussed in a later section.

Transperitoneal Approach

This approach shares the same steps to the retroperitoneal approach to the point of encountering the peritoneum and a generous incision is generally required (see Fig. 15). At this juncture, the

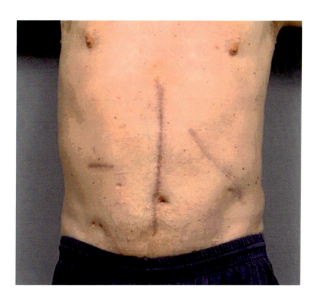

Fig. 15 Midline incision utilized for a L2-S1 transperitoneal approach in a patient with previous abdominal surgeries. Recall that safety and visualization, not the length of the scar, should be the first consideration

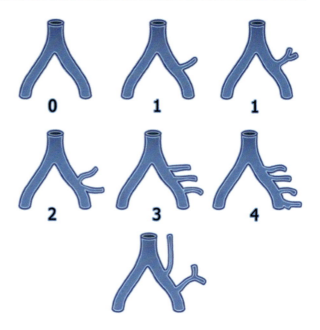

Fig. 16 Variances in the venous anatomy around the left common iliac vein. Notice increasing number of tributaries and sub-tributaries possible here as well as, in the bottom picture, the potential for the presence of an ascending iliolumbar vein

peritoneum is divided. Using the Thompson retractor, the descending colon and the sigmoid are packed into the left paracolic gutter. The transverse colon is lifted with a wide retractor and with displacement of the small bowel to the right, the duodenal-jejunal junction is defined. The major trunk vessels of the aorta and IVC are our main interest when approaching transperitoneally. Dissection nearby through the posterior peritoneum over the right aspect of the aorta allows progressive exposure of the aorto-caval space from the renal vein down aiming to remain to the right of the inferior mesenteric vein.

In the complex cases often requiring 4 level reconstruction or in revision surgery we have found it best to approach the superior aspect initially, defining the space between the aorta and vena cava dividing the segmental vessels and noting that there is often a large vein at the L3 level. In some cases, this is a retro-aortic renal vein. We progress inferiorly to the overlap of the right iliac artery over the left vein. Depending on the position of L4-L5 it may be approached by dissecting the right iliac artery free and mobilizing it and the aorta to the left and the IVC and the left vein to the right and inferiorly. Care with the superior aspect of the left iliac vein will allow control of the iliolumbar vein or neighboring branches to prevent tearing and problems with blood loss. While much attention is focused upon care for the vessel variations (see Fig. 16) and parasympathetics, attention should also be given to the nearby anterior neural structures such as the lumbosacral plexus and obturator nerve.

A lower disc may be better approached beneath the left iliac artery, displacing the vein inferiorly and to the right. In this case, it is sensible to control the vein branches along the superior border early. If the bifurcation of the main trunks is high, L4-L5 may be approached from below, similar to the L5-S1 approach with definition and ligation or clipping control of the middle sacral vessels. The aorta and the left iliac artery can also be displaced to the right, but this requires mobilization of the Iliac artery to limit the displacement forces while it is pinned. The IVC bifurcation is usually above the L5-S1 disc but may occur at the inferior edge of the disc. Adaptation of the standard approach will be required.

A vascular surgeon is required only when there are complex problems such as requirements for 3- to 4-disc level-based reconstructions and/or where there are definite arterial risks and venous anomalies or there is revision surgery.

In all of these cases a transperitoneal access is going to be safer by allowing easier treatment of complications if they occur and better controlled outcomes.

In the space between the aorta and IVC, segmental veins and arteries have to be defined, individually controlled and divided to enable freeing of enough length to allow adequate access of the disk space without tearing the major vessels. In this area, the surgeon should be aware of developmental anomalies and variations such as a 10% risk of a retro-aortic left renal vein, a horseshoe kidney, and left IVC variations (discussed below). Also one must consider abdominal aortic aneurysm and dense calcification of the arteries. These can all be recognized from the preoperative MRI or CT angiography scan and, conversely, can be a major problem if not anticipated at the time of surgery.

Tips on Approaching Operative Levels Below the Bifurcation

We choose to divide the peritoneum centrally over the sacral promontory to obtain access to L5-S1 and L4-L5 if the bifurcation is high enough. This area is usually clear of major structures and the areolar tissue can be progressively elevated until the anterior aspect of the left iliac vein is displayed. The medial edge is then clearly defined. Achieving this may take some time with slowly teasing away fibrous and vascular bands or may require diathermy and/or limited areas of sharp dissection. Once a satisfactory length of the vein edge can be visualized, attention is turned to elevating the vein and freeing it from the posterior surface. In general, it is attached to the anterior spinal ligament and surrounding fibrous tissue. Great care is required here and a combination of teasing the tissue with two pairs of forceps again with or without diathermy and sharp dissection with scissors, working to go deep into the ligament or periosteum, is used to achieve the separation. This is particularly important if this is revision surgery. The median sacral vessels can readily be isolated and controlled by dissection with the peanut dissector of any surrounding areolar tissue to allow clipping and diathermy.

Tips on Approaching Operative Levels Above the Bifurcation

Above the bifurcation the approach we usually use is between the aorta and IVC but a left lateral approach along the aorta can also be used. At the bifurcation level, the aorta can be displaced to the right with the veins. This latter area requires careful definition down the lateral edge of the vein behind the left iliac artery and requires progressive division of the iliolumbar vein, either with clips or ties, and the often multiple accompanying veins in this region, to enable de-tethering and mobilization.

As mentioned earlier, the L4-L5 level access depends on the position of the aortic bifurcation and, often, the easiest approach is between the left common iliac vein and the artery above. If the aorta is of normal size or small and of normal compliance, it is possible to dissect along the left side and retract it to the right. With this approach, care is necessary with the iliolumbar veins, which are best safely controlled early before they are avulsed from the iliac vessels. When dividing the veins attention to the local autonomic nerves is also required.

Revision and Previous Abdominal Surgery

As previously discussed, an approach to the spine in the revision setting will be complicated by adhesions close to the affected disc site affecting ability to access safe planes. The first decision again, with revision, is whether to proceed retroperitoneally or via a direct transperitoneal approach. The majority of the time for revision surgery we consider the direct transperitoneal approach to be safest with the best exposure, especially if access to more superior levels is required. When carefully performed with meticulous technique, blood loss with recent procedures has been of the order of 100 mls, even in a case where we had visualized large pressurized veins on the other side of the peritoneum (e.g., a patient with portal vein hypertension).

The retroperitoneal approach, if being made for the second time, involves tethering and scarring of the retroperitoneal space. The dissection can be difficult and the ureter is at risk in this tissue. It is our policy to have an urologist place a removable stent as a precaution as this allows easier recognition of the ureter and greatly reduces risk of damage to it. It needs to be recognized that direct trauma is not the only way that tissue damage can occur, with ischemia from repeated dissections being associated with damage to the ureter and the femoral nerve in this region. A resultant leak or neuropathy may result.

If using retroperitoneal dissection for revision, prosthetic replacement, or fusion, our approach has been to develop the dissection plane more superiorly. With access to L4-L5, the most difficult, we commence definition and mobilization of the aorta on its left side, then progress down to the bifurcation, and continue laterally to the left common iliac artery. We have found that the common iliac vein is attached posteriorly and it becomes exposed as the artery is displaced medially. This is likely to be the riskiest part of the operation. To get adequate access the vein has to be freed along 3–4 cm of its length so that it can be displaced medially and a little inferiorly.

Before the vessels can be freed up it is important to mobilize the ureter away from them. This is assisted by having a ureteric stent placed at the commencement of the operation which we do for initial recognition in the revision to the retroperitoneum setting. As the dissection progresses it allows easier definition between the ureter and the adhesions that need to be divided.

Once the lateral edge of the vein is visualized sharp dissection is used with scissors directed posterolaterally allowing the periosteum to be divided which protects the vein. This is continued progressively until the vein is adequately freed inferiorly.

In women it is also very important to understand their gynecological history. Past surgery laterally, e.g., around the tubes and ovaries or involving the lymphatics, can increase the problem. This is pertinent as a case has been reported to us where the iliac vein was totally avulsed as distraction was applied due to dense tethering in the pelvis. Obviously, this can occur no matter which direction of approach is used. Reports of direct iliac vein trauma probably only account for a limited number of cases that have occurred.

With all the endovascular options a skilled vascular surgeon can bring to the table, we have the ability to control hemorrhage with balloons or stent grafts if a major rent in the vessel occurs. We use this with either direction of approach and have the endovascular equipment in the theater. As the potential risks include a threat to life we have considered it useful to place a left iliac vein guidewire "0.035" at case commencement using duplex ultrasound via the common femoral vein when our team has considered the risks are significant. At the end of the case the wire is removed with less risk than a standard venepuncture.

If a prosthetic disc needs to be removed, it is important that the disc type be reviewed and information be sought on disassembly techniques and recommendations from the manufacturer. With the space opened at least partially, the metal plates can be freed with the careful use of a small osteotome between disk and vertebral body and prosthetic extraction facilitated by rotation of the plates. Sometimes a partially or complete vertebrectomy is required to explant the device. Careful assessment for and removal of any collections of metallosis or polyethylene debris should occur before proceeding with standard reconstruction.

We have found the vessels are adequately protected by a Gore-Tex™ prosthetic film allowing the dissection to proceed deep to the film when it has been implanted at the initial procedure.

Ability to Transfer Between Retro- and Transperitoneal Approaches

The decision to use one type of access does not rule out use of the alternative. It is straight forward to transfer from extraperitoneal dissection if problems arise and the upper extent of the approach is not reached. The peritoneum is opened and the bowel moved to expose the interval between the IVC and the aorta. This interval can be widened to expose the upper discs for repair.

Even if surgery has been planned based on a transperitoneal approach it can be switched to retroperitoneal should problems occur. This may require adjustment of what can be achieved.

Vascular Pathology, Injuries, and How to Manage Them

Venous Pathology and Handling Techniques

The iliac veins are at most risk in the pristine abdomen no matter whether the approach to the anterior lumbar spine is made retroperitoneally or transperitoneally. The first decision is whether to proceed retroperitoneally or via a direct trans-peritoneal approach.

In general, the approach is simplest from the retroperitoneal aspect. There is minimal gastrointestinal manipulation, a limited incision for the patient, and the approach to L5-S1, and L4-L5 is straightforward and L3-L4 often manageable.

The veins should be identified along their edges with progressive mobilization and elevation. The risk of damage and hemorrhage is mainly from unidentified branches. It is easy to get into this problem if the retraction is continuously maintained as the branches empty and remain emptied and thus can be divided without recognition until, with reduction of tension, the depth of the wound tends to allow the severed vein to rapidly fill with blood and into the wound.

The vascular problems are twofold: either hemorrhage from cuts, tears, or avulsions or obstruction with or without thrombus. The latter can occur as the result of trauma or from repairs after the iliac has been clipped/sewn to control hemorrhage or compressed by the operative area behind the right common iliac artery.

The frequent "injury" is really a part of the dissection in the majority that can be prevented by adopting a dissection technique of stretch and relax given that these are viscoelastic structures. Branch veins can be distracted and torn from the common iliacs and vena cava and it is often the smaller veins that are mainly responsible as under distraction they are difficult to visualize when distracted but if pulled or torn can result in a sizable hole in the main vessel with the potential of copious bleeding. Allowing tissue relaxation can give these vessels a chance to fill and, once defined, be isolated and appropriately controlled.

This problem is just part of the surgical approach and must be mastered to, firstly, limit injury by the technique of dissection and, secondly, to have the capacity to repair the vessel tear without significant further damage, i.e., enlarging hole or causing the result of critical stenosis or occlusion.

Usually it is the common iliac vein that is bleeding. A calm approach is required to not let this get out of hand. The bleeding needs to be controlled well enough to allow the placement of hemoclips or sutures for control without compromising the vessel lumen. If the lumen is compromised, the problem can be converted to an occlusion rather than a hemorrhage. The left common iliac can be tightly stenosed or occluded by taking large bites to control bleeding. At the time of surgery, secondary to this, all the pelvic veins will become congested as they develop an alternative pathway and feed into the right and on the ipsilateral side the iliolumbar region, this may well make further dissection more difficult and hazardous as all these small vessels will be increasingly pressurized and bleeding will be more likely to occur and more difficult to control. Postoperatively, the risk of extensive DVT is significant and adequate anticoagulation must be given. The patient should also be informed that they are likely to have a congested and swollen leg(s) and have the potential of a lifetime pressure/stasis problem.

Clusters of veins anterior to the sacrum if torn can be very difficult to control. When all else fails, osseous stapling of a mass of Surgicel™ or Gelfoam™ and Surgiflo™ or thrombin-based injectable onto the site to tamponade and coagulate the plexus can resolve a life-threatening situation (see Fig. 17).

If the solution appears very difficult, an endovascular option may be helpful. Recovery of the situation can be achieved by placing a guidewire through the common iliac vein and can allow it to be followed with a balloon to apply tamponade at the site until a formal repair can be carried out (see Fig. 18). The wire needs to be followed by imaging as it is common for the wire to travel up the iliolumbar branches.

If the bleeding is from a difficult or inaccessible site, e.g., posteriorly near the bifurcation, a covered stent may offer a means to solve the

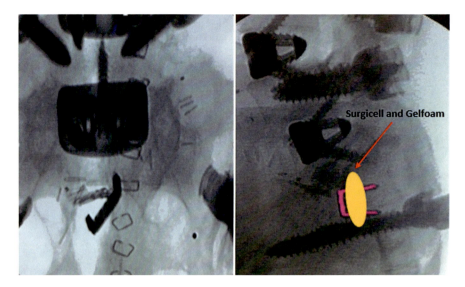

Fig. 17 Use of a Richards staple (usually found in a knee reconstruction instrument tray) for osseous purchase to secure a Gelfoam and Surgicel composite for tamponade in a case of difficult to control sacral venous bleeding

Fig. 18 Vascular balloon employed for hemorrhage control

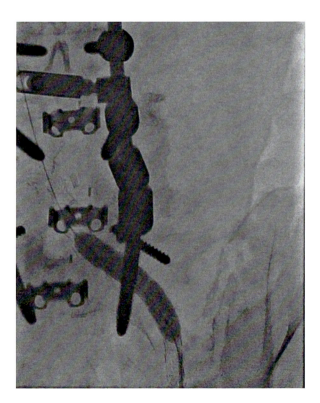

problem. These can also be used to recanalize an occlusion.

There are no *covered* stent grafts made especially for the major veins and the iliacs often require 14–18 mm stents to fit suitably. The internal method to seal venous injuries would require a type usually used for arterial aneurysm repair. If the prosthesis does not grip inside the iliac vein, it

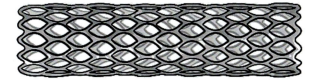

Fig. 19 Expandable stent graft

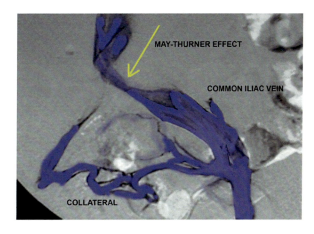

Fig. 20 May-Thurner effect of left common iliac vein compression by left common iliac artery

can end up being discovered as an embolus to the heart. These stent grafts need to be ordered as they are not usually "off the shelf." For a high-volume anterior spinal service it would be sensible to have a choice of stent grafts of different sizes to be suitable for any problems that may arise (e.g., grafts of 12, 14, 16, and 18 mm diameter and 5 cm long) on the shelf. This would generally enable suitable implants for any emergency. With experience, however, the chance of requiring this assistance with an initial approach should be very rare, but it would be useful in the more demanding setting of revision surgery.

The covered stents usually available for the larger sizes 12–14 mm are balloon expandable "Be" stent grafts or by "Atrium V12™." These covered nitinol self-expanding stent grafts maximize their size out at 10 mm but there are larger grafts usually used for iliac artery coverage in aneurysm repair that can be ordered (see Fig. 19).

Vascular surgeons have used these in situations where patients have been referred with a major vessel tear and with a repair attempted but complicated by a tight stenosis which have gone on to occlude, with ensuing serious results of massive limb swelling and stasis change. Thrombolysis with associated stenting has normalized the situation. For the case of simple stenosis (which is not all that uncommon) usually from right common iliac artery compression of the left iliac vein with swelling or a plate filling the space from behind, I prefer a bare sinus-Obliquus stent that conforms well at the IVC bifurcation.

The occlusive venous risk again is most common in the left common iliac vein. There is an anatomical variation associated with lumen compromise and flow restriction named as May-Thurner effect (see Fig. 20) which can be recognized in up to 25% of the population, particularly females.

Control and Repair of a Venous Hole

1. First, bleeding should be controlled by direct pressure on the area preferably by fingers or sponge sticks or, if necessary, placement of DeBakey vascular clamps. Help should be requested early if, upon assessment, the injury will likely require more than simple vascular repair techniques and, additionally, multiple skilled pairs of hands help the situation.
2. Then the vein should be mobilized and skeletonized above and below the site to allow the hole to be situated and visualized on the vein edge; then a series of small hemoclips can be

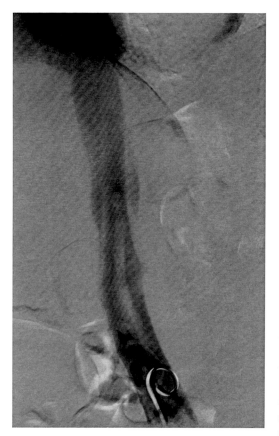

Fig. 21 IVC thrombus originating below the left renal vein after Floseal had been injected into a dural vein to control heavy bleeding during a posterior approach. Angiography at 6 months showed no residual

applied to the area to give satisfactory stable control. If necessary further stabilization can be achieved by using a 5/0 Prolene™ continuous suture to oversew the site without taking excessively large bites and tied with at least 7–8 throws.

3. Use of a sandwich repair can be a useful adjunct to close small venous pinholes without propagating further tearing. This employs a small Teflon or Dacron pledget skewered with the needle and thread twice and then slid down onto the holes for extra buttress and tamponade effect.

4. If extreme difficulty is encountered, then an endovascular wire, followed by a 10 cm angioplasty balloon, should be employed to allow control of iliac bleeding. A covered stent graft could be deployed in such instances if blood loss persists and there is a threat to the patient.

5. If the IVC is the problem, a side clamp or two small angled clamps can be applied until the site can be adequately secured.

6. The most important message is not to panic; a calm, well-considered approach is mandatory to achieve the best results.

Control of Venous Oozing

First-line management for nonspecific oozing is with judicious use of topical hemostatic agents (such as thrombin-based products) combined with an adjuvant volume filler such as Surgicel or Gelfoam. This is useful for controlling oozing from smaller veins, from vertebral body perforators, and from extradural veins. Generally, the topical hemostatic should be applied when there is no easily identifiable large bleeding source amenable to clipping, ligature, or diathermy and should then be assisted in its action by placement of the adjuvant volume filler, pressure agents (e.g., Surgicel gauze) with or without additional pressure from a packed surgical gauze. Adequate time should be given for this combination to take effect and repeating this sequence is often necessary until hemostasis is achieved. When used carefully, it will not upset the function of the cell saver. Spongostan™ can be used; however, it should not be sucked into the cell saver.

Excessive use of hemostatic agents is not without risks given its potential to be absorbed into local venous plexus, with the potential for forming and propagating clots. The vascular surgeon has been asked to treat a patient who developed a large floating clot in the IVC extending from the renal level up to the atrium (see Fig. 21). The clot developed after Floseal™ was injected into the lumen of a heavily bleeding epidural vein. He elected to manage the patient with full-dose low molecular weight heparin at a dose of 1 mg/kg twice daily, which in his experience is the most effective medication for large thrombus. It is possible that small emboli develop as the clot breaks up but no major functional problem developed and follow-up at 3 and 6 months angiography showed no residual thrombus. Furthermore, Floseal™ has been reported to have an association with inflammatory intraperitoneal adhesions. Fibrillar™ should be used with caution in conjunction with cell saver as thrombosis and emboli have been reported.

Another uncommon technique useful in extremis, particularly in the context of sacral plexus oozing, is for direct tamponade across a wide surface area by placement of an anterior sacral Gelfoam secured to a Richards staple and then impacted into the bleeding anterior sacrum.

Arterial Pathology

Although the veins are at most risk in the pristine abdomen, the arteries are also at risk no matter whether the approach to the anterior lumbar spine is made retroperitoneal or transperitoneal and, particularly, in the elderly patient. The most significantly affected arteries are the iliac arteries and the aorta. The two most common arterial problems are either related to hemorrhage from cuts, tears, and avulsions or related to obstruction. Neither problem is common in high-volume anterior lumbar practice where protocols anticipate significant pathology.

Bleeding events that we have encountered more often relate to the retraction forces required to displace the arteries. If the local segmental vessels or the mid sacral vessels are not appropriately located and ligated/clipped and divided prophylactically, an avulsion injury is possible and will require fine suturing to control.

The tension affecting the vessels can, on occasion, result in a split at the aortic bifurcation as the iliac arteries inferior to this are distracted. This is not able to be satisfactorily repaired until the tension is released, so it is best to pack and pressurize the spot and get on and do the spinal repair and come back and resolve this problem once the spinal repair is complete.

In a similar fashion, we have seen splits and tears develop in the iliac arteries at the site of pins or local retraction. Again, careful suturing is necessary especially in these smaller vessels where technical mistakes in the repair are likely to result in stenosis or occlusion. This is not an area to be controlled with hemoclips; attempts at

control with hemoclips have often required endovascular rescue for dissection and marked stenosis.

To check suspected postoperative problems a 3D CT angiogram is required and has to be assessed in multiple rotations. Careful assessment of the study is required as "normal" reports from radiology providers have been on occasion incorrect and proven to be high-grade postoperative stenosis (see Fig. 22).

The first decision in the presence of significant arterial disease is whether to proceed retroperitoneally or via a direct transperitoneal approach.

In general, the approach is simplest from the retroperitoneal aspect. As previously stated, there is minimal gastrointestinal involvement, a limited incision for the patient and the approach to L5,S1, and L4,5 is straightforward, and L3,4 can often be managed. This approach is associated with higher risk if using the ipsilateral side for revision surgery, discussed in detail in our earlier section of venous pathology. We, therefore, opt for a transperitoneal approach or, in some cases, a contralateral left retroperitoneal access (if a right-sided retroperitoneal was the previous access) in these cases. Getting to L3-L4 and above is managed by developing the space between the aorta and the IVC and, when there is extensive calcium plaques above, significant stress is built up in the process of retraction (see Fig. 23). We try and use areas relatively devoid of calcium for the main sites of displacement.

The risk of arterial thrombosis is related to arterial trauma, usually precipitated by dissection. This can occur in minimally diseased vessels but is much more likely when there is a significant degree of atherosclerosis with thick unstable areas of plaque. This is particularly so if there are areas of solid calcification (see Fig. 23). The degree of calcification will influence the risk and it is sensible to involve a vascular surgeon for the approach if the calcification is significant. We use CT 3D angiography with maximum intensity projection to provide the best assessment of the calcium effect.

Atheroma can remain undisclosed until the artery is compressed or retracted and, as such, preoperative identification provides additional

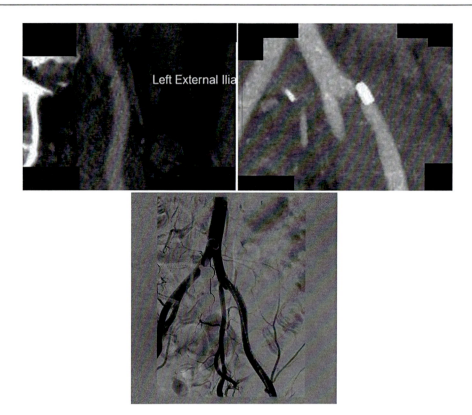

Fig. 22 Sutured tear assessed as satisfactory by radiology postoperatively but likely not assessed adequately in 3D. The problem was resolved with endovascular stent

Fig. 23 Different models of injury from arterial handling

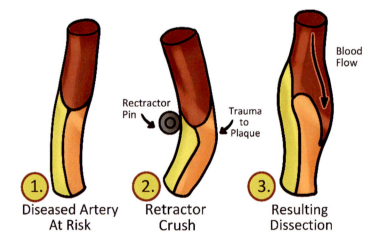

safety. While younger patients are at little risk of abdominal atheromatous disease, patients who are aged above 50 years and those who are smokers and diabetics and have other peripheral vascular disease risk factors such as hypertension or hyperlipidemia have an increasingly significant risk of involvement of the iliac artery and aorta. This includes solid plaques, ulcers, and unstable material. With extreme degrees of calcific atheroma now showing in the population, it is prudent to order a preoperative angiogram in any one aged over 50 or with multiple risk factors, particularly if multilevel surgery is required. If an inadvertent intraoperative thrombosis is diagnosed intraoperatively, it can be treated on table with thrombectomy and stenting (see Figs. 24, 25, and 26).

In some patients, this may require the decision not to proceed with an anterior approach but rather consider alternatives such as lateral or posterior approaches, accepting their relative limitations.

Fig. 24 Occluded iliac artery after retraction

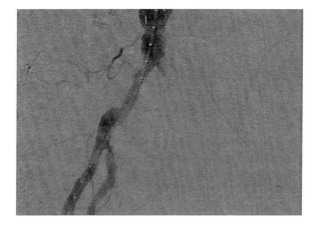

Fig. 25 Kissing stents to resolve occlusion

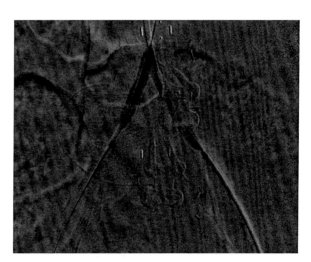

Fig. 26 Restored lumen and flow

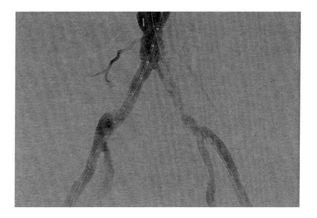

The full extent of their limitations and disadvantages are not discussed in detail here but include the inability to fully correct alignment and lordosis, higher risk of neurological injury, higher risk of inadequate disc clearance, and risk of displacement of the prosthesis laterally, posteriorly, or anteriorly. Recognition of such difficult situations may benefit from referral to a high-volume center of excellence for anterior, transperitoneal reconstruction. As discussed below, we have used combined synchronous spinal and vascular surgery to do open vascular and anterior column reconstruction in selected patients. Not only does this provide the opportunity to treat high-risk vascular lesions but also facilitates optimal spinal reconstruction, fusion, replacement, or hybrid procedures from an anterior approach, and we have subsequently published our experience and outcomes with this technique (Scott-Young et al. 2018b).

Another way to deal with lesser disease is to presume a significant risk of dissection and prepare the patient for an endovascular procedure to accompany the main spinal operation as required. When concerned by the vascular appearance perioperatively, a guidewire can be placed via the common femoral artery and placed to the diaphragm on the side considered to be at highest risk. This is usually on the left but depends on the planned approach and the atheroma distribution in the individual patient.

Occlusion occurring by dissection happens at the site of arterial compression and distortion (see Fig. 23). It is possible to check whether there is a high-risk atheroma by carefully palpating the artery before applying the high-grade force needed to manipulate the vessels away from the target intervertebral space.

In addition to dissection and progression to local occlusion, embolization of loose cholesterol or thrombus can occur. This will usually manifest as distal limb ischemia seen as pallor and coolness but often with patchy cyanosis. The emboli can be multiple, small, and fragmented and pulses may be normal at the ankle. There is often little that can be done with this material after the event but surgical intervention, either endovascular or open, may be required to prevent further progression which may be associated with progressive patient decline, possibly culminating with death if enough tissue is compromised.

The frequency of encountering this problem in the increasing number of senior patients with spinal disease has forced us to look at prevention rather than wait until confronted by a white leg with no femoral pulse and severe pain in the postoperative period. When operating on patients with arterial lesions it is critical to have balloons, stents, and stent grafts available to address any complication as soon as it is recognized (see Figs. 24, 25, and 26). This offers much better outcomes than the reported series, where serious limb arterial ischemia from dissection and occlusion has gone undiscovered for days after the index operation. Despite meticulous surgery and preoperative assessment, total unexpected occlusion of a main vessel can occur during surgery in some patients who have minimal vascular disease.

If a previously unrecognized vascular lesion is appreciated intraoperatively, it is advisable to assess further with an on-table angiogram under image intensifier. In this case, an arterial puncture is carried out using duplex ultrasound to minimize risk. A wire and diagnostic catheter can be placed in the aorta to allow general assessment with bilateral obliques as single plane studies may give an inaccurate impression. If the disease is severe, it is sensible to leave the 4F access sheath after the catheter has been withdrawn to allow a further angiogram after the formal spinal surgery has been completed. If the wire is left in situ, it also can serve as a railroad for the deployment of angioplasty balloons and stents if required to resolve dissection and occlusion at a later time. Removal of the wire at completion is a simple matter if intervention has not been required. If a sheath has been required, the site is managed in the usual way. Local pressure for a 4F sheath is used for diagnostics. If a larger sheath has been utilized for intervention, e.g., 6F or 7F, a StarClose™ device is used. Early in our experience with the very calcified vessels we decided to use guidewire placement routinely before opening the abdomen giving a railroad to fix dissection should it occur. With experience we have found we can manage these complex vessels with

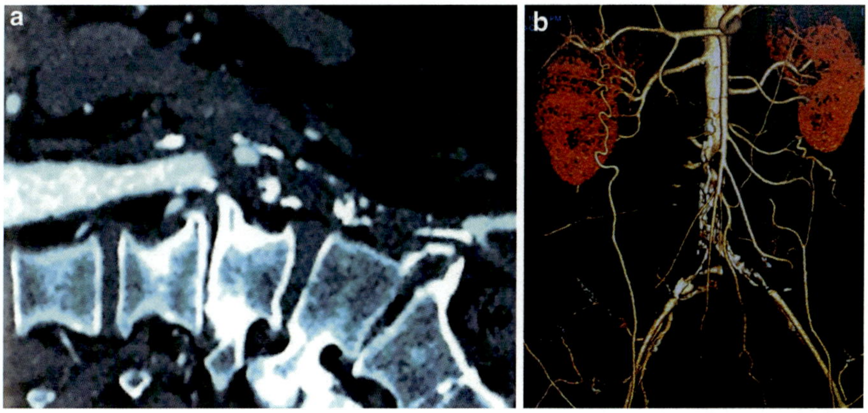

Fig. 27 CT Angiogram examples of an occluded aorta and iliac arteries in sagittal CT and 3D reconstruction modes

limited risk so the use of a pre-emptive wire has been discontinued and we treat if a problem is identified with blood flow obstruction.

Patients with substantial problems of aortic or iliac aneurysms at critical levels or functional occlusion require treatment as a priority above the spinal problem so this needs to be resolved (Fig. 27a, b). From our experience, the anterior approach is significantly superior to other methods we have undertaken, where possible, to proceed with combined synchronous surgical treatment of both problems at the same operation as detailed below (Scott-Young et al. 2018b).

The Rationale for Synchronous Treatment of Concurrent Spinal and Vascular Lesions with the Anterior Approach

When patients with combined pathology were identified, we as spinal and vascular surgeons faced problems from different aspects prior to developing a combined synchronous strategy. For example, the patient might initially present predominantly as a vascular problem with background symptoms related to spinal disease or as a primarily spinal problem with significant vascular disease. In the common case of claudication symptoms, it could be unclear whether it was mainly as a result of spinal pathology, vascular ischemia, or both processes.

Some of these patients had critical vascular problems and should we have had to make a decision for an individual operation, it would have been the vascular surgery which would have "overridden" the spine procedure in importance, in an era where endovascular treatment of AAA was increasingly common. Generally, because of the requirement for a reconstruction with either a complicated stent for the large aortic aneurysms or iliac stenosis or a full aortoiliac or aortofemoral bypass, there would have been great difficulty and significantly increased hazard in attempting an anterior spinal operation subsequently, due to the presence of the reconstruction and significant adhesions. This would leave the patient with residual symptoms from untreated spine pathology.

The other situation was that some patients, with major spinal problems indicating treatment in their own right, were found to also have moderate severity vascular pathology that warranted treatment for prevention of a critical event and the vascular treatment in a staged fashion would make future anterior spinal surgery unsafe. This subgroup of patients were challenging either because of moderate aneurysmal change (4 cm or greater) or extensive atheroma. In these situations, there is an extreme risk of dissection, occlusion, and emboli even in the primary anterior spinal approach setting.

As such, we have developed a multidisciplinary planning and review approach and this is necessary to inform all options for treatment to the patient, including our offering of an ability to control or repair the vascular disease along with lumbar spine reconstruction under the same anesthetic (Scott-Young et al. 2018b). Considering the need for major spinal reconstruction in someone also affected by moderate to severe vascular disease, we routinely explain the risks of a combined procedure, but that it is feasible to have the benefits of anterior spinal reconstruction if we were to resolve the vascular problem. Open anterior surgery was the best method for the spinal reconstruction notwithstanding the presence of a surgical vascular lesion and, given that vascular reconstruction was possible and indicated using the same exposure, this could potentiate treatment of the vascular lesions as well as excellent access for the spinal reconstruction. This approach also meant less trauma than if the surgeries were carried out independently of each other. We considered that overall there would be less physiological stress to the patient and less overall risk (i.e., from approach-related vessel injuries and adhesions) to the patient if we managed the problems in this way and also it would allow for repair in a group who would otherwise be excluded from the best option for reconstruction of their spine. It should be emphasized, however, that this approach is only undertaken after significant planning and discussion for suitable selected patients with combined disease.

In patients who we have assessed as moderate risk we prepare them for possible intervention at the time of surgery. This may require preparation for an open endarterectomy/thrombectomy (Figs. 28 and 29) if an occlusion occurs after the vessels are forcibly displaced or an angioplasty and stent in some of the higher-risk patients. The vascular surgeon has opted to place a guidewire at the commencement of the

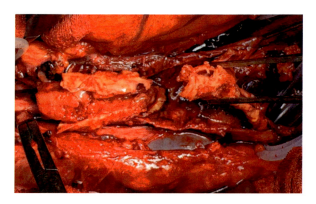

Fig. 28 Endarterectomy

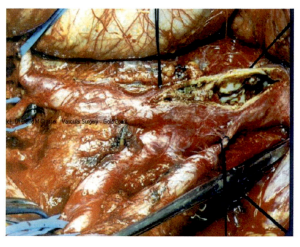

Fig. 29 Aortic iliac atheroma undergoing endarterectomy and thrombectomy

procedure so that we have immediate direct access into the artery.

As the patients required open access to the spine the decision on whether to use open repair for the aneurysms rather than endovascular depended on the fact that this was not more risky than endovascular procedures and, particularly, that in this younger group of patients, open reconstruction would be expected to give a long-lasting and durable result (in my own experience over 30 years) without the continued concern of "endoleak." "Endoleak" can occur in about 10% of endovascular AAA repairs and can present years after the intervention. Some of the vascular lesions we are presented with are extensive and circumferential or of the morphology that would make endovascular treatment difficult (see Figs. 30 and 31). The open approach would also be expected to reduce repeated exposure to CT irradiation and future risk. The same logic applies to management of the extensively diseased occluded aortas with iliac blockages also, where even with covered stent grafts, occlusive complications have been found to occur up to 10–15% with extended follow-up. The longevity with respect to the AAA reconstruction, if the patient survives 30 days postoperatively, is returned to normal.

Lesser vascular conditions require cover for the potential problems likely to occur at the time of surgery. We do this by having a vascular surgeon involved with the planning and risk stratification of selected cases and a plan for resolution if required. Be it an endovascular balloon and stenting of a stenosis or an occlusion associated with a dissection from ruptured plaque or an open endarterectomy or arterial thrombectomy, a vascular surgeon should be available for these contingencies. We are also

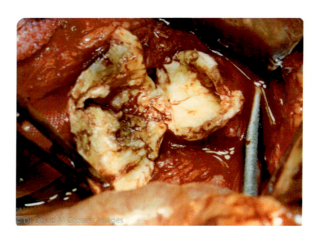

Fig. 30 Extensive circumferential atheroma

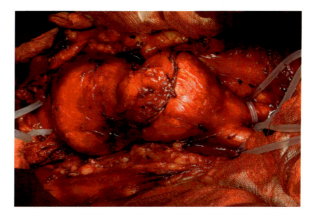

Fig. 31 Large saccular AAA prepared for reconstruction

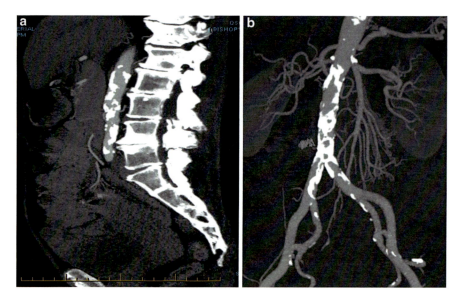

Fig. 32 The patient in this case was carefully approached for a major spine reconstruction but despite the care taken the plaque fractured and two iliacs occluded. This was corrected immediately by bilateral iliac endarterectomy and had no appreciable effect on recovery

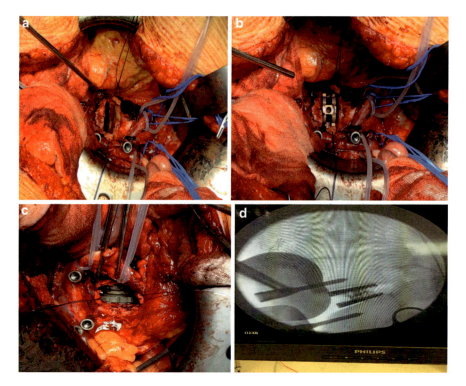

Fig. 33 Discectomy (**a**) and implantation (**b–d**) around open repair of aneurysm

prepared to perform an aorto-bi-Iliac or aorto-bi-femoral bypass with Dacron Gelsoft™ graft should a major problem occur (Figs. 32, 33, and 34).

As stated previously, when we are very concerned preoperatively, the vascular surgeon will place an endovascular wire or sheath via the femoral artery up to the diaphragm so that we have

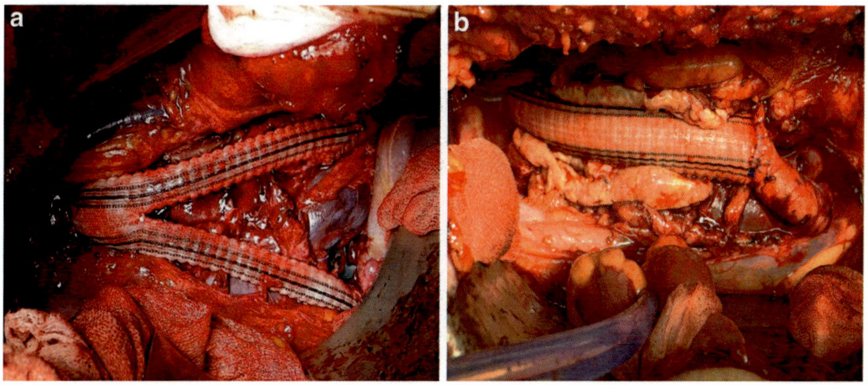

Fig. 34 Examples of aortoiliac treated openly via transperitoneal approach (**a–b**) combined with spinal reconstruction

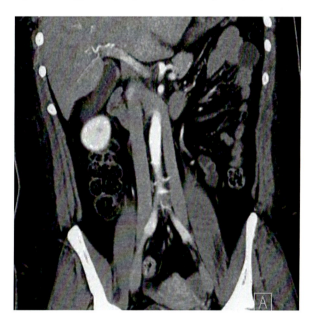

Fig. 35 Left IVC/duplication – if it is a duplication, it can be divided if the common iliac bridges to the IVC on the right. If it is the sole or main system, it will need to be protected in the same manner as a normal IVC

a direct line for repair. It is just as important for venous control when approaching around the bifurcation at the IVC in redo surgery as there is a significant risk of tears, damage, or compression, which could result in massive bleeding or thrombosis. We have managed these with covered stent grafts when it was not possible to resolve the problem intraoperatively.

Vascular and Renal Anomalies and Variations

Duplication or left inferior vena cava with or without left iliac connection and retro-aortic vein are well described:

- If an angiogram has not been carried out, you might be surprised by a dual IVC, with the left side arising in the same area as the iliolumbar, or by a retro-aortic renal vein.
- Duplication or left IVC (see Fig. 35) can cause problems if not recognized and a standard retroperitoneal approach is used. The bifurcation is such that the best approach to L4-L5 is to the left of the aorta and superior to the left iliac artery. Before ligating this vessel make sure that the right IVC exists or you may create a problem equivalent to an acute vena caval obstruction (Fig. 36).
- A retro-aortic renal vein can cross under the aorta in the L2 or even L2/3 region (Fig. 37a, b) making sideward traction hazardous if not

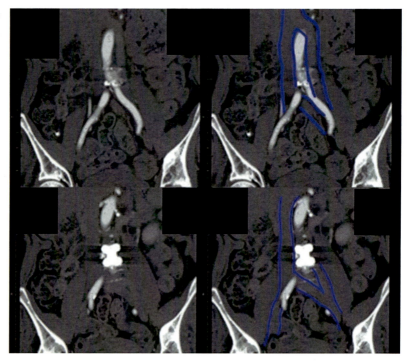

Fig. 36 Duplex IVC with a good left iliac vein; in this case the right IVC can be divided as long as adequate flow in left IVC. Not anticipated as problematic as discs requiring access are at L5-S1 and L3-L4

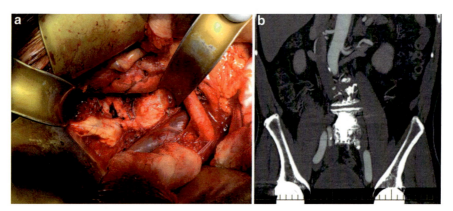

Fig. 37 Clinical (**a**) and coronal CT (**b**) images of a retro-aortic renal vein

recognized and anticipated. Consideration of whether to abandon planned treatment of a higher lumbar level needs to be given if it cannot be safely mobilized.

May-Thurner Effect: Severe Stenosis at Upper Common Iliac Vein Beneath Right Artery

Also known as iliocaval compression syndrome, the left common iliac artery compresses the left iliac vein against the L5 vertebral body. If this affected upper segment of the common iliac is traumatized requiring sutures or transverse clips, further compromise of the vein can occur. External pressure behind the vein after surgery that places plate and screws added to the anterior face of the spine can also limit the vein lumen by additional pressure. This can be resolved with stenting (Figs. 38 and 39).

Occlusion can also occur in this area secondary to emboli from the leg being trapped by the May–Thurner effect. The effect at revision

Fig. 38 Post-op external compression of iliac vein

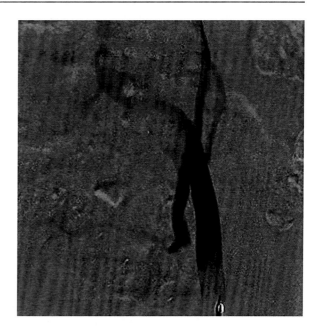

Fig. 39 Normal appearance of left iliac vein after stenting

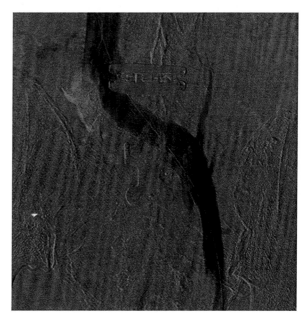

surgery is a pelvis widely congested by large collateral veins that have developed to divert the venous flow (Figs. 40 and 41). This will make the area between the iliacs much more difficult to dissect and quite hazardous with coils of highly pressurized fragile vessels.

Untreated, the outcome can be chronic stenosis with distal congestion, a swollen leg clinically and the development of large pelvic collaterals (Figs. 41 and 42). A complete occlusion may result with the prospect of permanent stasis effects (Fig. 43).

If recognized, it would be sensible to consider recanalizing and stenting the main iliac venous system prior to any planned anterior spinal surgery as this would resolve the pressure problems and reduce the vascular compromise risk (Fig. 44a, b). The diagnosis might require a 3D CT venous angiogram to adequately assess the venous problem (Figs. 40 and 41).

Fig. 40 3D venous angiogram demonstrates massive pelvic collateral shunts after left iliac thrombosis

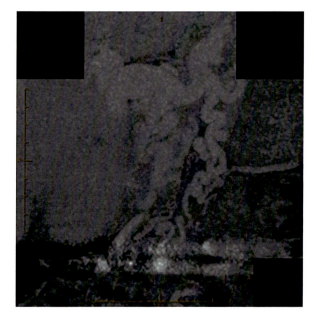

Fig. 41 Massive pelvic venous collaterals demonstrated on CT angiogram

Portal Vein Occlusion with Hypertrophy and Portal Hypertension

Other pathologies that cause significant intra-abdominal venous congestion include cirrhosis and portal vein occlusion and rare problems such as large portal vein aneurysms. These can substantially increase the hazard of the transperitoneal approach. If portal venous hypertension is recognized preoperatively it may be more appropriate to re-approach retroperitoneally in the revision setting.

Renal Anomalies: Low Lying or Horseshoe Kidneys

As well as the study of vessels on preoperative imaging, care should be taken to ensure that there are no renal anomalies that could make access dangerous or difficult. Examples of these are horseshoe kidneys that can obscure anterior access to the great vessels and low lying kidneys which often have associated anomalous vessels and pedicles that could be injured in the course of a standard retroperitoneal approach

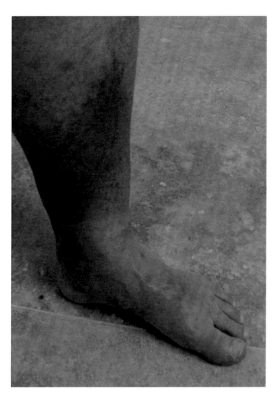

Fig. 42 Example of occluded iliac veins presenting with major leg swelling and stasis from iliac DVT 2 years after anterior spine surgery. The patient had been told nothing could be done to assist. Following treatment, the patient now has a patent vein system and normal functional leg

(see Fig. 44a, b). Preoperative consultation with a renal physician and determination of the differential function of each kidney is wise. In this case, a transperitoneal may be preferred to ensure that the renal pedicles do not undergo torsion as a result of the approach.

Postoperative Care

A team approach involving hospital medical staff and well-trained spinal nursing staff is important for prompt review of events related to anesthetic complications, hemodynamic instability, vascular events or neurological insult. Aggressive rehabilitation and specialty physiotherapy need to be involved preoperatively as well as postoperatively to guide and motivate the patients through appropriate perioperative conditioning, mobilization, and in-bed exercises to achieve optimal outcomes and restore functional motion. Judicious postoperative analgesic use is important to be run in conjunction with anesthesia, having regard from the often complex preoperative analgesia regimes these patients present with.

Ward care anticipating problems with anterior surgery is critical. Having an algorithm to manage ileus that commonly results with transperitoneal approach is helpful with regard to diet and fluid balance. For uncomplicated retroperitoneal surgery, a light diet is resumed the day following surgery. Close monitoring of the wound, patient pain control, and appropriate correction of any abnormalities in the patient's laboratory profile help to avoid complications. We aim to remove all drains and tubing (e.g., Painpump™, surgical drains) as soon as possible and complete only a short course of prophylactic intravenous antibiotics. An incentive spirometer is routinely provided as the pain from abdominal surgery can lead to splinting and atelectasis.

Regarding DVT prevention, we utilize low molecular weight heparin, mechanical calf measures and encourage high patient movement as an inpatient with outpatient aspirin and exercise prescribed until follow-up. 100 mg aspirin is prescribed as an outpatient until the first postoperative review.

We then monitor the patient routinely from a spinal point of view for follow-up then with close follow-up with a vascular surgeon for high-risk vascular patients. This includes not only radiological assessment to verify the efficacy of the correction performed but also clinical (outcome scores) and neurological (EMG) follow-up.

Conclusion

Anterior surgery for reconstruction of spine and related disorders is a vital approach to have in a surgeon's armamentarium. A variety of disorders such as degenerative, deformity, trauma, infection, and tumor can be best treated with the anterior exposure of the spine. The approach requires experience to achieve proficiency on the volume-performance curve and, in general, it is wise to have a close working relationship with a vascular surgeon.

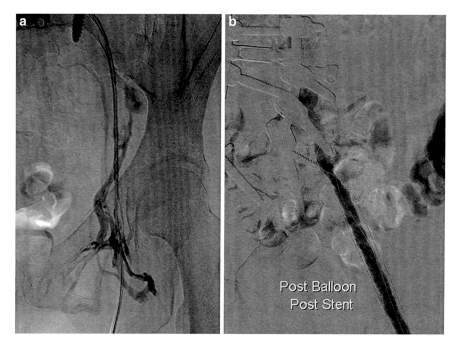

Fig. 43 (a) Left iliac vein occlusion. (b) Same case after recanalization and stent. Remains patent since, over 36 months

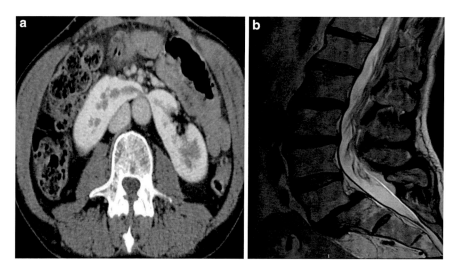

Fig. 44 Renal anomalies – (a) Horseshoe kidney overlying the great vessels (b) intrapelvic kidney lying in proximity to L5-S1 on a pre-op MRI of a patient to undergo correction of L5-S1 spondylolisthesis

References

Barrey C et al. (2011) Sagittal balance disorders in severe degenerative spine. Can we identify the compensatory mechanisms? Eur Spine J 20(Suppl 5):626–633

Chiriano J et al (2009) The role of the vascular surgeon in anterior retroperitoneal spine exposure: preservation of open surgical training. J Vasc Surg 50(1):148–151

Cunningham BW et al (2002) The use of interbody cage devices for spinal deformity: a biomechanical perspective. Clin Orthop Relat Res 394:73–83

Fraser RD (1982) A wide muscle-splitting approach to the lumbosacral spine. J Bone Joint Surg 64:44–46

Fritzell P, Hägg O, Wessberg P, Nordwall A, Swedish Lumbar Spine Study Group (2001) 2001 Volvo Award Winner in clinical studies: lumbar fusion versus nonsurgical treatment for chronic low back pain: a multicenter randomized controlled trial from the Swedish Lumbar Spine Study Group. Spine (Phila Pa 1976) 26(23):2521–2532. Discussion 2532–2534

Gao SG et al (2011) Biomechanical comparison of lumbar total disc arthroplasty, discectomy, and fusion: effect on adjacent-level disc pressure and facet joint force. J Neurosurg Spine 15(5):507–514. https://doi.org/10.3171/2011.6.SPINE11250

Gerber M et al (2006) Biomechanical assessment of anterior lumbar interbody fusion with an anterior lumbosacral fixation screw-plate: comparison to stand-alone anterior lumbar interbody fusion and anterior lumbar interbody fusion with pedicle screws in an unstable human cadaver model. Spine (Phila Pa 1976) 31(7):762–768

Hamdan AD et al (2008) Vascular injury during anterior exposure of the spine. J Vasc Surg 48(3):650–654. https://doi.org/10.1016/j.jvs.2008.04.028

Holt RT et al (2003) The efficacy of anterior spine exposure by an orthopedic surgeon. J Spinal Disord Tech 16 (5):477–486

Hsieh PC et al (2007) Anterior lumbar interbody fusion in comparison with transforaminal lumbarinterbody fusion: implications for the restoration of foraminal height, local disc angle, lumbar lordosis, and sagittal balance. J Neurosurg Spine 7(4):379–386

Ito H et al (1934) A new radical operation for Pott's disease. J Bone Joint Surg 16B:499–515

Jost B et al (1998) Compressive strength of interbody cages in the lumbar spine: the effect of cage shape, posterior instrumentation and bone density. Eur Spine J 7:132–141

Kim CH et al (2017) A change in lumbar sagittal alignment after single-level anterior lumbar interbody fusion for lumbar degenerative spondylolisthesis with normal sagittal balance. Clin Spine Surg 30(7):291–296

Labelle H et al (2005) The importance of spino-pelvic balance in L5-s1 developmental spondylolisthesis: a review of pertinent radiologic measurements. Spine (Phila Pa 1976) 30(6 Suppl):S27–S34

Lee Y, Zotti M, Osti O (2016) Operative management of lumbar degenerative disc disease. Asian Spine J 10(4):80

Modic MT, Steinberg PM, Ross JS et al (1988) Degenerative disk disease: assessment of changes in vertebral body marrow with MR imaging. Radiology 166(1 Pt 1):193–199

Panjabi MM (1988) Biomechanical evaluation of spinal fixation devices. I. A conceptual framework. Spine 13:1129–1134

Parvizi J et al (2017) Prevention of periprosthetic joint infection: new guidelines. Bone Joint J 99-B(4 Supple B):3–10. https://doi.org/10.1302/0301-620X.99B4.BJJ-2016-1212.R1

Pradhan BB et al (2002) Single-level lumbar spine fusion: a comparison of anterior and posterior approaches. J Spinal Disord Tech 15(5):355–361

Puvanesarajah V et al (2017) Morbid obesity and lumbar fusion in patients older than 65 years: complications, readmissions, costs, and length of stay. Spine (Phila Pa 1976) 42(2):122–127. https://doi.org/10.1097/BRS.0000000000001692

Regan JJ et al (2006) Evaluation of surgical volume and the early experience with lumbar total disc replacement as part of the investigational device exemption study of the Charité Artificial Disc. Spine (Phila Pa 1976) 31 (19):2270–2276

Roussouly P et al (2005) Classification of the normal variation in the sagittal alignment of the human lumbar spine and pelvis in the standing position. Spine (Phila Pa 1976) 30(3):346–353

Roussouly P et al (2011) Biomechanical analysis of the spino-pelvic organization and adaptation in pathology. Eur Spine J 20(Suppl 5):609–618. https://doi.org/10.1007/s00586-011-1928-x

Ruey-Mo L et al (2008) Mini-open anterior spine surgery for anterior lumbar diseases. Eur Spine J 17 (5):691–697

Sachinder H et al (2011) Iatrogenic arterial injuries of spine and orthopedic operations. J Vasc Surg 53(2):407–413

Scott-Young M et al (2018a) Concurrent use of lumbar total disc arthroplasty and anterior lumbar interbody fusion: the lumbar hybrid procedure for the treatment of multilevel symptomatic degenerative disc disease: a prospective study. Spine (Phila Pa 1976) 43(2):E75–E81

Scott-Young M, McEntee L et al (2018b, In press) Combined Aorto-iliac and anterior lumbar spine reconstruction: a case series. Int J Spine Surg 12:328–336

Sehgal N (2000) Internal disc disruption and low back pain. Pain Physician J 3(2):143–157

Steffen T, Gagliardi P, Hurrynag P (1998) Optimal implant-endplate interface for interbody spinal fusion devices. Meeting of the European Spine Society, 23–27 June 1998, Innsbruck, p 4

Watkins RG III, Watkins RG IV (2015) Surgical approaches to the spine. Springer, New York

Watkins RG et al (2014) Sagittal alignment after lumbar interbody fusion: comparing anterior, lateral and trans-foraminal approaches. J Spinal Disord Tech 27 (5):253–256. https://doi.org/10.1097/BSD.0b013e31828a8447

Part VII

Challenges and Lessons from Commercializing Products

Approved Products in the USA: AxiaLIF

65

Franziska Anna Schmidt, Raj Nangunoori, Taylor Wong, Sertac Kirnaz, and Roger Härtl

Contents

The Parabolic Phenomenon of a Surgical Technique 1212

AxiaLIF as an Example for the Rise and Fall of an Approved Product
in the USA 1212

Conclusions 1215

References 1215

Abstract

Axial lumbar interbody fusion (AxiaLIF) was a device used to treat instability and disc degeneration in L4-L5 and L5-S1. Through a para-coccygeal incision and presacral approach, muscular and ligamentous dissection could be avoided. The discectomy and instrumentation was performed through via trans-sacral rod. AxiaLIF was approved in 2004 through a 510 (k) clearance from a predicate device which was an anterior thoracolumbar plate for trauma and deemed a class 2 device with a moderate risk to the patient. While early studies had encouraging results in terms of fusion rates and improvement in clinical outcome measures, most also had a conflict of interest. As the technology was rapidly adopted, the first reports of complications which were visceral injuries emerged casting doubt onto its effectiveness. Only a handful of studies focused on the long-term fusion rate, restoration of lordosis, and indirect decompression. In retrospect, it is apparent that the 510 (k) clearance and classification of the device was incorrect because the predicate was not substantially equivalent to the existing device. The experience with AxiaLIF provides a cautionary tale about new technology – that it should be safe, clinically effective, and have long-term data prior to rapid adoption.

Keywords

AxiaLIF · Para-coccygeal approach · 510 (k) clearance · Class 2 · Fusion rates · Restoration of lordosis · Indirect

F. A. Schmidt (✉) · R. Nangunoori · T. Wong · S. Kirnaz
Department of Neurological Surgery, Weill Cornell Brain and Spine Center, Weill Cornell Medicine, New York, NY, USA
e-mail: Schmidt_Franziska@gmx.de; rkn587@gmail.com; taw4001@med.cornell.edu; sertackirnaz@gmail.com

R. Härtl
Department of Neurological Surgery, Weill Cornell Brain and Spine Center, New York–Presbyterian Hospital, Weill Cornell Medicine, New York, NY, USA
e-mail: roger@hartlmd.net

© Springer Nature Switzerland AG 2021
B. C. Cheng (ed.), *Handbook of Spine Technology*,
https://doi.org/10.1007/978-3-319-44424-6_43

decompression · Axial lumbar interbody fusion

The Parabolic Phenomenon of a Surgical Technique

Innovation is an essential part of advancing patient care (Riskin et al. 2006; Scott 2001). All innovations at some point in time are new; the challenge is to differentiate what is safe from unsafe. The parabolic phenomenon of technological advancement was characterized by Scott et al. and starts with widespread enthusiasm and media coverage on the basis of early case studies. This provides confidence and leads to the rapid adoption of new technology as the next standard of care. As negative reports emerge, the initial enthusiasm disappears as the new device is used in gradually limited circumstances and ultimately, falls into disuse. According to this phenomenon, as complications emerge, surgeons must react quickly to warrant patient safety (Hamilton et al. 2012). In conclusion, the challenge is to introduce new devices and technologies in a way that would flatten this parabolic phenomenon and identify unsafe devices early on with no or minimal exposure to patients.

AxiaLIF as an Example for the Rise and Fall of an Approved Product in the USA

Axial lumbar interbody fusion (AxiaLIF) (Fig. 1) was FDA-approved in 2004 through a 510 (k) clearance (Rapp et al. 2011). The predicate was the K-Centrum Anterior Spinal Fixation System, which was an anterior thoracolumbar plate and screw system for trauma and degenerative spinal disease and deemed substantially equivalent.

In contrast to the predicate system, however, AxiaLIF was a novel para-coccygeal approach to the L5-S1 disc space for interbody fusion through a paracoccygeal incision. The purported advantages over traditional surgery included avoidance of neural retraction and access to the disc space for interbody fusion without muscular and ligamentous dissection. This device was FDA approved for the treatment of degenerative disc disease and spondylolisthesis at levels L4-L5 and L5-S1 (Marotta et al. 2006). Importantly, this approach was seen by early adopters as an attractive option for fusion surgery in patients who had contraindications for traditional ALIF surgery at L4-5 and L5-S1. In addition, AxiaLIF was the first true minimally invasive option (MIS) for the L4-5 and L5-S1 levels as MIS approaches gained popularity with techniques such as lateral lumbar interbody fusion (LLIF). The advent of LLIF was in parallel with MIS approaches to spinal deformity correction; however, direct lateral access to L5-S1 had not yet been described. As a result, the presacral approach from AxiaLIF was an attractive alternative (Anand et al. 2008, 2010, 2014a, b; Boachie-Adjei et al. 2013). In retrospect, it is clear that the AxiaLIF had little in common with its predicate – a thoracolumbar anterior plate system. As described in Scott's parabolic model (Hamilton et al. 2012), the first studies were favorable – but most of them were retrospectively designed and included technical publications and retrospective case series touting the safety and efficacy of the approach (Fig. 2). The case series were promising because of the surgical corridor which obviated the need for an access surgeon, lack of critical structures in the presacral space, decreased operative time, and blood loss (Marotta et al. 2006; Aryan et al. 2008). This initial enthusiasm, driven by the results of several case studies, the presumed advantages, and the possibility for an MIS approach for fusion at the caudal lumbar levels in patients whom traditional approaches were contraindicated led to adoption of the technique by many surgeons. In 2010, the first case report describing a complication was published on a patient who suffered from a rectal injury during an AxiaLIF procedure at L5-S1, which led to an ileostomy (Botolin et al. 2010). In 2011, surgeons witnessed cases of failed multisegment AxiaLIF instrumentation. Revision strategies for AxiaLIF were, therefore, developed and were performed via anterior and posterior approaches (Hofstetter et al. 2011). In the same year, a group of surgeons published their experience after a 34-month follow-up with 68 patients who underwent AxiaLIF surgery. The overall complication rate was high at 26.5%, ranging from rectal perforation (2.9%), infection (5.9%) to pseudoarthrosis (8.8%). In contrast to the early case series, the authors had no conflict of interests (Lindley et al. 2011). In 2012 and 2013, numerous studies reported after AxiaLIF diminished disc height compared to the preoperative status as well as a loss of segmental lordosis. In these publications, pseudoarthrosis rates were high and mounting evidence from the spinal deformity literature was suggestive of the importance of lordosis, particularly at L4-5 and L5-S1 for sagittal balance, independent of fusion rates (Marchi et al. 2012; Hofstetter et al. 2013; Anand et al. 2014c). Due to improvement in some pain scale metrics (Oswestry Disability Index, Visual Analog Scale), the use of AxiaLIF continued despite a growing body of evidence demonstrating visceral complications, pseudoarthrosis, loss of lordosis, and technical reports describing revision strategies for failed AxiaLIF. For instance, some publications reported a decrease in back and leg pain with follow-up ranging from 2 to 6 years (Zeilstra et al. 2013; Tobler et al. 2013). Furthermore, a systematic literature review from 2015 demonstrated that AxiaLIF had a fusion rate of 93.15% (Schroeder et al. 2015). Interestingly, despite high fusion rates and improvement in pain scores, the authors of this systematic review concluded that AxiaLIF had a high complication rate (12.9%) and that the published literature on AxiaLIF was dominated by retrospective case series with many having conflicts of interest (Fig. 2). Follow-up of patients from the initial publications demonstrated that in long constructs for adult spinal deformity, AxiaLIF was not a good choice and an Anterior Lumbar Interbody Fusion (ALIF) was substantially better at L5-S1. The reported complication profile, morbidity of the approach, failure to achieve lordosis, and mounting evidence of need for revision surgery ultimately led to the demise of AxiaLIF (Anand et al. 2017, 2018) (Fig. 3).

Initially, surgeons looked at improvement in pain and fusion rates, but sagittal balance and other biomechanical metrics such as pelvic

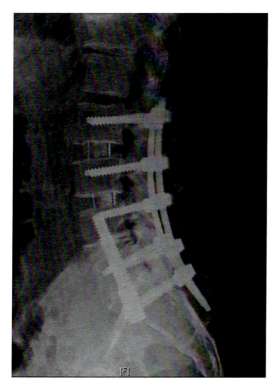

Fig. 1 X-ray showing a L2-S1 fusion with L4-S1 AxiaLIF implant

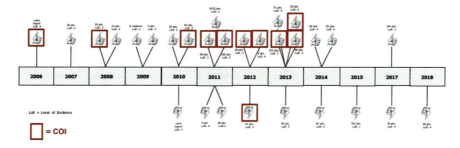

Fig. 2 Graph demonstrating the timeline of AxiaLIF publications and their overall, either positive or negative, conclusion and reported conflict of interest. The graph also illustrates the number of patients and the level of evidence

	Indirect Decompression (Volume change)	Restoration of Lordosis	Fusion rates	
2006	-	-	-	Marotta et al.
2007	-	-	-	Stippler et al.
2008	-	-	✓	Aryan et al.
	-	✓	-	Anand et al.
2009	-	-	✓	Luther et al.
	-	-	-	Erkan et al.
2010	-	✓	✓	Anand et al.
	-	-	-	Botolin et al.
	-	-	✓	Bohinski et al.
2011	-	-	-	Gundanna et al.
	-	-	✓	Tobler et al.
	-	-	-	Hofstetter et al.
	-	-	-	Lindley et al.
	-	-	✓	Gerszten et al.
2012	-	-	✓	Bradley et al.
	-	✓	✓	Marchi et al.
	-	✓	✓	Issack et al.
2013	-	✓	✓	Anand et al.
	-	-	✓	Tobler et al.
	-	-	✓	Zeilstra et al.
	-	✓	✓	Hofstetter et al.
	-	-	✓	Whang et al.
2014	-	✓	-	Anand et al.
	-	-	✓	Anand et al.
	-	✓	✓	Anand et al.
2015	-	✓	✓	Schroeder et al.
2017	-	-	✓	Zeilstra et al.
	-	✓	-	Anand et al.
2018	-	✓	✓	Anand et al.

L5/S1 Surgery Goals:
- Study without presenting data on fusion rates, restoration of lordosis, or indirect decompression.
✓ Study, which is presenting data on fusion rates, restoration of lordosis, or indirect decompression.

Fig. 3 Figure demonstrating whether AxiaLIF study presented data on fusion rates, restoration of lordosis, or indirect decompression

incidence, lumbar lordosis, sacral slope, and pelvic tilt were not primarily examined. In the early years of AxiaLIF, a series of company sponsored meetings were organized and the "Association of Presacral Spine Surgeons" was established with the idea of bringing together surgeons who were interested in this novel presacral approach. One of the authors of this present chapter (RH) remembers a meeting in 2009 when the presented cases clearly showed adequate fusion but the absence of lordosis and lack of restored disc height at L5/S1. This led to heated discussions among the contestants. Ironically, AxiaLIF emerged as a surgical technique at the same time that the influential work by Glassman et al. established the significance of lumbar lordosis to long-term patient outcomes, which began an era of investigation into global spinal alignment parameters (Glassman et al. 2005). Since 2012, and after several years of thorough follow-up, one of the initial authors of AxiaLIF (NA) utilized national seminars to repeatedly caution against using AxiaLIF at the bottom of long constructs for cases of adult spinal deformity. The limitations with the AxiaLIF technique were published by the same author (NA) in 2014 (Anand et al. 2014c). It is now understood that these global metrics may be substantially more important to patient outcomes than high fusion rates alone.

The rise and fall of an approved product in the USA such as AxiaLIF is a great example demonstrating the necessity for constant follow-ups, reflection, timely reporting, the recognition of a significant complication profile, and the failure to address sagittal alignment parameters, which were beginning to be understood as an increasingly important surgical goal, especially at L4/5 and L5/S1 (Schroeder et al. 2015). The predicate for AxiaLIF was the K-Centrum Anterior Spinal Fixation System, a device that was not equivalent to AxiaLIF in terms of the type of implant or access corridor, but received 510 (k) clearance anyways, demonstrating a loophole in the approval process. In retrospect, AxiaLIF should have been mandated as a class 3 device with an attention to safety, subsidence, indirect decompression, restoration of lordosis, and fusion rate.

Conclusions

The lesson to be learned from this failed product is that rapid adoption is not advisable unless independent groups can verify the data obtained by studies that may have conflicts of interest. For instance, Bisschop et al. described that the market approval process for all new spinal device implants should include at least one randomized controlled trial (Bisschop and van Tulder 2016). If favorable results are reported, this randomized controlled trial should then be repeated by different investigators and compared to the standard surgical device or product. In addition, Bisschop's publication also advocated for multi-institutional follow-up for at least 5 years to track long-term outcomes. In the case of AxiaLIF, one group of surgeons followed the recommendations outlined by Bisschop et al. and found that it was not an optimal device during long-term follow up with their patients, despite promising early results (Anand et al. 2008, 2010, 2013, 2014a, b, c, 2017, 2018).

The experience with AxiaLIF provides a cautionary tale about the enthusiastic adoption of new technology. New technology should be safe, clinically- and cost-effective, improve patient outcomes, and at least match the standard of care. Long-term data characterizing safety and effectiveness, without conflicts of interest, are essential. In addition, it also demonstrates that the goals of surgery may evolve and shift over time. When this occurs, the technology should undergo a thorough evaluation to determine whether it fits in the current clinical framework. In the case of AxiaLIF, our understanding of the importance of global sagittal alignment parameters and indirect decompression had significantly advanced by the time AxiaLIF was being utilized. The failure of AxiaLIF should not hamper innovative surgical techniques – but does provide lessons into the importance of long-term data and the duty of surgeons to ensure patient safety and outcomes (Bisschop and van Tulder 2016; Herndon et al. 2007) (Figs. 2 and 3).

References

Anand N, Baron EM, Thaiyananthan G, Khalsa K, Goldstein TB (2008) Minimally invasive multilevel percutaneous correction and fusion for adult lumbar degenerative scoliosis: a technique and feasibility study. J Spinal Disord Tech 21(7):459–467

Anand N, Rosemann R, Khalsa B, Baron EM (2010) Mid-term to long-term clinical and functional outcomes of minimally invasive correction and fusion for adults with scoliosis. Neurosurg Focus 28(3):E6

Anand N, Baron EM, Khandehroo B, Kahwaty S (2013) Long-term 2- to 5-year clinical and functional outcomes of minimally invasive surgery for adult scoliosis. Spine 38(18):1566–1575

Anand N, Baron EM, Khandehroo B (2014a) Does minimally invasive transsacral fixation provide anterior column support in adult scoliosis? Clin Orthop Relat Res 472(6):1769–1775

Anand N, Baron EM, Khandehroo B (2014b) Is circumferential minimally invasive surgery effective in the treatment of moderate adult idiopathic scoliosis? Clin Orthop Relat Res 472(6):1762–1768

Anand N, Baron EM, Khandehroo B (2014c) Limitations and ceiling effects with circumferential minimally invasive correction techniques for adult scoliosis: analysis of radiological outcomes over a 7-year experience. Neurosurg Focus 36(5):E14

Anand N, Cohen JE, Cohen RB, Khandehroo B, Kahwaty S, Baron E (2017) Comparison of a newer versus older protocol for circumferential minimally invasive surgical (CMIS) correction of adult spinal deformity (ASD)-evolution over a 10-year experience. Spine Deform 5(3):213–223

Anand N, Alayan A, Cohen J, Cohen R, Khandehroo B (2018) Clinical and radiologic fate of the lumbosacral junction after anterior lumbar interbody fusion versus axial lumbar interbody fusion at the bottom of a long construct in CMIS treatment of adult spinal deformity. J Am Acad Orthop Surg Glob Res Rev 2(10):e067

Aryan HE, Newman CB, Gold JJ, Acosta FL Jr, Coover C, Ames CP (2008) Percutaneous axial lumbar interbody fusion (AxiaLIF) of the L5-S1 segment: initial clinical and radiographic experience. Minim Invasive Neurosurg: MIN 51(4):225–230

Bisschop A, van Tulder MW (2016) Market approval processes for new types of spinal devices: challenges and recommendations for improvement. Eur Spine J 25(9):2993–3003

Boachie-Adjei O, Cho W, King AB (2013) Axial lumbar interbody fusion (AxiaLIF) approach for adult scoliosis. Eur Spine J 22(Suppl 2):S225–S231

Botolin S, Agudelo J, Dwyer A, Patel V, Burger E (2010) High rectal injury during trans-1 axial lumbar interbody fusion L5-S1 fixation: a case report. Spine 35(4):E144–E148

Glassman SD, Bridwell K, Dimar JR, Horton W, Berven S, Schwab F (2005) The impact of positive sagittal balance in adult spinal deformity. Spine 30(18):2024–2029

Hamilton D, Howie C, Gaston P, Simpson H (2012) Scott's parabola and the rise and fall of metal-on-metal hip replacements. BMJ: Br Med J 345:e8306

Herndon JH, Hwang R, Bozic KJ (2007) Healthcare technology and technology assessment. Eur Spine J 16(8):1293–1302

Hofstetter CP, James AR, Hartl R (2011) Revision strategies for AxiaLIF. Neurosurg Focus 31(4):E17

Hofstetter CP, Shin B, Tsiouris AJ, Elowitz E, Hartl R (2013) Radiographic and clinical outcome after 1- and 2-level transsacral axial interbody fusion: clinical article. J Neurosurg Spine 19(4):454–463

Lindley EM, McCullough MA, Burger EL, Brown CW, Patel VV (2011) Complications of axial lumbar interbody fusion. J Neurosurg Spine 15(3):273–279

Marchi L, Oliveira L, Coutinho E, Pimenta L (2012) Results and complications after 2-level axial lumbar interbody fusion with a minimum 2-year follow-up. J Neurosurg Spine 17(3):187–192

Marotta N, Cosar M, Pimenta L, Khoo LT (2006) A novel minimally invasive presacral approach and instrumentation technique for anterior L5-S1 intervertebral discectomy and fusion: technical description and case presentations. Neurosurg Focus 20(1):E9

Rapp SM, Miller LE, Block JE (2011) AxiaLIF system: minimally invasive device for presacral lumbar interbody spinal fusion. Med Devices (Auckland, NZ) 4:125–131

Riskin DJ, Longaker MT, Gertner M, Krummel TM (2006) Innovation in surgery: a historical perspective. Ann Surg 244(5):686–693

Schroeder GD, Kepler CK, Vaccaro AR (2015) Axial interbody arthrodesis of the L5-S1 segment: a systematic review of the literature. J Neurosurg Spine 23(3):314–319

Scott JW (2001) Scott's parabola: the rise and fall of a surgical technique. Br Med J 323:1477. https://doi.org/10.1136/bmj.323.7327.1477

Tobler WD, Melgar MA, Raley TJ, Anand N, Miller LE, Nasca RJ (2013) Clinical and radiographic outcomes with L4-S1 axial lumbar interbody fusion (AxiaLIF) and posterior instrumentation: a multicenter study. Med Devices (Auckland, NZ) 6:155–161

Zeilstra DJ, Miller LE, Block JE (2013) Axial lumbar interbody fusion: a 6-year single-center experience. Clin Interv Aging 8:1063–1069

Spine Products in Use Both Outside and Inside the United States

66

Tejas Karnati, Kee D. Kim, and Julius O. Ebinu

Contents

Introduction	1217
FDA Approval Process Compared to the European Process	1218
Case Studies of Failed Spinal Products in the United States	1220
Spine Products Approved Outside the United States	1223
Conclusion	1226
References	1226

Abstract

The spine product market in the United States and that of the rest of the world shares many similarities but also has significant differences. The FDA approval process of medical devices in the United States has a more stringent, often inconsistent, and prolonged pathway to final approval than when compared to the CE marking process in Europe. In fact, a large number of spinal implants have not yet been either approved or used as widely in the United States as compared to the rest of the world. There are three main spine product categories, namely lumbar artificial discs, interspinous spacers, and dynamic stabilization systems that can be identified as "new" and not as widely used in US markets. After analyzing why some of the products in these categories failed the FDA approval process, we present other unique spine products widely used in European and other international markets but not so commonly seen in US markets.

Keywords

Spine products · CE marking · FDA · Medical devices · Premarket approval · Dynesys · DIAM · Barricaid · LimiFlex · M6-L · Helifix

T. Karnati (✉) · J. O. Ebinu
Department of Neurological Surgery, University of California, Davis, Sacramento, CA, USA
e-mail: tkarnati@ucdavis.edu

K. D. Kim
Department of Neurological Surgery, UC Davis School of Medicine, Sacramento, CA, USA
e-mail: kdkim@ucdavis.edu

© Springer Nature Switzerland AG 2021
B. C. Cheng (ed.), *Handbook of Spine Technology*,
https://doi.org/10.1007/978-3-319-44424-6_54

Introduction

Spinal pathology is one of the most common health problems patients face in North America. Management of such pathology is associated with increasingly high healthcare costs (Raciborski et al. 2016). While the aging population is an important factor associated with these costs, another contributing factor includes the market approval process of spinal implants. According to a recent study in 2017, the global spinal product market is valued at over $14.4 billion USD and is projected to reach $18 billion USD by 2023 (https://idataresearch.com/product-category/spine/). The United States, with its vast medical resources and large population of spine surgeons, is undoubtedly the largest spinal product market, valued at $7.7 billion USD in 2017. In contrast, the European spinal market's total value is just over one-quarter the size of the US at $2.05 billion USD. Interestingly, China is the fastest growing market and, as of this writing, has surpassed the combined market value of 15 countries in Europe at $2.73 billion USD (https://idataresearch.com/product-category/spine/). The purpose of this chapter is to focus on the spine product market in the United States and highlight the differences with those of the European and international markets.

Given the enormous economic impact of spinal products in the healthcare systems of both the United States and Europe, it is imperative to understand the differing spinal products that are available for use in both continents. Additionally, an appreciation of the subtle differences in the approval processes of medical devices in the United States compared to Europe, and an analysis of why some products failed the US Food and Drug Administration (FDA) approval process, will provide further insight into the underlying differences between the two continents.

For the purposes of this chapter, we will focus on three specific spinal product categories: dynamic stabilization systems, artificial intervertebral discs, and interspinous and interlaminar stabilization/distraction devices. These categories were specifically chosen because of the relatively sparse usage of such devices in the US market when compared to the European and other international markets. Each of these product categories has specific examples of products that are either not widely used in the United States or not even approved at all by the FDA. In order to understand why certain products are not available in the US market whereas the same products have been widely used in Europe for many years, a thorough understanding of medical device regulations is important.

Therefore, we will first describe the FDA approval process of such spinal implants and compare and contrast it with respect to the European medical device regulatory process. We will then specifically discuss three case studies of spine products (the Dynesys® Dynamic Stabilization System, the Barricaid® Annular Closure Device, and the DIAM® Spinal Stabilization System) that failed to meet the rigors of the FDA approval process and shed light on the circumstances that led to their failure. Lastly, we discuss products that are not yet approved by the FDA in the United States but are widely used in the European and other international markets. Although there are a myriad of financial, regulatory, and clinical factors that contribute to the differing spine products in use among the US and European markets, we will attempt to highlight three case studies of spine products currently only approved in the European market: the M6®-L Lumbar Artificial Disc, the Helifix® Interspinous Spacer System, and the LimiFlex™ Spinal Stabilization System.

FDA Approval Process Compared to the European Process

From the standpoint of the FDA, medical devices are categorized into three regulatory classes (Sastry 2014). The medical devices that have the lowest risk of causing harm (i.e., thermometers, tongue depressors) are categorized as Class I medical devices. Devices that have some potential harm (i.e., powered wheel chairs, pregnancy test kits) are categorized as Class II medical devices and typically require premarket notification 510(k). Class III devices present significant risk and most require premarket approval (PMA) (Sastry 2014).

Most spinal products fall under Class II and Class III designation.

When a device manufacturer submits a 510(k) premarket notification, the manufacturer must demonstrate that the device is at minimum as safe and as effective ("substantially equivalent") to another already existing legally marketed device ("predicate device") (Rome et al. 2014). In their submission documents, the device companies must compare their device to the predicate device and demonstrate substantial equivalence. In order to meet the requirements of "substantial equivalence," the device must have the same intended use as the predicate device but not necessarily the same technological characteristics (as long as the device does not raise new questions of safety and effectiveness) (Rome et al. 2014).

The Premarket Approval (PMA) process is usually required for devices that pose a significant risk of illness or injury (i.e., Class III devices) or even devices that were found not substantially equivalent to Class I or Class II predicate devices through the 501(k) process (Lauer et al. 2017). Typically, the PMA process occurs in four steps: limited scientific review, in-depth review, advisory panel review, final deliberations and decisions for previously unapproved devices (Lauer et al. 2017). During the limited scientific review part of the process, the FDA will decide whether or not to file the submission at all; instance in which the FDA will refuse to file a submission are not uncommon. If information is incomplete, unclear, or would not stand up to scientific scrutiny, the FDA may refuse to even go forward with filing. During the in-depth review part of the process, the agency will evaluate the device's safety and effectiveness (French-Mowat and Burnett 2012). Safety and efficacy are typically demonstrated through clinical trials and scientifically validated research. The key difference of the European approval process with that of the US is the focus on "efficacy" (French-Mowat and Burnett 2012; Sorenson and Drummond 2014). As we will see later, the European approval process is predicated on safety but not necessarily "efficacy." The third step is the advisory panel review, which consists of a panel of experts independent of the FDA that hold a public meeting. After

conclusion of the meeting, the advisory committee submits a final report to the FDA. In the fourth and final step, the FDA will issue either an "approval order," an "approvable letter," a "not approvable letter," or an order denying approval (French-Mowat and Burnett 2012; Sorenson and Drummond 2014).

Often, in order to gather safety and effectiveness data to support a PMA application or a 501(k) submission, the FDA can grant an investigational device exemption (IDE) to a medical device. An IDE allows the investigational device to be used in a clinical study (Ament et al. 2017). Therefore, some of the products that we will discuss later on in the chapter are currently being used under an investigational device exemption until further data can be obtained.

In contrast to the FDA medical device approval process, the CE Mark is a certification mark that indicates conformity with health, safety, and environmental protection standards for products sold within the European Economic Area (EEA) (Mishra 2017). The key difference between the US and European approval processes lies in a device's "efficacy." In the US approach, the FDA assesses the device's effectiveness as well as its risk of harm; however, the CE mark indicates simply that the medical device satisfies certain high safety, health and environmental protection requirements (Mishra 2017). In short, the FDA approval process ensures that a device both poses no harm to consumers but also does what it claims to do. The US approach is not without its critics, who argue that this dual goal of "safety and efficacy" adds inordinate time and unpredictability to the approval process without in fact establishing the effectiveness of the device (Heneghan and Thompson 2012).

With its unpredictable, inconsistent, prolonged, and often expensive path, the FDA's approval process is widely considered more cumbersome than the CE marking process (Heneghan and Thompson 2012). Typically, there is a 1- to 3-year delay in launching new medical devices into general clinical practice in the US compared to in Europe according to a 2012 report by the Boston Consulting group after analyzing approvals from 2000 through 2011 of devices that were "the most innovative and potentially risky medical technologies" (those requiring PMA) (Kramer et al. 2012). Nevertheless, while the CE mark may be less arduous to obtain, some argue that it is a less powerful certification (Mishra 2017). Especially after the widely publicized breast implant scandal of the early 2000s, in which a French company sold silicone implants (which had CE mark approval but not approved by the FDA) that were later recalled after it was found they had been fraudulently manufactured with unapproved silicone gel, the FDA approval process is sometimes seen as safer for consumers (Lampert et al. 2012). Interestingly, the FDA approval means that the device is approved for use in many parts of the world, while the CE mark has restrictions, sometimes even within the EU. Importantly, even though a medical device has a CE marking, there is no guarantee that the device will be widely accepted by physicians or reimbursable by the government in each European country (https://www.ecnmag.com/article/2012/02/which-way-go-ce-mark-or-fda-approval).

Case Studies of Failed Spinal Products in the United States

Having established the important differences and similarities of the medical device approval process between the United States and Europe, we now turn our discussion to three specific spinal products that failed the FDA approval process: Dynesys® Dynamic Stabilization System (Zimmer Biomet), Barricaid® Annular Closure Device, and the DIAM® Spinal Stabilization System.

Before a discussion of Dynesys or any other pedicle screw-based dynamic stabilization systems can be had, the principle of dynamic stabilization must be described. Dynamic stabilization, also known as soft stabilization or flexible stabilization, involves insertion of flexible materials rather than the traditional rigid ones to allow some movement along the instrumented area of the spine (Tyagi et al. 2018). Essentially, dynamic stabilization devices place the posterior structures under tension and create a focal increase in

lordosis. This process may shift load transmission so that certain positions are more tolerable and may limit motion so that painful positions are not experienced (Gomleksiz et al. 2012). As of this writing, no dynamic stabilization devices have received approval from the FDA for use other than as an adjunct to spinal fusion. Once of these dynamic stabilization systems, the Dynesys® Dynamic Stabilization System by Zimmer Spine, was not granted a PMA initially by the FDA for standalone dynamic stabilization and then later was only approved as an adjunct to spinal fusion.

According to the FDA, the Dynesys® Dynamic Stabilization System is intended to "provide immobilization and stabilization of spinal segments as an adjunct to fusion in the treatment of and following acute and chronic instabilities or deformities and failed previous fusion" (https://www.accessdata.fda.gov/cdrh_docs/pdf6/K060638.pdf). The device consists of a titanium alloy pedicle screw and a spacer that consists of surgical polyurethane that holds the vertebrae in a more natural anatomical position; a nylon-like cord runs through the spacers and is pulled taut to limit flexion movements (see Fig. 1) (Dynesys Dynamic Stabilization System 2015).

Initially, Dynesys was granted a 501(k) clearance in March 2004 since it was considered to be "substantially equivalent" to the Silhouette® spinal fixation system. However, the Dynesys system as a standalone device for non-fusion stabilization was later recognized by the FDA as a new type of treatment, and consequently Zimmer Biomet had to apply for a PMA. On the 4th of November 2009, the PMA application for this device was rejected (https://medtech.pharmaintelligence.informa.com/MT028146/Panel-Rejects-Zimmers-Dynesys-Spine-Stabilization-Device).

The Dynesys system was rejected by the FDA despite initially being granted a 501(k) approval. Although the Dynesys system was considered to be substantially equivalent, it should have been regarded as a new dynamic stabilization technique as opposed to a new type of posterior technique (Bisschop and Van Tulder 2016). The Dynesys system thereby received its 510(k) approval long before the first and only randomized control trial was conducted (Bisschop and Van Tulder 2016).

Fig. 1 Dynesys device with titanium pedicle screws, polyurethane spacers and nylon cord the limits flexion (Dynesys Dynamic Stabilization System 2015)

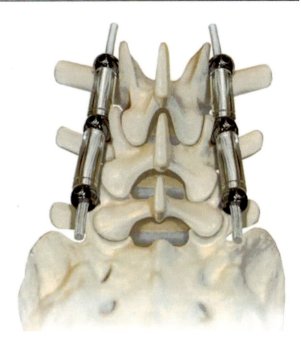

During this time period where there was little published research, the device was in clinical use in Europe since 1999 with over 40,000 patients receiving implants. Consequently, there was not enough evidence in the period between the 510(k) approval and the rejected PMA application to determine whether this device resulted in improved health outcomes compared to standard treatments (Bisschop and Van Tulder 2016). Indeed, short-term clinical results seemed favorable, while long-term complications arose: screw loosening, adjacent segment degeneration, late infection. In a 2010 study of 71 patients undergoing decompression using the Dynesys stabilization system, radiographic evidence of screw loosening occurred in 19.7% of patients (Ko et al. 2010). Of note, the radiographic findings of screw loosening failed to show any associated adverse effect on clinical improvement. The FDA reviewers also pointed to potential bias in the company's study, noting that a majority of patients were treated by researchers with a financial interest in the company. Admittedly, this bias could have been due to chance which would further question the relevance of such a bias claim (Ko et al. 2010).

Another example of a spinal product that failed the FDA's PMA process is the DIAM® Spinal Stabilization System by Medtronic. The device is an "H" shaped silicone and woven polyester device which is sandwiched between two adjacent spinous processes of the lumbar spine (see Fig. 2) (Phillips et al. 2006; DIAM™ 2019).

Once seated in the interspinous space, the device is secured by polyester cables and titanium crimps. This design helps stabilize movement in both flexion and in extension (Phillips et al. 2006). Medtronic described to the FDA that the device was intended to alleviate pain "through the reduction of stresses on the overloaded posterior disc and facet joints," while it "re-tensions the supraspinous ligament and other ligamentous structures" (https://www.fda.gov/downloads/AdvisoryCommittees/CommitteesMeetingMaterials/MedicalDevices/MedicalDevicesAdvisoryCommittee/OrthopaedicandRehabilitationDevicesPanel/UCM486374.pdf). It was first implanted in France in 1997 and has been used for more than 10 years outside the United States (Hrabálek et al. 2009). However, on February 19, 2016, the FDA's orthopedic and rehabilitation devices advisory panel unanimously recommended rejection of Medtronic's DIAM® spinal stabilization implant (https://www.fiercebiotech.com/medical-devices/fda-advisory-panel-votes-against-recommending-approval-medtronic-spine-implant).

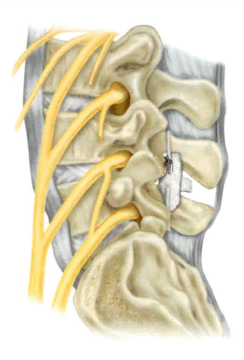

Fig. 2 DIAM® Spinal Stabilization System (DIAM™ 2019)

According to the analysis of the pivotal clinical trial, the FDA expressed significant concerns with the study design. The study population was too heterogenous; there were multiple, potential diagnostic subgroups included in clinical trial. Some of the investigational and nonoperative control subjects were subsequently treated with surgery at adjacent levels, or surgery involving more than one spinal level (https://www.fda.gov/downloads/AdvisoryCommittees/CommitteesMeetingMaterials/MedicalDevices/MedicalDevicesAdvisoryCommittee/OrthopaedicandRehabilitationDevicesPanel/UCM486692.pdf). Furthermore, the screening algorithm was also questioned. It was unclear how a symptomatic level was identified in subjects with multilevel degenerative spinal pathology, whether if subjects had subacute versus chronic degenerative spinal pathology, or if subjects were experiencing primary versus recurrent low back pain. Moreover, some patients randomized to the DIAM® group were allowed to undergo the same nonoperative treatments as the control group. Investigational and nonoperative control patients were also free to pursue non-prescription therapies such as massage and acupuncture. In addition, 60.8% of all of the nonoperative control patients crossed over to receive treatment with the DIAM® investigational device after at least 6 months of treatment lending itself to potential bias on the estimate of the treatment effect (https://www.fda.gov/downloads/AdvisoryCommittees/CommitteesMeetingMaterials/MedicalDevices/MedicalDevicesAdvisoryCommittee/OrthopaedicandRehabilitationDevicesPanel/UCM486692.pdf).

Perhaps the most controversial of the list of concerns that the FDA had were the observed radiographic spinous process erosions and fractures. In the pivotal study, a total of 44.0% (37/84) of subjects were observed to have had an erosion at either the superior or inferior spinous process (or both locations) at 36 months (Crawford et al. 2013). Superior spinous process fractures were observed by the core laboratory in 7.7% (14/181) of the DIAM® subjects at the 12 months timepoint (Crawford et al. 2013). Medtronic representatives pointed to the fact that published literature has documented that plain radiographs may lack sensitivity for the detection and diagnosis of spinous process fractures in subjects with interspinous process spacer devices (https://www.fda.gov/downloads/AdvisoryCommittees/CommitteesMeetingMaterials/MedicalDevices/MedicalDevicesAdvisoryCommittee/OrthopaedicandRehabilitationDevicesPanel/UCM486692.pdf). Despite numerous appeals by Medtronic, the DIAM® Spinal Stabilization System is currently only approved for use as an investigational device until further data can be obtained.

Finally, our discussion of case studies of FDA medical device approval failures will end on the Barricaid® annular closure device. The Barricaid® annular closure device is a permanent implant that is used after a limited lumbar discectomy is performed. The device consists of a polymeric mesh that sits in the posterior intervertebral disc space and is connected to a metallic anchor that is attached to the vertebral body and essentially blocks an opening in the anulus, thereby preventing re-herniation of the nucleus pulposus (see Fig. 3) (Parker et al. 2016; Hahn et al. 2014). The device is particularly effective in patients in whom a limited discectomy is performed with resulting large annular defects and a higher chance of re-herniation. More specifically,

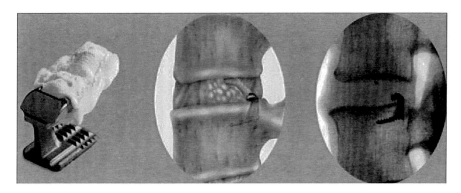

Fig. 3 A metal anchor is inserted parallel to the surface of the endplate and the polymeric mesh forms a barrier that prevents re-herniation of the intervertebral disc (Hahn et al. 2014)

it was indicated in patients with radiculopathy (with or without back pain), a posterior or posterolateral herniation, characterized by radiographic confirmation of neural compression using MRI, and a large annular defect (e.g., between 4 and 6 mm tall and between 6 and 12 mm wide) determined intraoperatively post discectomy, at one level between L4 and S1 (Hahn et al. 2014).

Barricaid's company, Intrinsic Therapeutics, initially submitted an Investigational Device Exemption (IDE) to start a clinical trial in the United States but the IDE was never approved due to safety concerns, so the company performed clinical trials outside of the United States (OUS), including a randomized clinical trial (RCT), to support initiation of a US clinical study (Ledic et al. 2015). FDA and Intrinsic therapeutics never reached consensus on OUS study design, associated protocols, or documents, but the RCT clinical data and nonclinical studies were submitted in a PMA (Ledic et al. 2015). The clinical trial included 554 randomized patients (Klassen et al. 2017). Its purpose was to determine whether a microdiscectomy with a bone-anchored annular closure device resulted in lower re-herniation and reoperation rates and increased overall patient clinical success, when compared to traditional lumbar discectomy without defect closure. The group treated with Barricaid had significantly lower rates of re-herniation (12% vs. 25%, $P < 0.001$), reoperations to address recurrent herniation (5% vs. 13%, $P = 0.001$), and index-level reoperations (9% vs. 16%, $P = 0.01$). However, one main criticism was the high rate of end-plate lesions in the Barricaid group. Eighty-eight percent of the patients receiving Barricaid had endplate changes versus 40% of control patients. The control patient's changes were smaller on average and appeared to stabilize sooner than the Barricaid patient's changes. The Barricaid endplate changes were larger and had a distinctive radiographic feature – according to the FDA's radiologist. Furthermore, some of these endplate changes were lytic in nature and were radiographically distinct from Schmorl's nodes and endplate changes. Furthermore, there were also device integrity issues in the clinical study. The study collected data on 63 implant and instrument retrievals during the study and commercial use and found that the average retrieval was 2.4 ± 1.8 years (Range = 0.1–5.8 years) after implantation (Ledic et al. 2015). The implant was removed primarily due to mesh detachment, migration, new or worsening pain, and/or instability (Ledic et al. 2015). For all these concerns, the FDA panel voted on December 2017 against approval.

Spine Products Approved Outside the United States

After having discussed those products that have failed the FDA approval process, we now turn our attention to those products that are widely used in the European market but not yet approved or widely used in the United States. There are three categories of spinal implants that are currently used widely in the European and other international markets but for a myriad of reasons are not so commonly seen in the US markets: lumbar artificial

Table 1 Lumbar artificial discs

Device name	FDA approval	Manufacturer
ProDisc®-L	Yes – single level	DePuy Synthes
INMOTION®	Yes – single level	DePuy Synthes
Activ-L™	Yes – single level	Aesculap®
Cadisc™-L	No	Rainier® Technology
FlexiCore®	No	Stryker®
Freedom® Lumbar Disc	No	AxioMed®
M6®-L	No	Spinal Kinetics™
XL TDR®	No	NuVasive
Maverick®	No	Medtronic®
Kineflex-L™	No	SpinalMotion®

Table 2 Interspinous/interlaminar devices

Device name	FDA approval	Manufacturer
Coflex® Interlaminar Stabilization Device	Yes	Paradigm Spine
Superion® Indirect Decompression System	Yes	VertiFlex, Inc.
DIAM™ Spinal Stabilization System	No (only IDE)	Medtronic Sofamor Danek
Aperius™-PercLID™ System	No	Medtronic
FLEXUS™	No	Globus Medical
Wallis® System	No (only IDE)	Zimmer Spine
In-Space	No	Synthes®

discs, interspinous devices, and dynamic stabilization systems. Although a substantial number of devices in all three categories exist and are currently manufactured, only a couple of these devices are approved for use in the United States. Although not an exhaustive list, Tables 1 and 2 list lumbar artificial discs and interspinous devices that are currently available and are used in different markets throughout the world. For the purposes of this chapter, we will discuss the M6® artificial disc systems by Spinal Kinetics, the Helifix® Interspinous Spacer System by Alphatec® Spine, and the LimiFlex Spinal Stabilization System by Empirical Spine, all of which have been in use in the European markets for many years but not yet approved for use in the United States.

Spinal Kinetics (Sunnyvale, CA) is the manufacturer of the M6 artificial disc systems and has, as of this writing, exceeded 50,000 total implantations of their M6-C Cervical and M6-L Lumbar artificial discs in international markets (https://www.odtmag.com/contents/view_ breaking-news/2017-07-25/spinal-kinetics-surpasses-50000-implants-of-its-m6-artificial-disc). M6-C was launched in 2006 and is now approved for use by the FDA in the US markets, while M6-L was launched in 2010 and is currently not FDA approved but used extensively in international markets and has the CE mark (Formica et al. 2017). The M6-L artificial disc has an artificial nucleus pulposus (made from polycarbonate urethane) and an artificial ring of fibrous material (made from polyethylene) and is designed to provide physiologic motion in 6 degrees of freedom (2 degrees in each axial, coronal, sagittal planes). As of this writing, this system is the only artificial disc with 6 degrees of freedom. The inner disc is sandwiched by two titanium outer plates with keels for anchoring the disc into the vertebral body (http://www.spinalkinetics.com/patients/ m6-l-artificial-lumbar-disc/). According the manufacturer, the M6-L disc is intended for use for treatment of symptomatic degenerative disc disease (DDD) at one or two adjacent levels between

L3 through S1 and may even have up to maximum 3 mm of spondylolisthesis at the intended level (http://www.spinalkinetics.com/patients/m6-l-artificial-lumbar-disc/). The M6-L disc is currently approved for use in the European Union countries, Australia, New Zealand, Russia, South Africa, Brazil, and United Arab Emirates.

There are a number of artificial lumbar discs that are approved for use in the United States and a wider array of discs that are not yet approved. Although it is by no means an exhaustive list, Table 1 contains a list of the currently used lumbar artificial discs in the United States and international markets (Food and Drug Administration 2012a, b, c).

In addition to lumbar artificial discs, interspinous and interspinous stabilization/distraction devices are also not as widely used in the US market as they are used in international markets. Before we discuss the various interspinous spacer devices used outside the United States, it is worth discussing the rationale behind interspinous distraction. There is no widely accepted term in spine literature, but lumbar interspinous process decompression (IPD) is also known as interspinous distraction, posterior spinal distraction, or interlaminar stabilization. These procedures have been proposed as minimally invasive alternative procedures to laminectomy and fusion (Landi 2014). By distracting the adjacent spinous processes and/or lamina and thereby restricting extension in patients with lumbar spinal stenosis and neurogenic claudication, these interspinous distractor devices can alleviate symptoms that arise from neural compression (Lee et al. 2015). These devices enlarge the neural foramen, decompress the causa equina, and act as spacers between the spinous processes to limit extension of the spinal interspace. Because of the minimally invasive nature of placement of these devices, proponents argue that such techniques lead to shorter hospital stay, preservation of local bone and tissue, reduced risk of cerebrospinal fluid leakage, and is reversible in such a way that it does not limit future treatment options. Some potential complications of these spacer devices are fracturing of spinous processes, incorrect positioning, and mechanical failure of the devices (Lee et al. 2015).

Of the devices that have been commercialized for placement between the spinous process or lamina of a motion segment in the US and international markets, only two such devices are approved for use in the United States: Coflex® Interlaminar Stabilization Device and the Superion® Indirect Decompression System. There is one interspinous spacer device not approved in the United States but widely used in the European markets that we will describe: the Helifix® Interspinous Spacer System. A list of some other interspinous devices can be found in Table 2 (Food and Drug Administration 2012a, b, c).

The Helifix® Interspinous Spacer System is essentially a PEEK (polyetheretherketone) implant that is self-distracting (Pintauro et al. 2017). It has a helical design that essentially can be "screwed in" between the spinous processes of adjacent vertebra. It has a percutaneous delivery mechanism through a posterolateral approach (https://atecspine.com/product-portfolio/thoracolumbar/helifix-interspinous-spacer-system/). After a 2–3 cm incision is made lateral to midline and, under fluoroscopy, a guidewire is inserted to find the interspinous space. Once another instrument, called the ligament splitter, dilates through the interspinous ligament, a dilator trial is positioned between the superior and inferior spinous processes (Gazzeri et al. 2014). Once proper fit is established, the Helifix device is inserted with a rotating movement of the self-distracting helical tip in the interspinous area. This device stretches the ligamentum flavum and the posterior fibers of the annulus fibrosus, thus enlarging the spinal canal (Gazzeri et al. 2014).

Lastly, the LimiFlex™ spinal stabilization system by Empirical Spine received its CE mark in 2009 and has been used to treat more than 2000 patients thus far, primarily in Europe (https://limiflex.com/healthcare-professionals/). The FDA has approved the device for investigational use and a clinical trial has been approved in patients with lumbar spinal stenosis with up to Grade I degenerative spondylolisthesis. The pivotal trial is a multicenter, prospective, concurrently controlled, non-blinded study in which patients will be randomized to receive either decompression and LimiFlex device implantation or decompression and posterolateral fusion. The LimiFlex device consist of two dynamic titanium rods each with a roller screw and a pre-attached ultrahigh molecular weight (UHMW) polyethylene textile band that straps around adjacent spinous processes. The two titanium rods sit on either side of adjacent spinous processes in a vertical position and the textile band is wrapped around the two spinous processes (LimiFlex Surgical Technique Manual: https://limiflex.com/wp-content/uploads/LB-10108.001.A-LimiFlex-STM_english-for-mail.pdf). By turning the screwing mechanisms on the roller screw part of the titanium rods, increased tension is created on the textile band that increases the angle of lordosis along two adjacent lumbar spinal levels. The pivotal study started in July 2017 in the United States and is expected to conclude in December 2023 (https://clinicaltrials.gov/ct2/show/NCT03115983).

Conclusion

The spine product market is replete with a wide variety of solutions for all of the different spinal pathologies, chief among them being back and neck symptoms. Although many spine products are concurrently used both in the United States and the wider international markets, there are quite a few spinal implants that have not yet been either approved or used as widely in the United States. There are three main spine product categories, namely lumbar artificial discs, interspinous spacers, and dynamic stabilization systems, which have a substantial selection of products that have been in use in the European market for many years. However, as a result of the FDA medical device approval process and its insistence on both safety and efficacy in evaluating spinal products, many of these devices are either still under investigation or outright rejected. In fact, an evaluation of some of the case studies of the FDA approval process of certain spine products has taught us valuable lessons. Ideally, new spinal devices should improve patient outcomes with increased safety at reasonable societal costs. As we have seen from studying the reasons behind the failure of some of the spine products, the main goal prior to dissemination of spinal implants or devices is to have a rigorous evidence-based evaluation. Only then can a successful product emerge from the FDA approval process that will ultimately both be safe for consumers and decrease healthcare costs in the long term.

At the same time, innovation must also be at the forefront of our minds, and spine surgeons in the United States must rapidly evolve and evaluate certain technologies that are widely used in the rest of the world. As we have seen from the extensive list of interspinous spacers, posterior dynamic stabilization systems, and artificial discs that are in use in the international markets, we must critically evaluate and employ the most effective devices available by conducting high-quality evidence-based studies so that our patients can have more surgical options when it comes to the vast array of spinal pathologies that they experience.

References

Ament JD, Mollan S, Greenan K, Binyamin T, Kim KD (2017) Understanding United States investigational device exemption studies-clinical relevance and importance for healthcare economics. Neurosurgery 80(6):840–846

Bisschop A, Van Tulder MW (2016) Market approval processes for new types of spinal devices: challenges and recommendations for improvement. Eur Spine J 25(9):2993–3003

Crawford RJ et al (2013) A prospective study of patient-reported outcomes for two years after lumbar surgery augmented with DIAM® Interspinous Implant. J Musculoskelet Res 15(3):2-222 to 29-235

DIAM™ Spinal Stabilization System Surgical Technique (2019) [ebook]. Medtronic, Memphis, p 12. http://www.arcos.com.uy/pdf/productos/194/437_953_diamst.pdf. Accessed 25 Mar 2019

Dynesys Dynamic Stabilization System (2015) [ebook]. Zimmer Biomet, Minneapolis, p 22. https://www.zimmerbiomet.com/content/dam/zimmer-biomet/medical-professionals/000-surgical-techniques/spine/dynesys-lis-less-invasive-surgery-surgical-technique.pdf. Accessed 25 Mar 2019

Food and Drug Administration (2012a) SUPERION® INTERSPINOUS SPACER. [cited June 12, 2018]. https://www.accessdata.fda.gov/cdrh_docs/pdf14/P140004D.pdf

Food and Drug Administration (2012b) Instructions for use: coflex® Interlaminar Technology. [cited June 12, 2018]. https://www.accessdata.fda.gov/cdrh_docs/pdf11/P110008C.pdf

Food and Drug Administration (2012c) Premarket approval application: coflex® Interlaminar Technology (P110008). [cited June 10, 2018]. https://www.accessdata.fda.gov/cdrh_docs/pdf11/P110008a.pdf

Formica M, Divano S, Cavagnaro L et al (2017) Lumbar total disc arthroplasty: outdated surgery or here to stay procedure? A systematic review of current literature. J Orthop Traumatol 18(3):197–215

French-Mowat E, Burnett J (2012) How are medical devices regulated in the European Union? J R Soc Med 105(Suppl 1):S22–S28

Gazzeri R, Galarza M, Alfieri A (2014) Controversies about interspinous process devices in the treatment of degenerative lumbar spine diseases: past, present, and future. Biomed Res Int 2014:975052

Gomleksiz C, Sasani M, Oktenoglu T, Ozer AF (2012) A short history of posterior dynamic stabilization. Adv Orthop 2012:629698

Hahn BS, Ji GY, Moon B et al (2014) Use of annular closure device (Barricaid®) for preventing lumbar disc reherniation: one-year results of three cases. Korean J Neurotrauma 10(2):119–122

Heneghan C, Thompson M (2012) Rethinking medical device regulation. J R Soc Med 105(5):186–188

Hrabálek L, Machác J, Vaverka M (2009) The DIAM spinal stabilisation system to treat degenerative disease of the lumbosacral spine. Acta Chir Orthop Traumatol Cech 76(5):417–423

Klassen PD, Bernstein DT, Köhler HP et al (2017) Bone-anchored annular closure following lumbar discectomy reduces risk of complications and reoperations within 90 days of discharge. J Pain Res 10:2047–2055

Ko CC, Tsai HW, Huang WC et al (2010) Screw loosening in the Dynesys stabilization system: radiographic evidence and effect on outcomes. Neurosurg Focus 28:E10

Kramer DB, Xu S, Kesselheim AS (2012) How does medical device regulation perform in the United States and the European union? A systematic review. PLoS Med 9(7):e1001276

Lampert FM, Schwarz M, Grabin S, Stark GB (2012) The "PIP scandal" – complications in breast implants of inferior quality: state of knowledge, official recommendations and case report. Geburtshilfe Frauenheilkd 72(3):243–246

Landi A (2014) Interspinous posterior devices: what is the real surgical indication? World J Clin Cases 2(9):402–408

Lauer M, Barker JP, Solano M, Dubin J (2017) FDA device regulation. Mo Med 114(4):283–288

Ledic D, Vukas D, Grahovac G, Barth M, Bouma GJ, Vilendecic M (2015) Effect of anular closure on disk height maintenance and reoperated recurrent herniation following lumbar discectomy: two-year data. J Neurol Surg A 76(3):211–218. https://doi.org/10.1055/s-0034-1393930

Lee SH, Seol A, Cho TY, Kim SY, Kim DJ, Lim HM (2015) A systematic review of interspinous dynamic stabilization. Clin Orthop Surg 7(3):323–329

Mishra S (2017) FDA, CE mark or something else?-Thinking fast and slow. Indian Heart J 69(1):1–5

Parker SL, Grahovac G, Vukas D et al (2016) Effect of an annular closure device (Barricaid) on same-level recurrent disk herniation and disk height loss after primary lumbar discectomy: two-year results of a multicenter prospective cohort study. Clin Spine Surg 29(10):454–460

Phillips FM, Voronov LI, Gaitanis IN, Carandang G, Havey RM, Patwardhan AG (2006) Biomechanics of posterior dynamic stabilizing device (DIAM) after facetectomy and discectomy. Spine J 6(6):714–722

Pintauro M, Duffy A, Vahedi P, Rymarczuk G, Heller J (2017) Interspinous implants: are the new implants better than the last generation? A review. Curr Rev Musculoskelet Med 10(2):189–198

Raciborski F, Gasik R, Kłak A (2016) Disorders of the spine. A major health and social problem. Reumatologia 54(4):196–200

Rome BN, Kramer DB, Kesselheim AS (2014) Approval of high-risk medical devices in the US: implications for clinical cardiology. Curr Cardiol Rep 16(6):489

Sastry A (2014) Overview of the US FDA medical device approval process. Curr Cardiol Rep 16(6):494

Sorenson C, Drummond M (2014) Improving medical device regulation: the United States and Europe in perspective. Milbank Q 92(1):114–150

Tyagi V, Strom R, Tanweer O, Frempong-boadu AK (2018) Posterior dynamic stabilization of the lumbar spine review of biomechanical and clinical studies. Bull Hosp Jt Dis (2013) 76(2):100–104

Trauma Products: Spinal Cord Injury Implants

67

Gilbert Cadena Jr., Jordan Xu, and Angie Zhang

Contents

SCI Background: Demographics, Economic Burden, and Grading Systems	1230
Introduction	1230
Demographics	1230
Economic Burden	1230
Grading Systems in SCI	1232
SCI Pathophysiology: Primary Versus Secondary Injury	1233
Primary Versus Secondary Injury	1233
Challenges for Recovery: Cervical Versus Thoracic SCI, Incomplete Versus Complete	1235
Cervical Versus Thoracic SCI	1235
Complete Versus Incomplete SCI	1236
Concepts: Functional Regeneration, Neuroprotection, Immunomodulation	1237
Results from Acute SCI Interventions: Stem Cells, Pharmacologics, Hemodynamic Strategies, CSF Drainage, Bio-Scaffolds, and Neuromodulation	1237
Stem Cell Therapies	1237
Pharmacologics	1238
Bio-Scaffolds	1241
Neuromodulation	1242
Challenges: Translational Studies, Timing of Intervention, Ethical and Economic Considerations, and Considerations in Clinical Trial Design	1244
Consistency of Results	1244
Risks	1245
Timing	1245
Ethical Challenges	1245
Costs	1246
Challenges and Considerations in Designing of Clinical Trials	1246
References	1247

G. Cadena Jr. (✉) · J. Xu · A. Zhang
Department of Neurological Surgery, University of
California Irvine, Orange, CA, USA
e-mail: gilbert.cadena@centracare.com; jcxu@uci.edu

© Springer Nature Switzerland AG 2021
B. C. Cheng (ed.), *Handbook of Spine Technology*,
https://doi.org/10.1007/978-3-319-44424-6_48

Abstract

The incidence of acute traumatic spinal cord injury (SCI) in the USA is approximately 27–81 cases per million people per year with cervical SCI being the most common site of injury. Despite early surgical decompression, secondary injury and the cascade of effects in the ensuing days and months remain one of the biggest barriers in achieving recovery in these patients. A host of pharmacologic, cellular, immunomodulatory, and rehabilitative interventions have been employed over the past several decades in an attempt to improve functional outcome in this population. Though no single intervention is likely to provide a cure, important information has been gained about the heterogeneity of this population and the myriad physiological processes underlying the acute and chronic phases of injury. Herein, we provide a broad overview of the underlying pathophysiology, discuss various cellular, structural, and pharmacologic therapies tested, and address the challenges and insights gained from completed SCI trials.

Keywords

Spinal cord injury · Spinal implants · Functional regeneration · Neuroprotection · Neuromodulation · Trauma products · Brain machine interface

SCI Background: Demographics, Economic Burden, and Grading Systems

Introduction

Trauma resulting in spinal cord injury (SCI) is often associated with significant morbidity. More than half of all SCI patients develop complications during both the initial hospital stay and after discharge, with long-term problems including veno-thrombo embolism and pressure ulcers from decreased mobility, as well as a variety of pulmonary and gastrointestinal complications (Eckert and Martin 2017). Despite advances in supportive care for SCI, technologies that can directly facilitate spinal cord recovery have been limited. In this chapter we review the various implants and therapies that have been used to improve functional outcome in the SCI population.

Demographics

The annual incidence of traumatic spinal cord injury (SCI) in the USA is estimated at 54 cases per one million people, translating to approximately 17,700 new cases each year (National Spinal Cord Injury Statistical Center Updated 2018). Motor vehicle accidents are the most common cause of traumatic SCIs (Fig. 1), accounting for 38% of all injuries between 2010 and 2014, followed by falls (31%) and acts of violence (13%). Traumatic SCI occurs predominantly in men (78%) and its prevalence is disproportionately high among non-Hispanic blacks (Fig. 2). Notably, the incidence of SCI from falls has nearly doubled since the 1970s (Chen et al. 2016). The average age at time of injury has also increased, from 29 years in the 1970s to 43 years old in 2018. Most injuries occur at the level of the cervical spinal cord (59%), followed by thoracic (32%) and lumbosacral (9%) spine (Ahuja et al. 2017b).

Economic Burden

SCI afflicts devastating physical, psychosocial, and economical consequences on its patients and their families. After discharge from acute hospitalization, the most common outcome is incomplete quadriplegia (Fig. 3). Approximately one-third of all patients with SCI are re-hospitalized in any given year, most often for infections of the skin and genitourinary system (National Spinal Cord Injury Statistical Center Updated 2018). This and other health ramifications lead to staggering costs, ranging from 2 to nearly five million dollars over the course of a patient's lifetime (Fig. 4). In addition, less than one-fifth of SCI patients are employed at 1 year after injury, adding

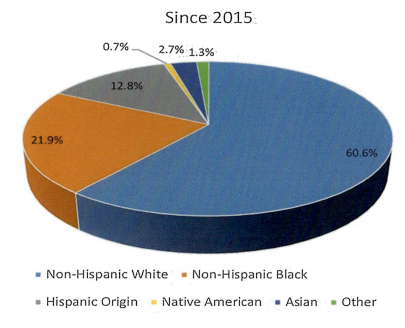

Fig. 1 Most common causes of SCI in the USA from 2015–2018. (National Spinal Cord Injury Statistical Center, Facts and Figures at a Glance. Birmingham, AL: University of Alabama at Birmingham, 2018. Redrawn with permission)

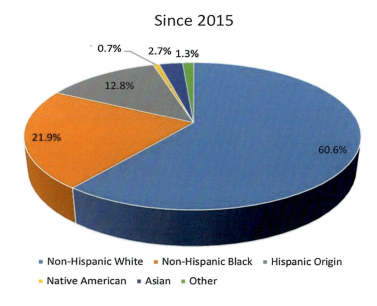

Fig. 2 Distribution of race/ethnicity in SCI cases in the USA from 2015–2018. (National Spinal Cord Injury Statistical Center, Facts and Figures at a Glance. Birmingham, AL: University of Alabama at Birmingham, 2018. Redrawn with permission)

on additional indirect costs from lost wages and benefits. Patients with SCI have persistently higher mortality rates when compared to age-matched controls (Fig. 5), with pneumonia and septicemia as the leading causes for the reduced life expectancy of these patients (Krause et al. 1997). The prognosis of restoring spinal cord function is largely dependent on the severity of injury (as measured by admission American Spinal Injury Association Impairment Scale [ASIA] grade, Frankel grade, or injury completeness) and the level of injury, as these two factors play important roles in the success of rehabilitation and regeneration (Wilson et al. 2012). Given the high morbidity and mortality associated with this population, there is a critical need for the development of new treatments and technologies to restore spinal cord function.

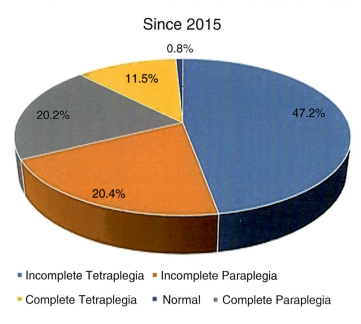

Fig. 3 Distribution of neurological injury in SCI cases in the USA from 2015–2018. (National Spinal Cord Injury Statistical Center, Facts and Figures at a Glance. Birmingham, AL: University of Alabama at Birmingham, 2018. Redrawn with permission)

	Average Yearly Expenses (in 2017 dollars)		Estimated Lifetime Costs by Age at Injury (discounted at 2%)	
Severity of Injury	First Year	Each Subsequent Year	25 years old	50 years old
High Tetraplegia (C1–C4) AIS ABC	$1,102,403	$191,436	$4,891,398	$2,688,229
Low Tetraplegia (C5–C8) AIS ABC	$796,583	$117,437	$3,573,960	$2,198,305
Paraplegia AIS ABC	$537,271	$71,172	$2,391,872	$1,569,714
Motor Functional at Any Level AIS D	$359,783	$43,700	$1,634,139	$1,153,420

Fig. 4 Costs of SCI patients in the USA. (National Spinal Cord Injury Statistical Center, Facts and Figures at a Glance. Birmingham, AL: University of Alabama at Birmingham, 2018. Redrawn with permission)

		Life Expectancy (years) for Post-Injury by Severity of Injury and Age at Injury									
		For Persons Who Survive the First 24 Hours					For Persons Who Survive the First 24 Hours				
Age at Injury	No SCI	AIS D—Motor Functional at Any Level	Para	Low Tetra (C5–C8)	High Tetra (C1–C4)	Ventilator Dependent Any Level	AIS D—Motor Functional at Any Level	Para	Low Tetra (C5–C8)	High Tetra (C1–C4)	Ventilator Dependent - Any Level
20	59.6	52.9	45.7	40.3	34.0	11.3	53.2	46.2	41.2	35.2	19.0
40	40.7	35.2	29.7	24.9	20.9	8.7	35.4	30.2	25.7	22.1	13.3
60	23.2	19.5	16.1	13.2	11.1	3.7	19.7	6.5	14.0	12.5	7.9

Fig. 5 Life expectancy of SCI patients in the USA. (National Spinal Cord Injury Statistical Center, Facts and Figures at a Glance. Birmingham, AL: University of Alabama at Birmingham, 2018. Redrawn with permission)

Grading Systems in SCI

To accurately characterize the level and extent of cord injury, five main grading systems have been developed and used (Tator 2006). In 1967, the Frankel system was the first to be developed and included five grades of severity. Although it was easy to use and understand, the Frankel system could not quantify recovery and definitions of each grade were imprecise. The Sunnybrook system was developed in 1982, which expanded the repertoire of severity grading and neurologic changes, allowing for increased precision in the clinical evaluation and quantification of recovery in a SCI patient. One major drawback of the Sunnybrook system is its lack of numerical scores of motor and sensory function. In 1982, the American Spinal Injury Association published the International Standards for Neurological Classification of Spinal Injury (NASCIS), a grading and classification system that would evolve into the current American Spinal Injury Association (ASIA) Impairment Scale (Roberts et al. 2017). Figures 6 and 7 depict the NASCIS scoring worksheet and ASIA Impairment Scale, respectively. Since then, the ASIA system has been refined many times, improving its precision and reproducibility in defining spinal cord injury and allowing for better understanding of the scale's therapeutic implications. The major advantages of the ASIA system include a more accurate definition of a complete SCI and improved methodology for determining motor and sensory scores. One of its major disadvantages is the ceiling effect of Grades C and D, in which patients seldom improve sufficiently to change to the next grade. Of note, the most recent classification system has been the Benzel system, but this system cannot be used in the acute phase of injury, rendering it inappropriate for use in clinical studies.

SCI Pathophysiology: Primary Versus Secondary Injury

Primary Versus Secondary Injury

Primary Injury

An understanding of the mechanisms underlying SCI is essential to develop new strategies for restoring spinal cord function. Traumatic SCI can be pathophysiologically divided into primary and secondary injuries and can be temporally divided into acute (<48 h), subacute (48 h to 14 days), intermediate (14 days to 6 months), and chronic (>6 months) stages (Ahuja et al. 2017b). The primary injury (i.e., the direct traumatic event) causes the immediate mechanical disruption and dislocation of the vertebral column, leading to compression or transection of the spinal cord. On the cellular level, this produces foci of glial and neuronal necrosis, alarmin release, disruption of the microvasculature, and breakdown of the blood–brain barrier (BBB) (Tran et al. 2018). These events together then directly trigger a prolonged secondary injury

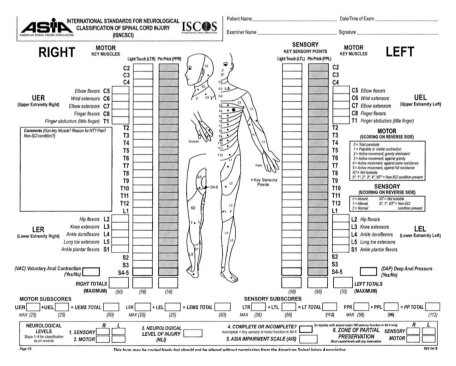

Fig. 6 International Standards for Neurological Classification of Spinal Cord Injury (ISNCSCI) and ASIA Impairment Scale (AIS) scoresheet

Fig. 7 International Standards for Neurological Classification of Spinal Cord Injury (ISNCSCI) and ASIA Impairment Scale (AIS) grading and classification guide. (https://asia-spinalinjury.org/wp-content/uploads/2016/02/International_Stds_Diagram_Worksheet.pdf. International Standards for Neurological Classification of SCI (ISNCSCI) Worksheet, Updated April 2019. © 2019 American Spinal Injury Association. Reprinted with permission)

cascade, which may viciously cycle for weeks until the glial scar is formed.

Surgical stabilization and decompression is the essential cornerstone of acute treatment for patients during the primary injury. Overall, the goal of surgery is to realign the spinal column and reestablish spinal stability as to relieve compression on the spinal cord. Typically surgery involves open reduction, decompression, and instrumented fusion to bring the spinal column back into anatomical position. By correcting the mechanical disruption, surgery aims to reverse or avert any further primary injury and prevent secondary injury. Although clinical class I randomized evidence supporting the efficacy of early surgical decompression is still needed, several prospective, nonrandomized studies have supported the safety and efficacy of early surgical intervention in traumatic SCI. The prospective, multicenter, cohort-controlled Surgical Timing in Acute Spinal Cord Injury Study (STASCIS; N = 313) compared patients with cervical SCI receiving either early (<24 h) or late (>24 h) decompression. They noted an increased odds of 2 grade improvement in the American Spinal Injury Association (ASIA) Impairment Scale with early decompression (within 24 h) as compared to late decompression (>24 h) (Fehlings et al. 2012). Another study showed that very early decompression (within 8 h) was associated with better motor recovery and improved ASIA grade at 1 year post-injury (Ahuja et al. 2017b). Additionally, several studies have shown an association between early decompressive surgery and greater motor scale recovery, reduced length of stay, lower complication rates, and shorter length of hospital stay.

Secondary Injury

The hallmark of secondary injury is cord damage beyond the initial site of injury. In the simple sense, cell death begets additional cell death in the secondary cascade. On the cellular level, disruption of the cord microvasculature leads to cell permeability, pro-apoptotic signaling, and ischemic injury, all of which potentiates additional cell death (Ahuja et al. 2017b). Incompetence of the BBB permits the unchecked entry of peripheral immune cells, toxic metabolic products, and other inflammatory substances into the central nervous system (CNS), called forth by alarmin release from necrotic cells. The ongoing necrosis of neurons and glial cells activates additional microglial cells, escalating the inflammatory response around the site of injury. This produces an overwhelming inflammatory response that inflicts damage to nearby cells by driving up oxidative stress, lipid peroxidation, and protein aggregation. Homeostatic imbalance, including the dysregulation of sodium and calcium channels and the overrelease of glutamate from dying cells, induces excitotoxic cell death, causing the release of additional glutamate and alarmin. As injured blood vessels lose their capability for autoregulation, the ensuing ischemia can persist for days to weeks after injury, extending further damage. All of these factors contribute to a harsh post-injury microenvironment that cyclically propagates and magnifies the original injury. In addition, disruption of the BBB in conjunction with the inflammatory response causes cord swelling, leading to further mechanical compression of the spinal cord.

Challenges for Recovery: Cervical Versus Thoracic SCI, Incomplete Versus Complete

Current strategies for neuroprotection and rehabilitation have focused primarily on preserving injured tissue and reducing the secondary injury. These approaches look to curb the inflammatory response and promote neural repair and regeneration after the primary insult. Without interventions to address the primary injury, inherent challenges in restoring SCI function vary depending on the level and completeness of cord injury.

Cervical Versus Thoracic SCI

Restoration of SCI function depends on the level of neurologic injury. A neurologic level of injury is defined as the most caudal spinal cord segment with intact sensation and Grade 3 or greater motor function (Roberts et al. 2017). For dermatomes not covered by muscles (C1-C4, T2-L1, and S1-S4-S5), the sensory level is used to delineate the level of injury.

Cervical SCI

For patients with cervical SCI, C6, and C7 are known to be the critical levels for achieving independence in most daily activities (Welch et al. 1986). The presence or absence of functional triceps is a critical determinant for functional independence in self-care tasks. A strong triceps permits stabilization of the elbow in extension, so that the shoulder depressors can act through the elbow in lifting the body weight, achieving independence in changing from lying to sitting position and in transferring to and from a wheelchair. Though patients with C6 injuries lack an intrinsic grasp, operating a motorized wheelchair is possible at this level through taking advantage of elbow flexion and shoulder abduction and flexion. Sparing of C7 allows for finger extension and flexion, affording even more independence with operating a wheelchair. Almost all cervical SCI patients will still need assistance with lower extremity dressing, bowel and bladder care, and driving. True functional ambulation is nearly impossible without truncal stability provided by thoracic musculature, though neuromodulation efforts have resulted in some success (discussed later in this chapter).

Compared to thoracic SCI, cervical SCI involves a mixture of damage to both central and peripheral nervous systems. Therefore, cervical SCI training approaches and chances of success is dependent on not only the injury of the spinal tracts but also on the damage of the motor neurons and roots, which can cause up to 70% of the paresis seen in C5 to C7 lesions (Dietz and Fouad 2014). The complexity and variability of cervical lesions makes the restoration of cervical cord function much more challenging. Patients with cervical injuries are also at a higher risk for lifelong ventilator dependency compared to patients with thoracic injuries, restricting rehabilitation capacity and long-term independence (Winslow and Rozovsky 2003). Cardiovascular dysfunction leading to orthostatic hypotension is also more prevalent in cervical injuries (West et al. 2012).

Thoracic SCI

With the assistance of leg braces and pelvic band, the ability to hold an erect position and stand upright is usually reached around T6. At these levels, truncal stability is achieved from innervation supplied to the long muscles of the back, upper intercostals, and transversus thoracis. To be able to ambulate, there must be sufficient pelvic control and at least grade $3/5$ strength in both hip flexors and one knee extensor (Branco et al. 2007). Though there is no set level for ambulation, most authors agree that lesions at or below T10 allow for ambulation due to innervation of the thoracic musculature and secondary hip flexors, including the external and internal obliques and latissimus dorsi.

Complete Versus Incomplete SCI

A complete SCI is defined as the absence of all motor and sensory functions in the sacral segments S4–S5. An incomplete injury is defined as the partial preservation of sensory or motor function in the lowest sacral segments (anal sensation, including deep anal pressure and voluntary external anal sphincter contraction). Incomplete SCI can then be further divided into sensory and motor incomplete subcategories. The presence of deep anal sensation may be the only indicator of an incomplete SCI. The determination of a complete or incomplete SCI requires resolution of spinal shock. Spinal shock is characterized by an initial depolarization of axonal tissue immediately after injury and is a physiologic response to trauma. Spinal shock involves a transient period of flaccid paralysis during which the patient is areflexic. Notably, this includes the absence of the bulbocavernosus reflex, a spinal-mediated reflex involving S2–S4. After return of this reflex, the patient can be assessed accurately for completeness of cord injury.

Complete SCI

The zone of partial preservation is used only with complete injuries. The zone of partial preservation documents dermatomes and myotomes caudal to the neurologic level of injury that remain partially innervated and is recorded for motor and sensation for the right and left sides. Presence of a zone of partial preservation is associated with improved neurologic recovery (Wilson et al. 2012). Otherwise, prognosis for neurologic recovery in a patient with a motor/sensory complete lesion (ASIA Impairment Scale A) is grim. Complete quadriplegics are unable to ambulate and only 5% of complete paraplegics will have sufficient motor recovery to permit ambulation (Branco et al. 2007). Waters et al. (1992) and Waters et al. (1993) found that of complete lesions assessed 1 month after injury, 90% of those causing complete quadriplegia and 96% of those causing complete paraplegia remained complete. Schönherr et al. (1999) found that approximately 1 month after injury, 100% of those causing complete quadriplegia and 96% of those causing complete paraplegia remained complete. The most recent data from the National Spinal Cord Injury Statistical Center (NSCISC) report a much higher rate of conversion from complete to incomplete status in quadriplegics, with 30% overall conversion and 15% to motor incomplete (Marino et al. 2011).

Incomplete SCI

As the amount of spared spinal cord tissue determines the effectiveness of rehabilitation training, the presence of an incomplete injury, as compared to a complete injury, is consistently associated with improved neurological recovery. This is because the success of facilitating meaningful plasticity during training depends on the presence of certain physiological prerequisites. In order to produce a locomotor electromyography (EMG) pattern in patients with SCI, afferent input from load receptors is needed (Dietz 2009). For example, no meaningful leg muscle activation occurs in individuals with complete SCI during supported stepping if there is no loading of the sole of the foot. This makes obtaining significant muscle activation in patients with complete SCI much more challenging than in patients with incomplete SCI. An additional factor that may also contribute to the smaller EMG amplitudes in patients with complete paraplegia is a loss of descending noradrenergic input to spinal locomotor centers (Dietz 2012).

Concepts: Functional Regeneration, Neuroprotection, Immunomodulation

Many promising strategies are being explored in an attempt to regain function after spinal cord injuries, with a strong focus on regeneration. In the literature, several comprehensive reviews have been conducted on the status of translational advances in spinal cord injury (Badhiwala et al. 2018; Venkatesh et al. 2019; Ahuja et al. 2017b). While there are a number of promising strategies for treating SCI, there are currently no effective and reliable treatments that have achieved regeneration in SCI. Two major principles in the field of regeneration have focused on neuroregeneration and neuroprotection. Neuroregenerative techniques have predominantly focused on stem cell lines that can restore lost neurons. For neuroprotection, research has focused on factors that both promote regeneration and limit inhibitory factors. Different stem cell lines have the potential to regenerate neural circuits, provide trophic support, modulate the inflammatory response, and remyelinate damaged axons. To support both endogenous and implanted stem cells, there has also been a strong focus on neurotrophic factors that promote neuron growth and reagents that block inhibitory factors such as the formation of scar tissue. These immunotherapies serve as an adjunct to promote spinal cord regeneration through increasing cell survival and engraftment (Ahuja et al. 2017a; Tsuji et al. 2019).

A parallel strategy focuses on biomaterials that can mimic the physiological extracellular matrix and provide support and structure to regenerating neural structures. Biomaterials can be used to make scaffolds that provide guidance for existing damaged neurons and allow implanted cells to propagate and differentiate. Another goal of scaffolds has been to minimize secondary injury after SCI through inhibiting apoptosis and necrosis, limiting the formation of glial scar and serve as a vehicle for the delivery of immunomodulators and cell therapies (Liu et al. 2019).

A separate strategy for restoring SCI function is through neuromodulation with electrical interfaces. Neuromodulation is currently successfully in use for many neurological disorders, including deep brain stimulation for Parkinson's disease and epidural stimulation for pain. In the setting of SCI neurostimulation has already been used to help with respiratory pacing, bladder control, and restoring volitional movements through peripheral stimulation. New research suggests that spinal cord stimulation may also promote regeneration. Neuromodulation can also be adapted to restore motor and sensory function by creating an interface that connects the central nervous system with peripheral extremities. Through brain–machine interfaces, a tetraplegic patient has been able to perform reach and grasp movements. As technological advances continue, the field of neuromodulation will continue to increase in potential (James et al. 2018).

While there have been significant steps toward restoring function in SCI patients, to date there is no widespread and effective therapy available for regeneration or technology that restores function. In the following sections of this chapter, we review the current advances made, as well as challenges encountered in the different focuses of ongoing research. As the field continues to advance, an effective therapy will likely emerge from a combination of the various strategies already in development.

Results from Acute SCI Interventions: Stem Cells, Pharmacologics, Hemodynamic Strategies, CSF Drainage, Bio-Scaffolds, and Neuromodulation

Stem Cell Therapies

A wide variety of stem cell types have been explored with the goals of neuroprotection and neuroregeneration. Extensive research has shown promise in restoring function in SCI animal models (Hong et al. 2018). Cell therapies have been proposed to provide functional benefits in SCI through several proposed mechanisms: neuroprotection, immunomodulation, axon sprouting and/or regeneration, neuronal relay formation, and myelin regeneration (Assinck et al. 2017). Stem cells for transplantation originally were developed from embryonic stem cells (ESCs), however given ethical and supply issues, induced pluripotent stem cells (iPSCs) have gained traction as favorable alternative. iPSCs have been successfully induced into specific neurons and glial cells, providing a more feasible source of cells to study. Studies have found success in integrating these stem cells into host circuitry in animal models. Neuronal stem cells (NSC) have been shown to enhance neurotrophic signaling, promote remodeling of neural circuitry, improve remyelination, and modify the extracellular matrix.

Supportive cell types are just as important as neurons themselves. Oligodendrocyte precursor cells (OPCs), mesenchymal stem cells (MSCs), olfactory ensheathing cells (OECs), and Schwann cells have thought to promote neuroprotection through several mechanisms. OPCs can secrete neurotrophic factors and remyelinate denuded axons. MSCs aim to regenerate damaged connective tissue and modulate inflammation. OECs have also been shown to induce regeneration and endogenous remyelination. Schwann cells are robust structural scaffolds in the peripheral nervous system that have been shown to modify astrogliosis in a way that facilitates remyelination and regeneration (Ahuja et al. 2017a; Badhiwala et al. 2018).

Cell therapy effectiveness will depend highly on the specific cell type, route of administration (IV, intrathecal and intraparenchymal), and timing. Ongoing trials explore and optimize each of these variables, with many different types of stem cells. The injection of MSCs, OPCs (AST-OPC-1), and autologous Schwann cells have all been been explored safely in phase I trials in SCI patients (Anderson et al. 2017; Priest et al. 2015; Jin et al. 2019). Despite successful preliminary trials, multiple late-stage trials have hit significant challenges, both scientific and financial. The Geron Corporation had launched a clinical trial of embryonic stem cells for SCI in 2010, citing "capital scarcity and uncertain economic conditions"(Lukovic et al. 2014). Another clinical trial by Stem Cells Inc. (Newark, CA) showed positive safety profiles in human fetal–derived NSC transplants into cervical and thoracic spinal cord injury sites. Further trials however did not show a significant response with clinically derived stem cell lines and studies have been terminated (Anderson et al. 2017). Despite failed trials, there continues to be a strong effort in clinical studies of cell therapies. The first human trial using iPSC has recently been approved to start (Tsuji et al. 2019).

Other than specific stem cells, other cells including macrophages and whole bone marrow have been thought to have potential to improve the microenvironment for neurological recovery. The bone marrow cells include hematopoietic stem cells, macrophages, lymphocytes, and marrow stromal cells. The bone marrow cells may secrete important cytokines as well as other essential factors for the survival and differentiation of neuronal cells.

While cell therapy is promising, there are also significant risks as with any new technology. A case of OEC implantation in the site of a chronic T10–11 complete SCI was found 8 years after implantation to have developed into an intramedullary spinal cord mass requiring resection (Dlouhy et al. 2014). This report raised concerns about the safety surrounding implantation of stem cells and highlights the importance of long-term monitoring. While the majority of stem cell trials have not reported significant adverse events, the follow-up time period for these trials has been limited to a few years. These reports should not deter the advancement of stem cell research but provide scientists with more information on how to safely proceed.

Pharmacologics

Steroids

Methylprednisolone sodium succinate (MPSS) was used in early spinal cord injury trials for its immunomodulatory and anti-inflammatory roles, both reducing neutrophil and macrophage migration while also playing an important role in reduction of membrane peroxidation. The (National Acute Spinal Cord Injury Study) NASCIS I was a double-blind, randomized controlled trial performed comparing low-dose and high-dose administration in acute SCI. Results showed no difference between groups and the trial was stopped early. The NASCIS II trial was conducted with higher dose MPSS and compared with naloxone and placebo. Analysis was stratified based on timing of administration (<8 versus >8 hrs) and adjusted for the severity of injury (Bracken et al. 1990). Results demonstrated improvement in motor and sensory function in those patients who received MP within 8 hrs and no significant differences in either treatment arm greater than 8 hrs from injury. Wound infections and GI bleeding was observed with greater frequency in the MP group. Lastly, the NASCIS III trial was designed to compare 24 h MPSS infusion to a 48-h MPSS infusion. These two dosing schedules were compared with a 48-h infusion of a third drug, tirilazad mesylate, an inhibitor of lipid peroxidation. Results suggested that patients who received MPSS infusion as a 48-h infusion had improved motor function at 6 weeks and 6 months. Preplanned subanalysis demonstrated those who received MPSS between 3 and 8 h post-injury were more likely to improve at 6 months, though with increased risk of severe sepsis and pneumonia (Bracken et al. 1997). As a result, published guidelines offer a Level I recommendation against the use of MPSS in acute SCI, though the debate continues and recommendations are to continue evaluation in future higher-quality studies (Donavan and Kirschblum 2018).

Hemodynamic Intervention

Prevention of secondary ischemia and spinal cord hypoperfusion is paramount in the management of acute SCI. Cervical and upper thoracic SCI carry a risk of spinal shock due to variable interruption of the sympathetic outflow from cell bodies residing in the thoracic spinal cord. As such, strict hemodynamic monitoring and blood pressure (BP) parameters have been advised, though the recommendations have been controversial. Loss of supraspinal sympathetic regulation permits hypotension, bradycardia, and arrythmias. Post-mortem histological analysis of SCI patients with severe cardiovascular dysfunction shows more severe white matter and axonal degeneration in those patients who have severe cardiovascular dysfunction after SCI (Furlan and Fehlings 2008). Maintenance of mean arterial pressure (MAP) 85–90 mm Hg for the first week after acute SCI currently carries a Level III recommendation, according to published guidelines (Walters et al. 2013). While there is no universal agreement as to the vasopressor of choice, some studies have indicated that norepinephrine may be preferable to dopamine (Altaf et al. 2017).

CSF Drainage

Similar to the principles of TBI management, swelling and edema after SCI can have deleterious effects on local tissue perfusion. Surgical decompression, followed by efforts to prevent hypotension, hypoxia, and other secondary insults, is paramount in the management of acute SCI. Prevention of spinal cord ischemia can be accomplished by increasing perfusion and/or reducing intrathecal pressure by way of CSF drainage. This has been well described in vascular surgery where it is routinely employed after thoracoabdominal aortic aneurysm repair. A prospective randomized trial of 22 patients (ASIA A-C, inclusive of C3-T11 injury levels) was performed to evaluate the effect of 72 h of CSF drainage on peak intrathecal pressure (ITP) and correlated with neurological recovery after acute SCI (Kwon et al. 2009a). Though the study was not powered to detect differences in neurological outcome, there were no significant associated adverse effects reported in the study arm and spinal cord perfusion pressure was consistently higher during the duration of the study. An ongoing multicenter Phase IIb randomized controlled trial (NCT02495545) comparing the efficacy of CSF drainage and MAP elevation to MAP elevation alone is currently enrolling patients through the December 2019.

Growth Factors

Several important growth factors, have been identified and gained attention as neurotrophic agents in the central nervous system. Conversely, inhibitory factors that inhibit neuronal growth have also been explored to understand the barriers to regeneration. All of these are being explored for their potential to be medications in treating SCI. By optimizing the microenvironment, these medications have the potential to enhance the regeneration of both endogenous cells and implanted stem cells.

Hepatocyte growth factor (HGF) is a c-Met receptor ligand that has been shown exert regenerative effects, including angiogenesis, after tissue injury in many epithelial organs. Studies in rodent models have shown that HGF in the central nervous system promotes angiogenic activity, prevents disruption of the blood–brain barrier, and promotes the survival of neurons after cerebral ischemia. Preclinical SCI studies investigating the use of intrathecal HGF in nonhuman primate models found improved ventral motor neuron survival and better motor outcomes, with significant improvements in upper limb recovery (Badhiwala et al. 2018; Kitamura et al. 2011). While most pharmacologics have been investigated through IV administration, a handful has been used intrathecally during the time of initial injury. In one human trial, HGF has administered as intrathecal injection at the time of injury and repeated weekly. Study results are still pending (Warita et al. 2019).

Granulocyte colony stimulating factor (G-CSF) is a small glycoprotein that has shown potential for enhancing cell survival and modulating inflammatory cytokine pathways. It is expressed on microglia and promostes expression of neurotrophic factors while inhibiting pro-inflammatory markers. G-CSF has been shown in mouse models to enhance neurogenesis and reduce apoptosis, with associated improvements in hind limb function (Koda et al. 2007). A randomized controlled trial in Iran showed that subcutaneous administration of G-CSF showed subtle but significant improvements in motor and sensory function compared to a placebo group in a population of incomplete subacute SCI (Derakhshanrad et al. 2018). Another group in Korea has experimented with G-CSF and isolated bone marrow cells injected during surgery at the site of contusion in six patients within 7 days of SCI. While the patients showed some neurologic

recovery, the study mainly demonstrated the safety of the procedure (Park et al. 2005). Further multicenter trials will be needed to reinforce these findings.

Fibroblast growth factor (FGF) protects against excitotoxic cell death and reduces free radical generation (Badhiwala et al. 2018). FGF has been shown in several animal models to promote regeneration of spinal tracts (Cheng et al. 1996). This was first translated to human studies in 2004 when surgeons applied sural nerve grafts and fibrin glue infused with a closely related FGF: acidic FGF (aFGF) in a SCI patient, with improvement from wheelchair bound to ambulating with a walker (Cheng et al. 2004). Two further trials of aFGF administered during laminectomy without the nerve grafting have been done with additional doses of aFGF/fibrin through lumbar puncture at 3 and 6 months post-surgery. No significant adverse events were reported and ASIA motor and sensory scores showed significant improvement 24 months after treatment (Wu et al. 2008, 2011). The studies however did not have control arms, and further large-scale randomized controlled trials are currently enrolling.

Rho Signaling

The Rho signaling pathway regulates axon growth and is upregulated after SCI. Activation of the GTPase Rho A leads changes in the actin cytoskeleton and collapse of regenerating axons, neurite retraction, and increasing apoptosis (Nori et al. 2017). Medications that inhibit the Rho pathway have the potential to relieve the inhibition on axonal growth. C3 transferase is a toxin produced by clostridium botulinum and can block the Rho pathway. VX-210 aka Cethrin is a C3 transferase that has been used in clinical trials dosed in a fibrin sealant and administered during planned decompression/stabilization surgery after SCI (Fehlings et al. 2018, 2011). However an interim analysis of phase IIb/III trial suggested futility, and studies have been halted.

Neurite growth inhibitor A (NOGO-A) is a myelin-associated protein that functions through NOGO receptors and forms co-receptor complexes with the TNF receptor family proteins to activate the GTPase Rho A. This activates the Rho pathway as discussed above and leads to inhibition of axon growth. A NOGO-A inhibitor (ATI355) has gone through a phase I clinical trial in 52 patients with SCI with acute SCI injury (4–60 days post-injury). The medication was administered via continuous intrathecal infusion over 24 h to 28 days. The published results confirmed safety and tolerability after 1-year follow-up with minimal adverse events, and no major adverse events attributable to ATI355. Future efficacy trials are still pending (Schwab and Strittmatter 2014; Kucher et al. 2018).

Excitotoxicity

Once the spinal cord is injured, the damaged cells release excess glutamate and leads to excitotoxic damage. Numerous medications have been explored in treating SCI with the targeted effect of limiting inflammation and excitotoxicity. Riluzole is a benzothiazole anti-epileptic that can reduce excitotoxicity though sodium blockade and reduction of presynaptic release of glutamate. This medication is already approved for neurodegenerative disorders including amyotrophic lateral sclerosis, and has been shown to reduce neurodegeneration and effects of traumatic injury by modulating excitotoxicity. The Riluzole in Spinal Cord Injury Study (RISCIS) trial has shown safety of oral administration of Riluzole in acute SCI. Phase II/III trials are now pending (Siddiqui et al. 2015).

Minocycline, a tetracycline antibiotic, also has been found to have anti-inflammatory, anti-oxidant, and anti-apoptotic properties. It has been shown to minimize N-methyl-D-aspartate (NMDA)-induced excitoxicity. Its effects may stem from its ability to modulate inflammatory effects of microglia (Shultz and Zhong 2017). Given its clinical availability and established safety as an antibiotic, it has been an attractive drug for clinical trials and is currently pending Phase III trials.

Gacyclidine is a noncompetitive NMDA recept antagonist that has targeted effects of reducing inflammation and excitotoxicity. Animal models showed possible attenuation of spinal cord damage. Early-stage SCI trials in humans have been completed but efficacy has yet to be shown (Donavan and Kirschblum 2018). Similarly, Magnesium also antagonizes NMDA receptors and has been explored for its potential to reduce inflammation and excitotoxicity. It has been used together with scaffolds, discussed in a later section, to treat SCI. To date, none of the medications have shown a clear benefit in the treatment of SCI.

Previously Explored Pharmacologics

A number of trials have shown negative results without further research. Thyroid releasing hormone (TRH) has been shown to antagonize many aspects of secondary injury. It was studied in a small clinical trial in 1995 in acute SCI with negative results (Lehrer 1996). Opioid antagonism has also been explored given the potential neurotoxic effects of endogenous opioids in SCI. The NASCIS II trial included a trial arm of naloxone which showed negative results. Gangliosides are glycolipids abundant in nervous tissue. In vitro experiments have shown potential benefit of gangliosides in enhancing axonal growth, and an experimental ganglioside GM-1 (mono-sialoate-trahexosylganglioside), also known as Sygen, had shown promise in animal models of SCI. Its efficacy however was not shown in a large RCT, and no future studies have been planned.

Bio-Scaffolds

Another strategy to facilitate regeneration focuses on providing support and guidance for damaged neurons through scaffolds that can both facilitate growth and inhibit antagonizing factors and the formation of glial scars. Many different natural materials have been used, including collagen, chitosan, alginate, and fibrin. Natural materials have advantages including biocompatibility, biodegradability, and low toxicity. Synthetic materials have also been developed, including various polymeric biomaterials, with the advantage of more controllable biodegradability and physicochemical and mechanical properties. Polyethylene glycol (PEG) is one synthetic biocompatible polymer that has shown potential to stimulate angiogenesis, reduce glial scar formation, and promote axonal regeneration through its ability to act as a fusogen and stabilize compromised neuronal membranes. In animal models of SCI, PEG has been shown to improve hind limb function.

The potential of biological scaffolds can be augmented by seeding with pharmacologics, immunomodulators, and stem cell lines, providing an environment that provides both neuroprotection and positive growth factors. The addition of magnesium to PEG has been used in rat models of SCI and shown to provide tissue protection and improved locomotor recovery (Kwon et al. 2009b). While promising results have been demonstrated in animal models, the evidence in humans is limited (Liu et al. 2019). Few biological scaffolds have been translated from the research lab to clinical trials. Complications ranging from seroma formation to meningitis have been reported in human subjects (Liu et al. 2019).

The neuro-spinal scaffold (InVivo Therapeutics Corp, Cambridge, Massachusetts) is a porous polymer scaffold composed of poly(lactic-co-glycolic acid) covalently conjugated to poly-L-lysine that has been one of the few scaffolds to be used in humans. A preliminary study of one patient who sustained a spinal cord contusion at T11–T12 had the neuro-spinal scaffold placed during surgical decompression and fusion for the acute injury (Theodore et al. 2016). The patient preoperatively had a T11 AIS grade A complete injury, and 12 months postoperatively was noted to be L1 AIS grade C. Preliminary results from a larger study of 16 patients (INSPIRE study) showed that 7 (44%) reached the primary endpoint of ASIA Impairment Scale conversion at 6 months. While one patient was lost to follow-up, the remaining six patients who converted retained their improvements at the 12-month mark. Further large-scale studies as part of INSPIRE 2.0 are currently undergoing (Anon n.d.).

Neuromodulation

An alternative approach to SCI focuses on the development of electrical stimulation to augment or modify neuronal function. As technological advances in materials science and computing are made, the possibility of neuromodulation to restore spinal cord function becomes closer to reality. Stimulation can be applied direction to the muscle, peripheral nervous system, or even central nervous system to drive motor output (Fig. 8). Technologies have been explored in brain stimulation, brain–machine interfaces, spinal cord stimulation, and peripheral stimulation.

Peripheral Stimulation

Functional electrical stimulation (FES) involves the application of electrical stimulus to generate muscle contractions. By targeting specific muscles of interest, FES can induce controlled movements in the limbs and body. FES has successfully been used to improve ambulation in patients with incomplete SCI through generating contractions in the limbs (Badhiwala et al. 2018). While originally envisioned as a permanently worn device, FES may also be used as a clinical intervention with a limited number of training session. One case report describes how FES training allowed a patient to regain meaningful movement in his upper extremities (Popovic et al. 2016). Through the help of stimulation patients can potentially gain muscle strengthening as well as neuroplastic changes and cortical reorganization. This concept will be compared with conventional occupational therapy in an ongoing phase III multicenter trial.

Brain Stimulation

Brain stimulation has also been explored as an option to help SCI and explores the possibility of increasing the activity of residual pathways. Transcranial direct current stimulation (tDCS) and transcranial magnetic stimulation (TMS) are two noninvasive techniques that have been explored in combination with motor training to augment existing neurological pathways. As noninvasive procedures, both tDCS and TMS are considered safe and low risk, and have been researched extensively in the treatment of neuropathic pain in SCI with mixed results (Fregni et al. 2006; Wrigley et al. 2013; Gao et al. 2017). Several small studies have used multiple sessions of tDCS in small groups of patients in the chronic phase of SCI with some improvement in upper and lower extremity function (Gunduz et al.

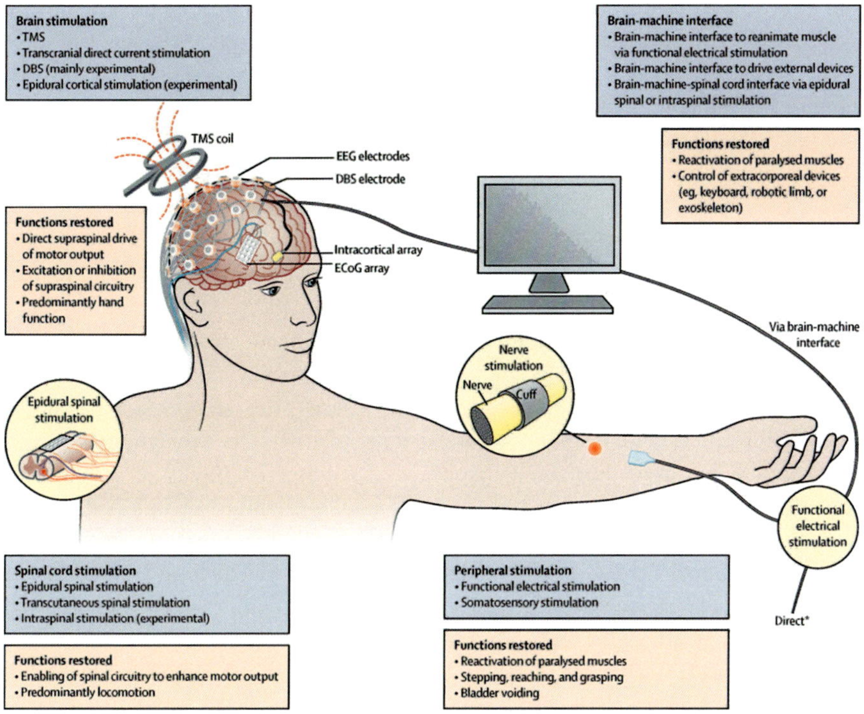

Fig. 8 Neuromodulation targets in SCI. https://ars.els-cdn.com/content/image/1-s2.0-S1474442218302874-gr1.jpg (Copyright requested through RightsLink). (Reprinted from Lancet Neurology Volume 17, Issue 10, Nicholas D James, Stephen B McMahon, Edelle C Field-Fote, Elizabeth J Bradbury, "Neuromodulation in the restoration of function after spinal cord injury," Pages 905–917, 2018, with permission from Elsevier)

2017). This therapy may be synergistic in conjunction with current rehabilitation therapies. tDCS in combination with direct motor cortex stimulation with epidurally implanted electrodes has also been shown to promote axonal sprouting and improve motor control in animal models. (Zareen et al. 2017)

Spinal Cord Stimulation

Epidural spinal cord stimulation has been implanted above the dorsal surface of the spinal cord extensively for chronic pain with a good safety record. This has allowed several initial SCI case studies to be completed where epidural stimulation of the lumbosacral spinal cord augmented recovery in chronic, motor-complete paraplegic SCI patients (Rejc et al. 2017; Grahn et al. 2017). In lumbosacral epidural electrical stimulation, a 16-electrode array is implanted over spinal segments L1–S1. Recovery of voluntary movement with epidural stimulation was observed in motor-complete patients 2 years post-injury (Angeli et al. 2014). There are two proposed mechanisms for this late recovery. First, there is activation of previously "silent" anatomical connections from epidural stimulation. These few remaining descending connections have been present since the time of injury but were insufficient to activate motor pools until epidural stimulation enhanced central excitatory drive by altering spinal cord circuitry. Stimulation may facilitate excitation of propriospinal interneurons, which support propagation of voluntary command to the lumbosacral spinal cord. Second, the continued improvement of voluntary movement with repetitive epidural stimulation and training suggests plasticity with the recruitment of novel neuronal pathways and synapses with time. These results suggest the ability of the spinal networks to learn with task-specific training and improve motor pool recruitment to promote force generation and accuracy.

Epidural stimulators have also been explored in chronic cervical injury and results suggest improvement in grip strength (Lu et al. 2016). In cervical epidural stimulation, two 16-contact percutaneous epidural leads are placed spanning C4–T1. Immediate improvements in maximal hand strength and control were seen in chronic, motor-complete tetraplegic SCI patients within one testing session and incremental, sustained improvements were observed after repeated stimulation. These results suggest that epidural stimulation facilitates the recruitment of viable, but previously nonparticipating, cervical interneuronal networks projecting to motor pools, similar to is observed with epidural stimulation to the lumbosacral spinal cord.

A number of clinical trials have also been investigating transcutaneous spinal cord stimulation (tcSCS) with a similar conceptual goal. tcSCS is less accurate but is noninvasive and inexpensive. However clinical trials of both direct SCS and aSCS are small, and larger clinical trials will need to be done to confirm initial findings.

Brain Machine Interfaces

Brain machine interfaces (BMI), also referred to as brain computer interface (BCI), have been a popular and exciting research topic. The technique allows the recording and decoding of brain activity and translates the information to generate functional output. This concept has shown promise in individual case studies (Bouton et al. 2016; Ajiboye et al. 2017). Recording devices range from noninvasive scalp surface EEG electrodes to invasive subdural or epidural implanted electrocorticography electrode arrays (ECoG) and intraparenchymal microelectrode arrays. The invasive devices have predominantly been implanted on the cortical surface gyri, often in the motor cortex. Neural activity is captured from these devices and decoded by a computer with the goal of interpreting movement plans. Noninvasive devices have a lower signal quality, while invasive recording devices are much more sensitive. The decoded signals are translated into commands that can be used to control an external device, such as prosthetic robotic limb or a cursor on a computer screen (Lee et al. 2013). One of the first implantations of BMI in humans was done by the BrainGate group at Brown University, where a small array implanted in the motor cortex allowed continuous control of a computer cursor in patients with severe SCI (Hochberg et al. 2006; Kim et al. 2008). Application of this technology goes beyond rehabilitation for SCI or stroke patients. Both the Department of Defense and commercial companies have funded efforts to master the ability to decode brain signal with the goal of human enhancement.

There are several barriers that will need to be addressed prior to widespread clinical applications of BMI. The technology currently requires a large amount of highly advanced and specialized technology, which has high costs that would make such a therapy inaccessible and unaffordable. Current participants are still required to be connected to a computer interface, which would not translate well to daily function. Implanted electrode arrays also are prone to gliosis and physical damage, resulting in loss of recordings. As the technology advances to become more portable, affordable, and durable, it will continue to expand in potential uses.

Challenges: Translational Studies, Timing of Intervention, Ethical and Economic Considerations, and Considerations in Clinical Trial Design

Consistency of Results

As with any research, the translation of research from the bench to patient care carries significant challenges. Animal models vary greatly from humans in body size, anatomy, and their response to severe SCI. While we try to control the variability in experimental models, patients carry significant heterogeneity in their individual characteristics and the type of SCI endured. This highlights the importance for early experiments to show a clear and significant benefit before moving on to higher-level trials, and likewise early-phase clinical trials will need to show the same to justify the cost of recruiting a larger population that is likely more heterogeneous.

Risks

While stem cells have been extensively studied in the laboratory, our experience with their use in humans is relatively limited, and we may not know the full range of potential side effects and complications associated with this therapy. As previously discussed, the use of olfactory nasal ensheathing cells for spinal transplantation led to the unwanted growth of a multicystic mass (Dlouhy et al. 2014). Conversely, immunotherapies that block the innate immune responses may prevent scar formation, but may also block protective processes and adversely affect spontaneous recovery (Putatunda et al. 2018). Government regulatory policies may help ensure the safety of translational research, but can also introduce further challenges by blocking potentially effective variations, such as adjustments in dose amount or number.

Timing

As the field of spinal cord injury moves into translational research, challenges surrounding clinical studies arise. One of the most important challenges is determination of timing of therapeutic intervention, which affects both recruitment strategy and potential treatment effects. A window of opportunity may exist for SCI patients in which a response to treatment is only possible for a limited time period after acute injury. This would limit studies to focus on patients in the acute or subacute phase of injury.

Studies that focus on acute injury face issues with accessing individuals within a few hours of presentation, requiring a developed infrastructure in trauma centers. In the acute setting, treatment effects are also more likely to be confounded by spontaneous neurologic recovery or other injuries and complications. The patients that are most likely to benefit from experimental treatments are sometimes also the most likely to spontaneously improve, which makes treatment effects very difficult to detect. While chronic SCI studies may be easier to recruit, patients showing high potential for recovery may not want to take on the risk an experimental approach carries. The burden of extensive follow-up and testing can also lead to difficulty in following the progress of studied patients (Blight et al. 2019; Blesch and Tuszynski 2009).

With respect to completed trials examining the effects of timing of surgical decompression, initial results from surgical trials (Vaccaro et al. 1997) showed no difference in ASIA motor score or grade in patients undergoing early ($<$72 hrs) versus late ($>$5 days) surgical decompression and 1-year average follow-up. All of these patients received high-dose methylprednisolone. The question of optimal surgical timing remained unanswered and gave rise to the Surgical Treatment of Acute Spinal Cord Injury Study (STASCIS) trial, which was intended to compare the efficacy of early ($<$24 hrs) versus late ($>$24 hrs) surgical decompression on 6-month outcomes in cervical SCI patients (Fehlings et al. 2012). All patients were aggressively medically managed with MAP goals $>$85 and 60% of enrolled patients received methylprednisolone as determined by each study team. Multivariate regression analysis, when controlled for methylprednisolone and type of injury (complete versus incomplete), demonstrated patients undergoing early surgical decompression were 2.8 times more likely to improve at least two grades in AIS (ASIA Impairment Scale) grade at 6 months. Several prospective cohort studies since have shown a benefit of early decompression. Subsequent retrospective studies have also shown significant improvement in 1-year SCIM (Spinal Cord Independence Measure) scores, AIS grade improvement, and AIS grade conversion in patients undergoing ultra-early decompression ($<$8–12 hr) after traumatic cervical SCI (Grassner et al. 2016; Burke et al. 2018).

Ethical Challenges

While a cure for SCI would transform the lives of affected patients, it is paramount to ensure that the research to achieve this goal is done ethically. The use of stem cells derived from embryos sparked intense ethical debate, with concern of whether embryos should be considered "persons." As a result of this concern governments worldwide have strictly regulated the use of embryonic stem cells in research. The advent of induced pluripotent stem cells (iPSCs) through genetic reprogramming of somatic cells provided a useful alternative to embryonic stem cells and has significantly minimized this ethical controversy. There are also ethical concerns surrounding animal models of SCI, as lesions of the spinal cord significantly impact the quality of life of an animal. Animal use protocols should ensure the well-being of any animals experimented on.

Once a promising therapy is demonstrated in an animal model, a difficult question is when this should be translated into human trials. Experimental interventions of the spinal cord carry significant risk, as their effects in animals do not always perform the same in humans. Given that recovery in the early stages of injury is difficult to predict, it is a concern that a new therapy could inhibit or negatively affect spontaneous recovery. It is important for researchers and governing research and ethics review boards to carefully ensure the scientific validity, favorable risk–benefit ratio, and respect for human rights. Additionally studies should be designed to maximize safety while ensuring the best care for the patient. An additional ethical challenge arises in patient selection. Patients that have the most severe injury need the most help; they are also the least likely to recover. Ultimately, the high costs of human trials limit the number of participants with strict inclusion criteria that maximize the potential to detect a possible therapeutic benefit (Rosenfeld et al. 2008).

Costs

It is predicted that by 2025, global spending on clinical trials will reach $68.9 billion a year (May 2019). The estimated costs for preclinical studies alone can costs hundreds of millions of dollars (Trounson and McDonald 2015). Stem cell lines can cost thousands of dollars to produce. While rodent models of SCI are widely available and accessible, they do not reflect humans in an anatomical and physiological way. Larger animal models, such as primates, greatly increase in costs as well as regulation. These costs are nominal compared to human clinical trials, and costs continue to increase for late-stage trials as the number of patients required increases and the study becomes more complex. The high financial burden makes careful planning and patient selection imperative to achieve the highest likelihood of identifying effective therapies.

Challenges and Considerations in Designing of Clinical Trials

Various surgical and therapeutic trials have been conducted over the last several decades with the goal of improving functional outcomes for the spinal cord–injured population. Some have been randomized controlled trials (RCT) while others have primarily been cohort studies. Extensive reviews of the trials in SCI can be found in the literature (Tator 2006; Hawryluk et al. 2008; Donavan and Kirschblum 2018). In order to conduct effective and efficient future trials, it is imperative to review the important lessons learned from previous experiences.

One must appreciate the unique challenges in studying this population as well as the areas of possible conflict that can impact results and affect the success of measured outcomes. As outlined eloquently in his review of human SCI trials, Dr. Tator addresses the important shortcomings in SCI research and trial design (Tator 2006). Spinal cord injury itself is a heterogenous disease process. Not only are cervical and thoracic SCI populations quite different in their injury pattern and propensity to regain function, but complete and incomplete populations are also crucially different. Incomplete SCI patients have more capacity to recover than those with complete injury but the myriad pathophysiological mechanisms at place are difficult to control for in any given trial. As alluded to earlier, the complete SCI population is more homogeneous and therefore may be a better population to study, despite the fact that potential for recovery is less than the former. Though it makes good sense to design trials based upon level of injury, thoracic ASIA grade A patients comprise less than 15% of the SCI population, for example, making trial accrual challenging. There are also limitations to our currently used ASIA grading system. For example, a ceiling effect exists for Grades C and D, making it difficult for patients to move out of one category to the next level of improvement. Spontaneous recovery presents a unique but omnipresent phenomenon. It was estimated by Burns et al. that approximately 60% of ASIA grade B through D patients move to a higher grade. Since these grades account for over 50% of patients with SCI, they must be included in trials. However, efficacy of a given intervention may be difficult to distinguish from the natural history in these patient populations. Further, combined upper and lower ASIA motor scores can be similar yet indicate vastly different cervical and thoracic SCI populations with varying degrees of injury and prognosis for recovery. As such, there has been an effort to stratify patients at trial entry into different levels and severities to more accurately quantify recovery. Lastly, and as is certainly the case for surgical trials, difficulty with blinding and timing of intervention provide unique challenges. It is of critical importance that examiners be blinded to treatment groups to minimize bias. This may require mandatory bandaging of surgical and "sham" sites or cooperation on the part of the patient to conceal their treatment group. Optimal timing of intervention has not been well established, though it has been hypothesized that neuroprotection be initiated early while subacute to chronic injury may be optimal for trials aimed at assessing regenerative capacity. These are only a few of the important points to consider when designing or evaluating results in SCI trials.

Currently no one effective therapy exists for restoring function in SCI. This emphasizes the need for continued research and translational studies. Despite the many challenges that exist in the research pathway, many translational human studies are underway. Both successes and failures will aid in our understanding of SCI and pave the road toward developing promising therapies. Throughout the process, it is equally important to ensure that ethical standards are adhered to. Through guidelines and careful review of each step, ethical challenges can be minimized to ensure the safety of all patients.

References

Angeli CA et al (2014) Altering spinal cord excitability enables voluntary movements after chronic complete paralysis in humans. Brain J Neurol 137(5):1394–1409

Ahuja CS, Nori S et al (2017a) Traumatic spinal cord injury-repair and regeneration. Neurosurgery 80(3S): S9–S22

Ahuja CS, Wilson JR et al (2017b) Traumatic spinal cord injury. Nat Rev Dis Primers 3:17018

Ajiboye AB et al (2017) Restoration of reaching and grasping movements through brain-controlled muscle stimulation in a person with tetraplegia: a proof-of-concept demonstration. Lancet 389(10081):1821–1830. Available at. https://doi.org/10.1016/s0140-6736(17)30601-3

Altaf F et al (2017) The differential effects of norepinephrine and dopamine on cerebrospinal fluid pressure and spinal cord perfusion pressure after acute human spinal cord injury. Spinal Cord 55:33–38

Anderson KD et al (2017) Safety of autologous human Schwann cell transplantation in subacute thoracic spinal cord injury. J Neurotrauma 34(21):2950–2963

Anon (n.d.) InVivo therapeutics announces presentation of twelve-month results from the INSPIRE Study of the Investigational Neuro-Spinal Scaffold™ in Acute Thoracic Complete Spinal Cord Injury – InVivo therapeutics. Available at: https://www.invivotherapeutics.com/press-releases/invivo-therapeutics-announces-presentation-of-twelve-month-results-from-the-inspire-study-of-the-investigational-neuro-spinal-scaffold-in-acute-thoracic-complete-spinal-cord-injury/. Accessed 17 May 2019

Assinck P et al (2017) Cell transplantation therapy for spinal cord injury. Nat Neurosci 20(5):637–647

Badhiwala JH, Ahuja CS, Fehlings MG (2018) Time is spine: a review of translational advances in spinal cord injury. J Neurosurg Spine 30(1):1–18

Blesch A, Tuszynski MH (2009) Spinal cord injury: plasticity, regeneration and the challenge of translational drug development. Trends Neurosci 32(1):41–47

Blight AR et al (2019) The challenge of recruitment for neurotherapeutic clinical trials in spinal cord injury. Spinal Cord 57(5):348–359

Bouton CE et al (2016) Restoring cortical control of functional movement in a human with quadriplegia. Nature 533(7602):247–250. Available at. https://doi.org/10.1038/nature17435

Bracken MB et al (1990) A randomized, controlled trial of methylprednisolone or naloxone in the treatment of acute spinal cord injury. NEJM 322(20):1405–1411

Bracken MB et al (1997) Administration of methylprednisolone for 24 or 48 hours or Tirilazad Mesylate for 48 hours in the treatment of acute spinal cord injury. JAMA 277(20):1597–1604

Branco F, Cardenas DD, Svircev JN (2007) Spinal cord injury: a comprehensive review. Phys Med Rehabil Clin N Am 18(4):651–679. v

Burke JF et al (2018) Ultra-early (<12 hours) surgery correlates with higher rate of American spinal injury association impairment scale conversion after cervical spinal cord injury. Neurosurgery 0(0):1–5

Chen Y, He Y, DeVivo MJ (2016) Changing demographics and injury profile of new traumatic spinal cord injuries in the United States, 1972–2014. Arch Phys Med Rehabil 97(10):1610–1619

Cheng H, Cao Y, Olson L (1996) Spinal cord repair in adult paraplegic rats: partial restoration of hind limb function. Science 273(5274):510–513

Cheng H et al (2004) Spinal cord repair with acidic fibroblast growth factor as a treatment for a patient with chronic paraplegia. Spine 29(14):E284–E288

Derakhshanrad N et al (2018) Granulocyte-colony stimulating factor administration for neurological improvement in patients with postrehabilitation chronic incomplete traumatic spinal cord injuries: a double-blind randomized controlled clinical trial. J Neurosurg Spine 29(1):97–107

Dietz V (2009) Body weight supported gait training: from laboratory to clinical setting. Brain Res Bull 78(1):I–VI

Dietz V (2012) Clinical aspects for the application of robotics in Neurorehabilitation. In: Dietz V, Nef T, Rymer WZ (eds) Neurorehabilitation technology. Springer, London, pp 291–301

Dietz V, Fouad K (2014) Restoration of sensorimotor functions after spinal cord injury. Brain J Neurol 137. (Pt 3:654–667

Dlouhy BJ et al (2014) Autograft-derived spinal cord mass following olfactory mucosal cell transplantation in a spinal cord injury patient. J Neurosurg Spine 21 (4):618–622. Available at. https://doi.org/10.3171/2014.5.spine13992

Donavan J, Kirschblum S (2018) Clinical trials in traumatic spinal cord injury. Neurotherapeutics 15:654–668. https://doi.org/10.1007/s13311-018-0632-5

Eckert MJ, Martin MJ (2017) Trauma: spinal cord injury. Surg Clin North Am 97(5):1031–1045

Fehlings MG et al (2011) A phase I/IIa clinical trial of a recombinant rho protein antagonist in acute spinal cord injury. J Neurotrauma 28(5):787–796

Fehlings MG et al (2012) Early versus delayed decompression for traumatic cervical spinal cord injury: results of the surgical timing in acute spinal cord injury study (STASCIS). PlosOne 7:2

Fehlings MG et al (2018) Rho inhibitor VX-210 in acute traumatic subaxial cervical spinal cord injury: design of the SPinal cord injury rho INhibition InvestiGation (SPRING) clinical trial. J Neurotrauma 35 (9):1049–1056

Furlan GC, Fehlings MG (2008) Cardiovascular complications after acute spinal cord injury:pathophysiology, diagnosis, and management. Neurosurg Focus 25(5):1–15

Grahn PJ et al (2017) Enabling task-specific volitional motor functions via spinal cord Neuromodulation in a human with paraplegia. Mayo Clin Proc 92(4):544–554. Available at. https://doi.org/10.1016/j.mayocp.2017.02.014

Grassner L et al (2016) Early decompression (<8 h) after traumatic cervical spinal cord injury improves functinoal outcome as assessed by spinal cord independence measure after one year. J Neurotrauma 33 (18):1658–1666

Gunduz A et al (2017) Non-invasive brain stimulation to promote motor and functional recovery following spinal cord injury. Regen Res 12(12):1933–1938

Hawryluk GWJ et al (2008) Protection and repair of the injured spinal cord: a review of completed, ongoing, and planned clinical trials for acute spinal cord injury. Neurosurg Focus 25(5):1–16

Hochberg LR et al (2006) Neuronal ensemble control of prosthetic devices by a human with tetraplegia. Nature 442(7099):164–171

Hong J, Rodgers CE, Fehlings MG (2018) Stem cell applications in spinal cord injury: a primer. Stem Cell Genetics Biomed Res:43–72. Available at. https://doi.org/10.1007/978-3-319-90695-9_4

James ND et al (2018) Neuromodulation in the restoration of function after spinal cord injury. Lancet Neurol 17 (10):905–917

Jin MC et al (2019) Stem cell therapies for acute spinal cord injury in humans: a review. Neurosurg Focus 46 (3):E10

Krause JS et al (1997) Mortality after spinal cord injury: an 11-year prospective study. Arch Phys Med Rehabil 78 (8):815–821

Kucher K et al (2018) First-in-man Intrathecal application of Neurite growth-promoting anti-Nogo-a antibodies in acute spinal cord injury. Neurorehabil Neural Repair 32 (6–7):578–589

Kwon BK et al (2009a) Intrathecal pressure monitoring and cerebrospinal fluid drainage in acute spinal cord injury: a prospective randomized trial. J Neurosurg Spine 10:181–193

Kwon BK et al (2009b) Magnesium chloride in a polyethylene glycol formulation as a neuroprotective therapy for acute spinal cord injury: preclinical refinement and optimization. J Neurotr 26(8):1379–1393. https://doi.org/10.1089/neu.2009.0884

Kim S-P et al (2008) Neural control of computer cursor velocity by decoding motor cortical spiking activity in humans with tetraplegia. J Neural Engin 5(4):455–476

Kitamura K et al (2011) Human hepatocyte growth factor promotes functional recovery in primates after spinal cord injury. PloS one 6(11):e27706

Koda M et al (2007) Granulocyte colony-stimulating factor (G-CSF) mobilizes bone marrow-derived cells into injured spinal cord and promotes functional recovery after compression-induced spinal cord injury in mice. Brain Res 1149:223–231

Lee B, Liu CY, Apuzzo MLJ (2013) A prime ron brain-machine interfaces, concepts, and technology: a key element in the future of functional neurorestoration. World Neurosurg 79(3-4):457–471

Lehrer N (1996) Treatment with thyrotropin-releasing hormone (TRH) in patients with traumatic spinal cord injuries. Neurol Rep 20(1):65

Liu S, Xie Y-Y, Wang B (2019) Role and prospects of regenerative biomaterials in the repair of spinal cord injury. Neural Regen Res 14(8):1352–1363

Lu DC et al (2016) Engaging cervical spinal cord networks to Reenable volitional control of hand function in Tetraplegic patients. Neurorehabil Neural Repair 30 (10):951–962

Lukovic D et al (2014) Perspectives and future directions of human pluripotent stem cell-based therapies: lessons from Geron's clinical trial for spinal cord injury. Stem Cells Dev 23(1):1–4. Available at. https://doi.org/10.1089/scd.2013.0266

Marino RJ et al (2011) Upper- and lower-extremity motor recovery after traumatic cervical spinal cord injury: an update from the national spinal cord injury database. Arch Phys Med Rehabil 92(3):369–375

May M (2019) Clinical trial costs go under the microscope. Nat Med. Available at. https://doi.org/10.1038/d41591-019-00008-7

National Spinal Cord Injury Statistical Center, Updated 2018. Spinal cord injury: facts and figures at a glance. Available at: https://www.nscisc.uab.edu/Public/Facts%20and%20Figures%20-%202018.pdf. Accessed 1 May 2019

Nori S, Ahuja CS, Fehlings MG (2017) Translational advances in the management of acute spinal cord Injury: what is new? what is hot?. Neurosurg 64 (CN_suppl_1):119–128

Popovic MR, Zivanovic V, Valiante TA (2016) Restoration of upper limb function in an individual with cervical Spondylotic myelopathy using functional electrical stimulation therapy: a case study. Front Neurol 7. Available at. https://doi.org/10.3389/fneur.2016.00081

Priest CA et al (2015) Preclinical safety of human embryonic stem cell-derived oligodendrocyte progenitors supporting clinical trials in spinal cord injury. Regen Med 10(8):939–958

Park HC et al (2005) Treatment of complete spinal cord injury patients by autologous bone marrow cell transplantation and administration of granulocyte-macrophage colony stimulating factor. Tissue Engineer 11 (5-6):913–922

Putatunda R, Bethea JR, Hu W-H (2018) Potential immunotherapies for traumatic brain and spinal cord injury. Chin J Traumatol = Zhonghua chuang shang za zhi/Chinese Medical Association 21(3):125–136

Rejc E et al (2017) Motor recovery after activity-based training with spinal cord epidural stimulation in a chronic motor complete paraplegic. Sci Rep 7:1. https://doi.org/10.1038/s41598-017-14003-w

Roberts TT, Leonard GR, Cepela DJ (2017) Classifications in brief: American spinal injury association (ASIA) impairment scale. Clin Orthop Relat Res 475 (5):1499–1504

Rosenfeld JV et al (2008) The ethics of the treatment of spinal cord injury: stem cell transplants, motor Neuroprosthetics, and social equity. Topics Spinal Cord Injury Rehabil 14(1):76–88

Schönherr MC et al (1999) Functional outcome of patients with spinal cord injury: rehabilitation outcome study. Clin Rehabil 13(6):457–463

Schwab ME, Strittmatter SM (2014) Nogo limits neural plasticity and recovery from injury. Curr Opin Neurobiol 27:53–60

Siddiqui AM, Khazaei M, Fehlings MG (2015) Translating mechanisms of neuroprotection, regeneration, and repair to treatment of spinal cord injury. Prog Brain Res 218:15–54

Shultz RV, Zhong Y (2017) Minocycline tarets multiple secondary injury mechanisms in traumatic spinal cord injury. Neural Regen Res 12(5):702–713

Tator CH (2006) Review of treatment trials in human spinal cord injury: issues, difficulties, and recommendations. Neurosurgery 59(5):957–987

Theodore N et al (2016) First human implantation of a Bioresorbable polymer scaffold for acute traumatic spinal cord injury: a clinical pilot study for safety and feasibility. Neurosurgery 79(2):E305–E312

Tran AP, Warren PM, Silver J (2018) The biology of regeneration failure and success after spinal cord injury. Physiol Rev 98(2):881–917

Trounson A, McDonald C (2015) Stem cell therapies in clinical trials: progress and challenges. Cell Stem Cell 17(1):11–22. Available at. https://doi.org/10.1016/j.stem.2015.06.007

Tsuji O et al (2019) Concise review: laying the groundwork for a first-in-human study of an induced pluripotent stem cell-based intervention for spinal cord injury. Stem Cells 37(1):6–13

Venkatesh K et al (2019) Spinal cord injury: pathophysiology, treatment strategies, associated challenges, and future implications. Cell Tissue Res. Available at 377:125. https://doi.org/10.1007/s00441-019-03039-1

Vaccaro AR et al (1997) Neurologic outcome of early versus late surgery for cervical spinal cord injury. Spine 22(22):2609–2613

Walters BC et al (2013) Guidelines for the management of acute cervical spine and spinal cord injuries: 2013 update. Clin Neurosurg 60:82–91

Warita H et al (2019) Safety, tolerability, and pharmacodynamics of Intrathecal injection of recombinant human HGF (KP-100) in subjects with amyotrophic lateral sclerosis: a phase I trial. J Clin Pharmacol 59(5):677–687

Waters RL et al (1992) Recovery following complete paraplegia. Arch Phys Med Rehabil 73(9):784–789

Waters RL et al (1993) Motor and sensory recovery following complete tetraplegia. Arch Phys Med Rehabil 74(3):242–247

Welch RD et al (1986) Functional independence in quadriplegia: critical levels. Arch Phys Med Rehabil 67 (4):235–240

West CR, Mills P, Krassioukov AV (2012) Influence of the neurological level of spinal cord injury on cardiovascular outcomes in humans: a meta-analysis. Spinal Cord 50(7):484–492

Wilson JR, Cadotte DW, Fehlings MG (2012) Clinical predictors of neurological outcome, functional status, and survival after traumatic spinal cord injury: a systematic review. J Neurosurg Spine 17(1 Suppl):11–26

Winslow C, Rozovsky J (2003) Effect of spinal cord injury on the respiratory system. Am J Phys Med Rehabil/Assoc Acad Physiatrists 82(10):803–814

Wu J-C et al (2008) Nerve repair using acidic fibroblast growth factor in human cervical spinal cord injury: a preliminary phase I clinical study. J Neurosurg Spine 8 (3):208–214

Wu J-C et al (2011) Acidic fibroblast growth factor for repair of human spinal cord injury: a clinical trial. J Neurosurg Spine 15(3):216–227

Zareen N et al (2017) Motor cortex and spinal cord neuromodulatino promote corticospinal tract axonal outgrowth and motor recovery after cervical contusion spinal cord injury. Experiment Neurol 297:179–189

Biologics: Inherent Challenges

68

Charles C. Lee and Kee D. Kim

Contents

Introduction	1252
Autologous Bone	1252
Allogeneic Bone	1253
Growth Factors	1254
BMP-2	1254
OP-1 (BMP-7)	1256
NELL1	1256
Growth Factor Combinations	1256
Stem and Progenitor Cells	1257
Gene Therapy	1258
Microenvironment	1260
Conclusions	1262
Cross-References	1262
References	1262

Abstract

A biologic product is derived from a biological system and may include osteogenic, osteoinductive, and/or osteoconductive properties that are required for efficient bone regeneration. Current biologics include autologous or allogenic bone products, growth factors, and bone graft substitutes. Each of these therapeutic approaches presents unique advantages and challenges that require further development. Due to known, inherent challenges of current biologics, a single approach may not be sufficient to address complex issues presented in patients with significant issues in the spine, especially those at high risk for nonunion. While a combinatory approach is certainly

C. C. Lee
Department of Cell Biology and Human Anatomy, School of Medicine, University of California, Davis, Davis, CA, USA
e-mail: cglee@ucdavis.edu

K. D. Kim (✉)
Department of Neurological Surgery, UC Davis School of Medicine, Sacramento, CA, USA
e-mail: kdkim@ucdavis.edu

© Springer Nature Switzerland AG 2021
B. C. Cheng (ed.), *Handbook of Spine Technology*,
https://doi.org/10.1007/978-3-319-44424-6_137

interesting, a new therapeutic modality that combines different aspects of bone repair and regeneration may be needed. Stem cell and gene-based approaches have been investigated to address unmet clinical needs, which may include systemic issues and/or suboptimal microenvironment at the site of implantation. Intricate interactions and molecular cross-talks between osteogenic, osteoinductive, and osteoconductive biologics that are required for successful bone fusion are still under investigation.

Keywords

Autograft · Allograft · Growth factors · Stem and progenitor cells · Gene therapy · Spine therapeutics · Spine biologics · Challenges · Microenvironment · Bone graft substitutes

Introduction

Two or more vertebrae may be fused to stabilize the spine for a variety of clinical conditions such as degenerative disc disease, vertebral fracture, scoliosis, and other conditions that cause instability of the spine. The height of an intervertebral disc may be up to 16 mm (Zhou et al. 2000) and represents the distance between two adjacent vertebrae that needs to be filled with new patient bone for a successful interbody fusion. In the absence of a true, effective osteogenic element in the treatment modality, this distance may present a significant challenge for the right type of cells on either side to migrate out from their native environment and bridge the bony gap, especially in patients with risk factors such as diabetes, aging, and smoking. As an example, a significant difference was observed in nonunion rates between non-smokers and smokers (14.2% vs. 26.5%) (Glassman et al. 2000). Thus, a proper understanding of various therapeutic modalities, size and type of bony gap (or defect), and microenvironment, where bone regeneration must take place, is critical for a successful fusion.

A biologic medical product (or biologic) refers to a therapeutic modality that is derived from a biological system (e.g., animals, humans) and used to address a clinical condition. For bone fusion, it includes any biological elements that have osteogenic, osteoinductive, and/or osteoconductive properties, which participate in regeneration of bone in a clinical setting. An intricate interaction or molecular cross-talk between these properties may be necessary for efficient bone regeneration. The osteogenic component may include stem and progenitor cells (e.g., mesenchymal stem cells [MSC]), bone marrow-derived products (bone marrow aspirate), and autograft/allograft. It is typically referred to viable cells that can directly contribute to repair and regeneration of bone at the affected site. Exogenous (transplanted) or endogenous (native) cells that can proliferate and differentiate into mature bone cells may be considered osteogenic. An osteoinductive factor participates in differentiation of bone precursors toward the terminal lineage. It can modulate local microenvironment to promote tissue healing post-injury. Osteoinductive factors include bone morphogenetic proteins (BMPs), fibroblast growth factor-2 (FGF-2), and platelet-derived growth factor (PDGF) (Khorsand et al. 2017; Kim et al. 2015). An osteoconductive element provides a scaffold that can support the function of osteogenic cells and osteoinductive factors structurally to regenerate, fill, and bridge bony void or defect with new bone. It may be derived from a biological system or produced synthetically. In this chapter, current clinical uses of each of these components that participate in bone repair and regeneration and their inherent challenges are discussed.

Autologous Bone

Currently, technologies addressing the need to repair bone defect or void are extensive, but none currently offer the benefit of a personalized approach that is safe, relatively noninvasive, and anatomically defined for autologous bone formation. As described above, an ideal bone product would be osteoconductive, osteoinductive, and osteogenic. Autologous bone (autograft) meets the above criteria and is considered the "gold standard" for bone repair and regeneration. Local autologous bone is often inadequate. Therefore, it is important to consider many factors such as potential morbidity due to the harvesting procedure, tissue condition at the harvest site, mechanical and structural implication of graft harvesting, and patient condition among others. Autologous bone is traditionally harvested from the iliac crest but can also be collected from other sites such as the rib, vertebral body, and fibula. The iliac crest has been considered the "gold standard" source since the amount of bone available for harvesting is relatively large compared to the other sites. A wide variety of harvesting techniques are available depending on the type of bone subject to harvesting and the approach used, including trapdoor (bone flap), window (removal of segmental bone), splitting (fissure in the iliac crest), and trephine extraction ("minimally invasive") techniques. Overall, there has not been consensus as to which technique results in a better clinical outcome, especially the risk of complication and morbidity associated with the harvest procedure of choice. While the use of autologous bone graft has shown exceptional efficacy, regardless of the type of harvesting methods or sites used, with a fusion rate greater than 94% (Sawin et al. 1998), donor site morbidity still presents a significant problem. Pain at the donor site was reported in 16% of patients who reported a worse pain at the donor site (iliac crest) than the primary surgical site in addition to difficulty walking (15.1%), employment (5.2%), recreation (12.9%), household chores (7.6%), sexual activity (7.6%), and irritation from clothing (5.9%) at 1 year follow-up (Kim et al. 2009). Similar findings were observed in other studies with a longer follow-up period (Sasso et al. 2005; Schwartz et al. 2009). Although there have been efforts to reduce donor site morbidity including restricting the area for harvesting and use of a suction drain (Kurz et al. 1989), the clinical outcome of using autologous bone is still less than optimal due to donor site morbidity, increased blood loss and frequent need for blood transfusion, prolonged operative time, risk of nerve damage, and a revision in many cases (Mulconrey et al. 2008).

Allogeneic Bone

Allogeneic bone (allograft) is considered an alternative source of graft material and eliminates the donor-site complications associated with autologous bone harvest procedures (Ehrler and Vaccaro 2000). It is harvested from cadaveric and living donors and is typically stored in a bone tissue bank. Fresh-frozen allografts, freeze-dried allografts, and demineralized bone matrix are the most commonly used forms of allogenic bone, and each preparation method brings certain advantages and disadvantages to a bone grafting procedure (An et al. 1995). Across all methods of preparation, allograft is more immunogenic and displays a decreased rate of graft integration in comparison to autograft (Ehrler and Vaccaro 2000). Fresh-frozen allografts are treated with antibiotics after harvesting and must be frozen below $-70\,^{\circ}C$ to preserve the quality and viability of the graft material (Laitinen et al. 2006). The composition and structure of fresh-frozen allograft is most similar to those properties of autologous bone. Fresh-frozen grafts retain the most structural integrity of any processed allograft, which provides an optimal matrix for osteogenesis at the graft site. This preparation method also preserves a significant amount of endogenous BMPs known to be osteoinductive. However, the presence of cellular debris activates a host immune response and leads to increased risk of graft-site inflammation, delayed bone regeneration, and higher failure rate (An et al. 1995; Laitinen et al. 2006). Freeze-dried allografts are prepared similarly to fresh-frozen grafts with the addition of a dehydration process. Dehydration renders the graft material shelf-stable at room temperature and thus more readily stored for future use (Ehrler and Vaccaro 2000). Freeze-dried allografts carry a decreased risk of infection and immunogenicity at the graft site. This allograft type contains no live cells and fewer viable growth factors, but the mineralized portion may still provide a significant osteoinductive effect (Mellonig 1995). A potential drawback of freeze-dried allograft is decreased mechanical stability and structural strength, rendering the graft more brittle and prone to fracture (Cornu et al. 2000).

Demineralized bone matrix (DBM) is more extensively processed than fresh-frozen and freeze-dried allografts. Acid treatments are used to remove the inorganic, mineralized components of bone leaving behind the organic collagen matrix composed of type I collagen, non-collagenous proteins, and various growth factors (Lee et al. 2005). DBM is sold in numerous commercial formulations and is subject to a great deal of variability in processing techniques. Differences in acid solutions, temperature, particle size, terminal sterilization, and demineralization time result in lot-to-lot variability, yielding inconsistencies in BMP content and osteoinductive potential (Lee et al. 2005; Wang et al. 2007). DBM does not provide structural support as the mechanical properties of bone are significantly diminished secondary to the demineralization process (Pacaccio and Stern 2005). Generally, DBM is indicated for use as a bone graft extender and is not thought to be sufficient as a stand-alone bone graft substitute for reconstructive orthopedic procedures. Allograft materials made of minimally or highly processed allogenic bone are relatively abundant in supply and are effective in regenerating bone. Overall, the structural integrity and osteoinductivity of fresh-frozen allograft is preferred over freeze-dried allograft and DBM products, despite the risk of immunogenic complications.

Growth Factors

BMP-2

Bioactive agents including growth factors with osteoinductive functions are administered for spinal fusion as a means to promote endogenous bone formation and healing. BMPs are a family of osteoinductive growth factors with important roles in development and are known to stimulate osteoblastic differentiation of stem cells. BMP-2, or recombinant BMP-2 (rhBMP-2) produced in vitro by mammalian or bacteria cells, is the most widely studied and has been approved by the FDA for anterior lumbar interbody fusion (ALIF) in 2002 (US Food and Drug Administration 2019a). rhBMP-2 is typically administered with a carrier such as demineralized bone matrix or biodegradable collagen sponges at doses in the range of 3.5–20 mg (Duarte et al. 2017). A commercially available product for ALIF, Infuse™ Bone Graft (Medtronic), contains rhBMP-2 with an absorbable collagen sponge carrier inserted in a titanium cage to provide mechanical support. Early clinical studies demonstrated the Infuse™ product to be superior to iliac crest bone graft with 94% versus 88.7% fusion, respectively (Burkus et al. 2002a, b). Radiographs and computed tomographic (CT) scans provided evidence of rhBMP-2 driven osteoinduction in the interbody cages within 6 months of surgery and new bone formation in the surrounding disc space by 24 months post-surgery in all patients (Burkus et al. 2003). Follow-up studies compared cortical allograft dowels with or without rhBMP-2 for single level ALIF. All patients receiving the allograft with rhBMP-2 showed radiographic evidence of fusion by 12 months which remained stable through 24 months ($n = 79$, 100% fusion) compared with controls ($n = 52$, 89% and 81.5% fusion at 12 and 24 months, respectively) (Burkus et al. 2005, 2006). In addition to improving bone formation, the use of Infuse™ is simple with reduced operation times, length of hospital stay, and volume of blood loss compared to bone graft therapy (Burkus et al. 2002a). FDA has granted additional approvals for Infuse™ in polyetheretherketone supports for ALIF and oblique lateral interbody fusion (OLIF, US FDA 2019b).

The use of rhBMP-2 with the Infuse™ product was widely adopted for ALIF (on-label) as well as many off-label applications. In 2010, off-label use of rhBMP-2 was reported to be as high 85% including primary cervical, primary thoracolumbar, posterior lumbar, transforaminal lumbar, and posterolateral spinal fusions (Ong et al. 2010). When used as directed for ALIF, industry-sponsored clinical study publications reported efficacy rates superior to allogenic or autologous bone with adverse events rarely observed. Follow-up publications and review of FDA data summaries focused attention on a number of complications and adverse events including implant displacement, subsidence, osteoclast-mediated bone resorption, ectopic bone formation, inflammation, life-threatening cervical swelling, urogenital complications, and possible increased risk of tumor formation (Carragee et al. 2011a; Savage et al. 2015; Faundez et al. 2016).

In addition to osteoinductive properties, rhBMP-2 enhances osteoclast activity which has been linked with subsidence or abnormal bone resorption. Localized areas of bone remodeling were noted at 12 months post-surgery by CT scan in 18% of patients receiving cortical allograft dowels with rhBMP-2 (Burkus et al. 2006). These regions resolved by 24 months in follow-up CT scans. Patients treated with rhBMP-2 within the disc space during transforaminal lumbar interbody fusion were observed with 69% vertebral bone resorption and 74% graft subsidence (McClellan et al. 2006). Radiologic analysis of interbody fusions in the cervical or lumbar spine performed with a polyetheretherketone cage and rhBMP-2 demonstrated endplate resorption in all cervical and most lumbar fusions with concomitant cage migration and disc space subsidence (Vaidya et al. 2008). Ectopic bone formation beyond the implant site has also been reported and is thought to be a consequence of leakage of rhBMP-2 from the collagen sponge due to pressure during the placement procedure or the use of irrigation or suction near the graft site (James et al. 2016). Leakage occurring in the epidural space may result in nerve root compression and is noted more frequently in transforaminal lumber interbody fusion (Rihn et al. 2009).

Use of rhBMP-2 in cervical fusions has resulted in severe, life-threatening swelling in the head and neck leading to dyspnea and dysphagia (Shields et al. 2006). Procedure-related swelling around the cervical spine contributes to constriction of airways and nerves that lie in close anatomical proximity. The life-threatening nature of these complications may require interventions such as intubation, anti-inflammatory medications, tracheotomies, or additional surgeries (Smucker et al. 2006; Fineberg et al. 2013). Seroma formation was also noted when the Infuse bone graft was used for occipitocervical fusion (Shahlaie and Kim 2008) with elevated levels of inflammatory cytokines including IL6, IL8, and TNFα reported in seroma fluid (Robin et al. 2010). The severity of inflammatory complications related to rhBMP-2 use in the cervical region prompted FDA to issue a public health warning regarding the use of this product in the cervical region in 2008 (Crawford et al. 2009). Use of rhBMP-2/Infuse™ products in pediatric spinal fusions is also contraindicated (Epstein 2013). No differences in infection rates were noted in spinal fusions with and without rhBMP-2 in a retrospective study by Williams et al. (2011) suggesting swelling events to be related to hypersensitivity rather than infection.

A number of urogenital complications have been reported including retrograde ejaculation and bladder retention with incidence rates up to 6.3% and 9.7% respectively, although in most studies these events were not statistically greater than observed in control groups (Carragee et al. 2011b). Further studies of 10-year outcomes following rhBMP-2/Infuse™ for ALIF suggest these complications may be related to concomitant prostatic disease with specific correlation to rhBMP-2 difficult to establish (Comer et al. 2012). Other studies suggest transperitoneal and laparoscopic approaches increase risk of urogenital adverse events in comparison to a retroperitoneal approach (Burkus et al. 2002a; Than et al. 2011).

BMPs have developmental and regenerative roles in many organ systems and are known to regulate cell growth and differentiation as well as stem and progenitor cell functions. Members of this family of proteins are frequently upregulated in tumors of diverse organ systems and the spine is a common site for tumor metastases leading to concerns that rhBMP-2 use in spinal fusions may increase tumor risk (Carragee et al. 2013). Retrospective analysis by Devine et al. (2012) did not find evidence of increased cancer risk when rhBMP-2 was used as approved. Higher doses of rhBMP-2 (40 mg) were linked to small increases in cancer risk, as was the use of rhBMP-7. In agreement with these findings, retrospective studies of cancer incidence in Medicare patients receiving rhBMP-2 over a 5-year period were not associated with an increase in cancer risk (Kelly et al. 2014; Beachler et al. 2016). Likewise, an analysis of database studies of rhBMP-2 in spinal fusion was unable to detect a significant increase in local or distant site tumor formation (Cahill et al. 2015).

OP-1 (BMP-7)

Osteogenic protein 1 (OP-1), also known as BMP-7, is another member of the family of BMPs with osteoinductive potential. In spinal fusion, rhBMP-7 is delivered in a collagen putty (Olyppus Biotech). When compared with autograft bone for noninstrumented posterolateral lumbar fusion, the OP-1 implant was similar, but not superior to autograft bone (Johnsson et al. 2002). FDA approval, under an investigational device exemption, was granted in 2004 for OP-1 putty in posterolateral spinal fusion (Vaccaro et al. 2003, 2004). Radiographic and clinical outcomes were at least equivalent to the autograft cohort over the 2-year follow up period, with no adverse events reported (Vaccaro et al. 2005a, b). The OP-1 putty was also assessed as an adjunct to autograft bone for intertransverse process lumbar fusion in a small clinical trial (Vaccaro et al. 2003). Inclusion of OP-1 putty with the autograft bone was not shown to improve spinal fusion, although product safety was demonstrated with no adverse events reported. In a study with 4-year follow up, OP-1 putty was compared with iliac crest autograft and shown to be equivalent in promoting fusion and advantageous from a procedural view due to less blood loss and shorter operative times (Vaccaro et al. 2008a, b). In a randomized, multicenter trial, there were fewer fusions in the OP-1 cohort (54%) than in the iliac crest autograft groups (74%) (Delawi et al. 2016). More recent clinical trials are focused on the use of OP-1 in uninstrumented posterolateral fusions (Olympus Biotech, http://www.clinicaltrials.gov/ct2/show/NCT00679107). The safety profile of OP-1 in terms of adverse events appears to be superior to rhBMP-2; however, this should be interpreted with caution until larger OP-1 trials are completed. In terms of effectiveness, OP-1 has not been shown to be superior to autograft bone and direct comparisons with rhBMP-2 have not been conducted.

NELL1

Other growth factors have shown potential in spinal fusion studies in animals. Neural EGFL like 1 (NELL1) is an osteoblast-specific growth factor which acts in bone maintenance and repair to promote expansion of a progenitor population (James et al. 2017). When delivered in a demineralized bone putty carrier, NELL1 was found to be effective for spinal fusion in rats (Li et al. 2010). In a sheep spinal fusion model, NELL1 delivered in demineralized bone matrix promoted 100% fusion by 3 months with advantages in bone volume and mineral density compared to demineralized bone alone (Siu et al. 2011). Studies by Yuan et al. (2013) compared NELL1 in demineralized bone matrix with the BMP-2/Infuse™ bone graft product in a rat spinal fusion model. Fusion rates were similar in both groups (100%). Histological analysis demonstrated the presence of adipocytes and cyst-like bone formation in the rhBMP-2 graft cohort and increases in bone formation, ossification, and vascularization in the NELL1 group (Yuan et al. 2013).

Growth Factor Combinations

Combinations of growth factors have also been tested in animal spinal fusion studies including angiopoietin 1 (ANG1) and cartilage oligomeric matrix protein (COMP). ANG1 is a secreted factor with important roles in vascular development and angiogenesis. COMP is a BMP-binding protein expressed during endochondral ossification and shown in vitro to enhance BMP-2 activity (Ishida et al. 2013). A recombinant chimeric protein composed of COMP and ANG1 delivered in a collage sponge for lumbar fusion was shown to be effective (89.5% vs. 38.5% in sham-treated controls) in rats (Park et al. 2011). Greater bone volume, mechanical strength, and vascularity were also reported in fusions with the COMP-ANG1 chimeric protein. Endogenous BMPs are soluble factors thought to be present in nanogram concentrations, but supraphysiological doses of 3.5–20 mg are commonly utilized in spinal fusion applications (Ishida et al. 2013). In an effort to reduce rhBMP-2 dosage and therefore improve product safety, co-administration of COMP and rhBMP-2 was shown to induce equivalent bone formation at lower doses than rhBMP-2 alone in a rat model of spinal fusion (Refaat et al. 2016).

Other growth factors studied in animal models for spinal fusion include AB204 (Activin A/BMP-2 chimera), calcitonin, FGF-2, growth differentiation factor 5, insulin-like growth factor 1, transforming growth factor β, noggin, peptide B2A, and secreted phosphoprotein 24 (Cottrill et al. 2019). In general, these studies suggest that growth factors alone or as an adjunct to bone autografts enhance spinal fusion efficacy. Adverse events associated with rhBMP-2 administration at pharmacological doses raise valid concerns about the safety of growth factor therapies with this family of proteins. Combinatorial approaches or methods to reduce dosage to physiological levels may provide safer, less expensive options. Growth factors such as NELL1, with greater specificity of action, may improve safety and efficacy but clinical trials have not yet been reported.

Stem and Progenitor Cells

An ideal population of cells for bone repair and regeneration must be able to self-renew and proliferate to provide a long-term reservoir of cells available to respond to biological cues. Also, they must be able to differentiate toward mature osteocytes without resulting in fibrosis. In a strict sense, a true stem cell must be able to self-review and differentiate toward multiple lineages clonally at the single-cell level. Progenitor cells have a limited capability to self-renew, proliferate, and differentiate. Stem and progenitor cells that participate in osteogenesis may reside in and near bone or at another location such as adipose tissue. Similar to the hematopoietic system, osteogenic stem and progenitor cells must also be capable of maintaining healthy, long-term homeostasis of the skeletal system. Single human bone marrow-derived high proliferative potential-mesenchymal colony-forming cells (HPP-MCFC) have been investigated for their ability to give rise to another generation of HPP-MCFC and differentiate into multiple lineages clonally (Lee et al. 2013). HPP-MCFC are present at a frequency of about 7% in typical bone marrow mesenchymal cell cultures and show a significant osteogenic activity when induced with osteogenic growth factors. Although controversies still exist, these proliferative and multipotent cell populations derived from different tissue sources are referred to as MSC. MSC have been identified in bone marrow, bone, fat, muscle, and umbilical cord (UC). However, most of studies on MSC are not performed clonally, and it is difficult to conclusively claim their "stem-cell-ness." MSC that have been isolated in culture show expression of different surface markers such as CD29, CD44, CD90, CD49a-f, CD51, CD73 (SH3), CD105 (SH2), CD106, CD166, and Stro-1 and lack of expression of CD45, CD34, CD14 or CD11b, CD79a or CD19, and HLA-DR in addition to plastic adherence and potential to differentiate toward multiple lineages (Dominici et al. 2006; Maleki et al. 2014). STRO-3+ mesenchymal precursor cells have shown an ability to partially reconstitute extracellular matrix (ECM) at 6 months following intradiscal administration into degenerate discs in sheep (Ghosh et al. 2012). Other cell types with similar phenotypic and functional properties have been described including adipose-derived stem cells (ASC) (Minteer et al. 2013), MSC isolated from different locations in umbilical cord (Nagamura-Inoue and He 2014), long bone (Toosi et al. 2017), and dental pulp (Kawashima et al. 2017).

Healthy bone is maintained through formation by osteoblasts and resorption by osteoclasts. An intricate balance between osteoblastic and osteoclastic activities determines the health of bone. The lifespan of osteoblasts (~2–100 days) and osteoclasts (~12 days) are relatively short, and these cells must continually be replenished by stem and progenitor cells. For example, bone marrow-derived MSC have been shown to proliferate and differentiate toward the osteogenic lineage, regulated by numerous signaling pathways such as β-catenin-dependent Wnt (D'Alimonte et al. 2013), Hedgehog (Salem et al. 2014), NELL1 (Lee et al. 2017b), and BMP signaling. Osteogenesis involves several steps that includes ECM production (organic phase) and mineralization by hydroxyapatite (inorganic phase), followed by differentiation of osteoblasts into osteocytes through embedding in bone matrix and subsequent apoptosis. Osteoclasts have also been shown to participate in osteogenesis by inducing expression of Wnt 1 through the release and activation of TGF-β (Ramasamy et al. 2014). Thus, the microenvironment for bone stem and progenitor cells would include several factors as described above. A deficiency in any component of the microenvironment may result in an aberration in osteogenesis, similar to the hematologic aberrations noted with hematopoietic microenvironment changes. Also, self-renewal and differentiation of stem cells that participate in bone regeneration can be affected directly by specific epigenetic changes, leading to aberration in gene expression and metabolic bone disease (Pérez-Campo and Riancho 2015). It is entirely possible that unnatural or adverse microenvironment provided by a bone graft substitute may have a negative impact in bone repair and regeneration, regulated by proliferation and differentiation of stem and progenitor cells. Thus, the stem and progenitor cell microenvironment need to be considered carefully when transplanting these cells for clinical applications.

MSC are very responsive to their microenvironment (Lander et al. 2012). However, a database of approximately 700 clinical trials on MSC registered in ClinicalTrials.gov as a therapeutic agent indicates that the majority of these trials either did not use any carrier by injecting MSC directly into patients or used ceramic-based carriers (unpublished observation). While an increasing number of preclinical studies focus on MSC and their microenvironment, no clinical trials have considered the effects of the microenvironment on stem cell function. The clinical effectiveness and efficacy of MSC remain inconclusive, although these cells are investigated extensively as shown above, and there have not been any MSC-based therapies approved by the US Food and Drug Administration. Another comprehensive review on the use of MSC in close to 500 clinical trials revealed that the beneficial effect of MSC-based treatment could be principally due to their immunomodulation and regenerative potential (Squillaro et al. 2016). However, issues such as MSC heterogeneity, lack of standard isolation methods, donor heterogeneity, immunogenicity, and cryopreservation bring complexity to the interpretation of clinical data. In addition, in most cases, the host microenvironment may not have been ideal for long-term survival and desired function of transplanted cells.

Gene Therapy

Gene therapy is one strategy to optimize the tissue environment for bone healing or regeneration through delivery of genes encoding for osteoinductive or osteogenic factors. Treatment with a therapeutic osteoinductive factor typically requires large doses of recombinant proteins. In contrast, the use of regional gene therapy to modify native or implanted cells to produce the same protein of interest is potentially more physiologically relevant and cost effective (Phillips et al. 2005). Typically, a gene of interest is delivered to cells which then produce and secrete the target protein into the extracellular environment in a sustained pattern mimicking physiological secretion. The transgene may be introduced in vivo via local injection into patient tissue or in conjunction with transplants of stem or progenitor cells with the gene of interest introduced ex vivo. Local in vivo injection methods are comparatively simple and may provide a readily available, off-the-shelf product, although precise control of transgene delivery to cells in vivo remains a major challenge. Advantages of the ex vivo transduction approach include the ability to expand cells in culture to sufficient numbers essential for bone repair and the capability to quantify target protein production. Some of the challenges include the time and labor necessary for cell expansion and transduction, and risk of inflammatory response due to viral proteins (Baltzer and Lieberman 2004).

Many studies have explored the use of BMPs in gene therapy applications for spinal fusion (Barba et al. 2014). Early attempts at direct percutaneous in vivo gene delivery demonstrated that adenoviral vectors containing BMP-2 promoted bone formation (Musgrave et al. 1999; Alden et al. 1999). Later studies explored various cell types for ex vivo gene modification followed by transplantation at the fusion site. An ideal cell type for this application should be readily harvested, easy to expand in culture, amenable to gene modification, and have inherent osteogenic and osteoinductive properties. MSC, which can be harvested from bone marrow and are known to differentiate toward osteogenic, chondrogenic, and adipogenic lineages, are a logical choice and have been widely studied as a source of autologous cells for this purpose. MSC genetically modified with BMP-2 were implanted by direct injection into thoracic disc spaces (Riew et al. 2003) or lumbar paraspinal muscles (Hasharoni et al. 2005) and resulted in bridging bone formation in pigs and mice, respectively. Biomechanical tests demonstrated BMP-2-modified MSC-mediated spinal fusion and were as effective as bilateral fusion with stainless steel pins in a mouse model (Sheyn et al. 2010). To improve efficacy, BMP-2-modified MSC were implanted on collagen sponges or demineralized bone carrier matrix and shown to promote abundant bone formation and fusion rates superior to administration of BMP-2 alone (Wang et al. 2003). Additional studies in the rat model of lumbar fusion drew similar conclusions (Peterson et al. 2005; Miyazaki et al. 2008a). In follow-up studies, Miyazaki et al. (2008b) found adipose-derived MSC, which are easier to isolate from patients, to be comparable with bone marrow MSC in this model. Adenoviral-mediated BMP-2 modification of fibroblasts prior to injection along the paraspinous musculature induced heterotopic ossification, new bridging bone, and greater than 90% fusion by 4 weeks in both immune-competent and immune-deficient mice ($n > 40$ per group, Olabisi et al. 2011). Other members of the BMP super-family explored in spinal fusion studies include BMP-4 (Zhao et al. 2007), BMP-6 (Laurent et al. 2004), BMP-7 (Hidaka et al. 2003), BMP-9 (Helm et al. 2000; Dumont et al. 2002), and combinations of BMP-2 and -7 (Zhu et al. 2004; Kaito et al. 2013) with most studies reporting high rates of fusion in rodent or rabbit studies. Despite promising preclinical results in small animal models, the translation of BMP-directed adenoviral mediated gene transfer for spinal fusion to human patients faces challenges including the presence of neutralizing antibodies from previous exposures to adenovirus (greater than 80% of adults are seropositive) which limit effectiveness (Kim et al. 2003), systemic toxicity, and regulatory barriers to clinical trial approval (Wang 2011).

Ideal viral vectors for gene delivery in spinal fusion applications should be capable of delivering the desired transgene efficiently to initiate the bone formation required for the fusion process, but also self-limiting to avoid excessive or abnormal bone formation. Non-integrating, adenoviral vectors induce high, yet transient, levels of gene expression and are the vector of choice for most spinal fusion studies as outlined previously. Lentiviral vectors, which may infect nondividing cells and are known to insert within the host cell genome, have also been studied for this application. Rat MSC transduced with a lentiviral-BMP-2 vector were implanted in hind limb muscle pouches and observed to induce robust bone formation (Sugiyama et al. 2005) in a proof-of-principle study. Bone marrow MSC transfected with lenti-BMP-2 were implanted in collagen sponges and shown to induce spinal fusion in rats (Miyazaki et al. 2008b). Direct comparisons of adenoviral or lentiviral BMP-2 ex vivo gene therapy in MSC seeded on collagen sponges for implantation suggest improved bone formation with lentiviral delivery (Miyazaki et al. 2008c) at 8 weeks post implantation. Longer studies will be necessary to assess safety and risks of insertional mutagenesis with lentiviral vector approaches. Nonviral gene transfer approaches including sonoporation, electroporation, and nucleofection are thought to be safer than viral delivery methods, but low transfection efficiency has limited their application in spinal fusion studies (Sheyn et al. 2008; Makino et al. 2018).

Other osteoinductive proteins studied for gene therapy applications in spinal fusion include NELL1, LIM mineralization proteins (LMPs), and SMAD family member 1 (SMAD1). Lu et al. (2007) utilized adenoviral-mediated NELL1 delivery in demineralized bone matrix to show improved bone quality and maturity at 6 weeks post implantation in a rat spinal fusion model. In comparison to BMP-2 administration, goat MSCs carrying the NELL1 transgene promoted less bone mass but greater trabecular and chondroid bone formation (Aghaloo et al. 2007). LMP1 encodes an osteoinductive intracellular protein which promotes bone growth and skeletal organization. Initial studies of LMP1 for gene therapy in spinal fusion utilized bone marrow cells transfected with LMP1 cDNA with results demonstrating successful fusion (Boden et al. 1998) in rats. Bone marrow or buffy-coat cells transduced with adenoviral LMP1 for 10 min and implanted in demineralized bone or collagen-ceramic composite sponges were shown to induce posterolateral lumbar fusion in rabbits (Viggeswarapu et al. 2001). To improve ease and efficiency of cell isolation, autologous dermal fibroblasts from skin biopsies were transduced with adenoviral-LMP3 and shown to induce ectopic bone formation in muscle (Lattanzi et al. 2008). SMAD1, a downstream target of BMPs, is a key intermediary in expression of genes driving osteoblast differentiation and thus proposed to induce osteogenesis more specifically than BMPs. MSC transduced with SMAD1C and implanted on gelatin sponges were shown to support efficient new bone formation in a rabbit model of lumbar spinal fusion (Douglas et al. 2010).

Despite the promising results of preclinical gene therapy studies for spinal fusion, translation of these results to therapies for human patients has shown little progress. Variables which will need to be optimized to move gene therapy approaches forward include choice of vector, therapeutic gene, delivery method (in vivo or ex vivo), source of cells for ex vivo modification, and implantation or injection strategies. The risks associated with the viral vector-mediated delivery including systemic toxicity, insertional mutagenesis, and genomic instability constitute significant barriers to bringing such therapies into the clinic (Wang 2011). Regulatory approval for gene therapy approaches for spinal fusion, a procedure which may influence quality of life but for which the underlying pathology is typically nonlethal, may be difficult to achieve.

Microenvironment

In a therapeutic situation, it may be critical to consider the effects of the implant site on the fate of transplanted stem and progenitor cells and growth factors. It is entirely possible for a damaged or diseased implant site to have an unintended consequence on incoming cells and growth factors. For example, as described above, growth factors may leak out of the implant site and result in unintended consequences locally and systemically, especially if growth factors are introduced at a dose significantly higher than their physiological levels (as seen with BMP-2). In addition, in a healthy intervertebral disc, numerous factors such as cytokines, growth factors, endogenous cells, enzymes, and mechanical stimuli regulate the balance between the anabolic and catabolic processes. However, decreased proteoglycans and collagen II, increased proteinases and cytokines, and decreased (acidic) pH have been observed in a degenerating intervertebral disc (Huang et al. 2013), caused by aging, disease, trauma, or mechanical stress. These factors may comprise an unfavorable microenvironment and present a significant challenge when stem and progenitor cells are considered as a therapeutic option. Thus, it is imperative that biologics must be delivered within a carrier that can retain therapeutic elements and maintain a healthy microenvironment for tissue repair and regeneration.

A hostile microenvironment at a disc impacts directly on the success of any attempted interbody fusion. Anterior cervical discectomy and fusion (ACDF) is the most common fusion surgery in cervical spine. Interbody fusion is also often utilized in lumbar spine: transforaminal lumbar interbody fusion (TLIF), posterior lumbar interbody fusion (PLIF), direct lateral interbody fusion (DLIF), and anterior lumbar interbody fusion (ALIF). Without successful fusion, the patients may have persistent or new pain and the implants utilized for fusion may fail, possibly leading to another surgery. Posterolateral fusion is very common in lumbar spine, and it may pose even more challenging microenvironment. Successful fusion depends on a solid bony growth spanning the transverse processes with adjacent ligament and musculature that may interfere with bone regeneration. As discussed previously, restoration of the microenvironment affected by various exogenous and endogenous, adverse physiological influences may be required prior to and/or at the time of stem and progenitor cell transplantation (Lee and Kim 2012).

Autologous bone harvesting and implantation result in transfer of not only stem cells that reside in the harvested bone but also their microenvironment. Undoubtedly, autologous bone harvested from a non-load-bearing site such as the iliac crest consistently results in a positive clinical outcome and is considered the "gold standard" for bone repair. As described above, transplanted bone marrow-derived MSC taken out of their native microenvironment and expanded do not result in clinical outcomes similar to implantation of bone. Thus, implantation of both stem and progenitor cells along with their native microenvironment ("autograft") is unarguably better than cells ("cell transplantation") alone. However, the clinical outcome when using the autograft approach is less than optimal due to donor site morbidity (especially pain), fracture, infection, increased blood loss, prolonged operative time, and risk of nerve damage. In addition, autograft is limited in quantity, and quality is suboptimal depending on the patient. Due to these risks and limitations, bone graft substitutes have been increasingly utilized instead of autograft. Bone graft substitutes are generally not indicated to be used with stem and progenitor cells but known to participate in bone regeneration, potentially reestablishing the microenvironment for bone stem and progenitor cells that are residing at the site of injury or transplanted.

Bone graft substitutes are discussed in ▶ Chap. 11, "Bone Grafts and Bone Graft Substitutes" in this book and broadly categorized into allografts, ceramics, polymers, and biologics. Several factors such as mechanical stress, vascularity, and surface characteristics of graft material have demonstrated to influence the stimuli and microenvironment of cells. These factors seem to have a collective influence on the osteogenic differentiation of stem cells through epigenetic/gene upregulation mechanisms. For example, mechanical stretch induced downregulation of GNAS (stimulatory G-protein alpha subunit) isoforms of mesenchymal cells and upregulation of osteogenic differentiation transcription in in vitro models (Vlaikou et al. 2017). In another study, mechanical stress in osteoblast precursor cells in 3D scaffolds experienced greater signaling through MAP kinase pathway (Appleford et al. 2007). Tissue engineered bone constructs of DBM and nanoscale self-assembling peptides provided with decreased pore size and increased charge field resulted in better enrichment of osteogenic cells (Hou et al. 2014).

Ceramic-based bone graft substitutes include calcium phosphate, calcium sulfate, hydroxyapatite (HA), β-tricalcium phosphate (β-TCP), and bioactive glass. HA scaffolds showed a greater degree of ectopic bone formation than β-TCP when implanted with MSC in rats (Denry and Kuhn 2016), but better attachment and spreading of MSC were observed with β-TCP while expressing G-protein coupled receptor (Barradas et al. 2013). Surface modification of calcium phosphate cement with arginine–glycine–aspartate (RGD) showed a significant improvement in attachment, survival, and proliferation of MSC (Chen et al. 2012), indicating that additional coating may be required to provide an optimal environment for MSC. Similarly, it has been postulated that the initial contact with blood primes the surface and prepares calcium phosphate ceramic scaffolds (CPS) for viable in situ cell seeding (Denry and Kuhn 2016). In addition, CPS show a very slow degradation rate (Bružauskaitė et al. 2016; Winter et al. 1981) and are not radiolucent, thus interfering with the visualization of new bone formation by radiographic evaluation.

Poly-lactic acid (PLLA) (Holderegger et al. 2015), polyglycolic acid (PGA) (Generali et al. 2017), and poly-DL-lactic-co-glycolic acid (PLGA) (Mendes Junior et al. 2017) have been proposed as a synthetic bone graft substitute. All of these synthetic materials show exceptional compatibility with MSC and support osteogenesis. However, PLLA, PGA, and PLGA degrade within 30 days (Generali et al. 2017) and may not be able to bridge a critical size bone defect, considering natural bone regeneration over several months. Growth factors such as BMP-2 has received a significant attention due to its potent osteogenic properties and safety issues including adverse effects (e.g., life-threatening inflammatory complications), ectopic bone formation, osteoclast activation, and induction of adipogenesis (James et al. 2016). Other growth factors such as fibroblast growth factor-2 and insulin-like growth factor-1 have been shown to induce osteogenesis (Nagayasu-Tanaka et al. 2015; Guntur and Rosen 2013). However, as shown with BMP-2, patient safety must carefully be considered before clinical utilization.

ECM is an essential regulator of stem cell function and a critical component of stem cell microenvironment. ECM is primarily comprised of proteins (e.g., collagen, laminin, fibronectin, elastin) and carbohydrates (polysaccharides). While various proteins in ECM have been investigated for bone regeneration, the clinical utility of the polysaccharide component has not yet been explored fully. Carbohydrate-based, polysaccharide materials have been used to regenerate bone in preclinical settings including cellulose (Park et al. 2015), alginate (Hung et al. 2016), chitosan (Levengood and Zhang 2014), and glycosaminoglycans (Mathews et al. 2014). Carbohydrate-based materials have a long history of use in various medical applications and are nontoxic, biocompatible, soluble, and biodegradable. These properties make carbohydrate-based polymers an excellent scaffold for tissue engineering. A recent study reported that supramolecular sulfated glycopeptide nanostructures with a tri-sulfated monosaccharide on the surface amplified signaling of BMP-2, resulting in enhanced bone formation (Lee et al. 2017a). Thus, carbohydrate-based polymers may play a key role in developing novel therapeutic approaches and addressing the microenvironment issue in near future as shown in recent studies on stem cells (Batchelder et al. 2015b), tumor heterogeneity (Batchelder et al. 2015a), and cardiac tissue regeneration (Baio et al. 2017).

These therapeutic approaches using bone graft substitutes aim to generate bone by providing an osteoconductive environment for endogenous cells and factors and/or inducing osteogenesis and migration of host cells into the target site, reestablishing the microenvironment conducive to tissue regeneration. Bone regeneration and repair are fairly successful in healthy individuals, but high-risk patient populations have been reported to show nonunion rates as high as 40% for bone fusion (Scott and Hyer 2013). In well-controlled preclinical studies, diabetes showed a significant negative effect on bone regeneration compared to healthy animals under the identical experimental conditions including bone graft substitutes (Camargo et al. 2017). Thus, it is clear that bone graft substitutes alone are not sufficient for bone regeneration in high risk patients.

Conclusions

Contrast to small defects that the body can heal spontaneously, critical bone defects (>25 mm) have a higher probability of nonunion (>50%) and require a therapeutic intervention (Haines et al. 2016). Current biologics available for bony fusion address one or two components of complete bone repair and regeneration: osteogenicity, osteoinductivity, and osteoconductivity. A product or combination product that possesses all of these properties may be needed to close or bridge a critical bone defect in a functionally meaningful manner, especially in patients with high risk factors, in addition to systemic or local environment (microenvironment) that may adversely influence the patient outcome.

Cross-References

▶ Bone Grafts and Bone Graft Substitutes

References

Aghaloo T, Jiang X, Soo C, Zhang Z, Zhang X, Hu J, Pan H, Hsu T, Wu B, Ting K, Zhang X (2007) A study of the role of NELL-1 gene modified goat bone marrow stromal cells in promoting new bone formation. Mol Ther 15:1872–1880

Alden TD, Pittman DD, Beres EJ, Hankins GR, Kallmes DF, Wisotsky BM, Kerns KM, Helm GA (1999) Percutaneous spinal fusion using bone morphogenetic protein-2 gene therapy. J Neurosurg 90:109–114

An HS, Lynch K, Toth J (1995) Prospective comparison of autograft vs. allograft for adult posterolateral lumbar spine fusion: differences among freeze-dried, frozen, and mixed grafts. J Spinal Disord 8:131–135

Appleford MR, Oh S, Cole JA, Carnes DL, Lee M, Bumgardner JD, Haggard WO, Ong JL (2007) Effects of trabecular calcium phosphate scaffolds on stress signaling in osteoblast precursor cells. Biomaterials 28:2747–2753

Baio JM, Walden RC, Fuentes TI, Lee CC, Hasaniya NW, Bailey LL, Kearns-Jonker MK (2017) A hyper-crosslinked carbohydrate polymer scaffold facilitates lineage commitment and maintains a reserve pool of proliferating cardiovascular progenitors. Transplant Direct 3:e153

Baltzer AW, Lieberman JR (2004) Regional gene therapy to enhance bone repair. Gene Ther 11:344–350

Barba M, Cicione C, Bernardini C, Campana V, Pagano E, Michetti F, Logroscino G, Lattanzi W (2014) Spinal fusion in the next generation: gene and cell therapy approaches. ScientificWorldJournal 2014:406159

Barradas AM, Monticone V, Hulsman M, Danoux C, Fernandes H, Birgani ZT, Barrère-de Groot F, Yuan H, Reinders M, Habibovic P, van Blitterswijk C, de Boer J (2013) Molecular mechanisms of biomaterial-driven osteogenic differentiation in human mesenchymal stromal cells. Integr Biol (Camb) 5:920–931

Batchelder CA, Martinez ML, Duru N, Meyers FJ, Tarantal AF (2015a) Three-dimensional culture of human renal cell carcinoma organoids. PLoS One 10:e0136758

Batchelder CA, Martinez ML, Tarantal AF (2015b) Natural scaffolds for renal differentiation of human embryonic stem cells for kidney tissue engineering. PLoS One 10: e0143849

Beachler DC, Yanik EL, Martin BI, Pfeiffer RM, Mirza DK, Deyo RA, Engels EA (2016) Bone morphogenetic protein use and cancer risk among patients undergoing lumbar arthrodesis: a case-cohort study using the SSER-Medicare database. J Bone Joint Surg Am 98:1064–1072

Boden SD, Titus L, Hair G, Liu Y, Viggeswarapu M, Nanes MS, Baranowski C (1998) Lumbar spine fusion by local gene therapy with a cDNA encoding a novel osteoinductive protein (LMP-1). Spine 23:2486–2492

Bružauskaitė I, Bironaitė D, Bagdonas E, Bernotienė E (2016) Scaffolds and cells for tissue regeneration: different scaffold pore sizes-different cell effects. Cytotechnology 68:355–369

Burkus JK, Gornet MF, Dickman CA, Zdeblick TA (2002a) Anterior lumbar interbody fusion using rhBMP-2 with tapered interbody cages. J Spinal Disord Tech 15:337–349

Burkus JK, Transfeldt EE, Kitchel SH, Watkins RG, Balderston RA (2002b) Clinical and radiographic outcomes of anterior lumbar interbody fusion using recombinant human bone morphogenetic protein-2. Spine 27:2396–2408

Burkus JK, Dorchak JD, Sanders DL (2003) Radiographic assessment of interbody fusion using recombinant human bone morphogenetic protein type 2. Spine 28:372–377

Burkus JK, Sandhu HS, Gornet MF, Longley MC (2005) Use of rhBMP-2 in combination with structural cortical allografts: clinical and radiographic outcomes in anterior lumbar spinal surgery. J Bone Joint Surg Am 87:1205–1212

Burkus JK, Sandhu HS, Gornet MF (2006) Influence of rhBMP-2 on the healing patterns associated with allograft interbody constructs in comparison with autograft. Spine 31:775–781

Cahill KS, McCormick PC, Levi AD (2015) A comprehensive assessment of the risk of bone morphogenetic protein use in spinal fusion surgery and postoperative cancer diagnosis. J Neurosurg Spine 23:86–93

Camargo WA, de Vries R, van Luijk J, Hoekstra JWM, Bronkhorst EM, Jansen JA, van den Beucken JJ (2017) Diabetes mellitus and bone regeneration: a systematic review and meta-analysis of animal studies. Tissue Eng Part B Rev 23:471–479

Carragee EJ, Hurwitz EL, Weiner BK (2011a) A critical review of recombinant human bone morphogenetic protein-2 trials in spinal surgery: emerging safety concerns and lessons learned. Spine J 11:471–491

Carragee EJ, Mitsunaga KA, Hurwitz L, Scuderi GJ (2011b) Retrograde ejaculation after anterior lumbar interbody fusion using rhBMP-2: a cohort controlled study. Spine J 11:511–516

Carragee EJ, Chu G, Rohatgi R, Hurwitz EL, Weiner BK, Yoon ST, Comer G, Kopjar B (2013) Cancer risk after use of recombinant bone morphogenetic protein-2 for spinal arthrodesis. J Bone Joint Surg Am 95:1537–1545

Chen W, Zhou H, Weir MD, Bao C, Xu HH (2012) Umbilical cord stem cells released from alginate-fibrin microbeads inside macroporous and biofunctionalized calcium phosphate cement for bone regeneration. Acta Biomater 8:2297–2306

Comer GC, Smith MW, Hurwitz EL, Mitsunaga KA, Kessler R, Carragee EJ (2012) Retrograde ejaculation after anterior lumbar interbody fusion with and without bone morphogenetic protein-2 augmentation: a 10-year cohort controlled study. Spine J 12:881–890

Cornu O, Banse X, Docquier PL, Luyckx S, Delloye C (2000) Effect of freeze-drying and gamma irradiation on the mechanical properties of human cancellous bone. J Orthop Res 18:426–431

Cottrill E, Ahmed AK, Lessing N, Pennington Z, Ishida W, Perdomo-Panoja A, Lo SF, Howell E, Holmes C, Goodwin CR, Theodore N, Sciubba DM, Witham TF (2019) Investigational growth factors utilized in animal models of spinal fusion: systematic review. World J Orthop 10:176–191

Crawford CH, Carreon LY, McGinnis MD, Campbell MJ, Glassman SD (2009) Perioperative complications of recombinant human bone morphogenetic protein-2 on an absorbable collagen sponge versus iliac crest bone graft for posterior cervical arthrodesis. Spine 34:1390–1394

D'Alimonte I, Lannutti A, Pipino C, Di Tomo P, Pierdomenico L, Cianci E, Antonucci I, Marchisio M, Romano M, Stuppia L, Caciagli F, Pandolfi A, Ciccarelli R (2013) Wnt signaling behaves as a "master regulator" in the osteogenic and adipogenic commitment of human amniotic fluid mesenchymal stem cells. Stem Cell Rev 9:642–654

Delawi D, Jacobs W, van Susante JL, Rillardon L, Prestamburgo D, Specchia N, Gay E, Verschoor N, Garcia-Fernandez C, Guerado E, Quarles van Ufford H, Kruyt MC, Dhert WJ, Oner FC (2016) OP-1 compared with iliac crest autograft in instrumented posterolateral fusion: a randomized, multicenter non-inferiority trial. J Bone Joint Surg Am 98:441–448

Denry I, Kuhn LT (2016) Design and characterization of calcium phosphate ceramic scaffolds for bone tissue engineering. Dent Mater 32:43–53

Devine JG, Dettori JR, France JC, Brodt E, McGuire RA (2012) The use of rhBMP in spine surgery: is there a cancer risk? Evid Based Spine Care J 3:35–41

Dominici M, Le Blanc K, Mueller I, Slaper-Cortenbach I, Marini F, Krause D, Deans R, Keating A, Prockop DJ, Horwitz E (2006) Minimal criteria for defining multipotent mesenchymal stromal cells. The International Society for Cellular Therapy position statement. Cytotherapy 8:315–317

Douglas JT, Rivera AA, Lyons GR, Lott PF, Wang D, Zayzafoon M, Siegal GP, Cao X, Theiss SM (2010) Ex vivo transfer of the Hoxc-8-interacting domain of Smad1 by a tropism-modified adenoviral vector results in efficient bone formation in a rabbit model of spinal fusion. J Spinal Disord Tech 23:63–73

Duarte RM, Varanda P, Reis RL, Duarte ARC, Correia-Pinto J (2017) Biomaterials and bioactive agents in spinal fusion. Tissue Eng Part B Rev 23:540–551

Dumont RJ, Dayoub H, Li JZ, Dumont AS, Kallmes DF, Hankins GR, Helm GA (2002) Ex vivo bone morphogenetic protein-9 gene therapy using human mesenchymal stem cells induces spinal fusion in rodents. Neurosurgery 51:1239–1244

Ehrler DM, Vaccaro AR (2000) The use of allograft bone in lumbar spine surgery. Clin Orthop Relat Res 371:38–45

Epstein NE (2013) Complications due to the use of BMP/INFUSE in spine surgery: the evidence continues to mount. Surg Neurol Int 4:S343–S352

Faundez A, Tournier C, Garcia M, Aunoble S, Le Huec JC (2016) Bone morphogenetic protein use in spine surgery – complications and outcomes: a systematic review. Int Orthop 40:1309–1319

Fineberg SJ, Ahmadinia K, Oglesby M, Patel AA, Singh K (2013) Hospital outcomes and complications of anterior and posterior cervical fusion with bone morphogenetic protein. Spine 38:1304–1309

Generali M, Kehl D, Capulli AK, Parker KK, Hoerstrup SP, Weber B (2017) Comparative analysis of polyglycolic acid-based hybrid polymer starter matrices for in vitro tissue engineering. Colloids Surf B Biointerfaces 158:203–212

Ghosh P, Moore R, Vernon-Roberts B, Path FRC, Goldschlager T, Pascoe D, Zannettino A, Gronthos S,

Itescu S (2012) Immunoselected STRO-3+ mesenchymal precursor cells and restoration of the extracellular matrix of degenerate intervertebral discs. J Neurosurg Spine 16:479–488

Glassman SD, Anagnost SC, Parker A, Burke D, Johnson JR, Dimar JR (2000) The effect of cigarette smoking and smoking cessation on spinal fusion. Spine 25:2608–2615

Guntur AR, Rosen CJ (2013) IGF-1 regulation of key signaling pathways in bone. Bonekey Rep 2:437

Haines NM, Lack WE, Seymour RB, Bosse MJ (2016) Defining the lower limit of a "critical bone defect" in open diaphyseal tibial fractures. J Orthop Trauma 30: e158–e163

Hasharoni A, Zilberman Y, Turgeman G, Helm GA, Liebergall M, Gazit D (2005) Murine spinal fusion induced by engineered mesenchymal stem cells that conditionally express bone morphogenetic protein-2. J Neurosurg Spine 3:47–52

Helm GA, Alden TD, Beres EJ, Hudson SB, Das S, Engh JA, Pittman DD, Kerns KM, Kallmes DF (2000) Use of bone morphogenetic protein-9 gene therapy to induce spinal arthrodesis in the rodent. J Neurosurg 92:191–196

Hidaka C, Goshi K, Rawlins B, Boachie-Adjei O, Crystal RG (2003) Enhancement of spine fusion using combined gene therapy and tissue engineering BMP-7-expressing bone marrow cells and allograft bone. Spine 28:2049–2057

Holderegger C, Schmidlin PR, Weber FE, Mohn D (2015) Preclinical in vivo performance of novel biodegradable, electrospun poly(lactic acid) and poly(lactic-co-glycolic acid) nanocomposites: a review. Materials (Basel) 8:4912–4931

Hou T, Li Z, Luo F, Xie Z, Wu X, Xing J, Dong S, Xu J (2014) A composite demineralized bone matrix – self assembling peptide scaffold for enhancing cell and growth factor activity in bone marrow. Biomaterials 35:5689–5699

Huang YC, Leung VY, Lu WW, Luk KD (2013) The effects of microenvironment in mesenchymal stem cell-based regeneration of intervertebral disc. Spine J 13:352–362

Hung BP, Naved BA, Nyberg EL, Dias M, Holmes CA, Elisseeff JH, Dorafshar AH, Grayson WL (2016) Three-dimensional printing of bone extracellular matrix for craniofacial regeneration. ACS Biomater Sci Eng 2:1806–1816

Ishida K, Acharya A, Christiansen BA, Yik JH, DiCesare PE, Haudenschild DR (2013) Cartilage oligomeric matrix protein enhances osteogenesis by directly binding and activating bone morphogenetic protein-2. Bone 55:23–35

James AW, LaChaud G, Shen J, Asatrian G, Nguyen V, Zhang X, Ting K, Soo C (2016) A review of the clinical side effects of bone morphogenetic protein-2. Tissue Eng Part B Rev 22:284–297

James AW, Shen J, Tsuei R, Nguyen A, Khadarian K, Meyers CA, Pan HC, Li W, Kwak JH, Asatrian G, Culiat CT, Lee M, Ting K, Zhang X, Soo C (2017)

NELL-1 induces Sca-1$^+$ mesenchymal progenitor cell expansion in models of bone maintenance and repair. JCI Insight 2:e92573. https://doi.org/10.1172/jci.insight.92573

Johnsson R, Strömqvist B, Aspenberg P (2002) Randomized radiostereometric study comparing osteogenic protein-1 (BMP-7) and autograft bone in human non-instrumented posterolateral lumbar fusion: 2002 Volvo Award in clinical studies. Spine 27:2654–2661

Kaito T, Johnson J, Ellerman J, Tian H, Aydogan M, Chatsrinopkun M, Ngo S, Choi C, Wang JC (2013) Synergistic effect of bone morphogenetic proteins 2 and 7 by ex vivo gene therapy in a rat spinal fusion model. J Bone Joint Surg Am 95:1612–1619

Kawashima N, Noda S, Yamamoto M, Okiji T (2017) Properties of dental pulp-derived mesenchymal stem cells and the effects of culture conditions. J Endod 43:S31–S34

Kelly MP, Savage JW, Bentzen SM, Hsu WK, Ellison SA, Anderson PA (2014) Cancer risk from bone morphogenetic protein exposure in spinal arthrodesis. J Bone Joint Surg Am 96:1417–1422

Khorsand B, Nicholson N, Do AV, Femino JE, Martin JA, Petersen E, Guetschow B, Fredericks DC, Salem AK (2017) Regeneration of bone using nanoplex delivery of FGF-2 and BMP-2 genes in diaphyseal long bone radial defects in a diabetic rabbit model. J Control Release 248:53–59

Kim HS, Viggeswarapu M, Boden SD, Liu Y, Hair GA, Louis-Ugbo J, Murakami H, Minamide A, Suh DY, Titus L (2003) Overcoming the immune response to permit ex vivo gene therapy for spine fusion with human type 5 adenoviral delivery of the LIM mineralization protein-1 cDNA. Spine 28:219–226

Kim DH, Rhim R, Li L, Martha J, Swaim BH, Banco RJ, Jenis LG, Tromanhauser SG (2009) Prospective study of iliac crest bone graft harvest site pain and morbidity. Spine J 9:886–892

Kim SE, Yun YP, Lee JY, Shim JS, Park K, Huh JB (2015) Co-delivery of platelet-derived growth factor (PDGF-BB) and bone morphogenic protein (BMP-2) coated onto heparinized titanium for improving osteoblast function and osteointegration. J Tissue Eng Regen Med 9:E219–E228

Kurz LT, Garfin SR, Booth RE Jr (1989) Harvesting autogenous iliac bone grafts. A review of complications and techniques. Spine (Phila Pa 1976) 14:1324–1331

Laitinen M, Kivikari R, Hirn M (2006) Lipid oxidation may reduce the quality of a fresh-frozen bone allograft. Is the approved storage temperature too high? Acta Orthop 77:418–421

Lander AD, Kimble J, Clevers H, Fuchs E, Montarras D, Buckingham M, Calof AL, Trumpp A, Oskarsson T (2012) What does the concept of the stem cell niche really mean today? BMC Biol 10:19

Lattanzi W, Parrilla C, Fetoni A, Logroscino G, Straface G, Pecorini G, Stigliano E, Tampieri A, Bedini R, Pecci R, Michetti F, Gambotto A, Robbins PD, Pola E (2008) Ex vivo-transduced autologous skin fibroblasts expressing human Lim mineralization protein-3

efficiently form new bone in animal models. Gene Ther 15:1330–1343

Laurent JJ, Webb KM, Beres EJ, McGee K, Li J, van Rietbergen B, Helm GA (2004) The use of bone morphogenetic protein-6 gene therapy for percutaneous spinal fusion in rabbits. J Neurosurg Spine 1:90–94

Lee CC, Kim KD (2012) Stem cell microenvironment as a potential therapeutic target. Regen Med 7:3–5

Lee KJ, Roper JG, Wang JC (2005) Demineralized bone matrix and spinal arthrodesis. Spine J 5:217S–223S

Lee CC, Christensen JE, Yoder MC, Tarantal AF (2013) Clonal analysis and hierarchy of human bone marrow mesenchymal stem and progenitor cells. Exp Hematol 38:46–54

Lee SS, Fyrner T, Chen F, Alvarez Z, Sleep E, Chun DS, Weiner JA, Cook RW, Freshman RD, Schallmo MS, Katchco KM, Schneider AD, Smith JT, Yun C, Singh G, Hashmi SZ, McClendon MT, Yu Z, Stock SR, Hsu WK, Hsu EL, Stupp SI (2017a) Sulfated glycopeptide nanostructures for multipotent protein activation. Nat Nanotechnol 12:821–829

Lee S, Wang C, Pan HC, Shrestha S, Meyers C, Ding C, Shen J, Chen E, Lee M, Soo C, Ting K, James AW (2017b) Combining smoothened agonist and NEL-like protein-1 enhances bone healing. Plast Reconstr Surg 139:1385–1396

Levengood SL, Zhang M (2014) Chitosan-based scaffolds for bone tissue engineering. J Mater Chem B 2:3161–3184

Li W, Lee M, Whang J, Siu RK, Zhang X, Liu C, Wu BM, Wang JC, Ting K, Soo C (2010) Delivery of lyophilized NELL-1 in a rat spinal fusion model. Tissue Eng Part A 16:2861–2870

Lu SS, Zhang X, Soo C, Hsu T, Napoli A, Aghaloo T, Wu BM, Tsou P, Ting K, Wang JC (2007) The osteoinductive properties of NELL-1 in a rat spinal fusion model. Spine J 7:50–60

Makino T, Tsukazaki H, Ukon Y, Tateiwa D, Yoshikawa H, Kaito T (2018) The biological enhancement of spinal fusion for spinal degenerative disease. Int J Mol Sci 19:2430

Maleki M, Ghanbarvand F, Reza Behvarz M, Ejtemaei M, Ghadirkhomi E (2014) Comparison of mesenchymal stem cell markers in multiple human adult stem cells. Int J Stem Cells 7:118–126

Mathews S, Mathew SA, Gupta PK, Bhonde R, Totey S (2014) Glycosaminoglycans enhance osteoblast differentiation of bone marrow derived human mesenchymal stem cells. J Tissue Eng Regen Med 8:143–152

McClellan JW, Mulconrey DS, Forbes RJ, Fullmer N (2006) Vertebral bone resorption after transforaminal lumbar interbody fusion with bone morphogenetic protein (rhBMP-2). J Spinal Disord Tech 19:483–486

Mellonig JT (1995) Donor selection, testing, and inactivation of the HIV virus in freeze-dried bone allografts. Pract Periodontics Aesthet Dent 7:13–22

Mendes Junior D, Domingues JA, Hausen MA, Cattani SMM, Aragones A, Oliveira ALR, Inacio RF, Barbo MLP, Duek EAR (2017) Study of mesenchymal stem cells cultured on a poly(lactic-co-glycolic acid)

scaffold containing simvastatin for bone healing. J Appl Biomater Funct Mater 15:e133–e141

Minteer D, Marra KG, Rubin JP (2013) Adipose-derived mesenchymal stem cells: biology and potential applications. Adv Biochem Eng Biotechnol 129:59–71

Miyazaki M, Zuk PA, Zou J, Yoon SH, Wei F, Morishita Y, Sintuu C, Wang JC (2008a) Comparison of human mesenchymal stem cells derived from adipose tissue and bone marrow for ex vivo gene therapy in rat spinal fusion model. Spine 33:863–869

Miyazaki M, Sugiyama O, Tow B, Zou J, Morishita Y, Wei F, Napoli A, Sintuu C, Lieberman JR, Wang JC (2008b) The effects of lentiviral gene therapy with bone morphogenetic protein-2-producing bone marrow cells on spinal fusion in rats. J Spinal Disord Tech 21:372–379

Miyazaki M, Sugiyama O, Zou J, Yoon SH, Wei F, Morishita Y, Sintuu C, Virk MS, Lieberman JR, Wang JC (2008c) Comparison of lentiviral and adenoviral gene therapy for spinal fusion in rats. Spine 33:1410–1417

Mulconrey DS, Bridwell KH, Flynn J, Cronen GA, Rose PS (2008) Bone morphogenetic protein (RhBMP-2) as a substitute for iliac crest bone graft in multilevel adult spinal deformity surgery: minimum two-year evaluation of fusion. Spine (Phila Pa 1976) 33:2153–2159

Musgrave DS, Bosch P, Ghivizzani S, Robbins PD, Evans CH, Huard J (1999) Adenovirus-mediated direct gene therapy with bone morphogenetic protein-2 produces bone. Bone 24:541–547

Nagamura-Inoue T, He H (2014) Umbilical cord-derived mesenchymal stem cells: their advantages and potential clinical utility. World J Stem Cells 6:195–202

Nagayasu-Tanaka T, Anzai J, Takaki S, Shiraishi N, Terashima A, Asano T, Nozaki T, Kitamura M, Murakami S (2015) Action mechanism of fibroblast growth factor-2 (FGF-2) in the promotion of periodontal regeneration in beagle dogs. PLoS One 10:e0131870

Olabisi RM, Lazard ZW, Heggeness MH, Moran KM, Hipp JA, Dewan AK, Davis AR, West JL, Olmsted-Davis EA (2011) An injectable method for noninvasive spine fusion. Spine J 11:545–556

Ong KL, Villarraga ML, Lau E, Carreon LY, Kurtz SM, Glassman SD (2010) Off-label use of bone morphogenetic proteins in the United States using administrative data. Spine 35:1794–1800

Pacaccio DJ, Stern SF (2005) Demineralized bone matrix: basic science and clinical applications. Clin Podiatr Med Surg 22:599–606

Park BH, Song KJ, Yoon SJ, Park HS, Jang KY, Zhou L, Lee SY, Lee KB, Kim JR (2011) Acceleration of spinal fusion using COMP-angiopoietin 1 with allografting in a rabbit model. Bone 49:447–454

Park S, Park J, Jo I, Cho S, Sung D, Ryu S, Park M, Min K, Kim J, Hong S, Hong BH, Kim B (2015) In situ hybridization of carbon nanotubes with bacterial cellulose for three-dimensional hybrid bioscaffolds. Biomaterials 58:93–102

Pérez-Campo FM, Riancho JA (2015) Epigenetic mechanisms regulating mesenchymal stem cell differentiation. Curr Genomics 16:368–383

Peterson B, Iglesias R, Zhang J, Wang JC, Lieberman JR (2005) Genetically modified human derived bone marrow cells for posterolateral lumbar spine fusion in athymic rats: beyond conventional autologous bone grafting. Spine 30:283–289

Phillips FM, Bolt PM, He TC, Haydon RC (2005) Gene therapy for spinal fusion. Spine J 5:250S–258S

Ramasamy SK, Kusumbe AP, Wang L, Adams RH (2014) Endothelial Notch activity promotes angiogenesis and osteogenesis in bone. Nature 507:376–380

Refaat M, Klineberg EO, Fong MC, Garcia TC, Leach JK, Haudenschild DR (2016) Binding to COMP reduces the BMP2 dose for spinal fusion in a rat model. Spine 41:E829–E836

Riew KD, Lou J, Wright NM, Cheng SL, Bae KT, Avioli LV (2003) Thoracoscopic intradiscal spine fusion using a minimally invasive gene-therapy technique. J Bone Joint Surg Am 85–A:866–871

Rihn JA, Patel R, Makda J, Hong J, Anderson DG, Vaccaro AR, Hilibrand AS, Albert TJ (2009) Complications associated with single-level transforaminal lumbar interbody fusion. Spine J 9:623–629

Robin BN, Chaput CD, Zeitouni S, Rahm MD, Zerris VA, Sampson HW (2010) Cytokine-mediated inflammatory reaction following posterior cervical decompression and fusion associated with recombinant human bone morphogenetic protein-2: a case study. Spine 35:E1350–E1354

Salem O, Wang HT, Alaseem AM, Ciobanu O, Hadjab I, Gawri R, Antoniou J, Mwale F (2014) Naproxen affects osteogenesis of human mesenchymal stem cells via regulation of Indian hedgehog signaling molecules. Arthritis Res Ther 16:R152

Sasso RC, LeHuec JC, Shaffrey C, Spine Interbody Research Group (2005) Iliac crest bone graft donor site pain after anterior lumbar interbody fusion: a prospective patient satisfaction outcome assessment. J Spinal Disord Tech 18:S77–S81

Savage JW, Kelly MP, Ellison SA, Anderson PA (2015) A population-based review of bone morphogenetic protein: associated complication and reoperation rates after lumbar spinal fusion. Neurosurg Focus 39(4):E13. https://doi.org/10.3171/2015.7.FOCUS15240

Sawin PD, Traynelis VC, Menezes AH (1998) A comparative analysis of fusion rates and donor-site morbidity for autogeneic rib and iliac crest bone grafts in posterior cervical fusions. J Neurosurg 88:255–265

Schwartz CE, Martha JF, Kowalski P, Wang DA, Bode R, Li L, Kim DH (2009) Prospective evaluation of chronic pain associated with posterior autologous iliac crest bone graft harvest and its effect on postoperative outcome. Health Qual Life Outcomes 7:49

Scott RT, Hyer CF (2013) Role of cellular allograft containing mesenchymal stem cells in high-risk foot and ankle reconstructions. J Foot Ankle Surg 52:32–35

Shahlaie K, Kim KD (2008) Occipitocervical fusion using recombinant human bone morphogenetic protein-2: adverse effects due to tissue swelling and seroma. Spine 33:2361–2366

Sheyn D, Kimelman-Bleich N, Pelled G, Zilberman Y, Gazit D, Gazit Z (2008) Ultrasound-based nonviral gene delivery induces bone formation in vivo. Gene Ther 15:257–266

Sheyn D, Rüthemann M, Mizrahi O, Kallai I, Zilberman Y, Tawackoli W, Kanim LE, Zhao L, Bae H, Pelled G, Snedeker JG, Gazit D (2010) Genetically modified mesenchymal stem cells induce mechanically stable posterior spine fusion. Tissue Eng Part A 16:3679–3686

Shields LB, Raque GH, Glassman SD, Campbell M, Vitaz T, Harpring J, Shields CB (2006) Adverse effects associated with high-dose recombinant human bone morphogenetic protein-2 use in anterior cervical spine fusion. Spine 31:542–547

Siu RK, Lu SS, Li W, Whang J, McNeill G, Zhang X, Wu BM, Turner AS, Seim HB 3rd, Hoang P, Wang JC, Gertzman AA, Ting K, Soo C (2011) NELL-1 protein promotes bone formation in a sheep spinal fusion model. Tissue Eng Part A 17:1123–1135

Smucker JD, Rhee JM, Singh K, Yoon ST, Heller JG (2006) Increased swelling complications associated with off-label usage of rhBMP-2 in the anterior cervical spine. Spine 3:2813–2819

Squillaro T, Peluso G, Galderisi U (2016) Clinical trials with mesenchymal stem cells: an update. Cell Transplant 25:829–848

Sugiyama O, An DS, Kung SP, Feeley BT, Gamradt S, Liu NQ, Chen IS, Lieberman JR (2005) Lentivirus-mediated gene transfer induces long-term transgene expression of BMP-2 in vitro and new bone formation in vivo. Mol Ther 11:390–398

Than KD, Wang AC, Rahman SU, Wilson TJ, Valdivia JM, Park P, La Marca F (2011) Complication avoidance and management in anterior lumbar interbody fusion. Neurosurg Focus 31(4):E6. https://doi.org/10.3171/2011.7.FOCUS11141

Toosi S, Naderi-Meshkin H, Kalalinia F, Pievandi MT, Hosseinkhani H, Bahrami AR, Heirani-Tabasi A, Mirahmadi M, Behravan J (2017) Long bone mesenchymal stem cells (Lb-MSCs): clinically reliable cells for osteo-diseases. Cell Tissue Bank 18:489–500

US Food and Drug Administration (2019a) Premarket approval database. https://www.accessdata.fda.gov/scripts/cdrh/cfdocs/cfpma/pma.cfm?id=P000058. Accessed 9 May 2019

US Food and Drug Administration (2019b) Premarket approval database. https://www.accessdata.fda.gov/scripts/cdrh/cfdocs/cfpma/pma.cfm?id=P000058S065. Accessed 9 May 2019

Vaccaro AR, Patel T, Fischgrund J, Anderson DG, Truumees E, Herkowitz H, Phillips F, Hilibrand A, Albert TJ (2003) A pilot safety and efficacy study of OP-1 putty (rhBMP-7) as an adjunct to iliac crest autograft in posterolateral lumbar fusions. Eur Spine J 12:495–500

Vaccaro AR, Patel T, Fischgrund J, Anderson DG, Truumees E, Herkowitz HN, Phillips F, Hilibrand A, Albert TJ, Wetzel T, McCulloch JA (2004) A pilot study evaluating the safety and efficacy of OP-1 putty (rhBMP-7) as a replacement for iliac crest autograft in posterolateral lumbar arthrodesis for degenerative spondylolisthesis. Spine 29:1885–1892

Vaccaro AR, Anderson DG, Patel T, Fischgrund J, Truumees E, Herkowitz HN, Phillips F, Hilibrand A, Albert TJ, Wetzel T, McCulloch JA (2005a) Comparison of OP-1 putty (rhBMP-7) to iliac crest autograft for posterolateral lumbar arthrodesis: a minimum 2-year follow-up pilot study. Spine 30:2709–2716

Vaccaro AR, Patel T, Fischgrund J, Anderson DG, Truumees E, Herkowitz H, Phillips F, Hilibrand A, Albert TJ (2005b) A 2-year follow-up pilot study evaluating the safety and efficacy of OP-1 putty (rhbmp-7) as an adjunct to iliac crest autograft in posterolateral lumbar fusions. Eur Spine J 14:623–629

Vaccaro AR, Lawrence JP, Patel T, Katz LD, Anderson DG, Fischgrund JS, Krop J, Fehlings MG, Wong D (2008a) The safety and efficacy of OP-1 (rhBMP-7) as a replacement for iliac crest autograft in posterolateral lumbar arthrodesis: a long-term (>4 years) pivotal study. Spine 33:2850–2862

Vaccaro AR, Whang PG, Patel T, Phillips FM, Anderson DG, Albert TJ, Hilibrand AS, Brower RS, Kurd MF, Appannagari A, Patel M, Fischgrund JS (2008b) The safety and efficacy of OP-1 (rhBMP-7) as a replacement for iliac crest autograft for posterolateral lumbar arthrodesis: minimum 4-year follow-up of a pilot study. Spine J 8:457–465

Vaidya R, Sethi A, Bartol S, Jacobson M, Coe C, Craig JG (2008) Complications in the use of rhBMP-2 in PEEK cages for interbody spinal fusions. J Spinal Disord Tech 21:557–562

Viggeswarapu M, Boden SD, Liu Y, Hair GA, Louis-Ugbo J, Murakami H, Kim HS, Mayr MT, Hutton WC, Titus L (2001) Adenoviral delivery of LIM mineralization protein-1 induces new-bone formation in vitro and in vivo. J Bone Joint Surg Am 83:364–376

Vlaikou AM, Kouroupis D, Sgourou A, Markopoulos GS, Bagli E, Markou M, Papadopoulou Z, Fotsis T, Nakos G, Lekka ME, Syrrou M (2017) Mechanical stress affects methylation pattern of GNAS isoforms and osteogenic differentiation of hAT-MSCs. Biochim Biophys Acta 1864:1371–1381

Wang JC (2011) Commentary: gene therapy for spinal fusion. Spine J 11:557–559

Wang JC, Kanim LE, Yoo S, Campbell PA, Berk AJ, Lieberman JR (2003) Effect of regional gene therapy with bone morphogenetic protein-2-producing bone marrow cells on spinal fusion in rats. J Bone Joint Surg Am 85–A:905–911

Wang JC, Alanay A, Mark D, Kanim LE, Campbell PA, Dawson EG, Lieberman JR (2007) A comparison of commercially available demineralized bone matrix for spinal fusion. Eur Spine J 16:1233–1240

Williams BJ, Smith JS, Fu KM, Hamilton DK, Polly DW Jr, Ames CP, Berven SH, Perra JH, Knapp DR Jr, McCarthy RE, Shaffrey CI, Scoliosis Research Society Morbidity and Mortality Committee (2011) Does bone morphogenetic protein increase the incidence of perioperative complications in spinal fusion? A comparison of 55,862 cases of spinal fusion with and without bone morphogenetic protein. Spine 36:1685–1691

Winter M, Griss P, de Groot K, Tagai H, Heimke G, von Dijk HJ, Sawai K (1981) Comparative histocompatibility testing of seven calcium phosphate ceramics. Biomaterials 2:159–160

Yuan W, James AW, Asatrian G, Shen J, Zara JN, Tian HJ, Siu RK, Zhang X, Wang JC, Dong J (2013) NELL-1 based demineralized bone graft promotes rat spine fusion as compared to commercially available BMP-2 product. J Orthop Sci 18:646–657

Zhao J, Zhao DY, Shen AG, Liu F, Zhang F, Sun Y, Wu HF, Lu CF, Shi HG (2007) Promoting lumbar spinal fusion by adenovirus-mediated bone morphogenetic protein-4 gene therapy. Chin J Traumatol 10:72–76

Zhou SH, McCarthy ID, McGregor AH, Coombs RR, Hughes SP (2000) Geometrical dimensions of the lower lumbar vertebrae – analysis of data from digitised CT images. Eur Spine J 9:242–248

Zhu W, Rawlins BA, Boachie-Adjei O, Myers ER, Arimizu J, Choi E, Lieberman JR, Crystal RG, Hidaka C (2004) Combined bone morphogenetic protein-2 and -7 gene transfer enhances osteoblastic differentiation and spine fusion in a rodent model. J Bone Miner Res 19:2021–2032

Robotic Technology

69

Kyle J. Holmberg, Daniel T. Altman, Boyle C. Cheng, and Timothy J. Sauber

Contents

Introduction	1270
Passive, Semi-Active, and Active Systems	1271
Supervisory-Controlled, Telesurgical, and Shared-Control Systems	1271
Image-Based Systems	1271
History of Robotics in Surgery	1272
Robotics and Computer-Assisted Navigation	1273
History of Robotics in Spine Surgery	1273
Mazor Robotics	1274
Mazor: SpineAssist	1274
Mazor: The Renaissance Guidance System	1275
Mazor: Mazor X	1275
ROSA Robot	1276
Excelsius GPS	1277
Benefits of Robotic-Assisted Spine Surgery	1277
Benefits: Improved Accuracy of Robotic-Assisted Pedicle Screw Placement	1277
Benefits: Decreased Radiation Exposure	1279

K. J. Holmberg (✉) · D. T. Altman · T. J. Sauber
Department of Orthopaedic Surgery, Allegheny Health
Network, Pittsburgh, PA, USA
e-mail: Kyle.Holmberg@ahn.org; Daniel.Altman@ahn.
org; Timothy.Sauber@ahn.org

B. C. Cheng
Neuroscience Institute, Allegheny Health Network, Drexel
University, Allegheny General Hospital Campus,
Pittsburgh, PA, USA
e-mail: Boyle.Cheng@ahn.org; bcheng@wpahs.org

© Springer Nature Switzerland AG 2021
B. C. Cheng (ed.), *Handbook of Spine Technology*,
https://doi.org/10.1007/978-3-319-44424-6_138

Costs	1280
Limitations	1280
Discussion	1280
References	1281

Abstract

With increasing numbers of patients requiring spine surgery, there has been an emphasis on technological advances designed to enhance surgical outcomes and improve patient safety. In particular, the number of elective spinal fusion surgeries in the USA continues to increase. Instrumentation of the spine with pedicle screws is frequently used for indications including deformity and instability. With neurovascular structures near the pedicle, accurate screw placement is of paramount importance to ensure good outcomes. Reported complications related to pedicle screw malposition range from 1% to 54% (Molliqaj et al., Neurosurg Focus 42(5): E14, 2017). Currently, there are several techniques described for inserting pedicle screws, including freehand manual insertion based on anatomic landmarks and fluoroscopy, manual insertion with navigation assistance systems, and robotic-assisted methods.

Keywords

Robot · Navigation · Spine · Surgery · Pedicle · Screw

Introduction

With increasing numbers of patients requiring spine surgery, there has been an emphasis on technological advances designed to enhance surgical outcomes and improve patient safety. In particular, the number of elective spinal fusion surgeries in the USA continues to increase. Instrumentation of the spine with pedicle screws is frequently used for indications including deformity and instability. With neurovascular structures near the pedicle, accurate screw placement is of paramount importance to ensure good outcomes. Reported complications related to pedicle screw malposition range from 1% to 54% (Molliqaj et al. 2017). Currently, there are several techniques described for inserting pedicle screws, including freehand manual insertion based on anatomic landmarks and fluoroscopy, manual insertion with navigation assistance systems, and robotic-assisted methods.

The goal of this review is to focus on robotic-assisted surgical platforms used for the insertion of pedicle screws by highlighting their history and reviewing the literature regarding the current state of robotic platforms in spine surgery. In addition to detailing the history of robotics in spine surgery we will discuss terminology and definitions as well as reported benefits and costs associated with their use.

To better understand the basic concepts of robotics, it is important to understand or define the terminology that is used to describe robotics pertaining to surgical platforms. By definition, a robot is a machine capable of automating one, or a series of actions or steps in a procedure (Urakov et al. 2017). According to the Robot Institute of America, a robot is defined as a reprogrammable, multifunctional manipulator designed to move material, parts, tools, or specialized devices through various programmed motions for the performance of a variety of tasks (Jacofsky and Allen 2016). Robots should play an integral role in surgery to be classified as "surgical robots." There are several different ways to categorize robotic platforms, but their actions are classified as being either direct or indirect. Robots act directly when they cut tissue or mill bone into a predetermined and final desired shape, or indirectly by holding cutting jigs or drill guides (Chen et al. 2018). Furthermore, both direct and indirect surgical robotic platforms are further classified as passive, semi-active, and active systems (Jacofsky and Allen 2016).

Passive, Semi-Active, and Active Systems

Passive systems are used to complete a portion of the surgical procedure under direct, continuous control of the operating surgeon. The robotic instrument acts as an extension of the surgeon's hand and is unable to function without an operator engaging the device. Examples of passive systems include the OMNIbotics (OMNI) and the da Vinci surgical robot (Intuitive Surgical) (Chen et al. 2018).

Semi-active systems require surgeon involvement but provide feedback in the form of haptics (tactile, auditory, or visual) that restrict what can be done surgically to prevent the user from operating outside of a predetermined boundary. Semi-active systems can theoretically enhance the surgeon's control of the robotic instruments and increase surgical safety by preventing the surgical instruments from entering a "no-fly zone" as determined by the surgical plan. Current examples of semi-active systems include the Mako robotic arm–assisted surgical platform (Stryker) for total knee arthroplasty and the Navio surgical system (Smith and Nephew) used in unicompartmental knee arthroplasty.

Finally, active systems can perform a task entirely independently through the use of pre-programmed algorithms and predetermined parameters for surgical resection of tissue without direct manipulation by a surgeon or operator. Active systems allow for surgeons to initiate and stop the robot's activity but do not allow for continuous control, or the ability to alter the robot's actions intraoperatively (Chen et al. 2018). Current examples of active systems which are fully autonomous include the ROBODOC Surgical System and TSolution One (Think Surgical) (Chen et al. 2018).

Supervisory-Controlled, Telesurgical, and Shared-Control Systems

Additionally surgical robots have been classified into three main categories of systems, according to Kochanski et al. which are supervisory-controlled systems, telesurgical systems, and shared-control systems (Kochanski et al. 2019).

Supervisory-controlled systems are designed so that the surgeon can plan the robotic portion of the case before beginning the procedure. The surgeon initiates the procedure, then allowing the robot to perform its portion of the case autonomously under the direct supervision of the surgeon, making it an active system. Telesurgical systems, which are passive, offer the surgeon complete control of the robot at all times. Examples of telesurgical systems include the Georgetown robot and the Spine Bull's-Eye robot, both designed for use in spine surgery and the da Vinci surgical robotic system (intuitive surgical) which is widely utilized in various surgical specialties. Lastly are the shared-controlled systems, which allow for the robot and surgeon to share control of surgical instruments concurrently. Notably, most spine surgical robots on the market today such as the Mazor X, Excelcius GPS, and the ROSA spine robot are classified as shared-control systems. These systems require either preoperative or intraoperative CT or fluoroscopic imaging and use proprietary software to plan trajectories for pedicle screw placement. Once the preoperative plan is confirmed, the surgeon manually inserts the pedicle screws using the robotic instruments as a guide.

Image-Based Systems

Currently, most orthopaedic and spine robotic systems require a preoperative plan and computer model on which to base the surgical procedure. Having a preoperative plan to execute is one aspect of orthopaedic and spine robotic systems that differentiates it from other surgical robots, such as the da Vinci. Having a specific reproducible plan, which allows the surgeon to preoperatively or intraoperatively analyze and predict the desired result before any tissue is resected is a significant advantage of these systems (Jacofsky and Allen 2016).

The patient's anatomy must be registered via mapping points on the bone with a navigated tool during the registration process or via intraoperative scan in both system types so that the robot knows where the instruments are in space relative to the patient's anatomy. In image-based systems, this

Fig. 1 Intraoperative CT scan used for imaged-based planning to map the patient's anatomy in the prone position and register instruments used in lumbar fusion

registration is referenced to the preoperative imaging. Currently, preoperative CT or an intraoperative CT scan is used (Fig. 1). Intraoperative CT limits the need for bony registration as the navigation arrays are typically included in the scan. Potential disadvantages of image-based systems include increased cost, patient inconvenience, and increased radiation exposure to the patient during CT (Jacofsky and Allen 2016). Potential advantages include extremely accurate mapping of the patient's anatomy, which increases the precision of implant placement.

History of Robotics in Surgery

Robotic surgical platforms are used in many surgical specialties, including gynecological, urological, thoracic, orthopaedic, and general surgery over the last two decades. According to early reports, the use of the UNIMATION PUMA 200 (Programmable Universal Manipulation Arm, Nokia) during a neurosurgical biopsy performed in 1985 marked the first documented robotic-assisted surgical procedure (Chen et al. 2018). That same system was repurposed and used for the transurethral resection of prostate tissue, which eventually lead to the development of the ProBot, which was designed explicitly for prostate surgery (Jacofsky and Allen 2016). With a growing body of evidence supporting their safe and predictable implementation, the field of robotic surgery began to proliferate. Since then, the integration of robotics in the operating room has steadily increased across a variety of surgical specialties.

Robotic surgery began to gain momentum in the year 2000 when the da Vinci surgical robotic system (intuitive surgical) was approved for use in urologic, gynecologic, and general laparoscopic procedures by the US Food and Drug Administration (Perez-Cruet et al. 2012). The da Vinci is a robotic system that offers surgeons the ability to perform a variety of soft tissue procedures with more minimally invasive technique (Jacofsky and Allen 2016).

The first applications of the da Vinci platform were directed towards soft tissue procedures such as laparoscopic prostatectomies and gynecological surgeries that were readily adopted. In fact, in 2012, 85% of prostatectomies in the USA were performed with robotic assistance (Chen et al. 2018). The da Vinci surgical robot is a telesurgical platform that allows the surgeon to control the docked surgical robot from a remote station that is outside the sterile field yet within the same operating room as the patient. This ability allows the surgeon to visualize the patient and surgical field in an augmented three-dimensional view, provided by endoscopic cameras, which allows for real-time decisions to be executed with unparalleled precision while minimizing soft tissue trauma typically experienced by the patient. The da Vinci articulating robotic arms serve as extensions of the surgeon's instruments and are under complete control of the operating surgeon at all times.

While the da Vinci robot has not been as widely adopted by spine surgeons for routine surgical techniques, it has been reported to be used for the resection of paraspinal tumors, transoral odontoidectomy, and in laparoscopic-assisted anterior lumbar interbody fusion (D'Souza et al. 2019). For the treatment of spine trauma, the da Vinci robot was recently used in a minimally invasive retroperitoneal approach for the treatment of an L3 fracture with corpectomy and expandable cage implantation with a good clinical outcome (Lippross et al. 2020).

Robotics and Computer-Assisted Navigation

Historically, robotics and image-guided navigation systems used in spine surgery have been considered mutually exclusive technologies. However, both technologies fundamentally rely upon radiographic imaging and stereotaxic principles in order to provide precision and accuracy during spine surgery (Ganz 2011). Stereotaxy refers to the use of instruments that relate anatomical targets within the body with radio frequency markers using a cartesian coordinate system. This can provide the surgeon with haptic feedback when accessing areas of the surgical field that are difficult to appreciate visually (Ganz 2011).

Stereotactic techniques have since been applied to the field of spine surgery and lead to the development of the intraoperative CT with Stealth Station navigation (Medtronic Inc.). The robotic systems described in this chapter involve the automated performance of various portions in the operation. However, image-guided systems have the benefit of providing real-time haptic feedback to the surgeon, who then uses that information to perform those portions of the case manually.

Both image-guided navigation and robotics systems that have historically been used in spine surgery are fundamentally based upon stereotactic principles coupled with pre/intraoperative patient imaging and registration (Fig. 2). Similar to navigation-assisted techniques, robotic systems also rely on radiographic imaging and stereotaxis for trajectory planning and drilling of pedicle screw bone tunnels and screw insertion (Kochanski et al. 2019). Historically robotic systems have lacked the real-time navigation that image-guided systems utilize. However, newer robotic systems such as the Excelsius GPS (Globus Medical), ROSA Spine (Zimmer-Biomet), and Mazor X Stealth (Medtronic) allow for real-time image guidance coupled with the precision and accuracy of an automated robotic arm to ensure a proper trajectory when inserting pedicle screws (Kochanski et al. 2019).

History of Robotics in Spine Surgery

Robotics have been implemented throughout the majority of surgical specialties, and spine surgery is no exception. To date, the use of robots in spine surgery has primarily been limited to using a planning platform to assist with pedicle screw placement which is often used to stabilize the vertebral column in spinal fusion surgery (Urakov et al. 2017). Metallic screws are inserted into each pedicle involved at each vertebral level involved in an operation, and then rods are inserted through the heads of all screws on each side in order to support the vertebral column. Accurate positioning of the pedicle screw is essential given the proximity of the neurovascular structures, including the spinal cord, spinal nerve roots, and major blood vessels. The goal of modern robotics systems in spine surgery is to assist surgeons in consistently placing pedicle screws with accuracy and precision based on advanced imaging and guidance software. Currently, there are only three manufacturers with five robots approved by the FDA indicated for instrumentation of the spine in the USA. These include Mazor's three generations of robots: the SpineAssist, the Renaissance Guidance System, and Mazor X (Mazor

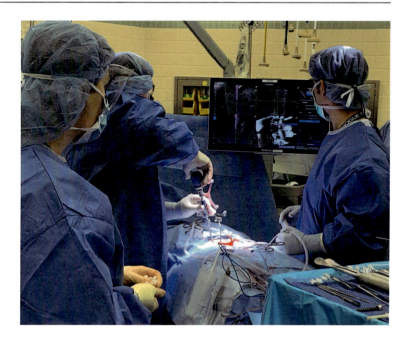

Fig. 2 Spine surgeon inserting pedicle screws under navigation with instruments registered via intraoperative CT scan

Robotics), the ROSA Spine robot (Zimmer Biomet), and the Excelsius GPS (Globus Medical).

Mazor Robotics

Mazor Robotics was founded in 2001 in Israel. The company brought the first commercially available robot for spine surgery to market. Starting with the FDA approval of SpineAssist in 2004, the company's technology continued to evolve with the launch of the Renaissance Guidance System in 2011, and eventually the Mazor X which was released in the USA in 2016. As Mazor is the longest standing company to produce approved spine surgical robots, the majority of the published literature has focused on these systems.

Mazor: SpineAssist

The first robotic-assisted platform to reach the US market, with applications specifically designed for use in spine surgery, arrived in 2001 with SpineAssist (Mazor Robotics). SpineAssist was approved for use in the USA in 2004 and became the first commercially available mechanical guidance system used for the placement of pedicle screws in thoracolumbar surgery (Theodore et al. 2018). SpineAssist, which is the spinal application of its predecessor SmartAssist (Mazor Robotics), is a shared-control semi-active platform that is rigidly attached to the patient's iliac crest or spinous process, unifying the patient and robot. This ensures that patient movement due to positional changes or respirations does not interfere with the position of the robot relative to the patient, thus potentially providing higher accuracy with the placement of pedicle screws. SpineAssist was designed to accurately guide and assist the surgeon in drilling and placing spinal implants such as pedicle screws, however, the actual instrumentation of the patient's spine is performed by the surgeon and not the robot itself.

The SpineAssist System consists of two units: a small hexapod robot that can move with six degrees of freedom and a separate computer workstation with a graphic user interface. Using the workstation, the surgeon plans the trajectory for pedicle screw placement at any time prior to the start of the case (Shoham et al. 2007). During the operation, the robot's base platform is mounted on the vertebrae using a disposable sterile clamp system. The coordinates of the desired screw insertion trajectory are obtained from preoperative plans made by the surgeon using a reconstructed CT image, which is obtained pre- or intraoperatively. Following a registration process that involves matching two intraoperative fluoroscopic images with the preoperative CT scan, the planned coordinates arc then translated to each of the robot's six articulators to create the required motion. The computer system controls the movement of the robot to the desired position where the guide arm is then activated to move into the preplanned trajectory. The surgeon performs the manual drilling and screw insertion via an open or percutaneous technique according to the patient's needs (Shoham et al. 2007).

Mazor: The Renaissance Guidance System

Mazor Robotics' second-generation robot and successor to SpineAssist is the Renaissance Guidance System. The Renaissance is a miniature robotic platform that acts as a mechanical guidance unit that can be mounted directly to the patient via bone-anchored attachments to the spine or pelvis or the operating table (Malham and Wells-Quinn 2019). A three-dimensional marker is attached to the platform before the intraoperative CT scan is performed using an intraoperative CT with the patient prone on the operating table. The images are uploaded into the Renaissance System's interface, and the surgeon uses the information to plan the procedure using the system's software. The patient's anatomy can be reviewed in coronal, axial, and transverse planes to ensure that the proper trajectory is planned before instrumentation. Once the plan is complete, the robot's guidance system is mounted to the platform and automatically sent to the preplanned trajectory where the surgeon can use the system to place K-wires that are then used to guide the insertion of cannulated pedicle screws. The Renaissance Guidance System has been used in approximately 30,000 cases.

Mazor: Mazor X

Newer robotic platforms such as the Mazor Robotics' third-generation spine surgical robot, the Mazor X (Fig. 3), which was released in the USA in 2016, utilize an automated mechanical arm fixed with a drill guide which assists in drilling and pedicle screw insertion, with or without K-wire placement. This feature requires surgeons to introduce instruments using the robotic arm, which provides haptic feedback if tools deviate from a preplanned trajectory. The surgical robot is either mounted to the operating table or the floor and is connected to the patient similarly using bone anchors as with the Mazor Renaissance System (Malham and Wells-Quinn 2019). There is an associated computer station with a user interface that contains the planning software that allows for planning of screw

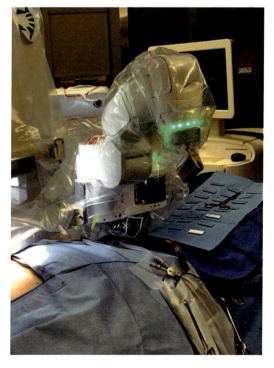

Fig. 3 Mazor X robotic arm used in posterior lumbar instrumented fusion

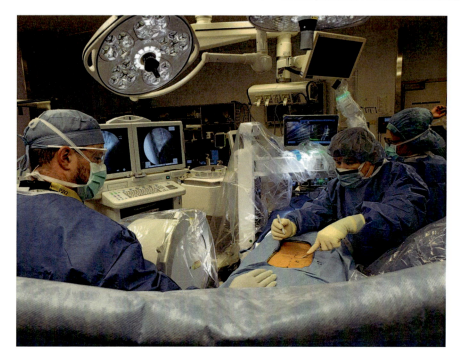

Fig. 4 Spine surgeons localizing appropriate levels using fluoroscopy for treatment of the lumbar spine with the Mazor X Robotic platform

trajectories based on the patient registration process. The automated robotic arm, which is controlled by the user interface and directed by the surgeon, can move between preplanned screw trajectories at various levels quite rapidly, allowing the surgeon to drill while providing tactile feedback to ensure adequate purchase as instruments enter the bone. Once the setup and registration process are performed, the automatic movements of the robotic arm between various levels allow for rapid and accurate insertion of pedicle screws potentially decreasing total surgical time. The Mazor X system retails at approximately $1000,000 with disposable costs ranging from $1000–$1500 per case according to recent market reports (Fig. 4).

Mazor Robotics' newest addition to their robotics line combines its navigation technology with Robotics to form the Mazor X Stealth robotic-assisted surgical platform. Mazor X Stealth received its FDA approval in 2018 and was used in its first cases in early 2019 just after the acquisition of Mazor Robotics by Medtronic (Minneapolis, MN). The Mazor X Stealth allows for the union of the computer navigation software, and the Mazor robotic-assisted surgical platform to improve accuracy and precision of pedicle screw placement in spine surgery. By combining these two technologies, surgeons can use software to formulate their preoperative plan based on a 3D analysis of the patient's anatomy. This allows surgeons to place pedicle screw trajectories accurately with the assistance of the robotic arm and receive navigation feedback in real time to ensure that trajectories match the preoperative plan.

ROSA Robot

The ROSA Robot, which stands for Robotic Stereotactic Assistance, was initially developed in 2007 by MedTech in Montpellier, France (now Zimmer-Biomet), for use in cranial surgery in Europe. The company expanded its applications for the platform and developed ROSA Spine which was cleared for use in the USA by the FDA in early 2016. ROSA Spine combines navigation technology with robotic assistance to aid surgeons in placing pedicle screws according to preplanned trajectories, and performance of transforaminal lumbar interbody fusions (Chen et al. 2018; Lonjon et al. 2016). With the patient in the prone position on the operating table, the ROSA Robot's floor-mounted platform is brought into position at the side of the table, while the separate navigation camera is placed at the foot of the patient. A referencing marker is rigidly mounted to the patient's iliac crest, which monitors the patient's movements, either respiratory or due to manipulation by the surgeon. Intraoperative fluoroscopic images are obtained using a specific registration pattern held by the robotic arm. The images are then uploaded directly to the user interface where the surgeon can begin using the planning software to plan the ideal trajectory for each screw. Once the trajectories have been planned, the articulating robotic arm can handle marked instruments that are confirmed by the navigation camera to enable accurate placement of instruments according to the preplanned trajectories. This allows for real-time adjustments to be made to account for the patient's respiratory movements, which are continuously monitored by the navigation cameras (Lonjon et al. 2016). The ROSA Spine Robot, which retails for just over $600,000, is the least expensive of the three leading systems with automated robotic arms available in the market.

Excelsius GPS

The Excelsius GPS released by Globus Medical (Audubon, PA) in 2018, received FDA approval for use in spine surgery in 2019. This system uses proprietary software for planning pedicle screw trajectories and an automated robotic arm to guide instruments simultaneously with navigation technology and has proved to be safe and effective as a means for inserting pedicle screws (Galetta et al. 2019). A recent single institution study demonstrated a 99% rate of successful pedicle screw placement out of 562 screws placed using the Excelsius GPS system in minimally invasive lumbar spine surgery using a variety of approaches (Huntsman et al. 2020). One advantage of this system is the ability for robotic-assisted insertion of pedicle screws without the use of K-wires. By using real-time image guidance, the surgeon is able to manipulate a variety of compatible surgical instruments according to the preoperative plan via the robotic–navigation interface. Excelsius GPS is indicated for spinal instrumentation from C1-pelvis and is compatible with most imaging systems using preoperative or intraoperative CT or fluoroscopy. The Excelsius GPS costs approximately $1,200,000, with an average cost of $1000 per procedure in disposables required for its use.

Benefits of Robotic-Assisted Spine Surgery

There have been many benefits reported in the literature attributed to the use of robotic-assisted spine surgery including minimally invasive applications, improved accuracy, reductions in radiation to OR staff, decreased blood loss, faster pedicle screw insertion time, reduction in human error due to tremor or fatigue, and decreasing the learning curve associated with pedicle screw insertion technique among resident and fellow trainees (Li et al. 2020; Siddiqui et al. 2019). Of which, pedicle screw placement accuracy and reduction in radiation exposure to surgeons and operating room staff have been the most widely reported.

Benefits: Improved Accuracy of Robotic-Assisted Pedicle Screw Placement

In many studies, the accuracy of pedicle screws is evaluated using postoperative CT scans, which are used to confirm the trajectory of the screw within the cortical bone of the pedicle. These images are graded based on the Gertzbein-Robbins classification, where the deviation of the screw is measured against the "ideal" trajectory of the screw. According to this classification system, Grade A is an interpedicular screw without a breach of the cortical layer, a Grade B screw breaches the cortex but is deviated <2 mm laterally from the pedicle, and Grade C and D are breaches of <4 and < 6 mm, respectively (Solomiichuk et al. 2017). While theoretically, a more displaced screw may have a higher chance of postoperative complications, the clinical significance of this radiographic definition in the absence of clinical symptoms is not well described (Kochanski et al. 2019). The majority of screws with a minor breach of the pedicle cortex may still maintain excellent biomechanical properties and are unlikely to require revision. Thus, in the absence of clinically apparent neurologic of vascular symptoms or biomechanical instability, a minimally displaced screw is unlikely to cause unwanted complications. However, clinically significant misplacement of pedicle screws can cause fractures, neurologic or vascular impairment, injury to the dura, and biomechanical instability (Solomiichuk et al. 2017).

A multicenter retrospective review evaluating the accuracy of pedicle screw placement using SpineAssist (Mazor Robotics) in the placement of 3271 pedicle screws was performed by DeVito et al. Pedicle screws were placed using the robotic platform and then evaluated and graded using post-op CT. They found 98% of robotic pedicle screws placed to be clinically acceptable when evaluated using fluoroscopic X-rays before leaving the operating room. One hundred and ninety-eight patients underwent postoperative CT analysis, and 98.3% of the screws placed were found to be within 2 mm of the pedicle cortex. Only 1.7% of the screws placed had a pedicle wall breach greater than 2 mm (Kochanski et al. 2019). This study was consistent with previously published literature demonstrating that robotic-assisted techniques are a safe and effective option for the insertion of pedicle screws.

In a retrospective series performed by Molliqaj et al., 439 thoracolumbar pedicle screws were inserted using SpineAssist (Mazor robotics), and 441 screws were inserted using a freehand fluoroscopic-guided technique by experienced spine surgeons. The accuracy of screw placement was determined by neuroradiologists who were blind to the treatment group. Each screw was independently evaluated and graded based on the Gertzbein-Robbins criteria. In the robot-assisted group, 366 (83.4%) of screws placed were found to be perfectly intrapedicular (Gertzbein-Robbins Grade A) vs. 335 (76%) in the freehand group. Additionally, 93.4% of robotic-assisted pedicle screws were classified as nonmisplaced, defined as Gertzbein-Robbins Grades A and B, compared to 88.9% of the screws in the freehand fluoroscopy (p = 0.005) (Molliqaj et al. 2017).

The Renaissance system (Mazor Robotics) was evaluated for accuracy and safety for use in lumbar spinal fusion surgery by Kim et al. The study compared robot-assisted posterior lumbar interbody fusion (Robot-PLIF) vs. a conventional freehand open approach (Freehand-PLIF). A total of 37 patients were treated using the robotic-assisted platform, and 41 patients were treated with the conventional freehand approach. Of the pedicle screws inserted using the Robot-PLIF approach, 93.7% were Grade A, with 5.7% Grade B, and 0.6% Grade C breaches, respectively. All breaches occurred in the lateral wall of the pedicle in the Robot-PLIF group. In the Freehand-PLIF group, 91.9%, 7.6%, and 0.6% of pedicle screws placed were classified as Grade A, B, and C breaches, respectively. The Grade C breach was an inferior wall violation resulting in a subsequent nerve root irritation that required revision. Additionally, it was noted that none of the 74 screws in the Robot-PLIF group violated the proximal facet joint, while 13 of the 82 Freehand-PLIF groups violated the proximal facet joint (P < 0.001) (Kim et al. 2017).

Hu et al. investigated the accuracy of robotic-assisted pedicle screw placement as well as learning curve associated with spinal instrumentation using the Mazor Renaissance platform. They found that over the course of 150 cases performed by a single surgeon there was an overall increase in the accuracy of pedicle screw placement that was associated with the increasing number of cases performed using the robot. By dividing patients into groups of 30, the investigators analyzed pedicle screw accuracy when placed by the robot, the rate at which pedicle screws had to be converted to manual insertion due to malposition, and the overall rate of malposition of screws placed robotically. They found that when analyzed by intraoperative fluoroscopy and post-op radiographs, there was a greater than 90% success rate in pedicle screw placement after the first 30 cases using the robot (82% in the first group of 30 cases, 91–95% in the following 120 cases). Additionally, the number of screws that had to be converted to manual insertion decreased from 17% in the first 30 cases and 7% in the last 30 cases in the study. Screw malposition using the robot also decreased with increased surgeon experience notably from 1.4% in the initial group and 0% in the last group. The authors concluded that improved accuracy, decreased rates of screw reinsertion, and decreased rates of screw malposition could be achieved after the surgeon completes the initial 30 cases using the robot. These findings were consistent with other published reports, and have been replicated with another surgeon evaluated outside of the study (Hu et al. 2014).

These studies suggest that robotic-assisted techniques can consistently, safely, and accurately be used to insert pedicle screws that are at least as acceptable as the current conventional techniques. Further research is needed to determine if robot-assisted techniques are both statistically and clinically significantly superior than the current standard in terms of safety and accuracy of pedicle screw insertion.

Benefits: Decreased Radiation Exposure

Intraoperative fluoroscopy has been widely used in orthopaedic surgery and is particularly helpful in minimally invasive spine surgery such as percutaneous pedicle screw placement. Radiation safety, including limiting dose and exposure, to patients, surgeons, and staff, is of significant concern and has been receiving increased attention in published literature. There are several ways to standardize and quantify radiation exposure. The Gray (Gy), which is used to express an absorbed dose of radiation, is the actual physical quantity of one joule of radiation energy per kilogram of matter. The sievert (Sv) represents the biological equivalent of an effective dose on tissue and can be used to measure the cumulative dose of radiation absorbed by tissue over time. For healthcare workers, an upper limit of 20 millisieverts (mSv) of radiation exposure per year has been established by the international commission of radiological protection (Hayda et al. 2018).

Radiation exposure among spine surgeons varies considerably and assessing radiation exposure in spine surgery can be challenging due to a lack of standardized reporting. For instance, some studies report radiation exposure in time (seconds) per screw, whereas others report total radiation generated throughout a case using an entirely different scale or measurement (Malham and Wells-Quinn 2019). As such, implementing new technologies such as robotic surgical systems that can potentially decrease the risk of this occupational hazard in spine surgery would likely be beneficial to OR staff.

One of the proposed benefits of robotic-assisted surgical systems is that they can potentially limit radiation exposure to operating room staff. CT imaging can be performed before the day of surgery or using an intraoperative CT scan with the surgical staff outside the operating room. In a study performed by Mendelsohn et al., patients undergoing intraoperative CT-assisted lumbar spine surgery were found to be exposed to 2.77 times more radiation (5.69 mSv) than patients in a fluoroscopic-assisted control group. However, the surgeon and OR staff were exposed to approximately 2.5 times less radiation in the intraoperative CT group when compared to the fluoroscopy group (Hayda et al. 2018). In another study by Costa et al., patients undergoing CT-based navigation-assisted lumbar spine surgery were exposed to a mean of 5.15 mSv of radiation, whereas the surgeon and staff who were outside the operating room at the time of the scan were reported as having no radiation exposure. Both studies confirm that the use intraoperative CT increases radiation exposure to the patient by approximately 7.5 mSv, which is less than that sustained during a routine lumbar CT. Additionally, radiation exposure to the surgeon and staff were reported to be decreased in both studies (Hayda et al. 2018).

A review of literature demonstrated promising results when evaluating radiation exposure measured in seconds of fluoroscopy time comparing robotic-assisted vs. traditional freehand insertion of pedicle screws. A study performed by Kandlehardt et al. demonstrated an average radiation exposure time of 34 s per screw in the robotic-assisted group vs. 77 s per screw in the fluoroscopic-assisted freehand group (Kochanski et al. 2019). Likewise, Schoenmayr and Kim reported a decrease in radiation exposure by 40% when comparing robotic systems to traditional techniques. Additionally, Roser et al. found an average fluoroscopy time of 31.5 s per screw for using traditional freehand technique vs. 15.98 s per screw in the robotic-assisted cohort in a prospective randomized study evaluating the two groups (Kochanski et al. 2019). These studies demonstrate positive results and support the proposed benefits of reductions in total radiation exposure associated with robotic-assisted spine surgery. However, further research is likely needed to determine the impact of robotics systems on total radiation exposure to both surgeons and patients due to differences in surgeon experience, operative technique, and imaging protocols for robotics systems used between these studies and others like them (Kochanski et al. 2019).

Costs

Introducing robotic-assisted techniques may potentially improve precision and accuracy in the operating room, but the ultimate acceptance of robotic surgery into mainstream surgical practice will be heavily dependent on its cost-effectiveness. According to a 2019 market analysis, the average upfront cost of a spine surgical robot can be anywhere from 600,000–1,300,000 dollars depending on the model and purchasing agreement for a health system. Additionally, there is typically an annual service contract that can cost as much as $100,000 to service and maintain the robot (Ahern et al. 2020). Over time the total costs to operate, service, and maintain each robot accumulate and thus increase the total cost of each surgical procedure in which the robot is used. As such, the added upfront costs, added surgical time, and training of the surgeon, and OR staff must be considered and weighed against the potential benefits of robotic-assisted surgery. According to a

review by Ahern et al., these added upfront costs could be offset if robotic-assisted spine surgery can continue to decrease operating room time, hospital length of stay, and rates of revision surgery (Ahern et al. 2020). Although the learning curve for robotic-assisted surgery is steep and volume dependent, once surgeons become familiar with their use, operating room times may decrease leading to more operating room efficiency (Ahern et al. 2020). Additionally, costs may be recouped by capturing a greater portion of patients who desire robotic-assisted procedures.

Limitations

Despite the reported benefits, there is still a lack of published data demonstrating conclusive evidence that suggests that long-term clinical outcomes in spine surgery are significantly improved with robotic-assisted techniques when compared to traditional methods. More extensive studies are needed to investigate further if robotic-assisted techniques in spine surgery are beneficial enough to justify the upfront and maintenance costs associated with their use.

Discussion

The goals of spine surgery are to improve patient functioning and reduce morbidity associated with disease of the spine. Over the last two decades, robotic-assisted technologies for use in spinal instrumentation have evolved rather quickly. The rapid growth and integration of imaging, navigation, and robotics in spinal surgery provide surgeons and health systems with an increasing number of options in spine surgery. Robotic techniques have been shown to be safe and effective in treating instability and deformity caused by degenerative spine disease or trauma as well as in other complex cases where patient anatomy may be distorted such as revision or tumor surgery. It is important to note that over the long term, most industries that have implemented robotic technology have demonstrated increased production capacity, improved precision, and decreased costs.

There are many reported advantages of robotic-assisted spine surgery including minimally invasive applications, improved accuracy of pedicle screw placement, and decreased radiation exposure to the surgeon, and operating room staff. While the reported advantages and potential benefits of using robotic-assisted techniques are promising, they must be weighed against the increased direct costs associated with purchasing and maintaining the surgical robot, added surgical time, and the ongoing education and training of the surgeon and operating room staff on the robot's use.

While it is not the goal of this review to argue in favor or against the use of robotics in spine surgery, robotics in the operating room are likely here to stay. In spine surgery, technologies such as computer navigation and robotic-assisted techniques continue to evolve in an effort to improve patient outcomes, decrease complications, and increase patient safety. As such, more high-quality studies are needed to investigate improvement outcomes, efficiency, and reductions in total costs of care in order to provide the information that spine surgeons and health systems will need to decide if they will implement robotics into their practice.

Acknowledgments The authors would like to thank Allegheny Health Network's Department of Orthopaedic Spine Surgery, the Allegheny Health Network Neuroscience Institute, Robert Pezzin, and all those who contributed to this review.

References

Ahern DP, Gibbons D, Schroeder GD, Vaccaro AR, Butler JS (2020) Image-guidance, robotics, and the future of spine surgery. Clin Spine Surg 33(5):179–184. https://doi.org/10.1097/BSD.00000000000000809. PMID: 31425306

Chen AF, Kazarian GS, Jessop GW, Makhdom A (2018) Robotic technology in orthopaedic surgery. J Bone Joint Surg Am 100(22):1984–1992. https://doi.org/10.2106/JBJS.17.01397

D'Souza M, Gendreau J, Feng A, Kim L, Ho A, Veeravagu A (2019) Robotic-assisted spine surgery: history, efficacy, cost, and future trends. Robot Surg Res Rev 6:9–23. https://doi.org/10.2147/RSRR.S190720

Galetta MS, Leider JD, Divi SN, Goyal DKC, Schroeder GD (2019) Robotics in spinal surgery. Ann Transl Med 7:S165. https://doi.org/10.21037/atm.2019.07.93

Ganz JC (2011) Gamma knife neurosurgery: principles of Sterotaxy, vol XXII. Springer, Vienna.

Hayda RA, Hsu RY, DePasse JM, Gil JA (2018) Radiation exposure and health risks for orthopaedic surgeons. J Am Acad Orthop Surg 26(8):268–277. https://doi.org/10.5435/JAAOS-D-16-00342

Hu X, Ohnmeiss DD, Lieberman IH (2014) What is the learning curve for robotic-assisted pedicle screw placement in spine surgery? Clin Orthop Relat Res 472:1839. https://doi.org/10.1007/s11999-013-3291-1

Huntsman KT, Ahrendtsen LA, Riggleman JR, Ledonio CG (2020) Robotic-assisted navigated minimally invasive pedicle screw placement in the first 100 cases at a single institution. J Robot Surg 14(1):199–203. https://doi.org/10.1007/s11701-019-00959-6

Jacofsky DJ, Allen M (2016) Robotics in Arthroplasty: a comprehensive review. J Arthroplast 31(10):2353–2363. https://doi.org/10.1016/j.arth.2016.05.026

Kim HJ, Jung WI, Chang BS, Lee CK, Kang KT, Yeom JS (2017) A prospective, randomized, controlled trial of robot-assisted vs freehand pedicle screw fixation in spine surgery. Int J Med Robot 13(3). https://doi.org/10.1002/rcs.1779

Kochanski RB, Lombardi JM, Laratta JL, Lehman RA, O'Toole JE (2019) Image-guided navigation and robotics in spine surgery. Neurosurgery 84:1179. https://doi.org/10.1093/neuros/nyy630

Li H-M, Zhang R-J, Shen C-L (2020) Accuracy of pedicle screw placement and clinical outcomes of robot-assisted technique versus conventional freehand technique in spine surgery from nine randomized controlled trials. Spine 45: E111. https://doi.org/10.1097/brs.0000000000003193

Lippross S, Jünemann K-P, Osmonov D, Peh S, Alkatout I, Finn J, Egberts J-H, Seekamp A (2020) Robot assisted spinal surgery – a technical report on the use of DaVinci in orthopaedics. J Orthop 19:50–53. https://doi.org/10.1016/j.jor.2019.11.045

Lonjon N, Chan-Seng E, Costalat V, Bonnafoux B, Vassal M, Boetto J (2016) Robot-assisted spine surgery: feasibility study through a prospective case-matched analysis. Eur Spine J 25(3):947–955. https://doi.org/10.1007/s00586-015-3758-8

Malham GM, Wells-Quinn T (2019) What should my hospital buy next?-guidelines for the acquisition and application of imaging, navigation, and robotics for spine surgery. J Spine Surg 5(1):155–165. https://doi.org/10.21037/jss.2019.02.04

Molliqaj G, Schatlo B, Alaid A, Solomiichuk V, Rohde V, Schaller K, Tessitore E (2017) Accuracy of robot-guided versus freehand fluoroscopy-assisted pedicle screw insertion in thoracolumbar spinal surgery. Neurosurg Focus 42(5):E14. https://doi.org/10.3171/2017.3.FOCUS179

Perez-Cruet MJ, Welsh RJ, Hussain NS, Begun EM, Lin J, Park P (2012) Use of the da Vinci minimally invasive robotic system for resection of a complicated paraspinal schwannoma with thoracic extension: case report.

Neurosurgery 71(1 Suppl Operative):209–214. https://doi.org/10.1227/NEU.0b013e31826112d8

Shoham M, Lieberman IH, Benzel EC, Togawa D, Zehavi E, Zilberstein B, Roffman M, Bruskin A, Fridlander A, Joskowicz L, Brink-Danan S, Knoller N (2007) Robotic assisted spinal surgery–from concept to clinical practice. Comput Aided Surg 12 (2):105–115. https://doi.org/10.3109/10929080701243981

Siddiqui MI, Wallace DJ, Salazar LM, Vardiman AB (2019) Robot-assisted pedicle screw placement: learning curve experience. World Neurosurg 130: e417. https://doi.org/10.1016/j.wneu.2019.06.107

Solomiichuk V, Fleischhammer J, Molliqaj G, Warda J, Alaid A, von Eckardstein K, Schaller K, Tessitore E, Rohde V, Schatlo B (2017) Robotic versus fluoroscopy-guided pedicle screw insertion for metastatic spinal disease: a matched-cohort comparison. Neurosurg Focus 42 (5):E13. https://doi.org/10.3171/2017.3.FOCUS1710

Theodore N, Ahmed AK (2018) The history of robotics in spine surgery. Spine 43:S23. https://doi.org/10.1097/BRS.0000000000002553

Urakov TM, Chang KH, Burks SS, Wang MY (2017) Initial academic experience and learning curve with robotic spine instrumentation. Neurosurg Focus 42(5): E4. https://doi.org/10.3171/2017.2.FOCUS175

Index

A

Abdominal obesity, 607
Abdominal organs, 698
Absolute contraindications, 576, 1170
Absorbable collage sponge, 232
ACADIA Facet Replacement System (AFRS), 851–852
Accelerated mechanical testing, 409
Access surgeons, 1167
Accuflex rod, 286
Accuflex system, 301
Acidic FGF (aFGF), 1240
Active systems, 1271
Activin A, 1257
Activities, 1065
ActivL prosthesis, 917
Acute cauda equina syndrome, 10
Adamkewicz, 1055
Additive manufacturing, 159
Adenoviral mediated gene transfer, 1259
Adhesion(s), 1185
    of amorphous calcium, 262
Adjacent-level degeneration, 868, 900
Adjacent level degenerative changes, 900
Adjacent level disease, 900, 916, 917
Adjacent segment, 808
Adjacent segment degeneration (ASD), 317–322, 762,
        767, 858, 874, 885, 901, 902, 912, 970
    and adjacent segment disease (ASD), 791
    arthroplasty, 887
    artificial disc replacement, 887
    biomechanical studies, 886
    cervical disc prosthesis, 887
    cervical fusion, 886
    intradiscal pressure, 886
    mechanical instability, 885
    motion preservation devices, 886
    motion preserving devices, 887
    myelopathy, 885
    natural progression, 886
    radiculopathy, 885
Adjacent segment disease (ASD), 295, 296, 311, 762,
        772, 782, 834, 835, 874, 901, 985, 989, 1003,
        1005, 1034
Adjacent segment facet joints, 543

Adjunctive treatments, 121
Adult deformity surgery, 709
Adult spinal deformity, 658, 970, 976, 978, 989
    early fixation constructs, 662–663
    minimally invasive surgery, 669–671
    non-operative management, 661–662
    operative management, 662
    pedicle subtraction osteotomy, 666–668
    prevalence of, 658
    proximal and distal extent of instrumentation, 663–664
    Smith-Petersen osteotomy, 664–666
    vertebral column resection, 668–669
Adverse local tissue reactions (ALTR), 460
Adverse reactivity to implant debris, 129
Aftersensations, 117
Aggrecans, 1114
Aggressive, 956, 958, 959
AHARA, 927
ALARA, 927
Allo-bone graft
    cortico-cancellous allograft, 214
    DBM-based product, 214–232
Allodynia, 117
Allogeneic bone, 1253–1254
Allograft(s), 159, 624, 1142, 1253
    in anterior column, 1021–1025
    bone banking, 1018–1019
    in cervical spine, 1021–1025
    description, 1017–1018
    lumbar spine, 1024–1025
    posterior elements, 1025
    safety, 1018
    in spine surgery, 1020–1021
    thoracolumbar, 1025
Allograft-based grafts, 1015
Ambulatory care, 119
American Academy of Orthopaedic Surgeons (AAOS)
        clinical guidelines, 575
American Association of Neurological Surgeons
        (AANS), 1033
American Society for Testing and Materials (ASTM), 298
American Society of Mechanical Engineers (ASME), 414
American Spinal Injury Association (ASIA) Impairment
        Scale, 1233

Amino-acid, 263
15-Amino acid peptide, 239
Amorphous calcium, 262
Amplification, 117
Analgesic, 1205
Anaplastic astrocytoma, 518–519
Anatomy, cervical spine, *see* Cervical spine anatomy
Aneurysms, 1179
Angiogram, 1172, 1176, 1181, 1193, 1194, 1196, 1197, 1201, 1204
Animal testing, 416–417
Ankylosing spondylitis, 12, 97, 99
Annex I, 1155
Annular defect area, 1125
Annular fissuring, 276
Annulotomy, 615, 696
Annulus fibrosus, 1113
Anorganic bone matrix (ABM), 239
Anterior and anterolateral plate, 618
Anterior approach, 998, 999, 1002, 1003
Anterior cervical decompression and fusion (ACDF), 750, 901, 1022
  procedures, 334, 335
  surgery, 772
Anterior cervical diskectomy and fusion, 322, 333, 418, 594, 808, 858, 1134–1137, 1260
  case study, 598–599
  complications, 597
  operative details, 596–597
  patient evaluation, 595–596
  patient positioning and monitoring, 596
Anterior cervical fusion, 597
Anterior cervical plate (ACP), 333
Anterior column, 1166–1168, 1171, 1196
Anterior Column Realignment (ACR), 979
Anterior column reconstruction, 979
Anterior column releases, 713
Anterior conformal shaped interbody, 396
Anterior longitudinal ligament (ALL), 723, 840
  resection, 968
Anterior lumbar interbody fusion (ALIF), 342, 343, 604, 743, 965, 970, 973–974, 976, 984, 985, 998, 1000, 1002, 1003, 1138, 1140, 1166, 1168, 1170, 1171, 1213, 1260
  advantages and disadvantages, 606–608
  AEGIS®, 1001
  anterior cages, 1000
  benefits, 646
  BMP-2, 1000, 1001
  case study, 611
  closure, 649
  complications, 649, 1001
  contraindications, 646
  device, 284
  dimensions, 623
  discectomy, 609–610, 649
  endplate preparation, 649
  exposure, 648
  failure, 1003

  graft/cage positioning, 649
  graft migration, 1002
  graft selection, 648
  H-shaped incision, 999, 1000
  incision and approach, 609
  indications, 646, 1001
  indications for plate usage, 616–617
  indirect decompression, 1001, 1002
  instruments, 648
  Integra Omni-Tract® retractor, 999
  integrated fixation cage, 624–625
  interbody design, 620–621
  material selection of plates, 619
  materials for, 621–623
  motion segmental stiffness, 967
  operative setup, 648
  plate design, 619–620
  positioning, 648
  positioning, draping, and mapping of incision site, 608
  vs. posterior approach, 605–606
  post-operative protocol, 650
  preoperative, 647
  Steinman pin retractors, 999
  surgical technique, 648
  variable *vs.* fixed-angle devices, 620
Anterior lumbar spinal reconstruction
  arterial pathology, 1193–1197
  benefits, 1167–1169
  biomechanical rationale, 1170–1171
  clinical examination, 1171
  contraindications, 1169–1170
  diagnosis, 1171
  indications, 1167
  medical and surgical history, 1169
  patient optimization, 1173
  personnel, 1174–1175
  postoperative care, 1205
  pre-operative evaluation, 1171
  retroperitoneal approach, 1176–1185
  revision and previous abdominal surgery, 1187–1188
  screening, 1171
  skin care and preparation, 1175–1176
  surgical planning, 1171–1173
  synchronous treatment, 1197–1201
  transperitoneal approach, 1185–1188
  vascular access, 1176–1177
  vascular and renal anomalies and variations, 1201–1206
  vascular clips and vascular/urological disposables, 1173–1174
  vascular repair, sutures and needle holders for, 1173
  venous hole, control and repair of, 1191–1192
  venous oozing, control of, 1192–1193
Anterior plate fixation, 966
Anterior plating, 604
Anterior retropharyngeal approach, 1074–1076
Anterior to psoas approach (ATP), 970
Anterolateral plating system, 618
Antibacterial material quality, 262

Antibio-fouling capability, 263
Antibiotics, 265, 266, 959
Anticoagulation, 1175
Anti-infection, 156
Anti-inflammatory agents, 1102
  celecoxib, 1104
  glucocorticoids, 1103–1104
  interleukin-6 receptor monoclonal antibody, 1102
Anti-TNF antibodies, 103
Aorta, 913
Architecture, 201, 266
Arterial occlusion, 1179
Arterial pathology, 1193–1197
Arterial thrombosis, 1193
Arteriogram, 1058
Arthrodesis, 333, 335, 342, 346, 621, 809, 1011, 1021, 1025, 1133
Arthropathy, 942
Arthroplasty, 762–767, 772–775, 780, 782, 783, 811, 820
  AS degeneration, 887
  devices, 790
Articular chondrocytes (AC), 178
Artifact, 440, 444–448, 453, 454, 456, 457
  measurements, 451, 452
  metallic implant, 454
  motion, 454
  quantitative measurement, 441
  reduction, 443
  size, 444, 448–450, 452, 457
  types, 454
Artificial cervical discs, 860, 865
Artificial disc, 902, 904
Artificial lumbar discs, 1225
ASARA, 927
Aseptic lymphocytic vasculitis-associated lesion (ALVAL), 461
ASR, 1157
Associated diffusion coefficient (ADC), 282
ASTM F1717, 408, 409
ASTM F 2119-01 testing protocol, 447
ASTM F2267, 410
ASTM F543, 410
ASTM International, 408, 410
Astrocytomas, 517
Atelectasis, 1205
Aterior cervical discectomy and fusion (ACDF), 634
Atheromatous vessels, 1178
Atherosclerosis, 1170, 1193
Atlanto-axial complex, 718–721
Augmentation, 556, 1172, 1184
Australian health expenditure, 1151
Australian Register of Therapeutic Goods, 1154
Australian Regulatory body, 1159
Australian Therapeutic Goods Administration, 1153
Autogenous bone graft, *see* Autograft
Autogenous iliac crest bone graft, 1142
Autograft, 201, 213, 238–240, 254, 263, 264, 624, 1015, 1252
  iliac crest, 213

limitations with autogenous iliac bone graft, 214
  local bone graft, 213
  low risk of disease transmission, 214
  reduction in the use, 214
Autologous bone, 198, 1252–1253
Autologous disc chondrocyte transplant (ADCT), 185
Autologous reinfusion, 1174, 1176
Automatic Class III designation (De Novo), 406
Avascular component, 1113
Axial back pain, 1033
Axial compression (AC), 333
Axial lumbar interbody fusion (AxiaLIF), 969, 972, 973, 983
  caution against using, 1215
  literature on, 1213
  parabolic phenomenon of surgical technique, 1212
  revision strategies, 1212
Axial rotation, 624, 967
Axillary roll, 694

B
B2A, 239–240
Back pain, 956
Bacterial adhesion, 263
Bactericidal carrier, 265
Baguera C, 767
BAK cage, 963
BalanC, 283
Ball-in-socket, 762, 763, 766, 767
Balloons, 1174, 1188–1192, 1196, 1199
Bandwidth (BW), 452, 454–456
Barricaid®, 1218, 1220, 1222, 1223
Basal ganglia, 475, 478, 479
Bedrest, 575
Beta tricalcium phosphate $Ca_3(PO_4)_2$ (TCP), 240
Bifurcation, 1177, 1178, 1181, 1186, 1187
Bilateral, 368, 369
Bilateral linear implants (BLL), 396
Bilateral pedicle screw (BPS), 339
Bioactive glass (BAG), 262–265
  antibacterial material quality, 262
  100% BAG, 264
  bone bonding, 262
  bone-forming cells migrate and colonize, 262
  clinical applications, 263
  direct connectivity for cell in-growth, 264
  dissolution products from implanted BAG, 264
  3D printing technology, 265
  formation of an HA layer, 262
  high pH, 262
  new bone like matrix, production of, 262
  overlapping and interlocking bioactive glass fibers, 264
  45S5, 263
  S53P4, 263
  scaffolds for regeneration of large bone defects, 264
  tissue engineered constructs for replacement of large bone defects, 265
Bioactive synthetic multi-domain peptide, 239
Biochemical and morphological characteristics, 1114

Biocompatibility, 129, 165, 416
Biocompatible three-dimensional scaffold, 201, 266
Biodegradable polymers, 157
Biofilms, 166, 952, 953
Bioflex system, 285, 301, 302
Biologic(s), 103
    bony union, 266
    challenges, 213
Biological scaffolds, 1241–1242
Biological therapy, 1092
    bone morphogenic protein-7, 1092–1093
    growth and differentiation factor 5, 1093
    growth and differentiation factor 6, 1094
    haematopoietic stem cells, 1097
    hepatocyte growth factor, 1094
    hypertonic dextrose, 1097–1098
    link-N, 1096
    statins, 1096–1097
Biomaterial(s), 165
    alloys, 460
    implant, 429
Biomechanical analysis, 283
Biomechanical functions in motion, 790
Biomechanical testing, 299, 300, 316, 332, 346
Biomechanics, 75, 295, 306, 350, 354, 355, 367–370, 379,
    381, 383, 387, 396, 1003
    artificial disc designs, center of rotation of, 322–323
    of interbody fusion, 963–965
    PDSS, 323–326
    spinal fusion, 317–322
Biomedical model, 111
Biomolecular treatment, degenerative disc disease,
    174–176
Biopsychosocial approach, 72
Biopsychosocial model, 111
Biplanar fluoroscopy, 709
Birbeck granules, 141
Bisphosphonates, 106, 575
Blood-brain barrier (BBB), 1233
Blood cultures, 957
Blood loss, 606, 690, 981
Blood serology testing, 957
BMP-2, 1254–1256
BMP-7, *see* Osteogenic protein 1 (OP-1)
Bone allografts, 1142
Bone anchoring, 279
Bone banking, 1018
Bone bonding, 262
Bone dowels, 634
Bone formation, 214
    cells, 213
    promoting, 214
    properties essential for, 213
Bone graft, 634–639, 641, 1010
    extenders, 240
    substitutes, 1254, 1258, 1261, 1262
    substitutes and extenders, 1142
Bone grafts and graft substitutes
    allografts, 214–232

    autograft, 213–214
    B2A, 239–240
    BMPs, 232–239
    cerapedics, 239
    ideal properties, 266
    material's mechanism of action, 266
    spinal surgery, 266
    structural and handling characteristics, 266
    synthetic materials and grafts, 240–265
    transverse processes, surgically placed between
        the, 207
    treatment of surgical site infection with/without bone
        destruction, 265
Bone healing, 212, 213, 266
Bone implant interface (BII), 279, 380–384, 387–389, 391,
    396, 398
Bone induction, 240
Bone ingrowth, 156
Bone marrow edema, 100
Bone mineral density, 105, 106, 412
Bone morphogenetic proteins (BMPs), 232, 1014, 1015,
    1020, 1024, 1033, 1143
    BMP-1, 232
    BMP-2, 232, 1000, 1001
    BMP-3, 232
    BMP-4, 232
    BMP-5, 232
    BMP-6, 232
    BMP-7, 232
    limitations for general use of, 239
    major and minor adverse effects, 239
    osteoinductive, 266
    rhBMP-2, 232, 238, 239
    rhBMP-7, 238
    in United States and Europe, 238
Bone morphogenic protein-7, 1092–1093
Bone regeneration, requirements of, 213
Bone remodeling, 201, 213, 1014–1015
Bone repair, 1011
Bone resorption, 213
Bone screws, 152
Bone tamp, 574
Bony component, 1113
Bony contact, 621
Bony healing, 1014
Bony ingrowth, 616
Bookwalter systems, 1173
Bowel perforation, 983
Bracing, 575
Brain machine interfaces (BMI), 1244
Brain stimulation, 1242–1243
Breast adenocarcinoma, 513
Breast cancer, 106
Bryan artificial disc, 753
BRYAN® Cervical Disc Arthroplasty study for the US
    FDA, 797
BRYAN® Cervical Disc Prosthesis, 792, 793, 803
Bryan CDA, 860
    vs. ACDF, 862–863

Bryan cervical disc, 751, 812, 860–863
Bryan disc, 824, 860
    prothesis, 863

C

Cadaver, 353, 355, 365
Cadaver biomechanical testing
    adverse events, 410
    bone-implant interface assessments, 410
    COV, 412, 413
    donor demographics, 412
    fatigue testing, 412
    high variability, 413
    macroscopic abnormalities, 413
    motion testing, 410
    non-destructive characterization, 413
    soft tissue and bone degradation, 410
    spinal implants, 414
    tissue-implant interface, 410
    variability, 413
Cadaver sources, 214
Cadisc-C, 766
Cadisc-L, 766
Cage, 154, 334
    materials, 639–641
Calcified, 738, 740, 741
Calcium phosphate materials, 240–254
Canadian Spine Surgery Outcomes Network (CSORN), 31
Cancellous bone, 159
Cancer, 104
Carbohydrate-based polymers, 1262
Carbonate ions, 262
Carbon-coated implants, 469
Carbon fiber, 639
Carbon fiber reinforced PEEK (CFRP), 442
Carotid artery, 907
Carotid sheath, 731–732
Cartilage derived morphogenic protein 1
    (CDMP-1), 1093
Cartilage resorption and mineralization, 1014
Cartilaginous component, 1113
Cauda equina syndrome, 30
Caudal injections, 941
CD Horizon Agile system, 302, 303
C5-6 disc space, 802
CDMP-2, 1094
Celecoxib, 1104
Cell-based grafts, 1016
Cell based therapy, degenerative disc disease, 176–179
Cell-free intradiscal implantation, 1101
    fibrin sealant, 1101–1102
    hyaluronate hydrogels, 1101
CE mark, 1155, 1157, 1158, 1219, 1220
Center for Biological Evaluation and Research
    (CBER), 402
Center for Devices and Radiological Health (CDRH), 402
Center for Drug Evaluation and Research (CDER), 402

Center for Food Safety and Applied Nutrition
    (CFSAN), 402
Center for Tobacco Products (CTP), 402
Center for Veterinary Medicine (CVM), 402
Center of rotation (COR), 762–764
    of artificial disc designs, 322–323
Central herniation, 32
Central nervous system (CNS), 1235
Central sensitization syndrome, 117
Ceramic(s), 156, 201, 232, 238, 240, 254, 262, 1143, 1261
    articulated surfaces, 450
Ceramic-based grafts, 1017
Certificate in neurophysiologic intraoperative monitoring
    (CNIM), 501
Cervical arthroplasty, 456, 790, 808
Cervical artificial discs
    Bryan, 751, 753
    M6-C, 758
    Mobi-C, 756–757
    nomenclature, 752
    porous coated motion prosthesis, 753–754
    Prestige LP, 755, 756
    Prestige ST, 755
    ProDisc-C, 754
    Secure-C, 757
Cervical disc designs, artificial, 811
Cervical disc devices
    artifact measurements, with BW kept constant, 452
    artifact measurements, with ETL kept constant, 451
    artifact size, 448, 449
    artificial cervical disc, implantation of, 451
    artificial cervical disc replacements, 447
    CCM alloy discs, 450
    cobalt-chromium alloy, 446
    MRI sequence and parameters, 448
    phantom grid, disc placement and artifact measurement
        on, 448
    post implantation MRI scans, 453
    stainless steel, 446
    titanium implants, 446
Cervical disc herniation, 866
Cervical disc replacement, 785
    Bryan Disc, 910–911
    complications, 911–912
    contraindications, 906
    indications, 905
    Mobi-C, 909–910
    nerves, 906–907
    outcomes, 912–913
    positioning and approach, 907–908
    postoperative protocol, 911
    Prestige LP, 908–909
    revision options, 912
    trachea and esophagus, 907
    vessels, 907
Cervical fusion, 886
Cervical interbody fusions, 634, 635, 637
Cervical levels, kinematics at, 791
Cervical myelopathy, 595, 596

Cervical pedicle screw fixation, 549–550
Cervical plate(s), 157, 335
  dynamic, 336
Cervical plating, 333–335
Cervical radiculopathy, 595, 776
Cervical SCI, 1235–1236
Cervical spine, 772–774, 776
  ACDF in, 798
  arthroplasty and, 799
  biomechanics of, 790
  extension kinematics, 802
  kinematics of, 796
  structural anatomy, 790
  subaxial, 801
Cervical spine anatomy
  atlanto-axial complex, 718–721
  cervical triangles, 734
  facet joint, 722–723
  fascia, 725–726
  intervertebral disc, 724–725
  lamina and spinous process, 723
  lateral mass, 723
  ligaments, 723–724
  muscles, 725–729
  neurovascular structures, 729–731
  pedicles, 722
  spinal canal, 723
  subaxial cervical spine, 720–721
  transverse process, 722
  ventral structures, 731–733
  vertebral body, 722
Cervical spine surgery
  anterolateral approach, 1083
  posterior approach, 1084–1088
  retropharyngeal approach, 1074–1076
  Smith and Robinson approach, 1078–1084
  surgical approach, 1070–1071
  surgical set-up, 1071
  surgical strategy, 1070
  transoral approach, 1071–1074
  Verbiest approach, 1076–1078
Cervical spondylosis, 751, 808, 858, 859, 863, 867, 868
Cervical total disc replacement (cTDR), 772–774,
    824, 830
  Baguera C, 767
  biomaterials, 774–776
  Cadisc-C, 766
  Class III medical device, 824
  complications, 780–782, 833–835
  CP ESP cervical disc, 765–766
  design considerations, 763
  discectomy, 817
  end plate preparation, 817
  evidence, 783–786
  FCD, 765
  footprint size, 818
  heterotopic ossification, 830–833
  hybrid treatment, 825
  implant insertion, 819

  indications, 824–826
  with level 1 evidence, 825
  M6-C artificial cervical disc, 764–765
  physiologic kinematics, 762–763
  placement of pins, 816
  planning, 813
  prosthesis, 833, 834
  Simplify cervical disc, 767
  surgical procedure and technical pearls, 776–780
  surgical technique, 814–820
  Synergy Cervical Disc, 766–767
  wound closure, 820
Cervical transforaminal epidural injections, 941
Cervical triangles
  dorsal, 734
  ventral, 733–734
Cervical vertebrectomy, 597
Cerviral disc arthroplasty, 809
Challenges, 1252, 1258–1260
Charité Artificial Disc Replacement, 1160
Charite III prosthesis, 903
CHARITE US IDE trial, 841, 842
Charlson comorbidity index, 954
Chemonucleolysis, 1098
  chymopapain, 1098–1099
  methylene blue, 1100
  oxygen-ozone, 1099
Chemonucleolysisgelified ethanol, 1099
Chest drain, 1059
Chest radiograph, 1063
Chest tube, 741
Chiropractors, 69
  benefits and risks of chiropractic care, 81
  biological rationale, 82
  biopsychosocial paradigm, 49
  bone setters, 70
  care of elderly patients, 57–58
  characteristics of chiropractic health care and
    practice, 75
  chiropractic and mainstream medicine, 74–75
  diagnosis and assessment methods, 76–77
  diagnostic procedures, 51–52
  education, 72–74
  education and training of, 46–48
  ethnographic observation studies of health
    encounter, 50
  examination procedures, 80
  frequency and duration of treatment, 80–81
  holistic approach, 69
  insurance coverage, 44–45
  management of chronic back pain, 82
  measurement of health outcomes, 81
  musculoskeletal conditions, treatment for, 54
  outcomes of care, 56–57
  patients with chronic back problems, 56
  radiographic procedures, 76
  re-evaluation and reexamination of patient
    condition, 81
  registration and licensing of, 71

Index

removal of legal or ethical barriers, 42–44
safety and efficacy, 40–42
scientific rationale of chiropractic profession, 77–78
spinal manipulation, 52–54, 75–76
systematic treatment plan, 69
things to know before talking to, 45–46
treatment of acute and chronic low back pain, 84–85
treatment of lower back pain in chiropractic, 79–80
treatment of spine, 48–57
use for spinal care, 40
Chronic inflammation, 165
Chronicity, 120
Chronic low back pain (CLBP)
activities, 114
affective and cognitive processing of pain, 117
afterthoughts, 120
barriers, 111
barriers to recovery, 120
behaviors, 121
beliefs and expectations, 113
biographical suspension, 116
clinical implications, 120
cognitive part, 121
coping, 115, 116
culpable, 115
education, 121
enmeshment, 115
financial loss, 117
follow-up, 120
inability to plan, 114
ingrained, 120
interactive tool, 120
invisible pain, 116
isolation, 114
Keele STarT Back, 120
lived experience, 116
loss of self, 115
low education, 114
mediated psychosocial factors, 112
mediators, 113
medical expenditure, 119
models of care, 119
modifiable psychosocial factors, 112
negative, 112
non-pharmaceutical treatment, 121
one-size fits all, 121
optimization, 121
patient confidence, 120
patient's acceptance, 113
pharmaceuticals, 119
psychosocial impacts, 120
psychosocial realm, 111
reassessment, 120
relationships, 114
remain active, 121
responsibility, 121
risk of chronicity, 112
second-line, 121
sickness/pain behaviors, 116

social roles, 115
socioeconomic impact, 117–120
socioeconomic status, 114
stigma, 115
support, 116
systematic reviews, 111
themes, 114
toolkit, 120
uncertain future, 116
work, 115
Chymopapain, 1098–1099
Circumferential fusion, 972, 973
Class III devices, 1215, 1219
Class II medical devices, , 1218
Class I medical devices, , 1218
Claudicant back pain, 31
Clinical efficacy, 286
Clinically Organized Relevant Exam (CORE) Back Tool, 7
Clinical outcomes, 316, 317, 321–323, 325, 326
Clinical testing, 332
Clinical uncertainty, 114
Clinical variability, 214
Clips, 1173–1174
c-met proto-oncogene product, 1094
Cobalt-chrome (Co-Cr), 153, 775
Cobalt chromemolybdenum alloys, 546
Cobalt chromium (CoCr), 440, 442
   alloys, 166, 416, 460
Cobalt-chromium-molybdenum (CoCrMo) alloy
   debris, 464
Cobb method, 660–661
Coccyx injections, 942
Code of Federal Regulations (CFR), 402
Coefficient of variation (COV), 412, 413
Coflex, 279, 281, 1225
   device, 855
   system, 896
Coflex® interlaminar stabilization, 309
Cognitive behavioral therapy (CBT), 121
Collagen sponge, 1254
Combination materials, 201
Combinations of growth factors, 1256–1257
Combined pathology, 1197
Comorbidity index, 954
Complementary and alternative medicine (CAM)
   therapies, 80
Complex vascular pathologies, 1167, 1170
Complications, 1001, 1254
Composite primary endpoint, 418
Compound muscle action potentials, 490
Compression, 606
   strength, 607, 619
   stress, 380
Computational modeling, 414–416
Computed tomography (CT), 419, 929–930, 957, 959,
   1056, 1057, 1060, 1133, 1171, 1172, 1176, 1187,
   1193, 1197, 1199, 1202–1204
   MRI artifact, 155
Computer stereotactic navigation techniques, 553

Conformity assessment, 1154
Congenital anomaly, 742
Congress of Neurological Surgeons (CNS), 1033
Conical screws, 544
Conjugation technology, 240
Conservative management, 575
Conservative therapy, 575
Constraint, 763
Contact healing, 1012
Contraindications, 576, 810, 1005
Controlled architecture, 201, 266
CORE Back Tool, 120
Core dislocation, 838
Coronal imbalance, 606
Corpectomy, 597
Corrosion, 130, 165
Cortical allograft, 214
Cortical amplification, 496
Cortical bone, 623
Cortical bone trajectory (CBT), 340, 392, 393
    screws, 1039
Corticosteroid, 929, 937, 1103
    injections, 937
Cosmic posterior dynamic system, 286, 302, 303
Costing-data sources, 118
Cost-of-illness, 1152
    studies, 118
Costotransversectomy, 740
CP ESP cervical disc, 765–766
CP ESP prosthesis, 766
Creatine kinase, 702
Critical bone defects, 1262
Crystallization of a bone-like HA, 262
CSF drainage, 1239
Cummins, 824
Cutaneous patch testing, 466
Cyclical loading, of spinal constructs
    basic science, 390–391
    interbody devices, 396–398
    pedicle screw constructs, 395–396
    pedicle screws, 391–395
Cytokine(s), 134, 277
    reaction, 840
Cytotoxic reactions, 619

D

Dallas Discogram Protocol, 936
Danger associated molecu lar patterns (DAMPs), 135
Danger signaling, 134
da Vinci surgical robot (Intuitive Surgical), 1271
Decompress, 1168, 1176
Decompression, 281, 972, 986, 988
Decorticator, 709
Deep brain stimulation (DBS), 474–480
Deep infections, 958
Deep SSIs, 950, 955, 957
Deep venous thrombosis (DVT), 1170, 1189, 1205
    prevention, 1205

Deformity, 691, 707, 991, 1167, 1168
    correction, 164, 166
Degeneration, 998, 1001, 1003, 1005, 1006
    arthritis, 742
    cascade, 279
    lesions, 1052, 1053
    listhesis, 33
    spondylolisthesis, 989
Degenerative disc disease (DDD), 31–32, 188, 280, 809,
        858, 871, 974–975, 981, 989, 998, 1001,
        1004–1006, 1167, 1168, 1171, 1224
    bioartificial total disc replacement therapies, 184–185
    biochemical changes, 1114–1119
    biological annulus fibrosus repair, 182–184
    clinical studies, 185–187
    differentiated cells, 178
    etiology, 1114
    gene therapy, 174–176
    morphological changes, 1115
    pathology, treatments and challenges, 172–174
    PRP, 176
    recombinant protein and growth-factor-based
        therapy, 174
    scaffold development, 182
    stem cells, 178–179
    unpublished clinical trials, 187–188
Demineralized bone matrix (DBM), 1020, 1142, 1254
Demineralized bone matrix (DBM) based product
    differentiation factors (donor bone), 232
    disease transmission, 232
    forms, 232
    manufacturers processing, 232
    morselized autografts, mixed with, 232
    organic phase of bone, 214
    promoting bone formation, 214
    as a stand-alone graft material, 232
    sterilization method, 232
    variable, 232
Depression, 112
Design requirements, for engineered
        biomaterials, 201
Developmental and regenerative roles, 1255
Developmental anomalies, 1187
Device class
    Class I, 403
    Class II, 403
    Class III, 403
Device for intervertebral assisted motion (DIAM), 308,
        309, 469
DEXA, 988
DHLSDNYTLDHDRAIH, 1096
Diagnosis
    blocks, 358
    SSIs, in spine surgery (*see* Surgical site infections
        (SSIs), in spine surgery)
    tests, 358
DIAM®, 279, 1218, 1220–1222
Diamond concept, 1011
Diaphragm, 1055, 1056, 1058, 1059, 1061, 1063

Diet, 1205
Differentiate toward multiple lineages, 1257
Diffuse idiopathic skeletal hyperostosis (DISH), 97
Diffusion weighted imaging, 279, 282
DiGeorge syndrome, 596
Direct bony healing, 1011–1013
Direct cost, 118
Direct lateral interbody fusion (DLIF), 1138, 1260
    See also Extreme lateral interbody fusion (XLIF)
Disability, 78
    score, 284
Disability-adjusted life years (DALYs), 110
Disc arthroplasty, lumbar, 32
Disc degeneration, 1114, 1115, 1118
Discectomy, 599, 609–610, 703–705, 817, 1062, 1118, 1167, 1168, 1183, 1184
Disc health, 282
Disc height, 962, 986, 998, 999, 1001–1003, 1005
Disc herniation, 858, 859, 867, 1115, 1117, 1125
Discogenic pain, 941, 970
Discography, 935, 1057
Discover artificial cervical disc, 813, 873
Discover CDA *vs.* ACDF, 874
DISCOVER™ artificial cervical discs, 447
Disc replacement, 468, 887
    artificial, 887
Disease modifying anti-rheumatic drugs (DMARDs), 469
Disease transmission, 214, 232, 1018
Disk height, 1171
Diskitis, 94, 95
Dissection, 1167, 1168, 1177–1179, 1181, 1182, 1185–1189, 1193, 1196, 1197, 1199
Dissimilar metals, 153
Distractor, 1182
    pins, 816
Donor site morbidity, 1253
Dorsal instrumentation, 542
Dorsal muscle
    deep layer, 728–729
    intermediate layer, 728
    superficial layer, 728–729
Dorsal pedicle screw systems, 547
Down syndrome, 596
3D printing technology, 265
Dual-functional graft, 265
Dual IVC, 1201
Duplication, 1201
Durable result, 1199
Dynamic lumbar fixation, 282
Dynamic neutralization system (Dynesys), 284
Dynamic soft stabilization (DSS) system, 896
Dynamic stabilization, 562, 1218, 1220, 1224, 1226
    artificial disc designs, center of rotation of, 322–323
    PDSS, 323–326
Dynesys, 323, 325, 1218, 1220, 1221
    pedicle screws and spacer, 895
    posterior dynamic stabilization, 297
    system, 283, 895
Dysfunction, 350, 357, 358, 360, 363, 369

E

Early postoperative contamination, 952
Easy procurement, 214
Echo train length (ETL), 442, 444, 448–452, 456
Ectopic bone formation, 239
Elastic modulus, 621
Elastomeric, 763, 767
Electromyography (EMG)
    anesthetic concern for, 494
    neurophysiology, 490
    pedicle screw threshold testing, 492–494
    recording methods, 490–494
    spontaneous, 491
    train of four test, 491
    triggered, 492–493
Electron beam (EB), 159
Elimination, 1176
Embolization, 1196
Embryonic stem cells (ESCs), 179, 1238
Emergency care, 119
En bloc resection, 740
Endarterectomy, 1198
Endoleak, 1199
Endoscopic techniques, 741
Endoscopy, 1031, 1044–1045
Endovascular, 1167, 1188, 1189, 1192, 1193, 1196, 1197, 1199, 1200
Endplate changes, 1223
Enhanced recovery after surgery (ERAS), 31, 1034
Enneking staging system, 507–508
Epidural abscess, 94, 95
Epidural fibrosis, 972
Epidural injections, 937, 940
Epidural spinal cord compression scale, 512
Equipment sterilization, 954
Eradication, 959
EU-AU MRA, 1154
Eucomed, 1157, 1158
EU healthcare expenditure, 1152
EU MDR, 1157, 1158
EU Medical Device Classification, 1153
European Economic Area (EEA), 1219
European market, 1155
European spinal market, 1218
European standards, 1155
Evidence based practice (EBP), 77
Evidence-based therapy, 79
Excelsius GPS, 1277
Excessive distances for the cells to migrate, 213
Exercise therapy, 121
Existing metrics, 289
Expandable cages, 638
Expandable lateral cages, 968
EXPEDIUM® Spine System, 441
Extenders, 1015
Extension, 1221
Extensive degeneration, 285
External jugular vein, 907
Extracellular matrix (ECM), 172, 178, 213, 239

Extreme lateral interbody fusion (XLIF), 343, 611, 978, 1140
  advantages, 611
  case study, 615–616
  disadvantages, 612
  indications for plate usage, 617–618
  patient positioning, 612–615
Extreme lateral lumbar interbody fusion, 691

F

Facet anatomy and biomechanics, 846–848
Facet arthroplasty systems
  history of, 848–849
  rationale and biomechanics of, 848
  TFAS, 849–850
Facetectomy(ies), 972, 1035
Facet joint, 722–723, 934, 935
  denervation, 939
  injection, 936, 938
Facetogenic pain, 938
Factor-based grafts, 1016
Failure, 1003
Familiarity, 608
Fascia
  investing layer, 725
  pre-tracheal layer, 725
  prevertebral layer, 725–726
Fast spin echo (FSE) protocols, 443, 448, 454
Fatigue failure, 380, 381, 383, 385, 387–390, 393, 397, 544
Fatigue testing, 380
  standardized cyclical loading protocols, 380–381
  in vitro cyclical loading, 381
Fat plane, 1173
Fat saturation, 443
FDA-approved devices, 750, 752
  See also Cervical artificial discs
FDA Modernization Act, 1158
Fear-avoidance, 112
Females, 355, 368, 369
Femoral nerve, 693
Femoral ring allografts (FRA), 974, 1024
Femoral stretch test, 14
Fernstrom, 824
Fibergraft® BG Morsels, 264
Fib-Gen, 1102
Fibre orientation, 1117
Fibrin sealant, 1101–1102
Fibroblast growth factor (FGF), 1240
Finger dissection, 609
Finite element, 366–369
Finite element analysis (FEA), 316, 322, 408, 414–416
Finite element modeling (FEM), 299, 300, 316, 317, 319, 320, 326, 795
First-line recommendations, 121
Fixed-angle, 624
Flexion, 1221
Flexion-extension, 624, 967

Fluoroscopic guided technique, 552
Fluoroscopy, 714, 926, 929
Food and Drug Administration (FDA), 402, 1218–1226
Food, Drug and Cosmetic Act (FD&C Act), 402
Foraminal height gain, 691
Foraminal stenosis, 706
Force nucleus, 543
Fracture(s), 1052, 1053, 1059
  liaison service, 526, 527, 536
Frankel system, 1232
FRAX score, 535
Freedom Cervical Disc (FCD), 765
Free-hand, 549
  pedicle screw placement, 548–549
Freehand open approach (Freehand-PLIF), 1278
Freeze-dried allograft, 1020
Frenchay artificial cervical joint, 865
Frenchay cervical disc, 811, 824
Frenchay prosthesis, 905
Fresh frozen allograft, 1019
Fretting corrosion, 133, 462
Friction-cost method, 118
Fulcrum Assisted Soft Stabilization (FASS), 285
Functional animal testing, 416
Functional electrical stimulation (FES), 1242
Fusion, 286, 350, 357, 360–365, 367–369, 625, 1132, 1254
  rates, 335, 607, 634, 1213, 1215
  time, 335

G

Gacyclidine, 1241
Gait, 474, 476–480
Galvanic corrosion, 154
Gap healing, 1013
Gear-shift probe, 548
Gelified ethanol, 1099
Gender, 355
Gene delivery, 1259
Gene therapy, 174–176, 1258–1260
Genipin, 1102
Genitofemoral nerve, 694
Geron Corporation, 1238
Gertzbein classification, 711
Gertzbein-Robbins Classification, 1277
Giant cell tumor, 575
Global Harmonization Task Force (GHTF), 1155
Global spinal devices market, 1153
Global spinal product market, 1218
Goal-oriented CBT, 121
Grade I degenerative spondylolisthesis, 1225
Gradient echo techniques, 454
Graf ligament, 283, 303, 304, 894, 895
Graf ligamentoplasty system, 283
Graft collapse, 337
Graft loading, 333
Graft material, 610
Graft migration, 1002

Index 1293

Graft system, 325
Grandfathering, 1157
Granulocyte colony stimulating factor (G-CSF), 1240
Granulomas, 138
Gravimetric wear, 462
Gross domestic product (GDP), 1150
Growth and differentiation factor 5, 1093
Growth and differentiation factor 6, 1094
Growth factors, 175, 201, 213, 232, 239, 254, 266
    BMP-2, 1254–1256
    combinations of, 1256–1257
    NELL1, 1256
    OP-1, 1256
    osteogenic, 238
Gynaecological history, 1188

H
Hard callus, 213
Hardware failure, 165
Harrington, 542
Healthcare expenditure, 119, 1150, 1151
Health expenditure, in Australia, 1150
Health related quality of life (HRQOL), 1142
Heart, 133
Helifix®, 1218, 1224, 1225
Hemangioblastoma, 517
Hemangioma, 574, 575
Hematogenous, 952
Hematoma formation, 213, 239
Hematopoietic marrow cavities, 213
Hematopoietic stem cells (HSC), 1097
Hemoclips, 1173, 1191
Hemodynamic intervention, 1239
Hemostasis, 511
Hemostatic agents, 1192
Hepatocyte growth factor (HGF), 1094, 1240
Herniations, 1182
Heterotopic ossification (HO), 811, 815, 818, 820, 841,
      859, 863, 874, 1259
    cervical total disc replacement (CTDR), 830–833
    clinical significance of, 832
    prevention of, 832
    risk factors for, 831–832
High endplate stresses, 969
High lumbar, 1055, 1056, 1059
High pH, 262
High signal intensity zone (HIZ), 931
High-touch low-tech health model, 77
Hip flexor weakness, 698
Histiocytosis, 138
HLA-B27, 100
Holistic approach, 69
Hollow modular anchorage screw, 683
Hook-rod, 542
Hooks, 152
Human-capital method, 118
Humanistic aspect, 77
Humanitarian device exemption (HDE), 405

Humanitarian use device (HUD), 405
Hyaluronate hydrogels, 1101
Hybrid cages, 641
Hybrid interbody cage, 638
Hybrid procedure, 713, 985, 1005–1006, 1169, 1196
Hydration, 1114
Hydroxyapatite (HA), 156, 159, 160, 546, 1143
    coating, 775
Hydroxyapatite $Ca_{10}(PO_4)_6(OH)_2$ (HA), 240
Hyperlordotic cage, 713
Hypersensitivity, 138–140
Hypertonic dextrose, 1097–1098
Hypertrophics, 278
Hypogastric nerve plexus, 707
Hypogastric plexus, 1177
Hypoglossal nerve, 733

I
IDEAL, 443
Ideal shape, 265
Identity, 115
IL-1β, 134
IL-6, 134
Iliac crest, 213, 691, 693, 971
    autograft, 634, 962
Iliac veins, 1178, 1179, 1181, 1186–1191, 1202, 1203,
      1205, 1206
Iliac vessels, 608, 609
Iliocaval compression, 1202
Iliohypogastric nerve, 691, 693, 698
Ilioinguinal nerve, 691, 693, 698, 707
Iliolumbar vein, 691, 914
Ilium, 351, 352, 354, 359, 361, 366
Image-based systems, 1271–1272
Image intensifier, 1058, 1059
Imaging, 957
Immunogenic, 1253
Immunomodulation, 1258
Implant(s), 951–953, 956, 958, 959
    breakage, 838
    failure, 838, 1171
    fatigability, 619
    materials, 456, 457
    migration, nerve root compression from, 842
Implant-allergy, 138
Implantation, 1182–1184, 1200
Inadequate bone graft volume, 213
Inadequate preparation of host bone, 213
Incidence, 951
Incision, 1054, 1055, 1058–1061, 1063
Incisional hernia, 698
Indirect costs, 117
Indirect decompression, 562, 690, 1001, 1002, 1215
Indirect fracture healing, 1013–1015
Induced pluripotent stem cells (iPSCs), 179, 1238, 1246
Infection, 213
    nidus, 619
Infectious lesions, 1052, 1057
Inferior articular process (IAP), 1035

Inferior (recurrent) laryngeal nerve, 733
Inferior vena cava (IVC), 913
  filters, 843
Inflammasome, 135
Inflammation, 212
Inflammatory cells, recruitment of, 213
Inflammatory response, 1258
Inflammatory state, 703
Innovation, 1153, 1156–1160, 1162
In silico, 365, 369
In-situ curing, 1120–1122
Instability, 277
Instantaneous axis of rotation (IAR), 306
Institutional review board (IRB), 406
Instructions for Use (IFU), 824
Integrated fixation cage, 624–625
Interbody, 152, 610, 1035–1036, 1042–1043
  devices, 384, 386–390, 396–398
  fusion, 339, 342, 343, 998–1003
  fusion devices, 343
  spinal fusion, 212
Interbody cage, 339
  allograft, 637
  design, 342, 343, 635
  standalone implantation, 635
Interlaminar injections, 941
Internal disc disruption (IDD), 998, 1001, 1005
International 10-20 system, 495
International Standardization Organization (ISO), 298
International Standards for Neurological Classification of
    Spinal Cord Injury (ISNCSCI), 1234
International Standards Organization (ISO), 408, 410
Interpedicular travel (IPT), 299, 300
Interprofessional knowledge, 120
Inter-professional Spine Assessment and Education
    Clinics (ISAEC), 7
Interspinous and interspinous stabilization/
    distraction, 1225
Interspinous devices (ISD), 306, 307, 854, 1224, 1225
  advantages, 563
  Coflex® interlaminar stabilization, 309
  complications, complication rates and reoperation
    rates, 568–569
  DIAM, 308, 309
  indications and contraindications, 564–565
  motion preservation, 562
  outcomes of, 570–571
  purpose of, 562
  representative, 563
  sources of complications, 569–570
  surgical technique, 565–568
  Wallis system, 307
  X-Stop device, 308
Interspinous fusion device, 562
Interspinous process decompression (IPD),1225
Interspinous process (ISP) devices, 326
Interspinous space,1221
Interspinous spacer, 280, 281
  devices, 279

Intervertebral disc (IVD), 172, 173, 182, 187, 724–725,
    774, 776, 777, 1218, 1222, 1223
  macroscopic degeneration in human, 1116
  structural defects in lumbar, 1117
  structure and function of, 1112–1113
Intervertebral height, 691
Intra-articular (distracting) cage, 683
Intradiscal pressure (IDP), 326, 328
Intradural spinal tumors
  intradural extramedullary tumors, 515
  intramedullary lesions, 517
  operative considerations, 513
Intramedullary lesions, 517
Intraoperative fluoroscopy, 1279
Intraoperative localization, 738, 742
Intraoperative neuromonitoring
    (IONM), 554
  amplitude, 486
  anesthetic effect on, 488–489
  baselines and documentation, 489
  delivery of electrical stimulation, 486–487
  electromyography, 490–494
  guidelines, 488
  history of, 484–485
  latency, 486
  morphology, 486
  motor evoked potentials, 497–500
  multimodality approach, 501
  protocol for intraoperative change, 489
  recording of electrical current, 486
  services, 501
  setup, 488
  SSEPs (see Somatosensory evoked potentials (SSEPs))
  troubleshooting, 487–488
  utility in spine procedures, 500–501
Intravenous drug, 94, 95
Investigational device, 1222
Investigational device exemption (IDE), 403, 406, 407,
    903, 1005, 1219, 1223
  trials, 887
In vitro, 365, 368, 369
  biomechanical studies, 618
In vitro cyclical loading, 381
  failure criteria, 388–390
  interbody devices, 384
  loading modality, 384–386
  loading rate, 382–383
  magnitude of loading, 386–388
  pedicle screw constructs, 383–384
  pedicle screws, 383
  recording frequency, 382
  specimen preparation, 381–382
  specimen selection and handling, 382
Ionising radiation, 926
Ions, 143
ISO 12189, 409
ISO 18192-1, 409
Isobar TTL system, 282, 283, 303, 304
Isthmic spondylolisthesis, 980, 989

Index

J
Jamshidi needle, 709
Joint arthroplasty, 772, 773, 781, 786

K
510(k), 1158
Kaneda system, 618
510k application, 1155
K-Centrum anterior spinal fixation system, 1212
510 (k) clearance, 1215
Kidney, 133
Kineflex-C artificial cervical disc, 872, 876
"Kinematically accurate" motion, 792
Kinematic data, 792
Kinematic loading profile, 277
Kinematic response, 279
Kinematic signature, 289
Kirschner wire, 1042, 1043
Klippel-Feil syndrome, 596
Knowledge translation (KT), 77
Kyphoplasty, 943
    acute fracture, 578
    acuteVCF, 576
    anesthesia, 580
    antibiotic impregnated cement, 587
    antibiotic prophylaxis, 587–588
    balloon tamp, 581
    barium sulfate, 582
    bedrest, 574
    bilateraltranspedicular, 587
    biopsy, 588
    bipedicular, 581
    biplanar fluoroscopy, 581
    bisphosphonate, 588
    bonescintigraphy, 578
    bone tamp, 581
    canal stenosis, 576
    cement extravasation, 583
    cement leakage, 576
    chemotherapy, 588
    chronic fractures, 578
    complications, 583
    computed tomography (CT), 578
    efficacy, 586
    embolization, 583
    extrapedicular approach, 581
    fluoroscopy, 581
    follow up, 583
    giant cell tumors, 588
    history, 576
    imaging, 578
    indication, 575
    kyphotic angle, 587
    kyphotic deformity, 580
    laboratory tests, 579
    local anesthetic, 581
    magnetic resonance imaging (MRI), 578
    metabolic bone disease, 583
    metastatic/primary bone tumors, 588
    midline position, 588
    morbidity, 574
    mortality, 574
    narcotics, 574
    neurologic deficits, 583
    neurologic injury, 576
    nonoperative management, 575
    outcomes, 586–587
    pain control., 583
    pedicle, 581
    physical examination, 576
    plain radiographs, 578
    posterior cortex, 576, 588
    postoperative care, 582–586
    preoperative testing, 579
    prone position, 580
    quality of life, 574
    radiation, 588
    radiation therapy, 588
    radiculopathy, 576
    radiopacifier, 582
    re-fracture, 585
    relative contraindications, 576
    sedation, 580
    sham procedure, 575
    Short-Form-36, 586
    single photon emission computed tomography, 578
    subsequent fracture, 583
    technique, 580–582
    timing, 589
    transpedicular approach, 581
    unhealedVCF, 577
    unilateral approach, 587
    unipedicular approach, 581
    viscosity, 582
    ziroconium dioxide, 582
Kyphosis, 709
Kyphotic deformity, 738
Kyphotic model, 320

L
L1-2 and T12-L1, 991
L4-5, 990
L5/S1 level, 606, 990
Lab-based biomechanical study, 316
Lack of fixation, 213
Lag screws, 624
Lamellar bone, 213
Lamina, 723
Laminectomy, 1034–1035
Laminoforaminotomy, 551
Laminotomy, 705
Langenbeck retractor, 1179
Laparoscopic lumbar discectomy, 704
Large portal vein aneurysms, 1204
Lateral approach, thoracolumbar junction,
    see Thoracolumbar junction (TLJ)

Lateral bending, 624, 967
Lateral canal stenosis, 284
Lateral cutaneous nerve, 693
Lateral decubitus position, 694
Lateral interbody fusion, 707
    role of, 712–713
Lateral lumbar interbody fusion (LLIF), 343, 690–691,
        966–968, 971, 978, 984, 985, 1140, 1212
    anatomy, 691–693
    complications, 697–698
    contraindications, 691
    implant placement, 696
    indications, 691
    patient surgical positioning, 694–695
    posterior instrumentation and fusion, 697
    preoperative planning and operative window, 693–694
    vascular anatomy, 693
Lateral mass, 723
Lateral migration, 280
Law of bone remodeling, 897
Leakage of rhBMP-2, 1255
Learning curve, 543, 1003
Least Burdensome Concept, 1158
Legitimacy, 71
Legitimate disability, 116
Leg length discrepancy (LLD), 358
Leg pain, 1124
Lentiviral delivery, 1259
Life-changing effects, 115
Life expectancy, 1151
Lifestyle factors, 111
Ligaments, 351–353, 357, 366–369
Ligamentum flavum, 705, 724
Ligamentum nuchae, 724
LimiFlex™, 1218, 1225
Linea alba, 609
Link-N, 1096
Liver, 133
Load, 351, 354, 355, 357, 358, 366, 367, 369, 370
Loading modality, 384
    interbody devices, 386
    pedicle screw constructs, 385–386
    pedicle screws, 385
Load sharing, 155, 283, 337, 1171
Load transfer, 277
Local autogenous bone graft, 1142
Local bone graft (LBG), 213
Localized bactericide, 265
Longisimus muscle, 703
Loosening, 281
Lordosis, 691, 962, 970, 973, 986, 1171
Lovastatin, 1096
Low angle laser light scattering (LALLS), 131, 132,
        462, 463
Low back pain (LBP), 172, 350, 356–358, 369, 370, 1031–
        1033, 1112, 1152
    abdominal muscles, 24
    Achilles tendon reflex, 16
    acute, 5

adherent, 19
adverse side effects of surgery, 87
analgesic medication, 21
asymmetrical, 23
back dominant pain, 10
boney encroachment, 20
burden, 110
chronic, 5
chronic condition (see Chronic low back pain (CLBP))
chronicity, 84
classification, 7
classification of, 78–79
constant, 10
core strengthening, 24
decompression, 24
degeneration, 4
direction-specific movements, 21
distinct trajectories, 111
early assessment, 120
etiology, 79
extension, 13
facet joint, 20
false positive, 6
fecal incontinence, 11
femoral nerve roots, 18
femoral stretch test, 14
flexion, 13
footstool, 24
hamstring pain, 14
history, 12
iatrogenic effect, 6
imaging technology by physicians, 84
impairment, 11
incidence, 4
individuals, 111
inflammation, 13
intermittent, 10
leg dominant pain, 23
lifestyle factors, 111
limitations, 118
lumbar roll, 21
mechanical, 7
medical approach to, 82–84
modality, 21
morbidity, 676
natural history, 111
nerve conduction deficits, 15
neurogenic claudication, 19
night roll, 22
non-specific, 6
opioids, 87
overmedicalization, 114
pain generator, 21
Pattern 1, 16
Pattern 1 PEN, 22
Pattern 1 PEP, 21
Pattern 2, 18
Pattern 3, 18
Pattern 4, 19

Index

Pattern 4 FA, 19
Pattern 4 FR, 19
patterns of pain, 7
pelvic tilt, 24
perineal sensation, 11
persistent LBP, 110
physical examination, 13
piriformis, 9
prevalence, 110
prognostic research, 111
prone, 22
prone passive extension, 14
psychological factors, 111
radicular pain, 9
radiculopathy, 16
recurrent LBP, 110
Red Flags, 7
referred pain, 9
risk factors, 111
saddle sensation, 16
sagittal, 23
scheduled rest, 22
sciatica, 18
sciatic nerve, 18
shopping cart sign, 19
sloppy push-up, 13
social factors, 111
societal cost, 118
spinal cord involvement, 15
spinal injections, 10, 86–87
spinal malignancy, 10
spinal manipulation, 5
spinal stenosis, 20
spontaneous onset, 12
STarT Back, 6
straight leg raising, 14
subgroup, 118
surgery, 85–86
Swiss exercise ball, 24
symptom related factors, 111
syndrome, 7
treatment of acute and chronic, 84–85
trigger points, 9
upper motor neuron tests, 15
urinary retention, 11
Z-lie, 22
Lower fusion rates, 214
Lower thoracic, 1052, 1055
Low risk of disease transmission, 214
LP ESP lumbar prosthesis, 765
316L stainless steel, 153
Lumbar arthroplasty, 904, 917
Lumbar artificial discs, 917, 1224–1226
Lumbar disc arthroplasty, 32
  aorta, 913
  complications, 915–916
  contraindications, 913
  iliac arteries and veins, 914
  indications, 913

IVC, 913
  outcomes, 916–917
  positioning and approach, 914–915
  postoperative protocol, 915
  ProDisc-L II, 915
  revision options, 916
  segmental vessels, 914
  sympathetic plexus, 914
  ureter, 914
Lumbar discectomy, 1123–1125
Lumbar disc herniation, 280, 941
Lumbar disc prolapse, 29–30
Lumbar facet denervation, 939
Lumbar facet injections, 938
Lumbar fusion, 317, 319, 604–605
Lumbar instability, 282
Lumbar lordosis (LL), 319, 742, 743
Lumbar plexus, 693
Lumbar spinal stenosis, 562, 570
Lumbar spine, 606, 741–743, 1132
Lumbar TDR design, 837
Lumbopelvic parameters, 987
Lumbosacral plexus, 743
Lung deflation, 1054, 1058
Luque, 542
  wiring, 662
Luquesublaminar wiring, 547
Lymphocyte proliferation test, 141
Lymphocyte transformation testing (LTT), 142–143, 466

M

M6 artificial disc, 1224
M6-C artificial cervical disc, 764–765
M6-C artificial disc, 758
M6-L, 1224, 1225
Macro-discectomy, 1182, 1183
Macrophages, 141, 213, 277
Magnetic resonance imaging (MRI), 153, 285, 413, 419, 442, 445, 929–933, 957, 1056, 1057, 1172, 1176, 1178, 1187, 1206
  artifact production, 442
  parameter abbreviations, 444
  scan sequence, 444
  1.5T MRI, 443–445
Magnitude of loading
  interbody devices, 388
  pedicle screw constructs, 387–388
  pedicle screws, 387
Mainstream medical practitioners, 69
Major vessel tear, 1191
Maladaptive behaviours, 112
Male, 355, 356, 368, 369
Manipulation, 53
Manufacturers, 1155
  processing, 232
Maptens, 141
Maximum von Mises stress, 320, 321

Mayer, H.M., 690
May-Thurner effect, 1191
Mazor Robotics, 1274
Mazor X, 1275
McKenzie, Robin, 13
McKenzie technique, 52
MDD 93/42/EEC, 1157
MDR compliance cost, 1158
Mechanical bench testing
    acceptance criteria, 409
    evaluation, 408
    non-fusion devices, 408
    pedicle screw system, 407
    performance evaluation, 409
    standards, 411
    standard test methods, 408
    vertebral body replacement devices, 409
Mechanical fixation, 212
Mechanical instability, 512
Mechanical perspective, 1168
Mechanical stability, 1011
Mechanical testing, 316
Mechanobiology, 277
Mechanotransduction, 277
Medial branch block, 938
Mediated, 112
Mediators, 113
Medical Device Directive (MDD), 1153, 1155
Medical Device Regulation, 1156
Medical devices, 402, 1218–1220, 1222, 1226
Meningioma, 515–517
Mental component scale (MCS), 360, 362, 363
Mesenchymal progenitor cells (MPCs), 185
Mesenchymal stem cells (MSCs), 178, 179, 187, 232, 264,
    1014, 1238
Metal-allergy, 138
Metal alloys, 460, 462
Metal-bone interface complications, 839
Metal hypersensitivity
    clinical presentation of, 467–468
    implant debris physical attributes and local
        physiological response, 463–464
    implant sources of particulate debris, 462–463
    physiology, 460–462
    risk factors for, 467
    spinal implant composition, 468–469
    systemic response to metal debris, 464–465
    testing for, 465–467
    treatment of, 469
Metal implants, 165
Metal ion toxicity, 876
Metal-on-metal bearing surfaces, alternative to, 794
Metal-on-metal devices, 794
Metal-on-metal facet replacements, 469
Metal sensitivities, 619
Metal sensitivity symptoms, 143
Metal serum levels, 132
Metastatic disease, 575
Metastatic lesions, 574

Metastatic spine disease
    case study, 512–513
    neurologic oncologic mechanical systemic framework,
        511–512
    predictive analytic scoring systems, 512
Methylene blue, 1100
Methylprednisolone sodium succinate (MPSS), 1238
Microbiology, 957
Micro-discectomy, 1053, 1065
Microendoscopic discectomy, 704
Microenvironment, 1252, 1258, 1260–1262
Midline incision, 441
Mineralization, 213
Minimally invasive, 360–363, 365, 368, 369
    lateral interbody fusion, 690
    procedure, 562, 563
    technologies, 1126
Minimally invasive lumbar fusion, 1040
    decompression for bilateral stenosis, 1043–1044
    endoscopy, 1044–1045
    fluoroscopy nuances, 1040–1041
    interbody, 1042–1043
    Jamshidi needle advancement, 1041–1042
    Kirshner wire management and screw placement, 1042
    rod passage, 1043
    tubular retractor, 1043
Minimally invasive option (MIS), 1212
Minimally invasive spine surgery (MISS), 702, 1030,
    1031, 1033–1034
    advantages and disadvantages of, 702–703
    deformity correction, 709–711
    discectomy, 703–705
    history of, 703
    laminectomy, 705–706
    lateral interbody fusions (*see* Lateral interbody fusion)
    limitations in ADS, 713
    navigation in, 714
    robotics, 714
    sacroiliac joint, 708–709
    transforaminal lumbar interbody fusion (MIS TLIF),
        706–707
Minimally invasive surgery (MIS), 669–671, 978, 1052,
    1054, 1055, 1058–1064
Minimally invasive TLIF procedure (MI-TLIF), 981
Minimally invasive transforaminal lumbar interbody
    fusion (MI-TLIF), 343
Minimal osteoinductive potential, 214
Minimum clinically important difference (MCID), 309
MIS, 361–363
MI-TLIF, 985
MOBI-C®, 805
Mobi-C CDA *vs.* ACDF, 872
Mobi-C cervical artificial disc, 870
Mobi-C cervical disc, 756–757, 813
Mobility, 351, 352, 355, 359, 368
Modic changes, 931
Modulus of elasticity, 164, 964
Monoaxial screws, 543
Mono-causal theory, 75

Monocytes, 141
Monopolar stimulation, 487
Morbidity, 1053, 1054
Morquio syndrome, 596
Motion artifacts, 454
Motion palpation, 52
Motion preservation, 32, 277, 562, 563, 565, 808, 860
  technology, 886
Motion sparing, 902, 912, 917
Motor evoked potentials
  anesthetic considerations, 498–499
  neurophysiologic system, 497
  obligate waveforms and warning criteria, 499
  parameters and techniques, 498
  recording method, 497
Movement preservation, 1003
MRSA carriage, 953
MR spectroscopy, 932
Mucopolysaccharidosis type IV, 596
Multi-disciplinary, 1197
Multidisciplinary approach, SSIs, in spine surgery,
  see Surgical site infections (SSIs), in spine surgery
Multifidus muscle, 702, 705
Multilevel lumbar, 1004
Multiple myeloma, 574, 575
Muscle(s), 351, 353, 368, 369, 1056, 1058, 1059,
  1061, 1062
  dorsal, 727–729
  reapproximation, 610
  ventral, 725–728
Mutual Recognition Agreement, 1154
Myelography, 929, 930, 935
Myelopathy, 751, 762, 764, 766, 767
Myxopapillary ependymoma, 515

N

Nano-crystalline hydroxyapatite (ncHA), as a surface
  layer, 262
Nanometer, 129
National Institute for Health and Care Excellence (NICE)
  guidelines, 84
Native anatomy, 623
Native growth factors/cytokines, release of, 213
Natural history, 958
Navigated optical technology, 554
Navigation, 714
Neck Disability Index (NDI), 418, 764–766
Negative local factors, 213
Neoplastic lesions, 1052, 1057
Nerve(s), 1054–1059, 1062
  root blocks, 936
root compression, from implant migration, 842
Neural EGFL like 1 (NELL1), 1256
Neural elements, 968
Neural foramen, 1225
Neural monitoring, 1059
Neural retraction, 972
Neurite growth inhibitor A (NOGO-A), 1240
Neurogenic claudication, 29, 562, 564, 565, 570

Neurological monitoring, 1176
Neurologic oncologic mechanical systemic (NOMS)
  framework, 511–512
Neuromodulation, 1235, 1237, 1242–1244
Neuromonitoring, 695, 971, 1058
Neuronal stem cells (NSC), 1238
Neurophysiology
  electromyography, 490
  principles of, 485
Neuroprotection, 1235, 1237, 1238, 1242, 1247
Neuro-Spinal Scaffold, 1242
Neurosurgical implants, 280
Neurovascular structures
  meninges and dura, 730
  nerve roots, 730
  spinal cord, 729–731
Neutral zone (NZ), 333, 894
Neutrophils, 141, 277
NFlex device, 304
Nitinol, 154
NLRP3 inflammasome, 134
N-methyl-D-aspartate (NMDA)-induced
  excitoxicity, 1241
Non-device related complications, 281
Non-fusion, 1220
Non-inferiority, 418
Noninvasive treatments, 69
Nonphysiologic loads, 276
Non-specific low back pain, 29
  central herniation, 32
  lumbar disc arthroplasty, 32
  spinal fusion, 31
  spondylolisthesis, 32–33
  surgical treatment, 31–33
Non-steroidal anti-inflammatory drugs (NSAIDs), 102,
  832, 1104
Non-surgical management (NSM), 358–360
Non-threaded cages, 637
Nonunion, 333, 625, 1252
Nonviable tissue, 214
North American Spine Society (NASS) coverage policy
  recommendations for interspinous devices, 564
Notching, 166
NovaBone®, 263
Novel implants, 767
NuBac, 1120
Nuclear medicine, 933
Nucleus pulposus, 1113
Nucleus replacement (NR) implants
  as adjunct to discectomy, 1125
  classification, 1120–1121
  clinical outcomes, 1120
  design criteria, 1118
  geometry and material properties, 1123
  in-situ curing, 1120–1125
  potential uses, 1126
  preformed elastomer, 1120–1124
Numerical rating scale (NRS), 362
Nutritional status, 1056

O

O-arm, 714
Obesity, 625
Oblique interbody fusions, 690
Oblique lateral interbody fusion (OLIF), 966, 970–971, 977, 1132, 1138, 1140
Oblique lumbar interbody fusion/anterior to psoas (OLIF/ATP), 343
Occlusion, 1191, 1196
ODI scores, 333
Odontoid, 1072, 1073, 1075
OECD, 1150, 1151
Ogzur, 707
Olfactory ensheathing cells (OECs), 1238
Oligodendrocyte precursor cells (OPCs), 1238
OMNIbotics (OMNI), 1271
Omnitract™, 1173
On table angiogram, 1196
Open lateral approach, TLJ, 1058–1059
Open lumbar fusion, 1034
    CBT screws, 1039
    facetectomy, 1035
    hybrid percutaneous screws, with mini-open interbody, 1040
    interbody, 1035–1036
    laminectomy, 1034–1035
    pedicle screw placement, 1036–1038
    pedicle screws, Wiltse approach, 1038
    positioning, 1034
    posterolateral fusion, 1038
Open operative biopsy, 958
Open posterior sacral-alar-iliac approach, 681–682
Open posterolateral iliosacral approach, 681
Open reconstruction, 1199
Open surgical fusion, 360
Open TLIF procedure (O-TLIF), 980
Open ventral-ilioinguinal approach, 680–681
Opiates, 31
Opioids, 69, 87
    use, 28
Opportunistic computed tomography, 528
Organic phase of bone, 214
Orthopedic spinal devices
    classifications, 402–404
    FDA and CDRH, 402
    premarket submission, 403–419
Orthophosphate, 130, 462
Osseoanchored Prostheses for the Rehabilitation of Amputees (OPRA), 405
Osseointegration, 426
Osseous fusion, 616
Osteitis condensans ilii (OCI), 97
Osteoblasts, 1257
Osteoblast-specific growth factor, 1256
Osteoclasts, 1257
Osteoconduction, 213, 214, 240, 262
Osteoconductive, 159, 1252
Osteoconductivity, 1020
Osteogenesis, 213, 232, 239, 262, 264

Osteogenic, 159, 1252
    cells, 201, 232
    growth factor, 238
Osteogenicity, 1011, 1020
Osteogenic protein 1 (OP-1), 174, 1092, 1256
Osteoinduction, 213, 232, 266, 1252
    potential, 214, 238, 24, 12540
Osteoinductive factor (OIF), 213, 232
Osteoinductivity, 1020
Osteointegration, 156
Osteolysis, 134–136, 144, 763, 767, 840
    mechanisms, 461
Osteomalacia, 137
Osteomyelitis, 94, 95
Osteopathic bone, 284
Osteopathy, 73
Osteoporosis, 105, 106, 574–576, 583, 625, 978
    complications of medical treatment, 533
    diagnosis, 527–529
    dual energy x-ray absorptiometry, 528
    education, 529–532
    epidemiology, 524–525
    mitigation of poor bone health, 534
    morbidity and mortality of fragility fractures, 524–525
    nutritional supplementation, 532
    pharmaceutical management, 533
    preoperative bone health program, 534–537
    preoperative optimization of spine surgery patients, 534–537
    secondary causes of, 533
    secondary fracture risk, 525–527
    treatment of, 529–534
Osteoporotic bone, 548
Osteoporotic fractures
    investigations, 105
    malignancy, 104
    spinal, 104
    vertebral, 104
Osteo-stimulative, 264
Osteotomy, 1055, 1062, 1063
Oswestry Disability Index (ODI), 284, 418, 713, 832, 916, 1142
    scores, 571
Oswestry Low Back Pain Disability Index (ODI), 360, 362, 363, 365
Overdistraction, 817
Own the bone quality improvement program, 527
Oxygen-ozone, 1099

P

P-15, 239
Pain, 350, 353–363, 365, 368–370
    catastrophizing, 112
    generators, 934
    hypersensitivity, 117
Pain Self-Efficacy Questionnaire (PSEQ), 120
Palmer, 73
Para-coccygeal approach, 1212

Paracoccygeal notch, 973
Paraspinal muscles, 972
Paraspinal sacrospinalis-splitting approach, 703
Parkinson's disease (PD), 474–480
Pars defect, 32, 742
Particulate polymethylmethacrylate (PMMA), 134
Passive systems, 1271
Patch testing, 141
Patient-based negative factors, 213
Patient reported outcome measures (PROMs), 299, 302, 304
PCM CDA *vs.* ACDF, 868
PCM® Device, 800, 801
Pedicle(s), 722, 737, 740–742
   anatomy, 543
   probe, 548
   screw-rod stabilization, 344
   subtraction osteotomy, 666–668
Pedicle screw(s), 152, 155, 286, 542, 969
   complications, 555–556
   design & anatomy of, 543–546
   fixation, 338, 546–548, 610, 966
   freehand technique, 548–549
   head, neck, body, 543
   indications for use, 548–549
   outcomes, 555
   rod-based devices, 279
   and rods, 281
   threshold testing, 492–494
Pedicle-screw based posterior dynamic stabilisers (PDS), 1126
Pedicle screw-rod stabilization, 344
Pedicle screw system, 407, 408
   cobalt chromium rods, 440, 444
   MR artifact production, 442
   MRI scan sequence, 444
   phantom construct, 442
   phantom setup, 3T MRI scanner with, 443
   phantom setup, with $CuSO_4$ solution, 442
   screw and rod combinations, 441
   stainless steel, 440, 444
   titanium screws, 440, 444
   1.5T MRI, 444, 445
   torso, implantation in, 441
Pedunculopontine nucleus (PPN), 474
Pelvic girdle pain (PGP), 356
Pelvic incidence (PI), 319, 742, 743
Pelvic parameters, 1172
Pelvic tilt (PT), 319
Pelvis, 350, 351, 353–356, 365, 367–369
Pentosan polysulfate (PPS), 185
Peptide-based materials, 201
   B2A, 239–240
   cerapedics, 239
Percudyn device, 306
Percutaneous pedicle screw, 1040, 1042, 1044
   fixation, 709
Percutaneous screw placement, 553
Percutaneous vertebral augmentation, 575

Perineural injections, 937, 940
Perioperative measures, 346
Peripheral endplate, 988
Peripheral stimulation, 1242
Peritoneum, 698, 1055, 1056, 1058, 1059
Personalized approach, 80
Personal relationships, 114
Pessimistic beliefs, 113
Phases of bone formation, 1012
Phosphate, 262
Physical, 951
   examination, 76, 80
   therapy, 350, 358, 359
Physical component scale (PCS), 360, 362, 363
Physiologic conditions, 332
Pilot holesin, 547
Placement, 365, 367, 368, 370
Plain radiographs, 926
Plate, 334, 616
Platelet rich plasma (PRP), 176, 185, 1095
Pleura, 1053, 1055, 1058, 1059, 1061
PLIF, *see* Posterior lumbar interbody fusion (PLIF)
Pneumothorax, 1053–1055, 1063, 1064
Pointillart prosthesis, 906
Point loading, 986
PolyArylEtherKetones, 430
Polyether ether ketone (PEEK), 155, 159, 166, 280, 416, 425, 426, 441, 442, 444, 450, 453, 468, 621, 640, 763, 767, 1036, 1225
Polyethylene, 763
   core fractures, 838
Polyethylene glycol (PEG), 1241, 1242
Polyglycolic acid (PGA), 182
Poly Implant Prosthesis (PIP), 1157
Poly-lactic acid (PLLA), 1261
Polymer
   advantages, 427
   implant, 428
   innovation in, 424
   mechanical properties, 429
   PolyEtherEtherKetone, 425
Polymer-based grafts, 1017
Polymethylmethacrylate (PMMA), 582, 634
   bone cement, 556
Polyurethane, 763
   block model, 620
Poor vascularity, 213
Porous coated motion (PCM), 813
   device, 868
Porous coated motion prosthesis
   device description, 753
   outcomes of, 753–754
Porous metals, 154
Porous PEEK, 156
Porous titanium, 166
Portal vein occlusion, 1204
Positioning, 1055, 1056
Positive feedback loop, 277
Posterior and posterolateral approaches, 741

Posterior cervical decompression and fusion (PCDF), 1137
Posterior dynamic stabilization (PDS), 281, 288, 297, 311, 846, 855
   Accuflex system, 301
   BioFlex system, 301, 302
   biomechanics, 894
   CD Horizon Agile system, 302, 303
   component and interface level static and dynamic testing, 298
   Cosmic posterior dynamic system, 302, 303
   Dynesys, 302
   Graf ligament, 303, 304
   Isobar TTL system, 303, 304
   NFlex device, 304
   Percudyn device, 306
   preclinical *in vitro* biomechanical testing, 298–299
   preclinical *in vitro* mechanical testing, 298
   Stabilimax NZ device, 304, 305
   in vivo performance, 299, 301
   Wallis® posterior dynamic stabilization system, 307
Posterior dynamic stabilization systems (PDSS), 323–326
Posterior fixation, 297, 966
   techniques, 338
Posterior longitudinal ligament (PLL), 724, 909
Posterior lumbar decompression and fusion, 1138
Posterior lumbar fusion, 319
Posterior lumbar interbody fusion (PLIF), 342, 343, 706, 968–969, 972, 981, 984, 1031, 1034, 1040, 1138, 1260
Posterior pedicle screw instrumentation, 1138
Posterior spinal instrumentation, 542
Posterior superior iliac spine (PSIS), 681
Posterior tension band, 607
Posterolateral fusion (PLF), 207, 212, 238, 264, 1025, 1038, 1138, 1139, 1256
Post-laminectomy kyphosis, 1086
Postoperative care, 1205
Postoperative patch testing, 466
Post-surgical complications, 284
Potts disease, 690
Preclinical tests, 316
Predicate device, 404
Predictive analytic scoring systems, 512
Preformed elastomer, 1120–1124
Pregnancy, 351, 353, 356, 357, 369
Pre-hospital, 954
Prelordosed plates, 619
Premarket approval (PMA), 404, 406, 1218–1221, 1223
Pre-Market Assessment, 1155
Premarket notification (510(k)), 403–405, 1218
Premarket submission
   device evaluations, 407–419
   types, 403–407
Pre-operative angiogram, 1194
Pre-operative assessment, anterior lumbar spinal reconstruction, *see* Anterior lumbar spinal reconstruction
Pre-psoas approach, 694–696
Pre-psoas L5-S1, 697

Presacral space, 973
Pressure hyperalgesia, 117
Prestige, 824
   artificial disc, 812
   CDA, 863, 864, 866
   cervical disc system, 811
PRESTIGE® Artificial Devices, 798
PRESTIGE® LP Device, 799
PRESTIGE® ST cervical joint, 800
Prestige LP, 867
   artificial disc, 755
   CDA, 876
   cervical disc, 756
Prestige ST, 866, 867
   cervical disc, 755
Prevention, of spinal SSIs, *see* Surgical site infections (SSIs), in spine surgery
Previous disc surgery, 982
Primary bone tumors, 575
Problem-focused CBT, 121
ProDisc-C, 812, 868, 870
   biomechanics of, 797
ProDisc-C CDA vs. ACDF, 870
ProDisc-C cervical disc, 754
ProDisc-L, 812
Productivity losses, 118
Prolapsed lumbar disc, 32
Proliferation index, 142
Prolotherapy, 1097
Prone Extension Negative (PEN), 17
Prone Extension Positive (PEP) patients, 17
Prophylactic antibiotics, 1175
Propofol, 489
Prostaglandins, 134
Prosthesis List, 1159
Prosthetic disc nucleus (PDN), 1123
Prosthetic extraction, 1188
Prosthetic inflammatory response, 840
Proteoglycan, 1114
Provocative discography, 1173
Proximal junctional kyphosis (PJK), 167
Pseuarthrosis, 296
Pseudarthrosis, 165, 619, 963, 969, 970, 984, 990, 1132, 1142
Pseudoarthrosis, 333, 335, 1167
Psoas muscle, 609, 707, 971
Psoriasis, 98
Psoriatic arthritis, 12, 97
Psychological distress, 78
Psychological factors, 111
Psychological inflexibility, 116
Psychotherapy, 121
Pubic angle, 355
Public Health Service Act (PHS Act), 402
Pullout resistances, 624
Pullout strength, 341, 544, 624
Pullout test, 341
Pulmonary function, 1056
Pure ethanol, 1099

Q

Quadriceps palsy, 971
Qualitative stability index (QSI), 299

R

Radial annular tear, 936
Radiation exposure, 1279
Radiation-induced osteosarcoma, 508–511
Radicular pain, 9, 14, 983
Radicular syndromes
    lumbar disc prolapse, 29–30
    pathology, 29
    spinal stenosis, 30
    surgical treatment, 29–31
Radiculitis, 941
Radiculopathy, 751, 762, 764–767, 932
Radiofrequency (RF), 350, 358, 359
    ablation, 939
    denervation of facet joints, 87
Radiographs, 1133
Radiology, 926
Radiolucent, 155
Radionuclide bone scanning, 933
Randomized clinical trial, 301, 1223
Randomized controlled trials (RCT), 1223, 1246
Range of motion (ROM), 297, 299, 301, 303, 306, 316, 324, 333, 354, 355, 365–369, 379, 381, 389, 395–398, 846, 894, 966, 1170
RANKL-OPG pathway, 1014
Receptor activator of nuclear factor-kappa B (RANK) production, 461
Recombinant human bone morphogenetic proteins (rhBMPs), 296
Rectal injury, 973
Recurrent back, 1124
Recurrent disc herniation, 1124
Recurrent laryngeal nerve, 906–907
Reductionism, 49
Referred pain, 9
Regulatory burden, 1156
Regulatory effect, 1153–1158
Regulatory reform, 1156
Rehabilitation, 1205
Reimbursement effect, 1159–1161
Reimbursement strategies, 1160
Relaxation techniques, 54
Release(s), 1182
    concentrations, 265
Remodeling, 213
Renaissance Guidance System, 1275
Renal anomalies, 1204–1206
Re-operation rates, 1122
Repair, 1191–1192
Repetition time (TR), 442, 448
Research utilization, 77
Resorption, 1024
Restoration of lordosis, 1215
Retractor, 1173, 1175, 1177–1180, 1182–1184, 1186
    pins, 816

Retro aortic renal vein, 1201
Retrograde ejaculation, 607, 743, 970, 1181
Retroperitoneal, 1053–1056, 1058, 1061, 1062
    approach, 609, 970, 998
    space, 707
Retroperitoneal approach, anterior lumbar spinal reconstruction, 1177–1179
    left sided retroperitoneal approach, L4-5 and L5-S1, 1179–1185
    non-vascular anatomy, 1176–1178
    revision retroperitoneal surgery, 1185
Revision discectomy, 1124
Revision retroperitoneal surgery, 1185
Revision surgery, 213, 283
rhBMP-2, 232, 238, 239, 254, 262, 625, 974
RhBMP-7, 238
Rheumatoid arthritis, 106, 141
Rheumatology, 100
Rho signaling pathway, 1240–1241
RI-ALTO, 365, 367, 368, 370
Ribs, 693
Rigid body devices, 288
Rigidity, 624, 964
Riluzole in Spinal Cord Injury Study (RISCIS) trial, 1241
Risk, 1053, 1055, 1056
Risk informed credibility assessment method, 415
Robot-assisted posterior lumbar interbody fusion (Robot-PLIF), 1278
Robotic(s)
    and image-guided navigation systems, 1273
    in spine surgery, 1273–1274
Robotic assisted techniques
    advantages, 1281
    costs, 1280
    history of, 1272–1273
    limitations, 1280
Rod(s), 152
    fatigue, 164
Roland Morris Disability Questionnaire (RMDQ), 55
ROSA Robot, 1276
Rotation, 351–355, 357, 359, 361, 363, 365, 367, 368
Roy-Camille, 542

S

S53P4 bioactive glass, 263
45S5 bioactive glass, 263
Sacralized, 739, 742, 744
Sacral slope (SS), 319
Sacrococcygeal joint, 943
Sacro-iliac injections, 942
Sacroiliac joint (SIJ), 100, 318, 350, 935
    anatomy, 350, 676
    biomechanical principles, 677
    clinical studies, 361–365
    conservative management strategies, 679
    diagnosis and evaluation, 678–679
    dysfunction, diagnosis of, 358
    etiology, 678

Sacroiliac joint (SIJ) (*cont.*)
    functional integrity, 677
    function and biomechanics, 351–354
    fusion, 708–709
    innervation, 677
    instrumentation options, 683
    in vitro and *in silico* studies, 365–369
    ligaments, 351
    minimally invasive approaches, 682–683
    minimally invasive SIJ fusion, 360–361
    muscles, 351
    non-surgical management, 358–360, 685
    open SIJ fusion, 360
    pain, causes of, 357–358
    pathology, 676
    range of motion, 354–355
    sexual dimorphism, 355
    surgical decision making, 679
    surgical options, 680–682
Sacroiliitis, 97, 100
Sacrospinalis muscle, 703
Sacrum, 351–355, 359, 361, 362, 366–370
Sagittal alignment, 962
Sagittal balance, 712, 1172
    restoration, 979
Sagittal correction, 713
Sagittal imbalance, 606, 987
Sagittal plane deformities, 713
Sagittal vertical axis (SVA), 319, 709
Saskatchewan Spine Pathway, 7
Scanning electron microscopy (SEM), 130
Schwann cells, 1238
Schwannomas, 515
ScientX, 289
Scoliosis, 707, 738, 978, 1025, 1167
    clinical evaluation, 659
    definition, 658
    etiology, 659
    imaging evaluation, 660
    spinal and spinopelvic parameters, 660–661
Scoliosis Research Society (SRS), 362, 1142
Screening, 1171
Screw loosening, 1221
Screw-plate fixation, SI joint, 683
Screw-rod fixation, 286
Secondary fracture prevention
    outcomes of, 526
    programs, 525–526
Secondary gain symptoms, 115
Secondary osteoporosis, 575
Second-generation lumbar cages, 963
Secure-C artificial cervical disc, 873
    vs. ACDF, 873
Secure-C cervical disc, 757
Segmental(s), 1055, 1058, 1059
    arteries, 691
    lordosis, 968, 990
Selective laser sintering (SLS), 159
Selective nerve root injections, 940

Self, 115
Self-efficacy, 112
Self-renew and proliferate, 1257
Semi-active systems, 1271
Sensitization, 140
Sensitizers, 138
Seroma formation, 1255
Serum analysis, 460
Sexual dimorphism, 355
Short Form Health Survey, 1032
Side clamp, 1192
SIFix, 365, 366
Signalling molecules, 1013
Signal-to-noise (S/N) ratio, 448, 452
Silica surface, 262
Silicate-substituted calcium phosphate
    (Si-CaP), 262
Silicon nitride, 156
SI-LOK, 364, 368, 370
SImmetry, 365, 370
Simvastatin, 1096, 1097
Single level disease, 607
Single photon emission computed tomography
    (SPECT), 578, 933
SLIP, 1032
Small interfering RNA (siRNA), 176
Small-medium sized companies, 1160
SmartAssist (Mazor Robotics), 1274
Smith and Robinson approach, 1078–1084
Smith-Petersen Osteotomy, 664–666
Social withdrawal, 114
Soft callus, 213
Soluble signals, 214
Somatosensory evoked potentials (SSEPs)
    anesthetic considerations, 496
    neurophysiologic system, 494–495
    obligate waveforms and warning criteria, 496
    parameters and technique, 495–496
    recording method, 495
Speciality pain, 118
Spinal arthroplasty, 809
Spinal canal, 723
Spinal column kinematics, 1118
Spinal cord, 729–730
    blood supply, 731
    stimulation, 1243–1244
Spinal cord injury (SCI)
    biological scaffolds, 1241–1242
    BMI, 1244
    brain stimulation, 1242–1243
    cervical SCI, 1235–1236
    clinical trials, 1246–1247
    complete SCI, 1236
    consistency of results, 1244
    CSF drainage, 1239
    demographics, 1230
    economic burden, 1230–1232
    ethical challenges, 1245–1246
    excitotoxicity, 1241

functional regeneration, neuroprotection,
  immunomodulation, 1237
grading systems, 1232–1234
growth factors, 1239–1240
hemodynamic intervention, 1239
incomplete SCI, 1236–1237
peripheral stimulation, 1242
primary injury, 1233–1234
Rho signaling, 1240–1241
risks, 1245
rodent models of, 1246
secondary injury, 1234–1235
spinal cord stimulation, 1243–1244
stem cell therapies, 1237–1238
steroids, 1238–1239
thoracic SCI, 1236
timing, 1245
Spinal cord stimulation (SCS), motor function in PD
  animal studies, 474–476
  human studies, 476–478
  mechanisms, 478–479
Spinal device(s), 1156, 1162
Spinal device clinical trials
  adverse events, 418
  de novo submissions, 417
  individual patient success, 418
  medical imaging, 418, 419
  neurologic assessments, 418
  overall study success, 418
  pain and function evaluations, 418
  patient satisfaction measures, 418
Spinal fracture, 104
Spinal fusion, 31, 201–213, 288, 317–322, 1258
  allografts, 214–232
  autograft, 213–214
  axial back pain, 1033
  B2A, 239–240
  BMPs, 232–239
  cerapedics, 239
  clinical and preclinical data, 266
  clinical use in, 201
  and complications, 295–296
  interbody, 212
  MISS fusion, 1033–1034
  open and minimally invasive spinal fusion, brief
    history of, 1031
  regional and global trends in, 1031–1032
  spondylolisthesis, 1032–1033
  and structural integrity, 294–295
  synthetic materials and grafts, 240–265
  thoracolumbar burst fractures, 1033
Spinal implants, 1218, 1223, 1226
Spinal infection
  clinical pearls and myths, 96
  diagnosis, 94
  diskitis, 95
  epidural abscess, 94
  initial antibiotic therapy, 95
  intravenous drug investigations, 95

osteomyelitis, 95
  symptoms, 94
Spinal injections, 86–87
Spinal instability, 318, 840
Spinal instability neoplastic score (SINS), 512
Spinal manipulative therapy, 69
Spinal movement disorders, 288
Spinal stimulators, 943
Spinal surgery, PDS, *see* Posterior dynamic
    stabilization (PDS)
Spinal tumors
  intradural, 513–519
  metastatic spine disease, 511–513
  multidisciplinary approach, 507
  patient evaluation, 506–507
  primary column tumors, 507–511
Spine, 350–352, 354, 357, 358, 367–370, 634, 639
  arthroplasty, 903
  biomechanics (*see* Biomechanics)
  fusion surgery, 86
  manipulation, 164
SpineAssist System, 1274
Spine Outcomes Research Trial, 30
Spine Patient Outcomes Research Trial (SPORT), 1032
Spine products, 1218, 1226
  Barricaid®, 1222, 1223
  DIAM®, 1221, 1222
  Dynesys®, 1220
  FDA approval process, 1218–1220
  interspinous and interspinous stabilization/
    distraction, 1225
  lumbar artificial discs, 1224
Spine surgery, 165, 166
  surgical site infections in (*see* Surgical site infections
    (SSIs), in spine surgery)
Spinopelvic parameters, 742
Spinous process, 567, 569, 570, 723, 1221, 1222,
    1225, 1226
Spleen, 133
Spondyloarthritis (SpA)
  ankylosing spondylitis, 103
  anti-TNF antibodies, 103
  bone marrow edema drug, 97
  classification, 97
  diagnosis, 97
  investigations, 100
  L4-L5 stenosis, 103
  MRI features, 100
  non-radiographic axial, 100
  non-steroidal anti-inflammatory drugs, 102
  psoriasis, 98
  subclinical, 100
  X-ray features, 100
Spondyloarthropathies, 12
Spondylolisthesis, 32–33, 287, 610, 742–744, 840, 975,
    978, 980, 982, 1032–1033
  degenerative, 989
  isthmic, 989
Spondylolysis, 840

Spondylolytic spondylolisthesis, 32
Spontaneous EMG, 491
SSSS, 1032
Stabilimax NZ device, 287, 289, 304, 305
Stability, 616, 1170
    to the spine, 201
Stabilization system, 1220–1222, 1225
Stable, 958
    arthrodesis, 962
Stainless steel (SS), 153, 416, 440, 442, 453, 456, 546, 775
Stainless steel alloys (SSA), 165
Stand-alone cage(s), 965, 1000
Standalone interbody devices
    ALIF, 998–1003
    lumbar TDR, 1003–1005
    TDR and fusion, 1005–1006
Standalone strategy, 1168
Standardized test methods, 408
Standards development organizations (SDOs), 408
Staphylococci, 587
Staphylococcus aureus, 265, 951
Static plate systems, 336
Statins, 1096–1097
Stem and progenitor cells, 1257–1258, 1260
Stem cell therapies, 1237–1238
Stenosed, 1189
Stenosis, 278
Stents, 1174, 1188, 1189, 1191, 1192, 1196–1199, 1201, 1206
Stereotaxy, 1273
Sterility assurance level (SAL), 1019
Sterilization method, 232
Sterilization process, 1019
Sternocleidomastoid (SCM), 725, 726, 734
Steroid(s), 940, 1238–1239
    perineural injections, 87
Stiffness, 166, 354, 366, 367, 964, 1170, 1171
Stigmatization, 115
Stimulating/supporting bone growth, 201, 266
Stimulation index, 142
Straight-forward technique, 547
Straight leg raising (SLR) test, 14
Streptococci, 587
Stress(es), 350, 353, 357, 367–369, 1113
    management techniques, 51
    strain, 380
Stress shielding, 158, 295, 964
    effect, 166
Stromal vascular fraction (SVF), 187
Structural abnormalities, 1116
Structural failure of disc, 1117
Structural instability, 213
Strut/morselized bone, 214
Subaxial cervical spine, 594, 720–721
Sublaminar wires, 542
Subluxation, 71, 82
Subsidence, 698
Substance use disorder, 69

Substitutes, 1015
Sunnybrook system, 1232
Superficial infections, 958
Superficial SSIs, 950
Superion®, 1225
Superion interspinous devices, 567
Superior articular process (SAP), 1035
Superior biocompatibility, 262
Superiority, 418, 419
Superior laryngeal nerve, 731–733, 906
Supervisory controlled systems, 1271
Supplementary fixation, 988
Supplementary pedicle screw instrumentation, 982
Support, 116
Surgical discectomy, 86
Surgical planning, 738
Surgical site infections (SSIs), in spine surgery, 954
    back pain, 956
    blood cultures, 957
    blood serology testing, 957
    CT, 957
    epidemiology, 950–951
    established risk factors, 953
    etiology, 951–953
    imaging, 957
    intraoperative, 955
    management, 958–959
    microbiology, 957
    MRI, 957
    open operative biopsy, 958
    postoperative, 955–956
    pre-operative, 954
    relative risk factors, 953
    sample preparation, 957
    ultrasound, 957
Surgical technical factors, 213
Surgical techniques, 369
Suspended future, 116
Suspended self, 116
Suspended wellness, 116
Swedish lumbar spine study, 32
Sympathetic chain, 733, 907
Sympathetic dysfunction, 970
Sympathetic plexus, 914
Symptom related factors, 111
Synchronous treatment, 1197–1201
Syndesmophytes, 100
Synergy Cervical Disc, 766–767
Synovial cyst, 940
Synthetic materials and grafts, 240
    advantageous properties of, 240
    BAG, 262–265
    calcium phosphate materials, 240–254
    silicate-substituted calcium phosphate, 262
Systemic lupus erythematosus, 106
Systemic response, 956
Systemic therapy, 958

T

Tamponade, 1189
Tap, 549
Tapping, 549
T cells, 138
Temporal summation, 117
1.5 Tesla MRI, 442
3 Tesla MRI, 442
TGA registration process, 1155
Theatre environment, 954
Therapeutic Goods Administration, 1159
Thoracic facet interventions, 939
Thoracic instrumentation, 738, 1044–1046
Thoracic kyphosis (TK), 319
Thoracic level, 474, 478
Thoracic SCI, 1236
Thoracic spine, 737–741
Thoracodorsal fascial, 705
Thoracolumbar burst fractures, 1033
Thoracolumbar fusion, 1031, 1046
Thoracolumbar junction (TLJ)
    anatomy, 1055–1056
    indications, 1052–1055
    MIS approach, 1059–1064
    open lateral approach, 1058–1059
    patient presentation, 1052
    postoperative, 1063–1065
    preoperative workup, 1056–1057
Thoracoscopic, 1052, 1054
Thoracotomy, 1054, 1055, 1058, 1059, 1063
Threaded cages, 636
Three-joint complex, 276
10th revision of the International Statistical
    Classification of Diseases and Related
    Health Problems, Clinical Modification
    (ICD–10–CM) codes, 5
Thrombectomy, 1198
Thyroid releasing hormone (TRH), 1241
Tissue engineering (TE), 180, 181, 1021
    degenerative disc disease, 179–185
Tissue regeneration, 1262
Tissue transplantation, 1010
Titanium (Ti), 133, 152, 156, 284, 440, 442, 619, 621, 640
    alloy(s), 165, 416, 546, 774
Titanium plasma spray (TPS), 160
T lymphocytes, 141
TNF-α, 134, 277
Tocilizumab, 1102
Tomography, 929
Total disc arthroplasty
    cervical disc replacement, 905–913
    lumbar disc arthroplasty, 913–917
Total disc replacement (TDR), 172, 174, 179, 182, 184,
    188, 322, 323, 751, 1166, 1168, 1169
    devices, 445, 453
    and fusion, 1005–1006
    lumbar, 1003–1005
    vascular revision strategies, 843–844

Total Facet Arthroplasty System (TFAS), 287,
    849–850, 852
Total facet replacement, 279
Total Posterior Arthroplasty System (TOPS), 287,
    848, 852–854
Total Posterior Element Replacement (TOPS) system, 896
Training, 1167
Train of four (TO4) test, 491
Trans-articular fusion rod, 683
Trans-articular threaded screw, 683
Transcranial direct current stimulation (tDCS), 1242
Transcranial magnetic stimulation (TMS), 1242
Transcranial motor evoked potentials (TcMEP), 484, 487
    See also Motor evoked potentials
Transcutaneous spinal cord stimulation (tcSCS), 1244
Transforaminal injections, 941
Transforaminal interbody fusion, 1138
Transforaminal lumbar interbody fusion (TLIF), 343, 968–
    969, 971–972, 980, 984, 985, 1031, 1033, 1034,
    1040, 1044, 1260
    complications, 654
    contraindications, 651
    discectomy, 653
    exposure, 652
    graft selection, 652
    indications, 650
    operative setup, 651
    positioning, 652
    post-operative protocol, 654
    procedure, 650
    surgical technique, 653
Translation per degree of rotation (TPDR), 299, 300
Transmission electron microscopy (TEM), 130
Transoral approach, 1071–1074
Transpedicular approach, 341, 738, 740
Transpedicular fixation, 557
    screws, 555
Transpedicular screw trajectory, 340
Transperitoneal approach, 970, 998
Transperitoneal approach, anterior lumbar spinal
    reconstruction, 1185–1187
    operative levels above the bifurcation, 1187
    operative levels below the bifurcation, 1187
Trans-psoas approach, 694, 695
Transpsoas technique, 1138
Transversalis fascia, 695
Trendelenberg position, 608
Trendelcnburg position, 1173
Triage pathway, 29
Triangular, 360–365, 367–370
    titanium implants, 360, 361, 363
Tricalcium phosphate, 1143
Triggered and free running EMG, 695
Triggered EMG, 492–493
T-scores, 527
Tubular retraction, 704
Tubular retractor, 1031, 1034, 1040, 1043
Tumor and deformity, 1024

U

Ultra-high molecular weight polyethylene (UHMWPE), 130, 380, 416, 767, 775, 838, 903
Ultrasound, 957
Unilateral, 368, 369
   oblique cage, 396
   pedicle screw, 339
   stabilization, 368
Unilateral laminotomy for bilateral decompression (ULBD), 1043
Union, 1011
Unrecognized vascular lesion, 1196
Upper instrumented vertebra, 663
Upper lumbar, 1055, 1056, 1063
Upper motor neuron tests, 15
Ureter, 691
   injury, 698
Ureteric stent, 1174, 1185
Urogenital complications, 1255
Urogynecologic surgical repair mesh, 1156
Urologist, 1174, 1188
US life expectancy, 1151
US market, 1218, 1223–1225
US national health expenditure, 1150

V

Valid scientific evidence, 417
Valsalva, 610
Vancomycin, toxic effects on hMSCs, 266
Variable-angle screw systems, 620
Varieties of structural and non-structural form, 214
Vascular access surgeon, 609
Vascular anatomy, 1055, 1167, 1172, 1177, 1181, 1185
Vascular injury(ies), 697, 971, 977, 1167, 1179, 1181
Vascular pattern, 932
Vascular repair, 1173–1175, 1191
Vasculogenesis and angiogenesis, 1014
Venogram, 1058
Venous oozing, 1192–1193
Venous thromboembolic (VTE) events, 1001
Ventral muscles, 725–728
Verbiest approach, 1076–1078
Verification and validation (V&V), 414
Vertebral artery, 731–732, 907
Vertebral body, 722
Vertebral body compression fractures (VCFs), 574
Vertebral column resection, 668–669
Vertebral compression fracture, 104, 575, 576, 585, 588
Vertebral endplates, 610
Vertebral fracture, 104
   assessment, 528

Vertebral functional segments, 212
Vertebrectomy, 1023, 1167, 1188
Vertebroplasty, 574, 575, 582–584, 586, 587, 943
Vertical ring (Cylinder) cages, 637–638
Vessel calcification, 1181
Vessel injury, 691
View angle tilt (VAT), 454
Viral vectors, 1259
Visceral pleura, 1055, 1058, 1061
Viscoelastic cervical disc, 762
Visual analog scale (VAS), 360–363, 365, 764–766
   scores, 333, 340
Visual analog scale (VAS)-back pain scores, 570
Visualize, 1189
Vitalistic concept, 49
Vitalistic theory, 68
Vitamin D, 532
   deficiency, 534
Volume performance, 1167

W

Wallis, 279
   interspinous implant, 280
   system, 307
Wear bebris, 840
Wear debris, 156, 840–842
Wide local clearance, 958
Wolff's law, 164, 295, 897
World Health Organization, 289

X

X-ray, 419, 926–929
X-Stop device, 279, 308, 854

Y

Years lived with disability (YLDs), 110
Yellow flags, 111
Yield strength, 623
Young's modulus, 164

Z

Zielke system, 618
Zirconium rods, 469
Zone of inhibition, 265
Z-scores, 527
Zygaphophyseal joint, 935